Essential Principles of Physics

P M Whelan MA · **M J Hodgson** MA

Sherborne School · King's School, Canterbury

Illustrated by **T Robinson**

John Murray Albemarle Street London

Essential Principles of Physics first published 1978
Reprinted 1979, 1982
Printed and bound in Great Britain by
J. W. Arrowsmith Ltd, Bristol

0 7195 3382 1

9.95

Essential Principles of Physics

The cover illustration is adapted from a holograph of vibration patterns in a thin-walled steel tube at one of its resonant frequencies, and is reproduced by courtesy of the National Physical Laboratory (Crown Copyright).

By the same authors

Questions on Principles of Physics

Preface

Aims

(*a*) To provide a rigorous one-volume reference book which discusses all the principles of physics that are needed by a student going on to read physics at the university.

(*b*) To present an account of physics up to GCE S level in which both the choice of subject matter and its emphasis are in line with current teaching trends, and in which the presentation, nomenclature and symbols follow the most recent practice. In accordance with the 1969 recommendations of the Royal Society and the Association for Science Education, SI units are used and are suitably stressed.

The layout of the book has been dictated by two considerations:

(*i*) We want to encourage the student to use the book *continuously*, and to make it the basis of his course. Every item has been fully indexed so that it can be found quickly.

(*ii*) It is intended that the book should be used for *learning* and revising (rather than for general reading).

The text is deliberately fragmented, and has been presented so that it makes an immediate visual impact. Throughout, the verbal explanations and descriptions have been reduced to a minimum, but their function has to some extent been replaced by 562 fully labelled diagrams.

Content

Deliberately, this book is *not* about *contemporary* physics: there is little or no treatment of topics such as fundamental particles, masers and lasers, holography or astrophysics. We feel that these developments contain ideas so difficult that they cannot be taught properly at this level, and certainly they merit no space in a book *of this type.*

Rather we have tried to stress those ideas on which the student can later build his understanding of contemporary physics. We have excluded material of secondary importance to make room for a thorough introduction of the fundamental principles.

The scope of the book is roughly that of the Special Paper of GCE A level, but in many topics there is enough material to satisfy the needs of candidates for Entrance Scholarships. Those parts of the text that go beyond A level are printed in closer type, and are marked with an asterisk (*).

The book assumes an understanding of the ideas developed during an O level course.

Details

(*a*) **Physical quantities** are defined by equations throughout the book, and the appropriate SI unit is indicated for each quantity involved in the definition.

(*b*) Great emphasis has been put on the part played by **molecules**. Intermolecular forces and energies are discussed at greater length than is usual at this level to pave the way for the study of materials science and the solid state, at the university.

(*c*) An attempt has been made to stress the wide application of the concept of **wave motion**. In Section III we develop the properties of waves in general, and then apply these ideas to electromagnetic waves and sound waves in Sections VII and VIII.

(*d*) In Section V on **thermal properties of matter** we have paid due attention to the recent considerable changes in terminology. The physical quantity *amount of substance*, and its unit the *mole*, have both been used freely. We have both here and elsewhere stressed the importance of being consistent about dimensions: for this reason we have avoided (as far as possible) the use of unit quantities such as 1 mol, or 1 kg. It is our experience that they often hide the true nature of a relationship.

Temperature is introduced through the thermodynamic scale, using the triple point of water and (by implication) absolute zero as the fixed points. The *Celsius* temperature is then developed from the thermodynamic scale.

(*e*) Sections IX and X on **electricity** follow *The 1966 Report on the Teaching of Electricity* published by John Murray for the ASE. At the time of writing we were not aware of any English book which used this treatment exclusively, and so we have given more detail here than in some other parts of the book.

(*f*) After some deliberation we decided to include no discussion of thermionic diodes and triodes, but have instead included **semiconductor devices**. The junction diode and transistor are used so much more widely that it is difficult to justify the time spent on obsolescent devices which perform similar functions. Enough of this material has been included to cater for the student who has no other solid state text.

Acknowledgements

(*a*) *C. J. Millar*, Senior Physicist of the King's School, Canterbury, generously put his own teaching notes at our disposal, and read part of the original manuscript.

(*b*) *Mrs. E. G. Hodgson*, *Miss Maureen Bryant* and *Miss Pam Card* cheerfully shared the unenviable task of preparing the typescript.

(*c*) *Dr. J. W. Warren*, of Brunel University, read the typescript. We are very grateful to him for correcting numerous minor errors, and for much constructive suggestion, particularly with regard to molecular behaviour. We are responsible for any errors that remain.

(*d*) *Tim Robinson* provided an unusual combination of artistic skill and understanding of physics to draw the diagrams. We are grateful for the way in which he interpreted our sketches and listened patiently to our suggestions, and we feel he has been most successful in making the diagrams a vital contribution to the book.

No book of this kind can ever claim originality. We have been influenced by many current publications, and particularly by the books listed in the Bibliography.

One of our hopes is that the reader will become sufficiently interested to seek them out for himself.

A companion volume, *Questions on Principles of Physics*, is also available. This is designed to develop and consolidate an understanding of the material presented in this book (rather than to test it) and aims to continue the teaching process.

August, 1970

P.M.W.
Sherborne, Dorset

M.J.H.
Canterbury, Kent

Preface to Second Edition
(*Essential Principles of Physics*)

This edition has been prompted both by the experience of using the first edition and by changing syllabuses. Since the first edition aimed to deal with fundamental principles we have not felt it necessary to prune much of the material it contained. The booklet *Entry to Physics Courses at Tertiary Level*, published by the Royal Society in March 1975, makes it clear that examining boards are by no means unanimous as to what areas of physics *are* essential to a student proceeding to university. The fact is that physics is a huge subject, and only facets of it can reasonably be examined after a sixth-form course. We therefore advise students to examine their own syllabuses carefully, and to omit from their programme any topic which they do not require.

We have rewritten some of the more subtle arguments, have made many small additions, and have completely reorganized certain chapters in order to incorporate a different approach or additional material. As far as possible this has been done in a way which does not affect the general plan of the book.

These are the principal changes:

(*a*) Section I contains more material on non-mechanical dimensions, physical quantities and differential equations.

(*b*) The chapter on **Gravitation** in Section **IV** has been completely rewritten in order to approach the subject through the concepts of field and potential.

(*c*) Section **V** has been reorganized to include an entirely new chapter on **Chance, Disorder and Entropy**. The chapter on kinetic theory now includes a full discussion of the *transport phenomena.*

(*d*) In Section **VII** the chapter on **Diffraction** has been expanded to include *X-ray diffraction* and the *Bragg equation.*

(*e*) Topics such as *semiconduction* and the *Hall effect*, previously discussed in Section **XI**, have been expanded and incorporated into their rightful places in Chapters **48** and **52** respectively. The former now deals with the *band theory* of conduction.

(*f*) Section **XI** has been reorganized into a more logical order. The chapters on **Electronics**, **X-rays and the Atom** and **Radioactivity** have all been rewritten in considerably more detail.

We would like to thank the Parkway Group for drawing the new diagrams in a style which matches so well that of the original book.

January, 1976 M.J.H. *Canterbury, Kent* P.M.W. *Sherborne, Dorset*

Publisher's note:

After a long illness, courageously borne, Paddy Whelan died in May 1980 at the age of 42.

A Note to the Student

Learning Physics

Acquiring proficiency in physics involves each of the following:

(*a*) *Complete familiarity* with the small number of *fundamental concepts* that constitute the real basis of the subject.

(*b*) *Access to detailed textbooks* which provide the thorough explanations necessary for a proper understanding, and which describe experimental work in some detail.

(*c*) *Ability to apply the basic laws and principles to both familiar and unfamiliar problems*, an ability which is developed by constant practice.

(*d*) A laboratory course of *practical work*.

The purpose of this book is to satisfy the first requirement. You are strongly recommended to make your own *summary of the book* in about a hundred pages, which can easily be done by emphasizing symbols and labelled diagrams, and economizing on words. This summary will help you to become familiar with the book, will demonstrate the *logical structure* of the subject, and will serve for last minute *revision*.

Revision

Well before the exam (say three months at least) draw up a *written timetable*, and keep to it.

(1) Set aside for each day a definite length of time (say *not more* than 30 minutes).
(2) Find a place where you will not be distracted.
(3) Have a pencil and paper to hand.
(4) Choose a particular topic to revise, and then devote *all* your attention to that topic.
(5) Pay particular attention to diagrams and their labels.
(6) Repeat to yourself the full meaning of a graph (e.g. emphasize which is the dependent, and which the independent variable, and whether some other quantity is by implication held constant).

(7) When you think that a topic is mastered, test yourself by *re-writing the main ideas*, as though explaining them to someone who has no previous knowledge of that topic.

Revision should be very much an active (not a passive) *occupation.*

Examination Technique

An examination is an exercise in efficient communication, and therefore also requires planning.

(1) Read the *rubric* at the top of the paper very carefully to see exactly what you have to do.
(2) See how much *time* can be spent on each question, and do not exceed it. If you run short of time, make your answer shorter, but make sure it includes the fundamentals.
(3) *Read the question* that you are answering very carefully to see *exactly* what is required. Do not spend time giving details that are not asked for, and do not repeat yourself. Work out the order in which the different parts of the question are best answered: it is usually better to tackle a problem before a lengthy description.
(4) Illustrate your answer by very large *clear labelled diagrams and graphs* wherever you can. Far more information can be conveyed by a quick sketch than in sentences of description. If you are asked to do a scale drawing, draw it accurately and as large as the paper will allow. State your scale.
(5) The following procedure is recommended in *numerical problems*:
(*i*) Read the question carefully to see *what has to be calculated.*
(*ii*) Summarize the information, making the *units* explicitly clear. This can frequently be done most clearly and concisely with a diagram.
(*iii*) State clearly the *principle* on which your method depends. Often an equation is enough.
(*iv*) Do the numerical computation in such a way that the examiner can easily find mistakes. Use units as recommended on p. 10.
(*v*) State your answer with a complete sentence, and make sure that it is of a *sensible order of magnitude.* If you think that it is not, then say so.
(6) When describing experiments, list events in the order in which they happen. Illustrate by symbols the measurements to be taken, and demonstrate how the result is to be calculated.
(7) *Good English and legible handwriting are essential*: without them you cannot communicate.

However good your examination technique, there is no substitute for knowing your subject.

Contents

I General Introduction

1 General Introduction

1.1 Physical Interaction and the Nature of Physics

The Structure of Physics

Natural Science gives descriptions of the natural world, but in the last resort it can only describe *what* happens rather than why it happens.

Some great themes persist throughout physics, and give the subject its unity. These include (for example):

(*a*) the concept of *energy*,
(*b*) the *conservation laws*,
(*c*) the concept of *wave motion*,
(*d*) the concept of *field*;

and it is possible to study the subject by grouping topics under those headings. However, it is more convenient (at pre-University level) to subdivide physics according to subject matter (Mechanics, Light, Electricity, etc.) and to coordinate during a degree course the ideas thus presented.

Some topics are too difficult to study at this elementary level. For example:

(*a*) **Mechanics**

(*i*) *Relativistic quantum mechanics* deals with the mechanics of all bodies.

(*ii*) *Quantum mechanics* treats the special cases where bodies have speeds that are very much less than that of light, but caters for all masses.

(*iii*) *Relativistic mechanics* is applied in the special cases where bodies have masses greater than (say) 10^{-27} kg, but caters for all speeds.

(*iv*) *Newtonian mechanics* (or *Classical mechanics*) is the only form discussed in detail in this book, and can be used in the double special case: (1) *relatively* small speeds, and (2) *relatively* large bodies. It is nevertheless applicable (as an excellent approximation) to all macroscopic bodies moving at these speeds. This is illustrated by fig. 1.1*a* and *b*.

(*b*) **Statistical mechanics** (which can be *classical* or *quantum*) applies the techniques of statistics to very large numbers of interacting particles in an attempt to explain the (macroscopic) properties of matter in bulk. Classical statistical mechanics then links *kinetic theory* and *thermodynamics* (chapter **26**).

(*c*) **Electromagnetism.** *Maxwell's equations* describe the behaviour of electromagnetic fields, and at the end of the nineteenth century provided a wonderful link between the theory of *light* and that of *electricity and magnetism.* The linking process is beyond the scope of this book.

The beginning student must always be prepared for some change of emphasis when he views the subject later with a more experienced eye.

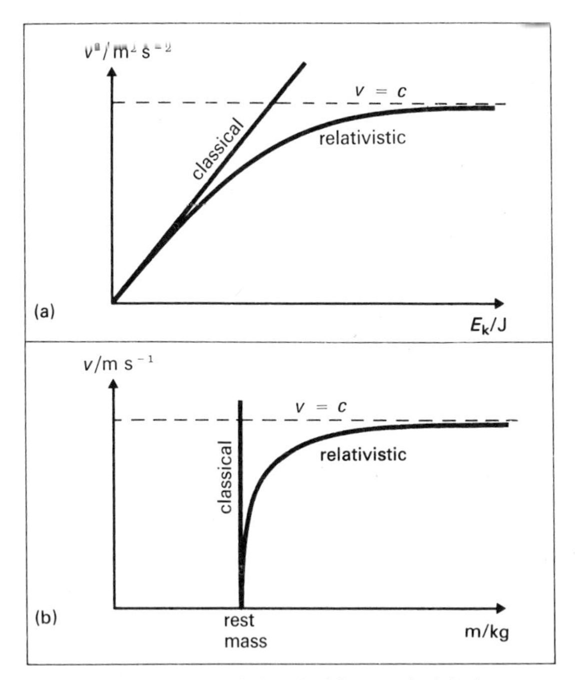

Fig. 1.1. Departure of classical from relativistic mechanics.

Interactions

An *interaction* in physics is a process by which one particle is enabled to influence (exert forces on) another. At present it seems that there are four interactions, which are summarized in this table:

*There is one further effect which could be called an interaction. According to the **Pauli exclusion principle** no two electrons in an atom may be described by the same set of quantum numbers, which is equivalent to saying that the electrons can exert some sort of repulsive force on each other (quite apart from the *Coulomb* interaction).

Fields

In this book we are mainly concerned with the electromagnetic and gravitational interactions. We visualize the following to happen:

(1) Body A establishes round itself a region in which it has the ability (or potential) to exert forces. This region is the **field**.

(2) If body B enters the field, then it experiences a *force* (as does A which is now in the field set up by B).

One can imagine the field to be the *cause*, and the force to be the *effect*.

In the past there has been a tendency to emphasize the field aspect to develop the theory of the electromagnetic interaction, and the force aspect for gravitation. In this book field ideas are used for both interactions—see chapters **20**, **43** and **52**.

Physical Laws

The aim of physics is to find some understanding of the physical world through the formulation of general principles. We look for statements that summarize ideas of importance with great precision and yet

Name	*Relative effectiveness*	*Comment*	*Example of interaction*
Strong or nuclear	1	large short-range forces between nuclear particles	the holding together of nucleons in a nucleus
Electro-magnetic	$\sim 10^{-2}$	account for the majority of the forces we meet in everyday life	repulsion between two solid bodies whose surfaces are brought into contact
Weak	$\sim 10^{-14}$	little understood at present	β-decay of nucleus (p. 475)
Gravitational	$\sim 10^{-40}$	only become significant when very large masses are involved (and then the force is dominant)	pull of Earth on Moon and Moon on Earth

simplicity: these statements should relate observations from widely different experiments, and should be capable of general application. *Such a summary is frequently called a* **law**. It is often in the form of an equation connecting symbols that represent operations which we carry out in the laboratory. We would like the equation to relate cause and effect, but laws are *not necessarily* obeyed. A law may represent the relation between cause and *probable* effect, particularly when quantum effects are significant.

In time it becomes apparent that some laws (such as *Newton's Law of Gravitation*) are inadequate: they do not represent *exactly* what happens, and may have to be modified before they can be applied in certain situations.

Some other Terms used in Physics

Unlike other terms in physics, these are *not* always used with exactly the same meaning. This list is a rough guide as to what is usually intended.

(*a*) A **fact** is an experimental observation which can be reproduced by repeating an experiment.

(*b*) **Law** (see above). Examples include *Snell's Law* (p. 95) (to which there are exceptions), *Newton's Second Law* (p. 34) (which in the form $\boldsymbol{F} = m\boldsymbol{a}$ only applies exactly in Classical mechanics), and the *Law of Conservation of Momentum*, to which there are no known exceptions.

(*c*) A **principle** is a general belief, and may often be called a law. *Heisenberg's Indeterminacy* or *Uncertainty Principle* is an example. The fact that the momentum of an isolated system remains constant is sometimes called a law, and sometimes a principle. The two terms are quite often interchanged.

(*d*) A **rule** is the name given to a series of working instructions, such as *Routh's Rules* (p. 56), and *Kirchhoff's Rules* (p. 362).

(*e*) A **theory** is a framework of ideas, based on experiment, which tries to relate previously disconnected observations by a single explanation. Examples include *Einstein's Theory of Relativity*, and the *Kinetic Theory of Matter* (p. 126). In the same way as a law, a theory may well be modified in the light of experience (e.g. *Newton's Corpuscular Theory* (p. 102)).

(*f*) An **hypothesis** is a more tentative idea proposed to explain observed phenomena, which has not yet been tested by a crucial experiment. If it survives the test it may then be called a theory.

Example. Avogadro's Law (p. 181) was for a long time referred to as *Avogadro's Hypothesis*.

Physical Models

Progress in physical science occurs in two ways:

(1) by experiment (the characteristic feature of science), and
(2) by appeal to models.

A **model** is a concept (rather than a material structure) proposed by a scientist as an aid to visualizing the possible processes within a particular physical situation. Such a model may give a highly simplified (or even misleading) picture of reality. If he then applies established principles to the hypothetical model, he may be led to make predictions which themselves lead to further experimental tests of that model. (See also the development of the history of gravitation (p. 151),)

Models have proved themselves especially useful in

(*a*) visualizing atomic structure (p. 465),
(*b*) developing the kinetic theory of matter (p. 126), and
(*c*) explaining the propagation of electromagnetic waves (p. 260).

1.2 Physical Quantities

We can classify physical quantities as

(*a*) **basic**, **primary** or **fundamental quantities**, which by agreement are treated as independent, and
(*b*) **derived** or **secondary quantities**.

(a) Basic Quantities

For convenience, seven quantities have been selected:

mass	*electric current*	*amount of substance*
length	*temperature*	
time	*luminous intensity*	

It is to be emphasized that these quantities are selected for *convenience*, and not through any necessity. Thus in mechanics one could equally well have chosen length, *force* and time, in which case mass would have been a derived quantity. Once these quantities are chosen, we can

(1) select an *arbitrary* unit, and then
(2) measure by direct comparison with that unit.

This is essentially a *counting* process, in which we follow a specified procedure.

(b) Derived Quantities

Any other physical quantity can be derived from the basic quantities by following the procedure implied in the defining equation (see below).

Defining Equations

A physical quantity is defined by a series of operations which are best summarized by an equation. The equation will relate the primary quantities by that series of operations, each of which involves only multiplication, division, integration or differentiation.

Examples

$$v = \frac{\mathrm{d}s}{\mathrm{d}t} \qquad \text{(p. 28)}$$

$$\Phi = \int \boldsymbol{B}\cdot\mathrm{d}\boldsymbol{A} \qquad \text{(p. 374)}$$

$$\frac{\mathrm{d}Q}{\mathrm{d}t} = -\lambda A \frac{\mathrm{d}\theta}{\mathrm{d}x} \qquad \text{(p. 216)}$$

These equations are more useful than verbal descriptions, since they can be applied immediately.

Notes. (*a*) Equations can only serve as complete definitions when each symbol used in the equation is explicitly defined.

(*b*) An equation involving only symbols for physical quantities does not specify or require any particular system of units.

(*c*) In this book each defining equation is repeated, and a recommended set of consistent units is indicated in *square* brackets after each term.

Example. If a normal force $\boldsymbol{F}$ acts on an area $\boldsymbol{A}$, then the average pressure p is given by

$$p = F/A$$

$$p[\mathrm{Pa}] = \frac{F[\mathrm{N}]}{A[\mathrm{m}^2]}$$

Strictly the units have no place in the equation.

Dimensions

The **dimensions** of a physical quantity indicate explicitly how the quantity is related, through its defining equation, to the basic quantities. If we write $[v] = [\mathrm{L\,T^{-1}}]$ we indicate that speed is measured by dividing a length by a time.

Some quantities (such as refractive index) are described by a number which is independent of the units of measurement chosen for the primary quantities: such quantities are said to be **dimensionless**.

Some quantities have units which are part dimensional, and part dimensionless. The name or abbreviation of the dimensionless unit is not always explicitly stated.

Example. Angular frequency (or pulsatance) ω is usually expressed in *radians per second* (rad $\mathrm{s^{-1}}$), but this is often written *per second* ($\mathrm{s^{-1}}$).

The uses of dimensions are discussed more fully in paragraph **1.10** (p. 19).

Conventions for Printing and Writing

This book adopts the convention in which

(*a*) symbols for physical quantities are in *italics* (e.g. mass m),

(*b*) symbols for units are in roman (upright) type (e.g. kg).

The value of any physical quantity can be expressed by

(physical quantity) = (numerical value)(unit)

This means that ordinary algebraic rules can be applied to

(*i*) physical quantities,
(*ii*) numbers, *and also*
(*iii*) units (see p. 10).

The chosen symbol represents both a *number* and a *unit*.

Examples
'$V = 5.0$ volts' is correct.
'Let the p.d. be V volts' is wrong.
'Let the p.d. be V' is correct.
We might later find

$$V = (5\times10^{-3})\,\mathrm{V} = 5\,\mathrm{mV}.$$

V is *not* just a number.

A symbol divided by a unit, such as $\boldsymbol{F}$/N, may represent a *number*: it is therefore the correct heading for tables and for labelling the axes of graphs (p. 25).

Example. This table summarizes the results of an experiment in which a force $\boldsymbol{F}$ stretched a spring through a displacement $\Delta\boldsymbol{x}$.

$\boldsymbol{F}$/N	$\Delta\boldsymbol{x}$/mm
0.0	0.0
9.8	25
19.6	49
	etc.

1.3 The Nature and Status of Physical Equations

Since physics is primarily an exact (quantitative) discipline, much of our work is concerned with equations. We can distinguish the following types.

(*a*) ***Defining equations.*** These were identified in the previous section. It helps if we all use technical terms with the same meaning, and so to reduce confusion *defining equations must be learned.*

(*b*) ***Laws.*** Equations representing laws were discussed in paragraph **1.1**. These too *must be learned*, since they often represent the result of experiment, and so cannot be simply deduced.

(*c*) ***Principles and useful results.*** There are many equations in physics that summarize the application of laws and defining equations to common situations. The following are well-known examples:

$$p + \tfrac{1}{2}\rho v^2 + h\rho g = \text{constant}$$

$$p = \tfrac{1}{3}\rho\overline{c^2}$$

$$V = \left(\frac{1}{4\pi\varepsilon_0}\right)\frac{Q}{r}$$

$$\mathscr{E} = I(R + r)$$

Since these equations can be derived from laws and definitions, it is *not* strictly *necessary* to learn any of them, but it is *useful* to learn them, as this will save a lot of time.

(*d*) ***Special results.*** You will often read in textbooks an equation which is the end-product of an analysis of a special situation which is not often encountered. *Such an equation should not be learned.* Rather, you should master the principles used to derive it.

Summary

When you meet a particular equation for the first time, decide whether it is

(*a*) a definition,
(*b*) a law,
(*c*) an important *idea* which you will want to use repeatedly, or
(*d*) a specialized result.

The equations in boxes in this book fall into the first three categories: they alone should be learned.

1.4 The Order in which Physical Quantities are Defined

The logical structure of physics, in some subject areas particularly, is based on a distinct hierarchy, and it is important to appreciate the order in which quantities have been defined. For example if we choose to define magnetic flux Φ by the scalar product $\Phi = \boldsymbol{B}\cdot\boldsymbol{A}$, then we cannot later define magnetic flux density $\boldsymbol{B}$ by the quotient $B = \Phi/A$. The latter equation can be regarded as a satisfactory definition of $\boldsymbol{B}$ only if both Φ and $\boldsymbol{A}$ have already been defined by some other means. (There exists an alternative approach to electromagnetism in which one defines Φ from $\mathscr{E} = -\mathrm{d}\Phi/\mathrm{d}t$, and then $B = \Phi/A$ becomes acceptable as the formal definition of $\boldsymbol{B}$.)

The schemes which follow show a logical order of defining physical quantities in five branches of physics. Every equation shown is a *definition*, and no attempt has been made to relate quantities by other equations. For simplicity the definitions are given in an elementary form. The boxes of the fundamental quantities (the logical starting points) are shown in grey.

Linear Dynamics

(See scheme 1 opposite.)

Notes

(*a*) No arrows are shown for mass ratio. The equation $\boldsymbol{F} = m\boldsymbol{a}$ defines the measure of *force*, and is also used to measure *inertial mass* through $\boldsymbol{F} = m_1\boldsymbol{a}_1 = m_2\boldsymbol{a}_2$.

(*b*) The definitions of rotational dynamics are not included—they follow a similar scheme.

Thermal Properties of Matter

(See scheme 2 opposite.)

Electrostatics

(See scheme 3 on page 8.)

Notes

(*a*) In this scheme quantities are introduced from mechanics only where they make a significant contribution to the understanding of the whole.

(*b*) Note carefully that electric charge is not a fundamental quantity, since it is derived from electric current through $Q = \int I\,\mathrm{d}t$.

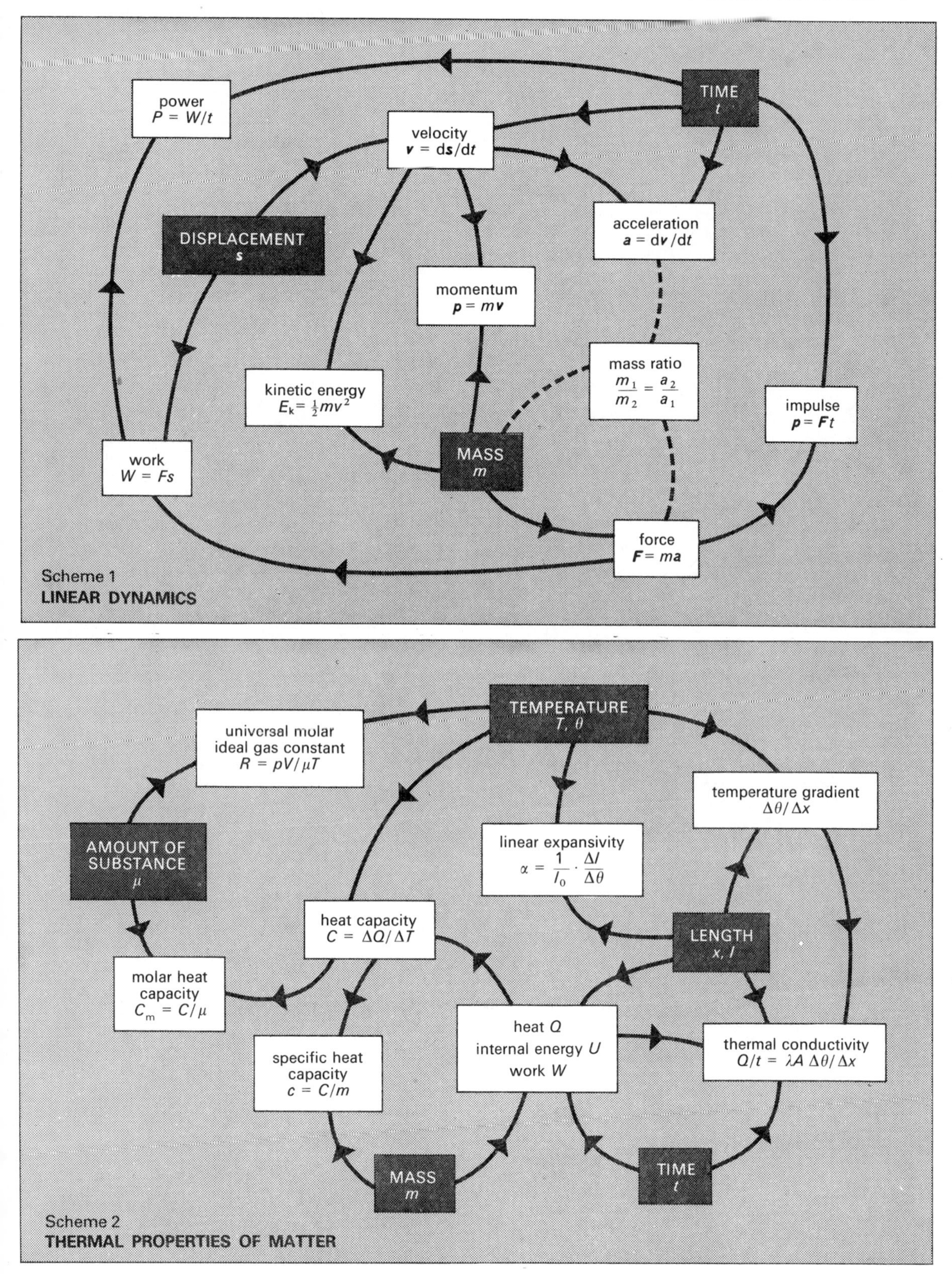

Scheme 1
LINEAR DYNAMICS

Scheme 2
THERMAL PROPERTIES OF MATTER

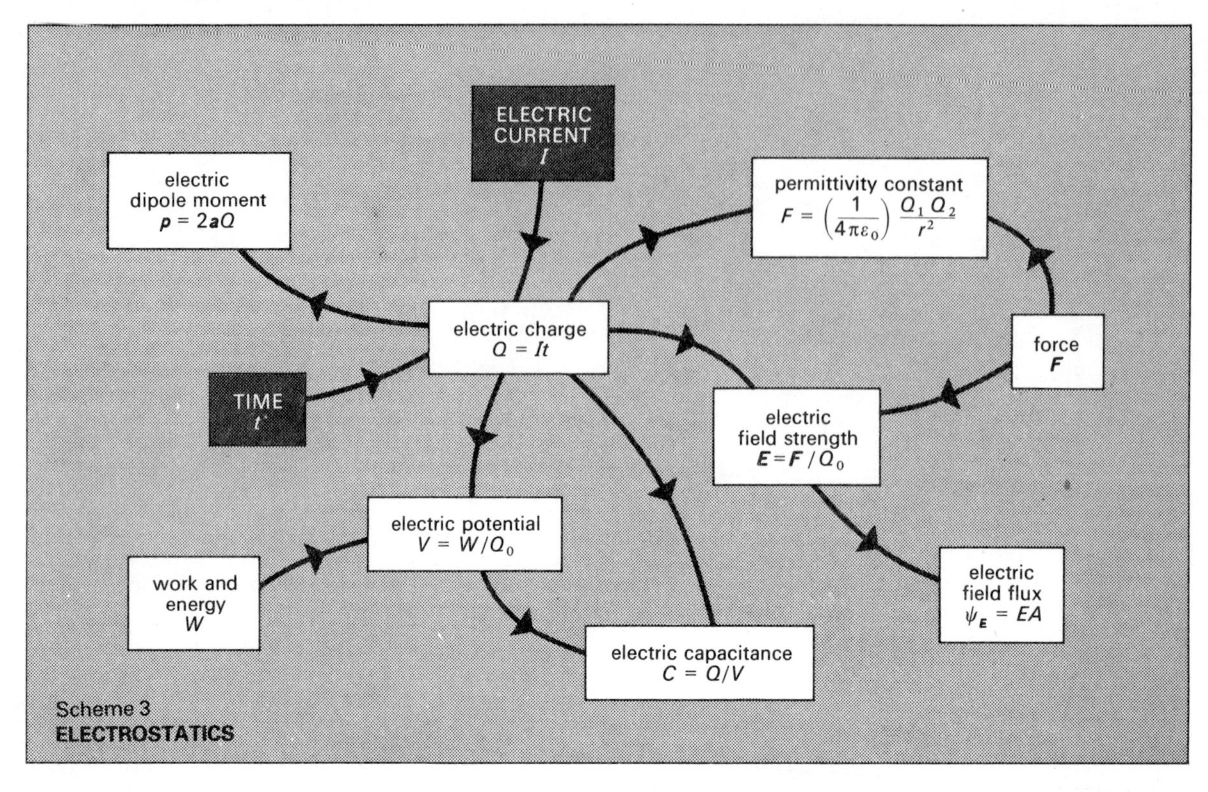

Scheme 3
ELECTROSTATICS

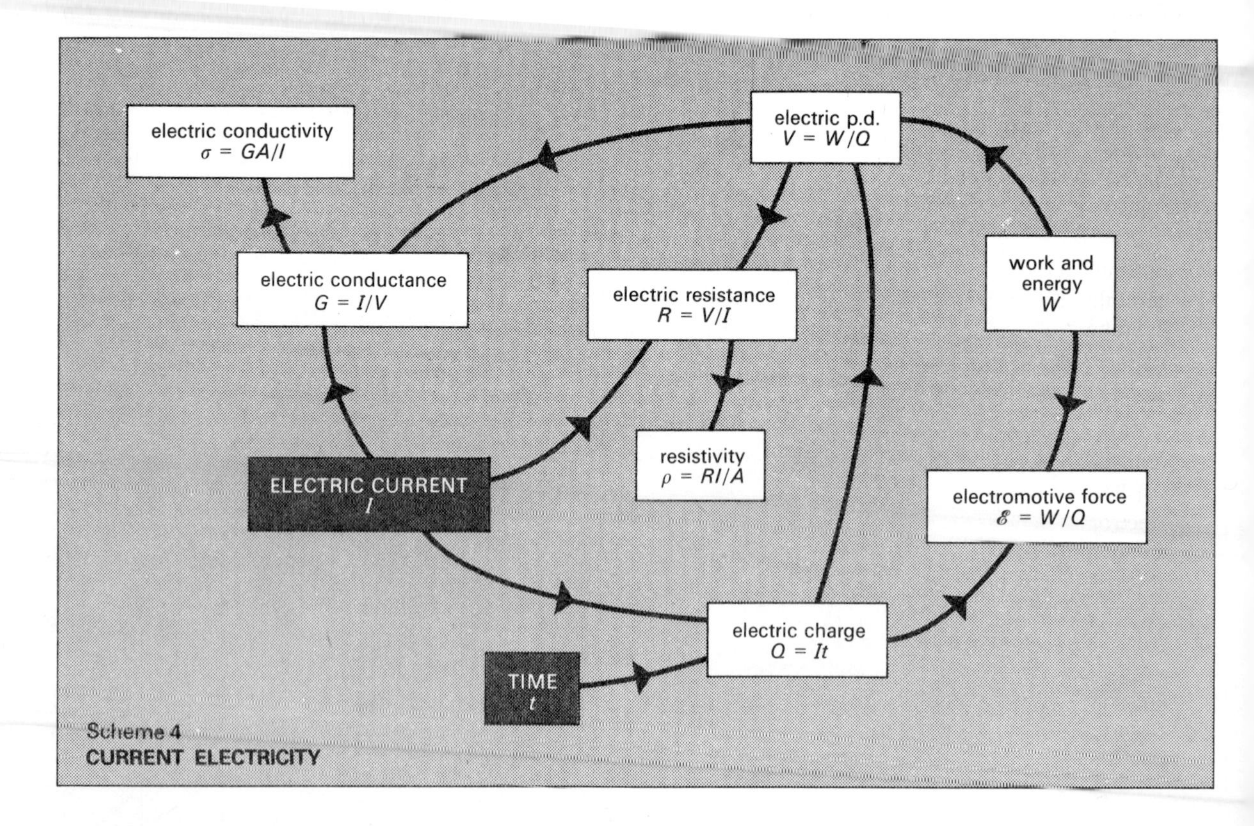

Scheme 4
CURRENT ELECTRICITY

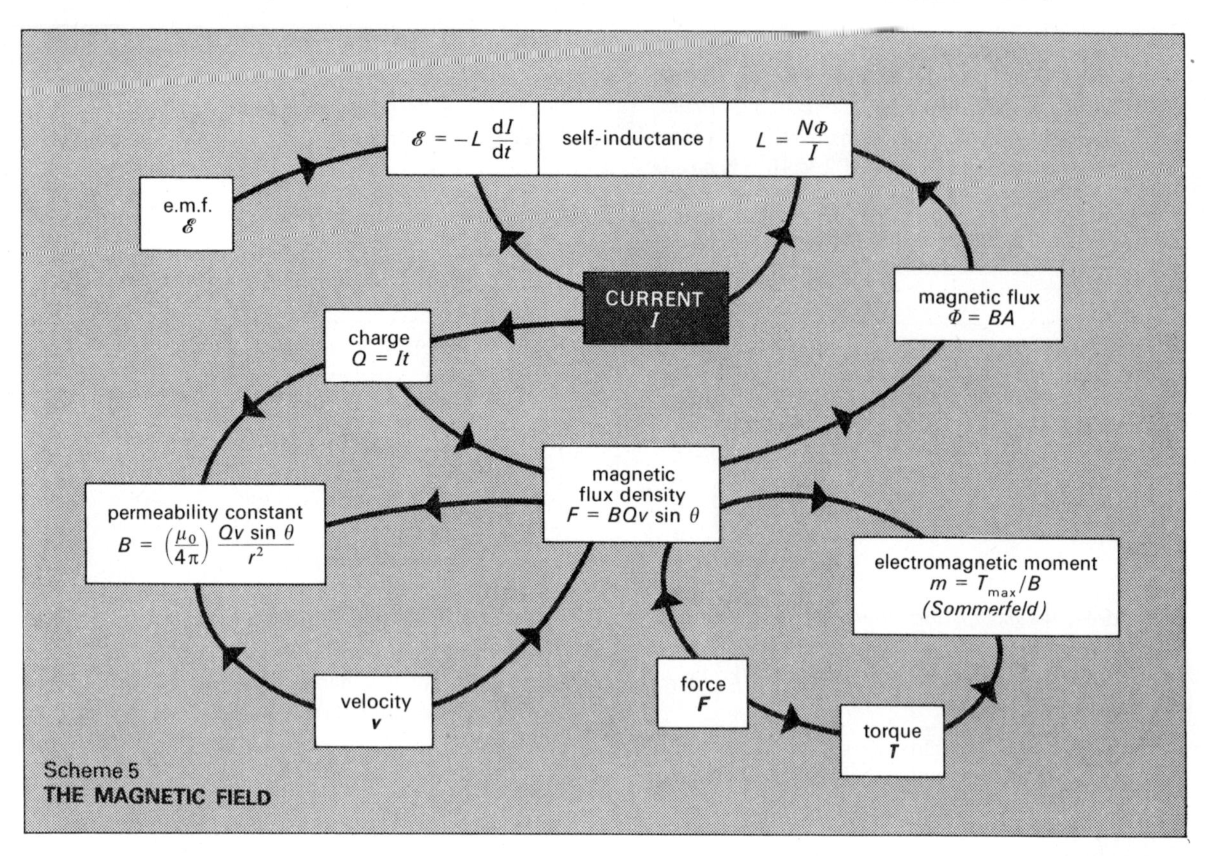

Scheme 5
THE MAGNETIC FIELD

Current Electricity

Notes on Scheme 4

(*a*) Mechanical quantities are introduced as in the previous scheme.

(*b*) The defining equations for $\mathscr{E}$ and V are symbolically identical, and it is important to identify carefully the quantity represented by W in each definition.

The Magnetic Field

Notes on Scheme 5

(*a*) There are several alternative approaches to electromagnetism. The one used here is that recommended by the *Association for Science Education*.

(*b*) The equations $F = BIl \sin\theta$ and

$$\delta B = \left(\frac{\mu_0}{4\pi}\right)\frac{I\delta l \sin\theta}{r^2}$$

are exactly equivalent to the two used above to define $\boldsymbol{B}$ and μ_0 respectively.

1.5 Units and the Realization of Standards

The Basic SI Units

SI Units are those in an **I**nternational **S**ystem of Units adopted in 1960 by the *Conférence Générale des Poids et Mesures*. The system

(*a*) is based on the metre, kilogram, second, ampere, kelvin, mole, and candela, and

(*b*) uses other units which can be obtained from the basic units by multiplication and division (i.e. without including *any* numerical factors).

Metre: *the metre is the length equal to* 1 650 763.73 *wavelengths in a vacuum of a specified radiation from an atom of krypton*-86.

Kilogram: *the kilogram is the mass of the international platinum-iridium prototype.*

Second: *the second is the duration of* 9 192 631 770 *periods of a specified radiation from an atom of caesium*-133.

Ampere: *the ampere is that constant current which, if maintained in two straight parallel conductors of infinite length, of negligible circular cross-section, and placed* 1 m *apart in a vacuum, causes each to exert a force of* 2×10^{-7} N *on one metre length of the other* (see p. 386).

Kelvin: *the kelvin is the fraction* $\frac{1}{273.16}$ *of the thermodynamic temperature of the triple point of water.*

To satisfy (*iii*), standards are frequently based on the properties of atoms: for this reason the present kilogram may be replaced at some future date by a property of the proton.

Standard Prefixes

The following prefixes attached to a unit abbreviation have the meanings shown:

Fraction	*Prefix*	*Symbol*	*Multiple*	*Prefix*	*Symbol*
10^{-3}	milli	m	10^{3}	kilo	k
10^{-6}	micro	μ	10^{6}	mega	M
10^{-9}	nano	n	10^{9}	giga	G
10^{-12}	pico	p	10^{12}	tera	T
10^{-15}	femto	f	10^{15}	peta	P
10^{-18}	atto	a	10^{18}	exa	E

Mole: *the mole is the amount of substance of a system which contains as many elementary units as there are carbon atoms in* 12×10^{-3} kg *of carbon*-12.

Thus a mole contains about $6.022\,169 \times 10^{23}$ specified elementary units, such as atoms, molecules, ions, electrons, etc.

Examples

(*i*) A mole of electrons has a charge of $(6.0 \times 10^{23}) \times (-1.6 \times 10^{-19})$ C, the size of which is called the *Faraday constant.*

(*ii*) A mole of protons has a mass of $(6.0 \times 10^{23}) \times (1.7 \times 10^{-27})$ kg.

(*iii*) A mole of any ideal gas at s.t.p. has a volume of 22.4×10^{-3} m^3.

One would refer respectively to the molar charge, molar mass or molar volume of the substance.

The term **molar** means 'divided by amount of substance'.

(*Candela* is not used in this book: it is a unit of luminous intensity).

Why these Units have been Chosen

Scientists continually exchange information, so their units must be understood with minimum explanation, and unambiguously. These ends are achieved by

(*a*) a small number of units which are
(*b*) internationally accepted, because they are

(*i*) easily compared
(*ii*) easily reproduced, and
(*iii*) invariant in time.

SI Units with Special Names

Some derived SI units are relatively complex, and are conveniently given a name whose abbreviation is simpler than the unit expressed in terms of the basic units.

Examples

The name *farad* (F) is given to m^{-2} kg^{-1} s^{4} A^{2} (p. 330).

The name *watt* (W) is given to m^{2} kg s^{-3} (p. 47).

The following units are defined at appropriate places in the text: hertz (Hz), newton (N), joule (J), pascal (Pa), coulomb (C), volt (V), ohm (Ω), siemens (S), weber (Wb), henry (H), tesla (T), becquerel, (Bq) and gray (Gy). They are also listed in appendix A (p. 480).

Units in Calculations

This book uses a **coherent** system of units, which means that any physical quantity has strictly only one correct unit. In using any equation it is possible to adopt one of two methods:

(1) one can replace each symbol by the numerical measure for each quantity *expressed in the basic units*, and *assume* the unit for the calculated quantity (which will be unambiguous), or

(2) one can replace each symbol by both its *number* and *unit abbreviation*: since the unit abbreviation can be treated algebraically, one can *calculate* the *unit* of the required physical quantity, as well as the number.

This method has two advantages.

(*i*) It indicates at once (by a wrong unit) if an incorrect equation has been used.

(*ii*) It causes the student to revise constantly the definitions of derived units, which leads to invaluable familiarity.

Method (2) *is strongly recommended.*

Laboratory measurements are frequently expressed in a non-SI unit (such as °C). This does not matter provided one converts to the SI unit before replacing a symbol in an equation by its number and unit abbreviation.

Example. Calculate the electrical energy converted to thermal energy when a current of 2.0 mA passes through a resistor of resistance 3.0 kΩ for a time of 10 minutes.

$$W = I^2Rt \qquad \text{(p. 357)}$$
$$= (2.0 \times 10^{-3}\ \text{A})^2 \times (3.0 \times 10^3\ \Omega) \times (600\ \text{s})$$
$$= 7.2\ \text{A}^2 \times \frac{\text{V}}{\text{A}} \times \text{s}$$
$$= 7.2 \frac{\text{C}}{\text{s}} \times \frac{\text{J}}{\text{C}} \times \text{s}$$
$$= 7.2\ \text{J}$$

7.2 joules of energy are converted.

The Realization of Standards

Metre: Realization involves marking off a length containing the stated number of wavelengths. Accurate reproduction of the krypton-86 orange line at 606 nm requires carefully specified conditions such as type of lamp, suitable temperature of operation to reduce *Doppler* broadening, current density etc. Measurements of particular distances can be made using interferometers or lasers, which determine the number of waves covering the distance.

Kilogram: Copies of the prototype are sent periodically to *Sèvres* for comparison and these can be adjusted easily after carefully weighing on appropriate balances. Masses of the order of 1 kg can be compared to 1 part in 10^9.

Second: The hyperfine transition of the caesium-133 atom is observed in an atomic beam apparatus. The magnetic deflection of the atoms is reversed when the frequency applied from a quartz oscillator equals that of the spectral line. Such equipment is required only rarely as many countries broadcast time signals which are accurate to 1 part in 10^{11}.

Ampere: The arrangement stated in the definition is not a practical one though it is consistent with electromagnetic theory. Current balances or dynamometers can be used to measure the forces exerted by currents very accurately. Good results can also be achieved using standard resistors and *Weston* cells, although such methods are not absolute. Nuclear magnetic resonance can also be utilized; one measures the frequency of precession of protons in a magnetic field produced by a current-carrying coil of suitable shape.

Kelvin: The International Practical Temperature Scale, I.P.T.S.-68, is a practical scale which is easily and accurately reproducible, and which gives thermodynamic temperatures. The defining fixed points are established by realizing specified equilibrium states between phases of pure substances (p. 164).

Mole: Although defined in terms of a number of entities, the mole is usually realized by weighing rather than counting. Volume comparison can be used for gases, and charge measurements can be used in electrolytic reactions.

1.6 Measurement

Measurement is that process by which we compare a physical quantity with the unit chosen to express that quantity. The statement 'The mass of the body is 2.0 kg' means that the body has a mass 2.0 times the mass of the standard kilogram: we arrive at the figure 2.0 by a specified procedure (detailed, in this instance, on p. 36).

To measure a particular physical quantity, we look for some physical phenomenon that depends on the quantity. Thus an electric current establishes a magnetic field; a change of temperature may cause a change of electrical resistance. We then choose an instrument whose design is based on a convenient phenomenon which

(*a*) can be applied in the range we are using, and
(*b*) gives the sensitivity that we require.

In this paragraph we discuss the measurement of some fundamental quantities.

(*a*) Measurement of Length

We choose an instrument appropriate to the order of magnitude, the nature of the length (e.g. it might be an internal tube diameter), and the sensitivity required.

(1) *Rulers* are used for lengths exceeding about 50 mm, and are probably accurate to about 0.5 mm.

(2) *Micrometer screw gauges* are frequently used for small lengths. If the screw has a pitch of 0.50 mm, and the sleeve has 50 divisions, each division corresponds to 0.01 mm (fig. 1.2).

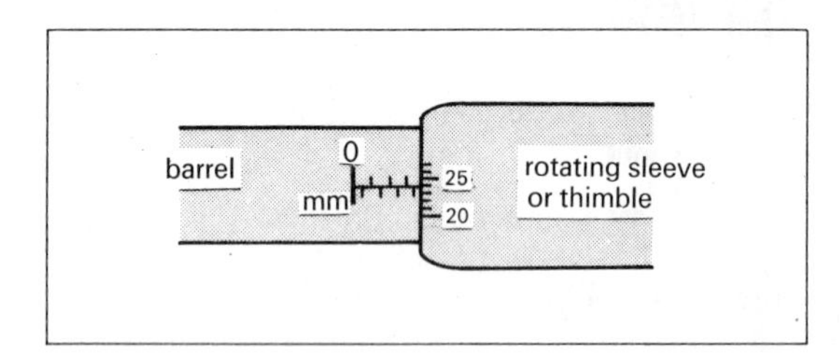

Fig. 1.2. A micrometer screw gauge indicating 3.74 mm.

(3) *A vernier scale* is a device which eliminates guesswork when we interpolate between scale divisions. Vernier scales are frequently used on calipers, barometers, travelling microscopes and spectrometers. Fig. 1.3 shows such a scale on a Fortin barometer.

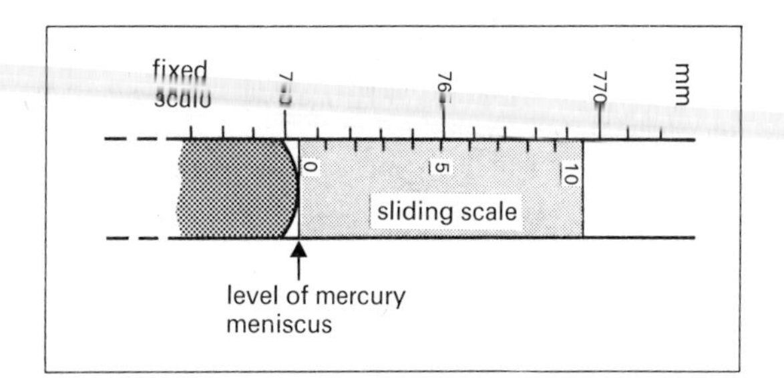

Fig. 1.3. A vernier scale indicating 760.4 mm (shown on its side for convenience).

(*b*) Measurement of Mass

Bodies possess both *inertial* mass and *gravitational* mass: experiment shows that these masses are equivalent to 1 part in 10^{12} (p. 84).

(1) *Inertial mass* can be measured by methods which do not require the existence of a gravitational field, such as

(*i*) the **inertial balance** of fig. 7.3 (p. 64),

(*ii*) trolley experiments, involving the mutual interaction of two masses (p. 36).

In each method one compares the mass to be measured with that of (a replica of) the standard kilogram.

(2) *Gravitational mass* is deduced from a measurement of the *weight* of a body (p. 38).

It is highly inconvenient to measure inertial mass *directly*, but relatively easy to measure or compare weights. The method adopted to find inertial mass is (usually)

(1) to measure (perhaps by direct comparison) a body's weight, then

(2) to calculate its gravitational mass, and so

(3) deduce its inertial mass, which is equivalent.

(*c*) Measurement of Time

Time is measured by a *clock*, a device which has some repetitive or periodic property, and which has been calibrated by comparison with a standard. The second used to be defined as 1/31 556 925.974 7 of the duration of the solar year 1900, but this time interval is *not reproducible*, and has therefore been replaced by the definition on p. 9.

One chooses a clock consistent with the requirements of the measurement to be made. Clocks have been based on the following properties:

(*i*) radioactive decay processes (p. 479),

(*ii*) the rotation of the Earth on its axis,

(*iii*) the oscillations of pendula and balance wheels (as in the wrist watch),

(*iv*) the oscillation of electrons in a circuit which carries an a.c. (as in the ticker-tape timer),

(*v*) the oscillation of a nitrogen atom within an ammonia molecule (an **atomic clock**).

The *quality* of a clock is determined by how well it keeps time with other clocks, and with the standard. An atomic clock is very little affected by any external influence (such as change of temperature), and is therefore very reliable.

(*d*) Measurement of Weight

Measurement of weight implies the comparison of a body's weight with the standard force unit (the newton).

(*i*) A **spring balance** is calibrated by known *forces*, and can then be graduated in *newtons*. Such a balance will then record the weight of a body to which it is attached provided the body and balance are in relative equilibrium. The balance registers the apparent weight of the body (not corrected for the Earth's rotation) *wherever it is used*.

(*ii*) A **beam balance** or a **lever balance** compares the pull of the Earth $\boldsymbol{G}_u$ on the **u**nknown mass m_u

with that G_k on **known** masses m_k (colloquially called 'weights'). Both masses are at the same place, so for a beam balance whose arms are of equal length

$$G_u = G_k$$
$$m_u g = m_k g$$
$$m_u = m_k$$

These balances compare unknown masses with standard masses, by comparing their weights. If the value of g is known *at that place*, then the weight can be calculated from $G = mg$.

(The **ballistic balance** (p. 52) compares inertial masses, but utilizes the gravitational field for the measurement of horizontal velocity.)

1.7 Orders of Magnitude

Developing a Feeling for Orders of Magnitude

In physics the phrase 'correct to an order of magnitude' means that the value quoted is reliable to within a factor of ten or so, and for many purposes the information may be sufficiently precise in this form. It is most important for a physicist to acquire a working knowledge of the size of typical quantities. This enables him

(*a*) to judge the plausibility of any given quantity, and

(*b*) to estimate the possible sizes of further quantities.

Even the best physicists make elementary errors in their working, but only a bad physicist is happy to present a solution whose size is patently absurd. It is vital to develop an awareness of reliable information sources from which orders of magnitude can be obtained.

Orders-of-Magnitude Tables

The tables opposite and overleaf show the number of particles to be found in various bodies, and, separately, the ranges of distances, masses and times found in the Universe. In both tables the numbers are presented on a logarithmic scale.

Parent body	*Number of particles*
Universe	10^{80}
	10^{70}
Sun	10^{60}
Earth	10^{50}
Earth's atmosphere	10^{40}
	10^{30}
human breath	10^{20}
	10^{10}
elementary particle	10^{0}

Energy

One of the principal interests of physicists is energy, and its conversion from one form to another. One reason for the concept of energy holding such an important place is that we can often use it to decide whether a particular physical change is likely to occur. A physical system achieves equilibrium by making its p.e. a minimum. Thus a liquid surface contracts, an electron in an atom moves as close as possible to the nucleus, and a radioactive nucleus disintegrates, each making an attempt to reduce the p.e. of the system.

In the SI *all* energies are expressed in *joules*, and this has the enormous advantage of revealing the true comparative values of energies previously expressed relative to other quantities (such as the electronvolt). The table below should be studied closely, since many useful conclusions may be drawn from it. For example one can predict that photons of violet light will cause photoelectric emission from caesium, but not from tungsten; or that air is ionized by photons from the middle ultra-violet, but not by those of lower frequency.

Description of energy W	*W/J*
binding energy of Earth-Sun system	10^{33}
translational k.e. of Moon	10^{28}
energy radiated by the Sun in 1 second	10^{26}
energy received per day on Earth from Sun	10^{22}
estimated Human energy requirement per year (1950)	10^{20}
energy associated with a strong earthquake	10^{20}
energy released by annihilation of 1 kg of matter	9×10^{16}
energy released by fission of 1 mol of ^{235}U	10^{13}
	1 TJ
energy dissipated by a lightning discharge	10^{10}
	1 GJ
energy released by combustion of 1 kg of petrol	10^{8}
energy converted by a 1 kW heater operating for 1 hour	3.6×10^{6}

continued on p. 15

Number representing l/m, m/kg, Δt/s	*Distance* l	*Body whose mass* m *is considered*	*Time interval* Δt
10^{28}		Sun	
10^{24}	distance to furthest photographed galaxy	Earth	
10^{20}	radius of our galaxy		age of Earth ≈ mean life of $^{238}_{92}U$
10^{16}	distance to nearest star		
10^{12}	radius of Solar system		
10^{8}		large ship	human life-span year
10^{4}	radius of Moon	car	day
$1 = 10^{0}$	man's height	cricket ball	time between heart beats
10^{-4}	thickness of Al foil		time period of highest audible note
10^{-8}	wavelength of visible light	oil-drop from atomizer	duration of photon emission
10^{-12}	diameter of molecule	blood corpuscle	
10^{-16}	diameter of nucleus		time period of visible light vibration
10^{-20}		protein molecule	
10^{-24}			time for light to cross an elementary particle
10^{-28}		proton	
10^{-32}		electron	

Description of energy W	W/J
energy required to charge a car battery	1 MJ
energy provided by a slice of bread	10^5
k.e. of a bowled cricket ball	10^1
energy stored in the magnetic field of a 2 H inductor carrying a current of 1 A	1
energy stored in the electric field of a 1 μF capacitor charged to 200 V	2×10^{-2}
	1 mJ
recommended maximum ionizing radiation dose to be absorbed per year by 1 kg of living tissue	4×10^{-4}
	1 μJ
maximum energy of proton in CERN synchrotron	10^{-8}
	1 nJ
rest-mass-energy of 1 unified atomic mass unit	1.6×10^{-10}
energy released by fission of 1 uranium nucleus	3×10^{-11}
typical binding energy per nucleon	1.3×10^{-12}
	1 pJ
minimum energy of γ-ray photon for pair production	1.6×10^{-13}
rest-mass-energy of 1 electron	8×10^{-14}
photon energy of typical X-ray	1 fJ
photon energy of middle u.v.	10^{-17}
average energy to create 1 ion-pair in air	5×10^{-18}
ionization energy of hydrogen atom	2×10^{-18}
width of forbidden region between energy bands in diamond crystal; change of binding energy per atom in a typical chemical change	1 aJ
work function energy of tungsten	7×10^{-19}
photon energy of visible light (violet)	5×10^{-19}
work function energy of caesium	3×10^{-19}
photon energy of visible light (red)	2.5×10^{-19}
width of forbidden region between energy level bands in germanium crystal	1×10^{-19}
photon energy of middle i.r.	2×10^{-20}
translational k.e. of ideal-gas molecule at 300 K	6×10^{-21}
photon energy of microwaves	10^{-23}
photon energy of medium band broadcast radio waves	10^{-27}

1.8 Treatment of Errors

We do not expect the result of a measurement to be the *true* value of the measured quantity. There always exists some uncertainty, which is usually referred to as *experimental error*. An experiment is successful if it is reliable.

Example. The results of two experiments to measure the specific heat capacity of water were quoted as

(1) $c = (4.0 \pm 0.2) \times 10^3\ \text{J kg}^{-1}\ \text{K}^{-1}$,
(2) $c = (4.10 \pm 0.01) \times 10^3\ \text{J kg}^{-1}\ \text{K}^{-1}$.

The second experiment is less successful, since the claim that the error was $10\ \text{J kg}^{-1}\ \text{K}^{-1}$ is obviously not justified (c for water $= 4.18 \times 10^3\ \text{J kg}^{-1}\ \text{K}^{-1}$).

In this paragraph we discuss (non-rigorously) causes of experimental error, and ways of estimating its size.

(a) How Errors Arise

The term *error* excludes *mistakes* which can be put down to the individual. Mistakes can arise through

(*i*) misreading scales,
(*ii*) faulty arithmetic,
(*iii*) trying to apply a theory where it does not hold,
(*iv*) faulty transcription, etc.

Experimental errors are of two main kinds (fig. 1.4).

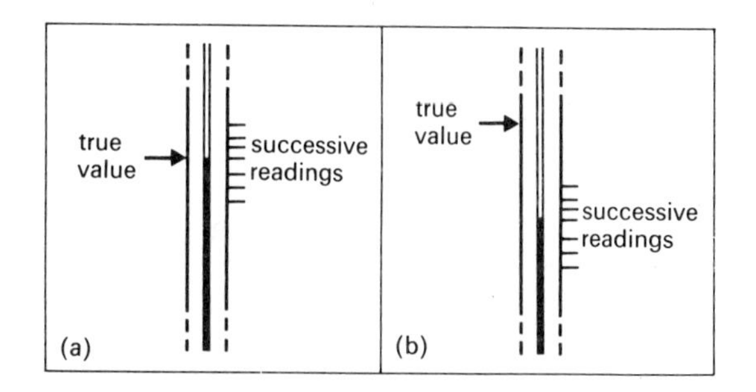

Fig. 1.4. Each bar represents a measurement taken with a thermometer, in which
(a) only random errors are present,
(b) a systematic error is superimposed on the random errors.

(1) A **random error** is one that has an equal chance of being negative or positive. It can be caused by

(*i*) lack of perfection in the observer,
(*ii*) the sensitivity of the measuring instrument,
(*iii*) a particular measurement not being reproducible.

The random error is revealed by repeated observation of a particular quantity: this procedure also helps to reduce its effect.

(2) A **systematic error** causes a random set of measurements to be spread about a value other than the true value. This effect can result from (for example)

(*i*) an instrument having a **zero error** (in which case the systematic error may be constant),

(*ii*) an instrument being incorrectly calibrated (such as a slow-running stopclock),

(*iii*) the observer *persistently* carrying out a mistimed action (e.g. in starting and stopping a clock).

Such systematic errors are *not* revealed by repeated measurement (fig. 1.5), but they can be eliminated or corrected by

(*i*) varying the conditions of measurement, or

(*ii*) suitable treatment of the observations (see the graph on p. 278).

In the remainder of this paragraph we assume that this has been done.

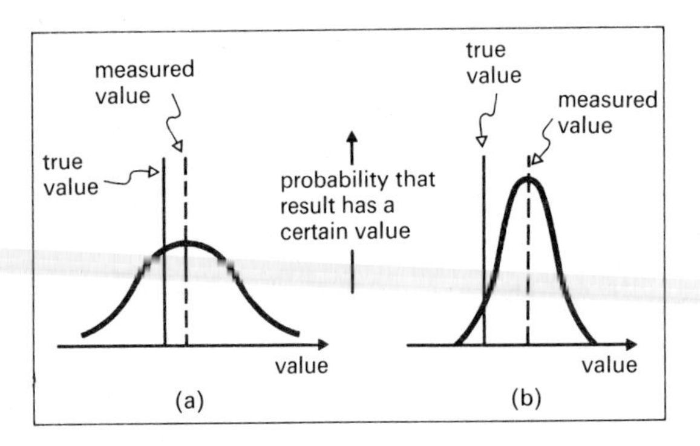

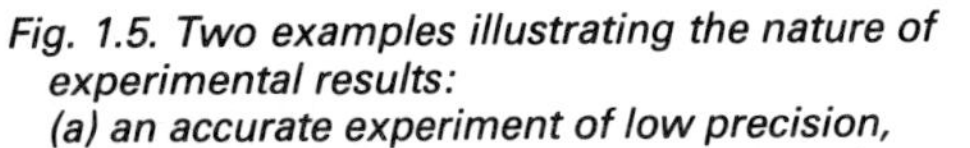

Fig. 1.5. Two examples illustrating the nature of experimental results:
(a) an accurate experiment of low precision,
(b) a less accurate but more precise experiment.

An **accurate** experiment is one for which the *systematic* error is relatively small. A **precise** experiment has a small *random* error. These points are illustrated by fig. 1.5.

(*b*) Estimating the Size of an Error

(1) Single Measurement

The error is guessed using commonsense.

(*i*) In experiments where a *setting* is required (e.g. using a jockey or an optical bench), the true position can be located by bracketing techniques. One might find, for an image distance,

$$v = +(316 \pm 10)\ \text{mm}$$

(*ii*) Where a scale has to be read, there will be a *sensitivity limit*. Thus we would write:

for an ammeter

$$I = (25.4 \pm 0.2)\ \text{mA},$$

or, for a metre rule

$$l = (371.0 \pm 0.5)\ \text{mm}$$

The *largest* error (which could be that resulting from the uncertainty of calibration) is the most important.

(2) Several Repeated Measurements

When n observations are taken of a single quantity, the likely difference between the mean value and the true value is reduced. When n is large, statistical techniques are used to find a *probable range of error*. They are not considered in this book, which deals only with the *maximum* error calculated in a non-rigorous way.

(*c*) Percentage Errors

Suppose a reading x is obtained of a quantity whose true value is X. Then

$$x = X \pm e$$

where e is the **absolute error**.

The **fractional error**

$$f = \frac{e}{X},$$

and the **percentage error**

$$p = \frac{e}{X} \times 100$$

Usually $f \ll 1$, and $p \ll 100$.

To find the *maximum error in compound quantities*, we proceed as follows:

(*i*) *Sum and difference*
We add the *absolute* errors.

If $\quad S = a + b$
then $\quad \Delta S = \Delta a + \Delta b$

where ΔS refers to the absolute error of the sum.

If $\quad D = a - b$
then $\quad \Delta D = \Delta a + \Delta b$

The error, which could be either side of the true value, is still *added*.

(*ii*) *Product and quotient*
We add the fractional or *percentage* errors.

If $$P = ab$$

$$\frac{\Delta P}{P} = \frac{\Delta a}{a} + \frac{\Delta b}{b}$$

$$p_P = p_a + p_b$$

where p_P refers to the percentage error of the product.

If $$Q = \frac{a}{b}$$

$$p_Q = p_a + p_b$$

Examples

(*i*) $$P = a^n$$

$$p_P = np_a$$

(*ii*) $$Q = \frac{x^2}{y}, \quad \text{and} \quad p_x = 2\%$$

$$p_y = 3\%$$

$$p_Q = (2 \times 2 + 3)\% = 7\%$$

(*d*) Estimating the Uncertainty in an Experiment

To investigate whether an experiment is successful, we can proceed as follows:

(1) Repeat the experiment several times, and calculate the required quantity independently for each set of readings.

For example, use

$$\eta = \frac{\pi r^4 \Delta p\, t}{8 l V} \qquad \text{(p. 149)}$$

several times, and obtain a set of values for η. Analyse them to obtain a maximum (or a probable) error.

(2) Estimate (as in (***b***)) the maximum errors in the individually measured quantities r, Δp, t, l and V, and calculate their percentage errors.

Since

$$\ln \eta = \ln\left(\frac{\pi}{8}\right) + 4 \ln r + \ln \Delta p + \ln t - \ln l - \ln V$$

$$\frac{\Delta\eta}{\eta} = 0 + \frac{4\,\Delta r}{r} + \frac{\Delta(\Delta p)}{\Delta p} + \frac{\Delta t}{t} - \frac{\Delta l}{l} - \frac{\Delta V}{V}$$

To find p_η, we *add* all the percentage errors, noting factors such as the 4 for p_r. (In fact the measurement of r, in this experiment, has more effect on the uncertainty in η than all the other readings put together.)

(3) Compare the errors calculated in (1) and (2). If they disagree significantly, then the experiment is not satisfactory.

(*e*) Treatment of Graphs

One purpose of a graph is to display results: it can also show an estimated error range for individual points (fig. 1.6).

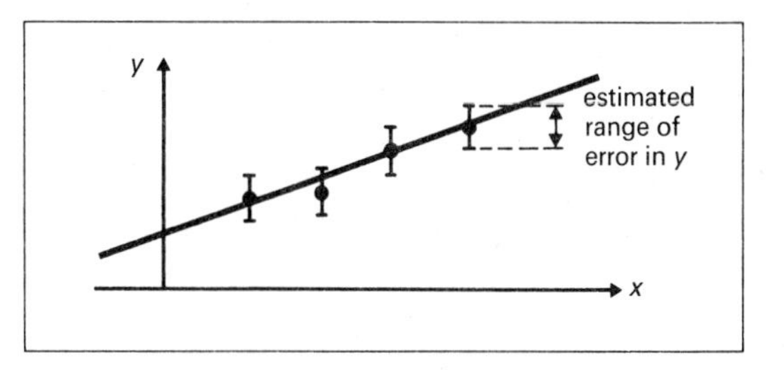

Fig. 1.6. Display of range of error on a graph.

Any line which fits the experimental points within their estimated ranges of error is in agreement with the experiment.

If a point is displaced significantly from the line, we can choose between

(*i*) the estimated limits of error being too optimistic,

(*ii*) there being a *mistake* (in which case the point can be disregarded), or

(*iii*) the theory under test not being applicable under the conditions of the experiment.

Estimating an Error from a Graph

(*i*) *Intercept.* See how far the line can be displaced without going seriously outside the error ranges.

(*ii*) *Slope.* Use the same techniques to measure the maximum and minimum slopes. If the graph is insufficiently sensitive, then we can pair the points, and evaluate the slopes analytically. Thus for 6 points we would calculate slopes:

m_1	from points	1 and	4
m_2		2	5
m_3		3	6

and then work out a mean slope.

It is important to keep the treatment of errors in proportion. For example it is futile to carry out a lengthy calculation of probable error, when the time could be better spent in reducing its main cause.

1.9 Scalar and Vector Quantities

Scalars and Vectors

Scalar quantities are those which are completely specified by a physical magnitude (expressed by the product of a positive or negative number and a unit) alone. They can be added *algebraically*.

Vector quantities require, in addition, a direction to be specified. The magnitude of a vector is inherently positive. In this book the symbol is printed in ***bold italic***, unless we refer to magnitude only.

Examples in Mechanics

Scalar quantity	*Vector quantity*
length l	displacement $\boldsymbol{s}$
mass m	area $\boldsymbol{A}$
time t	force $\boldsymbol{F}$
volume V	velocity $\boldsymbol{v}$
speed v	acceleration $\boldsymbol{a}$
energy W	momentum $\boldsymbol{p}$
pressure p	torque $\boldsymbol{T}$
density ρ	angular momentum $\boldsymbol{L}$

For equations in this book we adopt the following procedure:

(*i*) Vector quantities are printed in bold type if the *directions* of the quantities involved are *along the same straight line*.

(*ii*) They are represented by their magnitudes only (in italic) if, perhaps as a consequence of multiplication, the vector quantities involved lie along different directions.

Examples

(*i*) $\boldsymbol{F} = m\boldsymbol{a}$ (p. 36)

(*ii*) $F = BQv$ (p. 373)

A **vector** is a line drawn to represent a vector quantity in

(*a*) *direction*, by its direction, and

(*b*) *magnitude*, by its length.

When the answer to a problem is a vector quantity, its *direction* (not just its magnitude) must always be given, even if not asked for explicitly. Usually this is done best by a labelled diagram.

Adding and Subtracting Vectors

(*a*) Vectors must be *added* by the **parallelogram law** (fig. 1.7) which ensures that their directions, as well as their physical magnitudes, are taken into account. They may *not* be added algebraically. Vectors may be added in any order.

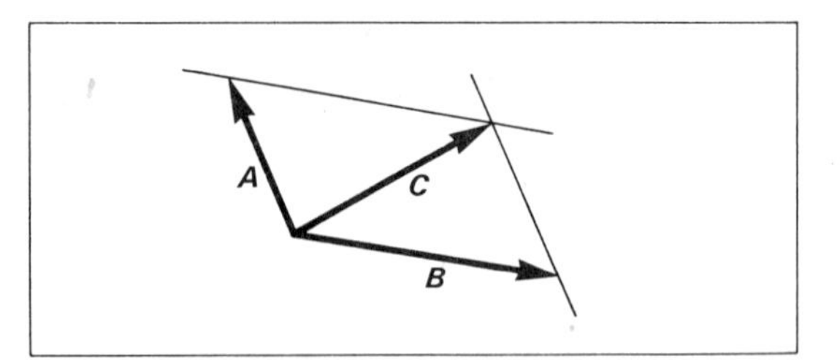

Fig. 1.7. Vector addition: $\boldsymbol{A} + \boldsymbol{B} = \boldsymbol{C}$.

(*b*) The *negative* of a vector is that vector with its direction reversed. So to *subtract* vector $\boldsymbol{E}$ from vector $\boldsymbol{D}$, we reverse the direction of $\boldsymbol{E}$ before adding by the parallelogram law (fig. 1.8). Example on p. 30.

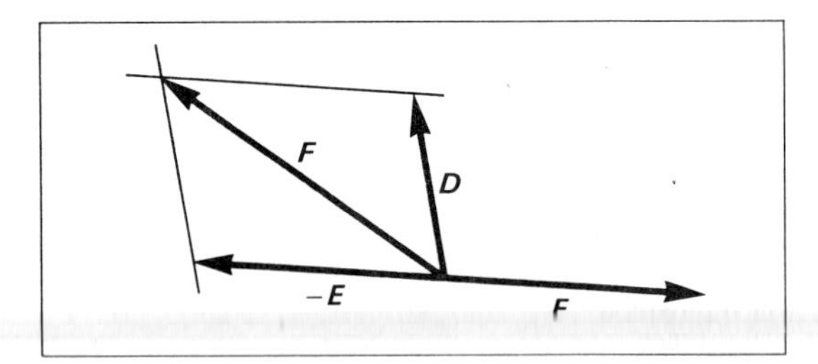

Fig. 1.8. Vector subtraction: $\boldsymbol{F} = \boldsymbol{D} + (-\boldsymbol{E}) = \boldsymbol{D} - \boldsymbol{E}$.

Resolving Vectors

In fig. 1.7 the vectors $\boldsymbol{A}$ and $\boldsymbol{B}$ have been compounded, or added, to give their **resultant** $\boldsymbol{C}$.

(*a*) The reverse process, in which $\boldsymbol{C}$ is replaced by two vectors, is called finding the **components** of the vector. There is an infinite number of parallelograms for which $\boldsymbol{C}$ could be a diagonal: there is an infinite number of ways in which $\boldsymbol{C}$ can be split into two components.

(*b*) When we find components $\boldsymbol{X}$ and $\boldsymbol{Y}$ which are *perpendicular*, then they are called **resolved parts** of the vector $\boldsymbol{Z}$ (fig. 1.9), and the process is called **resolving**.

The effect of $\boldsymbol{Y}$ along the direction Ox is zero, so $\boldsymbol{X}$ represents the *total effective value* of $\boldsymbol{Z}$ in that direction, where

$$X = Z\cos\theta$$

Similarly $$Y = Z\sin\theta$$

This process is particularly useful for forces (p. 38).

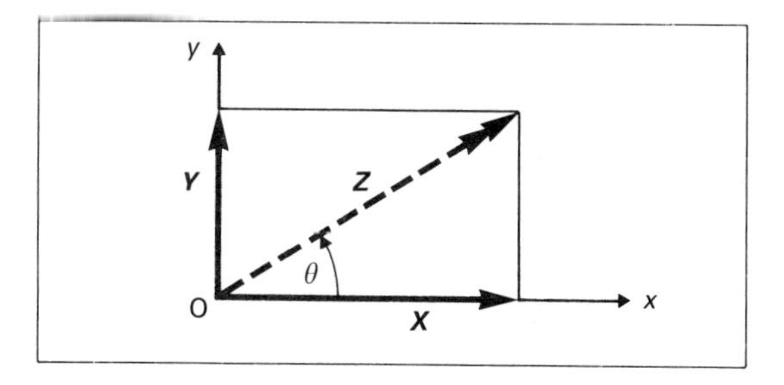

Fig. 1.9. ***X*** *and* ***Y*** *are the resolved parts of* ***Z****.*

*Multiplying Vectors

(a) Multiplication of a Vector by a Scalar

The result of multiplying the vector ***V*** by the scalar S is a *vector* of

(*i*) magnitude SV, and
(*ii*) the same direction as ***V***.

(b) Multiplication of a Vector by a Vector

When a vector multiplies a vector, the product can be

(*i*) a *scalar*, if the **dot product** ($\boldsymbol{A} \cdot \boldsymbol{B}$) is used, or
(*ii*) a *vector*, if the **cross product** ($\boldsymbol{A} \times \boldsymbol{B}$) is used.

In this case the order of multiplication (e.g. $\boldsymbol{A} \times \boldsymbol{B}$ or $\boldsymbol{B} \times \boldsymbol{A}$) determines the *sense* of the product. More properly such quantities are called **pseudovectors**, or **axial vectors**. The direction associated with them is that of an axis of rotation, and this enables them to be distinguished from ordinary (or **polar**) vectors.

We can choose (by convention) one of the two directions of the axis to represent the direction of the pseudovector—thus a pseudovector has a **sense** associated with it.

Example. The product of a *displacement* ***r*** and a *force* ***F*** can result in

(*i*) *work* (p. 42), in which we are interested in the resolved parts of ***r*** and ***F*** along the *same direction*, or
(*ii*) *torque* (p. 62), in which we are interested in their resolved parts along *perpendicular directions*.

Work is a scalar, but torque is a vector whose direction is perpendicular to the plane which contains ***r*** and ***F***, and whose sense is found conventionally by using a right-hand corkscrew rule.

1.10 The Uses of Dimensions

In a correct physical equation we can equate both number *and unit* for each term that appears. If we could not equate the unit, then a change of the system of units might result in the numbers changing by different factors, which would invalidate an equation whose numbers were previously equal.

Each term in a correct physical equation must have the same dimensions. Use of this fact is called the **method of dimensions**.

(*a*) Conversion of Units

When several systems of units were in common use, the method of dimensions gave a quick way of converting units for complex derived quantities (such as thermal conductivity) from one system to another. This procedure should no longer be necessary for the beginning student.

(*b*) To Check Equations

Since physical equations are dimensionally homogeneous, terms which are incorrect (other than by a dimensionless factor) can quickly be detected.

Examples

(***i***) $T = \frac{1}{5}\sqrt{l/10}$ is an approximate equation for the period T of a simple pendulum of length l. It is not dimensionally homogeneous, and can only be applied when T is in seconds and l in mm, i.e. for a *specified set of units*.

(***ii***) Suppose we are told that the speed c of transverse waves along a wire of tension F and mass m is given by

$$c = \sqrt{\frac{F}{m}}$$

$$[c] = [\mathrm{LT^{-1}}]$$

$$[F^{1/2}m^{-1/2}] = [\mathrm{MLT^{-2}}]^{1/2}[\mathrm{M}]^{-1/2}$$

$$= [\mathrm{L^{1/2}T^{-1}}]$$

Because the left- and right-hand sides have different dimensions, we detect a mistake: m is not mass, but mass *per unit length*, and has dimensions $[\mathrm{ML^{-1}}]$.

(***iii***) Consider the equation

$$\left(p + \frac{a}{V_\mathrm{m}^2}\right)(V_\mathrm{m} - b) = RT \qquad \text{(p. 213)}$$

Problems: (*a*) Is it dimensionally correct?
(*b*) If so, what is the unit for a?

(*a*) If the dimensions are right, each side of the equation will have the same unit. Taking the first term in each bracket, the left side has unit

$$(\mathrm{N\,m^{-2}})(\mathrm{m^3\,mol^{-1}}) = \mathrm{J\,mol^{-1}}$$

The right side has unit

$$(\mathrm{J\,mol^{-1}\,K^{-1}})(\mathrm{K}) = \mathrm{J\,mol^{-1}}$$

The equation is dimensionally correct.

(*b*) The unit for a/V_m^2 must be the same as that for p, since they occur in the same bracket. Therefore a has the unit of pV_m^2

$$= (\text{N m}^{-2})(\text{m}^3\,\text{mol}^{-1})^2$$
$$= \text{N m}^4\,\text{mol}^{-2}$$

(***iv***) Consider the equation

$$F' = \left(\frac{\mu_0}{2\pi}\right)\cdot\frac{I_1 I_2}{a} \qquad \text{(p. 386)}$$

What is the unit for $\boldsymbol{F}'$?

It has the same unit as the right side, namely

$$\left(\frac{\text{H}}{\text{m}}\right)\cdot\frac{\text{A}^2}{\text{m}} = \left(\frac{\text{N}}{\text{A}^2}\right)\cdot\frac{\text{A}^2}{\text{m}}$$
$$= \text{N m}^{-1}$$

$\boldsymbol{F}'$ tells us the force *per unit length* of the wire.

(*c*) Dimensional Analysis

This enables us to predict how physical quantities may be related. Some examples are given in the next paragraph.

1.11 Dimensional Analysis

Limitations

(*a*) The method cannot find the value of any dimensionless constants (such as 2π).

(*b*) Since in mechanics we have chosen to use only *three* fundamental quantities, [M], [L], and [T], we cannot in general determine how a given quantity depends upon more than *three* other quantities. Nevertheless problems in thermodynamics and electricity, for example, may need the introduction of further fundamental quantities, such as temperature [Θ], amount of substance [N] or electric current [I].

It will then be possible to relate a given quantity to more than three others. (See examples (3) and (4) below.)

(*c*) Even in mechanics it may be possible to relate a given quantity to more than three others if it is already known before the analysis how some of them *must* be related to one another. (See p. 147.)

(*d*) We must make a reasonable guess (based on physical intuition or experiment) as to all the dimensional quantities which may be involved in the equation (see p. 148).

(*e*) Dimensionless combinations of quantities will not be revealed.

(*f*) We can only discover relationships which involve the product of powers: no information, for example, can be obtained about exponential functions.

Applications

Dimensional analysis is particularly useful at this level for viscosity problems (p. 147), but can be applied in all branches of physics. Examples are found in the text, and below.

(1) *Speed of Ocean Waves*

Experiment indicates that the speed c is effectively independent of amplitude, and for long wavelengths is independent of surface tension.

Suppose $\qquad c = k g^x \lambda^y \rho^z$

where k is a dimensionless constant.

$$[\text{LT}^{-1}] = k[\text{LT}^{-2}]^x[\text{L}]^y[\text{ML}^{-3}]^z$$

By the method of dimensions, the powers of L, M and T must be the same on each side of the equation.

$$\begin{aligned} &(1)\ [\text{L}] & 1 &= x + y - 3z \\ &(2)\ [\text{T}] & -1 &= -2x \\ &(3)\ [\text{M}] & 0 &= z \end{aligned}$$

We conclude $c = k\sqrt{g\lambda}$, and is independent of ρ. (We expect this from physical intuition: the restoring force is gravitational pull, which is proportional to inertial mass.)

We cannot evaluate k by this method.

(2) *Oscillation of a Drop*

Suppose that a *small* liquid drop is disturbed from its spherical shape, and set oscillating. We assume that gravitational effects are negligible (p. 137). We consider the frequency f of oscillation to be controlled by the drop radius r, and the liquid surface tension γ and density ρ.

Put $\qquad f = k\gamma^x r^y \rho^z$

$$[\text{T}^{-1}] = k[\text{MT}^{-2}]^x[\text{L}]^y[\text{ML}^{-3}]^z$$

Comparing indices

$$\begin{aligned} &(1)\ [\text{L}] & 0 &= y - 3z \\ &(2)\ [\text{M}] & 0 &= x + z \\ &(3)\ [\text{T}] & -1 &= -2x \end{aligned}$$

We conclude $\qquad f = k\sqrt{\dfrac{\gamma}{\rho r^3}},$

a result which would be very difficult to obtain by any other method.

Physically, it is of the form

$$f \propto \sqrt{\left(\frac{\text{elasticity factor}}{\text{inertial factor (mass)}}\right)}$$

See the discussion of p. 78.

(3) *Thermal Conductivity of an Ideal Gas*

It is thought that the thermal conductivity k of an ideal gas will depend upon the mean molecular speed $\bar{c}$, the density ρ, the specific heat capacity at constant volume c_V, and the molecular mean free path λ. Find a possible relationship between these five quantities.

Suppose $k = (\text{constant})(\bar{c})^w \rho^x c_V^y \lambda^z$

$$[\text{MLT}^{-3}\Theta^{-1}] = [\text{LT}^{-1}]^w[\text{ML}^{-3}]^x[\text{L}^2\text{T}^{-2}\Theta^{-1}]^y[\text{L}]^z$$

Comparing indices

(1) [M] $\quad 1 = x$
(2) [Θ] $\quad -1 = -y$
(3) [T] $\quad -3 = -w - 2y$
(4) [L] $\quad 1 = w - 3x + 2y + z$

These equations have the solution

$$w = x = y = z = 1,$$

and the constant can be shown by other means to be 1/3. We conclude

$$k = \tfrac{1}{3}\bar{c}\rho c_V \lambda$$

The result is important for the correlation of the transport phenomena (p. 184).

Note that, since the problem involved *four* fundamental quantities, we were able to relate k to *four* other variables.

(4) *Energy Transfer by Electromagnetic Waves*

It is thought that the rate of flow of electromagnetic energy P through an area $\boldsymbol{A}$ placed normal to an electromagnetic wave will depend upon μ_0, and upon the instantaneous values of the electric and magnetic fields, $\boldsymbol{E}$ and $\boldsymbol{B}$. Find a possible relationship between these five quantities.

Suppose $\quad P = k\mu_0^w E^x B^y A^z$

$$[\text{ML}^2\text{T}^{-3}] = k[\text{MLT}^{-2}\text{I}^{-2}]^w[\text{MLT}^{-3}\text{I}^{-1}]^x \cdot [\text{MT}^{-2}\text{I}^{-1}]^y[\text{L}^2]^z$$

Comparing indices

(1) [M] $\quad 1 = w + x + y$
(2) [L] $\quad 2 = w + x + 2z$
(3) [T] $\quad -3 = -2w - 3x - 2y$
(4) [I] $\quad 0 = -2w - x - y$

These equations have the solution

$$w = -1, \quad \text{and} \quad x = y = z = 1.$$

We conclude

$$P = kEBA/\mu_0,$$

or that the intensity

$$\frac{P}{A} = k\frac{EB}{\mu_0}$$

The dimensionless constant can be shown by other means to be 1, and the vector intensity $(P/\boldsymbol{A})$ is referred to as the **Poynting vector**, $(\boldsymbol{E} \times \boldsymbol{B})/\mu_0$.

1.12 Some Useful Mathematics

Page numbers are references to places in the text where a particular point of mathematics is used.

(*a*) Algebra

Binomial Theorem

$$(1 + x)^n = 1 + nx + \frac{n(n-1)}{1 \times 2}x^2 + \frac{n(n-1)(n-2)}{1 \times 2 \times 3}x^3 + \cdots$$

If $x \ll 1$, then

$$(1 + x)^n \simeq 1 + nx$$
$$(1 + x)^{-n} \simeq 1 - nx \qquad \text{(p. 115)}$$

These useful approximations are valid when x^2 is negligible. They lead to

$$\frac{1+\alpha}{1+\beta} \approx (1+\alpha)(1-\beta) \approx 1 + (\alpha - \beta)$$

for small α and β.

Quadratic Equations

The equation

$$ax^2 + bx + c = 0$$

has the solution

$$x = \frac{-b \pm \sqrt{b^2 - 4ac}}{2a} \qquad \text{(p. 241)}$$

(*b*) Logarithms

Common logarithms are those to base 10.

Thus if $\quad x = 10^y$
then $\quad y = \lg x$
where lg implies $\log_{10}$.

Natural logarithms are to base e.

Thus if $x = e^y$

then $y = \ln x$

where ln implies $\log_e$. e is defined by

$$e = \lim_{n\to\infty}\left(1 + \frac{1}{n}\right)^n = 2.718\ldots$$

$$\ln(1 + x) \approx x - \tfrac{1}{2}x^2$$
$$\approx x$$

for very small x.

*Logarithmic and Exponential Relationships

$$\log_b(x \times y) = \log_b x + \log_b y$$

$$\log_b x^n = n \log_b x$$

$$\log_b\left(\frac{x}{y}\right) = \log_b x - \log_b y$$

If $p = a^x = b^y$

$x = \log_a p$, and $y = \log_b p$

Then $\log_a p = y \log_a b$

$= \log_a b \times \log_b p$

This shows $\log_e x = (\log_e 10)\log_{10} x$

or $\ln x = (2.303)\lg x$

and $\lg x = (0.434)\ln x$

(c) Trigonometry

Angles

Refer to fig. 1.10.

Angle θ is defined by

$$\boxed{\theta = s/r}$$

$$\theta[\text{rad}] = \frac{s[\text{m}]}{r[\text{m}]} \qquad \text{(p. 55)}$$

(*i*) When $s = r$, $\theta = 1$ **radian** (rad)

(*ii*) When $s = 2\pi r$ (one revolution)

$$\theta = \frac{2\pi r}{r}\text{rad} = 2\pi \text{ rad}$$
$$= 360°$$

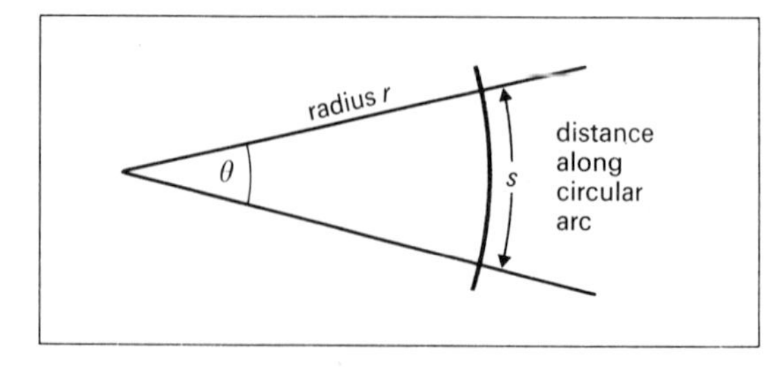

Fig. 1.10 Definition of angle θ.

(Note that the degree (°) is *not* an SI unit.)

$$\boxed{\pi \text{ rad} = 180°}$$

Thus 1 rad $\approx 57.3°$.

If $x = \sin\theta$,

then θ is the angle whose sine is x, and this would be abbreviated to

$$\theta = \arcsin x$$

Useful Relationships

The quadrants in which particular trig. functions are positive are shown in fig. 1.11.

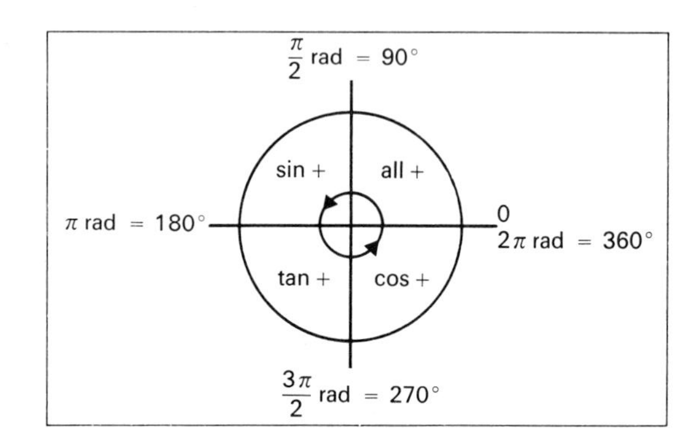

Fig. 1.11 Signs of trigonometric functions.

$$\tan\theta = \frac{\sin\theta}{\cos\theta}$$

$$\sin^2\theta + \cos^2\theta = 1 \qquad \text{(p. 80)}$$

$$\sin A + \sin B = 2\sin\left(\frac{A+B}{2}\right)\cos\left(\frac{A-B}{2}\right) \qquad \text{(p. 106).}$$

Sine Law. $\dfrac{a}{\sin A} = \dfrac{b}{\sin B} = \dfrac{c}{\sin C}$

Cosine Law. For any triangle

$$a^2 = b^2 + c^2 - 2bc\cos A$$

It can be shown that

$$\sin\theta = \theta - \frac{\theta^3}{3!} + \frac{\theta^5}{5!} - \cdots$$

So when $\theta \ll 1$ (say $\theta \sim 0.1$ rad)

$$\sin\theta \approx \theta$$

Similarly $\tan\theta \approx \theta$ but

$$\cos\theta \approx 1$$

These approximations are used frequently in s.h.m. problems (p. 83) and in optics (p. 231).

(d) Calculus

Derivatives and Integrals

$y = f(x)$	$\frac{dy}{dx} = f'(x)$	$\int y\,dx$
$y = x^n$	nx^{n-1}	$\frac{x^{n+1}}{n+1} + C\begin{pmatrix}\text{except when}\\ n = -1\end{pmatrix}$
$y = \frac{1}{x}$	$-\frac{1}{x^2}$	$\ln x + C$
$y = \ln x$	$\frac{1}{x}$	$x \ln x - x + C$
$y = \sin x$	$\cos x$	$-\cos x + C$
$y = \cos x$	$-\sin x$	$\sin x + C$

C is the constant of integration, whose value will be determined by the limits of integration.

Logarithms and Exponentials

$$\text{If} \quad \frac{dx}{dt} = cx$$

$$\ln x = ct + C$$

giving

$$x = A\,e^{ct}$$

$$\frac{d}{dx}(\exp x) = \exp x$$

$$\int e^x \cdot dx = e^x + C$$

Average Value of a Function

The **mean** or **average** value $\langle y\rangle$ of $y = f(x)$ over the interval $x = a$ to $x = b$ is given by

$$\langle y\rangle = \frac{1}{b-a}\int_a^b y\,dx \qquad \text{(p. 411)}$$

Similarly

$$\langle y^2\rangle = \frac{1}{b-a}\int_a^b y^2\,dx$$

(pp. 179 and 411)

The **root mean square** (r.m.s.) value of y

$$y_{\text{r.m.s.}} = \sqrt{\langle y^2\rangle}$$

For a *periodic* function, the interval $(b - a)$ is understood to be taken over an integral number of periods or half periods.

(e) Geometry

In fig. 1.12 the intersecting chords theorem gives

$$(2R - x)x = r^2$$

When x is small

$$x \approx r^2/2R \qquad \text{(p. 278)}$$

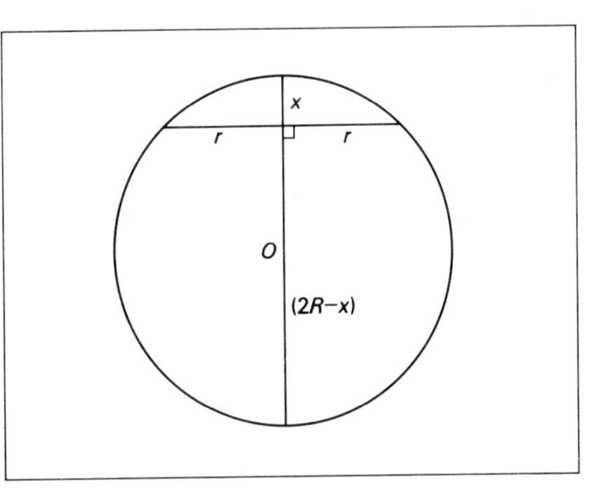

Fig. 1.12. The sagitta relationship.

Use of Symbols

Δx represents some *finite change* in the value of x. The change might be positive or negative: thus $\Delta V = -3.0\ \text{m}^3$ would signify a volume reduction. (We do *not* write $-\Delta V$, and in so doing antipicipate the negative sign.)

δx represents a *variation* in the value of x, a *small* increment whose size is later to be considered as vanishingly small. (Refer to p. 42 for an example.)

*1.13 Differential Equations in Physics

Their Origin

Differential equations arise in many branches of physics, as the following examples show.

(*a*) ***Mechanics.*** The solution of many problems in dynamics depends upon the equation

$$\boldsymbol{F} = m\boldsymbol{a} = m(d^2\boldsymbol{x}/dt^2)$$

$\boldsymbol{F}$ may be a function of t, $\boldsymbol{x}$ and/or $\dot{\boldsymbol{x}}$, and so we might have to integrate to find $\boldsymbol{x}$ as a function of t.

(*b*) ***Oscillations and waves.*** When $\boldsymbol{F}$ can be written in the form $-k\boldsymbol{x}$, the equation becomes

$$m(d^2\boldsymbol{x}/dt^2) + k\boldsymbol{x} = 0$$

which is referred to as the **differential equation of s.h.m.** Its solution is discussed below. Wave motions have a similar important differential equation: in one dimension

$$\frac{\partial^2 y}{\partial x^2} = \frac{1}{c^2}\frac{\partial^2 y}{\partial t^2}$$

(*c*) ***Heat flow.*** In one dimension the rate of heat flow through a material is described by the differential equation

$$dQ/dt = -\lambda A(d\theta/dx)$$

If, in a given situation, dQ/dt is constant, and both λ and A depend upon x, then we would have to integrate to find the variation of θ with x.

(*d*) ***D.C. circuits.*** The growth of current in a d.c. *LR* circuit is described by the equation

$$\mathscr{E} = L(dI/dt) + IR$$

The decay of charge on the plates of a capacitor in a d.c. *CR* circuit is described by the equation

$$IR + Q/C = 0$$
$$R(dQ/dt) + Q/C = 0$$

(*e*) ***A.C. circuits.*** The application of the law of energy conservation in an a.c. circuit leads to the differential equation

$$\mathscr{E}_0 \cos \omega t = L(dI/dt) + Q/C + IR$$

Since $I = dQ/dt$, we may write

$$\mathscr{E}_0 \cos \omega t = L\frac{d^2Q}{dt^2} + R\frac{dQ}{dt} + \frac{1}{C}Q$$

a differential equation which can be solved to find Q (and I) in terms of t.

(*f*) ***Radioactive decay.*** The disintegration of a radioactive nucleus is an entirely random process. This means that $-dN/dt$, the rate of decay of a particular sample, is proportional to the (large) number N of active nuclei that it contains. We write

$$dN/dt = -\lambda N$$

If a radioisotope is being created in a nuclear reactor at the constant rate C, but at the same time the new nuclide disintegrates at the rate $-\lambda N$, then we can calculate the net rate of increase of nuclei from

$$dN/dt = C - \lambda N$$

The number of nuclei present after a given time interval must be found by solving this differential equation. (Note the formal similarity between

$$dN/dt + \lambda N - C = 0, \text{ and}$$
$$dI/dt + (R/L)I - (\mathscr{E}/L) = 0 \quad \text{(above)}$$

Similarities of this kind enable us to solve many differential equations by comparing them with others whose solutions are already known.)

The Solution of Differential Equations

There exists a considerable scheme for classifying differential equations as to their order and degree, and also a formal set of rules for their solution. Most of the differential equations which we meet can be solved by one of two methods.

(*a*) ***By separation of variables.*** If the variables can be separated onto the two sides of the equation, then the solution will be obtained by direct integration. Several examples are given below.

(*b*) ***By inspection.*** Because the equations that we deal with tend to treat a small number of distinct physical situations, previous experience and physical intuition will often enable us to *guess the form of the solution*. We can then test our proposed solution by substituting it into the original differential equation.

Example: *the LC circuit.* Experience teaches us that an equation of the form

$$\frac{d^2Q}{dt^2} = -\left(\frac{1}{LC}\right)Q$$

describes an oscillatory situation, since it closely resembles

$$\frac{d^2x}{dt^2} = -\left(\frac{k}{m}\right)x$$

Suppose we write $(1/LC) = \omega^2$, and then suggest

$$Q = A \sin \omega t + B \cos \omega t$$

as our solution of

$$d^2Q/dt^2 = -\omega^2 Q$$

A and B are arbitrary constants. Differentiating twice, we have

$$(1) \qquad dQ/dt = A\omega \cos \omega t - B\omega \sin \omega t$$
$$(2) \qquad d^2Q/dt^2 = -A\omega^2 \sin \omega t - B\omega^2 \cos \omega t$$
$$= -\omega^2 Q$$

Since this is the original equation, we have verified that

$$Q = A \sin \omega t + B \cos \omega t$$

is a solution.

Constants

The above solution can also be expressed in the form

$$Q = C \sin(\omega t + \delta)$$

where C and δ are two further arbitrary constants related to A and B. To eliminate *two* such constants, and hence arrive back at the original equation, it was necessary to differentiate *twice*. This illustrates a general rule:

> *The* ***general solution*** *of an nth order equation (involving* $d^n y/dx^n$*) will have n constants.*

A **particular solution** is one in which the arbitrary constants have been given specified values. Thus to obtain the *particular* solution of a differential equation describing s.h.m. we need to have *two* separate items of information, or **boundary conditions**. For example we might be told that $\dot{x} = 0$ and $x = +a$ when $t = 0$, and this would enable us to write a particular solution involving no arbitrary constants.

1.14 Graphs

The Usefulness of Graphs

(*a*) They demonstrate pictorially how a pair of varying quantities are related. They can indicate, for example, the point at which a relationship is no longer linear. Graphs used like this are visual aids.

(*b*) If we know how two quantities are related, we can determine graphically the constants of the equation which relates them. (We use the graph as a way of judging the best line through the points.)

Examples

Equation	*Plot along*		*Slope m*	*Intercept c*
	y-axis	x-axis		
$T^2h = Bh^2 + A$	T^2h	h^2	B	A
$T = AI^{1/2}$	T	$I^{1/2}$	A	zero
$1/n = Bx + A$	$1/n$	x	B	A

(*c*) A graph is frequently an effective substitute for a table of values when an instrument is calibrated. For a thermocouple thermometer one could plot a graph of galvanometer deflection in divisions against temperature in kelvins.

Straight and Curved Graphs

(*a*) For pictorial purposes a curved graph may be acceptable, or even preferable to a straight line.

The equation

$$y = ax^2 + bx + c$$

represents a **parabola**, of which examples are given on p. 30 and p. 315.

(*b*) *Equation of a straight line* (fig. 1.13).

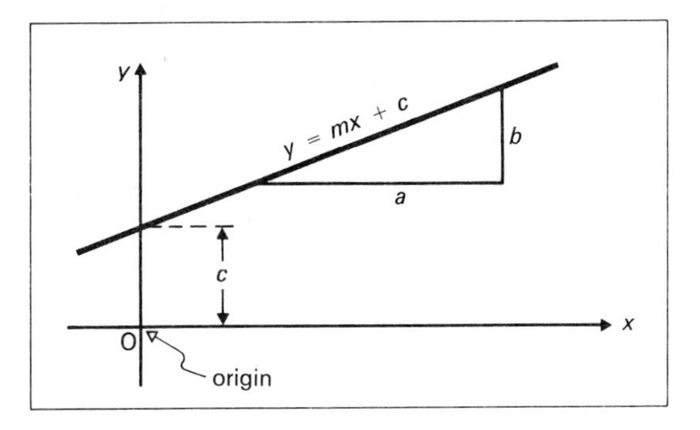

Fig. 1.13. A typical straight line.

In the equation $y = mx + c$, $m = b/a$ is the **slope** or **gradient** of the line.

When $x = 0$, $y = c$, which is the **intercept** on the y-axis. Similarly when $y = 0$, $x = -c/m$.

x is the **independent** variable, and y the **dependent** variable.

Choosing What to Plot

If a graph is to be used for calculation, one nearly always chooses to plot a *straight* line from which the slope m and/or the intercept c can be measured.

Log Graphs

Suppose we assume that two quantities y and x are related by

$$y = Ax^B$$

where A and B are constants.

Then

$$\lg y = \lg A + \lg (x^B)$$
$$= \lg A + B \lg x$$

If we plot $\lg y$ along the vertical axis, and $\lg x$ along the horizontal axis, then

(1) a *straight line graph* would confirm that the original assumption was correct, and

(2) its slope would be B, and intercept $\lg A$.

Hence *A and B could be evaluated.*

How to Plot the Graph

(*a*) (*i*) Plot the quantity that is controlled during the experiment (the *cause*) along the x-axis (as *abscissa*).

(*ii*) Plot the quantity which changes as a result (the *effect*) along the y-axis (as *ordinate*).

(*b*) Essentially a graph relates *numbers* (rather than physical quantities). The axes should be labelled in the same way as the headings for a column in a table (fig. 1.14 overleaf).

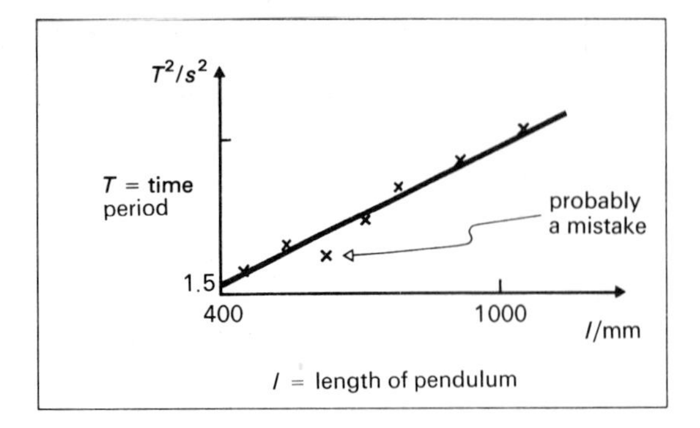

Fig. 1.14. A graph drawn to reach a conclusion from a single pendulum experiment.

(*c*) *The scale* chosen is inevitably a compromise. Consider the following.

(*i*) The scale should be simple to prevent mistakes.

(*ii*) The paper should be filled, and experimental points distributed as widely as possible.

(*iii*) If an intercept is required, the origin may have to be included.

(*iv*) If the slope of a straight line is to be measured, it should be inclined to the x-axis at an angle between 30° and 60°.

(*d*) *Points* can be marked ×, ⊙ or ⌶ if it is required to show the range of likely error. (An isolated · is easily lost.)

(*e*) A *smooth* curve or *straight* line should be drawn, even if this does emphasize a random error.

(*f*) Straight line graph for which $c = 0$ is expected should *not* be drawn through the origin automatically. Non-coincidence may reveal a systematic error (see p. 278 for example).

Fig. 1.14 illustrates some of these points.

The graph should be plotted before apparatus is dismantled. This has two advantages.

(*i*) Mistakes will become obvious, and can be checked.

(*ii*) Further measurements can be taken for crucial regions of the graph. These can then be replotted on a larger scale.

II Principles of Mechanics

2 Kinematics

2.1 Definitions

Kinematics is the study of the motion of points (or massless particles): we do not consider the agent that causes the motion. The following types of motion occur frequently and can be described by simple mathematics:

(*a*) linear motion
(*b*) circular motion
(*c*) parabolic motion
(*d*) simple harmonic motion.

The **displacement** s of a particle is the length and direction of the line drawn to the particle from some fixed point (frequently called the **origin**). Displacement is a *vector* quantity.

Velocity

Suppose that during some motion, the displacement $\boldsymbol{s}$ of a particle changes by $\Delta\boldsymbol{s}$ during a time interval Δt.

The average velocity $\boldsymbol{v}_{av}$ during the interval is defined by the equation

$$\boldsymbol{v}_{av} = \frac{\Delta\boldsymbol{s}}{\Delta t}$$

$$\boldsymbol{v}_{av}\left[\frac{\text{m}}{\text{s}}\right] = \frac{\Delta\boldsymbol{s}[\text{m}]}{\Delta t[\text{s}]}$$

Average velocity is a *vector* quantity. The vector $\Delta\boldsymbol{s}$ is found by subtracting the initial displacement from the final displacement. Suppose that $\Delta\boldsymbol{s}$ and Δt become vanishingly small, and are written $\delta\boldsymbol{s}$ and δt.

The instantaneous velocity $\boldsymbol{v}$ (at the instant contained by δt) is defined by the equation

$$\boldsymbol{v} = \lim_{\delta t \to 0}\left(\frac{\delta\boldsymbol{s}}{\delta t}\right) = \frac{d\boldsymbol{s}}{dt}$$

Instantaneous velocity is a *vector* quantity which may be measured in m s^{-1}.

Speed

The *average speed* of a particle is defined by

$$\frac{\text{distance travelled along actual path}}{\text{time taken } \Delta t}$$

The *instantaneous speed* is the value of this expression when the time interval Δt becomes vanishingly small. From these definitions it follows that

(*a*) the instantaneous speed equals the magnitude of the instantaneous velocity (speed being a scalar quantity), but
(*b*) the average speed is *not* necessarily equal to the magnitude of the average velocity (although it can be).

Acceleration

Suppose that the velocity changes by $\Delta\boldsymbol{v}$ during some time interval Δt. Then we define

The average acceleration $\boldsymbol{a}_{av}$ by

$$\boldsymbol{a}_{av} = \frac{\Delta\boldsymbol{v}}{\Delta t}$$

and

***The* instantaneous acceleration a by**

$$\boxed{a = \lim_{\delta t \to 0} \frac{\delta v}{\delta t} = \frac{dv}{dt}}$$

$$a\left[\frac{m}{s^2}\right] = \frac{\delta v[m/s]}{\delta t[s]}$$

Acceleration is a *vector* quantity, since Δv is calculated by vector subtraction. A body moving with uniform velocity has zero acceleration: this means that neither its speed nor its directions of motion is changing with time.

2.2 Linear Motion

(*a*) With Constant Acceleration

Suppose that $v = u$
and $s = 0$
when $t = 0$.

Then after time t

$$a = \frac{\Delta v}{\Delta t} = \frac{v - u}{t}$$

so $$v = u + at \quad (1)$$

Because

displacement = (average velocity)(time)

we write $$s = \frac{u + v}{2} \cdot t \quad (2)$$

(1) and (2) may be combined so as to eliminate v, u or t. The results are listed for reference.

$$v = u + at$$
$$s = \frac{u + v}{2} \cdot t$$
$$s = ut + \tfrac{1}{2}at^2$$
$$s = vt - \tfrac{1}{2}at^2$$
$$v^2 = u^2 + 2as$$

Uniformly accelerated linear motion

Comments

(i) If we know any *three* of v, u, a, s and t, then the remainder can be found.

(*ii*) s represents *displacement* (not distance travelled by particle).

(*iii*) These results can be found quickly by integration.

Integration of

$$\frac{dv}{dt} = a$$

gives $$v = u + at$$

(since a is to be constant).

$$a = \frac{dv}{dt} = \frac{ds}{dt} \cdot \frac{dv}{ds} = v \cdot \frac{dv}{ds}$$

which gives $$\tfrac{1}{2}v^2 = \tfrac{1}{2}u^2 + as$$

The constants of integration have been inserted.

(*b*) With Variable Acceleration

If a depends on t in some simple way, then calculus can be used. If a varies in an irregular manner, the problem is best solved by plotting a graph.

(i) The Displacement–Time Graph

Since $v = ds/dt$, the *slope* of the graph represents the *instantaneous velocity*.

The area under the graph has no particular significance.

(ii) The Velocity–Time Graph

(1) Since $a = dv/dt$, the *slope* of the graph represents the *instantaneous acceleration*.

(2) Since $$v = \frac{ds}{dt}$$

$$s = \int v \cdot dt$$

the *area* under the graph represents the *displacement*.

Areas above the time-axis are counted as positive, to ensure that a velocity in a given direction corresponds to a displacement in the same direction. The area under a graph can be calculated by counting squares on graph paper.

A point moving with simple harmonic motion has a variable acceleration: graphs which describe the motion are plotted on p. 81.

2.3 Relative Velocity

The velocity of a point A relative to a point B is the velocity which A *appears* to have to an observer who is moving with B. The velocity of B relative to A is a vector of equal magnitude in the opposite direction.

To calculate the velocity of A relative to B we adopt this procedure (fig. 2.1).

A velocity $(-\boldsymbol{v}_B)$ equal and opposite to that of B has been superimposed on both A and B, so that B has been brought to rest. $\boldsymbol{v}_{AB}$ is A's new velocity, which is found by vector addition.

$$\boldsymbol{v}_{AB} = \boldsymbol{v}_A + (-\boldsymbol{v}_B)$$
$$= \boldsymbol{v}_A - \boldsymbol{v}_B$$

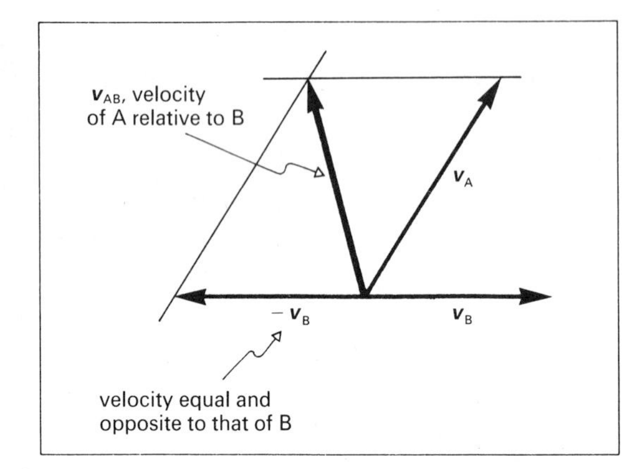

Fig. 2.1. Calculation of relative velocity.

Similarly $\boldsymbol{v}_{BA} = \boldsymbol{v}_B - \boldsymbol{v}_A = -\boldsymbol{v}_{AB}$

Thus to find the velocity of A *relative to* B, *we subtract (vectorially) the velocity of* B *from that of* A.

2.4 Parabolic Motion

Suppose a point is made to move in two dimensions as shown in fig 2.2.

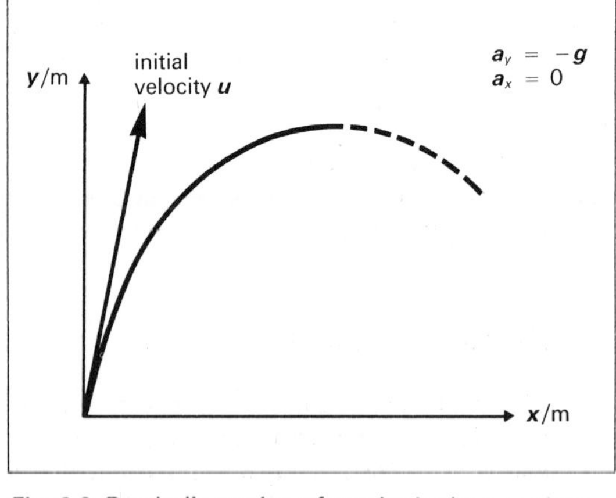

Fig. 2.2. Parabolic motion of a point in the x–y plane.

Suppose that for this motion

$$\text{acceleration in the } y\text{-direction } \boldsymbol{a}_y = -g$$
$$x\text{-direction } \boldsymbol{a}_x = 0$$
$$\text{initial velocity in the } y\text{-direction} = \boldsymbol{u}_y$$
$$x\text{-direction} = \boldsymbol{u}_x$$

It is convenient to treat the x- and y-motions independently. Thus using

(*a*) $v^2 = u^2 + 2as$, we have
$v_y^2 = u_y^2 - 2gy$ and
$v_x^2 = u_x^2$.

(*b*) $\boldsymbol{s} = \boldsymbol{u}t + \frac{1}{2}\boldsymbol{a}t^2$, we have
$\boldsymbol{y} = \boldsymbol{u}_y t - \frac{1}{2}\boldsymbol{g}t^2$ and
$\boldsymbol{x} = \boldsymbol{u}_x t$.

We can put $t = x/u_x$ to eliminate t from the last two equations, giving

$$y = u_y \frac{x}{u_x} - \tfrac{1}{2}g\left(\frac{x}{u_x}\right)^2$$

This relation between x and y is called the **trajectory equation**, and shows that the point describes a *parabolic path*. The motion is described by the graphs of fig. 2.3 (opposite).

The graphs should be read together with the comments on p. 29.

Examples of Parabolic Motion

A point describes a parabolic trajectory when it experiences a uniform acceleration in one direction while having a component of velocity in some other direction. This may occur

(*a*) in motion under gravity, as implied in this section,

(*b*) when an electrically charged particle is subject to a *uniform electric field* (p. 315).

2.5 Angular Motion

Suppose that a line is changing its orientation with time, and that it rotates through an angle $\Delta\theta$ in a time interval Δt.

The average angular velocity ω_{av} during the interval is defined by the equation

$$\omega_{av} = \frac{\Delta\theta}{\Delta t}$$

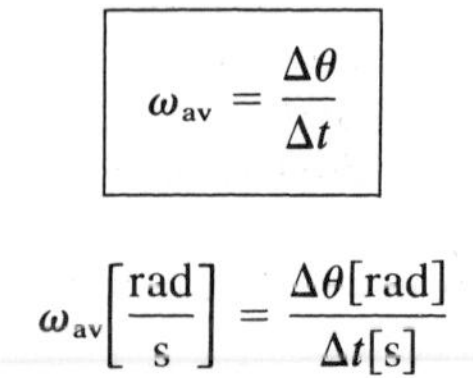

$$\omega_{av}\left[\frac{\text{rad}}{\text{s}}\right] = \frac{\Delta\theta[\text{rad}]}{\Delta t[\text{s}]}$$

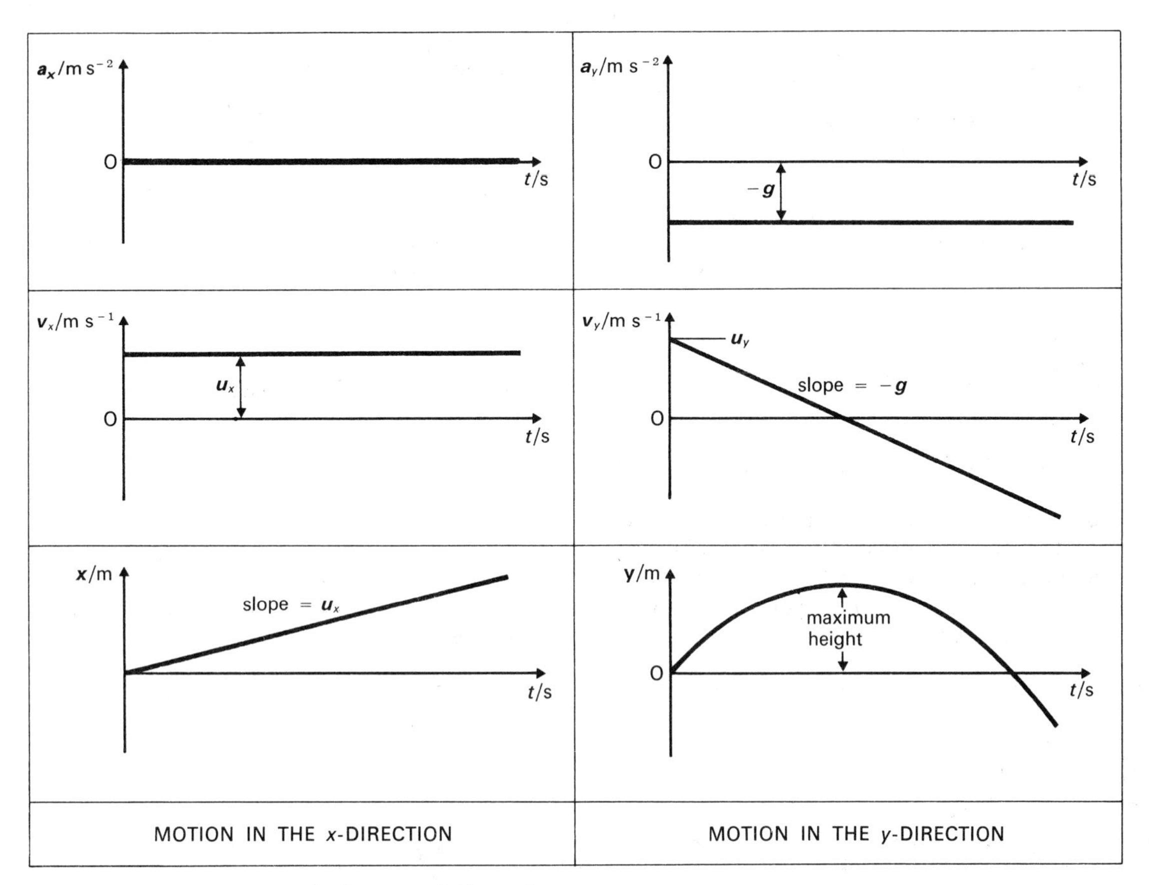

*Fig. 2.3. **a**–t, **v**–t and **s**–t graphs for a parabolic motion.*

The instantaneous angular velocity ω is defined by the equation

$$\boldsymbol{\omega} = \lim_{\delta t \to 0}\left(\frac{\delta \boldsymbol{\theta}}{\delta t}\right) = \frac{d\boldsymbol{\theta}}{dt}$$

Instantaneous angular velocity is a *pseudo-vector* quantity† which is usually measured in rad s^{-1}, and whose dimensions are [T^{-1}].

Frequently we are concerned with points moving at a fixed speed in a circular path (fig. 2.4). When we refer to the angular velocity of P, we mean the angular velocity of the radius vector OP drawn to P. If the linear speed v is fixed, then so is the angular velocity $\boldsymbol{\omega}$; we can write

$$\boldsymbol{\omega} = \frac{d\boldsymbol{\theta}}{dt} = \frac{\Delta\theta}{\Delta t}$$

† In fig. 2.4 the angular velocity of P is (by convention) directed into the paper, so as to obey a right-hand corkscrew rule.

The **period** T for one revolution is the time taken to describe an angle 2π radians, thus using

$$\Delta t = \frac{\Delta\theta}{\omega}$$

we have

$$T = \frac{2\pi}{\omega}$$

Since the point P covers a distance $2\pi r$ in time T at speed v

$$T = \frac{2\pi r}{v}$$

from which

$$v = \omega r$$

$$v\left[\frac{\text{m}}{\text{s}}\right] = \omega\left[\frac{1}{\text{s}}\right] r[\text{m}]$$

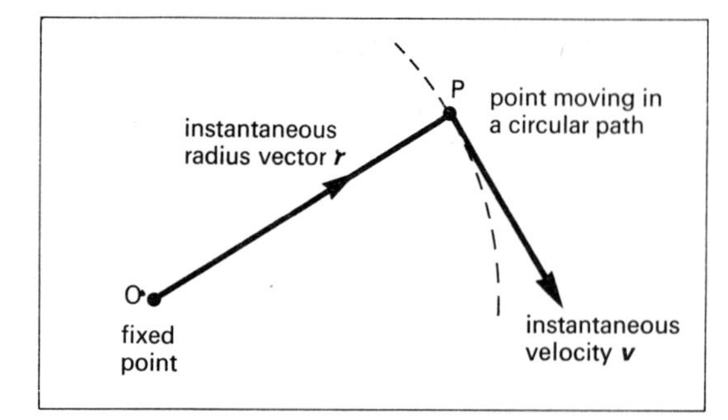

Fig. 2.4. The angular velocity of a point.

2.6 Uniform Motion in a Circle

Fig. 2.5 shows a point moving with instantaneous velocity $\boldsymbol{v}$ and constant speed v in a circle of radius r. The angular speed $\omega = v/r$. Since the direction of motion changes with time, the point is accelerated. We will calculate its acceleration.

Suppose the successive positions P_1 and P_2 are separated by an angular displacement $\Delta\boldsymbol{\theta}$. This corresponds to a time interval

$$\Delta t = \frac{\text{length of path}}{\text{speed}} = \frac{r\,\Delta\theta}{v}$$

(a) Average Acceleration $\boldsymbol{a}_{av}$

Since (from the diagram)

$$\Delta \boldsymbol{v} = 2v \sin\left(\frac{\Delta\theta}{2}\right)$$

towards the centre of the circle,

$$\boldsymbol{a}_{av} = \frac{\Delta \boldsymbol{v}}{\Delta t} = \frac{2v \sin(\Delta\theta/2)}{r\Delta\theta/v}$$

$$= \frac{v^2}{r} \cdot \frac{\sin(\Delta\theta/2)}{(\Delta\theta/2)}$$

and is also directed towards the centre of the circle, i.e. is **centripetal**.

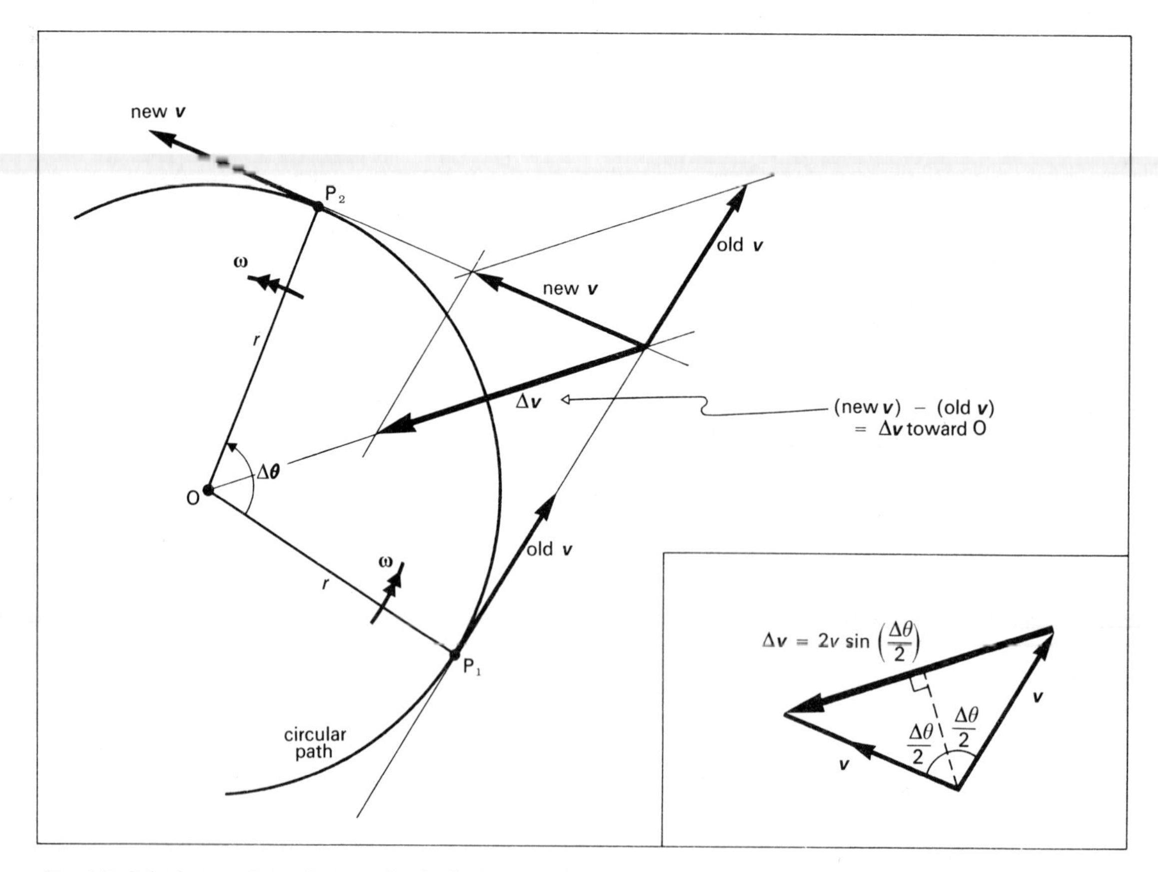

Fig. 2.5. Calculation of the change of velocity $\Delta\boldsymbol{v}$ over the time interval Δt.

(b) Instantaneous Acceleration $\boldsymbol{a}$

$$\boldsymbol{a} = \lim_{\Delta t \to 0} \left(\frac{\Delta \boldsymbol{v}}{\Delta t}\right)$$

$$= \frac{v^2}{r} \cdot \lim_{\Delta\theta \to 0} \frac{\sin(\Delta\theta/2)}{(\Delta\theta/2)}$$

$$= \frac{v^2}{r}$$

towards the centre of the circle.

$\boldsymbol{a}$ is the *limiting value* of the average acceleration.

$$a = \frac{v^2}{r} = \omega^2 r = v\omega$$

directed towards the centre of the circle

This is not uniformly accelerated motion, because the value of $\boldsymbol{a}$ is fixed only in magnitude, and not in direction. It is shown by a particle moving under a centrally directed force of constant size (see, for example, p. 39).

3 Newton's Laws

3.1 Introduction

Dynamics is the branch of mechanics which relates the properties of a body (such as its *mass*) and the effects of the body's environment (through an applied *force*) to describe the changes of motion of the body.

Classical mechanics confines itself to situations where the speeds of bodies relative to one another are small compared to c, the speed of electromagnetic radiation. Under these conditions (which are usually assumed in this book) **Newtonian mechanics**, as it is called, is known to be a very good approximation to **relavistic mechanics.** (See p. 2.)

A **particle** is an object without extent. In practice large bodies are treated as particles when they are small relative to their environment: thus when we analyse the dynamics of a car rounding a corner, we frequently ignore the fact that the mass of the car is distributed over a finite volume.

Momentum p

Suppose a body of mass m has an instantaneous velocity $\boldsymbol{v}$.

The momentum p of the body is defined by the equation

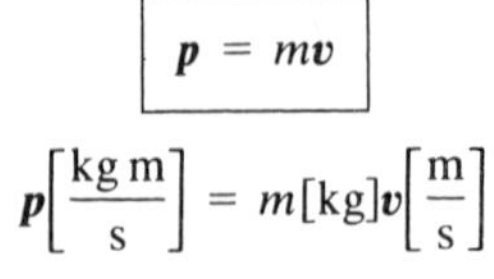

$$\boldsymbol{p} = m\boldsymbol{v}$$

$$p\left[\frac{\text{kg m}}{\text{s}}\right] = m[\text{kg}]v\left[\frac{\text{m}}{\text{s}}\right]$$

Notes. (*a*) 1 N s is a unit equivalent to $1\ \text{kg m s}^{-1}$, since $1\ \text{N} = 1\ \text{kg m s}^{-2}$ (p. 36).

(*b*) Because mass is a scalar and velocity is a vector, momentum is a *vector* along the direction of the velocity.

(*c*) Changes of momentum are calculated by *vector subtraction*, thus:

$$\Delta \boldsymbol{p} = \left(\begin{matrix}\text{final}\\ \text{momentum}\end{matrix}\right) - \left(\begin{matrix}\text{initial}\\ \text{momentum}\end{matrix}\right)$$

The concept of momentum is used in stating *Newton's Second Law*.

3.2 Newton's Laws of Motion

Law I *Unless a resultant force acts on a body, its velocity will not change.*

Law II *The rate of change of momentum of a body is proportional to the resultant force that acts on it.*

Law III *If body* A *exerts a force* $\boldsymbol{F}$ *on body* B, *then body* B *exerts a force* $\boldsymbol{F}$ *on body* A *of the same size and along the same line, but in the opposite direction.*

In symbols

Law I	If $\boldsymbol{F} = 0$, $\Delta \boldsymbol{v} = 0$
Law II	$\boldsymbol{F} \propto \frac{d}{dt}(m\boldsymbol{v})$
Law III	$\boldsymbol{F}_{AB} = -\boldsymbol{F}_{BA}$

Comments on Newton's Laws

(*a*) Law I gives us an intuitive meaning of force.

A **resultant force** *is that agent which changes the velocity (and momentum) of a body.*

(*b*) Laws I and II together enable us to define the meanings of, and measures for, both mass and force (p. 36).

(*c*) Law I is a special case of Law II.

(*d*) Law III refers to a pair of forces which must always act on two *different* bodies. These two forces have the same size at every instant of time.

(*e*) A pair of forces acting on the same body *cannot* be the 'action–reaction' pair referred to in Law III. Possible difficulty is avoided by referring to any force in full detail, such as 'the upward push exerted *by* the table *on* the book'.

(*f*) We cannot *prove Newton's Laws*, because they are assertions. *Newton* himself used the words '*Axiomata Sive Leges Motus*'—'The axioms or laws of motion'. Nevertheless we are justified in using them because (with the qualifications mentioned on p. 2) everything which they predict agrees with experimental observation.

3.3 Forces

Force is a *vector* quantity: this is an experimental result obtained by the following procedure.

(*a*) Apply several different forces in turn to a given body, and note the acceleration produced by each.

(*b*) Apply the same forces simultaneously. Their resultant produces the same acceleration as the vector sum of the independent accelerations.

This leads to the **superposition principle** for forces:

When several forces act on a body simultaneously, their single equivalent resultant can be found by vector addition.

The Origin of Forces

The *interactions* of nature are discussed on p. 3. A force experienced by one body shows us that it is interacting with another. In mechanics we are concerned with two types of force.

(*a*) *Gravitational forces* are so weak that we usually ignore them unless a massive body (such as the Earth) is part of the environment. The gravitational forces we discuss are always attractive.

(*b*) *Electromagnetic forces* are responsible for most of the effects that we observe (e.g. the push of a table on a book). They may be attractive or repulsive.

We say that bodies are *separated* when they do not exert appreciable contact forces on each other: these forces are electromagnetic in origin. When the separation becomes large, the interaction between two bodies takes place through a *long-ranged force*, such as a gravitational or electrostatic force obeying an inverse square law. (The *range* of a force is discussed fully on p. 120.)

Description of Forces

A force is described by specifying the following:

(*i*) its magnitude and direction,

(*ii*) the body on which it acts, and which part of the body,

(*iii*) the body that exerts the force, and

(*iv*) the nature of the force.

The gravitational forces that we discuss are always pulls, and are usually referred to as *weights*. The electromagnetic forces can be pushes or pulls, and are referred to variously as normal *contact* forces, *frictional* forces, fluid *upthrusts*, *lift* and *drag* forces, *electrostatic* and *magnetic* forces.

(*a*) *Weights* are discussed on p. 37.

(*b*) *Normal contact forces* are brought into play when the molecules of one body are so close to those of another that their respective electrons and atomic nuclei repel each other. The bodies are then said to *touch*. When the force is always normal to the bodies' common surface, then one of the surfaces is said to be *smooth*. If the bodies are *rough*, the resultant force is the vector sum of the normal force and the frictional force.

(*c*) *Tension* is the condition of a body subjected to equal but opposite forces which attempt to increase its linear dimensions along their line of action. Tension forces are then the *pulls* exerted by stretched strings or rods on the bodies to which they are attached, and act along the direction of the string or rod. If frictional forces are negligible the magnitude of a tension force is not altered by a change of direction of a string.

Compression is the opposite condition: a rod, but not a string, can exert a compressive force, which we think of as a *push*.

(*d*) *Frictional forces* act along the common surface between two bodies in such a direction as to oppose the *relative* movement of the two bodies.

Thus a car is accelerated by the frictional push of the road on the tyres, a force which prevents the tyres from sliding on the road. The *moving* car experiences a drag force from the air through which it moves.

As with all forces, frictional forces occur in pairs, and one must be careful to specify which force one is considering.

The Coefficient of Friction

Fig. 3.1 is a free-body diagram (p. 39) for a book resting on a table.

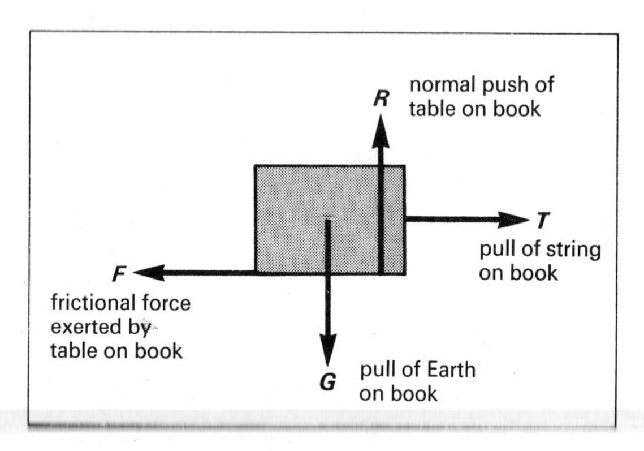

Fig. 3.1. Frictional forces.

It is found that

(1) $\boldsymbol{F}$ and $\boldsymbol{T}$ are in opposite directions,

(2) $F \leqslant T$,

(3) $F \leqslant F_{max}$, the largest value that $\boldsymbol{F}$ can take,

(4) $F_{max} = \mu R$, where μ is a number (without unit) called the **coefficient of friction** for that pair of surfaces.

Comments

(*i*) $\boldsymbol{F}_{max}$ is controlled by the material and cleanliness of the surfaces in contact, but not by the common surface area.

(*ii*) The value of F is only as large as is necessary to maintain equilibrium (when this is possible), and will frequently be less than μR.

(*iii*) When sliding occurs the frictional force may become smaller, but we usually approximate by setting it equal to μR.

(*iv*) If either surface is *smooth*, then $\mu = 0$ and $\boldsymbol{F} = 0$.

3.4 Discussion of Newton's Second Law

On p. 34 Newton's Second Law was stated in the form

$$\boldsymbol{F} \propto \frac{\mathrm{d}}{\mathrm{d}t}(m\boldsymbol{v})$$

or

$$\boldsymbol{F} = k\frac{\mathrm{d}}{\mathrm{d}t}(m\boldsymbol{v})$$

where k is a non-dimensional constant. In classical mechanics, if we have a *body* (of constant mass)

$$\boldsymbol{F} = km\frac{\mathrm{d}\boldsymbol{v}}{\mathrm{d}t} + k\boldsymbol{v}\frac{\mathrm{d}m}{\mathrm{d}t}$$

$$= km\frac{\mathrm{d}\boldsymbol{v}}{\mathrm{d}t} \qquad \left(\text{Since } \frac{\mathrm{d}m}{\mathrm{d}t} = 0\right)$$

$$= km\boldsymbol{a}$$

We have units for m [kg] and $\boldsymbol{a}$ [m s^{-2}] but not for $\boldsymbol{F}$. We then *choose* $k = 1$, and in so doing we also *define our unit for force.*

$$\boldsymbol{F}[\mathrm{N}] = m[\mathrm{kg}]\,\boldsymbol{a}\left[\frac{\mathrm{m}}{\mathrm{s}^2}\right]$$

1 **newton** (N) *is that force which accelerates a mass of* 1 kg *at* 1 m s^{-2}.

$$1\ \mathrm{N} = 1\ \mathrm{kg\,m\,s^{-2}}$$

The newton is an **absolute** force unit.

$\boldsymbol{F} = m\boldsymbol{a}$	*One form of Newton's Second Law*

Notes. (*a*) $\boldsymbol{F}$ is the *resultant* force experienced by the body, i.e. the vector sum of all the forces acting on the body.

(*b*) Writing $\boldsymbol{a} = \boldsymbol{F}/m$, in which $\boldsymbol{a}$ and $\boldsymbol{F}$ are vectors, we see

(*i*) $\boldsymbol{a} \propto \boldsymbol{F}$ for a given m,

(*ii*) $\boldsymbol{a}$ occurs along the same direction as $\boldsymbol{F}$, and

(*iii*) $\boldsymbol{a} \propto 1/m$ for a given $\boldsymbol{F}$.

(*c*) $\boldsymbol{F} = m\boldsymbol{a}$ defines the procedure for the *measurement of force*: we measure the acceleration it produces in a known mass (such as the standard kilogram).

The Principle of the Measurement of Mass

The **inertia** of a body is its ability to resist a change of motion.

*The **mass** of a body is the quantitative measure of that inertia (for linear motion).*

Suppose the same force $\boldsymbol{F}$ acts on two masses, m_0, the standard kilogram, and m, a mass to be measured. ($\boldsymbol{F}$ can be produced by extending a given spring by a fixed amount.)

Name	Physical quantity	Unit for measurement	Is a measure of	Measured by	Variation
mass	mass (scalar)	kilogram	inertia	comparison with standard mass	does not vary
weight	force (vector)	newton	gravitational attraction of the Earth	calibrated spring balance	variable

We define the ratio of the two masses by

$$\frac{m_0}{m} = \frac{a}{a_0}$$

where $\boldsymbol{a}$ and $\boldsymbol{a}_0$ are the accelerations produced in the unknown and standard masses respectively.

Since $\boldsymbol{a}$ and $\boldsymbol{a}_0$ can be measured directly, and m_0 is known (by definition), the value of m can be found.

Notes. (*a*) If a different common force is used, and gives accelerations $\boldsymbol{a}'$ and $\boldsymbol{a}'_0$, then *experiment* shows

$$\frac{m_0}{m} = \frac{a}{a_0} = \frac{a'}{a'_0}$$

The ratio of the masses does not depend on the common force.

(*b*) The procedure can be used to compare any (non-standard) masses:

$$\frac{m_1}{m_2} = \frac{a_2}{a_1}$$

(*c*) The experiment can be used to show that *mass* is a *scalar* quantity. Two bodies of masses m_1 and m_2, when fastened together behave as would a mass $(m_1 + m_2)$.

(*d*) In this section we show how masses can be compared *in principle*; *in practice* (p.12) different methods are simpler.

3.5 Mass and Weight

In this book the **weight** of a body is defined to be the gravitational attraction of a massive body (such as the Earth) for that body. *Weight* is thus the name given to a particular type of force, and must be distinguished from mass (see table).

(Note that the term *weight* is not always defined in this way: some authors define it to be the downward force exerted by the body on its support.)

The pull of the Earth on a body is controlled by the *gravitational mass* (p.152): the resistance to a change of motion depends on the *inertial mass*. Because the gravitational and inertial masses are proportional (p. 84), inertial mass is closely related to weight, *but is a different physical quantity*.

Variation of Weight and Constancy of Mass

(*a*) To obtain an intuitive feeling for the *mass* of a body, we frequently apply by hand a force to it, and sense the resistance it offers when undergoing a particular acceleration. This force *can* be applied *horizontally*.

(*b*) To measure the *weight* of a body, we frequently apply by hand a force to it which maintains it in approximate equilibrium. This force *must* be applied *vertically*.

Suppose we do this for a body of mass 6 kg, both at the surface of the Earth and at the surface of the Moon.

(*a*) To accelerate the body at $10\,\mathrm{m\,s^{-2}}$ horizontally, we apply a force

$$\begin{aligned} F &= ma \\ &= (6\,\mathrm{kg}) \times (10\,\mathrm{m\,s^{-2}}) \\ &= 60\,\mathrm{N} \end{aligned}$$

both on the Earth, and on the Moon.

The inertia (mass) is the same in both cases.

(*b*) To hold the body in approximate equilibrium, we apply an upward vertical force.

(*i*) On Earth $F = 60\,\mathrm{N}$.

(*ii*) On the Moon we would find by experiment that $F = 10$ N.

The weight on the Moon is 1/6 its value on Earth.

This difference between weight and mass would be emphasized if one were to kick a massive body at a place where the gravitational field was vanishingly small.

Measurement of Weight

Suppose that $\boldsymbol{g}$ is the acceleration, relative to the surface of the Earth, of a body in a state of free-fall. Then if we ignore the effect of the Earth's rotation about its axis, we can in principle use

$$\boldsymbol{F} = \text{ma}$$

in the form

$$\boldsymbol{G} = m\boldsymbol{g}$$

Measurement of m and $\boldsymbol{g}$ would enable us to calculate $\boldsymbol{G}$. In practice $\boldsymbol{G}$ can be measured directly by using a calibrated spring balance. Nevertheless the interpretation of the balance reading is complicated, because

(*a*) $\boldsymbol{g}$ varies from point to point on the Earth, and

(*b*) the surface of the Earth is an accelerated frame of reference.

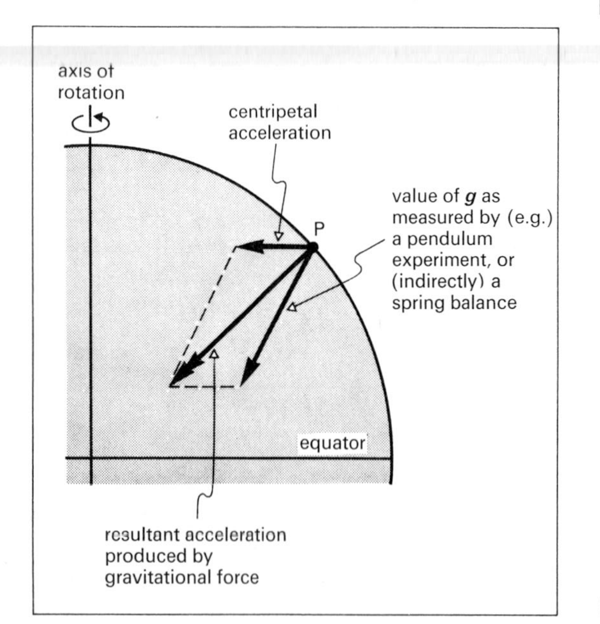

*Fig. 3.2. The effect of the Earth's rotation on the measured value of **g** (greatly exaggerated).*

Fig. 3.2 shows that observed values for $\boldsymbol{g}$ do not give an exact measure for the pull of the Earth on a body: $m\boldsymbol{g}$ and the true value for $\boldsymbol{G}$ differ in both magnitude and direction.

Gravitational Force Units

These are force units related to the *apparent* pull of the Earth on a standard mass.

1 **kilogram weight** (kgwt) *is the apparent pull of the Earth on a mass of* 1 kg.

1 **kilogram force** (kgf) *is the apparent pull of the Earth on a mass of* 1 kg *at a place where* $g_0 = 9.806\,65\ \text{m s}^{-2}$ (*standard gravitational acceleration*).

It is seen that the kilogram weight has a variable magnitude, so a physicist chooses to use a newton, which has a constant magnitude.

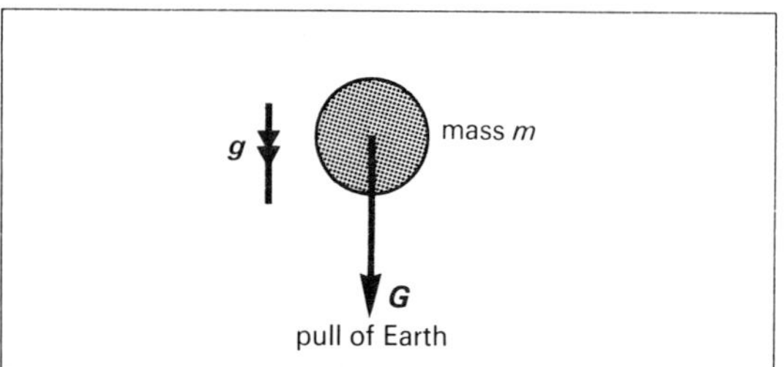

Fig. 3.3. A body in a state of free-fall in a vacuum.

Refer to fig. 3.3, in which $\boldsymbol{G}$ can be expressed in two ways:

(*a*) using $\boldsymbol{F} = m\boldsymbol{a}$

$$G[\text{N}] = m[\text{kg}]\ g\left[\frac{\text{N}}{\text{kg}}\right], \quad \text{or}$$

(*b*) G is numerically equal to m, if G is expressed in kgwt, and m in kg.

Thus we can relate gravitational and absolute force units:

1 kwgt is about 9.8 N, whereas
1 kgf = 9.806 65 N.

It should be noted that *neither* of these units is an SI unit, and their use should be avoided.

3.6 Examples of the Use of $\boldsymbol{F} = m\boldsymbol{a}$

In

$$\boldsymbol{F} = m\boldsymbol{a}$$

$\boldsymbol{F}$ is the sum of the resolved parts of the forces in a particular direction, and $\boldsymbol{a}$ is the acceleration of the body in that direction.

To apply the law, adopt the following procedure.

(1) Draw the diagram that represents the general situation, marking all bodies in the environment.

(2) Select *one body* from the situation whose motion is to be analysed, and draw a **free-body diagram** for that body. For this the body is removed from its environment, together with *all the forces* exerted *on* it by the bodies with which it interacts. These forces should be marked in absolute units.

(3) Select and orientate convenient coordinate axes, marking the acceleration $\boldsymbol{a}$ in the direction to be considered.

(4) Calculate the value of $\boldsymbol{F}$ defined above, and apply the law.

One common mistake is to mark the forces exerted *by* the chosen body: these are not relevant.

(Learning these rules is no guarantee of facility in applying them: practice is essential.)

(*a*) Linear Motion

For the situations (*b*) and (*c*) of fig 3.4, *Newton's Second Law* is written

$$(\boldsymbol{R} - \boldsymbol{w}) = m\boldsymbol{a} \quad \text{for load}$$

$$(\boldsymbol{T} - \boldsymbol{R} - \boldsymbol{W}) = M\boldsymbol{a} \quad \text{for lift}$$

where m, M are masses, and $\boldsymbol{w}$, $\boldsymbol{W}$ weights.

(*b*) Uniform Circular Motion

Every body moving in a circular path is accelerated (p. 32). For the special case of a body moving at speed v in a circle of radius r, the centripetal accceleration is always perpendicular to the instantaneous velocity, and has magnitude v^2/r. Using

$$\boldsymbol{F} = m\boldsymbol{a}$$

we have

$$F = \frac{mv^2}{r}$$

as the constant magnitude of the inward force that acts on such a body of mass m.

$$F = m\frac{v^2}{r} = m\omega^2 r = mv\omega$$

centripetal force

Notes. (*a*) *Newton's Third Law* indicates that there is an action–reaction pair in such a situation. The body under study *experiences a centripetal force* (towards the centre)—in some situations it may *exert a* **centrifugal** *force* (away from the centre) on some body in the environment. Since this centrifugal force is not acting *on* the body under study, it is not relevant to our analysis.

(Note that in general there is *no* centrifugal force. Two bodies such as the Sun and a planet attract

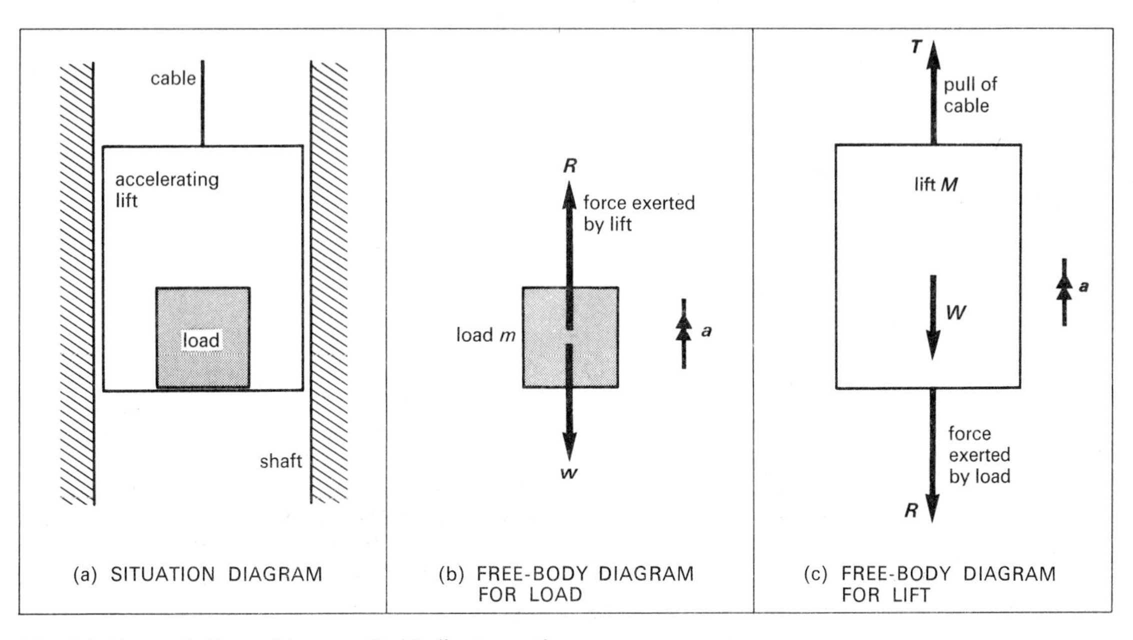

Fig. 3.4. Newton's Second Law applied to linear motion.

mutually through a field of force: the force that each experiences and exerts is centrally directed.)

(*b*) The centripetal force is perpendicular to the instantaneous velocity. Sudden removal of this force would allow the body to assume uniform motion, which would be along the tangent to the circle.

(*c*) *Newton's Laws* are correctly applied by an observer in a *non-accelerated* (or **inertial**) *frame of reference*. A passenger in a car turning a corner is *not* such an observer, whereas (if we ignore the Earth's rotation) an observer standing on the pavement *is*. The passenger *seems* to experience an outward force. The passenger actually *experiences* a resultant *inward* force *from* the car upholstery: he *exerts* an equal *outward* force *on* the upholstery.

Examples of Centripetal Forces

Situation	*Description of centripetal force*
(1) Conker being whirled in a horizontal circle	(1) Force exerted by string (which is in tension)
(2) Earth in orbit around the Sun	(2) Gravitational force exerted by Sun
(3) Car turning a corner	(3) Frictional force exerted by road on tyres
(4) Aircraft banking	(4) Normal push of air
(5) Electron in hydrogen atom	(5) *Coulomb* attraction exerted by proton in nucleus
(6) Electron describing a circular path in a magnetic field	(6) Magnetic force exerted by agent that sets up magnetic field

(7) *The banking of a bend on a road*

Problem: at what angle should a road be banked at a bend to avoid the need for a centripetal frictional force? (fig. 3.5).

Treat the car as a particle (p. 34) and resolve parallel to the line of greatest slope.

$$\boldsymbol{F} = m\boldsymbol{a}$$

becomes
$$mg\sin\theta = m\left(\frac{v^2}{r}\cos\theta\right)$$

$$\tan\theta = \frac{v^2}{rg}$$

The equation (which should *not* be learned) shows that a bend of given radius r can only be banked in this way for cars of a given speed v.

(c) Simple Harmonic Motion

Further examples of the technique of using $\boldsymbol{F} = m\boldsymbol{a}$ are given in the section on s.h.m. (p. 78). In these examples a body is subjected to a force which varies with displacement and with time.

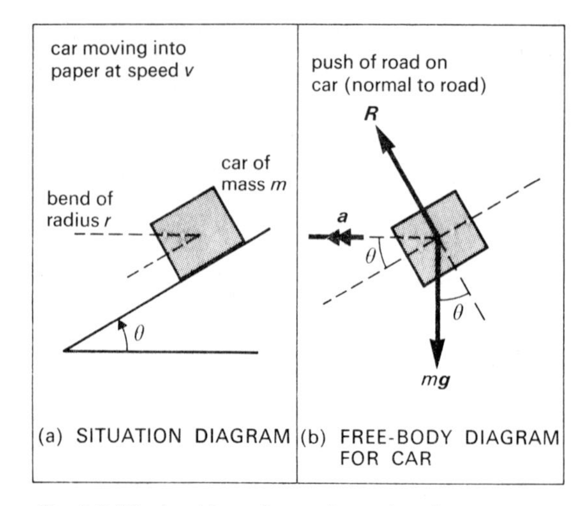

Fig. 3.5. The banking of a road on a bend.

3.7 Skidding and Toppling

Skidding

If a car of mass m travelling at speed v is to round a bend of radius r, it must experience a centripetal force mv^2/r. The *maximum* value of the centripetal frictional push F_{max} exerted by the road on the tyres is given by

$$F_{max} = \mu R = \mu mg \qquad \text{(p. 36)}$$

If the necessary centripetal force cannot be provided by this frictional push, then the car will skid.

So skidding occurs if

$$\frac{mv^2}{r} > \mu mg$$

or
$$v > \sqrt{\mu gr}$$

Toppling

Under certain conditions the car will topple before it skids. A quantitative treatment of this situation is complicated by the spin angular momentum (p. 59). (We usually ignore the rotation of the car about its own centre of mass by treating it as a *particle*, but here we must apply forces to *different parts* of the car.)

The tendency to topple can be reduced by

(*a*) lowering the centre of mass of the car, and
(*b*) increasing the width of the axle.

Thus a racing car holds the road better if it has large tyres, and a low-slung frame with widely-spaced wheels.

*3.8 The Tension in a Rotating Hoop

Fig. 3.6 shows a hoop of radius r rotating at angular speed ω about an axis through its own centre, and perpendicular to its plane.

The net centripetal force acting on the element of mass shown is

$$F = 2T\cos(90° - \Delta\theta)$$
$$\approx 2T\Delta\theta$$

if $\Delta\theta$ is small.

Apply $\boldsymbol{F} = m\boldsymbol{a}$ parallel to CO:

$$2T\Delta\theta = m\omega^2 r$$
$$= (2\mu r\,\Delta\theta)\omega^2 r$$
$$\therefore \quad T = \mu r^2\omega^2$$

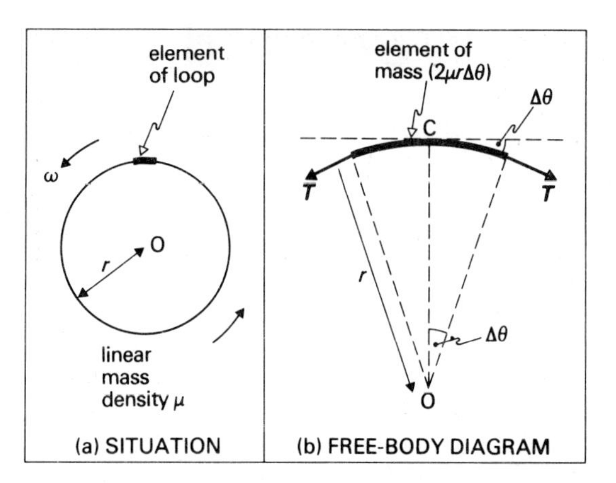

Fig. 3.6. A rotating hoop is in a state of tension.

The tension T in (for example) a car tyre is therefore proportional to the square of the speed. The stretching associated with it may cause the tyre to leave the rim of the wheel.

4 Work and Energy

4.1 Work

In physics the term **work** always has this meaning:

Work is done when a force moves its point of application so that some resolved part of the displacement lies along the direction of the force.

(a) Work Done by a Constant Force

Refer to fig. 4.1, which shows a constant force $\boldsymbol{F}$.

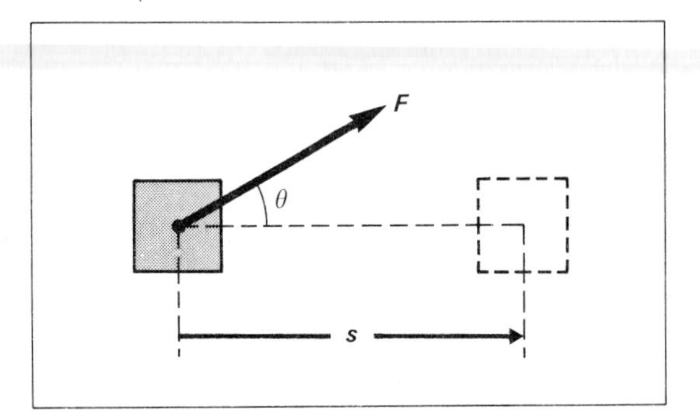

Fig. 4.1. Definition of work.

We define the work W done by the force $\boldsymbol{F}$ by the equation

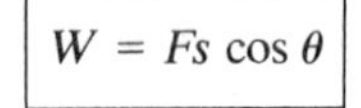

$$W = Fs \cos\theta$$

$$W[\mathrm{N\,m}] = F[\mathrm{N}]\, s[\mathrm{m}] \cos\theta$$

Notes (a) W may be visualized as

(i) (force)$\cdot$(resolved part of displacement in direction of force), or

(ii) (resolved part of force in direction of displacement)$\cdot$(displacement).

In each case we obtain the same result.

(b) Work is a *scalar* quantity (obtained by a particular way of multiplying two vectors).

(c) The unit *newton metre* (N m) occurs so frequently it is conveniently called a **joule** (J): thus

$$1 \text{ joule} = 1 \text{ newton metre}$$

*(b) Work Done by a Variable Force

The equation

$$W = Fs\cos\theta$$

is more correctly written

$$\Delta W = F\Delta s \cos\theta$$

If the magnitude of $\boldsymbol{F}$ varies with s, the work δW done over a very small displacement δs is given by

$$\delta W = F\cos\theta\,\delta s$$

The work done over a finite displacement can be calculated from

$$W = \int_{s_1}^{s_2} F\cos\theta\, \mathrm{d}s$$

Examples are given on pp. 157 and 325.

Examples of work done by forces

(a) If the *force does not move*, the work done is *zero*; e.g. a man supporting a beam. (There will, nevertheless, be some conversion of chemical energy to thermal energy by muscular action.)

(b) If the *force is perpendicular to the movement* of the body, the work done is *zero*; e.g. the centripetal push of a road on a car, or the magnetic force acting on a charged particle made to move in a circular path. (The force does not change the body's speed.)

(c) If the *force has a resolved part in the same direction as the displacement*, then by our definition, the work is *positive*; e.g. the work done by a mass of gas when it increases its volume by pushing back its surroundings (p. 186). (This is consistent with the convention in thermodynamics which agrees to call the work done *by* the substance positive.)

(*d*) If the *force* has a resolved part along the line of the displacement which is *oppositely directed*, then the work done by that force is negative. (Since $\theta > 90°$, $\cos\theta$ is negative.) Thus when sliding occurs between two surfaces in contact, the work done by the frictional forces is always negative, since the forces oppose relative motion.

4.2 Energy

Suppose body A exerts a contact force on body B, and does positive work: it follows that body B does negative work. Body A, having done positive work, is said to lose energy, while body B, having done negative work, is said to gain energy. *The work done measures the interchange of energy between A and B.*

***A body which can do work has* energy.**

Orders of Magnitude

Description	*Amount of energy W/J*
binding energy of Earth–Sun system	10^{33}
energy radiated by Sun during 1 second	10^{26}
energy released by annihilation of 1 kg of matter	10^{17}
energy released by fission of 1 mole of $^{235}_{92}U$	10^{13}
electrical energy dissipated in a lightning discharge	10^{10}
energy released by combustion of 1 kg of petrol	10^{8}
energy required to charge a car battery	10^{6}
energy provided by a slice of bread	10^{5}
kinetic energy of a bowled cricket ball	10^{1}
energy stored in the electric field of a 10 μF capacitor, charged to 20 V	10^{-3}
maximum energy of proton in CERN synchrotron	10^{-8}
work function of tungsten	10^{-18}
energy of visible light photon	10^{-19}

Energy and work are both scalars, and have the same dimensions and unit (joule). Our concept of energy is closely tied to our concept of force. In this book we are mainly concerned with gravitational and electromagnetic forces, and most of the energies we consider are derived from these two interactions.

Nevertheless, as with forces (p. 35) it is usually convenient to discuss energies in a descriptive way, and the following terms are used:

(*a*) kinetic energy
(*b*) potential energy—gravitational
—elastic
—electrostatic
(*c*) thermal and internal energy
(*d*) radiant energy
(*e*) chemical energy
(*f*) nuclear energy
(*g*) mass energy.

On a microscopic scale, *all* forms of energy can be classified as either kinetic or potential.

Changes occur between different forms of energy, and the amounts possessed by different bodies, but, if we take all forms into account, we find

There is no change in the total energy in the Universe. This is the **Law of the Conservation of Energy**.

If we *appear* to lose energy in some process, then rather than abandon this law, we look for a different form of energy.

4.3 Forms of Energy

(a) Kinetic Energy E_k

Suppose that a number of forces acting on a body of mass m have a resultant $\boldsymbol{F}$, and that by acting over a displacement $\boldsymbol{s}$ (in the direction of $\boldsymbol{F}$), $\boldsymbol{F}$ does work on the body, and thereby changes its velocity from $\boldsymbol{u}$ to $\boldsymbol{v}$.

The acceleration $\boldsymbol{a}$ produced by $\boldsymbol{F}$ is uniform, and is therefore related to $\boldsymbol{u}$, $\boldsymbol{v}$ and $\boldsymbol{s}$ by

$$v^2 = u^2 + 2as \qquad \text{(p. 29)}$$

Using $\quad \boldsymbol{F} = m\boldsymbol{a}$

and $\quad W = Fs$

we have

$$\begin{aligned} W &= mas \\ &= m\left(\frac{v^2}{2} - \frac{u^2}{2}\right) \\ &= \tfrac{1}{2}mv^2 - \tfrac{1}{2}mu^2 \end{aligned}$$

The work done by the force equals the change of the quantity $\frac{1}{2}m(\text{velocity})^2$.

The kinetic energy E_k of a body is defined by the equation

$$\boxed{E_k = \tfrac{1}{2}mv^2}$$

$$E_k[\text{J}] = \tfrac{1}{2}m[\text{kg}]\, v^2\left[\frac{\text{m}^2}{\text{s}^2}\right]$$

The kinetic energy (k.e.) is a positive *scalar* quantity that represents the energy associated with the body *because of its motion*. It is equal to either

(*i*) the work done *by* the resultant force in accelerating the body from rest to an instantaneous speed v, or

(*ii*) the work done *by* the body on some external agent which brings it to rest.

(*i*) and (*ii*) are equivalent.

The relationship

$$\boxed{W = \tfrac{1}{2}mv^2 - \tfrac{1}{2}mu^2}$$

is sometimes called the **work-energy theorem**.

In words

$$\left(\begin{array}{c}\textit{the work done by}\\ \textit{the forces acting on}\\ \textit{the body}\end{array}\right) = \left(\begin{array}{c}\textit{the change of the}\\ \textit{kinetic energy of the}\\ \textit{body}\end{array}\right)$$

It is a very powerful principle, and is frequently used in solving problems.

(*b*) Potential Energy

Potential energy (p.e.) *is the energy possessed by a system by virtue of the relative positions of its component parts.*

Examples (*i*) *Gravitational p.e.*

(For a fuller discussion refer to p. 156.) Suppose we exert forces on a body of mass m and on the Earth, and thereby push the body to a rest position a vertical distance h above its initial position.

We do work $\quad W = \boldsymbol{F} \cdot \boldsymbol{s}$

$$E_p = mgh,$$

and yet there is no gain of k.e. The pulls of the Earth on the body and the body on the Earth have done *negative* work: we say the system has gained *gravitational p.e. mgh*.

When the system is released the two gravitational forces both do positive work on the body and on the Earth. Both, in principle, acquire k.e., but that gained by the Earth is negligible. The p.e. is associated with the relative positions (i.e. separation) of the two masses making up the system.

(*ii*) *Elastic p.e.* (*stretching*)

Suppose we stretch a spring of force constant k from its natural length until the extension is x. Then we do work W given by

$$\begin{aligned} W &= (\text{average force})\cdot(\text{distance}) \\ &= (\tfrac{1}{2}kx)\cdot(x) \\ E_p &= \tfrac{1}{2}kx^2 \end{aligned}$$

The spring also exerts a force of equal magnitude which is oppositely directed: this force does *negative* work, and we say the spring has gained elastic p.e. $\frac{1}{2}kx^2$.

If there were a body attached to the spring, and the spring released, the spring force would do positive work on the body, which would cause a gain of k.e. The spring force does this work by virtue of the change in the spring's *shape* during deformation. The components of this system are the molecules that make up the spring, and the shape of the spring determines their relative positions.

(*iii*) *Elastic p.e.* (*twisting*)

Suppose that a pair of equal but oppositely directed torques $\boldsymbol{T}$ cause a rod or wire to be twisted through an angular displacement of θ. Then the equation

$$T = c\theta$$

defines the torsion constant c of that system (p. 134). The external agent that caused the twist has done work

$$\begin{aligned} W &= (\text{average torque})\cdot(\text{angle}) \qquad \text{(p. 57)} \\ &= (\tfrac{1}{2}c\theta)\cdot(\theta) \\ \therefore \quad E_p &= \tfrac{1}{2}c\theta^2 \end{aligned}$$

This equals the negative work done by the rod. There is no gain of k.e., and so it represents the gain of the elastic p.e. of the system.

The components of the system are again the molecules that make up the rod. (The energy is electromagnetic in origin.)

(c) Internal Energy

A frictional force always opposes relative motion, and when surfaces slide over one another, such a force always does negative work. This work represents energy being transferred to random molecular potential and kinetic energy (**internal energy**). (Frequently a negligible quantity of sound energy is produced, and this is eventually dissipated as internal energy.)

In fig. 4.2, B exerts a frictional force on A to the right, which moves its point of application to the left, and so does negative work. Macroscopically we see that A experiences a force which reduces its speed. Microscopically work is being done on a molecular scale that results in an increase of the random kinetic and potential energies of individual molecules. We observe a temperature increase along the common surface.

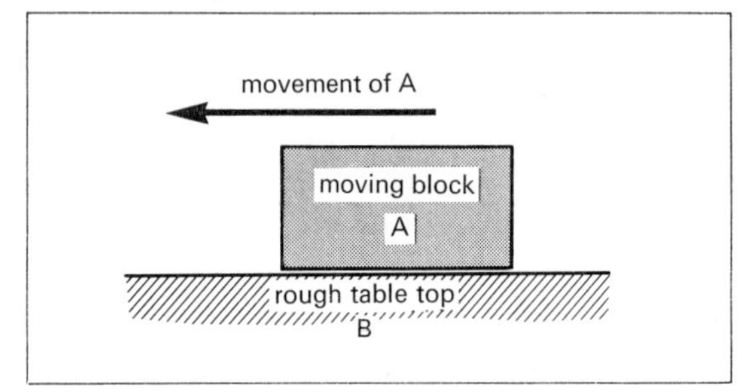

Fig. 4.2. Conversion of ordered energy to internal energy.

4.4. Energy Interchanges

Consider a block sliding down a rough surface in the Earth's gravitational field. We can view the problem in three ways:

(1) The *total* energy of the system is constant (energy conservation). This is not helpful for problem solving because of the difficulty of keeping account of the internal energy.

Suppose now that *frictional forces are absent.*

(2) We can apply the work-energy theorem to the *block*

$$W = \text{change of body's k.e.}$$

where W is the work done by the environment *on* the body.

(3) We can view the *system* (Earth and block) as a whole, and say that any change of k.e. is accounted for by a reduction of the p.e. of the Earth-block system. Thus

$$(\text{gain of k.e.}) = (\text{loss of p.e.})$$

The definition of gravitational potential energy has led to a useful conservation law.

Law of conservation of mechanical energy

In the absence of forces other than gravitational forces the sum of the kinetic energies and mutual potential energy of the bodies in an isolated system remains constant.

This law is true by the definition of p.e.

Some Examples of Energy Interchange

Situation	*Energy converted*		*Is mechanical energy conserved?*
	From	*Into*	
Bullet strikes a block and becomes embedded	k.e. of bullet	internal energy of block and bullet	no
Ball thrown into the air	k.e. of ball	p.e. of ball	yes, if we ignore air resistance
Pendulum swings in a vacuum	alternately k.e. of bob and p.e. of bob	alternately p.e. of bob and k.e. of bob	yes
Loudspeaker cone vibrates	(1) k.e. of cone (2) ordered energy of sound wave	(1) ordered energy of sound wave (2) internal energy of environment	(1) yes, but difficult to account for (2) no

*Potential Energy and Conservative Forces

Potential energy can be defined for those forces which are **conservative**, such as the gravitational and electrostatic forces (p. 323).

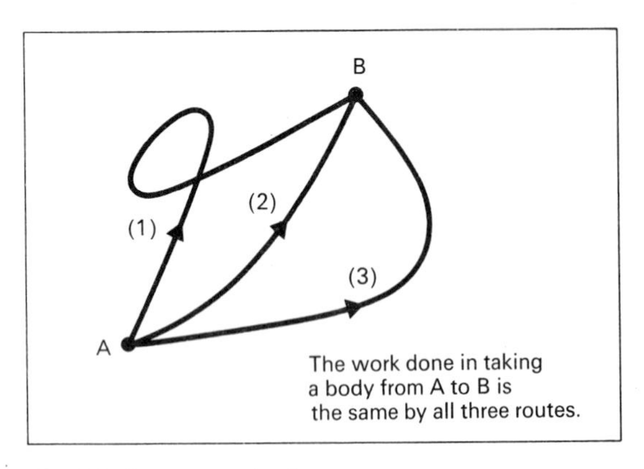

Fig. 4.3. A conservative force.

A conservative force is one such that the work done in taking a body (or charge) from one point to another does *not* depend on the route taken (fig. 4.3). This implies

(*a*) that the difference of potential between two points depends only on the location of those points, and

(*b*) if a conservative force takes a body round a closed path (i.e. back to the starting point) then the total work done is zero.

The force of kinetic friction depends on the direction of a body's velocity, not just its position, and is therefore a **non-conservative** force. We cannot define a potential energy for such a force.

Notes. (*a*) Potential energy is really a property of a *system* of interacting particles: it depends on their relative positions.

(*b*) Because we are nearly always concerned with *differences* of potential energy, we are free to choose our own *zero*. For gravitational p.e. we usually choose the Earth's surface, or that of a body at infinity. (See also p. 323.)

4.5 Circular Motion in a Vertical Plane

Fig. 4.4 shows a small mass m describing a circle in a vertical plane. The only forces acting on it are the pull mg of the Earth, and the centripetal pull $\boldsymbol{F}$ of the string: these forces are conservative, and so we may apply the law of conservation of mechanical energy.

Since
$$E_p + E_k = \text{constant}$$
$$mgy + \tfrac{1}{2}mv^2 = \text{constant}$$

At any instant the pull of the Earth may have a resolved part along the tangent to the circle, and this will cause a gain or loss of speed. The instantaneous

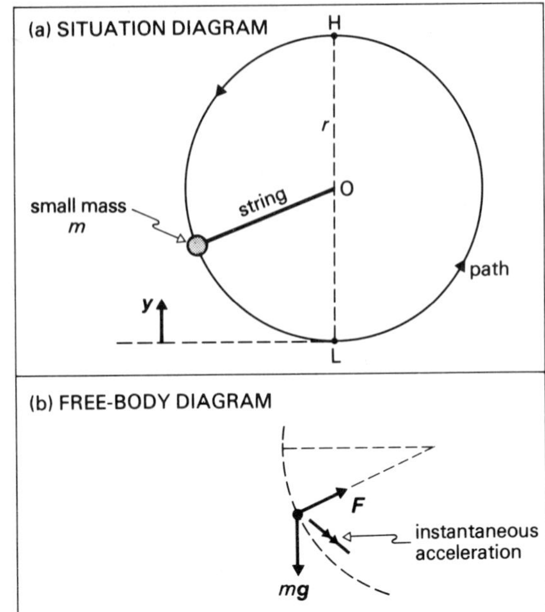

Fig. 4.4. Motion in a vertical circle.

speed can be found from the equation above, in which the constant will be determined by the starting conditions.

Minimum Speed for a Complete Circle

When the body reaches H the net downward force on it is

$$\boldsymbol{F} - m\boldsymbol{g}$$

Newton's second law applied along the line HO gives

$$F + mg = m\left(\frac{v^2}{r}\right)$$

The minimum speed the body may have at this point if it is to describe a complete circle is obtained by putting $\boldsymbol{F} = 0$.

Then
$$m\frac{v_{\min}^2}{r} = mg$$

If $v < \sqrt{rg}$ at this point then the path would (temporarily) become parabolic.

4.6 Power P

The power of a machine, or of the force exerted by a machine, tells us how fast it can transfer energy from itself to some other body.

Suppose a machine transfers energy ΔW over a time interval Δt.

Then the average power P_{av} over the time interval Δt is defined by

$$P_{av} = \frac{\Delta W}{\Delta t}$$

$$P_{av}\left[\frac{J}{s}\right] = \frac{\Delta W[J]}{\Delta t[s]}$$

The instantaneous power P is defined by

$$P = \lim_{\delta t \to 0} \frac{\delta W}{\delta t} = \frac{dW}{dt}$$

Notes. (*a*) Power is a *scalar* quantity.

(*b*) The unit *joule* (*second*)$^{-1}$ occurs so often it is called a **watt** (W).

Thus $$1\ J\ s^{-1} = 1\ W$$

and it follows that

$$1\ J = 1\ W\ s$$

The *watt second* is a unit of *energy*.

(*c*) Power has dimensions $[ML^2T^{-3}]$.

Calculation of Instantaneous Power

Suppose a machine moves its point of application of a force $\boldsymbol{F}$ at velocity $\boldsymbol{v}$ in the direction of $\boldsymbol{F}$.

Then since $$\delta W = F\delta s$$

and $$P = \lim_{\delta t \to 0} \frac{\delta W}{\delta t}$$

its instantaneous rate of working is given by

$$P = \lim_{\delta t \to 0} F\frac{\delta s}{\delta t}$$

$$= F\frac{ds}{dt}$$

$$= Fv$$

Thus the instantaneous power of a motor car depends not only on the force applied to it, but also on its instantaneous velocity.

5 Momentum

5.1 Impulse and Momentum

On p. 34 the **momentum** $\boldsymbol{p}$ of a body was defined by

$$\boxed{\boldsymbol{p} = m\boldsymbol{v}}$$

and it was seen that we could measure **force** $\boldsymbol{F}$ from the equation

$$\boxed{\boldsymbol{F} = m\frac{\mathrm{d}\boldsymbol{v}}{\mathrm{d}t}}$$

when it acted on a body of constant mass.

(a) Constant Force Applied

Suppose a constant force $\boldsymbol{F}$ acts over a time interval Δt.

Then

$$\boldsymbol{F} = m\frac{\Delta \boldsymbol{v}}{\Delta t}$$

$$\boldsymbol{F}\Delta t = m\,\Delta\boldsymbol{v}$$

$$= \Delta\boldsymbol{p}$$

where $\Delta\boldsymbol{v}$ and $\Delta\boldsymbol{p}$ are the velocity change and momentum change, respectively.

The impulse** of a constant force **is defined by the equation

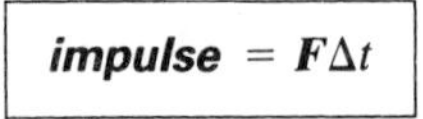

$$\boxed{\textit{impulse} = \boldsymbol{F}\Delta t}$$

$$\text{impulse [N s]} = \boldsymbol{F}[\text{N}]\cdot\Delta t\,[\text{s}]$$

Notes. (*i*) Impulse and momentum have the same dimensions $[\mathrm{MLT}^{-1}]$. Both may be measured in N s or kg m s^{-1}.

(*ii*) Impulse is a *vector* quantity, whose direction is that of $\boldsymbol{F}$.

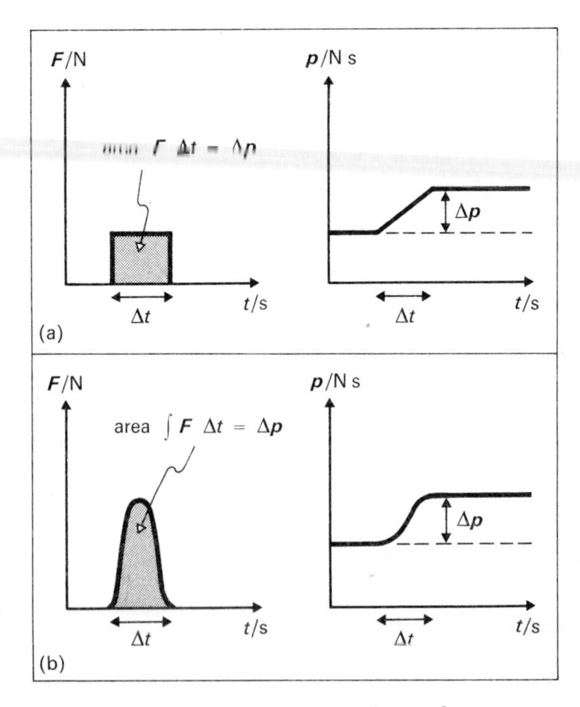

Fig. 5.1. Impulse and momentum change for (a) constant, and (b) variable forces.

(*iii*) The impulse of a steady force is illustrated by fig. 5.1*a*.

(b) Variable Force Applied

From $\boldsymbol{F} = m(\mathrm{d}\boldsymbol{v}/\mathrm{d}t)$, integrating

$$\int \boldsymbol{F}\,\mathrm{d}t = m\int \mathrm{d}\boldsymbol{v}$$
$$= \Delta \boldsymbol{p}$$

The* impulse *of a* variable force *is defined by

$$\textit{impulse} = \int \boldsymbol{F}\,\mathrm{d}t$$

The impulse of a variable force is illustrated by fig. 5.1*b*.

The Impulse-Momentum Theorem

$$\int \boldsymbol{F}\,\mathrm{d}t = \Delta \boldsymbol{p}$$
$$= m\boldsymbol{v} - m\boldsymbol{u}$$

where the impulse changes the velocity of a body of mass m from $\boldsymbol{u}$ to $\boldsymbol{v}$.

The relationship

$$\int \boldsymbol{F}\,\mathrm{d}t = m\boldsymbol{v} - m\boldsymbol{u}$$

is sometimes called the **impulse–momentum theorem**.

In words

$$\left(\begin{matrix}\textit{impulse applied}\\ \textit{to a body}\end{matrix}\right) = \left(\begin{matrix}\textit{change of momentum}\\ \textit{experienced by a body}\end{matrix}\right)$$

This should be compared with the *work–energy theorem* (p. 44).

The term *impulsive force* is applied to a force which is variable in size, and which acts for a very short time only, such as that applied by a golf club to a ball. The impulse of the force *can* be measured from the vector change of momentum of the ball, even though the values of Δt, and that of $\boldsymbol{F}$ at each instant, are uncertain.

5.2 Conservation of Momentum

Effects of External Forces

(i) Single Particle

Fig. 5.2 shows a mass m acted upon by forces $\boldsymbol{F}_1$, $\boldsymbol{F}_2$, etc.

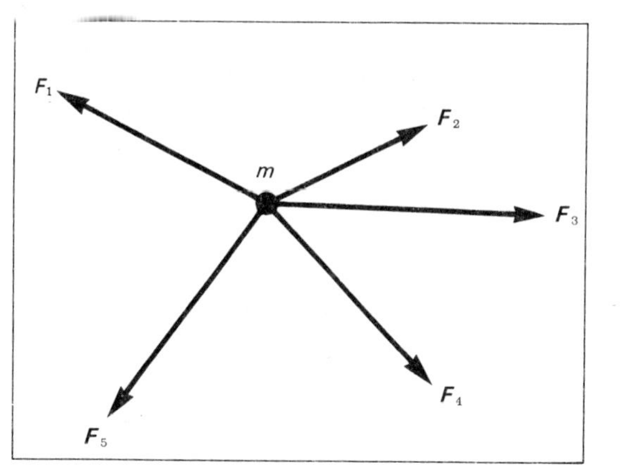

Fig. 5.2. Momentum conservation for a single particle.

According to *Newton's Second Law, if* there is a direction along which the sum of the resolved parts of the forces $\boldsymbol{F}$ is zero, *then* because there can be no acceleration in that direction, neither can there be any change of velocity or momentum.

(ii) System of Particles

As far as changes of momentum are concerned, the centre of mass of a system of interacting particles (defined on p. 64) behaves as though the whole mass of the system were concentrated at the centre of mass, and as though all the external forces were applied at the centre of mass. Thus the argument of (*i*) above, applied to a single particle, can be extended (as far as external forces are concerned) to systems of particles and bodies.

Effects of Internal Forces

These are forces exerted by one particle within the system on another *within the system.* (Such forces might be contact forces, or we could imagine separated masses to be connected by springs, etc.) *At every instant,* the instantaneous forces are an action-reaction pair of *Newton's Third Law.* Since change of momentum $= \int \boldsymbol{F}\,\mathrm{d}t$, a change of momentum experienced by one particle is accompanied by an equal but oppositely directed change of momentum experienced by the other particle. Since the *net* momentum change is found by *vectorial* addition, it is *zero.* This argument can be extended to all the pairs of *internal forces,* which thus produce *zero momentum change for the system.*

The Law of Conservation of Linear Momentum

This is a fundamental law of physics, and is applicable even in non-Newtonian mechanics.

The momentum of an isolated system is conserved.

An isolated system is one which experiences zero interaction with its environment, but of course in practice it is not easy to isolate a system. The following statement is a special case of the more general law which will be found very useful for solving problems.

If there is a direction in which no external forces act on a system, the total vector momentum of that system in that direction is constant (even if the bodies making up the system act on each other).

Using the Law

(1) Select the *system* to which it is to be applied.

(2) Draw *before* and *after* diagrams of the situation, marking the masses and velocities of the bodies that constitute the system.

(3) Marking a line along which the law is applicable, choose one direction as positive, and equate the total momentum before the interaction to that after.

The rocket and the ballistic balance are given as examples.

*5.3 The Rocket

Refer to fig. 5.3. The exhaust gases have a fixed speed u *relative to the rocket,* and Δm is the *change* in the mass of the rocket (which will, of course, be negative).

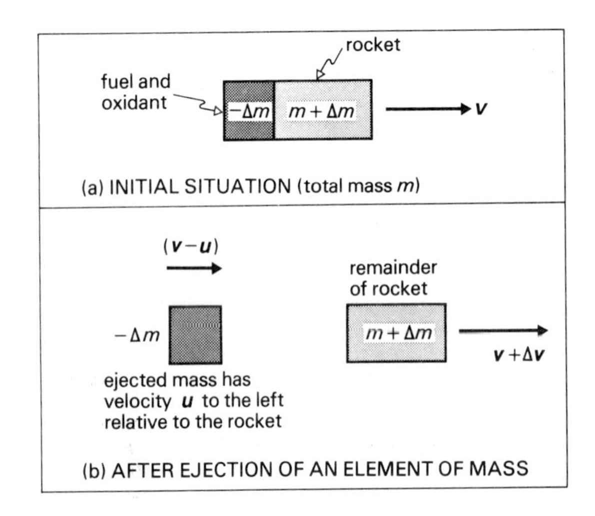

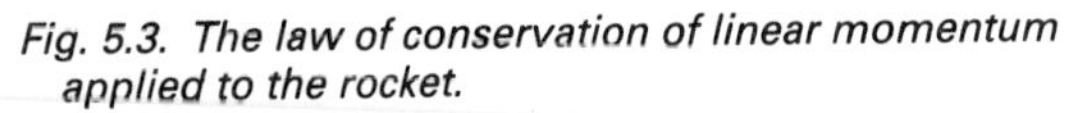
Fig. 5.3. The law of conservation of linear momentum applied to the rocket.

In (a) the linear momentum $\boldsymbol{p}$ to the right is given by

$$\boldsymbol{p} = m\boldsymbol{v}$$

In (b) it may be written

$$p = (m + \Delta m)(v + \Delta v) + (-\Delta m)(v - u)$$

Applying the law of conservation of linear momentum

$$mv = (m + \Delta m)(v + \Delta v) + (-\Delta m)(v - u)$$

Neglecting the term $\Delta m\, \Delta v$, this gives

$$m\,\Delta v = -u\,\Delta m$$

or

$$dv = -u\left(\frac{dm}{m}\right)$$

Suppose the rocket starts from rest with an initial mass m_0. If the speed has become v by the time the mass has been reduced to m,

$$\int_0^v dv = -u\int_{m_0}^{m} \frac{dm}{m}$$

$$\therefore \quad v = -u \ln (m/m_0)$$

$$\therefore \quad m = m_0 \exp(-v/u)$$

This result can also be deduced from *Newton*'s second law (p. 34).

5.4 Force equals Rate of Change of Momentum

Suppose we write *Newton's Second Law* in the equation

$$\boldsymbol{F} = \frac{d}{dt}(m\boldsymbol{v})$$

$$= m\frac{d\boldsymbol{v}}{dt} + \boldsymbol{v}\frac{dm}{dt}$$

We can use the equation conveniently in three forms.

(a) Constant Mass, Variable Velocity

Since

$$\frac{dm}{dt} = 0$$

we use

$$\boldsymbol{F} = m\frac{d\boldsymbol{v}}{dt} = m\boldsymbol{a}$$

This is the commonest use: an example is given on p. 39.

(b) Constant Velocity, Variable Mass

Since

$$\frac{d\boldsymbol{v}}{dt} = 0$$

we use

$$\boldsymbol{F} = \boldsymbol{v}\frac{dm}{dt}$$

dm/dt is the rate at which a body changes its mass while moving at velocity $\boldsymbol{v}$ relative to the observer.

Examples

(*i*) ***The conveyor belt.*** If sand falls vertically at a mass rate of flow of dm/dt onto a conveyor belt which moves at a horizontal velocity $\boldsymbol{v}$, the horizontal resolved part of the force exerted by the belt on the sand is $\boldsymbol{v}(dm/dt)$.

(*ii*) ***The hovering helicopter.*** Suppose the helicopter imparts a downward velocity $\boldsymbol{v}$ to air of density ρ which it collects over an area $\boldsymbol{A}$.

In time Δt

$$\begin{aligned}\Delta m &= \rho \times (\text{volume}) \\ &= \rho A v\, \Delta t \\ \Delta m/\Delta t &= \rho A v\end{aligned}$$

The downward force exerted by the helicopter on the air

$$\begin{aligned}F &= v(\Delta m/\Delta t) \\ &= \rho A v^2\end{aligned}$$

The air exerts an equal but opposite force on the helicopter. $\boldsymbol{v}$ can be adjusted for this to equal the pull of the Earth on the helicopter.

(*iii*) ***A molecular beam.*** Consider a molecular beam of cross-sectional area $\boldsymbol{A}$, containing a number density n of molecules, and moving at velocity $\boldsymbol{c}$, to be incident on a wall which absorbs the molecules without rebound. The mass of molecules hitting the wall in time Δt will be

$$\Delta M = nmAc\, \Delta t$$

where m is the mass of each molecule.

The force exerted on the wall will be to the right, and of size

$$\begin{aligned}F &= c\frac{dM}{dt} \\ &= c(nmAc) \\ &= nmAc^2\end{aligned}$$

The pressure $p = F/A = nmc^2$.

The reader should consider the effect on the pressure of

(*a*) the molecules rebounding elastically,

(*b*) the molecular beam being directed obliquely,

or

(*c*) the molecules having random velocities (p. 179).

It should be noted that the viewpoints (*a*) and (*b*) (opposite) are equivalent. Consider an *element* of sand from situation (*b*)(*i*), whose mass is Δm, and whose velocity is changed from 0 to $\Delta \boldsymbol{v}$ over a time interval Δt. (For convenience suppose the acceleration to be uniform, though this is not necessary.)

We could write

$$(a) \qquad \boldsymbol{F} = m\frac{d\boldsymbol{v}}{dt} \quad \text{as} \quad \boldsymbol{F} = (\Delta m)\left(\frac{\Delta \boldsymbol{v}}{\Delta t}\right)$$

$$(b) \qquad \boldsymbol{F} = \boldsymbol{v}\frac{dm}{dt} \quad \text{as} \quad \boldsymbol{F} = (\Delta \boldsymbol{v})\left(\frac{\Delta m}{\Delta t}\right)$$

If (*b*) we effectively apply $\boldsymbol{F} = m\boldsymbol{a}$ to an *element of mass* Δm.

*(c) Variable Mass and Variable Velocity

Refer back to the diagram of the rocket, fig. 5.3. The force exerted on the rocket (the thrust) has a size $m(dv/dt)$, and is to the right. The force exerted on the exhaust gases has a size $u(dm/dt)$, and is to the left. Since the total force exerted on the system is zero

$$m\frac{dv}{dt} = -u\frac{dm}{dt}$$

or

$$dv = -u\left(\frac{dm}{m}\right)$$

This is the same result as that found from the law of conservation of linear momentum (p. 50).

5.5 Collisions

A **collision** is a process in which the time interval during which bodies interact is small relative to the time for which we can observe them: we can then make a clear distinction between *before* and *after*. If such a collision takes a small time interval, large impulsive forces are brought into play. We can usually ignore the relatively small change of momentum brought about by an external force compared to that imposed on the bodies that collide by the impulsive (internal) forces.

We use the law of conservation of momentum in the form:

The total momentum of a system just before collision is equal to that just after collision if the time of collision is sufficiently small.

Classification of Collisions

Energy is always conserved (when all forms of energy are considered), but when mechanical energy is converted to internal energy it is not always possible to apply this principle usefully.

(*a*) **Elastic collisions** are those in which kinetic energy *is* conserved. Truly elastic collisions can only occur in practice on an atomic scale. (Even then they are not *always* elastic.)

(*b*) **Inelastic collisions** are those in which kinetic energy is not conserved: it may be converted to internal energy (as usually happens), or perhaps elastic potential energy of deformation. On a macroscopic scale this is the most common type of collision.

A **completely inelastic** collision is one in which two bodies stick together after impact (as a bullet being embedded in a target). The loss of kinetic energy is large but not complete.

(*c*) **Explosive collisions** are those in which there is an *increase* of kinetic energy. This could occur if potential energy were released by the impact, as when the collision of one trolley with another releases a compressed spring. (Then the action-reaction forces of separation would exceed those of approach.) These collisions are sometimes called **superelastic**—see p. 179.

Note carefully:

(1) momentum *is* always conserved, but

(2) kinetic energy is *not* always conserved, as it is usually converted to some other form of energy.

The Coefficient of Restitution

By definition

$$e = \frac{\text{relative speed of separation}}{\text{relative speed of approach}}$$

for two bodies that collide. The speeds are measured along their common normal. For a given pair of bodies e is roughly constant. This is sometimes called *Newton's experimental law of impact.*

e gives us an alternative way of classifying collisions:

e	*Type*
>1	explosive or superelastic
1	elastic
<1	inelastic
0	completely inelastic

5.6 The Ballistic Balance

This device is discussed here to illustrate the conservation laws of momentum and mechanical energy. Refer to fig. 5.4.

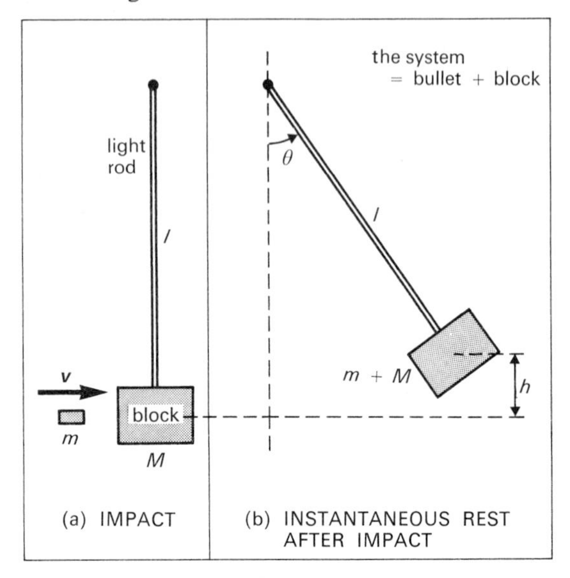

Fig. 5.4. The ballistic balance.

(a) The Impact

During this process

(*i*) mechanical energy is *not* conserved (because of the non-conservative internal frictional force), and internal energy is produced, but

(*ii*) linear momentum *is* conserved in the horizontal direction, along which there is no external force.

Put $\boldsymbol{V}$ = the block's velocity immediately after impact.

Using (*ii*), and taking the horizontal direction to the right as positive, we have

$$mv = (M + m)\boldsymbol{V}$$

(b) The Swing after Impact

During this process

(*i*) mechanical energy *is* conserved: the conservative gravitational force causes the conversion of k.e. to p.e., but

(*ii*) momentum is *not* conserved, as an external resultant force (the pull of the Earth) acts on the bullet-block system for a significant time interval.

Using (*i*)

$$\tfrac{1}{2}(m + M)V^2 = (m + M)gh$$

$$V = \sqrt{2gh}$$

Uses of the Balance

(1) As described, the balance can be used to measure the speed of the bullet.

(2) With modification the balance can be used to *compare masses*. m could be a second pendulum which is made to interact with the block M.

Instead of
$$\frac{m_1}{m_2} = \frac{a_2}{a_1} \qquad \text{(p. 37)}$$

we use
$$\frac{m}{M + m} = \frac{V}{v}$$

in which both v and V are measured. This can bc done directly (photographically) or indirectly (by measuring θ).

Note in fig. 5.4*b*, that $h = l(1 - \cos\theta)$.

6 Rotational Dynamics

6.1 Introduction

A **rigid body** is one that suffers a negligible deformation when subjected to external forces. Such a body undergoes motion which is made up of *translation* and *rotation*. Motion is purely

(*a*) **translational** if every particle on a body has the same instantaneous velocity,

(*b*) **rotational** if every particle moves in a circle about the same straight line (the axis of rotation).

(*a*) To describe translational motion we use the equation $\boldsymbol{F} = M\boldsymbol{a}$, in which $\boldsymbol{F}$ is the resultant force applied to a *body*, M is the mass of the body, and $\boldsymbol{a}$ is the acceleration of the *centre of mass* (p. 64).

(*b*) For the rotational motion we relate quantities that are analogous to $\boldsymbol{F}$, M and $\boldsymbol{a}$.

6.2 Rotational Kinematics

Angular velocity $\boldsymbol{\omega}$ is defined on p. 30.

Average *and* instantaneous angular accelerations α *are defined by the equations*

$$\boldsymbol{\alpha}_{av} = \frac{\Delta\boldsymbol{\omega}}{\Delta t}$$

$$\boldsymbol{\alpha} = \lim_{\delta t \to 0} \frac{\delta\boldsymbol{\omega}}{\delta t} = \frac{d\boldsymbol{\omega}}{dt}$$

$$\boldsymbol{\alpha}_{av}\left[\frac{\text{rad}}{\text{s}^2}\right] = \frac{\Delta\boldsymbol{\omega}[\text{rad/s}]}{\Delta t[\text{s}]}$$

Angular acceleration $\boldsymbol{\alpha}$ is a *pseudo-vector* quantity (see p. 19), measured in rad s^{-2}. It has dimensions $[\text{T}^{-2}]$.

Linear and Rotational Kinematics Compared

Linear motion		*Rotational motion*	
quantity	*dimensions*	*quantity*	*dimensions*
s[m]	[L]	θ[rad]	none†
$\boldsymbol{v} = \frac{ds}{dt}$	$[\text{LT}^{-1}]$	$\boldsymbol{\omega} = \frac{d\boldsymbol{\theta}}{dt}$	$[\text{T}^{-1}]$
$\boldsymbol{u} = (\boldsymbol{v})_{t=0}$	$[\text{LT}^{-1}]$	$\boldsymbol{\omega}_0 = (\boldsymbol{\omega})_{t=0}$	$[\text{T}^{-1}]$
$\boldsymbol{a} = \frac{d\boldsymbol{v}}{dt}$	$[\text{LT}^{-2}]$	$\boldsymbol{\alpha} = \frac{d\boldsymbol{\omega}}{dt}$	$[\text{T}^{-2}]$

† Although it has been argued that the assignment of no dimensions to angle can lead to inconsistencies.

Provided the angular acceleration is constant, the equations of uniformly accelerated motion (p. 29) follow:

$$\boldsymbol{\omega} = \boldsymbol{\omega}_0 + \boldsymbol{\alpha}t$$
$$\theta = \frac{\omega_0 + \omega}{2} \cdot t$$
$$\theta = \omega_0 t + \tfrac{1}{2}\alpha t^2$$
$$\theta = \omega t - \tfrac{1}{2}\alpha t^2$$
$$\omega^2 = \omega_0^2 + 2\alpha\theta$$

Uniformly accelerated angular motion

θ represents the angular *displacement* (not the total angle). These equations can be obtained by integration of the defining equations $\boldsymbol{\omega} = \mathrm{d}\boldsymbol{\theta}/\mathrm{d}t$, and $\boldsymbol{\alpha} = \mathrm{d}\boldsymbol{\omega}/\mathrm{d}t$. (Note that if a point describes several revolutions, then θ is of the form $[2n\pi \text{ rad} + \phi]$.)

Linear and Rotational Kinematics Related

Suppose we have a particle distant r from the axis of rotation of a rigid body, and that there is no translational motion of the body.

(*a*) *Displacements* are related by

$$s = r\theta \qquad \text{(p. 22)}$$

(*b*) *Speeds*

$$\frac{\mathrm{d}s}{\mathrm{d}t} = \theta\frac{\mathrm{d}r}{\mathrm{d}t} + r\frac{\mathrm{d}\theta}{\mathrm{d}t}$$

$$v = r\omega$$

($\mathrm{d}r/\mathrm{d}t = 0$ since r is constant for a particle on a rigid body.)

v is the instantaneous linear *tangential* speed.

(*c*) *Accelerations*

The total acceleration of the particle is made up of two components.

(*i*) *Tangential* linear acceleration $\boldsymbol{a}_t$

$$\frac{\mathrm{d}v}{\mathrm{d}t} = r\frac{\mathrm{d}\omega}{\mathrm{d}t}$$

$$a_t = r\alpha$$

This will be zero when $\boldsymbol{\alpha} = 0$.

(*ii*) *Radial* (centripetal) linear acceleration $\boldsymbol{a}_r$

$$a_r = \frac{v^2}{r} = \omega^2 r \qquad \text{(p. 33)}$$

This will be zero only when $\boldsymbol{\omega} = 0$.

The total acceleration must be found by vector addition: it has a magnitude

$$a = r\sqrt{\omega^4 + \alpha^2}$$

An important distinction follows from these equations: *all* points on a rigid body undergoing rotation about a fixed axis *necessarily* have the *same angular* displacement, speed and acceleration. Their *linear* values, however, depend on r.

6.3 Torque and Rotational Motion

The *moment of a force*, or its **torque**, ***T***, is defined on p. 62 by $T = Fr$.

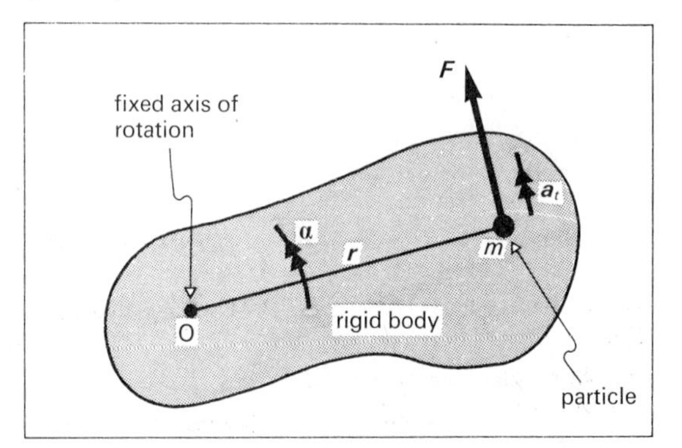

Fig. 6.1. Newton's second law applied to a particle of a rigid body.

In fig. 6.1, ***F*** is the *tangential* component of the resultant force that acts on the particle. (The radial component cannot produce angular acceleration.) Using *Newton's Second Law*

$$F = ma_t$$
$$= mr\alpha$$

Thus torque

$$T = (mr\alpha)r$$
$$= mr^2\alpha$$

Summing for the whole body

$$\sum T = \sum mr^2\alpha$$

or

$$\sum T_{\text{ext}} = \alpha \sum mr^2$$

since

(*a*) $\boldsymbol{\alpha}$ has the same value for all such particles m on the body, and (*b*) the sum of *internal torques* is zero, in the same way that it was shown on p. 49 that the sum of the *internal forces* is zero.

Moment of Inertia (m.i.)

The moment of inertia I of a body about a particular axis is defined by the equation

$$I = \sum mr^2$$

where r is the distance of a particle of mass m from the axis of rotation.

$$I[\text{kg m}^2] = \sum m[\text{kg}]\, r^2[\text{m}^2]$$

Notes. (*a*) The moment of inertia (m.i.) of a body is a *scalar* quantity.

(*b*) The dimensions of I are $[\text{ML}^2]$.

(*c*) I depends on

(*i*) the mass of a body,

(*ii*) the way the mass is distributed,

(*iii*) *the axis of rotation* (and is therefore not a constant for a rigid body).

Newton's Second Law for rotation about a fixed axis can now be written in the form

$$\boldsymbol{T} = I\boldsymbol{\alpha}$$

where $\boldsymbol{T}$ represents the *resultant* torque acting on the body.

Example. **The simple pendulum** (fig. 6.2)

Choose the anticlockwise sense of rotation as positive. For the bob

$$I = \sum mr^2 = ml^2$$

$$T = -(mg \sin\theta)l$$

(since $\boldsymbol{P}$, a radial force, makes zero contribution),

$$\boldsymbol{\alpha} = \ddot{\boldsymbol{\theta}}$$

so

$$\boldsymbol{T} = I\boldsymbol{\alpha}$$

becomes

$$-mg \sin\theta\, l = ml^2\ddot{\theta}$$

Rearranging

$$\ddot{\theta} = -\left(\frac{g}{l}\right)\sin\theta$$

$$\approx -\left(\frac{g}{l}\right)\theta \quad \text{for small } \theta$$

We conclude (p. 79) that the motion is simple harmonic for small amplitudes, and of period $2\pi\sqrt{l/g}$.

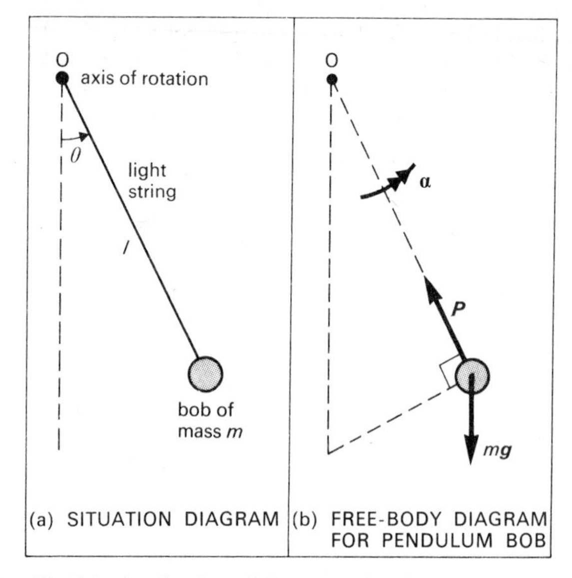

Fig. 6.2. Application of $\boldsymbol{T} = I\boldsymbol{\alpha}$ to the simple pendulum.

The analysis of the torsional pendulum follows the same pattern (p. 134).

6.4 Calculation of Moment of Inertia

Suppose a body of mass M has a moment of inertia I.

The radius of gyration k is defined by the equation

$$I = Mk^2$$

A point mass M placed a distance k from the axis of rotation would have the same value of I. (This is *not* the location of the centre of mass, which may coincide with the axis of rotation.)

*Calculation of I by Calculus

$I = \sum mr^2$ enables us to calculate I for a body composed of discrete point masses.

For a continuous distribution of matter the summation is replaced by the integration

$$I = \int r^2\, \mathrm{d}m$$

where the integration is taken over the whole body.

For some simple bodies it is more convenient to use

Routh's Rules:

Suppose a body of mass M has three mutually perpendicular axes of symmetry. Let $2a$ and $2b$ be lengths of two of these axes. Then the m.i. I_0 about the third axis (which will be through the centre of mass) is given by

$$I_0 = M\left(\frac{a^2 + b^2}{n}\right)$$

where $n = 3$ for rectangular bodies,
$n = 4$ for elliptical bodies,
$n = 5$ for ellipsoidal bodies.

The rules have no theoretical foundation.

Examples

(*a*) *Rectangular lamina*, length l and breadth h

$$I_0 = M\left(\frac{l^2 + h^2}{12}\right)$$

for an axis perpendicular to the plane of the lamina.

(*b*) *Circular disc*, radius r

$$I_0 = \frac{Mr^2}{2}$$

for an axis as in (*a*).

(*c*) *Sphere*, radius r

$$I_0 = \frac{2}{5}Mr^2$$

The rules calculate I_0, the m.i. about an axis *through the centre of mass* for a homogeneous body. For other axes we use

The Theorem of Parallel Axes

Suppose M is the mass of a body,

I_0 is the m.i. about a particular axis through the centre of mass,
I is the m.i. about some other *parallel* axis distance d from the first.

Then the theorem of parallel axes states

$$I = I_0 + Md^2$$

An example of its use is given on p. 59 (the compound pendulum). The theorem shows that a body will have its minimum moment of inertia about an axis through the centre of mass.

6.5 Energy of Rotational Motion

(a) Work Done by a Constant Torque

Refer to fig. 6.3 in which $\boldsymbol{F}$ is the tangential component of the resultant force acting at P.

Suppose $\boldsymbol{F}$ is applied while the body undergoes an angular displacement θ. Then $\boldsymbol{F}$ does work W, where

$$W = F\text{ (distance along arc)}$$
$$= Fr\theta$$
$$= T\theta \quad \text{since } T = Fr$$

$$W = T\theta$$

$$W[\text{J}] = T[\text{Nm}]\theta[\text{rad}]$$

When $\boldsymbol{T}$ and θ are in the same sense, W is positive. The force transfers energy to the body, whose (rotational) k.e. increases. (When the direction of $\boldsymbol{T}$ and the axis of rotation are perpendicular, $W = 0$.)

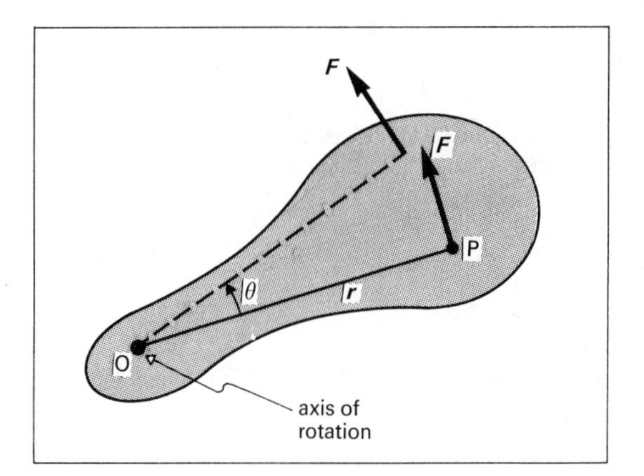

Fig. 6.3. Calculation of work done by a torque.

*(b) Work done by a Variable Torque

The equation $W = T\theta$ is more correctly written $\Delta W = T\,\Delta\theta$. Then if the magnitude of $\boldsymbol{T}$ varies with θ, the work δW done over a very small displacement $\delta\boldsymbol{\theta}$ is given by

$$\delta W = T\delta\theta$$

The work done over a finite displacement can be calculated from

$$W = \int_{\theta_1}^{\theta_2} T\,\mathrm{d}\theta$$

An example is given on p. 316.

Rotational Kinetic Energy

Suppose that the angular velocity of a rigid body is increased from $\boldsymbol{\omega}_0$ to $\boldsymbol{\omega}$ when a constant resultant torque $\boldsymbol{T}$ acts through a displacement θ. Since $\boldsymbol{\alpha}$ is constant, we can apply

$$\omega^2 = \omega_0^2 + 2\alpha\theta \qquad \text{(p. 55)}$$

in which $\boldsymbol{T} = I\boldsymbol{\alpha}$

Using $W = T\theta$

we have

$$W = I\alpha\theta$$

$$= I\left(\frac{\omega^2}{2} - \frac{\omega_0^2}{2}\right)$$

$$\boxed{W = \tfrac{1}{2}I\omega^2 - \tfrac{1}{2}I\omega_0^2}$$

This represents the **work-energy theorem** for rotational motion. In words

$$\begin{pmatrix}\textit{The work done by}\\ \textit{the torques acting}\\ \textit{on the body}\end{pmatrix} = \begin{pmatrix}\textit{The change of the}\\ \textit{rotational kinetic}\\ \textit{energy of the body}\end{pmatrix}$$

The* rotational k.e. *of a body can be calculated from

$$\boxed{E_k = \tfrac{1}{2}I\omega^2}$$

$$E_k[\text{J}] = \tfrac{1}{2}I[\text{kg m}^2]\ \omega^2\left[\frac{\text{rad}^2}{\text{s}^2}\right]$$

$\frac{1}{2}I\omega^2$ is a convenient way of writing $\sum \frac{1}{2}mv^2$, since

$$\sum \tfrac{1}{2}mv^2 = \tfrac{1}{2}\sum mr^2\omega^2$$

$$= \tfrac{1}{2}\omega^2 \sum mr^2$$

$$= \tfrac{1}{2}I\omega^2$$

It is more convenient to use $\boldsymbol{\omega}$ than $\boldsymbol{v}$, because all points on the rigid body have the *same value of* $\boldsymbol{\omega}$ (for a given axis of rotation).

A flywheel is designed so that most of its mass is concentrated at the rim. Thus the flywheel has a high moment of inertia and, for a given angular speed, high kinetic energy. This finds application in piston engines, steam rollers, friction-drive toy cars, etc.

Total Kinetic Energy of a Rigid Body

The total kinetic energy is made up by

$$\text{(translational k.e.)} + \text{(rotational k.e.)}$$

$$= \tfrac{1}{2}Mv^2 + \tfrac{1}{2}I_0\omega^2$$

in which M is the total mass,
$\boldsymbol{v}$ is the velocity of the centre of mass,
I_o is the m.i. about an axis through the centre of mass, and
$\boldsymbol{\omega}$ is the angular velocity about that axis.

Measuring Moment of Inertia

Refer to fig. 6.4.

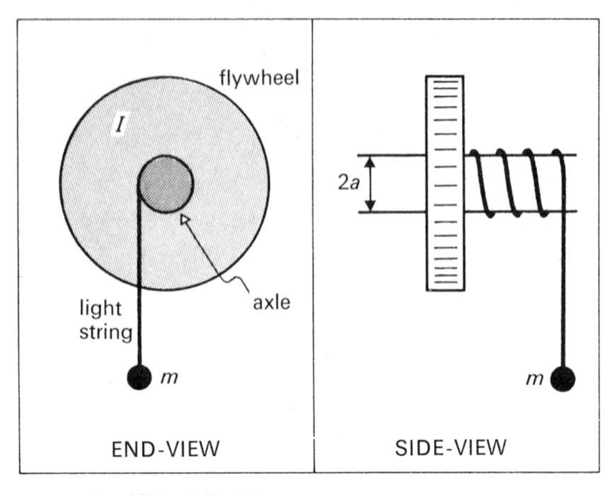

Fig. 6.4. Measurement of the moment of inertia of a flywheel.

Suppose the mass m falls a vertical distance h from rest, and acquires a speed v.

$$\text{Loss of gravitational p.e.} = mgh$$

$$\text{Gain of k.e.} = \tfrac{1}{2}I\omega^2 + \tfrac{1}{2}mv^2$$

If only conservative forces act, these energies are equal:

$$mgh = \tfrac{1}{2}I\omega^2 + \tfrac{1}{2}mv^2$$

The values of m, $\boldsymbol{g}$ and h are known. $\boldsymbol{\omega}$ and $\boldsymbol{v}$ can be found from a knowledge of t, the time of fall, h, the distance fallen, and a, the radius of the axle:

$$v = 2\left(\frac{h}{t}\right)$$

$$\omega = \frac{v}{a}$$

We can correct for the frictional torque by a separate experiment in which we measure the energy dissipated during each revolution.

*6.6 Angular Momentum and its Conservation

Angular Momentum

(a) Angular Momentum of a Particle

Refer to fig. 6.1 (p. 55) in which the particle m has tangential speed v. Its linear momentum is $m\boldsymbol{v}$: its *angular* momentum is the *moment* of the linear momentum about the axis of rotation.

$$\text{Angular momentum } L = (mv)r$$
$$= (m\omega r)r$$
$$= (mr^2)\omega$$

(b) Angular Momentum of a Body

The angular momentum L of a body about an axis of rotation is defined by

$$\textbf{angular momentum} = \sum (mr^2)\boldsymbol{\omega}$$

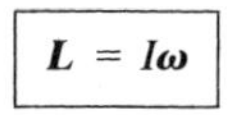

$$\boldsymbol{L} = I\boldsymbol{\omega}$$

where I is the moment of inertia about that axis

Notes. (*a*) Angular momentum is a *pseudo-vector* quantity, its sense and direction for a symmetrical body being that *of* $\boldsymbol{\omega}$ (p. 31).

(*b*) It may be measured in $(\text{kg m}^2) \times (1/\text{s})$

$$= \text{kg m}^2\,\text{s}^{-1} = \text{N m s}$$

(*c*) The total angular momentum of a body (such as the Earth) may be made up by

(*i*) its *spin angular momentum* about an axis through its centre of mass, and

(*ii*) its *orbital angular momentum* about some external axis.

This is an important idea in atomic physics.

The Law of Conservation of Angular Momentum

In paragraph 5.1 (p. 48) we discuss *linear* impulse and momentum: the same arguments can be developed for *angular* impulse and momentum. They give these results:

(*a*) The **angular impulse** of a variable torque is equal to $\int \boldsymbol{T}\,dt$.

(*b*) The **angular impulse-momentum theorem** states

$$\int \boldsymbol{T}\,dt = I\omega - I\boldsymbol{\omega}_0$$

In words

$$\begin{pmatrix}\textit{angular impulse}\\ \textit{applied to a}\\ \textit{body}\end{pmatrix} = \begin{pmatrix}\textit{change of angular}\\ \textit{momentum experienced}\\ \textit{by body}\end{pmatrix}$$

An example is given on p. 390.

(*c*) *The* **law of conservation of angular momentum** *states that if no resultant external torque acts on a system, then the total angular momentum of that (isolated) system remains constant.*

This is a fundamental principle of physics.

Examples of its Application

(*i*) *A skater* can reduce his moment of inertia by drawing in his arms. A reduction of I from I_1 to I_2 causes $\boldsymbol{\omega}$ to increase from $\boldsymbol{\omega}_1$ to $\boldsymbol{\omega}_2$ where

$$I_1\boldsymbol{\omega}_1 = I_2\boldsymbol{\omega}_2 \quad (\text{since } \boldsymbol{T} = 0)$$

Since the energy of rotation $= \frac{1}{2}I\omega^2$, his kinetic energy has increased: he does work in pulling in his arms.

(*ii*) *A gyroscope* mounted so that it experiences no resultant external torque points in a fixed direction, and can be used as a compass.

(*iii*) *In the Earth–Moon system,* tidal forces are reducing the Earth's *spin* angular momentum: the consequent increase in *orbital* angular momentum occurs as the Earth–Moon separation increases.

6.7 The Compound Pendulum

Fig. 6.5*a* shows a rigid body of arbitrary shape free to make oscillations about a horizontal axis through S.

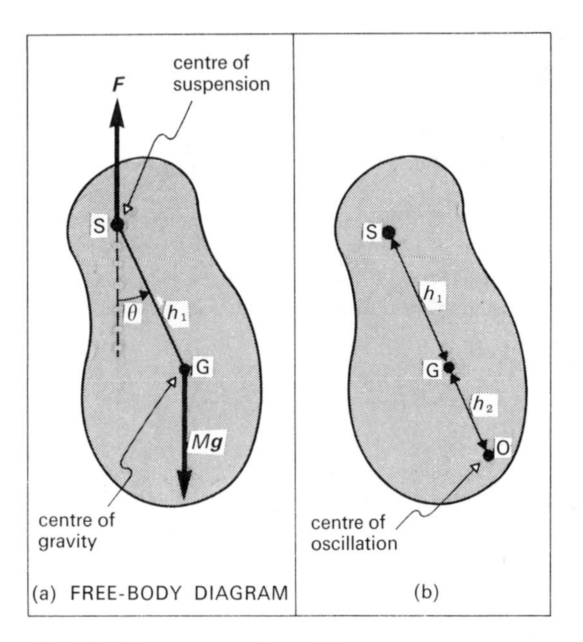

Fig. 6.5. The compound pendulum.

Put	$I_G = Mk^2$
then	$I_S = Mk^2 + Mh_1^2$
Using	$T_S = I_S\alpha$

and choosing the anticlockwise sense to be positive we have

$$-Mgh_1 \sin\theta = M(k^2 + h_1^2)\ddot{\theta}$$

or, for small amplitudes of oscillation,

$$\ddot{\theta} \approx \frac{-(gh_1)}{(k^2 + h_1^2)}\theta$$

We conclude (p. 79) that the motion is simple harmonic for small amplitudes, and of period

$$2\pi\sqrt{\frac{k^2 + h_1^2}{h_1 g}}$$

Minimum Time Period

If suspension from O (fig. 6.5*b*) gives the same time period as from S, then O and S are *centres of oscillation* and *suspension*, respectively, in fig. 6.5*a*.

$$\frac{k^2 + h_2^2}{h_2} = \frac{k^2 + h_1^2}{h_1}$$
$$k^2(h_1 - h_2) = h_1 h_2(h_1 - h_2)$$

We deduce either

(*a*) $h_1 = h_2$, or (*b*) $k^2 = h_1 h_2$, or
(c) $k = h_1 = h_2$ (special case).

In general four axes through a compound pendulum (such as a metre rule) will give equal time periods. When $k = h_1 = h_2$, we have two coincident pairs, and the period is a minimum: the smallest value of

$$\frac{k^2 + h_1^2}{h_1} = \frac{(k - h_1)^2 + 2kh_1}{h_1} = \frac{(k - h_1)^2}{h_1} + 2k$$

occurs when $k = h_1$, and is $2k$.

The *minimum period* is $2\pi\sqrt{2k/g}$.

The Simple Equivalent Pendulum

We have shown above that $k^2/h_1 = h_2$.

$$\therefore \quad \frac{k^2 + h_1^2}{h_1} = \frac{k^2}{h_1} + h_1 = (h_1 + h_2)$$

Then time period $= 2\pi\sqrt{\dfrac{(h_1 + h_2)}{g}} = 2\pi\sqrt{\dfrac{l}{g}}$

where $l = (h_1 + h_2)$ is the length of the simple equivalent pendulum: it is the distance between the centres of oscillation and suspension.

The Metre Rule as a Compound Pendulum

Suppose the time period T is measured for an oscillating metre rule for different values of h measured from the centre of gravity. Then

$$T^2 = 4\pi^2\left(\frac{k^2 + h^2}{hg}\right)$$

Fig. 6.6 shows T^2 plotted against h.

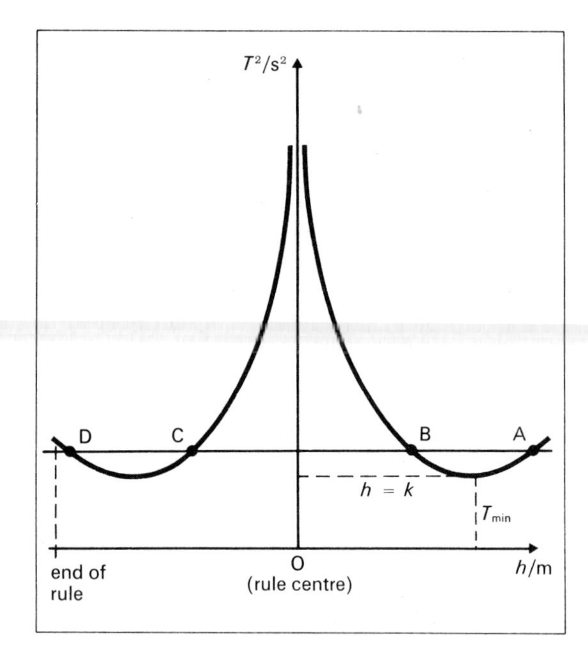

Fig. 6.6. Measurements with a metre rule as compound pendulum.

Notes. (*a*) T_{min} occurs when $h_1 = h_2 = k$.

(*b*) $T_{max} = \infty$ occurs when $h = 0$.

(*c*) The horizontal line (for a given T) may cut the graph at four points A, B, C and D.

Then AC = BD = $(h_1 + h_2)$ and is the length of the simple equivalent pendulum.

(*d*) The *reversible* compound pendulum was used by *Kater* for an early accurate measurement of $\boldsymbol{g}$.

6.8 Comparison of Linear and Rotational Dynamics

Analogous Quantities

For kinematic analogues, see p. 54.

Linear motion		*Rotational motion*	
dimensions	*quantity*	*quantity*	*dimensions*
[M]	m	$I = \sum mr^2$	[ML^2]
[MLT^{-2}]	$\boldsymbol{F}$	$T = Fr$	[ML^2T^{-2}]
[ML^2T^{-2}]	$E_k = \frac{1}{2}mv^2$	$E_k = \frac{1}{2}I\omega^2$	[ML^2T^{-2}]
[MLT^{-1}]	$\boldsymbol{p} = m\boldsymbol{v}$	$\boldsymbol{L} = I\boldsymbol{\omega}$	[ML^2T^{-1}]

Note carefully the respective dimensions.

Analogous Equations

Linear motion	*Rotational motion*
$\boldsymbol{F} = m\boldsymbol{a}$	$\boldsymbol{T} = I\boldsymbol{\alpha}$
$W = Fs$	$W = T\theta$
$W = \frac{1}{2}mv^2 - \frac{1}{2}mu^2$	$W = \frac{1}{2}I\omega^2 - \frac{1}{2}I\omega_0^2$
power $P = Fv$	$P = T\omega$
$\boldsymbol{F}t = m\boldsymbol{v} - m\boldsymbol{u}$	$\boldsymbol{T}t = I\boldsymbol{\omega} - I\boldsymbol{\omega}_0$

7 Equilibrium

7.1 Equilibrium of a Particle

A particle is in **equilibrium** if its acceleration is zero. The equilibrium is **static** if, in addition, the particle's velocity is also zero, so that it remains continuously at rest relative to the observer.

Since $\boldsymbol{F} = m\boldsymbol{a}$, in which $\boldsymbol{F}$ and $\boldsymbol{a}$ are vectors, if $\boldsymbol{a} = 0$, then $\boldsymbol{F} = 0$. This gives the **condition for the equilibrium** of a particle:

> *the vector sum of all external forces acting on the particle must be zero.*

Coplanar forces, which we meet most frequently, are those whose lines of action are confined to one plane. Suppose this plane is the x–y plane: then

(*a*) *To prove that a particle is in equilibrium,* we must show that the sums of the resolved parts of the forces on the particle in *any two directions* are each zero. This is called a **sufficient condition**.

(*b*) *Given that a particle is in equilibrium,* we know that the sums of the resolved parts of the forces on the particle in *all directions* are each zero. This is called a **necessary condition**.

Procedure for Solving Problems

(Refer also to p. 39.)

(1) Draw a situation diagram.

(2) Draw a free-body diagram for the particle in equilibrium.

(3) Resolve *twice* in *any* two convenient directions:

$$\text{e.g. put} \sum F_x = 0, \quad \sum F_y = 0.$$

This will result in two equations, which can be used to find *two* unknowns.

Notes. (*a*) Since a particle is a point mass, forces acting on it must pass through a single point.

(*b*) If the particle moves with constant velocity, the same procedure can be followed.

(*c*) When the forces are not co-planar then we obtain *three* (rather than two) independent equations.

7.2 Equilibrium of a Body

The Moment of a Force, or Torque T

The moment of a force is a measure of its ability to rotate a body about a given axis. Refer to fig. 7.1.

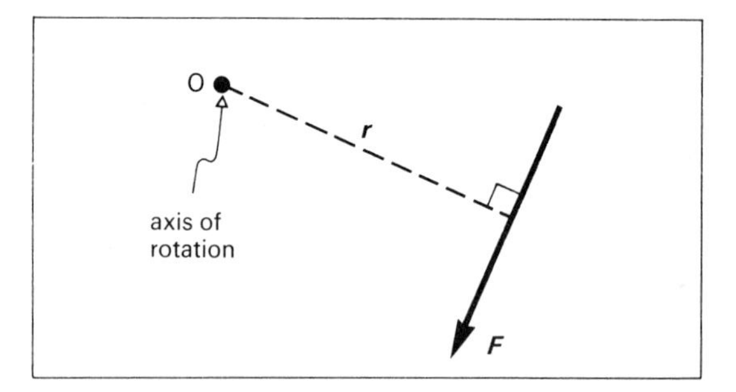

Fig. 7.1. Definition of torque.

The moment or torque T of the force F about the axis through O is defined by the equation

$$\boxed{T = Fr}$$

$$T[\text{N m}] = F[\text{N}]\, r[\text{m}]$$

Notes. (*a*) Torque has dimensions $[\text{ML}^2\text{T}^{-2}]$ (the same as energy, but this does *not* imply any direct relationship).

(*b*) Torque is a *pseudo-vector* quantity: the direction of the torque in fig. 7.1 is perpendicular to the paper, and its sense is into the paper (so as to obey a right hand corkscrew rule).

(*c*) 1 N m is a unit which here has no special name. (Be careful not to confuse the unit for *torque* with that for *work*—the joule.)

(*d*) The length r is sometimes called the **moment arm**.

(*e*) For co-planar forces we speak loosely of taking *moments about a point* (rather than an axis), since the axis is represented as a point in a two-dimensional diagram.

Equilibrium of a Body

(a) Translational

The centre of mass of the body must be non-accelerated. The conditions of paragraph 7.1 apply.

(b) Rotational

A body is in rotational equilibrium if its angular acceleration $\boldsymbol{\alpha}$ about every axis is zero.

Since $\boldsymbol{T} = I\boldsymbol{\alpha}$, in which $\boldsymbol{T}$ and $\boldsymbol{\alpha}$ are vectors, if $\boldsymbol{\alpha} = 0$, then $\boldsymbol{T} = 0$.

The condition is that the vector sum of all the external torques acting on the body must be zero.

Suppose we confine ourselves to forces in the x–y plane: then

(*i*) *To prove that a body is in equilibrium*, we must show that the sum of the torques acting on the body about *any one axis* is zero. This is a *sufficient* condition.

(*ii*) *Given that a body is in equilibrium*, we know that the sums of the torques acting on the body about *all axes* are each zero. This is a *necessary* condition.

In (*i*) and (*ii*), the axes will be drawn perpendicular to the x–y plane, and therefore along the z-direction. When the forces are not co-planar, then we have *three* independent directions about which we can take moments.

The necessary condition (*ii*) is called the **principle of moments**.

If a body is in equilibrium, the algebraic sums of the torques acting on the body about all axes are each zero.

7.3 Equilibrium Conditions

Summary of Necessary Conditions

(a) Forces in Three Dimensions

(*i*) *Translation* $\quad \boldsymbol{a} = 0$

$$\therefore \quad \sum \boldsymbol{F} = 0$$

which implies

$$\sum F_x = \sum F_y = \sum F_z = 0$$

(*ii*) *Rotation* $\quad \boldsymbol{\alpha} = 0$

$$\therefore \quad \sum \boldsymbol{T} = 0$$

which implies

$$\sum T_x = \sum T_y = \sum T_z = 0$$

(b) Forces Confined to $x - y$ Plane

(*i*) *Translation* $\quad \sum \boldsymbol{F} = 0$

which implies $\quad \sum F_x = \sum F_y = 0$

(ii) Rotation $\quad \sum \boldsymbol{T} = 0$

which implies $\quad \sum T_z = 0$

Thus in (*a*) we have six independent conditions, and in (*b*) we have three. When solving problems involving forces which are co-planar, we can only establish three independent equations, and find three unknowns: a fourth equation (found by resolving in a new direction, or taking moments about a new axis) contains no new information.

Theorems about Equilibrium

(*a*) *If three forces maintain a rigid body in equilibrium, their lines of action are concurrent.* (Suppose three forces $\boldsymbol{P}$, $\boldsymbol{Q}$ and $\boldsymbol{R}$ act on a rigid body. If $\boldsymbol{R}$ does not pass through the intersection of $\boldsymbol{P}$ and $\boldsymbol{Q}$, there will be a resultant torque which gives the body an angular acceleration about an axis through that point.)

(*b*) (*i*) *The Triangle of Forces Theorem*

If three forces acting on a particle can be represented in size and direction by the sides of a triangle taken in order, they will maintain the particle in equilibrium.

(*ii*) *Its converse*

If three forces maintain a body in equilibrium, they can be represented in size and direction by the sides of a triangle taken in order.

(*i*) and (*ii*) are consequences of the parallelogram law of addition (p. 18). The converse is the more useful for solving problems.

7.4 Couples

The resultant of two co-planar forces is usually found by vector addition using the parallelogram law, but this method cannot be used for *parallel* forces.

(*a*) *Like parallel forces* ***P*** and ***Q*** have a resultant along the same direction of magnitude $P + Q$. Its line of action is such that it produces the same torque as ***P*** and ***Q***, and is between their lines of action.

(*b*) *Antiparallel forces* ***P*** and ***Q*** (***P*** > ***Q*** say) have a resultant magnitude $(P - Q)$ whose line of action lies outside those of ***P*** and ***Q***.

(*c*) *Equal antiparallel forces* cannot be replaced, for all effects, by a single force, and constitute an example of a **couple**.

Couples

A **couple** *is a system of forces which produces a turning effect only (i.e. zero translational effect).*

A couple often consists of a pair of equal antiparallel forces acting along a straight line, as in

(*i*) the system of forces applied to a car steering wheel, or

(*ii*) the forces experienced by two sides of a rectangular coil which carries a current while in a magnetic field (p. 378).

We cannot always distinguish two *distinct* forces, as in

(*i*) the forces applied by one car clutch-plate on the other, or

(*ii*) the forces exerted by a torsion fibre on the moving-coil of a galvanometer.

The Torque of a Couple

The couple shown in fig. 7.2 illustrates

(*a*) the moment of a couple is the *same about any axis* drawn perpendicular to the plane it defines (since O is an arbitrary point), and

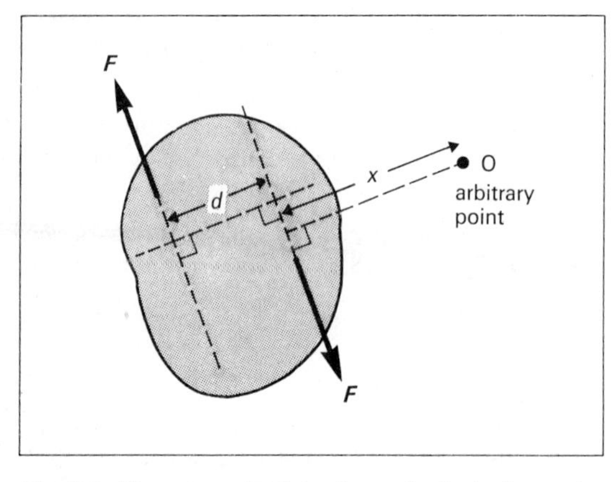

Fig. 7.2. The torque ***T*** *of the forces is clockwise and of size* $F(x + d) - Fx = Fd$.

(*b*) *the torque of a couple is calculated from*

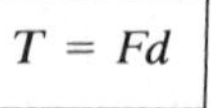

$$T = Fd$$

where F is the magnitude of either force, and d is the *moment arm*, the perpendicular distance between the forces' lines of action.

7.5 Centre of Gravity

Centre of Mass

Suppose a force is applied to the rod of fig. 7.3 at the point C, which accelerates the rod and both masses.

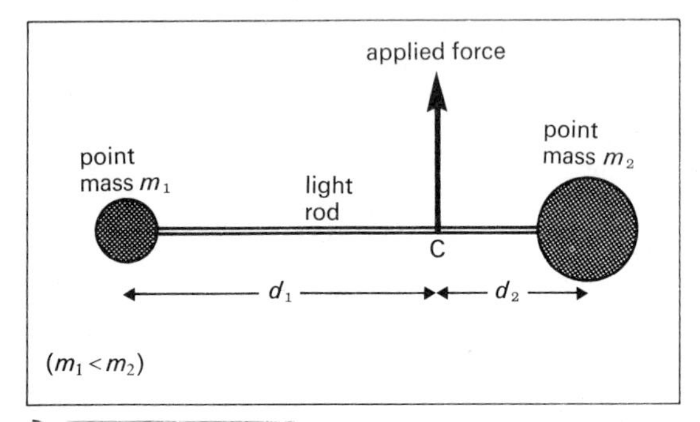

Fig. 7.3. The concept of centre of mass.

It is found by experiment that the rod *does not have angular acceleration if*

$$m_1 d_1 = m_2 d_2$$

C is the *centre of mass* of the two point masses. Suppose that relative to an arbitrary origin, m_1, m_2 and C have coordinates (x_1, y_1), (x_2, y_2) and $(\bar{x}, \bar{y})$ respectively. Then

$$\bar{x} = \frac{m_1 x_1 + m_2 x_2}{m_1 + m_2}, \qquad \bar{y} = \frac{m_1 y_1 + m_2 y_2}{m_1 + m_2}$$

We define the centre of mass $\bar{x}$ ***of a number of particles by the equation***

$$\bar{x} = \frac{\sum (m_1 x_1)}{\sum m_1}$$

where the particle m_1 ***has coordinate*** x_1 ***relative to the origin.***

* For a continuous distribution of matter (i.e. for a body not composed of discrete point masses) we would calculate the location of C from

$$\bar{x} = \frac{\int x \, dm}{\int dm}$$

Centre of Gravity

The weights of the individual particles of which a body is composed form a system of forces which are effectively parallel. Their single resultant constitutes the **weight *G*** of the body.

> *By the* **centre of gravity** *of a body we mean that single point through which the line of action of the weight passes however the body is oriented.*

It follows that the resultant gravitational torque on the body about an axis through that point will always be zero.

Suppose the individual particles of a body have weight $\boldsymbol{w}_1$ etc. Then the centre of gravity is located at $\bar{x}$, where

$$G\bar{x} = \sum (w_1 x_1)$$

$$\boxed{\bar{x} = \frac{\sum (w_1 x_1)}{\sum w_1}}$$

*For a continuous distribution of matter we again write

$$\boxed{\bar{x} = \frac{\int x \, dw}{\int dw}}$$

Since $\boldsymbol{w}_1 = m_1 \boldsymbol{g}$, it follows that where $\boldsymbol{g}$ can be assumed constant, the centre of mass and centre of gravity coincide. This is true on the the surface of the Earth.

Location of the Centre of Gravity

(*a*) Use obvious physical homogeneity and geometrical *symmetry*.

(*b*) By *calculation*, using

(*i*) for simple bodies, the principle of moments,

or

(*ii*) in more complex situations

$$\bar{x} = \frac{\int x \, dw}{\int dw}$$

(which incorporates the principle of moments).

(*c*) *By experiment.* Suspend the body freely from two different points on it. In each case the centre of gravity must lie vertically below the point of suspension. Its location is the intersection of the lines on which it must lie.

7.6 Stability of Equilibrium

Equilibrium or Not?

Distinguish carefully between a body in *unstable* equilibrium, and a body which is *not* in equilibrium (accelerated). For a body to be in equilibrium when resting on a surface, the vertical line passing through its centre of gravity must pass within the (implied) boundary of contact with the surface.

In fig. 7.4, (*a*) and (*c*) are possible equilibrium positions, but (*b*) and (*d*) are not.

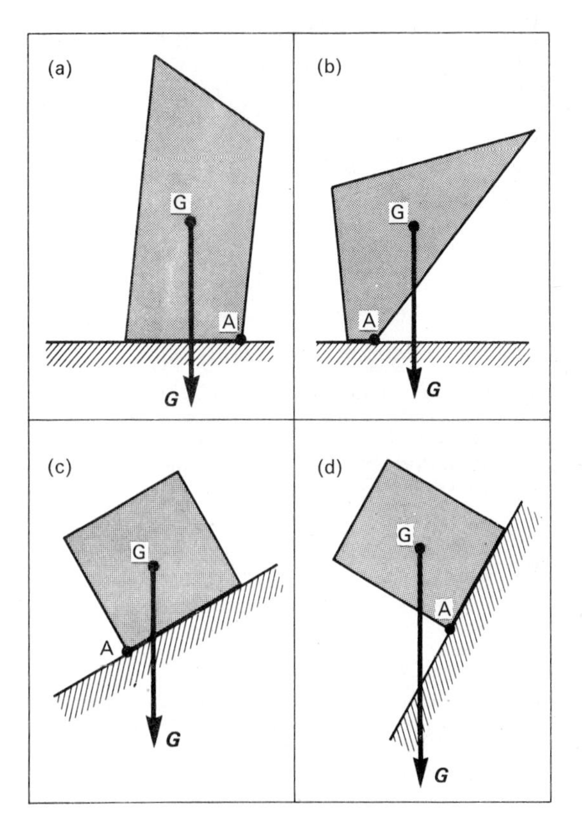

Fig. 7.4. Equilibrium, and line of action of weight.

For each example, consider the torque acting on the body about A, the corner on which pivoting is possible. In (*a*) and (*c*) the gravitational torque is balanced by the torque of the contact force: the body is *in equilibrium.* In (*b*) and (*d*) the gravitational torque is unbalanced: the body is *not in equilibrium.*

Stability of Equilibrium

The table overleaf shows how to distinguish **stable**, **unstable** and **neutral** equilibrium situations.

The potential energy has a stationary value at a position of equilibrium. It is a *maximum* for unstable equilibrium, and a *minimum* for stable equilibrium.

Situation	*Displace-ment*	*What happens to* c.g.	*What happens to* p.e.	*Subsequent behaviour*	*Name of original equilibrium*
marble in saucer	displace from centre	raised	increased	marble returns to original position	stable
marble on upturned saucer	displace from centre	lowered	reduced	marble rolls off saucer	unstable
marble on flat table	displace horizontally	stays at same level	no change	marble remains in displaced location	neutral

*Analysis of Equilibrium Stability

The magnitudes of the changes in potential energy in gravitational and electrostatic interactions are given by $\Delta E_p = mgh$, and $\Delta W = QEy$, respectively. If $\boldsymbol{F}$ represents the force of interaction, and $\boldsymbol{x}$ the displacement vector, we write

$$\delta E_p = -\boldsymbol{F} \cdot \delta \boldsymbol{x}$$

The negative sign is necessary because potential increases in a direction opposite from that of field. The result

$$F = -\frac{dE_p}{dx}$$

is of general application, and is used in this book on p. 124 to analyse the interaction between a pair of molecules.

(*a*) For a body to be in *equilibrium* $\boldsymbol{F} = 0$, so $dE_p/dx = 0$. The E_p–x curve must show a *stationary value*.

(*b*) For the equilibrium to be *stable* a small displacement must lead to an increase in E_p. The E_p–x curve must be at a *minimum* value. d^2E_p/dx^2 must be positive.

(*c*) For the equilibrium to be *unstable* the curve must show a *maximum* value. d^2E_p/dx^2 must be negative.

(*d*) When $d^2E_p/dx^2 = 0$ the equilibrium is apparently neutral, but it is wise to investigate the situation more fully. These points are illustrated in fig. 7.5.

The graph can be visualized simply in terms of a ball-bearing rolling along a curved track in a vertical plane. Then E_p is proportional to height above an arbitrary level.

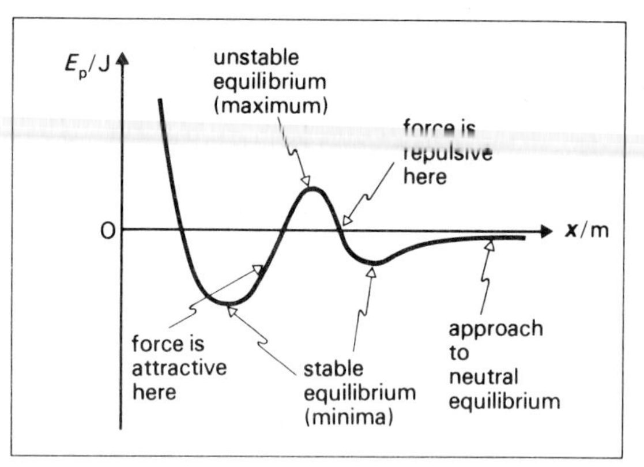

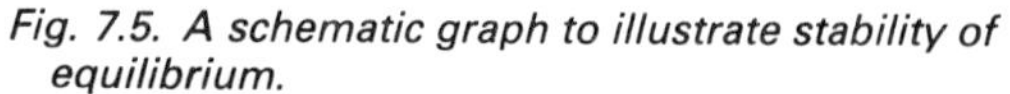

Fig. 7.5. A schematic graph to illustrate stability of equilibrium.

8 Fluids at Rest

This chapter summarizes concisely what is taught as **Hydrostatics** in O level courses.

8.1 Density and Pressure

Density

Suppose the sample of a particular *substance* has a mass m and volume V.

The average density ρ of the substance is defined by the equation

$$\rho = \frac{m}{V}$$

$$\rho\left[\frac{\text{kg}}{\text{m}^3}\right] = \frac{m[\text{kg}]}{V[\text{m}^3]}$$

The table indicates the extent to which densities can vary.

Material	$\rho/\text{kg m}^{-3}$
interstellar space	10^{-20}
best laboratory vacuum	10^{-16}
air	1
water	10^3
dense metal	10^4
atomic nucleus	10^{17}

The **relative density** of a substance is defined by

$$\text{relative density} = \frac{\text{density of substance}}{\text{density of water}}$$

For example, the density of mercury is 1.36×10^4 kg m^{-3}, and its relative density is 13.6.

Pressure

Suppose an area $\Delta \boldsymbol{A}$ totally immersed in a fluid experiences a normal force $\Delta \boldsymbol{F}$.

The average pressure p_{av} over the area is defined by the equation

$$p_{av} = \frac{\Delta F}{\Delta A}$$

$$p_{av}\left[\frac{\text{N}}{\text{m}^2}\right] = \frac{\Delta F[\text{N}]}{\Delta A[\text{m}^2]}$$

The pressure p at a point in the fluid is defined by the equation

$$p = \lim_{\delta A \to 0} \frac{\delta F}{\delta A}$$

where the area δA contains the point.

Notes. (*a*) The unit N m^{-2} is, for convenience, called a *pascal* (Pa). Thus

$$1 \text{ Pa} \equiv 1 \text{ N m}^{-2}$$

(*b*) Pressure is a *scalar* quantity. $\Delta \boldsymbol{F}$ is a *vector*, and $\Delta \boldsymbol{A}$ is a *pseudo-vector*, the direction of $\Delta \boldsymbol{A}$ being that of the normal to the surface. The *force* that we associate with a pressure has a direction, but the *pressure* does not.

(*c*) It is found experimentally that if the orientation of $\Delta \boldsymbol{A}$ is altered, then the magnitude of $\Delta \boldsymbol{F}$ (and hence the value of p) remains fixed, even though $\Delta \boldsymbol{F}$ remains normal to $\Delta \boldsymbol{A}$.

The Origin of Pressure Forces

(a) *The pressure caused by a gas* is discussed fully on p. 180.

(b) *In a liquid* molecules have mainly vibrational energy (kinetic and intermolecular potential energy), and a small proportion of translational k.e. When a liquid molecule strikes a surface, it will suffer a much smaller momentum change than would an analogous gas molecule because its speed is (on average) much smaller at the instant of impact. Nevertheless there may be 10^3 times as many molecules in unit volume, and the pressures in the gas and liquid may have the same order of magnitude.

(c) *In a solid* the mechanism is similar, but differs in that the solid's molecules have only vibratory (and no translatory) kinetic energy. When vibration causes overlap between the electron clouds of adjacent molecules, they will repel each other (p. 119).

8.2 Facts about Fluid Pressure

(a) *Pressure increase with depth*
Imagine two points separated by a *vertical* distance h in a fluid of constant density ρ. The pressure difference Δp between the points is given by

$$\boxed{\Delta p = h\rho g}$$

where h is measured as positive downward.

$$\Delta p\left[\frac{\mathrm{N}}{\mathrm{m}^2}\right] = h[\mathrm{m}]\,\rho\left[\frac{\mathrm{kg}}{\mathrm{m}^3}\right]\cdot g\left[\frac{\mathrm{m}}{\mathrm{s}^2}\right]$$

(b) It follows from (a) that *the pressures at two points at the same horizontal level in the same liquid are the same*, provided that the points are connected by that liquid only, and that the liquid is static. (If this were not so, fluid would move from the place of higher pressure to that of lower until the condition was satisfied.)

(c) The change of pressure Δp depends on the vertical distance h, ρ and g *only*: it is therefore *independent of the cross-sectional area* and shape of any containing vessel.

(d) *For a gas* (whose density is very low) Δp is small unless h is large. *The gas pressure is the same at all points in volumes of reasonable size.*

*For volumes of large size (such as the Earth's atmosphere) the gas density ρ will not be constant because the lower layers are compressed. One cannot apply $\Delta p = h\rho g$ under these conditions: one must use the more rigorous form

$$\frac{dp}{dh} = -\rho g,$$

in which h is measured vertically upward, and the negative sign is inserted because p decreases as h increases. ρ depends upon p.

(e) **Pascal's Principle** *states that pressure applied to an enclosed liquid is transmitted without change to every part of the liquid, whatever the shape of the liquid.*

If the pressure at a liquid surface is p_0 (e.g. atmospheric pressure), then the pressure at a point in the liquid can be found from

$$p = p_0 + \Delta p$$

where Δp is caused by changes of vertical height in the liquid.

8.3 Archimedes's Principle and Flotation

Archimedes's Principle *states that when a body is immersed in a fluid* (liquid or gas) *it experiences an upthrust equal in magnitude to the* weight *of fluid displaced.*

We can justify the principle by experimental verification, a pressure argument, or a force argument.

Force Argument
Consider the equilibrium of that part of the fluid which has the same volume and shape as the part of the body to be immersed. It experiences a force, *through its centre of gravity*, of the same magnitude as its weight, but oppositely directed. When that part of the fluid is replaced by the body to be immersed, then the body will experience the same pressure forces. This proof, although not analytical, is rigorous.

Notes. (a) The upthrust still acts when the body rests on some support (such as the bottom of a vessel) provided the space between the body and the support is not evacuated.

(b) *Newton's Third Law* requires that a body which experiences a buoyant upthrust should exert an equal but downward thrust on the fluid responsible.

Flotation

The Principle of Flotation *states that a floating body displaces its own weight of fluid.*

It is a consequence of the condition for equilibrium and *Archimedes's Principle*. It is applied in situations with

(*a*) *variable weight and constant upthrust* (such as the submarine), and

(*b*) *constant weight and variable upthrust*, such as

(*i*) the Balloon, and

(*ii*) the **Hydrometer**. When it floats freely, the *hydrometer experiences an upthrust equal in magnitude to its weight* (a constant): it therefore reaches equilibrium with a particular volume under the surface depending on the density of the liquid in which it is immersed.

8.4 Measurement of Density and Pressure

(*a*) Density and Relative Density

(*i*) Direct measurement of m and V.

(*ii*) Measurement of V by displacement methods.

(*iii*) Use of *relative density bottle* for liquids, or finely divided solids.

(*iv*) Use of *Archimedes's Principle*:

(1) *Solid.* If it floats use a sinker. If it dissolves in water use some other non-solvent liquid.

(2) *Liquid.* Use a solid (such as a glass stopper) of constant volume.

(*b*) Pressure

Absolute pressure is the actual pressure at a point in a fluid.

Gauge pressure is that recorded by a pressure gauge, and is frequently the difference between absolute pressure and atmospheric pressure.

The equation $\Delta p = h\rho g$ indicates why it convenient to refer to pressures by **heads of liquid**.

A unit commonly used for gas pressures is the **atmosphere** (atm), which is *defined* to be 101 325 Pa. It is equivalent (but not *exactly* equal) to that exerted by 760 mm of mercury (mmHg) of specified density under **standard gravity** ($g_0 = 9.806\,65\ \text{N kg}^{-1}$). Thus

$$\begin{aligned} 1\ \text{atmosphere} &= 760\ \text{mmHg} \\ &= (0.760\ \text{m}) \times (1.36 \times 10^4\ \text{kg m}^{-3}) \\ &\quad \times (9.81\ \text{N kg}^{-1}) \\ &= 1.013 \times 10^5\ \text{N m}^{-2}\ (\text{Pa}). \end{aligned}$$

[Note that the conventional millimetre of mercury exerts a pressure of 133 Pa. Neither the atm nor the mmHg is an SI unit.]

Atmospheric pressure can be measured by

(*i*) a **simple barometer** (normally using mercury),

(*ii*) a **Fortin** barometer, which incorporates refinements such as a *Vernier* scale,

(*iii*) an **aneroid** barometer, which consists of an evacuated flexible metal chamber.

The simple barometer is an example of a closed-tube **manometer**.

Other pressure-measuring devices include

(*i*) the **U-tube** manometer,

(*ii*) the **Bourdon** gauge,

(*iii*) the **McLeod** gauge (a device for measuring very low pressures).

9 The Bernoulli Equation

9.1 Description of Streamline Flow

In general, fluid dynamics is too complex to analyse, but we can proceed if we make the following *assumptions.*

(*a*) *The fluid is non-viscous* (see p. 144). No mechanical energy is converted to internal (thermal) energy.

(*b*) *The fluid is incompressible.* This means that its density remains constant.

(*c*) *The fluid angular momentum plays no part.* Flow patterns for which this is true are said to be **irrotational**.

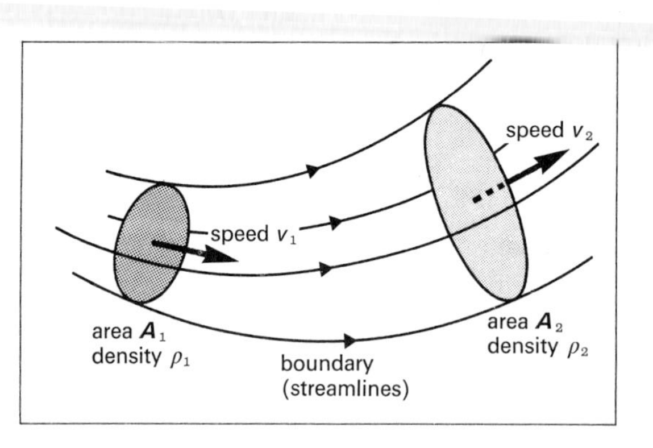

Fig. 9.1. A tube of flow.

(*d*) *The flow is* **steady**. Flow is said to be *laminar, orderly* or **streamline** if successive particles, while passing a given point have the same velocity. (Flow is **turbulent** or *disorderly* if the velocity of particles at the given point depends on the instant of observation.)

Although *irrotational steady* flow of a *non-viscous incompressible* fluid does not occur in practice, the analysis which follows is a reasonable description of some real physical situations. (See also the equation that describes an adiabatic change, p. 190.)

For steady flow, we define

(*i*) A **streamline** to be the path taken by a fluid particle. The instantaneous velocity of a fluid particle at a point lies along the tangent to the streamline at that point. Like electric field lines (p. 314), streamlines can never cross.

(*ii*) A **tube of flow** to be an imaginary boundary defined by streamlines drawn so as to enclose a tubular region of fluid. No fluid crosses the *side* boundary of such a tube (see fig. 9.1).

9.2 The Equation of Continuity

This equation follows from an application of the law of conservation of *mass* to ideal fluid flow.

In time Δt, a mass m_1 enters the tube of flow (fig. 9.1) across the cross-section $\boldsymbol{A}_1$, where $m_1 = \rho_1 A_1 v_1 \Delta t$. A mass m_2 leaves at $\boldsymbol{A}_2$ in the same time, where $m_2 = \rho_2 A_2 v_2 \Delta t$.

Equating $m_1 = m_2$

$$\boxed{\rho_1 A_1 v_1 = \rho_2 A_2 v_2} \quad \textit{The equation of continuity}$$

Since we are to consider incompressible fluids only, $\rho_1 = \rho_2$, so

$$\boxed{A_1 v_1 = A_2 v_2}$$

Av is called the *flow rate*, or **volume flux**. The fact that it is constant demonstrates that v is large where A is small, and vice-versa.

9.3 The Bernoulli Equation

This equation follows in a similar way from the law of conservation of *energy*.

(a) Change of Speed (fig. 9.2)

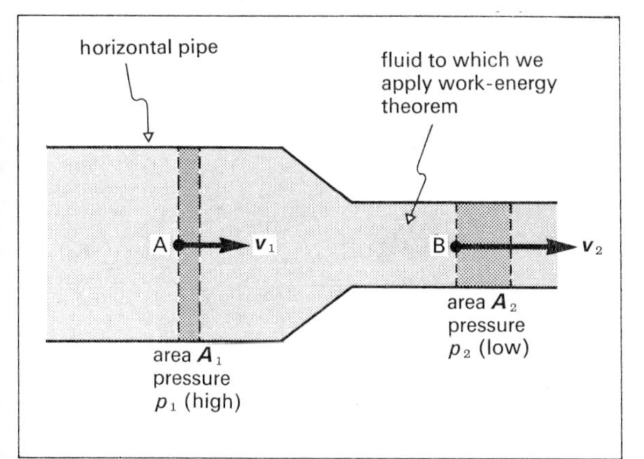

Fig. 9.2. Change of pressure that accompanies change of speed.

Consider a time interval Δt.

(*i*) The work done *on* the fluid during this time

$$W = \begin{pmatrix}\text{work done by} \\ \text{forces at A}\end{pmatrix} + \begin{pmatrix}\text{work done by} \\ \text{forces at B}\end{pmatrix}$$
$$= (p_1 A_1)(v_1 \Delta t) - (p_2 A_2)(v_2 \Delta t)$$
$$= (p_1 - p_2)(A_1 v \ \Delta t)$$

since $A_1 v_1 = A_2 v_2$. Note that the forces at B do *negative* work on the fluid.

(*ii*) The gain of kinetic energy is that of a mass

$$(A_1 v_1 \Delta t)\rho \qquad [\text{or } (A_2 v_2 \Delta t)\rho]$$

whose speed changes from v_1 to v_2.

$$\text{Gain of k.e.} = \tfrac{1}{2}(A_1 v_1 \Delta t \rho)(v_2^2 - v_1^2)$$

We now apply the *work-energy theorem*

$$W = \text{change of k.e.}$$
$$(p_1 - p_2)(A_1 v_1 \Delta t) = \tfrac{1}{2}(A_1 v_1 \Delta t \rho)(v_2^2 - v_1^2)$$
$$(p_1 - p_2) = \tfrac{1}{2}\rho(v_2^2 - v_1^2)$$
$$p_1 + \tfrac{1}{2}\rho v_1^2 = p_2 + \tfrac{1}{2}\rho v_2^2 \qquad (1)$$

(b) Change of Vertical Height (fig. 9.3)

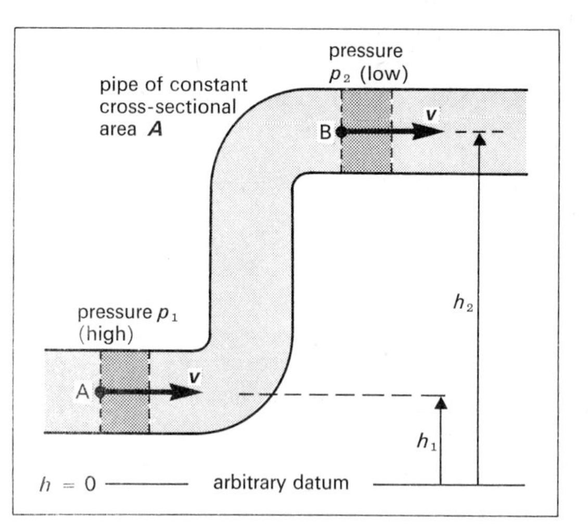

Fig. 9.3. Change of pressure that accompanies change of vertical height.

Consider a time interval Δt. Because there is no change of kinetic energy, the net work done *on* the system is *zero*. The (positive) work done by the pressure forces has the same magnitude as the (negative) work done by the gravitational forces:

$$(\text{work done by pressure forces}) = (\text{gain of p.e.})$$
$$(p_1 - p_2)(Av \Delta t) = (Av \Delta t \rho)g(h_2 - h_1)$$
$$(p_1 - p_2) = \rho g(h_2 - h_1)$$
$$p_1 + h_1 \rho g = p_2 + h_2 \rho g \qquad (2)$$

(c) Change of Speed and Height

If (*a*) and (*b*) happen simultaneously, so that both kinetic energy and gravitational potential energy change, then (1) and (2) are combined as

$$p_1 + \tfrac{1}{2}\rho v_1^2 + h_1 \rho g = p_2 + \tfrac{1}{2}\rho v_2^2 + h_2 \rho g$$

More generally, since A and B are arbitrary points, we have

$$\boxed{p + \tfrac{1}{2}\rho v^2 + h\rho g = \text{constant}}$$

which is the **Bernoulli equation**, and *which applies to*

all points along a streamline for steady, non-viscous incompressible flow. When the flow is irrotational, the constant has the same value for all streamlines.

Notes. (*i*) Each of the terms has the unit N m^{-2} (Pa) and dimensions [ML^{-1}T^{-2}] of pressure, and can be regarded as energy per unit volume.

(*ii*) Suppose $v_1 = v_2 = 0$, so that the fluid is at rest. Then

$$(p_1 - p_2) = -(h_1 - h_2)\rho g \quad \text{or} \quad \Delta p = h\rho g$$

This is the familiar result of p. 68.

$p + h\rho g$ is called the **static pressure** and $\frac{1}{2}\rho v^2$ is called the **dynamic pressure**.

9.4 Applications

In all real situations, the conditions for the *Bernoulli* equation to be exact are violated. (We can seldom escape turbulence; nearly every fluid is viscous, and gases are easily compressed.) Nevertheless the qualitative argument holds: where streamlines are drawn together, as at a constriction, *the speed of the fluid increases* (equation of continuity). If this happens at a horizontal level, then the *pressure of the fluid falls* (*Bernoulli*). This general result has many applications.

(*a*) The Venturi Meter

This device (fig. 9.4) measures the rate of flow of fluid (its *volume flux*) through a pipe.

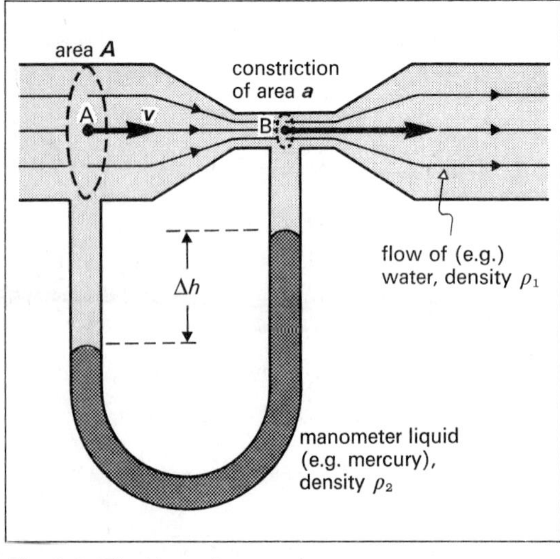

Fig. 9.4. The Venturi meter.

The pressure difference between A and B is recorded by the attached manometer. By measuring ***A***, ***a***, ρ_1, ρ_2 and Δh, one can find

(*i*) the fluid speed v at A (by using the *Bernoulli* equation), and hence

(*ii*) the volume flux $= Av$ (by using the equation of continuity).

(*b*) Dynamic Lift

This force is exerted on a body which is moving relative to a fluid in such a way that the streamlines are close on one side and further apart on the other—the pattern is not symmetrical.

(*i*) The Aerofoil

The aerofoil of fig. 9.5 is shaped so that air flows faster over the top than underneath.

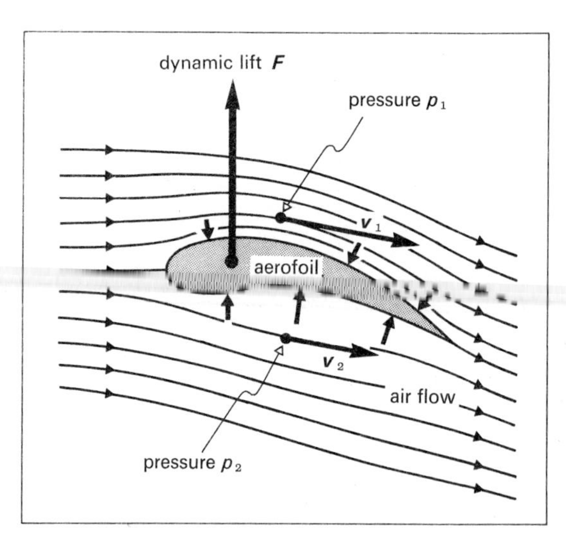

Fig. 9.5. An aerofoil experiences dynamic lift.

Since $v_1 > v_2$, $p_1 < p_2$ (*Bernoulli*).

The fact that the pressure below an aeroplane wing exceeds that above accounts for most of the force which supports the plane. (Note that fig. 9.5 is drawn for an observer *travelling with the aerofoil.*) When the angle of attack is too great the flow over the upper surface becomes turbulent, which reduces the pressure difference, and leads to stalling.

(*ii*) The Spinning Ball

This time the dynamic force on the ball (fig. 9.6) is caused, not by its shape, but because it spins.

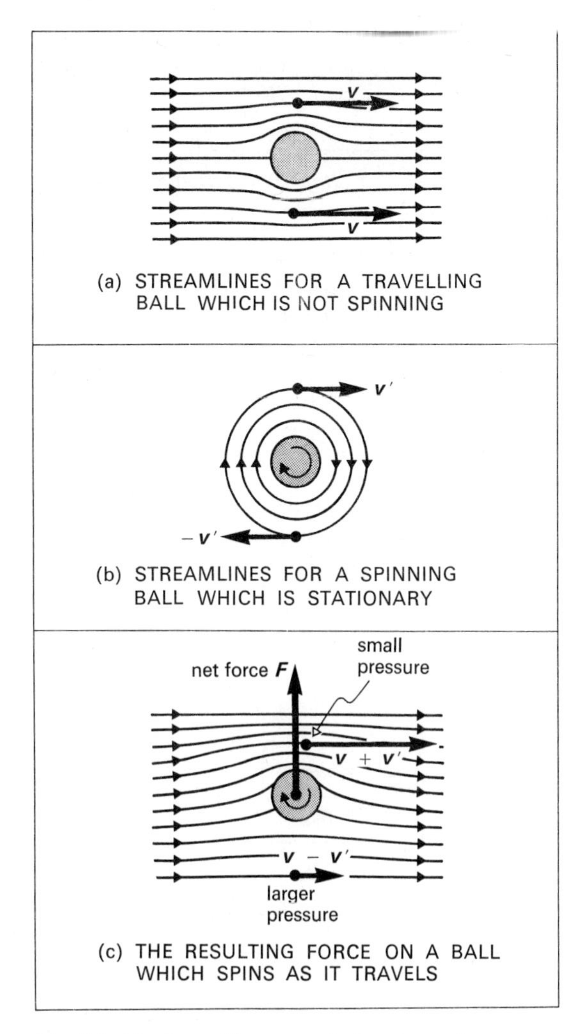

Fig. 9.6. The lateral force exerted on a moving spinning ball. The streamlines are drawn for an observer travelling with the ball.

The resulting lateral force causes the ball to have a curved trajectory, such as that of a sliced golf ball.

Note that a *viscous* medium is needed: in (b) air is dragged round by the ball.

(c) Torricelli's Theorem

Assume that the fluid of fig. 9.7 is non-viscous, and apply the *Bernoulli* equation to the points A and B on the streamline shown.

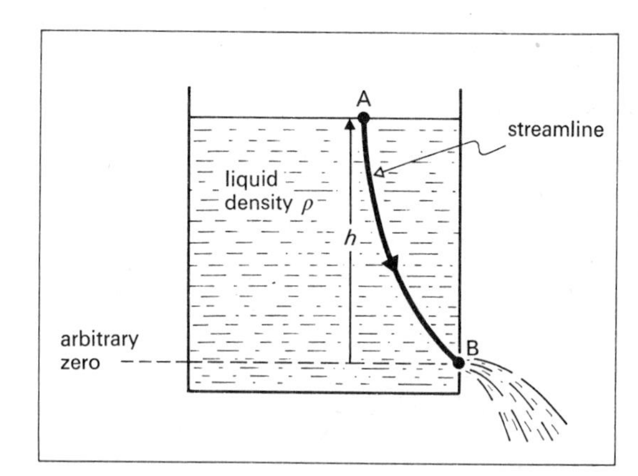

Fig. 9.7. Liquid escapes from a hole punched in a container.

At A p = atmospheric, $v = 0$.
At B p = atmospheric, v is to be found.
Then $p + \frac{1}{2}\rho v^2 + h\rho g = \text{const.}$ becomes

$$\text{(A)} \quad p + 0 + h\rho g = p + \tfrac{1}{2}\rho v^2 + 0 \quad \text{(B)}$$

$$v = \sqrt{2gh}$$

The horizontal speed of efflux equals the vertical speed acquired by free fall from a height h.

The *Bernoulli* equation also finds application in the action of filter-pumps, bunsen burners, carburettors, boomerangs, atomizers, sailing, propellors, etc.

Summary of Mechanics

	Defining equations	*Laws*	*Important derivations*
$\boldsymbol{v}$	$\boldsymbol{v} = \frac{\mathrm{d}\boldsymbol{s}}{\mathrm{d}t}$		$\boldsymbol{v} = \boldsymbol{u} + \boldsymbol{a}t$
$\boldsymbol{a}$	$\boldsymbol{a} = \frac{\mathrm{d}\boldsymbol{v}}{\mathrm{d}t}$		$\boldsymbol{s} = \frac{\boldsymbol{u} + \boldsymbol{v}}{2} \cdot t$
$\boldsymbol{\omega}$	$\boldsymbol{\omega} = \frac{\mathrm{d}\boldsymbol{\theta}}{\mathrm{d}t}$		$T = \frac{2\pi}{\omega}$ $v = \omega r$ $a = v^2/r = \omega^2 r = v\omega$
$\boldsymbol{p}$ $\boldsymbol{F}$	$\boldsymbol{p} = m\boldsymbol{v}$ (*i*) defined by *Newton* I (*ii*) measured from *Newton* II	NEWTON I If $\boldsymbol{F} = 0, \Delta\boldsymbol{v} = 0$ II $\boldsymbol{F} \propto \frac{\mathrm{d}}{\mathrm{d}t}(m\boldsymbol{v})$ III $\boldsymbol{F}_{AB} = -\boldsymbol{F}_{BA}$	
mass ratio	$\frac{m_0}{m} = \frac{a}{a_0}$		$\boldsymbol{F} = m\boldsymbol{a}$ $\boldsymbol{G} = m\boldsymbol{g}$
W E_k P	$W = Fs \cos \theta$ $E_k = \frac{1}{2}mv^2$ $P = \frac{\mathrm{d}W}{\mathrm{d}t}$	Law of conservation of energy	$W = Fs$ $= \frac{1}{2}mv^2 - \frac{1}{2}mu^2$
	impulse $= \boldsymbol{F}\,\Delta t$	Law of conservation of momentum: if $\boldsymbol{F} = 0, \Delta\boldsymbol{p} = 0$ for an isolated system.	$\boldsymbol{F}\,\Delta t = m\boldsymbol{v} - m\boldsymbol{u}$
$\boldsymbol{\alpha}$ $\boldsymbol{T}$ I	$\boldsymbol{\alpha} = \frac{\mathrm{d}\boldsymbol{\omega}}{\mathrm{d}t}$ $T = Fr$ $I = \sum mr^2$		$a_t = r\alpha$ $a_r = \frac{v^2}{r} = \omega^2 r = v\omega$ $\boldsymbol{T} = I\boldsymbol{\alpha}$ (from *Newton* II) $I = I_0 + Mh^2$ $\Delta W = T\,\Delta\theta$ $E_{k,rot} = \frac{1}{2}I\omega^2$

Summary of Mechanics—cont.

	Defining equations	*Laws*	*Important derivations*
$\boldsymbol{L}$	$\boldsymbol{L} = I\boldsymbol{\omega}$	Law of conservation of angular momentum: if $\boldsymbol{T} = 0$, $\Delta\boldsymbol{L} = 0$ for an isolated system.	$W = \frac{1}{2}I\omega^2 - \frac{1}{2}I\omega_0^2$ $\boldsymbol{T}\,\Delta t = I\boldsymbol{\omega} - I\boldsymbol{\omega}_0$
$\bar{x}$ $\bar{x}$	$\bar{x} = \frac{\int x\,\mathrm{d}m}{\int \mathrm{d}m}$ $\bar{x} = \frac{\int x\,\mathrm{d}w}{\int \mathrm{d}w}$		Equilibrium conditions $\sum \boldsymbol{F} = 0$ $\sum \boldsymbol{T} = 0$
ρ p	$\rho = \frac{m}{V}$ $p = \lim_{\delta A \to 0}\left(\frac{\delta F}{\delta A}\right)$		$\Delta p = h\rho\boldsymbol{g}$
			Equation of continuity $\rho_1 A_1 v_1 = \rho_2 A_2 v_2$ $A_1 v_1 = A_2 v_2$ *Bernoulli* $p + \frac{1}{2}\rho v^2 + h\rho\boldsymbol{g} = \text{constant}$

III Oscillations and Wave Motion

10 Simple Harmonic Oscillation

10.1 Introduction

A **periodic** or **cyclic** motion is one during which a body continually retraces its path at equal time intervals.

Examples

(*a*) planetary motion

(*b*) the motion of an air molecule disturbed by a sound wave (this being superimposed on its random motion).

Oscillations

Before a system can be set into oscillation, it must satisfy two conditions:

(1) it must be able to store *potential energy* (so a mechanical system would need some springiness or elasticity), and

(2) it must have some inertia (or mass) which enables it to possess *kinetic energy*.

These two concepts (*elasticity* and *inertia*) are very useful for developing an intuitive understanding of any oscillation.

An oscillation represents the continual interchange of potential and kinetic energy.

Examples

Nature of oscillating system	p.e. *stored as*	k.e. *possessed by moving*
mass on helical spring	elastic energy of spring	mass
cantilever	elastic energy of bent rod	rod
single pendulum	gravitational p.e. of bob	bob
vertical rod floating in liquid of zero viscosity	gravitational p.e. of rod or liquid	rod

Simple Harmonic Oscillations

Dimensional Argument

Problem: for what type of oscillation is the time period independent of the amplitude?

Put T = time period,

$\boldsymbol{F}$ = restoring force at displacement $\boldsymbol{x}$,

m = mass that oscillates.

Suppose

$$T = k(F)^a(x)^b(m)^c$$

Equating dimensions for M, L and T (as on p. 20), we find

$$a = -\tfrac{1}{2}$$
$$b = \tfrac{1}{2}$$
$$c = \tfrac{1}{2}$$

so $$T = k\left(\frac{mx}{F}\right)^{1/2}$$

The motion is **isochronous** (has T independent of amplitude) provided x/F is constant, since m is constant.

Then $$\boldsymbol{x} \propto \boldsymbol{F}$$

Definition of s.h.m.

(*a*) *Linear.* If the acceleration $\ddot{x}$ of a point is always directed toward, and is proportional to its distance from, a fixed point, then the motion is simple harmonic (fig. 10.1).

$$\ddot{\boldsymbol{x}} = -(\text{const})\boldsymbol{x}$$

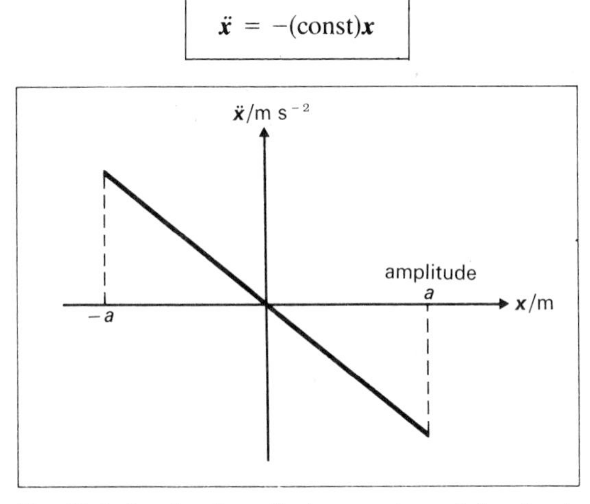

Fig. 10.1. Acceleration–displacement graph for simple harmonic motion.

(*b*) *Angular.* Suppose θ represents the angular displacement. Then

$$\ddot{\theta} = -(\text{const})\,\theta$$

The Importance of s.h.m.

(1) It occurs very frequently (see below).

(2) T does *not* depend on amplitude.

(3) It can be described by relatively simple mathematics.

(4) Any *periodic* motion which is not simple harmonic can be analysed into its simple harmonic components: it can then be regarded as the superposition of the separate components, which *can* be analysed.

s.h.m. in Practice

s.h.m. will be shown by any system subjected to a force which obeys *Hooke's Law*, since if

$$\boldsymbol{F} = -k\boldsymbol{x} \qquad \text{(p. 129)}$$

then $$\ddot{\boldsymbol{x}} = -\omega^2\boldsymbol{x}$$

where ω^2 is some positive constant, and ω is called the **pulsatance**.

In practice *Hooke's Law* is nearly obeyed for most *small* distortions (see p. 129): hence many particles and systems will show at least approximate s.h.m. for *small* oscillations about a point of stable equilibrium.

The value of T will increase if

(*i*) the inertia factor increases, or

(*ii*) the elasticity factor decreases.

Note that in some systems (such as the simple pendulum) the two may be related, since mass and gravitational force are proportional.

10.2 The Kinematics of s.h.m.

Definitions

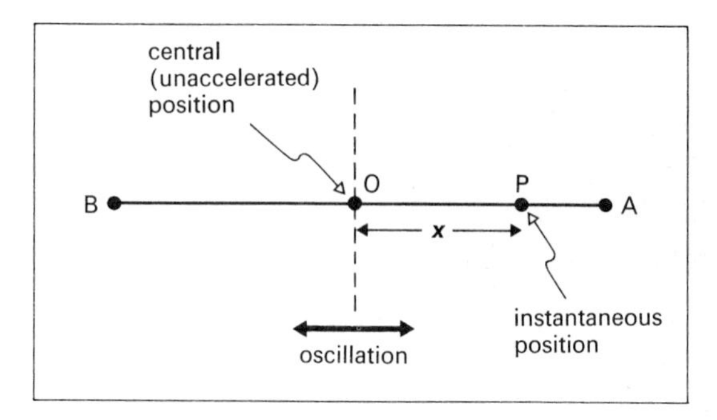

Fig. 10.2. Description of linear s.h.m.

Refer to fig. 10.2. in which P describes s.h.m.

Displacement $\boldsymbol{x}$ is the vector $\overrightarrow{OP}$: it takes negative values when P lies between O and B.

Amplitude a is the magnitude of the maximum displacement from the central position, i.e.

$$\text{OA} = \frac{\text{AB}}{2}$$

(It is always positive.)

Period T is the time taken to describe a complete oscillation (cycle), such as the path OAOBO.

Frequency f is the number of oscillations performed in each second. Thus

$$f = \frac{1}{T}$$

$$f[\text{Hz}] = \frac{1}{T[\text{s}]}$$

The unit *cycle per second* is called the **hertz** (Hz).

The auxiliary circle

Suppose (fig. 10.3) that N describes a circle of radius a at steady angular speed ω. We will show that P, under these conditions, describes s.h.m. according to our earlier definition.

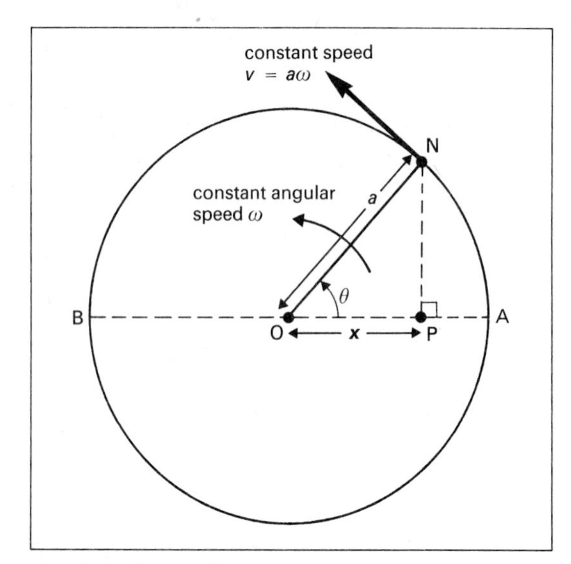

Fig. 10.3. The auxiliary circle.

Acceleration of N $= \omega^2 a$ toward O.
Resolved part of this acceleration parallel to AB

$$= \omega^2 a \cos\theta \quad \text{toward O.}$$

Acceleration of P $= \omega^2 a \cos\theta$ toward O.

$$\ddot{x} = -\omega^2(a\cos\theta)$$
$$\therefore \ddot{\boldsymbol{x}} = -\omega^2 \boldsymbol{x}$$

where $\ddot{\boldsymbol{x}}$ is measured in the same direction as $\boldsymbol{x}$.

Since ω^2 is a positive constant, P describes s.h.m.

The *time period* T of P is the time period for one revolution of N.

$$T[\text{s}] = \frac{\text{angle through which N turns [rad]}}{\text{angular speed}\left[\dfrac{\text{rad}}{\text{s}}\right]}$$

$$\boxed{T = \frac{2\pi}{\omega}}$$

which gives $\qquad \omega = 2\pi f.$

Variation of x, $\dot{x}$ and $\ddot{x}$ with Time t

(a) Displacement x

From fig. 10.3, $x = a\cos\theta$

$$= a\cos\omega t \quad (\theta = \omega t)$$

(b) Velocity $\dot{x}$

Velocity of P in direction OA is that of the resolved part of N's velocity in that direction.

Thus $\qquad \dot{x} = (a\omega)\sin\theta \quad \text{along AO}$

$$= -a\omega\sin\theta \quad \text{along OA}$$
$$= -a\omega\sin\omega t$$
$$= \pm a\omega\sqrt{1-\cos^2\omega t} \qquad \text{(p. 22)}$$
$$= \pm a\omega\sqrt{1-\left(\frac{x}{a}\right)^2}$$
$$= \pm\omega\sqrt{a^2-x^2}$$

Special cases:

When $x = 0$, $\quad \dot{x} = \pm a\omega$ (its largest value)

When $x = \pm a$, $\quad \dot{x} = 0$.

(c) Acceleration $\ddot{x}$

As was shown above

$$\ddot{x} = -a\omega^2\cos\theta$$
$$= -a\omega^2\cos\omega t$$

Special cases:

When $x = 0$, $\quad \ddot{x} = 0$

When $x = a$, $\quad \ddot{x} = -\omega^2 a$ } its largest value

$x = -a$, $\quad \ddot{x} = +\omega^2 a$ }

These results can also be found using calculus. Since $x = a\cos\omega t$

$$\frac{\mathrm{d}x}{\mathrm{d}t} = \dot{x} = -a\omega\sin\omega t$$

$$\frac{\mathrm{d}^2x}{\mathrm{d}t^2} = \ddot{x} = -a\omega^2\cos\omega t$$

These results assume $\theta = 0$ (and so $x = a$) when $t = 0$. If we had chosen $\theta = 3\pi/2$ rad (and so $x = 0$) when $t = 0$, we would have found

$$x = a\sin\omega t$$
$$\dot{x} = a\omega\cos\omega t$$
$$\ddot{x} = -a\omega^2\sin\omega t$$

Summary

Quantity	*Variation with* $\boldsymbol{x}$	*Variation with* t
$\boldsymbol{x}$	—	$x = a\cos\omega t$
$\dot{\boldsymbol{x}}$	$\dot{x} = \pm\omega\sqrt{a^2-x^2}$	$\dot{x} = -a\omega\sin\omega t$
$\ddot{\boldsymbol{x}}$	$\ddot{\boldsymbol{x}} = -\omega^2\boldsymbol{x}$	$\ddot{x} = -a\omega^2\cos\omega t$

It is instructive to examine the graphs (fig. 10.4) which illustrate these equations. Note particularly the phase relationships.

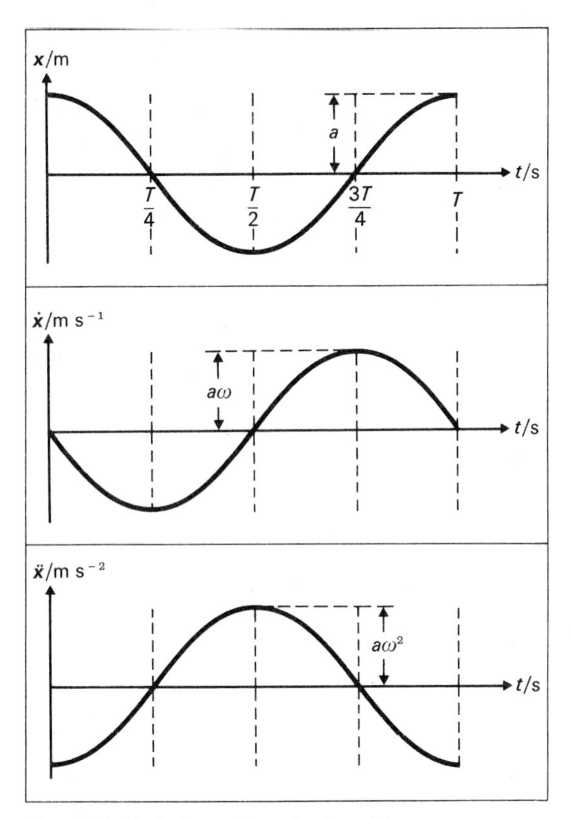

Fig. 10.4. Variation with t of x, $\dot{x}$ and $\ddot{x}$.

See also figs. 10.2 and 10.3.

Note. The equations of uniformly accelerated motion (p. 29) can **never** be applied to s.h.m., since the acceleration changes with time.

Use of the Auxiliary Circle

Any linear s.h.m. can be regarded as having associated with it a point moving with constant speed in a circle whose diameter is the path of the object. Such a circle greatly simplifies the calculation of the time taken by the object to cover a given distance, without resort to algebra. If we relate $\ddot{x}$ and x by the equation

$$\ddot{x} = -\omega^2 x,$$

then

(*a*) For the *object performing s.h.m.*, ω^2 is simply a positive constant, of dimensions $[\mathrm{T}^{-2}]$, from which T can quickly be found.

$$T = \frac{2\pi}{\omega}$$

(*b*) For the *point in the auxiliary circle*, ω is an angular speed, of dimensions $[\mathrm{T}^{-1}]$, and unit rad s^{-1}.

Beware not to confuse ω and $\dot{\theta}$ for an angular s.h.m. described by

$$\ddot{\theta} = -\omega^2\theta.$$

Here ω does *not* represent an angular speed, but is merely a constant of the motion. It is here called the **pulsatance**.

10.3 The Dynamics of s.h.m.

To Discover whether a Body Executes s.h.m.

(a) Linear Motion

(1) Draw the body an *arbitrary* displacement x from the mean position O.

(2) Mark in *all* the forces that act on the body, and evaluate the restoring force $\sum \boldsymbol{F}$.

(3) Mark the acceleration $\ddot{x}$ in the chosen x-direction (that in which x increases).

(4) Apply the equation of motion

$$\sum \boldsymbol{F} = m\ddot{x}$$

in a specified direction.

(5) The motion is s.h.m. if the equation of motion can be written

$$\ddot{x} = -(\text{positive constant})\, x$$

The positive constant is usually written as ω^2 for convenience, since time period $T = 2\pi/\omega$.

Examples are given below.

(b) Angular Motion

Repeat the procedure for an arbitrary angular displacement θ.

Apply torque $\boldsymbol{T} = I\boldsymbol{\alpha}$.

An equation of the form

$$\ddot{\theta} = -(\text{positive constant})\ \theta$$

means that the body executes angular s.h.m.

Examples are given on pp. 83 and 380.

10.4 Examples of Exact s.h.m.

(*a*) Mass on a Spiral Spring

Suppose that in fig. 10.5 the spring exerts a force $\boldsymbol{F}$ *on* the mass, where

$$\boldsymbol{F} = -k\boldsymbol{x}$$

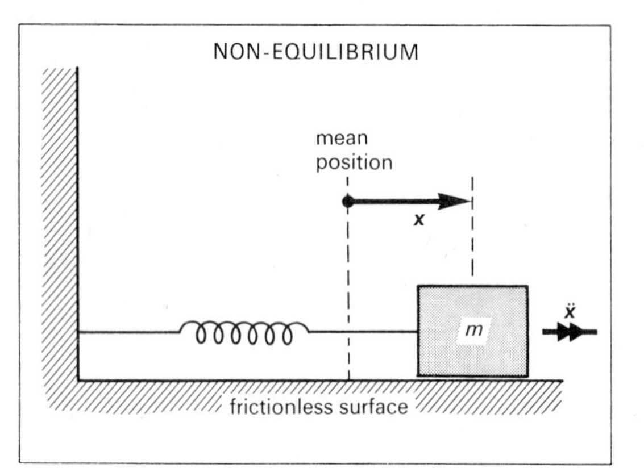

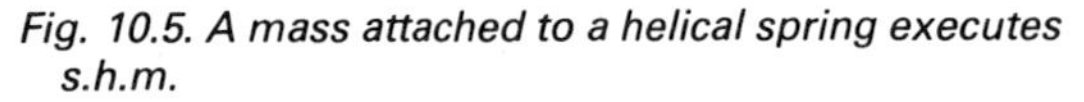

Fig. 10.5. A mass attached to a helical spring executes s.h.m.

k is the **spring constant**, measured in N m^{-1}. The negative sign means that the force is to the *left*.

Apply the equation of motion *to the right*. Then $\boldsymbol{F} = m\boldsymbol{a}$ becomes

$$-k\boldsymbol{x} = m\ddot{\boldsymbol{x}}$$

$$\ddot{\boldsymbol{x}} = -\left(\frac{k}{m}\right)\boldsymbol{x}$$

This is s.h.m., so long as *Hooke's Law* is obeyed, of period

$$T = 2\pi\sqrt{\frac{m}{k}}$$

The motion is illustrated by fig. 10.6.

Notes

(*a*) T is independent of

(1) amplitude

(2) $\boldsymbol{g}$ (the weight $m\boldsymbol{g}$ of the mass m is not relevant).

(*b*) $T = 2\pi\sqrt{\dfrac{\text{mass}}{\text{force at unit displacement}}}$

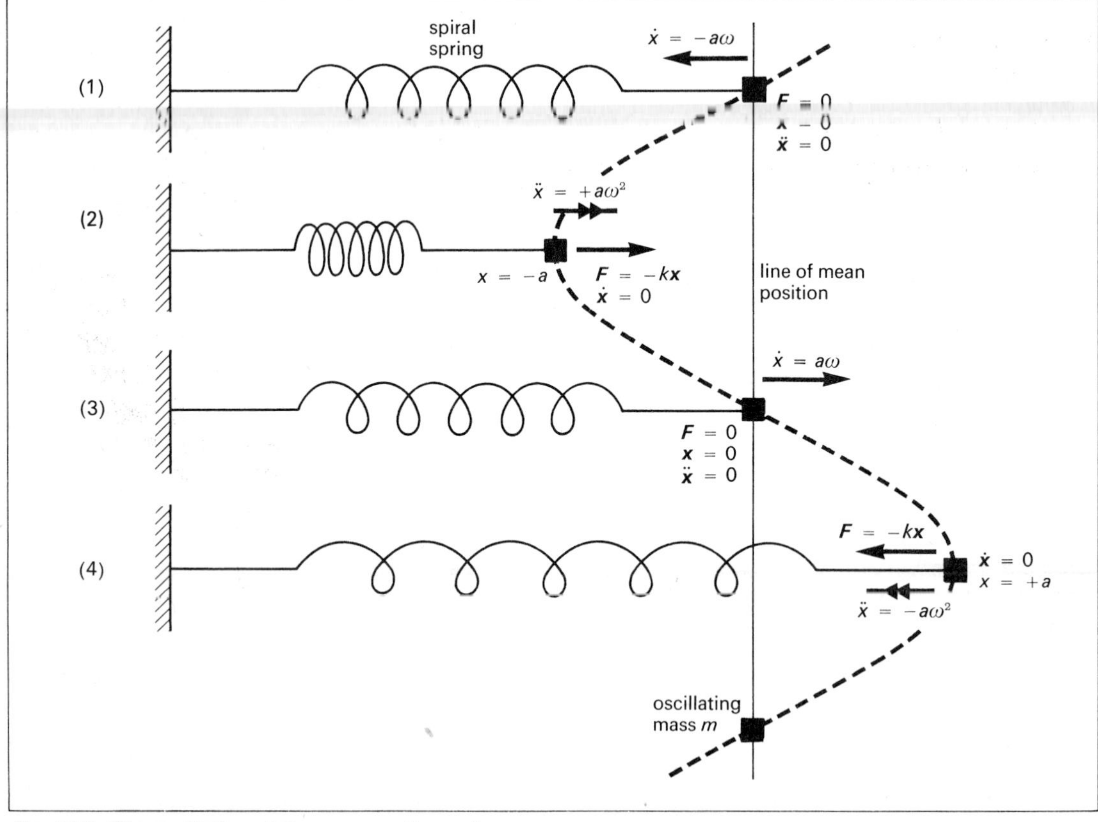

Fig. 10.6. The oscillation of the mass on the spring.

(*b*) Cylindrical Hollow Tube Floating in a Liquid

In fig. 10.7 the test-tube is not in equilibrium, because it has been displaced x from the mean position.

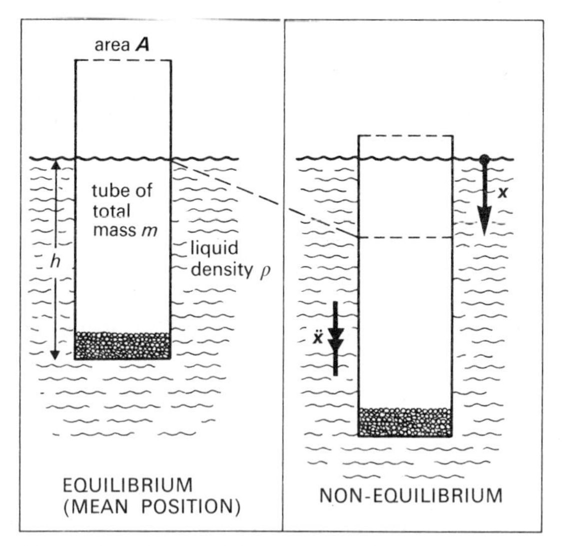

Fig. 10.7. A floating tube executes s.h.m.

The Archimedean *upthrust* exceeds the tube's weight by $(Ax)\rho g$.

Applying $\boldsymbol{F} = m\boldsymbol{a}$ in the *downward* direction, we have

$$-(A\rho g)x = m\ddot{x}$$

$$\ddot{x} = -\left(\frac{A\rho g}{m}\right)x$$

We deduce that the motion is s.h.m. (in the absence of friction).

For the equilibrium situation

$$(Ah)\rho g = mg$$

so

$$\ddot{x} = -\left(\frac{g}{h}\right)x$$

$$\therefore T = 2\pi\sqrt{\frac{h}{g}}.$$

h is a measure of the system's inertia, and g of its springiness. In the absence of a gravity field there would be no method of storing p.e., and so no oscillation could occur.

(In practice the liquid's viscosity produces two effects:

(*i*) it causes heavy damping (p. 88), and

(*ii*) it causes a significant mass of liquid to be set into oscillation. This liquid has taken its k.e. from the rod.)

(*c*) Liquid Oscillating in a U-tube

The motion is s.h.m. because the restoring force has a magnitude proportional to the length of displaced liquid, and so to the displacement. The force always acts so as to restore the liquid to the mean position. Viscous effects cause the motion to be damped (p. 87), but in their absence

$$T = 2\pi\sqrt{\frac{l}{2g}}$$

where l is the length of the liquid column.

(*d*) The Torsional Pendulum

On p. 134 it is shown that the time period T of a torsional pendulum consisting of a body of moment of inertia I suspended by a system of torsion constant c is given by

$$T = 2\pi\sqrt{I/c}$$

10.5 Examples of Approximate s.h.m.

(*a*) The Simple Pendulum

(An angular treatment is given on p. 56.)

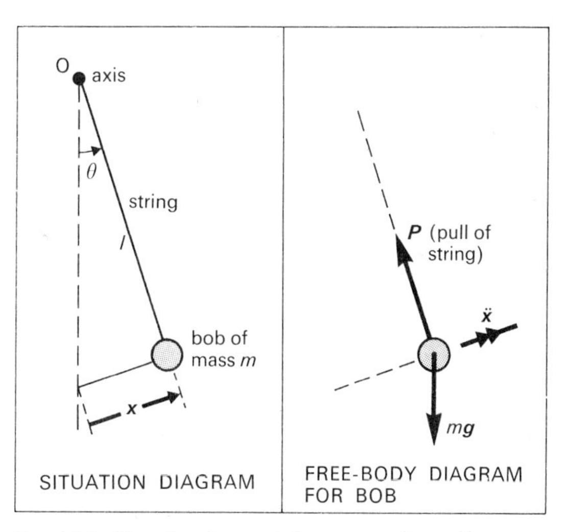

Fig. 10.8. The simple pendulum treated as a linear motion

The bob can move only along the tangent. Apply $\boldsymbol{F} = m\boldsymbol{a}$ to the *right*. Then

$$-mg \sin\theta = m\ddot{x}$$

where $mg \sin\theta$ is the magnitude of the resolved part of mg along the tangent.

When $\theta < \frac{1}{6}$ rad (10°) or so,

$$\sin\theta \approx \theta \approx \frac{x}{l}$$

if θ is measured in radians.

Then $$-m\left(\frac{g}{l}\right)\boldsymbol{x} \approx m\ddot{x}$$

$$\ddot{\boldsymbol{x}} \approx -\left(\frac{g}{l}\right)\boldsymbol{x}$$

So the motion is *approximate* s.h.m., and

$$T = 2\pi\sqrt{\frac{l}{g}}$$

T is independent of amplitude, *provided a is small.*

The angular treatment on p. 56 demonstrates that ml^2 is a measure of the system's inertia (its m.i.), while $\boldsymbol{mgl\theta}$ is a measure of the springiness (the restoring torque).

Because inertial mass $\propto$ gravitational mass, m cancels: T is independent of m. (Expressed differently, different masses always give the same T, demonstrating that all masses experience the same acceleration $\boldsymbol{g}$ at a given point.)

(b) Heavy Piston Inside Frictionless Cylinder

In fig. 10.9 suppose that atmospheric pressure can be ignored, and that all dissipative effects (p. 87) are negligible.

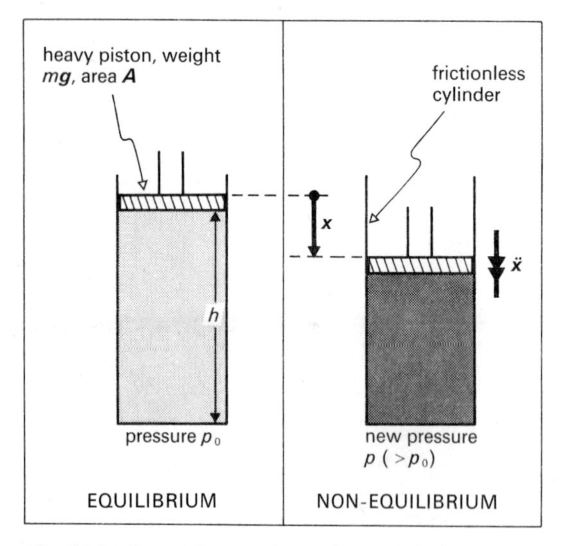

Fig. 10.9. Approximate s.h.m. shown by a heavy piston.

If the displacement $\boldsymbol{x}$ is small, then the pressure changes will be so small that they are approximately proportional in size, but opposite in sign, to the volume changes, and hence to the displacements $\boldsymbol{x}$. (Expressed differently, we can approximate the *Boyle's Law* hyperbola to a straight line over small pressure ranges: the line's slope is negative.) The force resulting from the pressure change is then proportional to $\boldsymbol{x}$, but oppositely directed: s.h.m. occurs.

It can be shown that

$$T = 2\pi\sqrt{\frac{h}{g}}$$

for small displacements.

The inertial factor here is m, the mass of the piston, and the springiness factors the pressure of the gas (since a compressed gas can store p.e.) and the gravitational p.e. of the piston.

(c) The Compound Pendulum

On p. 60 it is shown that the time period T of a compound pendulum is given by

$$T = 2\pi\sqrt{\frac{k^2 + h^2}{hg}}$$

10.6 Energy in s.h.m.

When an oscillation is **undamped**, the system does no work against resistive forces, and so its total energy remains constant in time. Fig. 10.10 demonstrates

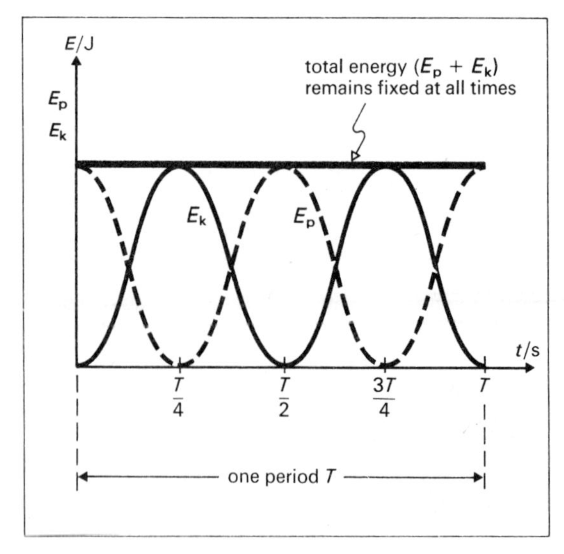

Fig. 10.10. Time variation of the energy of an s.h.m.

that the p.e. and k.e. oscillate with double the frequency of the motion. (See also a.c., on p. 411.)

Consider the spring of fig. 10.5.

(*a*) *Potential energy*

$$\text{at extension } x\ E_p = \int_0^x F\,dx$$
$$= \int_0^x kx\,dx$$
$$= \tfrac{1}{2}kx^2 \qquad \left(\omega^2 = \frac{k}{m}\right)$$
$$= \tfrac{1}{2}m\omega^2x^2$$

(*b*) *Kinetic energy*

$$\text{at extension } x\ E_k = \tfrac{1}{2}mv^2 \quad = \tfrac{1}{2}m\omega^2(a^2 - x^2)$$

(*c*) *Total energy*

$$\text{Total energy} = E_p + E_k$$
$$= \tfrac{1}{2}m\omega^2x^2 + \tfrac{1}{2}m\omega^2(a^2 - x^2)$$
$$= \tfrac{1}{2}m\omega^2a^2$$

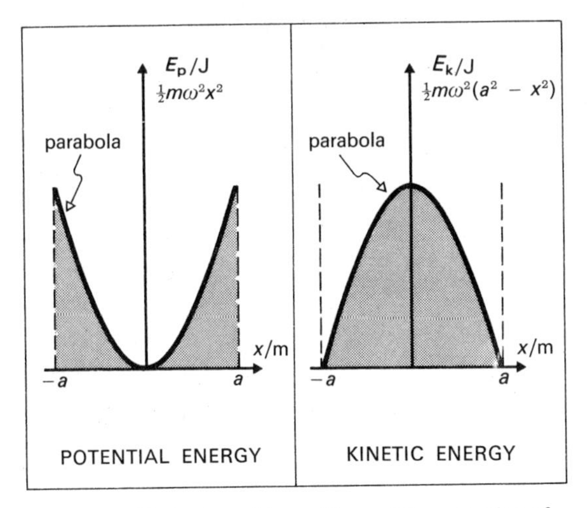

Fig. 10.11. Variation with position of the energies of an s.h.m. As in fig. 10.10, the total energy remains constant.

Since the total energy is independent of the variable x, it remains constant. Note that the energy of an oscillation is proportional to

(*i*) mass,
(*ii*) (frequency)^2,
(*iii*) (amplitude)^2.

*10.7 Energy Approach to s.h.m.

Suppose that a system capable of oscillation can possess energy in the forms

$$E_p = \tfrac{1}{2}ax^2,$$
and
$$E_k = \tfrac{1}{2}b\dot{x}^2,$$

where a and b are positive constants for the system, and x is a variable.

If the system loses no energy (is undamped) the total energy

$$E = (E_p + E_k)$$

is constant, so

$$E = \tfrac{1}{2}ax^2 + \tfrac{1}{2}b\dot{x}^2$$

Differentiating w.r.t. time

$$ax\dot{x} + b\dot{x}\ddot{x} = 0$$
$$\therefore \ddot{x} = -\left(\frac{a}{b}\right)x$$

(if $\dot{x} \neq 0$).

The system is capable of s.h.m. of time period

$$T = 2\pi\sqrt{\frac{b}{a}}$$

Examples

Oscillation	E_p	E_k	T
mass on a spiral spring	$\tfrac{1}{2}kx^2$	$\tfrac{1}{2}m\dot{x}^2$	$2\pi\sqrt{\frac{m}{k}}$
torsional pendulum	$\tfrac{1}{2}c\theta^2$	$\tfrac{1}{2}I\dot{\theta}^2$	$2\pi\sqrt{\frac{I}{c}}$
charge in *LC* circuit	$\tfrac{1}{2}\left(\frac{1}{C}\right)Q^2$	$\tfrac{1}{2}L\dot{Q}^2$	$2\pi\sqrt{LC}$

*10.8 Superposition of Two s.h.m.

In this paragraph we discuss the behaviour of a particle which is subjected to two simultaneous simple harmonic motions in the same plane.

(a) Same Line, Same Frequency

The resultant motion is an s.h.m. of the same frequency. The particle's resultant displacement is found by using the *principle of superposition* (p. 103), and a special case of this behaviour is discussed under *stationary waves* (p. 104).

(b) Same Line, Different Frequencies

The resultant motion is not *simple* harmonic because the amplitude oscillates with time—it is said to be **modulated**. The frequency of oscillation of the amplitude equals the difference of the frequency of the superposed motions. The behaviour is analogous to the sound emitted by two tuning forks of close but different frequencies—the variation with time of the sound intensity is called **beating** (p. 301). Fig. 10.12 (overleaf) shows what happens when two motions have different amplitudes a_1 and a_2, ($a_1 > a_2$).

See p. 109 for a situation where $a_1 = a_2$.

(c) Perpendicular Directions

(*i*) *Same frequency*. In general the path of the particle is an ellipse. When the amplitudes are equal, and the phase

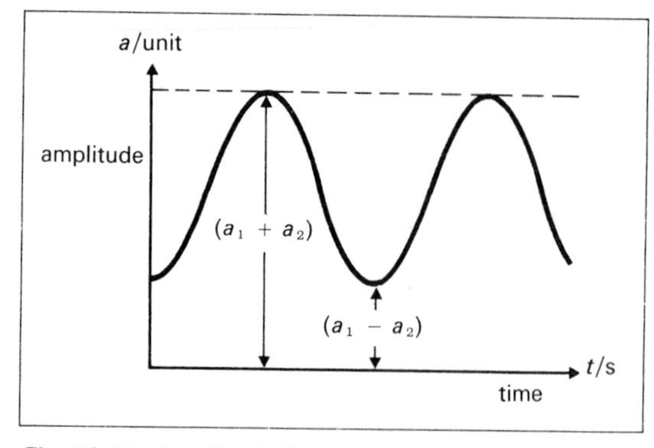

Fig. 10.12. Amplitude fluctuation (beats).

differences are 0, $\pi/2$ rad and π rad, we have a straight line, circle and straight line respectively (p. 97).

(*ii*) *Different frequencies.* The path of the particle is a continuous curve whose general form is controlled by the phase difference and frequency ratio of the superposed motions. Fig. 10.13 shows the **Lissajous figures** for a phase difference of $\pi/2$ rad between two motions whose frequency ratio f_y/f_x takes some simple values.

If the two simple harmonic motions have different amplitudes, the *Lissajous* figures have different shapes in detail, but their general form is not altered.

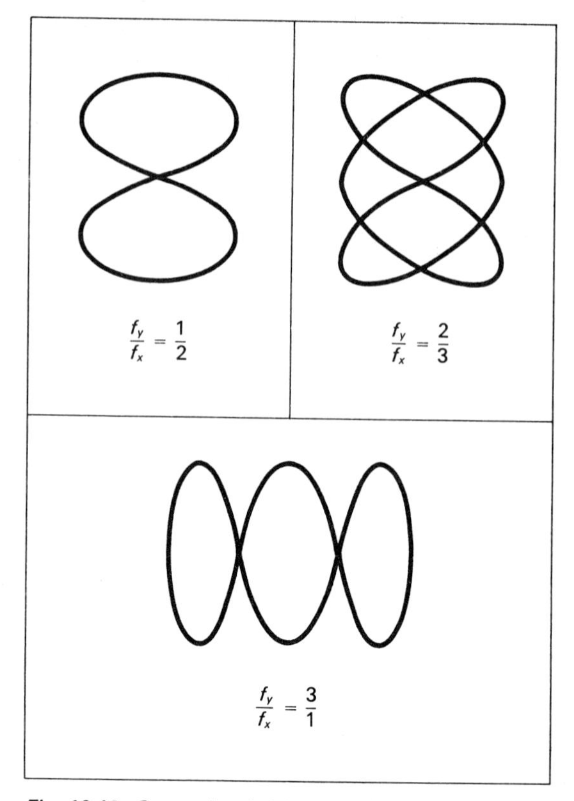

Fig. 10.13. Some simple Lissajous figures.

11 Damped and Forced Oscillation

11.1 Damped Oscillations

When a system executes *true* s.h.m., then

(*a*) its period is independent of its amplitude,
(*b*) its total energy remains constant in time.

In practice many bodies execute *approximate* s.h.m. because

(*a*) $\boldsymbol{F}$ is *not* proportional to $-\boldsymbol{x}$, and so the oscillations are not isochronous, and
(*b*) the energy of the system decreases in time (fig. 11.1).

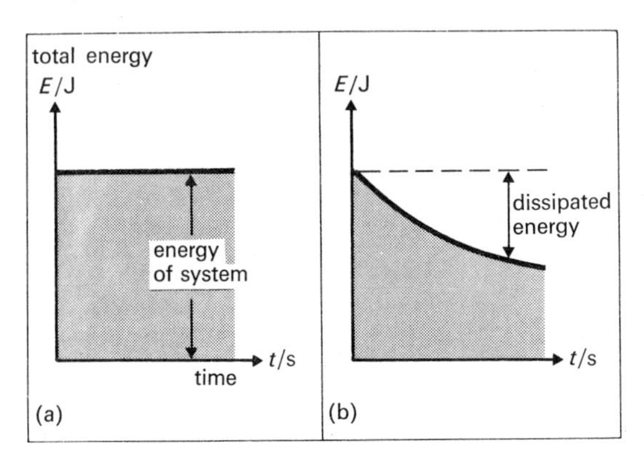

Fig. 11.1. Energy comparisons for (a) ideal and (b) approximate (damped) s.h.m.

A real oscillating system is opposed by dissipative forces, such as air viscosity. The system then does positive work: the energy to do this work is taken from the energy of oscillation, and usually appears as thermal energy. **Damping** is the process whereby energy is taken from the oscillating system.

Examples of Damping

(a) Natural

(*i*) The internal forces of a spring can be dissipative, and prevent the spring being perfectly elastic.

(*ii*) Fluids exert viscous forces that prevent relative movement of fluid layers, and hence oppose movement through the fluid. The internal energy of the fluid increases.

(b) Artificial

(*i*) Galvanometers utilize either electromagnetic damping (p. 400) or dashpots to prevent tiresome oscillation. (See also below.)

(*ii*) The panels of cars are coated to reduce vibration.

(*iii*) The shock absorber of a car provides a damping force which prevents excessive oscillation.

Examples of Damped Vibrations

(a) Natural

(*i*) A percussive musical instrument (such as a bell) gives out a note whose intensity decreases with time.

(*ii*) The paper cone of a loudspeaker vibrates, but is heavily damped so as to lose energy (as sound wave energy) to the surrounding air.

(b) Demonstration

Refer to fig. 11.2 overleaf.
Note three characteristics:

(*i*) the mass, which can possess k.e. because of its inertia,

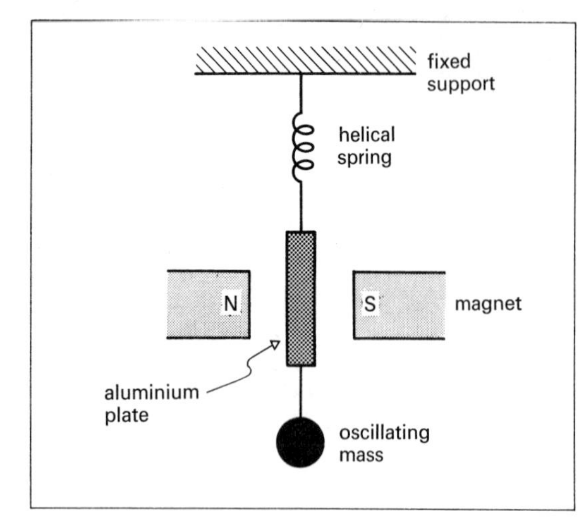

Fig. 11.2. Apparatus to demonstrate the electromagnetic damping of a mass oscillating on a helical spring.

(*ii*) the spring, which can store p.e. because of its elasticity, and which provides a restoring force $\boldsymbol{F} = -k\boldsymbol{x}$, and

(*iii*) the aluminium plate which experiences a resistive force

$$\boldsymbol{R} = -(\text{constant})\,\boldsymbol{v}$$

because of eddy current damping (p. 400). This force causes *Joule* heating in the plate.

The resistive forces in damping are frequently roughly proportional to the speed of movement (p. 148). An assumption that they are so, greatly simplifies the mathematical analysis.

Notes. (*a*) When the mass is at rest, $\boldsymbol{R} = 0$. $\boldsymbol{R}$, on its own, would never stop the body from reaching the equilibrium position.

(*b*) The frequency of a damped motion is slightly less than that of an undamped motion.

11.2 Amplitude of Damped Oscillations

Fig. 11.3 shows displacement-time graphs for three categories of damping.

(*a*) **Slight damping** results in definite oscillation, but the amplitude of oscillation decays exponentially with time.

The **decrement**

$$\delta = \frac{a_1}{a_2} = \frac{a_2}{a_3} = \frac{a_3}{a_4} \cdots \text{etc.}$$

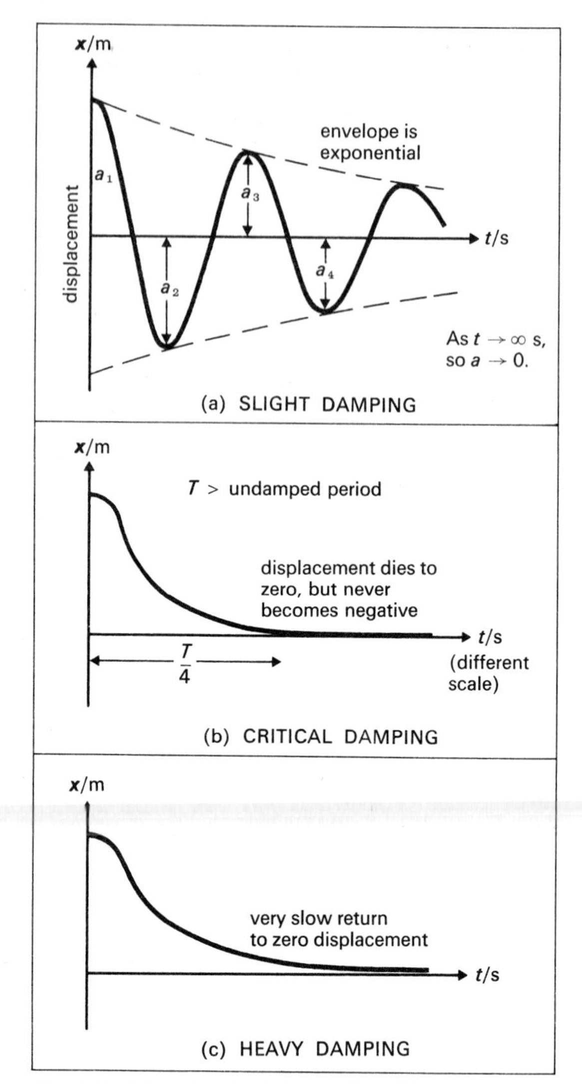

Fig. 11.3. Three kinds of damped oscillation.

For exponential decay δ is a constant. The **logarithmic decrement** $\ln \delta$ is used as a criterion of the amount of damping, and for making corrections.

(*b*) **Critical damping** results in no real oscillation as such: the time taken for the displacement to become effectively zero is a *minimum*.

(*c*) **Heavy damping** is damping in which the resistive forces exceed those of critical damping. The system returns very slowly to the equilibrium position.

The Moving-Coil Galvanometer

A galvanometer is

(*i*) **dead-beat** when it is critically damped,

(*ii*) used **ballistically** when the damping is as small as possible (See p. 390.)

Fig. 11.4 should be studied carefully.

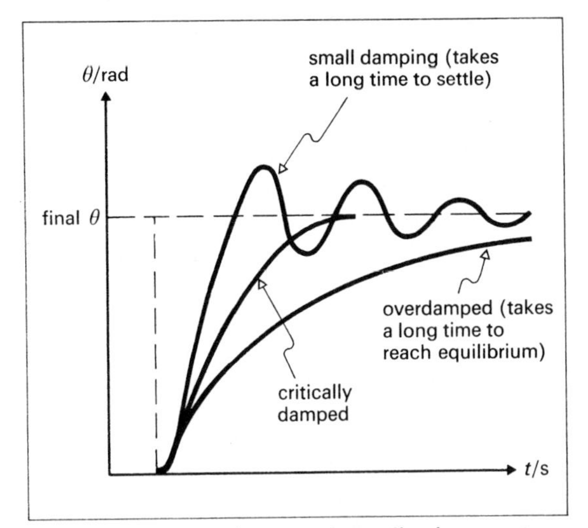

Fig. 11.4. θ-t curves for a moving-coil galvanometer being used to measure an electric current which becomes steady.

11.3 Forced Vibration

Suppose that in fig. 11.2, the support is set into continuous oscillation by some external agent, so that a periodic driving force is imposed on the oscillating mass. We will consider what happens when conditions are *steady*. (In the early stages beats (p. 109) will occur between the forced and natural vibration, giving rise to **transient** oscillations which we will ignore. The frequency of the natural free vibration is determined by the inertia and elasticity factors of the system.)

One observes the following:

(a) Energy

Under steady conditions the amplitude of vibration is fixed for a fixed driving frequency. We deduce that the driving force does work *on* the system at the same rate as the system loses energy by doing work against dissipative forces. The power of the driver is controlled by damping.

(b) Amplitude

The amplitude of vibration depends on

(*i*) the relative values of the natural frequency of free oscillation, and the frequency of the driving force, and also

(*ii*) the extent to which the system is damped.

These observations are summarized by fig. 11.5.

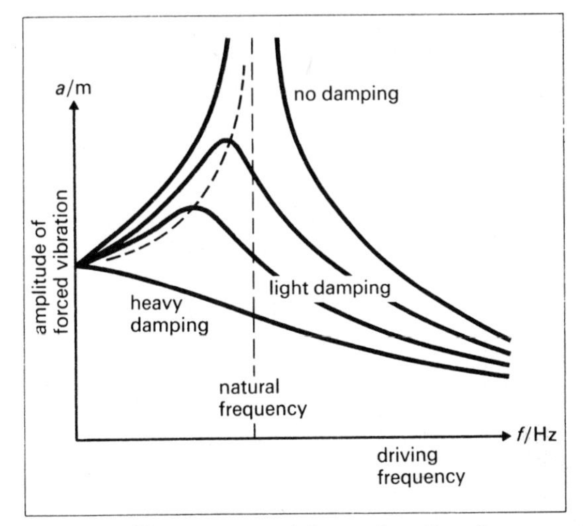

Fig. 11.5. The response of damped systems to varying driving frequencies.

Damping shows two effects:

(1) It causes the maximum amplitude to be reached when the driving frequency is a little less than either the damped or undamped natural frequencies.

(2) It reduces the response of the forced system over a range of frequencies.

(It cuts down the sharp peak of fig. 11.5.)

(c) Phase

Whatever its natural frequency, the forced vibration takes on the frequency of the driving force, but it shows the *phase lag* illustrated by fig. 11.6.

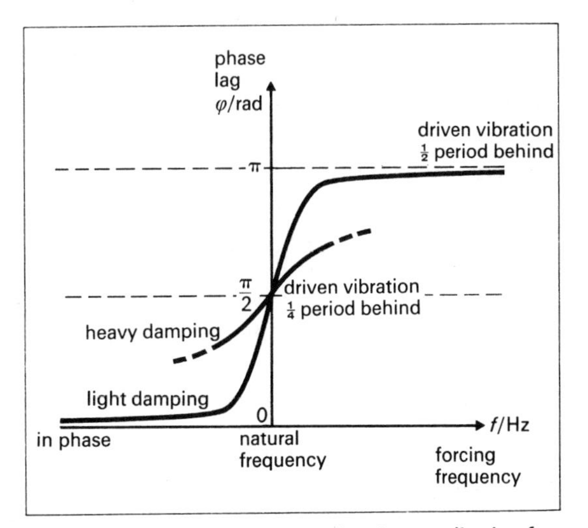

Fig. 11.6. Phase relationships for the amplitude of a forced vibration.

Barton's Pendulums

The apparatus of fig. 11.7 can be used to demonstrate these observations.

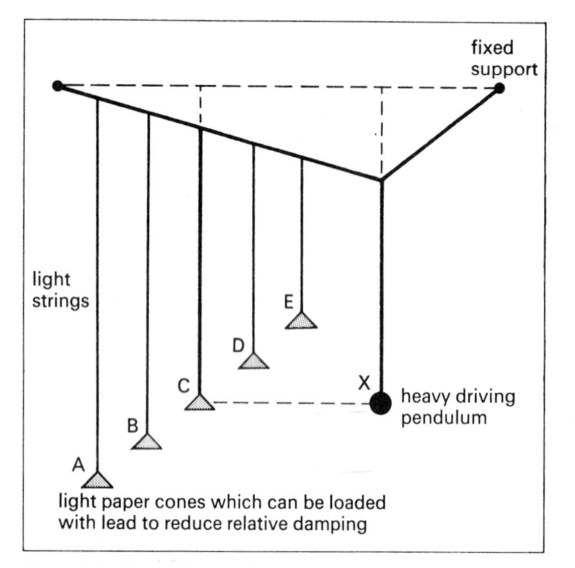

Fig. 11.7. Barton's pendulums.

When X oscillates perpendicular to the plane of the paper, the steady oscillations of pendulums A to E demonstrate figs. 11.5 and 11.6. While A and B may be almost $\frac{1}{2}$ period behind X, C will be $\frac{1}{4}$ period behind, and D and E will be nearly in phase.

C will show the largest amplitude.

11.4 Resonance

Fig. 11.5 illustrates that the *amplitude* of forced vibration reaches a maximum when the driving frequency is just less than the natural frequency of free oscillation. Several definitions of resonance are possible.

Energy resonance (velocity resonance) *occurs when an oscillator is acted upon by a second driving oscillator whose frequency equals the natural frequency of the system.*

When this happens the driving oscillator (some external agent) most easily transfers its energy to the oscillating system: *the energy of the system becomes a maximum.*

For

(*a*) *zero damping*, the energy would (in theory) become infinite,

(*b*) *light damping*, the driving frequency is close to that required for amplitude resonance.

Examples of resonance are widespread in all branches of physics.

(*a*) *Mechanics*

(*i*) The oscillation of a child's swing.

(*ii*) The destruction of the *Tacoma* bridge (*Washington* 1940).

(*b*) *Sound*

(*i*) Resonance tube (p. 307),

(*ii*) *Kundt's* tube (p. 307).

(*c*) *Electricity*. Tuning a radio circuit— the natural frequency of the circuit is made equal to that of the incoming electromagnetic wave by changing its capacitance. (See p. 422.)

(*d*) *Light*. Maximum absorption of infra-red waves by an NaCl crystal occurs when their frequency equals that of the vibration of the Na^+ and Cl^- ions.

*(*e*) *Modern Physics*. Reference should be made to more detailed books for **paramagnetic resonance** and the concept of **scattering cross-section.**

Resonant Systems

(*a*) Consider a mass on a helical spring. The inertia of the system and the restoring force are both *concentrated*: there is *one value* of the driving frequency for resonance.

(*b*) In a system such as the resonance tube (p. 307), the inertia and the restoring force are *evenly distributed* in the system: there are *several* (related) *driving frequencies* which give resonance.

*11.5 Quality Factor or *Q*-Factor

The Q-factor of a system undergoing forced oscillation is one way of indicating the sharpness of resonance. It is defined by

$$Q = 2\pi \frac{\text{(maximum energy contained in system)}}{\text{(average energy dissipated per oscillation)}}$$

Under steady conditions the energy dissipated by the system equals the work done on it by the external agent. A system of high Q (lightly damped) would be expected to continue vibrating for a long time as the energy taken from the system during each oscillation is small. Some typical values of this dimensionless quantity Q are given below:

test tube in water	10^1
typical inductor	2×10^2
simple pendulum in air	4×10^2
quartz crystal in air	3×10^4

It can be seen from fig. 11.5 that the smaller the damping (i.e. the higher the Q-factor) the sharper the resonance. This is particularly important in the tuned circuit of a radio receiver where a highly selective response is required.

12 Wave Motion

Some familiarity with elementary wave motion is assumed in this chapter and in chapter **14**.

12.1 Progressive Waves

A **progressive** *or* **travelling wave** *is the movement of a disturbance from a source which transfers energy and momentum from the source to places around it.*

Waves are of two fundamental types:

(*a*) *Mechanical waves*, which require a material medium for their propagation, and

(*b*) *Electromagnetic waves*, which can travel through a vacuum. They are discussed in detail in chapter **37**, p. 260.

Nevertheless all types of wave motion have the same basic properties, and can be treated analytically by equations of the same form. Wave phenomena

Examples of Progressive Waves

Wave type	*Nature of disturbance*	*Possible result of energy transfer*
water waves	water molecules describe a periodic motion in a closed path	oscillation of a floating body
sound waves	vibratory motion superposed on random motion of air molecules	diaphragm of a microphone set oscillating
secondary earthquake waves	shaking motion (transverse movement) of material of Earth's crust	buildings collapse
electromagnetic waves	transverse variation of electric and magnetic fields	could stimulate retina of human eye

occur in many branches of physics, and an understanding of the general *principles* underlying their behaviour is very important.

Wave Characteristics

(*i*) A wave transfers *energy*: there is no net translation of matter. (Waves breaking on a beach are *not* typical of the waves that interest us.)

(*ii*) In a mechanical wave the medium has particles which have inertia and elasticity (some means of interacting with their neighbours). These particles execute oscillations of small amplitude about their equilibrium positions.

(*iii*) Each particle oscillates in the same way as its neighbour, but shows a time lag if it is further from the source of energy.

Longitudinal and Transverse Waves

A wave is said to be

(*a*) **longitudinal** if the displacement of mechanical particles is *parallel* to the direction of translation of energy (as in sound waves), or

(*b*) **transverse** if the wave has associated with it some (vector) displacement which is *perpendicular* to the direction of translation of energy (as in electromagnetic waves).

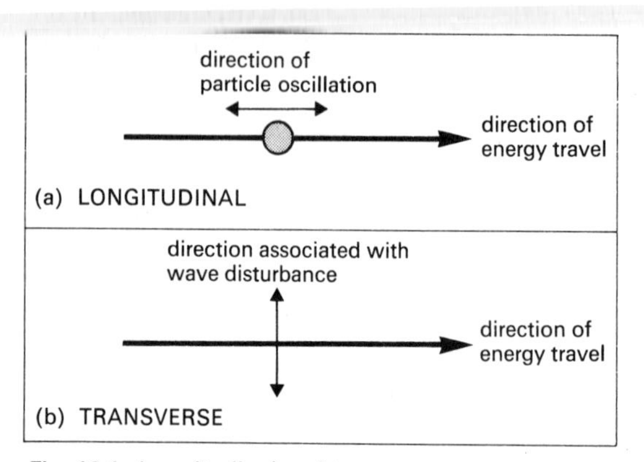

Fig. 12.1. Longitudinal and transverse waves.

12.2 Analytical Description of Waves

Definitions

The **wavefront** is a surface (reducing to a line and point respectively in two and one dimension(s)) over which the disturbance has the same phase at all points.

Examples for an **isotropic** medium (one in which the wave speed is the same in every direction):

Nature of source	*Shape of wavefront*
point	point (in one dimension) circle (in two dimensions) sphere (in three dimensions)
line	line (in two dimensions) cylinder (in three dimensions)
plane	plane
any source	plane at a distant point

A **ray** is the path taken by the wave energy in travelling from the source to the receptor. A ray is usually perpendicular to a wavefront, and is then a **wave normal.**

The **speed** c of a *mechanical* wave motion is that with which an observer must move so as to see the wave pattern apparently stationary in space. (We will not distinguish between **phase** and **group velocity** in this book.) According to the theory of relativity, *electromagnetic* waves have the same speed in vacuo relative to all observers. The speed of the wave motion is then defined as that at which the wavefront passes the observer.

The **wavelength** λ is the spatial period of the wave pattern at a fixed instant, and is therefore the distance between two consecutive particles which have the same phase. (It could, for example, be the distance between adjacent crests.) The wavelength of most waves remains constant, but when waves have a small radius of curvature, the wave crests may not be evenly spaced.

The **frequency** f of the wave is the number of vibrations performed in each second by the source, and is thus the number of crests that pass a fixed point in each second. Unit: hertz (Hz), (see p. 79).

(But see also *Doppler* effect. p. 112.)

Hence in time Δt a number $f\,\Delta t$ waves, each of length λ, pass a fixed point.

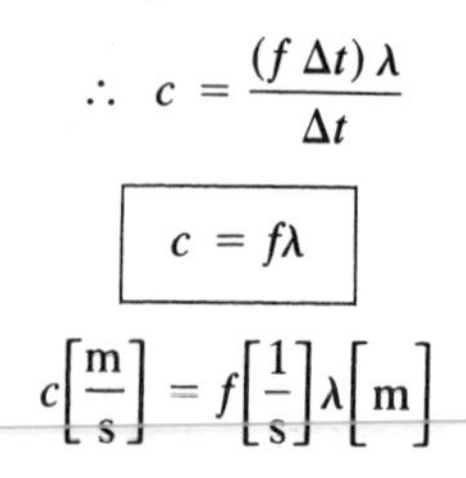

$$\therefore\quad c = \frac{(f\,\Delta t)\,\lambda}{\Delta t}$$

$$\boxed{c = f\lambda}$$

$$c\left[\frac{\text{m}}{\text{s}}\right] = f\left[\frac{1}{\text{s}}\right]\lambda\left[\text{m}\right]$$

Wave Diagrams

We will use the following symbols:

c for the *wave speed* (see below).
x for the *location* of a particular point in the medium.
y for the *disturbance* transmitted by the wave.

Thus for

(*a*) string waves, y is the transverse displacement of a string particle,

(*b*) sound waves, the longitudinal displacement of an air molecule,

(*c*) electromagnetic waves, the electric field strength $\boldsymbol{E}$ (p. 260).

Two types of graph are valuable.

(1) Displacement-Position Curves

A plot of y against x can be called a **wave profile**: it represents what is happening *at an instant of time.*

On the graph the axes for x and y are drawn perpendicular to each other. *In practice* x and y are only perpendicular when we consider a transverse wave: thus fig. 12.2 can never represent *directly* the displacements of a longitudinal wave (as is done by a photograph).

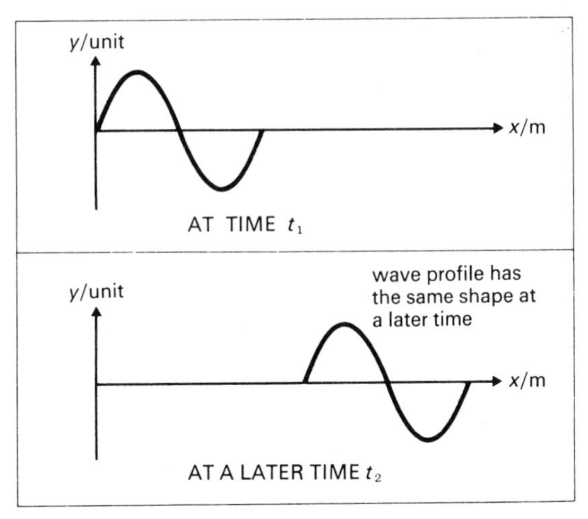

Fig. 12.2. Propagation of a wave profile.

Notes. (*a*) The wave profile retains its shape when there is no **distortion.**

(*b*) A **harmonic wave** is one whose wave profile is sinusoidal. (Refer to the comment on p. 97.)

(*c*) Be careful not to confuse

(*i*) $\dot{y}$, which might represent an instantaneous *particle speed*, with

(*ii*) $\dot{x} = c$, which is the speed at which the wave profile propagates (the *wave speed*).

(2) Displacement-Time Curves

A plot of y against t shows the nature of the oscillation of a particle at a *particular location.*

These ideas are developed in paragraph **12.8** on p. 97.

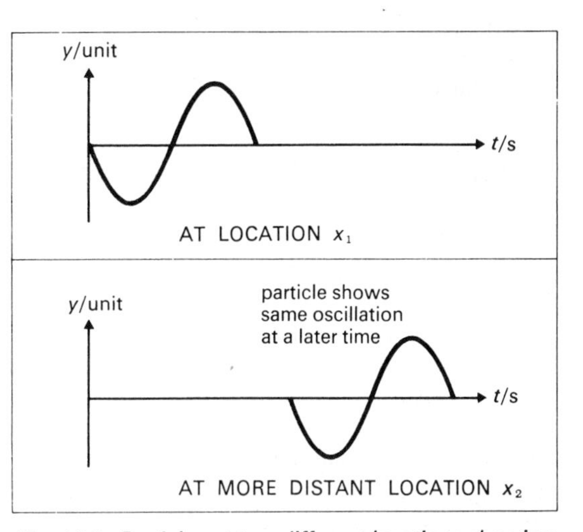

Fig. 12.3. Particles at two different locations showing the same oscillations at different times.

12.3 Wave Classification

Waves can be classified by

(1) Periodicity

(*a*) A *periodic wave* is generated by a periodic source which gives a **wavetrain**, a succession of disturbances.

(*b*) A *non-periodic wave* consists of a single wavefront (as is shown by a shock wave, p. 115).

(2) Energy Movement

(*a*) A *progressive or travelling wave* transfers energy from one location (the source) to another.

(*b*) A *stationary or standing wave* shows no *net* energy transfer in space (but in time it shows an interchange of energy between kinetic and potential).

(3) Direction of Displacement

(*a*) *Longitudinal* (p. 92)

(*b*) *Transverse* (p. 92)

(*c*) *Torsional waves* are those propagated along a rod when one end is given a rapid twist.

(4) Nature of Disturbance

Some examples are given on p. 91.

Thus a hand-clap gives rise to a wave which is
(1) non-periodic,
(2) progressive,
(3) longitudinal, in which
(4) the disturbance is a compression caused by the longitudinal displacement of air molecules.

12.4 Summary of Wave Properties

These properties are brought together for reference.

(a) Properties Controlled by (change of) Medium

(1) Wave *speed* (depends on inertial and elastic characteristics).

(2) *Reflection* occurs at an interface where wave speed changes.

(3) *Transmission* into the second medium occurs in (2), and may be accompanied by (*i*) *refraction*, and (*ii*) *dispersion*.

(4) *Polarization* can occur with transverse waves only.

These are properties which can be shown by *particles* under appropriate conditions.

(b) Properties which can be explained by superposition (chapter **14**)

(1) *Interference*, *beats*, and *stationary waves* appear when several waves coincide in the same region of space.

(2) *Diffraction* and *scattering* occur when different obstacles obstruct the path of a wave, and distort the wavefront.

Apart from scattering, these properties are *not* shown by particles *in classical physics*.

It is again to be emphasized that, apart from dispersion and polarization, *all waves should exhibit these properties*: obviously some wave types lend themselves more easily to demonstration.

12.5 Reflection

(a) One Dimension

(*i*) Fig. 12.4 shows diagrammatically the phases of reflected and transmitted pulses for *transverse* waves.

These phase changes can be demonstrated when shaken pulses cross from one stretched spring to another of different physical properties.

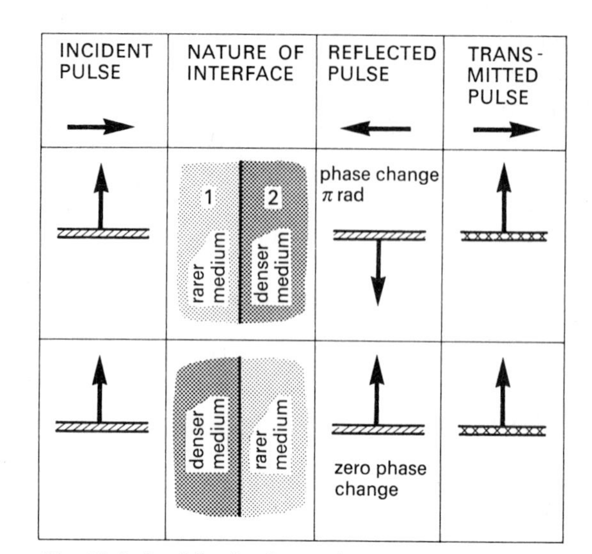

Fig. 12.4. Partial reflection and transmission of transverse waves.

Notes. (1) Energy (not amplitude) is conserved.

(2) There is a phase change of π rad when the wave is reflected at the denser medium: we will meet this again in physical optics (p. 270).

(3) For an infinitely massive boundary there is no transmission and total reflection. For a string there is total reflection at a free end.

(*ii*) Fig. 12.5 illustrates total reflection for a *longitudinal* pulse on a spring.

We will use these ideas when we discuss sound reflection (p. 303).

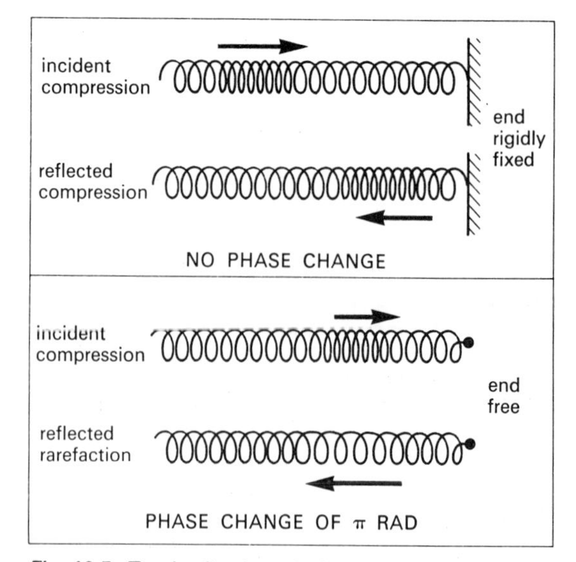

Fig. 12.5. Total reflection of a longitudinal pulse.

(b) Two Dimensions

[A knowledge of the behaviour of waves in a ripple tank is assumed.]

(*i*) *Plane waves* readily demonstrate the law of reflection $\theta = \theta'$ (p. 230). Note that an angle is measured either

(1) between a wavefront and a boundary, or

(2) between a ray and a normal.

(*ii*) *Circular waves* illustrate the ideas of

(1) *object* (centre of circular waves sent out by the source), and

(2) *image* (centre of circular waves after reflection or refraction).

Fig. 12.6 shows the formation of a virtual image (p. 234) by reflection of a circular wave at a plane surface.

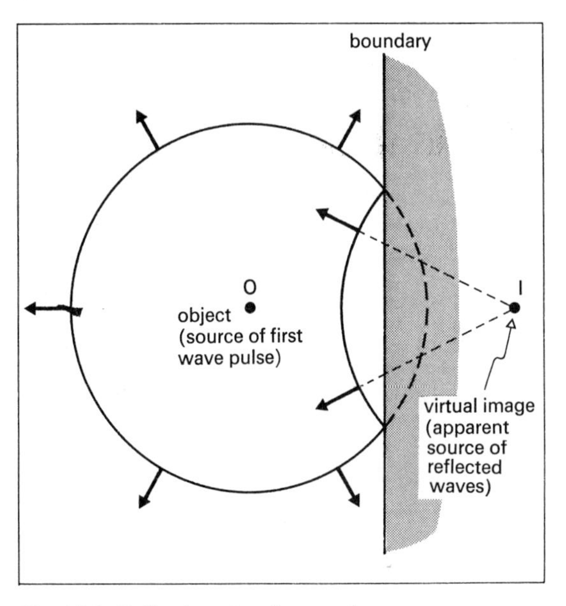

Fig. 12.6. Reflection at a plane surface.

12.6 Refraction and Dispersion

Refraction

Fig. 12.7 shows in diagram form what happens when a wave impinges on an interface between two media.

When speed $c_1 \neq c_2$, some energy is nearly always reflected back into the original medium. At the same time the direction of travel of the transmitted energy deviates from that of the incident energy. This process (change of wavefront direction) is called **refraction**.

Since $c = f\lambda$, and f is fixed by the source, a decrease in c causes a decrease in λ.

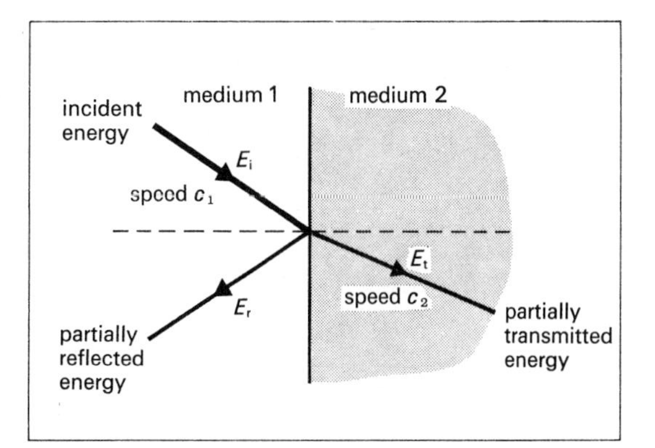

Fig. 12.7. Partial reflection and transmission.

Snell's Law

The **relative refractive index**, ${}_1n_2$, for waves passing from medium 1 to 2 is given by

$$ {}_1n_2 = \frac{c_1}{c_2} \qquad \text{(See p. 231)} $$

Refer to fig. 12.8.

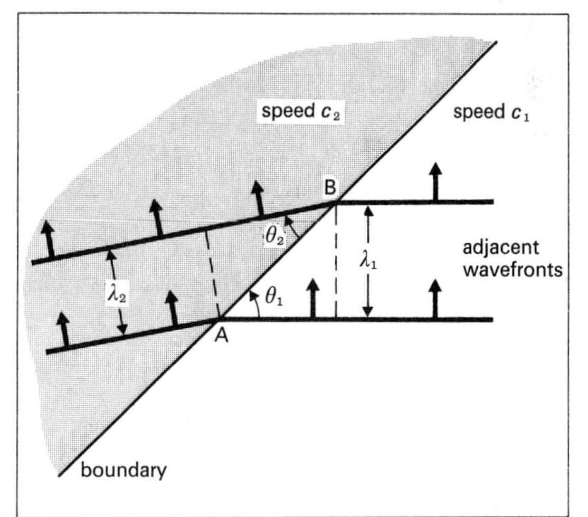

Fig. 12.8. 'Photograph' of a pair of adjacent wavefronts crossing a boundary.

$$ {}_1n_2 = \frac{c_1}{c_2} = \frac{f\lambda_1}{f\lambda_2} = \frac{\lambda_1/\text{AB}}{\lambda_2/\text{AB}} = \frac{\sin\theta_1}{\sin\theta_2} $$

Since c_1 and c_2 are constants,

$(\sin\theta_1)/(\sin\theta_2)$ *always has the same value for waves of a given frequency passing from medium 1 into medium 2.*

This is **Snell's Law**.

We have demonstrated that this constant value is the relative refractive index (defined as c_1/c_2).

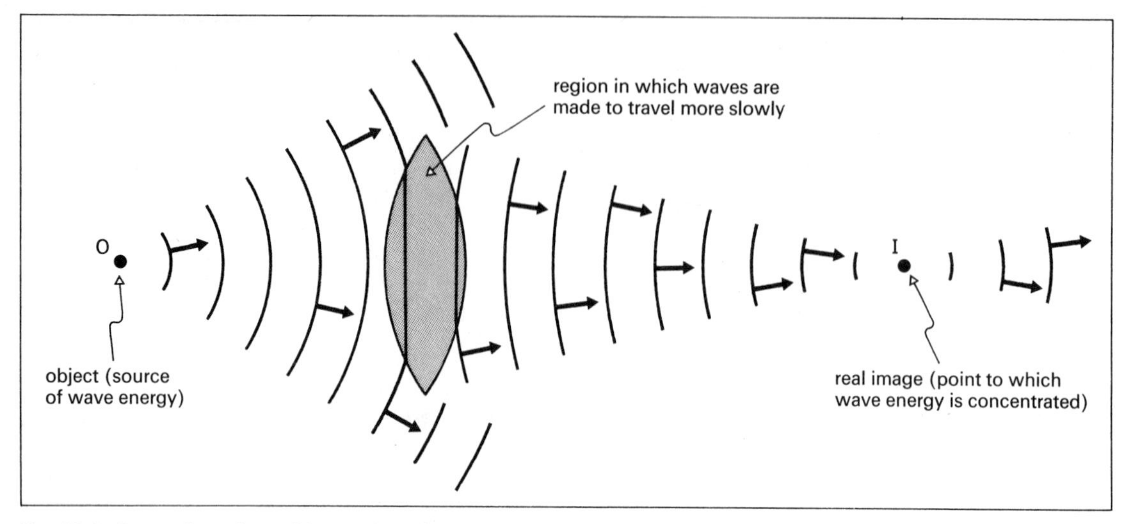

Fig. 12.9. Formation of a real image by refraction (ignoring diffraction).

Image Formation

The general idea is shown in fig. 12.9, in which diffraction effects are ignored. An image is formed only if the shaded region is correctly shaped.

The circular wavefront diverging from O is reshaped by refraction to one which converges on I. The converging agent would be concave in shape if the medium of which it was made speeded up the wave motion.

Examples of Media that Cause Refraction

Wave type	*Medium in which wave travels more quickly ('rarer')*	*Medium in which wave travels more slowly ('denser')*
ripple waves on water	deep water	shallow water
visible light	air	glass
sound	hydrogen	carbon dioxide
transverse waves on spring	spring of small mass per unit length	spring of greater mass per unit length
earthquake	dense rocks below surface	less dense surface rocks

Dispersion

Dispersion is that phenomenon which occurs when the speed of propagation of a wave motion depends on the wave frequency (and thus wavelength). The medium is called a **dispersive medium**.

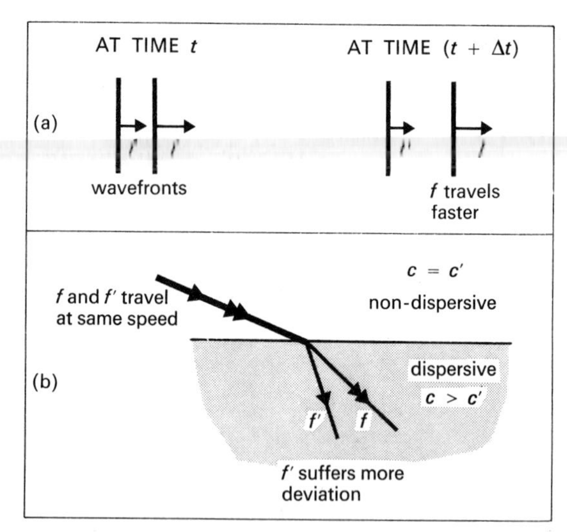

Fig. 12.10. Dispersion of two frequencies f and f′, (a) in one medium, (b) at an interface.

Waves of different frequency may become separated

(*a*) by travelling different distances through the same medium in the same time, and/or

(*b*) by being deviated through different angles when refracted at an interface between two media.

Fig. 35.1 on p. 246 illustrates (*b*) for light waves.

Examples of Dispersion

Wave type	Dispersive medium
light	glass: higher frequencies (such as blue) travel at lower speeds
surface water waves	at large λ, $c = \sqrt{g\lambda/2\pi}$ (gravity waves)
surface water waves	at small λ, $c = \sqrt{2\pi\gamma/\lambda\rho}$ (ripples)
electro-magnetic	ionosphere

Air is *not* a dispersive medium for sound waves at audible frequencies.

*Dispersion and Distortion

A genuine harmonic wave (which must in theory extend to infinity) has a unique frequency, and so remains undistorted when traversing different media. Any other waveform can be made by the superposition of harmonic waves of *different frequencies*: since these frequencies travel at different speeds in dispersive media, non-harmonic waves will then show distortion.

12.7 Polarization

Refer to fig. 12.11.

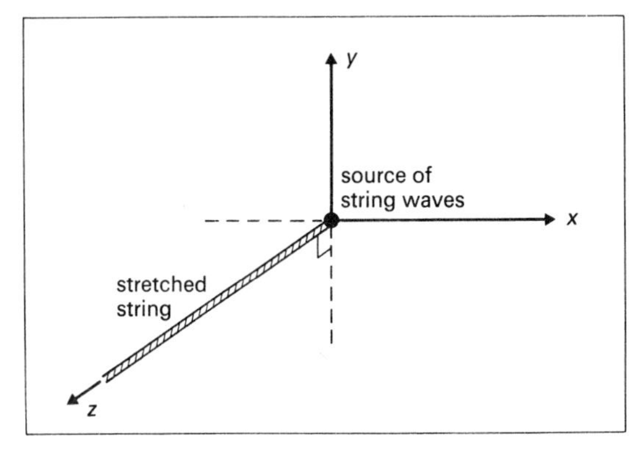

12.11. *Situation to describe polarization.*

(*a*) Suppose that the source oscillates parallel to the x-axis. Then any string particle is made to oscillate along a straight line, and the wave is **linearly** (plane) **polarized**. The **plane of polarization** is that containing the direction of energy travel, and the string displacement (the x–z plane). It is sometimes called the **plane of vibration**.

(*b*) Suppose that the source oscillates parallel to the x- and y-axes simultaneously, and that the two motions are *in phase*. Then we can superpose these two wave motions; their resultant is another linearly polarized wave.

(*c*) Suppose that in (*b*) the two vibrations *differ in phase by* $\pi/2$ rad but have the same amplitude. Now superposition gives a **circularly polarized** wave. At any given instant the string is *helical*.

(*d*) When the two motions of (*c*) have different amplitudes, the wave is **elliptically polarized.**

(*e*) When the motions have the same amplitude, and any phase difference other than 0, $\pi/2$ rad or π rad, then the wave is again **elliptically polarized**.

The points are illustrated by fig. 12.12.

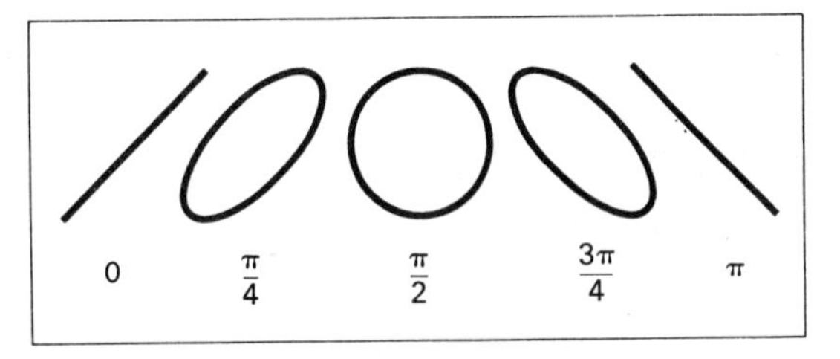

Fig. 12.12. End-on views of the string for superposed x- and y-motions of the source of the same amplitude, but which differ in phase.

**Note* that whereas a transverse wave on a a string transmits *linear* momentum, so a circularly polarized wave transmits *angular* momentum.

Example of Polarization

Television and v.h.f. transmissions are polarized, and so receiving aerials are oriented so as to be parallel to the electric field vector of the electromagnetic wave.

*12.8 The Equation of a Sinusoidal Travelling Wave

Importance of Sinusoidal Waves

(*a*) They are generated by sources that describe s.h.m., and are therefore very common (p. 79).

(*b*) They propagate without distortion (other than that introduced by dissipation of energy).

(*c*) Any wave shape can be regarded as the superposition of sinusoidal waves of appropriate frequencies, amplitudes and phase relationships.

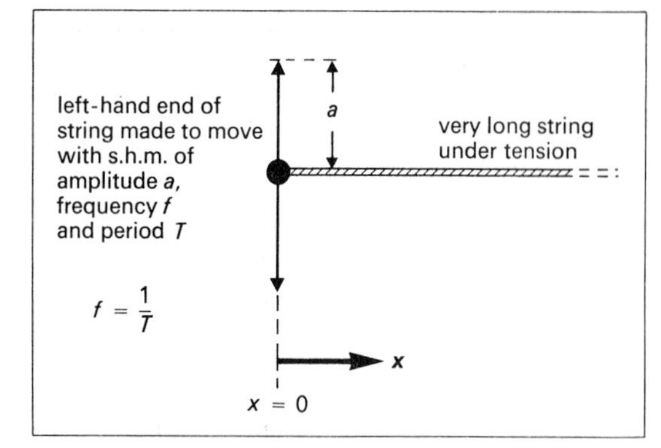

Fig. 12.13. To derive a wave equation.

Development of the Equation

We write the transverse displacement as $y(x, t)$. Thus at the left end (where $x = 0$)

$$y(0, t) = a \sin \omega t$$

where $$\omega = 2\pi f = \frac{2\pi}{T}$$

At some location $x (x > 0)$, the displacement y

(*i*) varies with t in the same way, but

(*ii*) shows a **phase lag** δ.

For this location we write

$$y(x, t) = a \sin (\omega t - \delta)$$

Now if the wave speed c is constant

$$\delta \propto x$$

$$\delta = kx$$

where k is the circular wave number (see below).

The equation of a one-dimensional wave travelling *to the right* is

$$\boxed{y(x, t) = a \sin (\omega t - kx)}$$

It is important to note that the x and t terms have *opposite signs*.

For a wave travelling *to the left*, the displacement at x would show a **phase lead** of δ. We would write

$$y(x, t) = a \sin (\omega t + kx)$$

in which the x and t terms have the *same sign*.

Useful Terms and Relationships

The **wavelength** λ is the distance between points having the same phase (same displacement, velocity and acceleration).

If x changes by λ, the phase changes by 2π rad.

$$\therefore k\lambda = 2\pi$$

$$\boxed{k = \frac{2\pi}{\lambda}}$$

The circular **wave number** k is the number of wavelengths in a length 2π units.

$$k[\text{m}^{-1}] = \frac{2\pi}{\lambda[\text{m}]}$$

Since $\omega = 2\pi/T$, we can write the equation in the alternative form

$$y = a \sin 2\pi\left(\frac{t}{T} - \frac{x}{\lambda}\right)$$

When t changes by a period T, the phase is unchanged so the location has changed by one wavelength λ. The wave profile advances by a distance λ in time T.

$$c = \frac{\lambda}{T} = f\lambda$$

It follows that $$c = (2\pi f)\left(\frac{\lambda}{2\pi}\right)$$

$$= \frac{\omega}{k}$$

The **phase constant** ϕ is introduced for situations where

$$y \neq 0 \quad \text{when } x = 0, t = 0$$

Thus

$$y(x, t) = a \sin [(\omega t - kx) - \phi]$$

The ideas of this section are developed in chapter **14**, p. 103.

*12.9 The Power of a Wave Motion

In this paragraph we consider string waves to deduce a result which is of more general application. Refer to fig. 12.14.

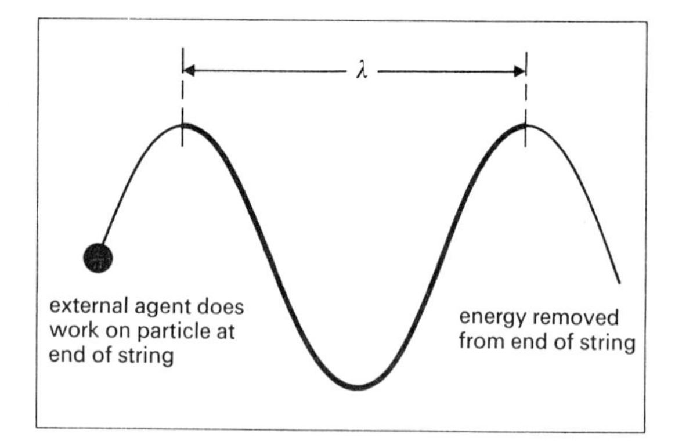

Fig. 12.14. Calculation of wave power.

Momentum and energy are transferred from particle to particle as the state of motion is propagated. The energy of each oscillating particle of mass m is $\frac{1}{2}m\omega^2 a^2$, being partly k.e., and partly p.e.

If the string has mass/unit length ρ, a length λ has energy E where

$$E = \tfrac{1}{2}(\rho\lambda)\omega^2 a^2$$

The energy of length λ is taken steadily from the string in time T, so the power P is given by

$$P = \frac{E}{T}$$

$$= \frac{\frac{1}{2}\rho\lambda\omega^2 a^2}{T} \qquad \left(c = \frac{\lambda}{T}\right)$$

$$= \tfrac{1}{2}\omega^2 a^2 \rho c$$

Note. The power of a wave motion is proportional to
(*a*) (frequency)^2
(*b*) (amplitude)^2
(*c*) wave speed c.

Suppose a wave motion causes energy E to cross an area A at right angles to the wave velocity in time t.

The wave intensity I is defined by

$$I = \frac{W}{At} = \frac{P}{A}$$

$$I\left[\frac{\mathrm{W}}{\mathrm{m}^2}\right] = \frac{P[\mathrm{W}]}{A[\mathrm{m}^2]}$$

Care must be taken to distinguish variations of *intensity* and variations of *amplitude* with distance r from the source.

Shape of wavefront	*Dependence of a on r*	*Dependence of I on r*
plane	independent	independent
cylindrical	$a \propto \frac{1}{\sqrt{r}}$	$I \propto \frac{1}{r}$
spherical	$a \propto \frac{1}{r}$	$I \propto \frac{1}{r^2}$

The student should note the close similarity between these results and those on p. 317 (*Gauss's Law* in electrostatics).

$I \propto 1/r^2$ is simply the well-known **inverse square law**. Note the conditions to which it can be applied.

13 Huygens's Construction

13.1 Statement of Huygens's Hypotheses

The Hypotheses

Each point on an existing wavefront can be considered as a source of secondary wavelets.

(1) *The wave amplitude at any point ahead can be obtained by superposition of these wavelets.*

(2) *From a given position of an uninterrupted wavefront, a later position can be determined as the envelope of the secondary wavelets.*

Validity of the Hypotheses

Huygens's construction is not a fundamental physical principle, but a useful mathematical device for the convenient solution of some problems in optics. It applies (in diffraction)

(*a*) to obstacles large relative to the wavelength, and

(*b*) at large distances for small angles of diffraction.

It takes no account of the physical interaction of a wave with an obstacle.

Uses of the Construction

It can be applied to all types of wave motion, but is particularly useful for diffraction problems in physical optics (p. 280). In geometrical optics it is frequently clumsier than the use of the ray concept.

Explanation of Propagation of Wave Motion

The following procedure has been adopted in fig. 13.1:

(1) Several arbitrary points on the existing wavefront have been chosen as centres of the secondary (spherical) wavelets.

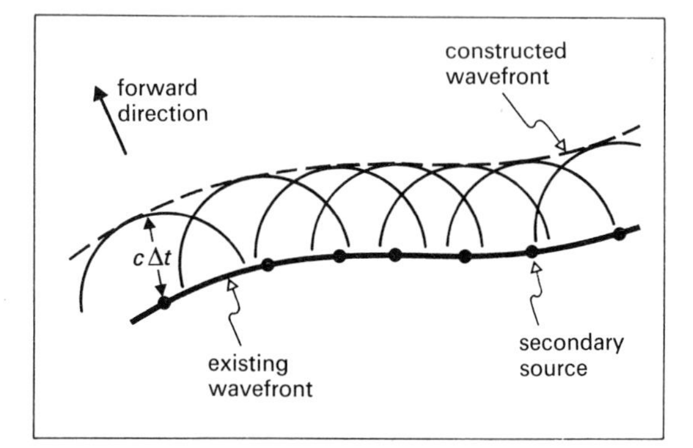

Fig. 13.1. Propagation of an arbitrary wavefront described by Huygens's construction.

(2) The radius of the secondary wavelets has been taken to be $c\ \Delta t$.

(3) The envelope of these wavelets has been taken to be the new wavefront after the time interval Δt.

[We *assume* (so that our result agrees with observation) the intensity of the spherical wavelets to vary continuously from a maximum in the forward direction to zero in the backward. *Fresnel* showed this assumption to be justified.]

Malus's Theorem

The discussions of the next two paragraphs are based on

(*a*) *Huygens's construction*, together with

(*b*) **Malus's Theorem**. This states:

The time taken for one point on a wavefront to travel to its corresponding point (at a later time) is the same for all such pairs of points.

13.2 Application to Reflection

Two situations serve as examples.

The Law of reflection

Refer to fig. 13.2.

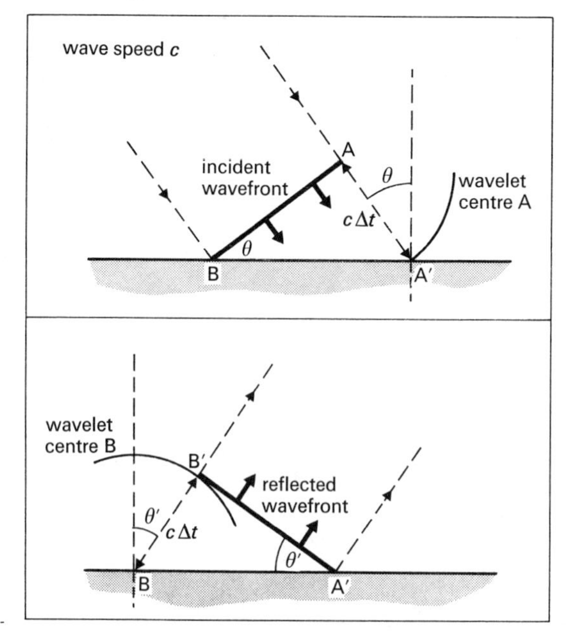

Fig. 13.2. The reflection of a plane wave at a plane surface explained by Huygens's construction.

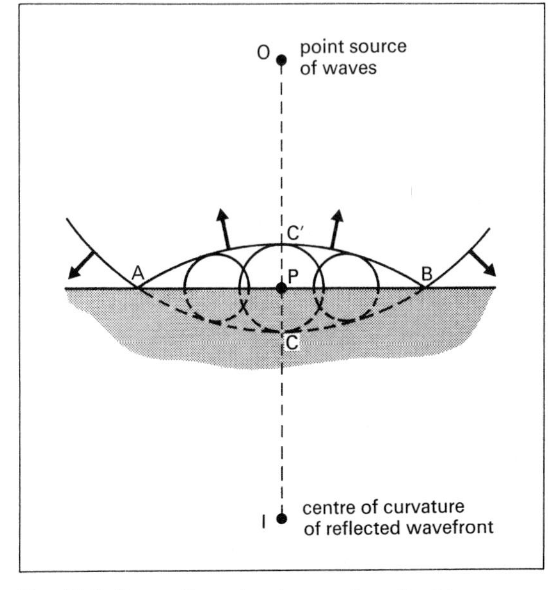

Fig. 13.3. Formation of an image described by Huygens's construction.

Suppose the wavefront at A takes time Δt to reach A′. Then $AA' = c\,\Delta t$.

The radius of secondary wavelet centre B is then also $c\,\Delta t$: the tangent to this wavelet from A′ gives B′, the point to which the wavefront previously at B has advanced in time Δt.

It follows that $\theta = \theta'$.

Image Formation

Fig. 13.3 shows the reflection of a spherical wave at a plane surface.

The secondary wavelets have been drawn as they originate along the reflecting boundary: it follows that they have different radii.

But $$CP = C'P$$
$$\therefore \quad OP = IP$$
a familiar result of elementary optics.

13.3 Application to Refraction

Huygens's construction can be used to derive results such as the lens-maker's formula (p. 239), but ray optics is usually much neater. These examples illustrate some important ideas.

The Law of Refraction

Refer to fig. 13.4, which has been drawn by the method used for fig. 13.2.

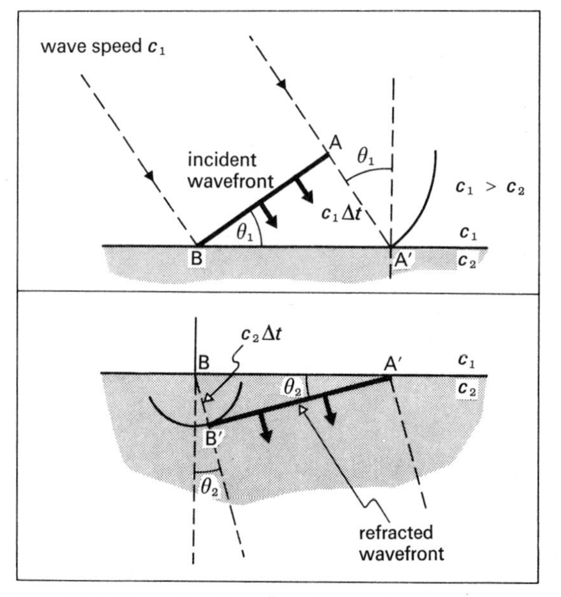

Fig. 13.4. The refraction of a plane wave at a plane surface explained by Huygens's construction.

Dispersion

If white light is used, in a dispersive medium the radius BB′ ($= c_2\,\Delta t$) will depend on the frequency of the waves. Then θ_2 will be different for different colours: this is the phenomenon of **dispersion**.

The *wave theory* can explain observed phenomena on the assumption that light travels *more slowly* in the medium in which it makes a smaller angle with the normal. *Newton's corpuscular theory* required that the light should travel *more rapidly* in such a medium if the corpuscular mass did not change as it crossed the boundary. Experiment (p. 267) verifies the wave assumption, and so for a time the wave explanation for refraction was preferred.†

Total Internal Reflection

Refer to fig. 13.5.

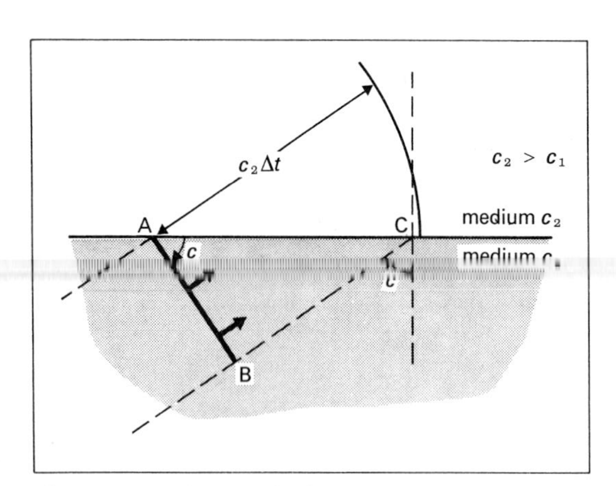

Fig. 13.5. Total internal reflection described by Huygens's construction.

If the radius of the secondary wavelet centre A exceeds AC, no refracted wavefront can be drawn in medium 2 (since C would be inside the circle). *Grazing emergence* occurs when

$$\begin{pmatrix}\text{time to travel}\\ \text{AC in medium 2}\end{pmatrix} = \begin{pmatrix}\text{time to travel}\\ \text{BC in medium 1}\end{pmatrix}$$

$$\frac{\mathrm{AC}}{c_2} = \frac{\mathrm{BC}}{c_1}$$

$$\frac{\mathrm{BC}}{\mathrm{AC}} = \frac{c_1}{c_2} = \frac{n_2}{n_1} = \sin c.$$

c is the *critical angle* for that pair of media.

Prism at Minimum Deviation

In fig. 13.6 wavefronts have been drawn for the ray that passes symmetrically.

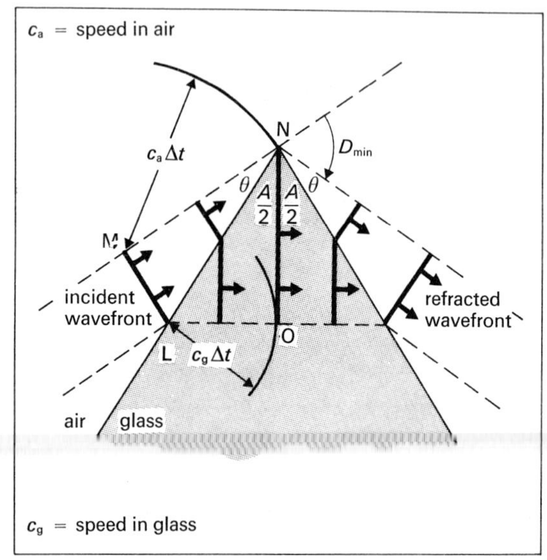

Fig. 13.6. Wavefronts drawn for a prism at minimum deviation.

The wavelets centred on M and L are drawn for the same interval Δt. Thus NO is a new wavefront which bisects the refracting angle at N.

Since

$$\mathrm{MN} = c_a\,\Delta t \quad \text{and} \quad \mathrm{LO} = c_g\,\Delta t$$

$$\frac{n_g}{n_a} = \frac{c_a}{c_g} = \frac{\mathrm{MN}}{\mathrm{LO}}$$

$$= \frac{\mathrm{NL}\cos\theta}{\mathrm{NL}\sin(A/2)}$$

But $$2\theta + A + D_{\min} = 180°$$

so $$\frac{n_g}{n_a} = \frac{\sin[(A + D_{\min})/2]}{\sin(A/2)}$$

a result which is proved for ray optics on p. 235.

† Nowadays we can combine the two since the theory of relativity suggests that the *mass* of the photon will change at the interface. *Momentum* is conserved parallel to the boundary, and in the denser medium it increases normal to the boundary. We must be careful when drawing conclusions about changes to the photon *velocity*.

14 The Principle of Superposition

14.1 The Principle of Superposition

It is observed experimentally that when simple water waves of small amplitude collide, each wave has afterwards the same shape, momentum and energy as it had before the collision. We deduce that such waves do not interact with each other. (In more advanced physics it is observed that a collision between say a photon and an electron *does* affect the waves concerned: nevertheless our general conclusion still holds good.)

The **principle of superposition** follows:

The net disturbance at a given place and time caused by a number of waves which are traversing the same space is the vector sum of the disturbances which would have been produced by the individual waves separately.

$$y = y_1 + y_2 + y_3 \text{ etc.}$$

This should be compared with the superposition principle for forces (p. 35). Travelling *particles* cannot be superposed in this way: a collision between two particles alters the velocity of both.

The principle can be applied

(*a*) to all electromagnetic waves,

(*b*) to elastic waves in a deformed medium so long as the deformation is proportional to the restoring force, i.e. so long as *Hooke's Law* applies.

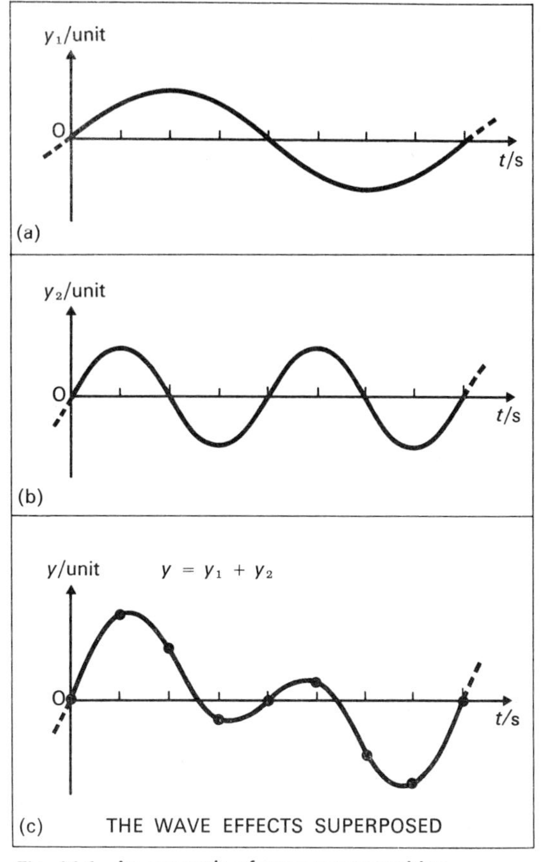

Fig. 14.1. An example of wave superposition.

One importance of the principle is that it enables us to analyse complicated wave motions by representing them in terms of simple harmonic motions (p. 79).

An example of Superposition

Fig. 14.1 shows the variation with time of the disturbance caused at a particular point by

(*a*) a sinusoidal wave,

(*b*) a second sinusoidal wave of the same amplitude but with twice the frequency, and

(*c*) both waves together.

One can draw similar curves for *y* plotted against *x*. These would be for a fixed time instant (i.e. would be 'photographs').

Application of the Principle

Wave superposition results in the phenomena classified in this table:

Phenomenon	*Waves that are superposed*
stationary waves	two wavetrains of same amplitude and frequency travelling at the same speed in opposite directions (a special case of interference)
interference	coherent waves from identical sources
beats	two wavetrains of close frequency travelling in the same direction at the same speed
diffraction	secondary wavelets originating from coherent sources on the same wavefront

The word *interference* is sometimes used where in this book we use *superposition*.

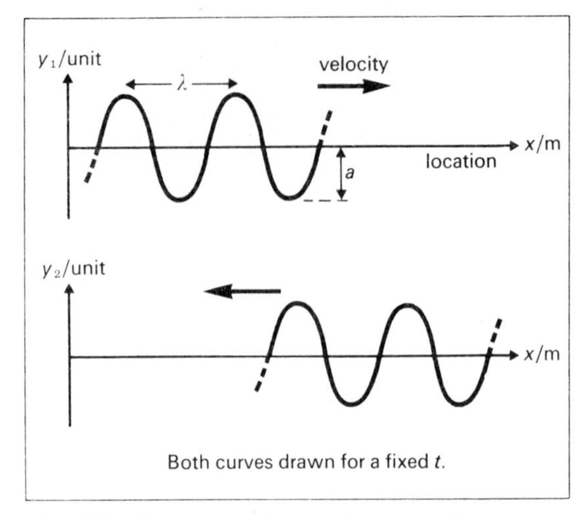

Fig. 14.2. Two wavetrains which give stationary waves when superposed.

It is convenient to describe the result of superposition in terms of what would be observed using transverse waves on a stretched string (fig. 14.3), and the table opposite compares and contrasts this with a progressive wave motion.

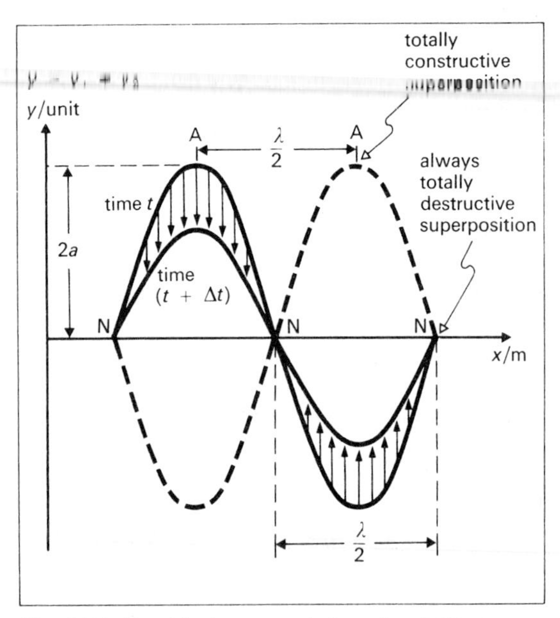

Fig. 14.3. Graphical representation of a stationary wave motion drawn at times t and (t + Δt).

The *eye* sees a *time-average*, and cannot distinguish between the phases of adjacent segments, whereas a *photograph* can.

14.2 Stationary or Standing Waves

Stationary or **standing** waves result from superposition of two wavetrains which have

(*a*) the same amplitude and frequency (and therefore wavelength), but

(*b*) equal but opposite velocities.

Two such wavetrains (supposed sinusoidal for convenience) are shown in fig. 14.2. The formation of stationary waves is thus a special case of interference (p. 106) which will be observed along the line joining the two sources in a *Young's* experiment.

COMPARISON OF STATIONARY AND PROGRESSIVE WAVE MOTION

	Stationary	*Progressive*
amplitude	Varies according to position from zero at the **nodes** (permanently at rest) to a maximum of $2a$ at the **antinodes**.	Apart from attenuation, is the same for all particles in the path of the wave.
frequency	All the particles vibrate in s.h.m. with the same frequency as the wave (except for those at the nodes which are at rest).	All particles vibrate in s.h.m. with the frequency of the wave.
wavelength	$2 \times$ distance between a pair of adjacent nodes or antinodes.	Distance between adjacent particles which have the same phase.
phase	Phase of all particles between two adjacent nodes is the same. Particles in adjacent segments of length $\lambda/2$ have a phase difference of π rad.	All particles within one wave-length have different phases.
waveform	Does not advance. The curved string becomes straight twice in each period.	Advances with the velocity of the wave.
energy	No translation of energy, but there is energy associated with the wave.	Energy translation in the direction of travel of the wave.

Formation of Stationary Waves

Frequently one wavetrain is derived from the (nearly) total reflection of the other. The foregoing description applies, but one is also concerned with **boundary conditions**.

The phase of the reflected wave can be found by
(*a*) experimental observation (p. 94), or
(*b*) working out the boundary condition appropriate to the type of wave and boundary being considered.

Examples

Wave	*Boundary*	*Boundary condition*	*There exists at the boundary a(n)*	*Example on page*
transverse string wave	rigid wall	displacement always zero	displacement node	305
longitudinal spring wave	rigid wall	displacement always zero	displacement node	—
sound wave	rigid boundary	displacement always zero	displacement node, pressure antinode	303
sound wave	open boundary	pressure change effectively zero	displacement antinode, pressure node	303
electromagnetic wave	conducting metal surface	electric field zero (p. 319)	electric field vector node	266

*14.3 Analytical Treatment of Stationary Waves

Suppose the waves of fig. 14.2 are represented by

$$y_1 = a \sin(\omega t - kx) \qquad \text{(to the right)}$$

and

$$y_2 = a \sin(\omega t + kx) \qquad \text{(to the left)}$$

Using the principle of superposition, the equation of the standing wave is given by

$$\begin{aligned} y &= y_1 + y_2 \\ &= a \sin(\omega t - kx) + a \sin(\omega t + kx) \\ &= 2a \cos kx \sin \omega t \end{aligned}$$

which may be written

$$y = \begin{pmatrix} \text{location} \\ \text{dependent} \\ \text{term} \end{pmatrix} \cdot \begin{pmatrix} \text{time} \\ \text{dependent} \\ \text{term} \end{pmatrix}$$

Deductions from the equation

The equation represents fig. 14.3.

(*a*) Every particle oscillates at a pulsatance

$$\omega = 2\pi f = 2\pi/T$$

(*b*) The oscillation amplitude varies with location according to $\cos[(2\pi x)/\lambda]$, since $k = 2\pi/\lambda$.

(*c*) At the *nodes*

$$y = 0 \quad \textit{always}$$

$$\cos\frac{2\pi x}{\lambda} = 0$$

$$\frac{2\pi x}{\lambda} = n\left(\frac{\pi}{2}\right), \qquad n = 1, 3, 5, \text{etc.}$$

Distance between nodes $= \lambda/2$.

(*d*) At the *antinodes* y reaches a maximum amplitude $2a$. This happens when

$$\cos\frac{2\pi x}{\lambda} = \pm 1$$

$$\frac{2\pi x}{\lambda} = m\pi, \qquad m = 0, 1, 2, \text{etc.}$$

Distances between antinodes $= \lambda/2$.

(*e*) There is no net energy transfer.

14.4 Interference

The term **interference** is best applied to the result of superposing waves from *identical sources*.

When a *pair* of sources is used, we frequently refer to **Young's experiment** (p. 272).

If an interference pattern (or a system of **fringes**) is to be seen, the sources must be coherent. They must also produce waves of about the *same amplitude*.

Sources are **coherent** if they have

(*a*) a constant phase relationship (i.e. are **phase-linked**), which implies

(*b*) the *same frequency*.

If the waves are transverse, then they must be either unpolarized, or polarized in the same plane.

What happens when these conditions are not observed is discussed on p.

The interference *at a particular point* is said to be

(*i*) **constructive** when the wave disturbances reinforce each other, or

(*ii*) **destructive** when they are self-cancelling.

Except where one wave is superposed on another, there is no trace of the other's presence.

Since (amplitude)$^2 \propto$ intensity, the energy associated with a wave takes no account of phase. Thus for surface water waves the same energy is associated with a double trough as with a double crest.

Description of Young's Experiment

Fig. 14.4 represents two-source interference for *any type* of wave motion: the reader should be able to associate it with a ripple tank experiment.

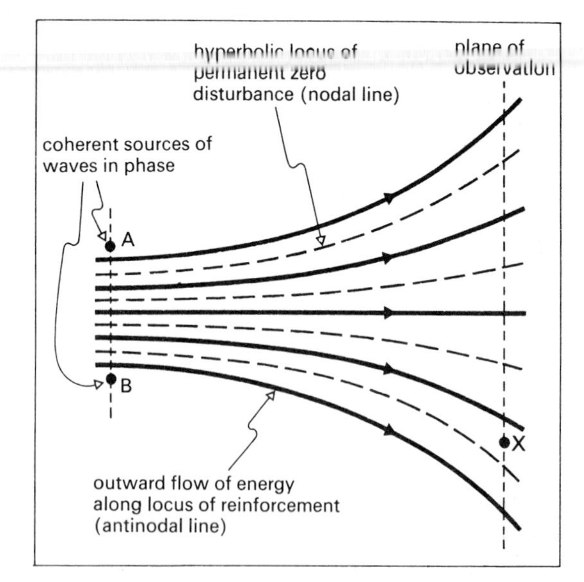

Fig. 14.4. A time-average of the interference pattern obtained in a Young's experiment.

An arbitrary point X will lie on

(*a*) a **nodal** line (cancellation) if

$$(AX - BX) = (2m + 1)\frac{\lambda}{2}$$

since the disturbance will then be *permanently* zero,

(*b*) an **antinodal** line (reinforcement) if

$$(AX - BX) = (2m)\frac{\lambda}{2}$$

The disturbance at X will vary with time: the amplitude will be double that produced by A or B separately. (In each case m is an integer.)

There is a *net translation of energy* away from the sources along the antinodal lines (fig. 14.5).

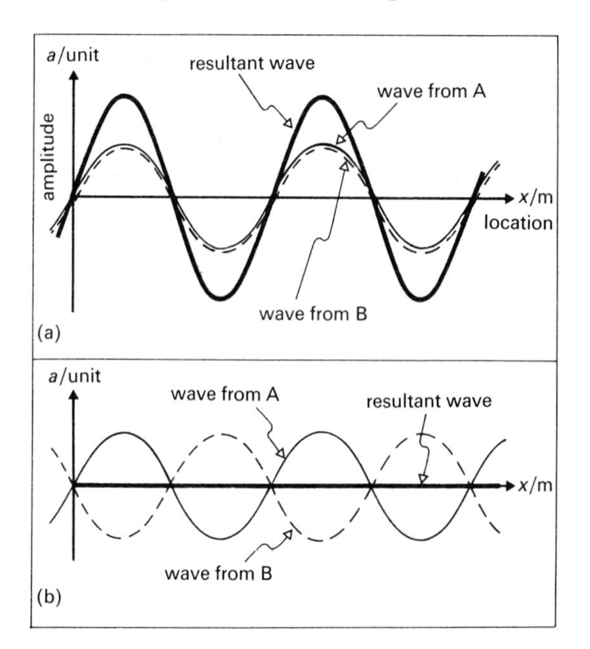

Fig. 14.5. Diagram (drawn for transverse waves) showing the result of superposition (a) along an antinodal line, (b) along a nodal line, for one instant of time.

A 'photograph' of the wave pattern would show that neighbouring antinodal lines consist of travelling waves in antiphase.

The student should work out what happens to the intensity of the waves and the distance between antinodal lines (the **fringe separation**) when the plane of observation is

(1) parallel to AB,
(2) perpendicular to AB, using

(*i*) different source to plane-of-observation distances,
(*ii*) A at different distances from B,
(*iii*) sources A and B with various phase relationships (such as a phase difference of π rad),
(*iv*) sources A and B of higher or lower frequencies,
(*v*) a source A of different frequency from that of B.

*14.5 Analytical treatment of Young's Experiment

In this paragraph we will calculate the variation of intensity with position on a screen when the screen is placed parallel to AB.

Suppose the two waves (of equal amplitude) are represented by

$$y_1 = a \sin(\omega t - kx)$$
$$y_2 = a \sin(\omega t - kx - \phi)$$

where ϕ is the phase difference between the disturbances at a particular point (produced in this example because the waves come by different routes).

By the superposition principle

$$y = y_1 + y_2$$
$$= 2a \cos\frac{\phi}{2} \sin\left(\omega t - kx - \frac{\phi}{2}\right)$$

The $[2a \cos(\phi/2)]$ term shows the variations of y's *amplitude* with ϕ, i.e. with position on the screen.

The $\sin[\omega t - kx - (\phi/2)]$ term shows how y varies with t for each value of x (the distance from the plane through AB to the plane of the screen).

Intensity Variation

$$\cos\frac{\phi}{2} = 0 \quad \text{when} \quad \phi = (2m + 1)\pi$$

Then $$y_{max} = 0$$

$$\cos\frac{\phi}{2} = \pm 1 \quad \text{when} \quad \phi = (2m)\pi$$

Then $$y_{max} = 2a$$

The amplitude (*not* displacement) $= 2a \cos(\phi/2)$, so

$$\text{intensity} \propto 4a^2 \cos^2\frac{\phi}{2}$$

$$I \propto \cos^2\frac{\phi}{2}$$

This is illustrated by fig. 14.6 overleaf. *Young's* fringes are sometimes called '**$\cos^2$ fringes**'.

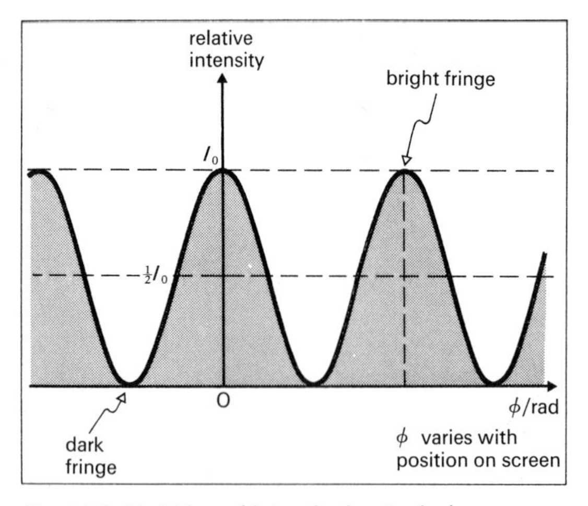

Fig. 14.6. Variation of intensity in a typical interference pattern.

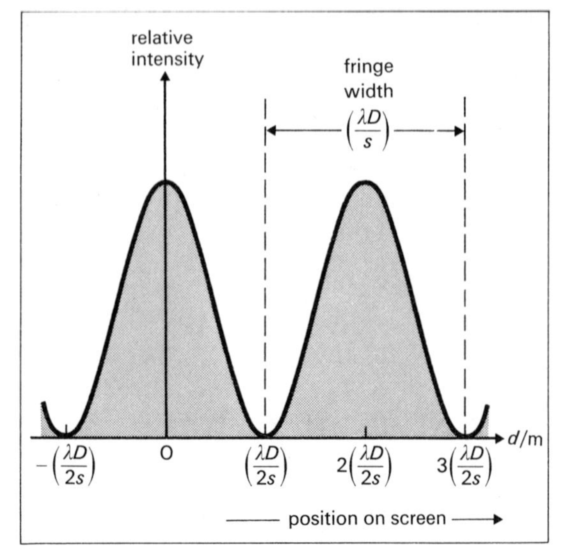

Fig. 14.8. Fringe maxima related to location on the screen.

Energy Conservation

Fig. 14.7 suggests how energy is conserved.

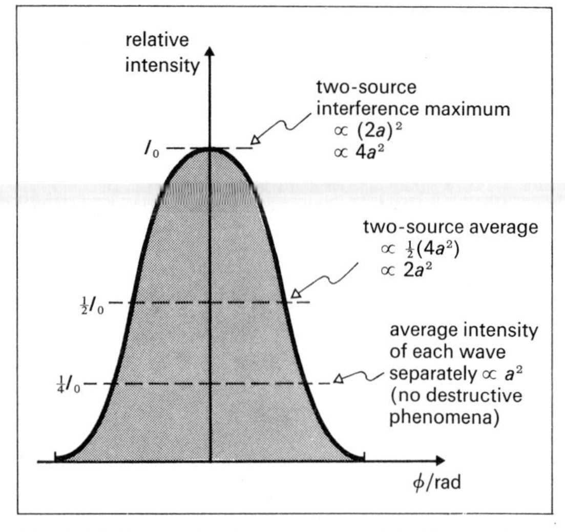

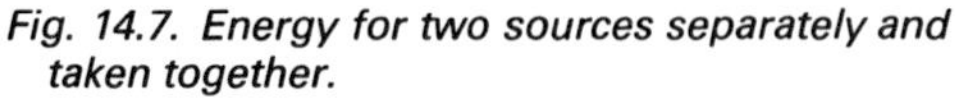

Fig. 14.7. Energy for two sources separately and taken together.

Intensity and Fringe Width

Suppose ϕ is the result of a path difference p between waves that originate from sources in phase.

Then
$$\frac{\phi}{2\pi} = \frac{p}{\lambda}$$

$$\therefore \ \phi = \frac{2\pi p}{\lambda}$$

and
$$I \propto \cos^2\left(\frac{\pi p}{\lambda}\right)$$

It is shown on p. 272 that

$$p = \frac{ds}{D} \quad \textit{under certain conditions}$$

where d is the distance along the screen from the central maximum, s is the source separation, and D is the screen-source distance.

Thus
$$I \propto \cos^2\left(\frac{\pi ds}{\lambda D}\right)$$

Maxima will occur where

$$\cos^2\left(\frac{\pi ds}{\lambda D}\right) = 1$$

so
$$\frac{\pi ds}{\lambda D} = m\pi$$

and
$$d = \frac{m\lambda D}{s}$$

Thus the fringe width w is given by

$$d_{m+1} - d_m = \frac{\lambda D}{s}$$

$$\boxed{w = \frac{\lambda D}{s}}$$

Note the conditions (p. 272) under which we can use this expression: it *cannot*, for example, be used in ripple-tank and microwave experiments. It is illustrated by fig. 14.8.

14.6 Beats

Suppose two wavetrains of slightly different frequencies f_1 and $f_2 (f_1 > f_2)$, but of equal amplitudes are superposed while travelling in the same direction at the same speed.

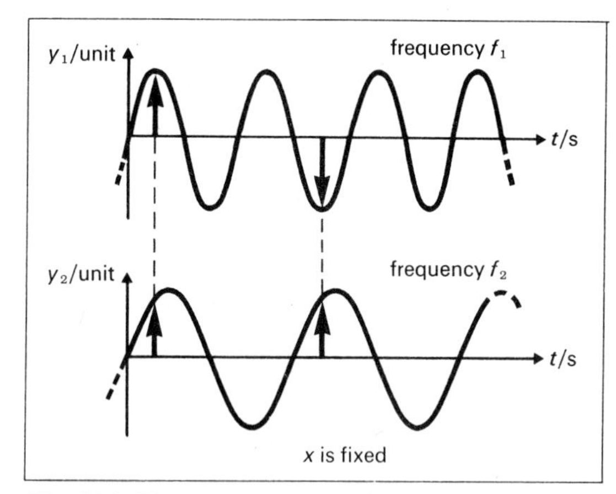

Fig. 14.9. Two wavetrains which give beats when superposed.

Consider their superposition *over a period of time at a fixed point.* Fig. 14.9 shows how the result is sometimes reinforcement, and sometimes cancellation, because the sources are not coherent. Fig. 14.10 shows the result of applying the principle of superposition.

Notes.

(*a*) The *amplitude* varies with time with a frequency

$$\left(\frac{f_1 - f_2}{2}\right)$$

(*b*) The **beat frequency** f_{beat} is the number of times the *magnitude* of the amplitude reaches a maximum in each second.

Thus $f_{\text{beat}} = 2 \times$ (amplitude frequency)

$= (f_1 - f_2)$

More generally

$$f_{\text{beat}} = (f_1 \sim f_2)$$

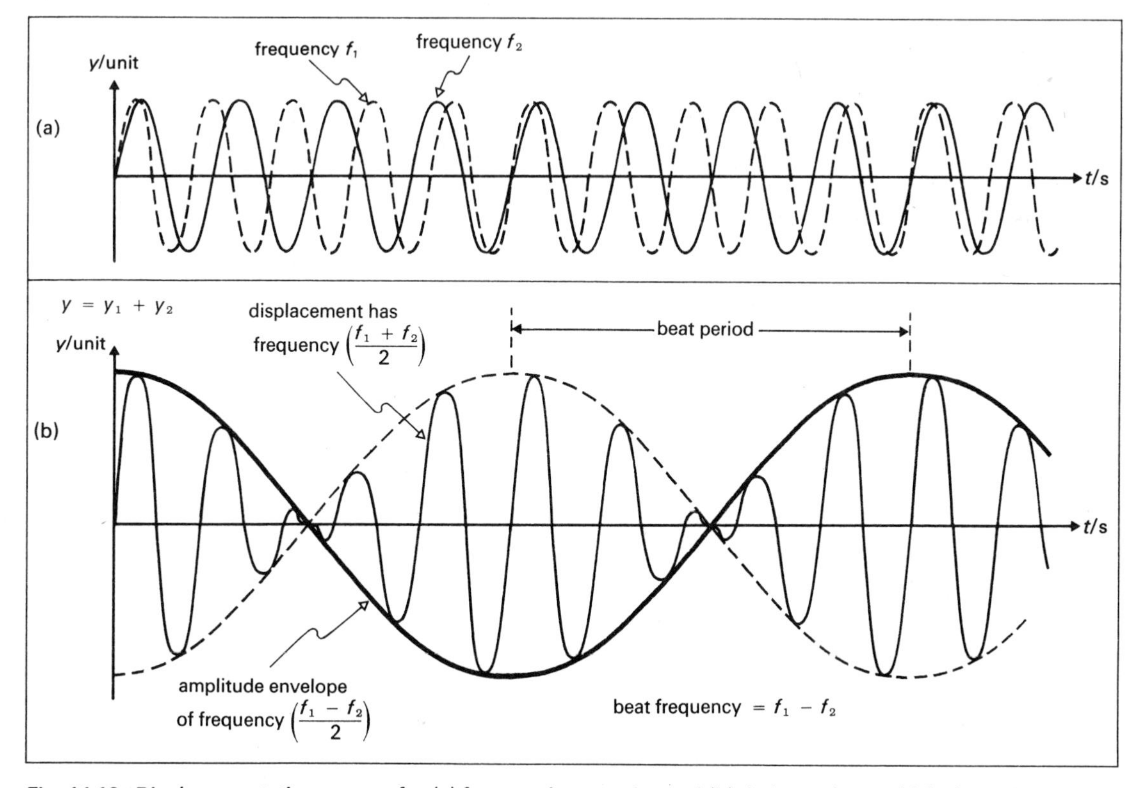

Fig. 14.10. Displacement–time curves for (a) frequencies f_1 and f_2 and (b) their resultant which shows beats.

This result is proved below, and in the next paragraph.

(*c*) The resultant wave has a (displacement) frequency

$$f = \frac{f_1 + f_2}{2}$$

the average of the superposed waves. Usually f_1 is close to f_2: then f is roughly equal to either.

Calculation of the Beat Frequency

Suppose that in time T (the **beat period**), the wave of frequency f_1 completes one cycle more than wave f_2.

The number of cycles they complete are f_1T and f_2T respectively.

$$\therefore \quad f_1T - f_2T = 1$$

So

$$f_{\text{beat}} = \frac{1}{T} = f_1 - f_2$$

Beats in Sound

The ear responds to the energy (intensity) variation.

Thus (loudness) $\propto$ (amplitude)^2

The ear does not distinguish the two halves of fig. 14.10, and hears *two* sounds of maximum intensity in each cycle of period $2/(f_1 - f_2)$. Thus the loudness variation has a frequency $(f_1 - f_2)$.

Beats in sound are discussed further on p. 301.

*14.7 Analytical Treatment of Beats

We aim to find the variation of y with t while x is constant. It is *not* a special case if, for convenience, we choose $x = 0$. (The time variations elsewhere are of the same form.)

Using $y = a \sin(\omega t - kx)$ where $\omega = 2\pi f$, we wish to superpose

$$y_1 = a \sin 2\pi f_1 t$$

and

$$y_2 = a \sin 2\pi f_2 t$$

So

$$\begin{aligned} y &= y_1 + y_2 \\ &= a\,(\sin 2\pi f_1 t + \sin 2\pi f_2 t) \\ &= \left[2a \cos 2\pi \left(\frac{f_1 - f_2}{2}\right) t\right] \sin 2\pi \left(\frac{f_1 + f_2}{2}\right) t \end{aligned}$$

We conclude that the resultant disturbance

$$y = \left\{\begin{array}{c}\text{an amplitude which} \\ \text{varies at frequency} \\ \left(\dfrac{f_1 - f_2}{2}\right) \\ \text{i.e.} \\ \text{slowly in time}\end{array}\right\} \times \left\{\begin{array}{c}\text{a displacement} \\ \text{which has frequency} \\ \left(\dfrac{f_1 + f_2}{2}\right) \\ \text{i.e. varies} \\ \text{rapidly in time}\end{array}\right\}$$

This time variation of y is shown in fig. 14.10.

14.8 Diffraction

Diffraction is the name given to the phenomena occurring when a wave motion passes an obstacle, or passes through an aperture. The change of wave velocity, or its spreading out, results from the superposition of wavelets that originate from points along the same wavefront.

Suppose the arrangement of fig. 14.11 is established in a ripple tank.

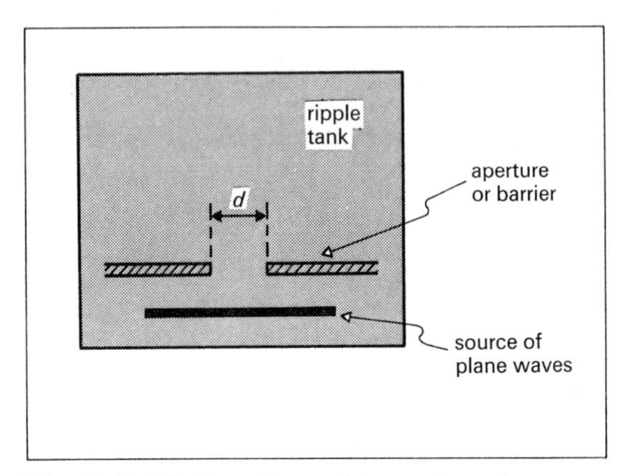

Fig. 14.11. Ripple tank experiment to investigate diffraction.

One can adopt one of two procedures:

(*a*) vary d while using waves of fixed wavelength, or

(*b*) vary λ while using an aperture or obstacle of fixed size.

Some results are given in fig. 14.12. These are simplified diagrams: in practice the pattern is complicated by subsidiary maxima and minima (p. 285).

Conclusions

(*i*) Diffraction is relatively more important when λ becomes large in comparison with d.

(*ii*) When d is very small, the wavelets are nearly semicircular (or, in three dimensions, hemispherical): this corresponds to the idea of a *Huygens* wavelet.

Application: the **diffraction grating**, the principles of which are discussed on p. 280. Such gratings can be made for all wave-types—the wavelength controls the choice of grating spacing.

Diffraction in Light and Sound

(a) Sound

The wavelength of sound for a man's voice is about

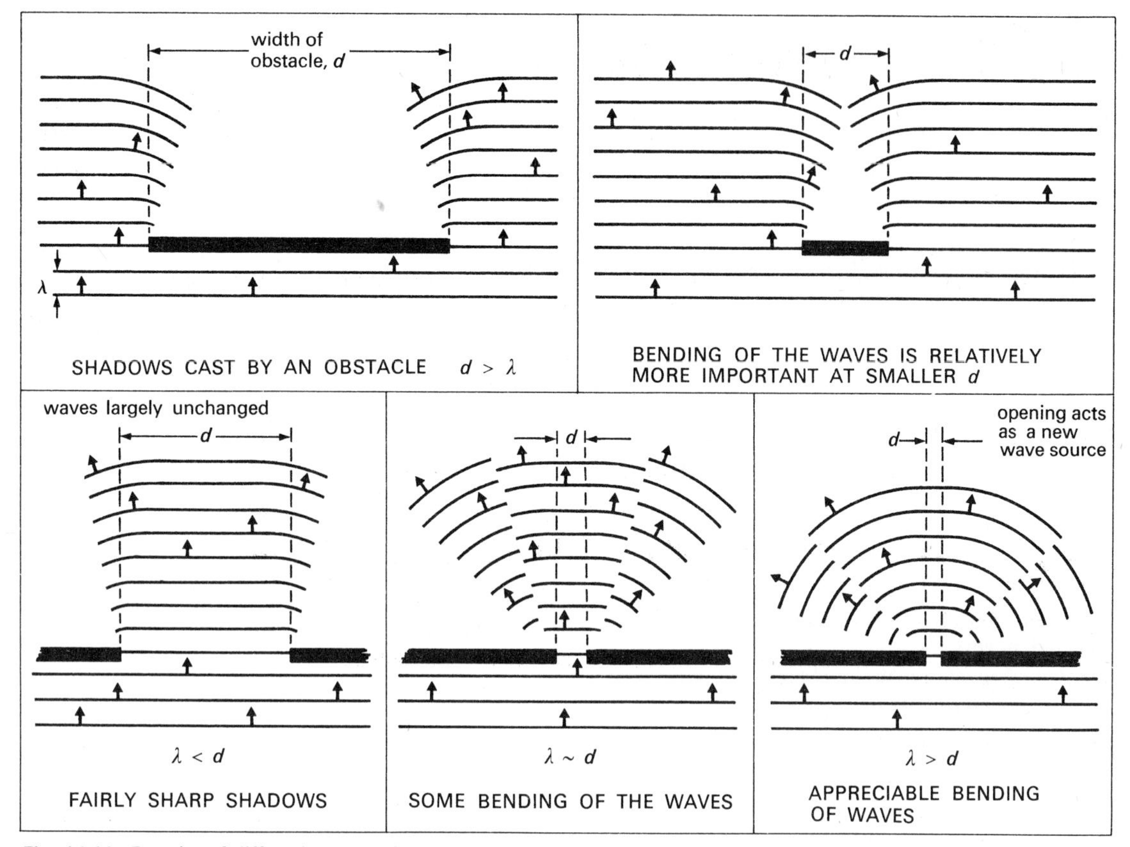

Fig. 14.12. Results of diffraction experiment.

2 m, and common objects (such as furniture) have smaller linear dimensions than this.

Diffraction (and reflection) enables us to hear sounds from hidden sources.

(b) Light

λ for visible light $\sim 10^{-7}$ m which is a small fraction of the diameter of (say) a window. Diffraction effects are not frequently observed unless deliberately promoted. They are nevertheless very important, and are fully discussed in chapter **39**, p. 279.

14.9 The Impact of Waves on Obstacles

Consider two extreme cases.

(a) Object Large Relative to λ

Then fig. 14.12 indicates two of the three effects:

(1) a shadow is produced behind the obstacle,
(2) the wave bends round the corners,
(3) a reflected wave moves back from the obstacle.

For example under the correct conditions a ball bearing causes light waves from a point source to leave a circular shadow when they are reflected, but there is also the possibility of a bright spot on the axis (p. 283).

(b) Object Very Small Relative to λ

Two processes operate:

(1) Diffraction enables the wavefront effectively to reform beyond the object.

(2) **Scattering** causes the object to act as a new source of secondary wavelets.

This second process accounts for the colour of the sky. Small particles in the atmosphere scatter blue light (small λ) much more than they do red light (large λ).

15 Doppler Effect

15.1 Calculation of Frequency Change

When there is movement of source or observer relative to the medium in which a wave is propagated, then although

(*i*) the wave speed c through the medium is *not* changed, nevertheless

(*ii*) *the frequency f_0 measured by an observer differs from f_s, that emitted by the source.*

This is called the **Doppler effect** (1842).

Notation c for wave speed,
v for source or observer speed,
u for speed of medium.

(*a*) Observer Moving

Refer to fig. 15.1. The basic effect here is a change in the *number of wave crests received per second.* The wavelength λ is unaltered.

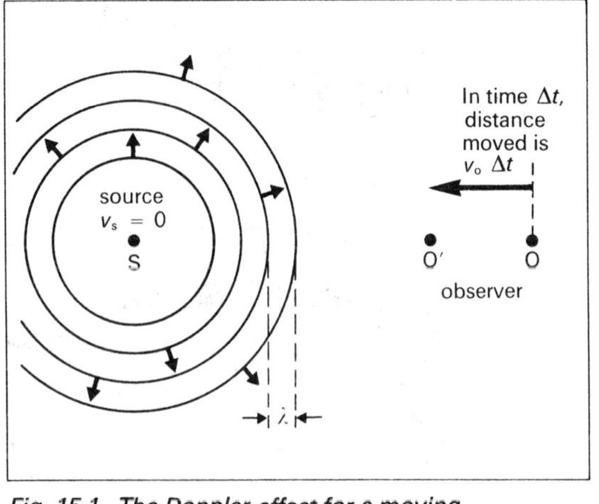

Fig. 15.1. The Doppler effect for a moving observer (relative approach).

Consider a time interval of Δt.

$$f_0 = f_s + \left(\frac{\text{number of waves in distance travelled in } \Delta t}{\Delta t}\right)$$

$$= f_s + \left(\frac{v_0\,\Delta t/\lambda}{\Delta t}\right) = f_s + \left(\frac{v_0}{\lambda}\right)$$

but $f_s\lambda = c$,

so $$f_0 = f_s + v_0\frac{f_s}{c}$$

$$= f_s\left(1 + \frac{v_0}{c}\right) \quad \textit{Observer approaches source}$$

or $$f_0 = f_s\left(1 - \frac{v_0}{c}\right) \quad \textit{Observer recedes from source}$$

Then $$\boxed{f_0 = f_s\left(1 \pm \frac{v_0}{c}\right)}$$

The equation applies for a receding $v_0 < c$. When $v_0 > c$, no signal is received.

(*b*) Source Moving

Fig. 15.2 might represent a source moving through a still medium (e.g. a trailing ripple tank vibrator) or a medium moving past a still source (e.g. a paddling swan on a river): *relative movement* is important.

The source movement changes the wave speed relative to the source, which causes

(*i*) a concentration of wave crests (and consequent decrease in wavelength) in front of the source,

(*ii*) a corresponding increase in wavelength behind.

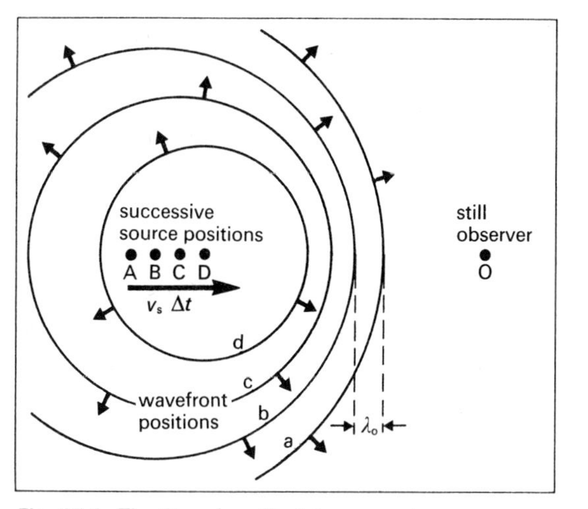

Fig. 15.2. The Doppler effect for a moving source (relative approach).

The basic effect now is a *change of wavelength*.

Consider a time interval of Δt.

$f_s\,\Delta t$ waves are compressed into a length $(c - v_s)\,\Delta t$.

Hence the apparent

$$\lambda_0 = \frac{\text{distance}}{\text{number of wave crests}} = \frac{(c - v_s)\,\Delta t}{f_s\,\Delta t}$$

So $f_0 = \dfrac{c}{\lambda_0} = f_s\left(\dfrac{c}{c - v_s}\right)$

$= f_s\left(\dfrac{1}{1 - (v_s/c)}\right)$ *Source approaches observer*

or $f_0 = f_s\left(\dfrac{1}{1 + (v_s/c)}\right)$ *Source recedes from observer*

$$f_0 = f_s\left(\frac{1}{1 \mp v_s/c}\right)$$

The equation applies for an approach $v_s < c$ (see p. 115).

(c) Medium Moving

(*i*) *For sound* the effective velocity of the wave motion relative to the source or observer is changed from $\boldsymbol{c}$ to $(\boldsymbol{c} \pm \boldsymbol{u})$, where $\boldsymbol{u}$ is the resolved part of the medium's velocity along the line joining source and observer.

(*ii*) *For light* the problem is complicated by the fact that no medium is involved. (See p. 115.)

(d) Source and Observer Moving

Suppose both movements cause relative approach. Since the source moves

$$\lambda_0 = \frac{c - v_s}{f_s}$$

Since the observer moves in addition

$$f_0 = \frac{c}{\lambda_0} + \frac{v_0}{\lambda_0} = \left(\frac{c + v_0}{c - v_s}\right) f_s$$

(e) Summary

(1) In General

$$f_0 = \left(\frac{c \pm v_0}{c \mp v_s}\right) f_s$$

in which *upper* signs indicate relative *approach*, and *lower* signs relative *recession*.

(2) For Sound

$$f_0 = \left(\frac{\text{relative speed of sound w.r.t. observer}}{\text{relative speed of sound w.r.t. source}}\right) f_s$$

This takes into account the velocity of the medium.

(3) An Approximate Relationship

Suppose $v_0 = 0$. Then for approach of source

$$f_0 = \frac{f_s}{1 - v_s/c}$$

$$\therefore\ f_0 - f_0\left(\frac{v_s}{c}\right) = f_s$$

$$(f_0 - f_s) = \Delta f = \left(\frac{v_s}{c}\right) f$$

where $f \approx f_0 \approx f_s$ if Δf is small.

Since $f\lambda = c$, and c is constant,

$$\ln f + \ln \lambda = 0$$

Differentiating,

$$\frac{\Delta f}{f} = -\frac{\Delta \lambda}{\lambda}$$

$$\frac{\Delta f}{f} \approx -\frac{\Delta \lambda}{\lambda} \approx \frac{v_s}{c}$$

This is a useful non-relativistic expression which is applicable when $v_s \ll c$. It is therefore particularly useful for electromagnetic waves.

(4) Special Cases

Suppose	*Then if*	*f_0 becomes*
$v_0 = 0$	$v_s = c$ (approach)	∞ (shock wave)
	$v_s = c$ (recession)	$f_s/2$
$v_s = 0$	$v_0 = c$ (approach)	$2f_s$
	$v_0 = c$ (recession)	0 (no signal)

The general ideas of this paragraph apply to all types of wave motion: in detail they apply to waves through material media.

15.2 Examples in Practice

(*a*) Train Passing through a Station

The value of v_s to be used is the resolved part of the train's velocity along the line joining source to observer. Then f_0 varies from

$$f_s\left(\frac{1}{1-(v_s/c)}\right)$$

through f_s to

$$f_s\left(\frac{1}{1+(v_s/c)}\right)$$

as the train recedes.

The frequency change is most abrupt as the train passes.

(*b*) Radar

Detection

The source and receiver are together, the receiver being tuned for frequencies other than f_s. The distance to the moving reflector is found from the time delay before the signal is received.

To Measure Speed

There are two problems:

(1) The moving object of speed v receives a frequency

$$f' = f_s\left(1 + \frac{v}{c}\right)$$

(2) The moving object is now a moving source emitting at frequency f'.

The detector observes a frequency

$$f_0 = f'\left(\frac{1}{1-(v/c)}\right)$$

$$= f_s\left(\frac{c+v}{c-v}\right)$$

(*c*) Rotating Bodies

(*i*) A body like the Sun shows equal but opposing frequency shifts for *Fraunhöfer* lines received from the opposite ends of a diameter.

(*ii*) Double stars show a similar effect. (One of the pair may not emit visible light—but the other still shows alternate red and blue shifts.)

(*iii*) *Doppler* effect observations of Saturn's rings show that they are made of discrete bodies (not rigid discs), since different parts have different angular velocities.

(*d*) The Red Shift

A retreating light source causes $f_0 < f_s$ so a particular spectral line is shifted toward the red part of the spectrum (fig. 15.3). The red shift observations are fundamental to our understanding of the universe. Considered as *Doppler* effects, the red shifts show that the universe of galaxies is expanding in all directions at a rate that increases in proportion to the distance r from the Earth. The speed of recession is given by $v = Hr$, where H is known as the **Hubble constant**.

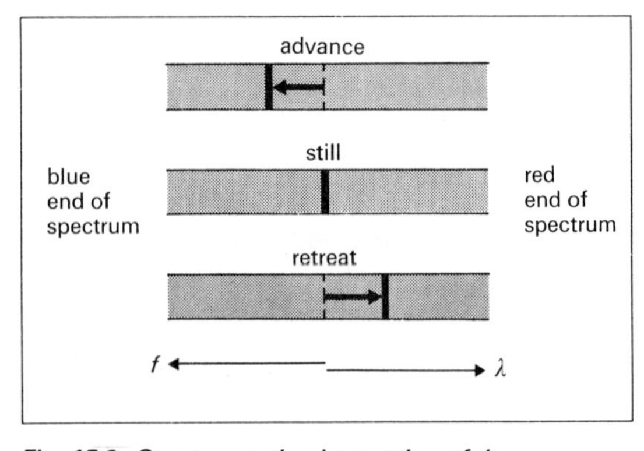

Fig. 15.3. Spectroscopic observation of the expanding universe. (According to relativistic theory, a red shift is also shown by a purely tranverse motion.)

(e) The Broadening of Spectral Lines

Gas molecules emitting light have variable velocities resolved along the line of observation. This results (fig. 15.4) in a broadening of spectral lines.

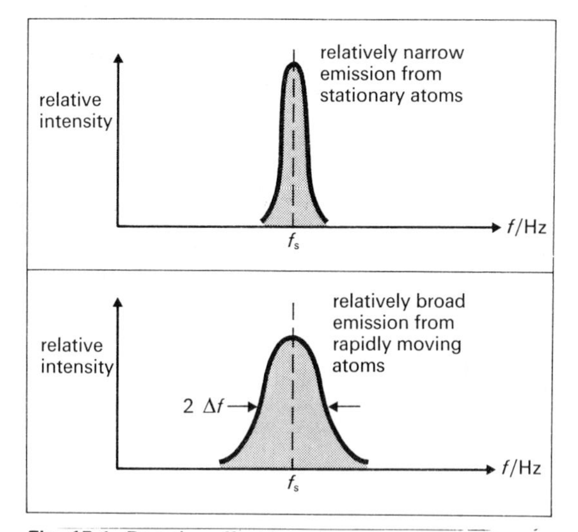

Fig. 15.4. Doppler effect broadens the spectral lines from a discharge tube.

Since

$$\tfrac{1}{2}m\overline{v^2} = \tfrac{3}{2}kT \qquad \text{(p. 192)}$$

and

$$\frac{\Delta f}{f} = \frac{v_s}{c},$$

high temperatures T result in large values of v_s and so of Δf.

Notes.

(*i*) A substance should therefore be cooled before accurate spectroscopic measurements are made on emitted light.

(*ii*) Measurement of Δf enables T to be calculated. Δf is thus a *thermometric property* which can be used to measure the temperature of a thermonuclear gas plasma when there is present a gas which is not totally ionized.

*15.3 The Doppler Effect in Light

Suppose an observer approaches a stationary source.

Then

$$f_O = f_s\left(1 + \frac{v_O}{c}\right)$$

If the source approaches a stationary observer

$$f_O = f_s\left(\frac{1}{1 - (v_s/c)}\right)$$

$$= f_s\left[1 + \left(\frac{v_s}{c}\right) + \left(\frac{v_s}{c}\right)^2 + \cdots\right], \quad \text{etc.}$$

When $v_s < c$, then in either case,

$$f_O \approx f_s\left(1 + \frac{v}{c}\right)$$

where $\boldsymbol{v}$ is their *relative* velocity.

We conclude

(*a*) when v is small, there is no distinction between source and observer movement, but

(*b*) when v is not negligible compared to c, we *seem* to have a method of distinguishing source movement from observer movement.

Explanation

(1) For *mechanical waves* (which require a material medium) v_s and v_O can be measured *relative to the medium, and so can c.*

(2) For *electromagnetic waves* no such medium exists, and *c is the same* (in a vacuum) *for any observer* whether he and/or the source is moving. (*This is a basic postulate of the* **special theory of relativity**.)

The analysis above is not applicable, and one should use

$$f_O = f_s\frac{1 + (v_R/c)}{\sqrt{1 - (v/c)^2}}$$

in which v_R is the radial component of the velocity of the source as determined by the observer, and v is the magnitude of the velocity. v_R is taken to be positive when the source–observer separation is decreasing.

When $(v/c)^2 \ll 1$, one is justified in using

$$f_O \approx f_s\left(1 + \frac{v_R}{c}\right)$$

for electromagnetic waves.

*15.4 Shock Waves

Suppose the observer is at rest, but that $v_s > c$. In any time interval the source moves further than the wavefront, and a cone (the **Mach cone**) encloses all the waves emitted by the source (fig. 15.5 overleaf). A, B, C, etc., are successive source positions.

The envelope of the secondary wavelets is a cone of semi-vertical angle ϕ, where

$$\sin\phi = \frac{\text{DP}}{\text{DE}} = \frac{cT}{v_sT} = \frac{c}{v_s}$$

$$\operatorname{cosec}\phi = \frac{v_s}{c}$$

The ratio (v_s/c) is called the **Mach number** (Ma).

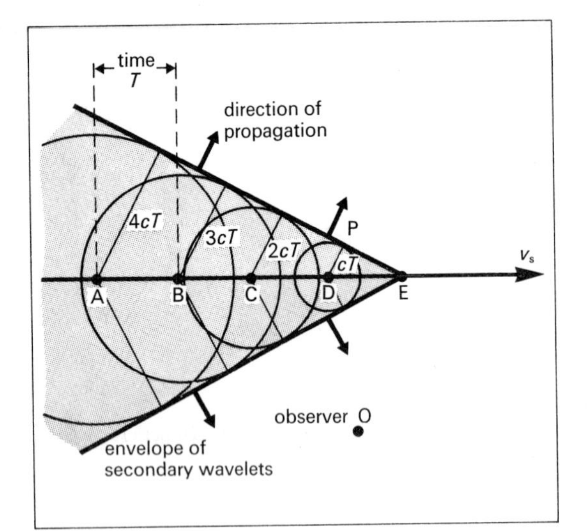

Fig. 15.5. Shock wave constructed by the Huygens method.

For a sound wave, an observer at O hears

(1) a shock wave from the large pressure difference across the conical wavefront (which has a *high energy intensity*), then

(2) later whistling sounds emitted from distant parts.

Shock Waves in Practice

(a) Sound

The waves (from e.g. a bullet) can be photographed because the large pressure difference causes a significant change of refractive index.

(b) Water

The speed of a fast boat can be calculated by photographing the conical envelope of the bow wave pattern, and measuring ϕ.

(c) Light

When an electrically charged particle moves through a medium at high speed $v_s > c_n$, the speed of light *in that medium*, a bluish light (**Čerenkov radiation**) is emitted. *Čerenkov* radiation can be used to measure the speed of the particle.

IV Structure and Mechanical Properties of Matter

16 Structure of Matter

16.1 Atoms

An **atom** is the smallest neutral particle that represents an *element.*

We can visualize atoms as being made up of electrons, protons and neutrons, some of whose properties are given below. (In fact there exist a very large number of other sub-atomic particles whose behaviour does not concern us here.)

Particle	*Mass*	*Charge*
electron	9.1×10^{-31} kg	-1.6×10^{-19} C
proton	1.7×10^{-27} kg	$+1.6 \times 10^{-19}$ C
neutron	1.7×10^{-27} kg	0

$$\frac{m_p}{m_e} \approx 1840$$

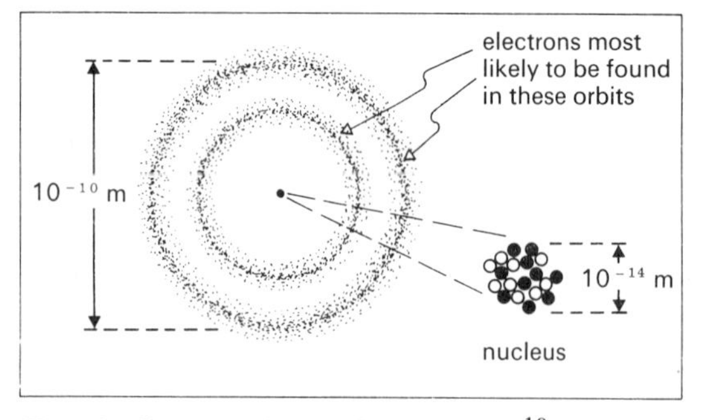

Fig. 16.1 One model of the fluorine atom $^{19}_{9}$F *(not drawn to scale).*

Atomic diameters lie in the range 1 to 3×10^{-10} m.

Nuclear diameters lie in the range 1 to 7×10^{-15} m.

The number of electrons in the outermost **shell** (group of imagined orbits) determines the chemical behaviour of an atom. An **ion** is any charged particle, but usually it represents an atom or group of atoms that has lost or gained one or more electrons from its outermost shell.

Most *gases* are ionized to a small extent. Their ionization can be increased by

(*a*) heating,

(*b*) fast-moving charged particles (such as those emitted by radioactive substances),

(*c*) exposure to electromagnetic radiation of high energy (e.g. X-rays and γ-rays).

Many *solutions* are ionized, particularly those of acids, bases and salts. The fact that water has a high electric permittivity (p. 338) helps to bring this about.

Several *solid* materials are ionized (p. 122).

16.2 Molecules

A molecule of a *substance* is the smallest particle of it which can exist under normal conditions. For example

(*a*) an argon molecule is an argon atom,

(*b*) an oxygen molecule is a pair of oxygen atoms,

(*c*) fig. 16.2 shows a model of a water molecule in the gaseous phase.

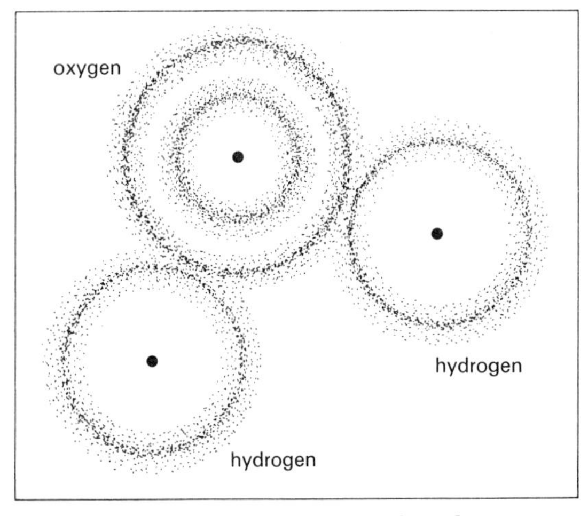

Fig. 16.2. A model of a water molecule H_2O.

For several purposes we can imagine a molecule *as though* it were *spherical*, as we will do in this section. On occasions we must be more exact: on p. 192 and elsewhere we treat diatomic molecules as though they were *dumb-bell-shaped*.

The idea of a molecule frequently breaks down. Some solids consist of ions and electrons, and it may not be possible to attribute a particular ion to a particular molecule (p. 122). The word *molecule* is frequently used loosely when referring to these ions.

16.3 Intermolecular Forces and Energies

Forces

The four *interactions* of nature, which together are the sole cause of all the forces that exist, are discussed on p. 3. It should be stressed that the **electromagnetic interaction**, and *not* the gravitational interaction, is responsible for *intermolecular* forces. (The gravitational forces are too weak to be significant on this scale.) A full understanding can only be obtained from wave mechanics.

Fig. 16.3 contains a great deal of information and should be studied carefully.

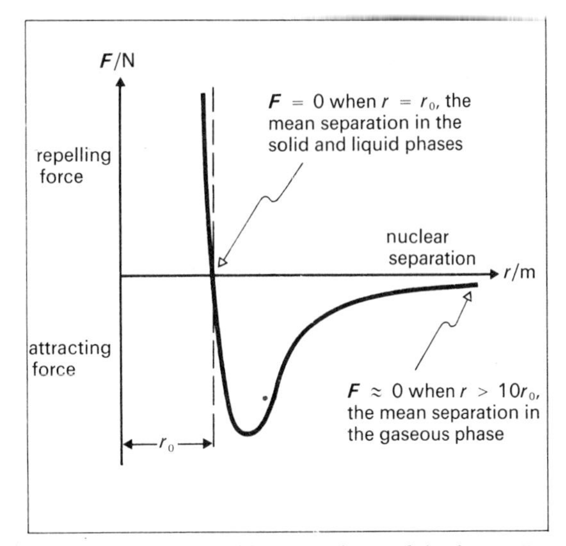

Fig. 16.3. Variation with separation r of the forces F which molecules exert on each other.

A simple explanation for the intermolecular force

Imagine the molecules of fig. 16.4 to be

(1) Distant: they are electrically neutral, and there are no electromagnetic forces.

(2) Closer: molecular distortion results in attraction.

(3) Closer still: interpenetration of electron shells introduces a repulsive force. The resultant force becomes zero (equilibrium).

(4) Too close: there is a resultant repulsive force.

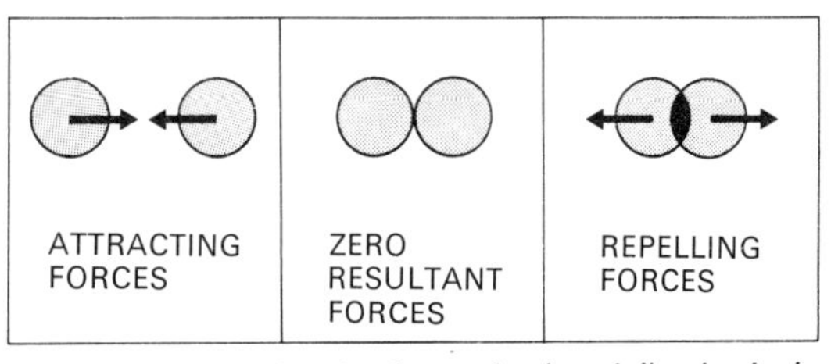

Fig. 16.4. Intermolecular forces for (model) spherical molecules.

Any displacement from the equilibrium position results in attraction or repulsion and thus a continuing *vibration* of amplitude about 10^{-11} m. The vibration is accompanied by a periodic interchange of k.e. and p.e., as each molecule is a simple harmonic oscillator.

Energies

We can classify the phases of matter according to the relative sizes of **the intermolecular attraction energy** ε (fig. 16.5) and the mean thermal energy of the molecules (which is a function of the temperature T):

(a) When the thermal energy is a minimum, then r

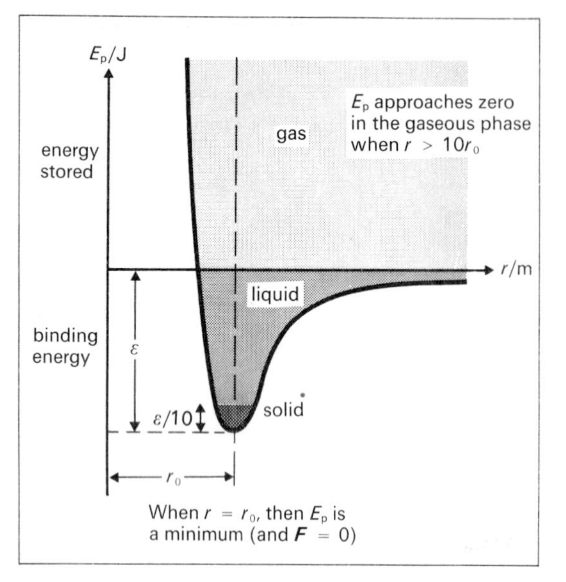

Fig. 16.5. Variation with separation r of the potential energy E_p of a system consisting of a pair of molecules.

is very close to r_0. As the thermal energy increases, then so long as it is less than about $\varepsilon/10$, the molecular pattern remains *ordered* and we have a **solid**. The thermal motion is restricted to a *vibrational* motion about a *fixed* mean position.

(*b*) As the thermal energy is increased to about $\varepsilon/10$, some bonds are broken and the molecules escape from their fixed positions, as they can now exchange places with their neighbours. Their motion is no longer entirely vibrational (as clusters of molecules migrate) and melting occurs: we have a **liquid**.

(*c*) When the thermal energy exceeds ε, a molecule has enough energy to leave a parent cluster, and escapes with some translational kinetic energy; we now have a **gas**.

The way in which we evaluate ε is discussed in paragraph **16.5.**

* The Effective Range of a Force

Consider a particle interacting with other identical particles by which it is surrounded. If we exclude short distances the number of interacting particles between the range r and $r + \delta r$ is $\propto r^2$.

Suppose that the potential energy of the interaction falls off as $k \cdot r^{-n}$ (again excluding very small distances), where k is a constant.

(*a*) If $n > 2$, then the total interaction p.e. between the particle considered and those lying between r and $r + \delta r$ is $\propto r^{2-n}$, and this *decreases* as r increases. Since the particle interacts mostly with near neighbours, the force is said to be **short-ranged**.

(*b*) If $n < 2$, then the same argument shows that the p.e. of interaction *increases* as r increases (provided the medium continues). The force is said to be **long-ranged**

Phase	*Macroscopic behaviour*	*Molecular order and behaviour*	*Mean molecular separation*	*Number per* m^3	*Volume of molecules relative to that of substance*	*Mean number of 'collisions' per second at s.t.p.*	*Thermal energy E of the molecules*
gas near s.t.p.	has variable shape and volume	totally random translational motion	$10r_0$	10^{25}	10^{-3}	10^{10}	$E > \varepsilon$
liquid	has variable shape but volume is nearly constant	molecules vibrate within ordered clusters which have some restricted translational motion	r_0	10^{28}	1	10^{14}	$\varepsilon > E > \dfrac{\varepsilon}{10}$
solid	both shape and volume are nearly constant	molecules vibrate about a fixed site in the lattice	r_0	10^{28}	1	10^{14}	$E < \dfrac{\varepsilon}{10}$

Examples

(a) When we are dealing with short-ranged forces the scale is relatively unimportant: thus, the specific latent heat of vaporization is the same for two samples of the same liquid, even if they have different volumes. Since l is determined by the intermolecular forces, it follows that they must be short-ranged.

(*b*) The properties of material bound by long-ranged forces depend radically on scale, and this is illustrated by the behaviour of stars and planets of different size.

16.4 Phases of Matter

There are three principal phases of matter whose characteristics are compared in the table opposite:

(*a*) The Gaseous Phase

Intermolecular forces are short-range forces (they are only effective at a range of a few molecular diameters). Since the average molecular separation is about $10r_0$, the molecules only interact significantly on impact or very close approach. Different molecules at any instant have different speeds, according to the *Maxwellian distribution* (p. 179), but a typical average random speed is $10^2\ \mathrm{m\,s^{-1}}$. (A quantitative account of the dynamics of gas molecules is given on p. 180.)

A **plasma** is made by heating gas atoms and molecules to a high temperature. At a temperature of (say) 10^4 K, some of the originally neutral particles will have acquired enough thermal energy (about 10^{-18} J) to cause them to ionize. The plasma consists of the resulting mixture of neutral particles, positive ions and negative ions (many of which are electrons).

(*b*) The Liquid Phase

The molecules *vibrate* about a *moving* centre of vibration, because thermal motion causes a constant change of relative positions. There are no permanent linkages, and although there is *no* **long-range order**, the molecules gather in clusters.

(*c*) The Solid Phase

A solid is characterized by a fixed shape (which means that it has a modulus of rigidity) as it can *permanently* resist a shear stress (p. 130). This is because each molecule or ion has a permanent linkage with its neighbours, and so retains a fixed position in the overall pattern (there *is* long-range order).

The molecular or ionic pattern can be one of two types:

(*i*) **Ordered patterns** exist in crystalline substances, and have a periodic repetition of the **unit cell** (specified group of particles) in three dimensions.

(*ii*) **Disordered** arrangements exist in supercooled liquids, such as glass. They are less stable, and therefore rare.

The ordered pattern has the lower potential energy, and so substances tend to be crystalline (since this is more stable).

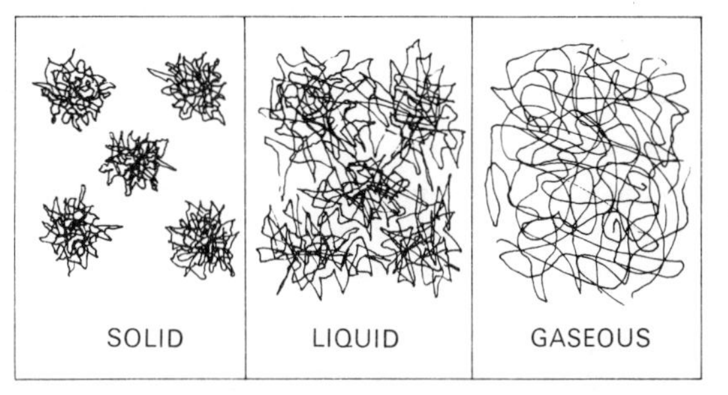

Fig. 16.6 Models of particle behaviour in solid, liquid and gaseous phases.

Fig. 16.6 indicates the behaviour of particles in the different phases, as predicted from data fed into a computer.

16.5 Phase Changes

A discussion in terms of the intermolecular attraction energy ε is given on p. 203, while the macroscopic view leads to a definition of specific latent heat of transformation.

Condensation

In a molecule makes an *elastic* collision with a surface, it is bound to escape. If the collision is *inelastic*, the molecule may lose sufficient translational k.e. to prevent subsequent escape. The probability that this will happen increases as the temperature decreases (since the molecular k.e. is then reduced), and molecules will then be attracted into a vibratory motion on the surface.

Solidification

As a liquid is cooled, the molecular vibrational energy is reduced until the molecules can no longer change relative positions. Removal of energy (the latent heat) has converted the disordered pattern into an ordered one, and the potential energy of the new structure is smaller.

When the solid is heated the breaking up of the ordered lattice can be detected near the melting point by a sharp change of slope in the curve which plots density against temperature. This indicates an increase in the number of sites in the lattice which are empty (these are called **vacancies**).

Sublimation

This is the process that occurs when, on gaining heat, the particles of a solid evaporate directly without passing through the intermediate liquid phase.

The Estimation of ε

Consider a system in which each particle makes effective contact with n nearest neighbours. n is called the **coordination number**.

(*a*) ***Sublimation.*** Assume the molecule in a solid interacts with nearest neighbours only. Then the binding energy per molecule is $\frac{1}{2}n\varepsilon$. The $\frac{1}{2}$ factor ensures that we do not count each bond twice. The molar latent heat of sublimation

$$L_{s,m} = \tfrac{1}{2}N_A n\varepsilon$$

The molar latent heat of vaporization $L_{v,m}$ would be slightly less.

(*b*) ***Surface tension.*** A similar argument shows that surface energy should be about $\frac{1}{2}L_v$, so

$$(\sigma A) \approx \tfrac{1}{2}(\tfrac{1}{2}N\varepsilon)$$

where N molecules are contained in a surface area $\boldsymbol{A}$.

The value of ε can be estimated from these expressions by assuming values for n and (N/A). Its size is determined by the nature of the interparticle bond. Typical values might be 10^{-19} J for an ionic or covalent bond, but only 10^{-22} J for a *van der Waals* bond.

* 16.6 The Nature of Bonding in Matter

In this paragraph we discuss the types of binding interactions which cause atoms and molecules to attract at relatively large distances, and which join atoms together to form molecules and other structures.

(*a*) Strong Bonds

These bonds result from the exchange or sharing of electrons, and occur because the bonded system has a lower p.e. than the separate atoms.

(*i*) In the **covalent bond** *electrons are shared* (at least in part) between different atoms. The bond energy is electrostatic in origin. In diamond (fig. 16.7) the individual carbon atoms are linked by covalent bonds to form a rigid crystal lattice.

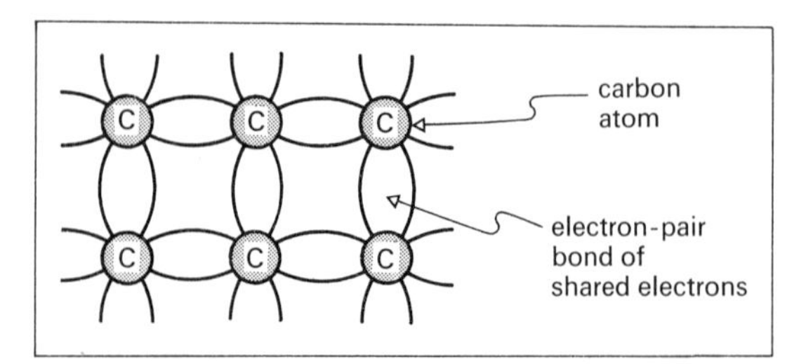

Fig. 16.7. Representation in two dimensions of the covalent bonds between carbon atoms in a diamond crystal.

(*ii*) In the **ionic bond** *electrons are transferred* from one neutral atom to another, and the resulting positive and negative ions attract each other by *Coulomb* forces. In sodium chloride (fig. 16.8) Na^+ and Cl^- ions are arranged so that the ions of each type form a cubic pattern.

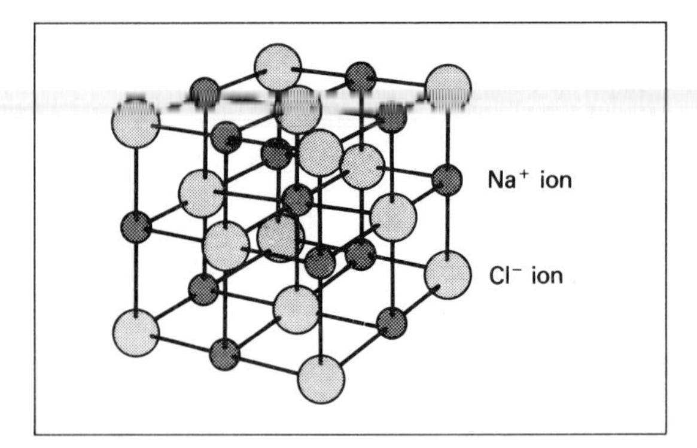

Fig. 16.8 The cubic lattice of Na^+ *and* Cl^- *ions in an* NaCl *crystal.*

In some substances (such as dry HCl) the bonds are *partly ionic* and *partly covalent* (the electron largely belongs to one 'ion', but is to some extent shared by both).

(*iii*) **Metal bonds** and **π-bonds** occur where electrons have greater freedom of movement, i.e. are less obviously tied to their parent atoms than they are in (*i*) and (*ii*). The lower energy which the electrons then possess enables them to contribute to binding over a larger region of the structure.

Examples. (1) The special bonding in *metals* (fig. 16.9*a*) results from the extended orbits of the conduction electrons.

(2) The stability of the *benzene molecule* (fig. 16.9*b*) comes from the extended orbits of the π-bond electrons.

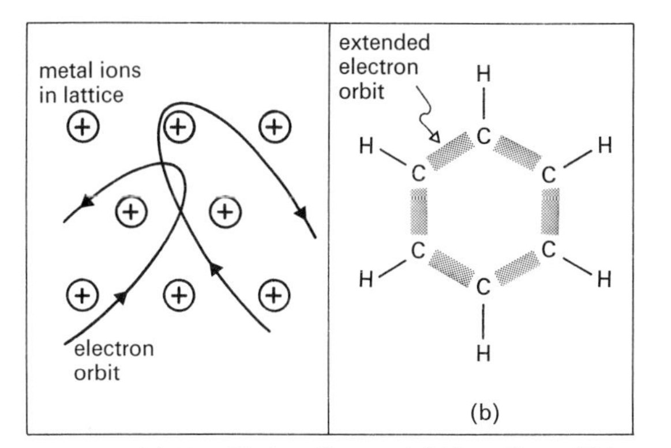

Fig. 16.9 (a) Metallic bonding and (b) bonding within the benzene molecule.

(*b*) Weak Bonds

These bonds result from the redistribution of electrons *within a molecule*, and enable it to interact with other molecules. The molecules retain their chemical identity.

(*i*) Van der Waals Forces

These are weak attractive forces acting over short distances outside a molecule and enable it to attract both polar and non-polar neighburs. The forces which link the chains of polymers such as polyethylene are of this type.

There are three effects which produce attraction.

(1) **Orientation effect**. Fig. 16.10*a* shows how *two permanent* electric dipoles will attract one another if their relative orientations are favourable. In practice, orientations do produce an attractive force, but its size decreases with increasing temperature as the molecular distribution becomes more random.

(2) **Induction effect**. *One* molecule which has a *permanent* electric dipole moment will **polarize** (distort) a neighbouring molecule, causing it to have an *induced* anti-parallel dipole moment. The attractive force which results is independent of temperature.

(3) **London effect**. This effect causes an attraction between a pair of molecules *both* of which are *non-polar*. These forces, which have been called **dispersion forces**, are for most gases more important than those resulting from the orientation and induction effects.

In fig. 16.10*b* molecule A sets up the electric field typical of a dipole because of a *random* displacement of its electron cloud. (All molecules have a fluctuating dipole moment of this kind.) Molecule B, finding itself in the field set up by A, acquires an *induced* moment (as in effect (2)).

Notes.

(*a*) The *pair* of dipole moments always cause *attraction*, whatever the direction of A's random moment.

(*b*) The average moment of either molecule is zero over a time which is long relative to the orbital period of an electron.

(*ii*) Hydrogen Bonding

Hydrogen bonds are basically electrostatic interactions between molecules whose surface regions have different electron densities. These bonds are particularly important

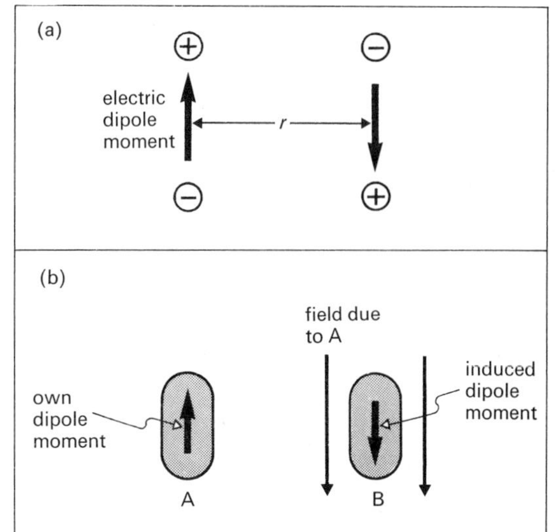

Fig. 16.10. Origin of van der Waals forces: (a) illustrating attraction between permanent electric dipoles and (b) illustrating attraction between temporary electric dipoles.

in determining the properties of water (fig. 16.11) and the structure of biological substances such as DNA (deoxy-ribonucleic acid).

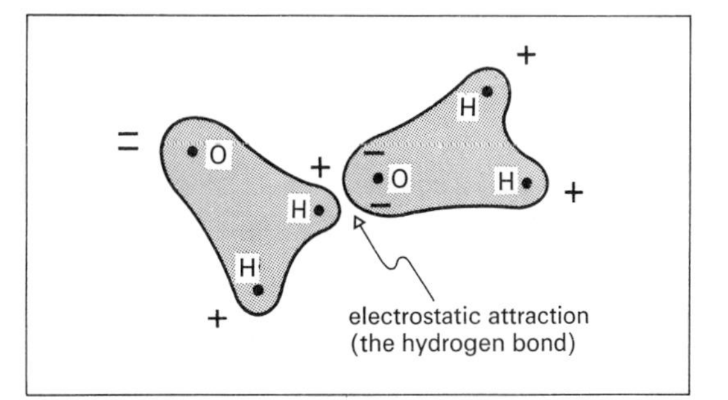

Fig. 16.11. Hydrogen bonding between water molecules.

We would expect the value of the specific heat of sublimation to be closely related to the strength of bond in the substance concerned. Experiment shows that it is highest for ionic and covalent solids, is very high for most metals, but comparatively small for solids bound by *van der Waals* forces.

Repulsive forces

The interactions discussed so far in this paragraph all lead to attractive forces, and these *attractive* forces draw atoms and molecules together. Matter does not contract indefinitely because when there is interpenetration of different atoms' electron clouds they exert very strong *repulsive* forces on each other.

16.7 Crystals and Their Structures

The Crystalline State

The vital distinction between a crystal and an amorphous solid lies in the former's long-range order, that is its *regularly repeated* arrangement of atoms, ions or molecules. This essential inner regularity is partly reflected in the external habit, and although the shapes of a particular crystal species are not exactly reproduced from crystal to crystal, the *angle* between specified faces is.

A crystal can be made from single atoms (as in diamond), ions (as in NaCl), or molecules (as in ice). The **unit cell** of a particular lattice is the smallest and simplest unit from which the three-dimensional periodic pattern can be built. It may contain several atoms or molecules. The choice of the unit cell is not unique, but is usually made so that its symmetry corresponds to that of the crystal. Its importance to X-ray crystallography is discussed on p. 287. The physical properties of solids are determined by the nature of the unit cells and their geometrical arrangements. The degree of imperfection in the crystal is also an important factor.

Solid specimens (e.g. most metals) are usually **polycrystalline**, consisting of numbers of small crystals or *grains* of arbitrary shape packed close together, and separated by grain boundaries. These boundaries are regions having more energy than the grain interiors.

Some Crystal Structures

The structure of a crystal is determined by

(*i*) the kind of bond(s) between its particles, and
(*ii*) the size and shape of those particles.

Thus in metals the non-directional nature of the π-bonds results in a close-packed structure, whereas a more open structure is found in covalent solids, since the bonding is directional.

(*a*) ***Face-centred cubic*** packing is shown in fig. 16.8. The cube has a particle at each corner, and in addition one at the centre of each face. (This pattern can be made in fig. 16.12 by positioning the top layer such that each of its spheres lies above a *gap* in the bottom layer.)
Examples: Al and Cu.

(*b*) ***Hexagonal close-packing*** is shown in fig. 16.12. The structure is built up from layers of hexagons, with particles in alternate layers in identical positions.

Examples: Mg and Zn.

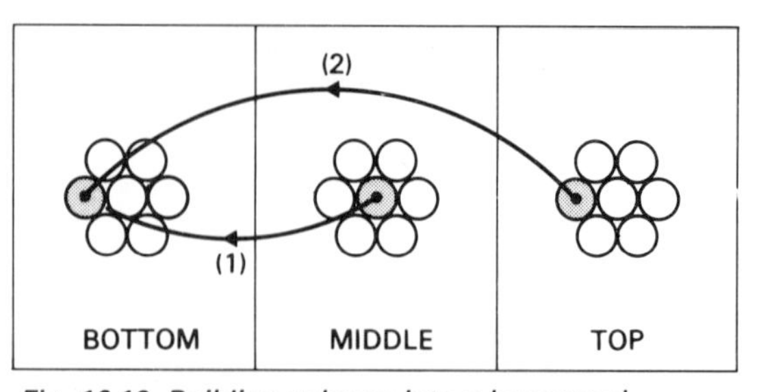

Fig. 16.12. Building spheres into a hexagonal close-packed structure: (1) place middle layer to superpose dot on dot, and (2) place top layer so as to superpose bottom layer exactly.

(*c*) ***Body-centred cubic*** packing has a particle at each corner of a cube, and one at the centre.
Examples: Fe and Na.

(*d*) ***Tetrahedral structures*** have one particle at each of the corners of a regular tetrahedron, and one at the centre. The structure is exemplified by the covalent bonding of diamond.

In each case a structure is aiming at maximum stability by minimizing the crystal's energy density. This is done by

(*i*) keeping the crystal electrically neutral,
(*ii*) preserving the directional nature of covalent bonds,
(*iii*) minimizing the strong *Coulomb* repulsion between like ions and, as far as is consistent with these requirements, by
(*iv*) packing the ions as densely as possible.

* The Energies of Crystals

A *single* crystal is an ideal solid in that its particles are in a well-defined environment, and this enables the energy to be computed relatively simply.

(*a*) ***Ionic crystals.*** The interionic forces are (*i*) long-range electrostatic (*Coulomb*) repulsion and attraction, and (*ii*) the shorter-ranged repulsive forces between the electron clouds. If the charges are $\pm e$, and a structure contains N ions, the total p.e. is given by

$$E_p = -\left(\frac{Ne^2}{4\pi\varepsilon_0}\right)\frac{A}{x} + \frac{B}{x^9}$$

in which x is the side of the unit cell, and A and B are constants. (A is called the **Madelung constant**, a characteristic of a particular ionic structure.) The binding energy is dominated by the *Coulomb* term, and is approximately equal to it.

(*b*) ***Molecular crystals***. Fig. 16.5 shows, for a pair of molecules, the p.e. curve resulting from the long-range attractive forces and the shorter-ranged repulsive forces.

The total energy E_p of the structure is the sum of the energies of each molecular pair. It can be expressed in the form

$$E_p = -\frac{A}{r^6} + \frac{B}{r^{12}}$$

where A and B are constants. The equilibrium value of E_p is a minimum, and so the equilibrium separation r_0 of the particles can be found by putting

$$\frac{dE_p}{dr} = 0$$

This, in turn, enables E_p to be calculated.

16.8 Brownian Motion

The concept of pressure (developed for an ideal gas in paragraph **24.5**) breaks down for very small areas because of the randomness of the molecular bombardment. This is illustrated in the **Brownian movement**.

We deduce that the smoke *particles* of fig. 16.13 (typically about 10^{-6} m across) are being bombarded by the *molecules* of the air. We cannot see the air molecules, so they must be very small. To produce the observed effects they must, if small, be moving at high speeds.

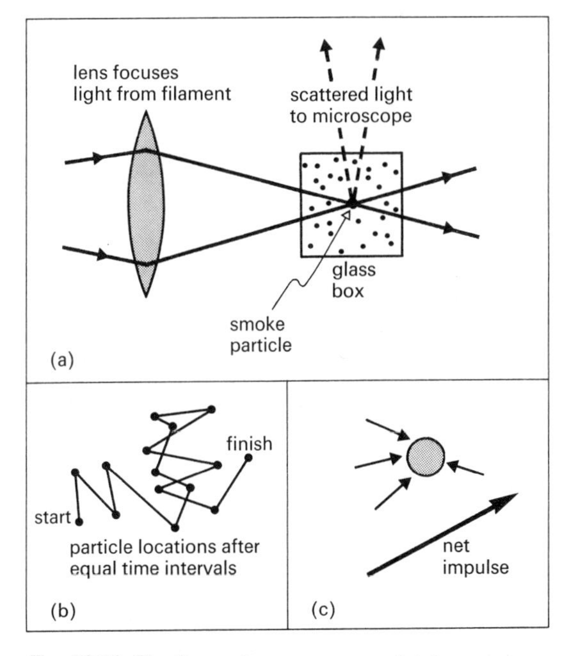

Fig. 16.13. The Brownian movement: (a) the experimental arrangement, (b) the observation, and (c) the explanation (not to scale).

Orders of Magnitude

(*i*) A typical smoke particle might move about 10^{-6} m (its own diameter) during one second.

(*ii*) The size and direction of the resultant force on the particle changes significantly during a time interval of about 10^{-10} s.

According to the principle of equipartition of energy (p. 192)

$$\tfrac{1}{2}m\overline{c^2}\text{ (molecule)} = \tfrac{1}{2}M\overline{C^2}\text{ (particle)}$$

so if $M \gg m$, it follows that $\overline{c^2} \gg \overline{C^2}$.

Because these are *mean* square speeds the speed of a particular molecule at a particular instant *may* be very small.

Notes.

(*a*) The study of the path taken by a particular particle is called the **random (drunkard's) walk problem**, and was carried out by *Einstein*. It leads to a rough method of measuring the *Avogadro constant* N_A.

(*b*) The average *displacement* of a particle starting from an origin on successive journeys is zero, even though the average *distance* travelled from the origin is not zero. This is because all directions of travel are equally likely.

(*c*) On a macroscopic scale *Brownian* motion explains

(1) the quivering movement of the suspension of a delicate mirror galvanometer, which sets a limit on the attainable sensitivity,

(2) the random noise of an amplifier (the **Johnson noise**) caused by the varying voltages which accompany random electron movement within a metal.

Both effects are reduced by lowering the temperature.

(*d*) We regard *Brownian* movement as indirect *experimental confirmation* that fluids consist of small particles moving at high speeds.

16.9 Measurement of the Avogadro Constant N_A, and Atomic and Molecular Diameters

(a) The Avogadro Constant N_A

(i) X-ray diffraction

The ions or atoms in crystals are arranged in regular planes separated by equal distances. This causes them to diffract X-rays in the same way that a ruled grating diffracts visible light (p. 280).

The wavelength of X-rays ($\sim 10^{-10}$ m) can be measured by a separate experiment in which they are diffracted by a ruled grating of known spacing.

The spacing of the planes in the crystal is calculated from a knowledge of the angles of diffraction and the wavelength of the X-rays, and from this N_A can be deduced simply.

(ii) Radioactivity

Radium emits α-particles whose rate of emission can be measured by a scintillation counter. The volume of helium that collects in a given time can be measured at a known temperature and pressure. This enables the mass of the helium to be calculated, and so the mass m_α of an α-particle can be found.

Since the molar mass of helium is 4.0026 $\times 10^{-3}$ kg mol^{-1}, N_A is calculated from

$$N_A[\text{mol}^{-1}] = \frac{4.0026 \times 10^{-3}\ \text{kg mol}^{-1}}{m_\alpha\ [\text{kg}]}$$

The value obtained is only approximate.

(iii) Electrolysis

The **Faraday constant** F is numerically equal to the charge required to liberate one mole of singly charged ions in electrolysis, and is measured in coulombs mole^{-1}. It's value (p. 10) can be calculated from measurements on the electrical deposition of copper. Then N_A is calculated from

$$N_A = \frac{F}{e}$$

in which the electronic charge e can be measured separately by *Millikan's oil drop experiment* (p. 440). Historically, this experiment was used to measure e (rather than N_A) many years before *Millikan's* oil drop experiment. Nowadays, the diffraction method for N_A, combined with the electrolysis determination of F, gives a value of e which is many orders of magnitude more precise than *Millikan's* value.

(iv) Brownian Motion

The *Boltzmann constant* k can be determined from a study of *Brownian motion*. We use it to find N_A from

$$N_A = \frac{R}{k}$$

where R is the universal molar gas constant (p. 178).

(b) Atomic Diameters

These can be found by direct measurement using X-rays (above). The values obtained agree with the orders of magnitude predicted from a knowledge of N_A combined with the densities of the substances being used.

(c) Molecular Diameters

Approximate Methods

(*i*) Let a drop of a fatty acid (such as oleic acid) whose volume is known fall gently onto a water surface dusted with lycopodium powder. The drop spreads out into a layer, the extent of which is shown by the powder. The film thickness t cannot be less than one molecular diameter, where

$$t = \frac{\text{volume of drop}}{\text{area of layer}}$$

In practice the acid is usually dissolved in alcohol, which evaporates or dissolves in the water.

(*ii*) When *mica* is cleaved the steps of cleavage can be measured by an interferometer. They are found always to be multiples of a particular distance.

(*iii*) The thickness of a *soap film* can be measured, and shown to be as small as 10^{-8} m.

More Exact Methods

(*i*) The molecular diameter r_0 can be estimated from measurements on *deviations from the gas laws*, and from critical data. The constant b in *van der Waals's equation* (p. 212) can be written

$$b = \tfrac{2}{3}\pi N_A r_0^3,$$

so that if b is measured by experiment, then r_0 can be evaluated.

(*ii*) The *kinetic theory* of gases relates measurements of viscosity, diffusion and thermal conductivity (the *transport phenomena* (p. 184)), from which it is possible to calculate r_0.

(*iii*) r_0 can also be estimated from the *dielectric properties* of liquids whose molecules are permanent electric dipoles.

16.10 Kinetic Theory

Kinetic theory is that branch of Physics based on the assumption that matter consists of molecules. It treats the following topics, to which reference should be made for details:

(a) Change of Phase
Evaporation (p. 205)
Vapour Pressure (p. 205)
Latent Heat (p. 203)

(b) Pressure
Liquid and Solid Pressure (p. 68)
Gas Pressure, and deductions from

$$p = \frac{1}{3}\rho \overline{c^2}$$ (p. 180)

(c) Change of Shape and Volume
Elasticity (p. 128)
Hooke's Law (p. 129)

(d) Transport Phenomena (p. 182)
Diffusion—transport of mass (p. 182)
Viscosity—transport of momentum (p. 184)
Electrical conduction—transport of electric charge (p. 349)
Thermal conduction—transport of heat (p. 215)

(e) Thermal Expansion (p. 169)

(f) Surface Energy and Surface Tension (p. 136)

In each example the predictions of the theory are in satisfactory agreement with experiment. Sometimes the agreement is quantitative. Thus the assumptions of the kinetic theory are indirectly confirmed.

17 Elasticity

17.1 Elasticity, Stress and Strain

When a pair of opposed external forces is applied to a body, it changes the relative positions of the body's molecules. This calls restoring forces into play, and accounts for the **elasticity** of a body—its ability to return to its original form after the distorting forces have been removed.

Fluids possess only **volume elasticity**.

Solids possess in addition **rigidity**, since they can also resist a change of shape.

Solids also possess **tensile elasticity**, ability to resist longitudinal elongation and compression. (This property is a consequence of a solid having volume and rigidity elasticities.)

***Stress** is a measure of the cause of a deformation and is defined by*

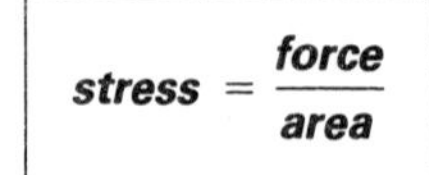

$$\textbf{stress} = \frac{\textbf{force}}{\textbf{area}}$$

There are three kinds of stress, each of which has dimensions $[\mathrm{ML^{-1}T^{-2}}]$, and unit Pa ($\mathrm{N\,m^{-2}}$).

***Strain** is a measure of the extent of deformation, and is defined by*

$$\textbf{strain} = \frac{\textbf{change of dimension}}{\textbf{original dimension}}$$

There are also three kinds of strain each of which is a pure number, with neither dimensions nor a unit.

17.2 The Elastic Moduli and Hooke's Law

***An elastic modulus** is defined by the equation*

$$\textbf{modulus} = \frac{\textbf{stress}}{\textbf{strain}}$$

$$\text{modulus [Pa]} = \frac{\text{stress [Pa]}}{\text{strain [no unit]}}$$

Three kinds of moduli are defined to correspond to the three kinds of stress and strain: each has dimensions $[\mathrm{ML^{-1}T^{-2}}]$.

Name of elastic modulus	*Nature of stress*	*Nature of deformation*
bulk modulus K	change of pressure	change of size but not shape
shear modulus G	$\dfrac{\text{tangential force}}{\text{area}}$	change of shape but not size
Young's modulus E	$\left(\dfrac{\text{tensile or compressive force}}{\text{area}}\right)$	change (usually) of both size and shape

These moduli are discussed in the following paragraphs, but for convenience a table of comparative values is given here.

Substance	*Modulus*		
	Bulk modulus K/Pa	*Shear modulus* G/Pa	*Young's modulus* E/Pa
typical metal	1×10^{11}	4×10^{10}	1×10^{11}
water	2×10^{9}	—	—
gas at pressure 1×10^{5} Pa	1×10^{5}	—	—

Hooke's Law

Hooke's Law was first enunciated for the deformation of a spring. It is a macroscopic (measurable) result of the fact that the graph of fig. 16.3 is a straight line in the region r close to r_0.

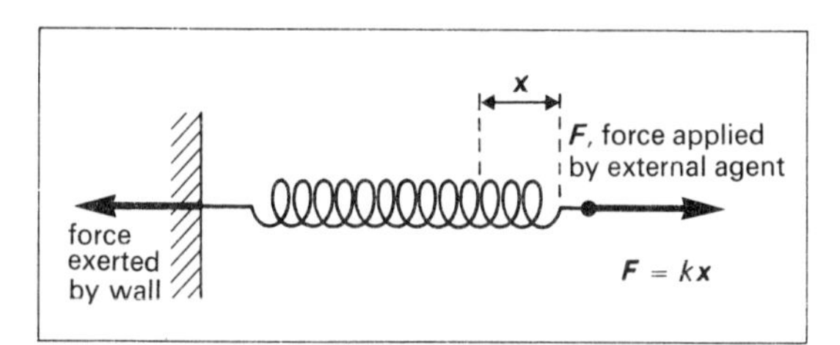

Fig. 17.1 Hooke's Law for a spring.

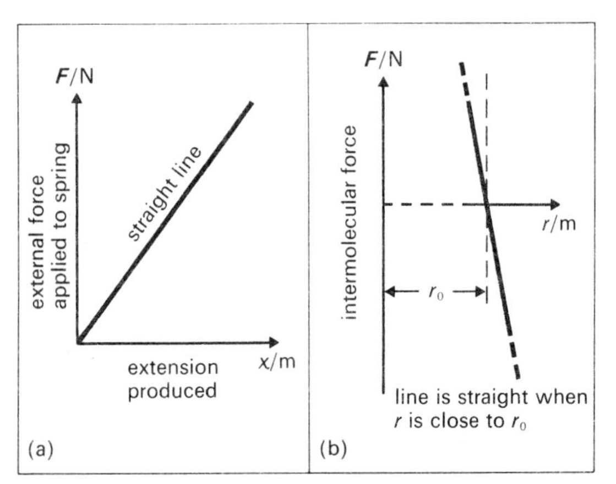

Fig. 17.2. (a) Macroscopic representation, and (b) microscopic cause, of Hooke's Law.

We now extend **Hooke's Law** to elastic deformation in other situations by writing

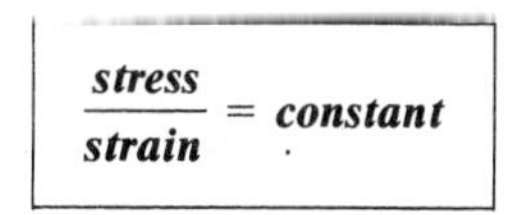

$$\frac{\textbf{\textit{stress}}}{\textbf{\textit{strain}}} = \textbf{\textit{constant}}$$

provided the deformation is small.

Thus under some conditions we can treat the elastic moduli as constants: when *Hooke's Law* does not apply they remain as useful concepts which can still be used (as average values).

It is important to note that the behaviour described by *Hooke's Law* makes it possible for deformed bodies to vibrate with s.h.m.

17.3 Young's Modulus E

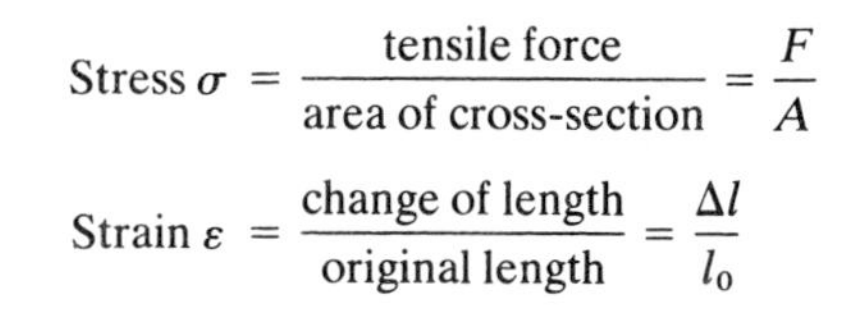

$$\text{Stress } \sigma = \frac{\text{tensile force}}{\text{area of cross-section}} = \frac{F}{A}$$

$$\text{Strain } \varepsilon = \frac{\text{change of length}}{\text{original length}} = \frac{\Delta l}{l_0}$$

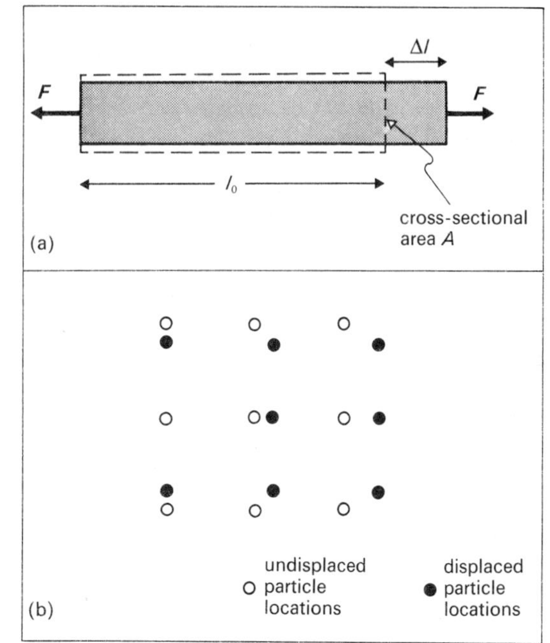

Fig. 17.3 Illustration of Young's modulus on (a) a macroscopic, and (b) a microscopic scale (both exaggerated).

Young's modulus E *is defined by the equation*

$$E = \frac{\sigma}{\varepsilon} = \frac{F/A}{\Delta l/l_0}$$

17.4 Bulk Modulus K

$$\text{Stress}\ \sigma = \frac{\text{change of normal force}}{\text{area}} = \Delta p$$

(the change of hydrostatic fluid pressure)

$$\text{Strain}\ \theta = \frac{\text{change of volume}}{\text{original volume}} = \frac{\Delta V}{V_0}$$

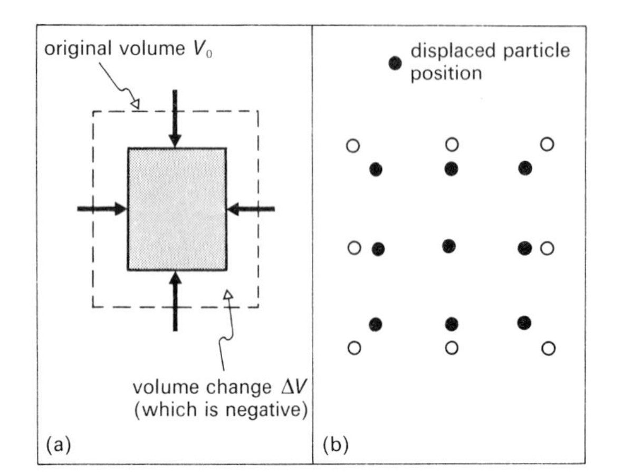

Fig. 17.4. To illustrate bulk modulus on (a) a macroscopic, and (b) a microscopic scale (both exaggerated).

Bulk modulus K ***is defined by the equation***

$$K = \frac{\sigma}{\theta} = \frac{-\Delta p}{\Delta V/V_0}$$

Notes.

(*a*) An increase in p produces a diminution in V. The negative sign is introduced to make K positive.

(*b*) $1/K$ *is called* ***compressibility*** (κ). A material is easily compressed if it has a *small* bulk modulus.

(*c*) K is approximately constant for the condensed phases (*Hooke's Law*).

(*d*) We can *calculate* K for gases, as shown on p. 133.

17.5 Shear Modulus G

Shear modulus is sometimes called **rigidity** modulus, and is possessed only by *solids*, since the shape of a fluid is determined by its container.

$$\text{Stress}\ \tau = \frac{\text{tangential force}}{\text{area over which it is applied}} = \frac{F}{A}$$

$$\text{Strain}\ \gamma = \text{angle of shear}\ \gamma = \frac{\Delta x}{y},$$

and has no dimensions.

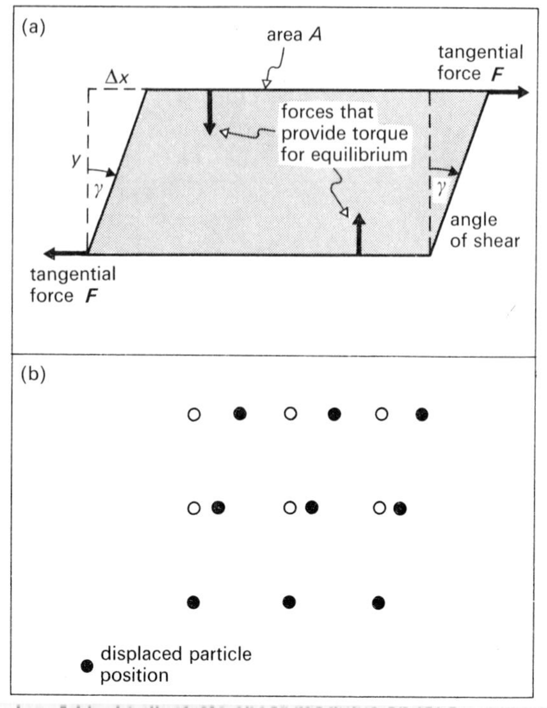

Fig. 17.6. To illustrate shear modulus on (a) a macroscopic, and (b) a microscopic scale (both exaggerated).

Shear modulus G ***is defined by the equation***

$$G = \frac{\tau}{\gamma} = \frac{F/A}{\gamma}$$

$$G[\text{Pa}] = \frac{F[\text{N}]/A[\text{m}^2]}{\gamma[\text{rad}]}$$

(This equation should be compared with the defining equation for the *coefficient of viscosity* (p. 146)).

Examples of Shearing

(*a*) The restoring torque $\boldsymbol{T}$ exerted by a wire twisted through angle θ is given by

$$T = c\theta$$

The **suspension** (or **torsion**) **constant** c is numerically equal to the restoring torque per unit angular displacement. See

(*i*) moving-coil galvanometer (p. 389).

(*ii*) *Boys's* experiment (p. 159).

(*b*) The torsional pendulum (p. 133).

(*c*) The extension of a helical spring involves the *twisting* of segments of the wire.

17.6 Measurement of Young's Modulus

In a measurement of E we might use quantities of the following orders of magnitude:

$\boldsymbol{F} \sim 10^2$ N (**F** is *not* expressed in kgf)
$\boldsymbol{A} \sim 10^{-1}$ mm² (the mean value from several micrometer readings)
$\Delta l \sim 2$ mm
$l_0 \sim 2$ m

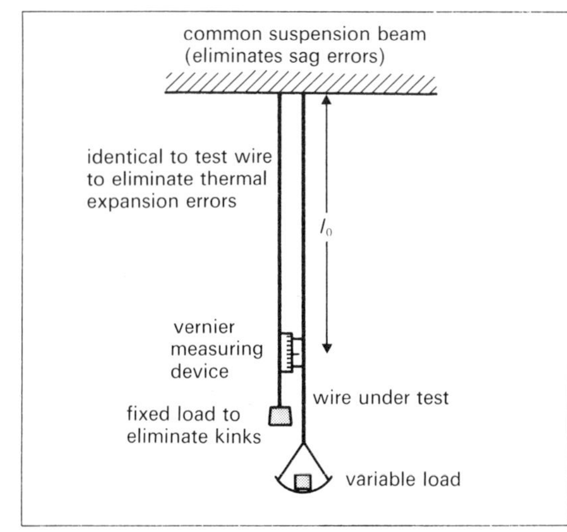

Fig. 17.6. Experimental arrangement for the measurement of E.

Obviously the values of **A** and Δl need especially careful measurement.

Δl should be measured during loading and unloading, using the same time delays, to reduce *hysteresis* (see fig. 17.7, in which the effect is exaggerated).

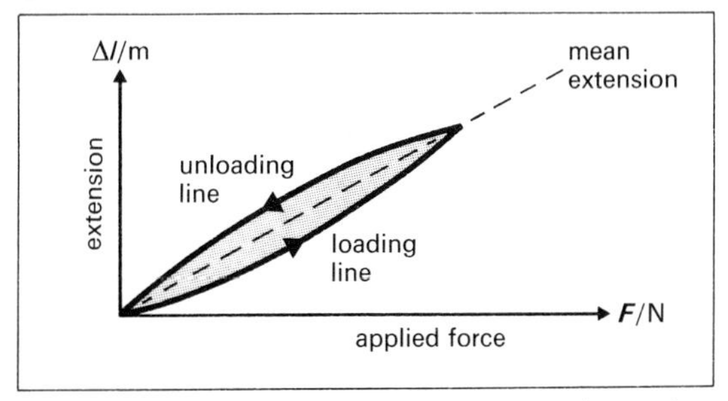

Fig. 17.7. Calculation of mean values of Δl for a wire taken through a hysteresis loop.

In
$$E = \frac{l_0}{A(\Delta l/F)}$$

l_0 and **A** are known, and the mean value of $(\Delta l/F)$ can be found from the graph.

17.7 Behaviour of a Wire under Stress

Fig. 17.8 shows a load-extension curve: the stress–strain curve is similar, but not identical, because l and **A** change during the experiment. (See p. 132.) The extension, rather than the load, is usually plotted along the x-axis.

(*a*) After the **limit of proportionality** the wire no longer obeys *Hooke's Law*, but still returns to its original form when the load is removed. It remains *elastic*.

(*b*) After the **elastic limit**, the wire shows a **permanent set**, and never regains its original shape and size. The deformation has been *dissipative* (p. 185).

(*c*) At the **yield point** there is a marked change in internal structure brought about by the slipping of crystal planes.

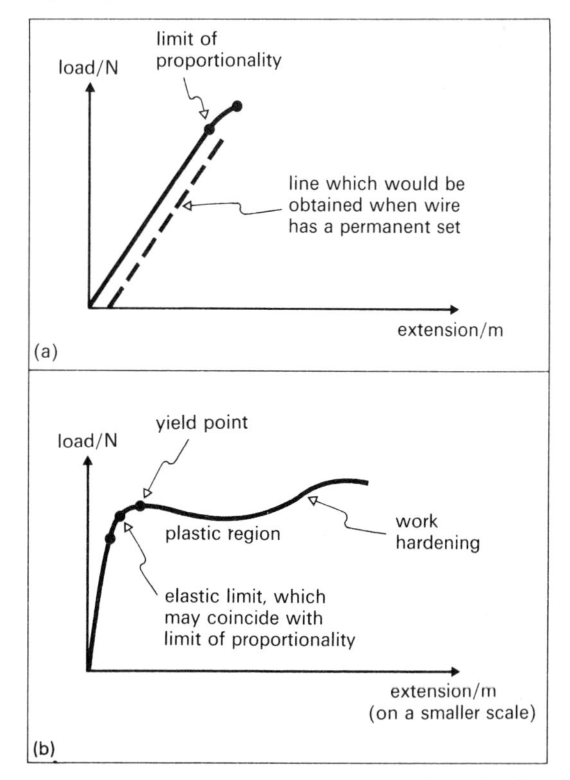

Fig. 17.8. Typical load-extension graphs for a ductile material.

(*d*) In the **plastic region** small increases in load produce marked increases in extension because of *flow* processes.

(*e*) Eventually a *constriction* or neck forms at a weak point, and the tensile forces pull the rod apart (see below).

(*f*) Before the constriction forms, further rearrangement of the crystal imperfections (**dislocations**) allows the stress to increase and finally reach its maximum value (the **breaking stress**). This process is called **work-hardening**.

Similar processes take place when a rod is compressed. At the breaking point the rod is *fractured* by compressive forces.

Strength of Solids

Types of Fracture

(*a*) **Brittle fracture** (fig. 17.9*a*) occurs when the applied force is greater than the maximum molecular attraction over a particular small area. When that part breaks the stress is transferred to adjacent areas.

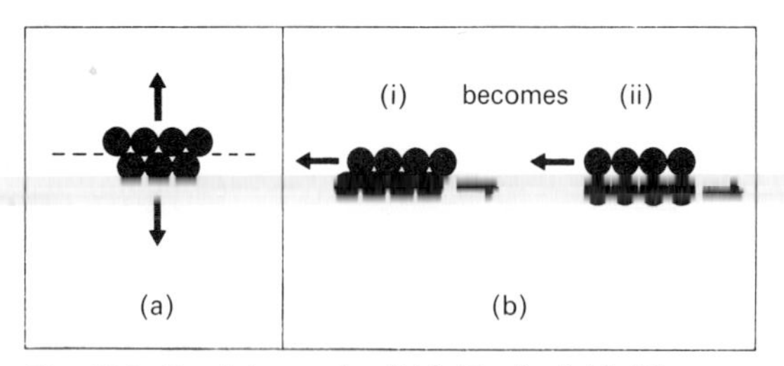

Fig. 17.9. Breakdown of solid lattice by (a) brittle fracture, and (b) plastic flow.

(*b*) **Ductile fracture** (fig. 17.9*b*) occurs when the elongation and narrowing that accompanies the sliding of the molecular planes results in a neck of small cross-sectional area A. The breaking stress σ is then exceeded, since $\sigma = F/A$.

(*c*) **Fatigue fracture** occurs when a fluctuating stress is applied to a material. Repeated work-hardening causes the material to become more inhomogeneous locally as plastic regions are introduced. If the same stress were applied steadily the material would be safe from fracture.

(*d*) **Creep fracture** occurs when metals are stressed for considerable periods of time at temperatures high enough to allow easy movement of dislocations or sliding at grain boundaries. The strain increases with time, and the creep rate accelerates just before fracture owing to the growth of cracks.

In practice the strength of materials is *very much less* than that predicted by theory because of minute surface cracks and other imperfections of the lattice.

The work done in stretching the specimen up to the point of brittle fracture is closely related to the **heat of sublimation** of the substance, and so we use it to calculate the theoretical strength.

In fact the best way of evaluating the molecular attraction energy ε (p. 120) is from the heat of sublimation.

17.8 Stressed Wires

(*a*) Energy Stored in a Stretched Wire

If we apply a force $\boldsymbol{F}$ *to* a wire, then

$$\boldsymbol{F} = \frac{EA}{l_0}\cdot\boldsymbol{x}$$

when the wire is extended $\boldsymbol{x}$. The energy W stored in the wire is the work done *on* the wire during extension, where

$$W = \int_0^{\Delta l} F\mathrm{d}x$$

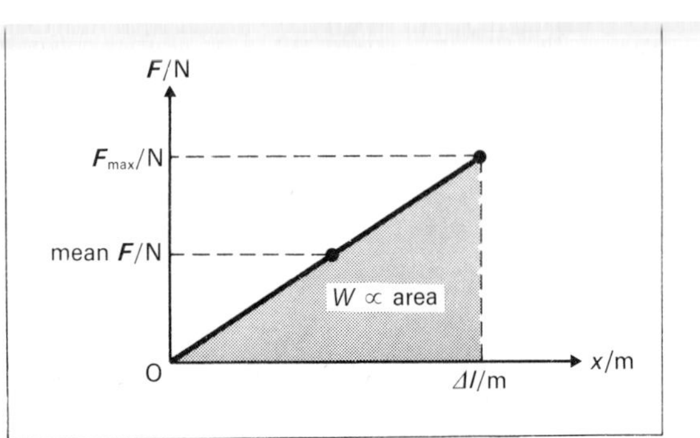

Fig. 17.10. Potential energy stored by a stretched wire.

W is represented by the area under the line of fig. 17.10, so

$$W = \tfrac{1}{2}F_{\max}\Delta l$$

$$= \tfrac{1}{2}\left(\frac{F_{\max}}{A}\right)\cdot\left(\frac{\Delta l}{l_0}\right)\cdot(Al_0)$$

$$= \tfrac{1}{2}\text{stress} \times \text{strain} \times \text{volume}$$

When a wire is deformed, the energy W stored elastically in volume V is given by

$$\frac{W}{V} = \tfrac{1}{2}\,stress \times strain$$

$$\frac{W}{V}\left[\frac{\mathrm{J}}{\mathrm{m}^3}\right] = \tfrac{1}{2}\,\text{stress}\left[\frac{\mathrm{N}}{\mathrm{m}^2}\right] \times \text{strain}\left[\frac{\mathrm{m}}{\mathrm{m}}\right]$$

If the deformation of a wire is not perfectly elastic, the energy dissipated when the specimen is taken through a cycle such as that of fig. 17.7 is represented by the shaded area. (See also p. 429.)

(*b*) Tension in a Cooling Wire

Suppose a wire of natural length l_0 is heated through $\Delta\theta$, and prevented from contracting, as is the steel rim of a locomotive wheel when it is being fitted. The heating produces an extension Δl, where

$$\Delta l = l_0\alpha\,\Delta\theta \quad \text{(def. of } \alpha\text{)}$$

but

$$F = EA\frac{\Delta l}{l_0} \quad \text{(def. of } E\text{)}$$

so the tension in the wire, when thus stretched, would be given by

$$F = EA\alpha\,\Delta\theta$$

17.9 Poisson's Ratio μ

When a rod is stretched, its cross-sectional area decreases as the length increases, and there is usually a change of volume.

Poisson's ratio μ is defined by the equation

$$\mu = \frac{-\textit{transverse strain}}{\textit{longitudinal strain}}$$

Thus

$$\mu = \frac{-\Delta r/r_0}{\Delta l/l_0}$$

where r might represent the radius of a cylindrical rod. The negative sign is inserted so that μ should be a positive number. It has no dimensions and no unit. For a cylindrical rod

$$V = \pi r^2 l$$

$$\Delta V = \pi r^2\,\Delta l + 2\pi r l\,\Delta r$$

For *zero volume change* on stretching,

$$\Delta V = 0$$

so

$$\mu = \frac{-\Delta r/r_0}{\Delta l/l_0} = \frac{1}{2}$$

It should be noted that the three moduli of elasticity used in this book are *not independent*. When one analyses waves travelling through elastic media, μ is a convenient parameter to relate these three moduli, and express the modulus appropriate to a particular situation.

* 17.10 Isothermal and Adiabatic Bulk Modulus

(*a*) Isothermal Bulk Modulus

From p. 130

$$K_{\text{iso}} = \frac{-\Delta p}{\Delta V/V_0}$$

$$= -V_0\frac{\mathrm{d}p}{\mathrm{d}V}$$

Under isothermal conditions an ideal gas obeys *Boyle's Law* pV = constant. Differentiating

$$-V\left(\frac{\mathrm{d}p}{\mathrm{d}V}\right)_{\text{iso}} = p$$

Thus the *isothermal* bulk modulus of an ideal gas is equal to its pressure.

(*b*) Adiabatic Bulk Modulus

Under certain conditions (p. 190) the pressure and volume of an ideal gas are related by

$$pV^\gamma = \text{constant}$$

$$p = \text{const } V^{-\gamma}$$

Differentiating

$$\left(\frac{\mathrm{d}p}{\mathrm{d}V}\right)_{\text{ad}} = -\gamma \text{ const } V^{-\gamma-1}$$

$$-V\left(\frac{\mathrm{d}p}{\mathrm{d}V}\right)_{\text{ad}} = \gamma p$$

The *adiabatic* bulk modulus of an ideal gas

$$K_{\text{ad}} = \gamma p$$

K_{ad} controls the speed of audible sound waves through the gas (p. 302).

* 17.11 The Torsional Pendulum

(*a*) An Elemental Cylinder

In fig. 17.11 (overleaf), angle of shear $= \gamma$

Effective area $\boldsymbol{A}$ over which tangential force is applied is half the shaded area

$$A = \pi r\delta r$$

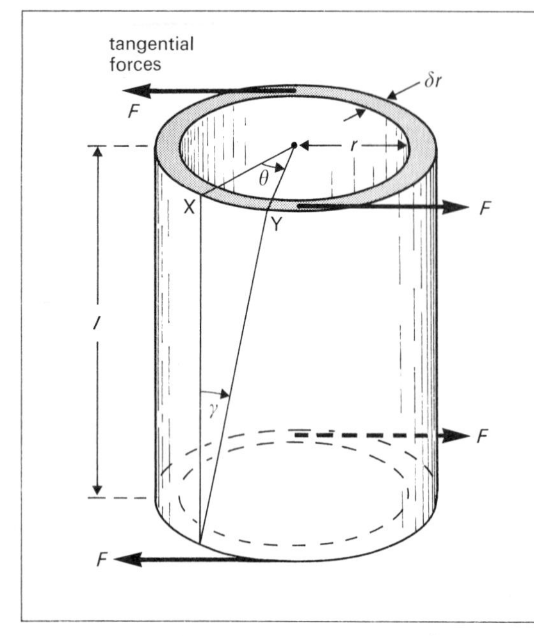

Fig. 17.11. Equal but opposite torques applied to a cylindrical shell cause an angle of shear γ.

By geometry XY $= r\theta \approx l\gamma$ if γ is small. Then

$$G = \frac{F/A}{\gamma} = \frac{F/\pi r\,\delta r}{r\theta/l}$$

The torque of the couple is

$$F2r = \frac{G\theta 2\pi r^3\,\delta r}{l}$$

Note that the shear is zero at the centre ($r = 0$) and increases to a maximum at the surface.

(*b*) A Solid Cylinder

Total torque to produce a twist θ in fig. 17.12.

$$T = \int_0^a \frac{G\theta 2\pi r^3\,\mathrm{d}r}{l} = \frac{\pi G a^4}{2l}\theta = c\theta$$

The torque exerted *by* the cylinder is given by $T = -c\theta$, because it is a restoring couple and acts so as to decrease θ. c is the **torsion constant**.

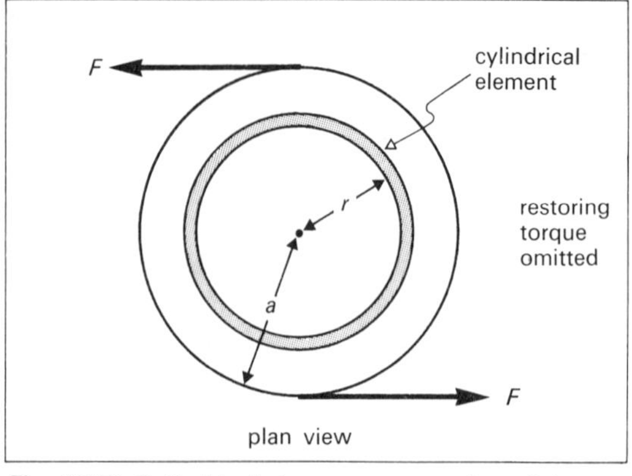

Fig. 17.12. Cylindrical element as part of a solid cylinder.

(*c*) Application to the Pendulum

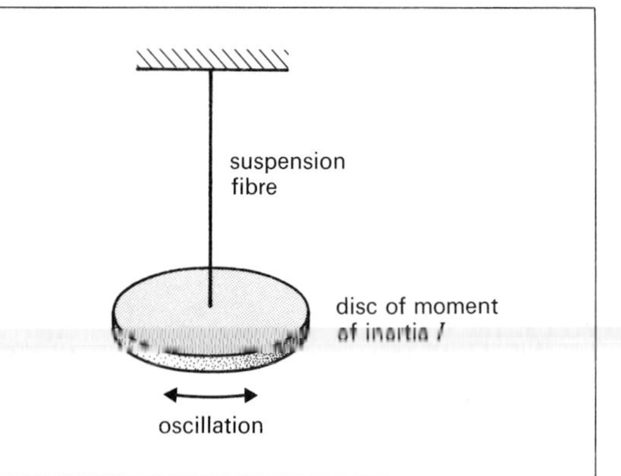

Fig. 17.13. The torsional pendulum.

Use

$$\boldsymbol{T} = \boldsymbol{I}\ddot{\boldsymbol{\theta}} \qquad \text{(p. 56)}$$

$$\frac{-\pi G a^4}{2l}\theta = I\ddot{\theta}$$

$$\ddot{\theta} = \frac{-\pi G a^4}{2lI}\theta$$

Thus the motion is simple harmonic, and has a time period

$$2\pi\sqrt{\frac{2lI}{\pi G a^4}}$$

This result can be used for the *measurement of G*, the shear modulus.

18 Surface Tension

18.1 Definitions

Surface Tension γ

Refer to fig. 18.1.

The surface tension γ is defined by the equation

$$\gamma = \frac{F}{\Delta l}$$

$$\gamma\left[\frac{\text{N}}{\text{m}}\right] = \frac{F[\text{N}]}{\Delta l[\text{m}]}$$

γ is a force per unit length.

$$[\gamma] = \frac{[\text{MLT}^{-2}]}{[\text{L}]} = [\text{MT}^{-2}]$$

γ decreases as the temperature increases, approaching zero near the critical temperature. At this temperature the liquid and vapour become indistinguishable (see p. 210).

Substance	γ/N m^{-1}	*T*/K
di-ethyl ether	1.7×10^{-2}	293
soap solution	2.5×10^{-2}	290
water	5.9×10^{-2}	373
water	7.3×10^{-2}	290
mercury	47×10^{-2}	293

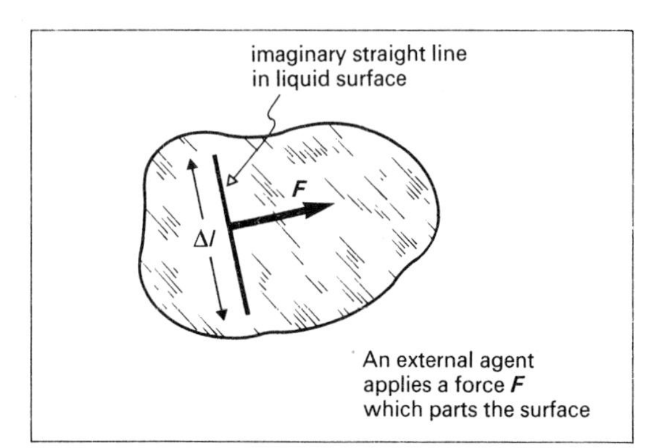

*Fig. 18.1. Definition of surface tension γ. (Note that **F** is drawn in the plane tangential to the surface.)*

Free Surface Energy σ

Suppose that under *isothermal* conditions mechanical energy W is required to create an additional area $\boldsymbol{A}$ of surface.

The free surface energy σ is defined by the equation

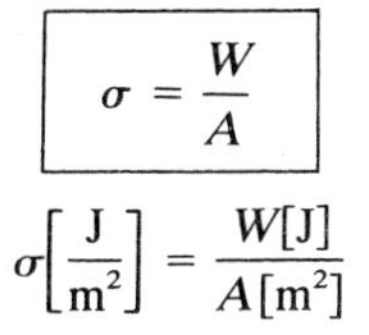

$$\sigma = \frac{W}{A}$$

$$\sigma\left[\frac{\text{J}}{\text{m}^2}\right] = \frac{W[\text{J}]}{A[\text{m}^2]}$$

σ represents work done per unit area and is numerically and dimensionally equal to γ.

$$[\sigma] = \frac{[\mathrm{MLT^{-2}L}]}{[\mathrm{L^2}]} = [\mathrm{MT^{-2}}]$$

*Under *adiabatic* conditions (such that no heat enters or leaves the system) the temperature decreases. Under *isothermal* conditions heat is added to the surface to restore the original temperature. Thus

$$\begin{pmatrix}\text{additional energy}\\ \text{of film}\end{pmatrix} = \begin{pmatrix}\text{work}\\ \text{done}\end{pmatrix} + \begin{pmatrix}\text{added thermal}\\ \text{energy}\end{pmatrix}$$

so total surface energy $> \sigma$.

The decrease of temperature under adiabatic conditions leads to an increased γ, and thus an increased force to oppose the increase of area—an example of *Le Chatelier's Principle*.

To show $\gamma = \sigma$ numerically

The mechanical work done by the agent in fig. 18.2 is

$$F\,\Delta x = (2\gamma a)\,\Delta x$$

where the 2 takes account of the two surfaces of the film.

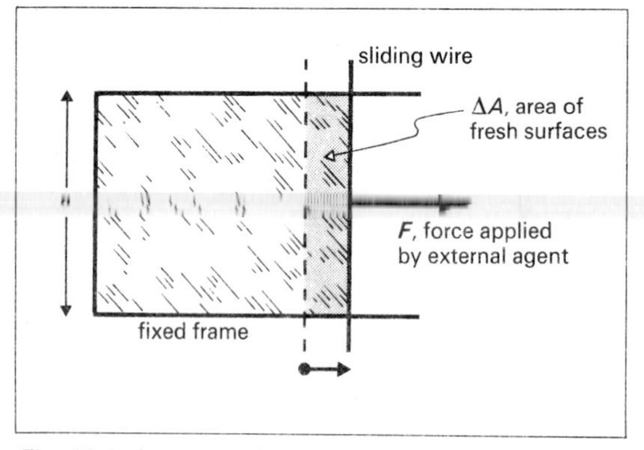

Fig. 18.2. An external agent increases the area of a soap film.

Increase in free surface energy is

$$\sigma\,\Delta A = \sigma(2a\,\Delta x)$$

Under *isothermal* conditions we equate

$$2\gamma a\,\Delta x = \sigma 2a\,\Delta x$$

so $$\sigma = \gamma$$

The Angle of Contact θ

θ is the angle between the tangent planes to the two surfaces, measured *through the liquid* (fig. 18.3). Even for a given liquid and solid, the value of θ is not constant (p. 141).

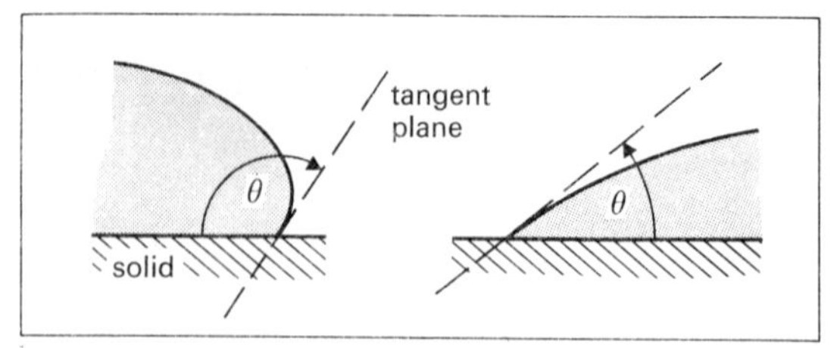

Fig. 18.3. To define angle of contact θ.

18.2 Discussion of Molecular Behaviour

For our microscopic examination of surface tension, we can consider either forces or energies, as we did for the macroscopic definitions.

(a) Discussion in Terms of Forces

A liquid molecule experiences short-ranged forces which are both attractive and repulsive, but the repulsive forces have the shorter range. Fig. 18.4 brings out these points:

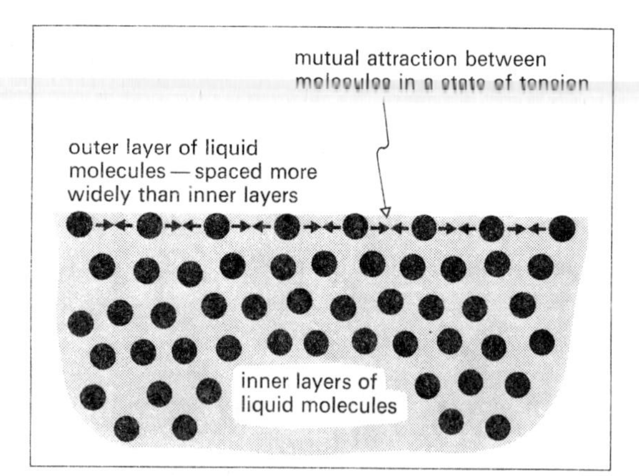

Fig. 18.4. Attractive forces between surface liquid molecules which are in a state of tension.

(*i*) The top layer of molecules does not experience any interaction from above. Molecules which try to leave this layer by moving upwards experience a downward force which may prevent them from doing so.

(*ii*) The individual molecules in the top layer are more widely spaced than those in the bulk of the liquid. Therefore the attractive forces that they experience from either side from their close neighbours cause them to be in a *state of tension* (p. 35). If an attempt is made to part the liquid surface, and thereby to break *on one side only* the bonds that cause the tension force, then the molecules at that point experience a resultant attractive force from the molecules on *the other side.*

The effect is appropriately called **surface tension**, since the increased lateral spacing of the surface molecules causes the surface to behave like a stretched membrane. (But note that because we are dealing with a liquid, an increase in area brings more molecules into the surface layer, so γ does *not* increase as the area increases.)

(*b*) Discussion in Terms of Energy

Fig. 18.4 shows that surface molecules have an increased separation: fig. 16.5 then indicates that their mutual intermolecular p.e. has become *less negative*, i.e. has increased. This point is illustrated in a different way by fig. 18.5.

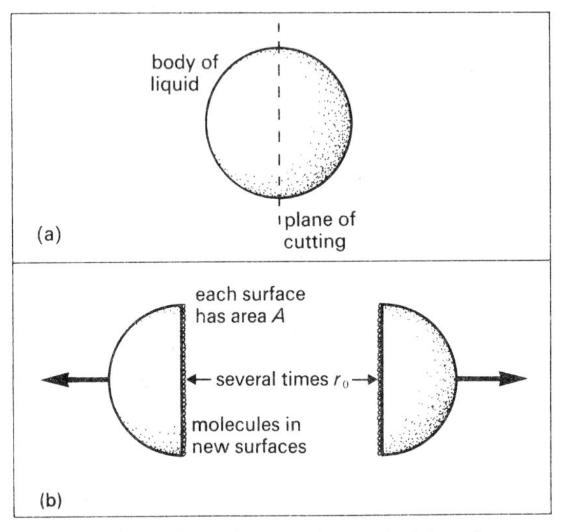

Fig. 18.5. A method of estimating σ: in (a) we have a spherical drop, which in (b) has been cut into two halves.

The parting of the two hemispheres involves work being done by an external agent to part molecules which attract one another during separation. The work done by the agent appears as the molecular p.e. associated with molecules in the new surface, i.e. as **surface energy**.

*Suppose each molecule in the new surface has lost about three near neighbours. Then the work W done by the agent is given by

$$W = \begin{pmatrix}\text{number of}\\ \text{lost}\\ \text{neighbours}\end{pmatrix} \times \begin{pmatrix}\text{energy for}\\ \text{separation}\\ \text{of each}\\ \text{molecular}\\ \text{pair}\end{pmatrix} \times \begin{pmatrix}\text{number}\\ \text{of}\\ \text{molecular}\\ \text{pairs}\\ \text{involved}\end{pmatrix}$$

$$\approx 3\varepsilon\left(\frac{A}{r_0^2}\right)$$

where ε and r_0 were defined in fig. 16.5 and the area occupied by a molecule in the surface has been taken as r_0^2. The new area created is $2A$.

Using typical values $r_0 \sim 10^{-10}$ m and $\varepsilon \sim 10^{-21}$ J, we find

$$\sigma = \frac{W}{2A} = \frac{3\varepsilon}{2r_0^2} \sim 10^{-1}\ \text{J m}^{-2}$$

The fact that the result is of the right order of magnitude lends support to the kinetic theory (p. 126).

A liquid surface has p.e. Any system takes up, in equilibrium, that configuration for which the p.e. becomes a minimum. This is a useful tool, provided one remembers that other types of p.e. may be involved.

18.3 The Shape of Liquid Surfaces

Some useful ideas

(*a*) The surface of a liquid in equilibrium is parallel to the tension forces established in the surface, which are perpendicular to the resultant attractive force experienced by a molecule displaced above the surface. If this were not so the attractive forces would have a resolved part along the surface which would accelerate the surface molecules.

(*b*) The gravitational force on an individual molecule is entirely negligible compared to the intermolecular forces.

(*c*) The pull of the Earth on a liquid drop assumes greater importance as the drop size increases.

$$\text{The gravitational p.e.} \propto \text{mass} \propto r^3$$

whereas

$$\text{surface p.e.} \propto \text{area} \propto r^2$$

for a spherical drop.

(1) Absence of Solids and Gravitational Forces

In two dimensions a cross-section of the liquid surface will be a circle (fig. 18.6).

In three dimensions the surface will be a sphere. Since this is the shape for which a given volume has the smallest area, it follows that a given liquid mass has smallest surface p.e. This situation holds approximately for

(*i*) a soap bubble (whose mass is negligible),

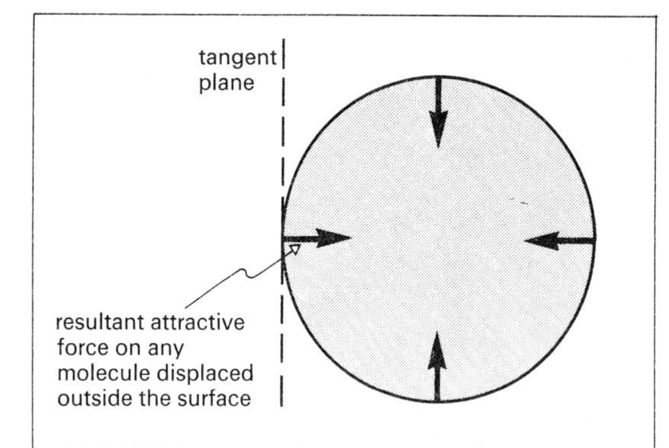

Fig. 18.6. A liquid surface shown in two dimensions. (Repulsive forces not shown.)

(*ii*) very small droplets, and

(*iii*) globules suspended by a fluid of density equal to their own, or in a state of free fall.

(2) When Gravitational Forces are Appreciable

In fig. 18.7 the drop p.e. is made up by

(*i*) surface p.e., $A\sigma$, plus

(*ii*) gravitational p.e., mgh.

The centre of gravity is pulled down until, in the first instance, the decrease of gravitational p.e.

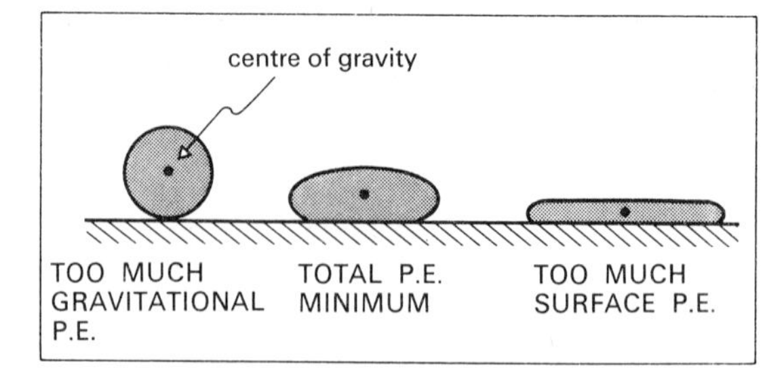

Fig. 18.7. Possible shapes for a mercury drop.

exceeds the increase of surface p.e. Equilibrium is attained when any *further* movement would make the decrease of one equal the increase of the other. The total p.e. is then a minimum.

(3) The Effect of a Solid

When a liquid surface adjoins a solid surface, we consider, for a particular liquid molecule,

(*i*) the attractive forces exerted by the neighbouring liquid molecules—the **cohesive** force,

(*ii*) the attractive forces exerted by the molecules of the solid—the **adhesive** force.

Fig. 18.8 needs careful study.

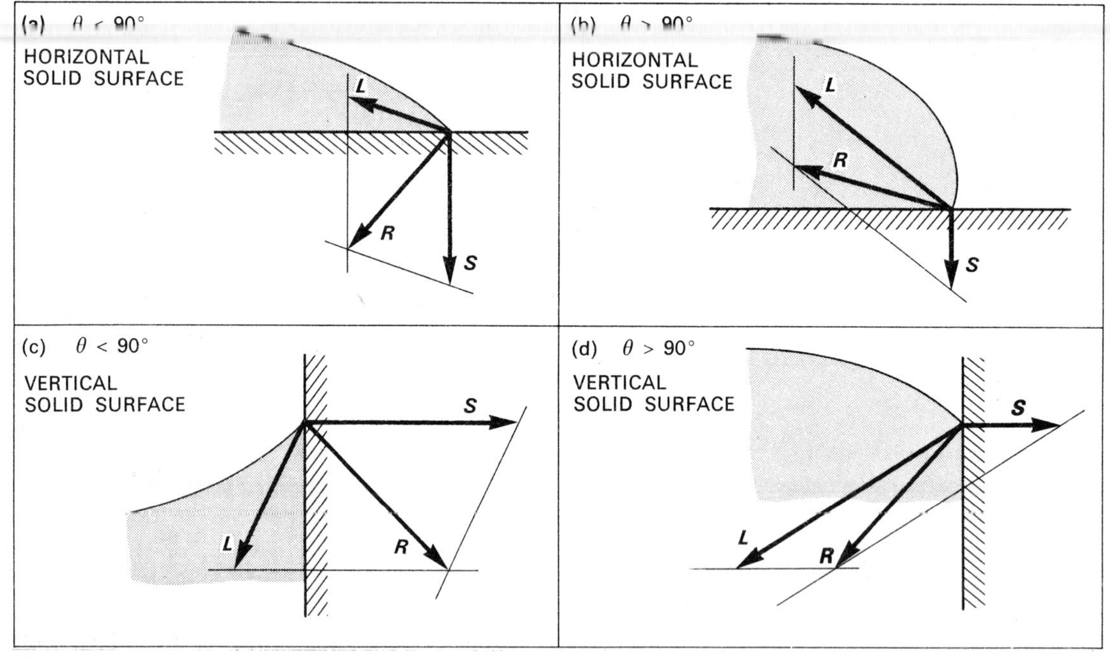

*Fig. 18.8. Shape of a liquid surface close to a solid, **L** is the attractive force exerted by liquid molecules (the cohesive force), **S** is the attractive force exerted by solid molecules (the adhesive force), and **R** is the resultant of **L** and **S**. The liquid surface has a tangent plane perpendicular to the direction of **R**. Note that **R** is opposed by appropriate repulsive forces.*

The *relative* magnitudes of **L** and **S** control the shape of the liquid surface, and its behaviour.

The Angle of Contact θ

θ was defined on p. 136. A liquid for which θ is small is said to **wet** the particular surface it adjoins, and **spreads** over that surface.

A **wetting agent** (such as a detergent) greatly reduces θ.

A **waterproofing agent** aims to prevent wetting by *increasing* θ.

A **meniscus** is a consequence of θ having a particular value. The gravitational p.e. becomes relatively less significant when the vessel's dimensions are *small*. A *fine* tube shows greater **capillary effects** than a wide tube.

18.4 Excess Pressure Inside a Spherical Liquid Surface

The diagram shows a spherical cap in equilibrium under the action of two sets of forces.

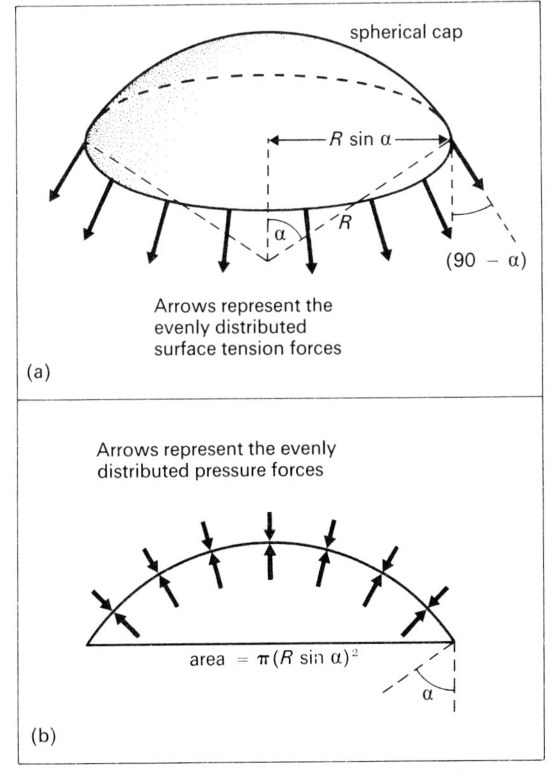

Fig. 18.9. Forces acting on a spherical cap: (a) due to surface tension, and (b) due to (excess) pressure.

(a) The Surface Tension Forces

The force at a particular point is tangential to the surface. The resultant of all these forces

(*i*) has a zero resolved part in the plane of the base, but

(*ii*) has a resolved part perpendicular to this plane

$$\gamma(2\pi R \sin\alpha)\cos(90° - \alpha) = \gamma 2\pi R \sin^2\alpha$$

(b) The Pressure Force

To maintain equilibrium the pressure inside the surface exceeds that outside, by Δp, say.

This pressure difference causes a resultant force which

(*i*) has a zero resolved part in the plane of the base, but

(*ii*) has a resolved part perpendicular to this plane of

$$\Delta p \times (\text{area of base}) = \Delta p \times \pi(R \sin\alpha)^2$$

Equating

$$\Delta p \times \pi R^2 \sin^2\alpha = \gamma 2\pi R \sin^2\alpha$$

so

$$\Delta p = \frac{2\gamma}{R}$$

Notes. (1) Δp is independent of α: therefore this result holds for all surfaces which form part of a sphere, e.g. a hemisphere.

(2) A *film* has *two* surfaces, and there is a pressure difference across *each*. Therefore a spherical soap film shows an excess pressure of $4\gamma/R$.

(3) Note that small soap bubbles have a larger Δp than big bubbles.

18.5 General Result for Excess Pressure

Suppose a curved surface has principal radii of curvature R_1 and R_2. Then it can be shown by using the principle of virtual work that

$$\boxed{\Delta p = \gamma\left(\frac{1}{R_1} + \frac{1}{R_2}\right)}$$

The pressure on the concave side of the surface exceeds that on the convex side.

(*a*) A **synclastic** surface is one for which the principal centres of curvature are found on the *same* side of the surface, e.g. a sphere.

(*b*) An **anticlastic** surface is one for which they are on *opposite* sides, e.g. a saddle-shaped surface, such as that formed by separating two coaxial funnels joined by an initially cylindrical film.

For a synclastic surface, R_1 and R_2 have the same sign. For an anticlastic surface R_1 and R_2 are allotted *opposite signs*.

Special Cases

(1) The soap bubble and liquid drop.
(2) The cylindrical film

$$\Delta p = 2\gamma\left(\frac{1}{R} + \frac{1}{\infty}\right) = \frac{2\gamma}{R}$$

and liquid jet

$$\Delta p = \gamma\left(\frac{1}{R} + \frac{1}{\infty}\right) = \frac{\gamma}{R}$$

(3) For an anticlastic surface showing no pressure difference, such as that in (*b*), $\Delta p = 0$ means that R_1 and R_2 are numerically equal, but of opposite sign.

* 18.6 Surface Energy and the Spherical Soap Bubble

Suppose the radius of a spherical soap bubble be increased from r_0 to $(r_0 + \delta r)$.

If r_0 represents the *equilibrium* radius, the net energy change is zero. Thus, using the **principle of virtual work,**

$$\begin{pmatrix}\text{work done by} \\ \text{pressure} \\ \text{forces}\end{pmatrix} + \begin{pmatrix}\text{work done by} \\ \text{surface tension} \\ \text{forces}\end{pmatrix} = 0$$

$$\begin{aligned}\Delta p\,\delta V &= \text{increase in surface energy} \\ &= \sigma\,\delta A\end{aligned}$$

assuming the change to be reversible (p. 185).

Using

$$A = (4\pi r^2) \times 2 \qquad V = \tfrac{4}{3}\pi r^3$$

$$\delta A = (8\pi r\,\delta r) \times 2 \qquad \delta V = 4\pi r^2\,\delta r$$

$$\Delta p 4\pi r_0^2\,\delta r = \sigma 16\pi r_0\,\delta r$$

whence

$$\Delta p = \frac{4\sigma}{r_0}$$

18.7 Capillary Action

This is typical of a situation which may be analysed by a discussion of

(*a*) forces (indicated on p. 136),
(*b*) energy, or
(*c*) pressures, as is done here.

In fig. 18.10 we suppose the liquid surface to be part of a sphere and the sides of the tube to be nearly vertical so that the angles marked can be taken as equal.

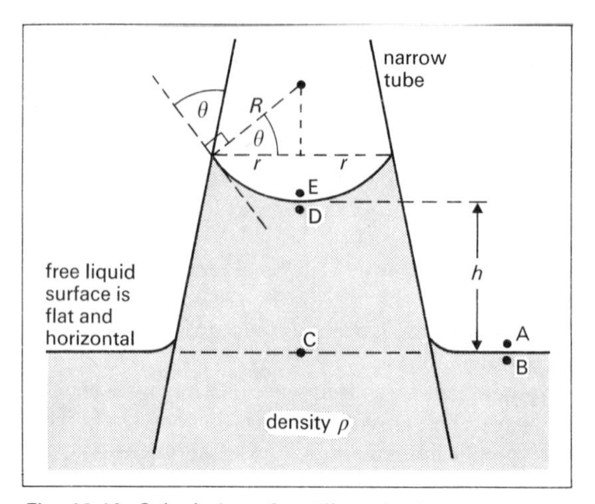

Fig. 18.10. Calculation of capillary rise by a pressure argument.

$$\begin{aligned}\text{Pressure at A} &= \text{pressure at B} \quad \text{(flat surface)} \\ &= \text{pressure at C} \\ &= \text{pressure at D} + h\rho g\end{aligned}$$

$$\begin{aligned}\text{Pressure at E} &= \text{pressure at D} + \frac{2\gamma}{R} \\ &= \text{pressure at D} + \frac{2\gamma\cos\theta}{r}\end{aligned}$$

$$\begin{aligned}\text{Pressure at A} &= \text{pressure at E} + hg \times \text{(gas density)} \\ &\approx \text{pressure at E}\end{aligned}$$

so

$$\frac{2\gamma\cos\theta}{r} = h\rho g$$

Notes. (*a*) $h \propto \gamma$, and $h \propto 1/r$. (*b*) If h and r are measured, then γ can be found.
(*c*) When $\theta < 90°$, h is +ve
$\theta = 90°$, h is 0
$\theta > 90°$, h is −ve.

A negative h corresponds to **capillary depression** as is shown by mercury, for which θ can be 140°.

(*d*) Since we assume *here* that the value of θ is constant, the orientation of the liquid surface changes according to the shape of the containing tube.

18.8 Some Surface Tension Phenomena

In this paragraph we mention briefly surface phenomena, some of which the reader should try to explain using basic principles.

(*a*) If two *parallel plates* are pressed together so as to sandwich a thin film of liquid between them, a large force is needed to pull them apart. (Think in terms of the *small* pressure that results *within* the liquid.)

(*b*) If a *narrow tube* contains a large number of liquid drops, it may not be possible to clear the tube by blowing at one end. (Think in terms of the pressure difference across *each* drop when distorted.)

(*c*) Imagine *two bodies floating* in a liquid with which they make angles of contact of θ_1 and θ_2 respectively.

Then if

(1) θ_1 and θ_2 are both less than 90°, or

(2) θ_1 and θ_2 are both more than 90°, then the bodies are drawn together. But if

(3) $\theta_1 > 90° > \theta_2$, or vice-versa, then the bodies are forced apart.

(Think in terms of the *pressure differences* across the respective meniscuses.)

(*d*) **Dependence of equilibrium vapour pressure on surface curvature**

Compare situations (*a*) and (*b*) in fig. 18.11. On attempted escape, A and B experience attraction from the molecules within the spherical surface drawn. The net attractive force on B exceeds that on A. The relative ease of A's escape means there will be more molecules in the vapour, and so the vapour pressure above the convex surface will exceed that above a comparable plane surface, which in turn exceeds that above a concave surface.

Alternatively one may consider the excess pressure within the convex surface as being responsible for a force which encourages A's escape, and vice-versa for B.)

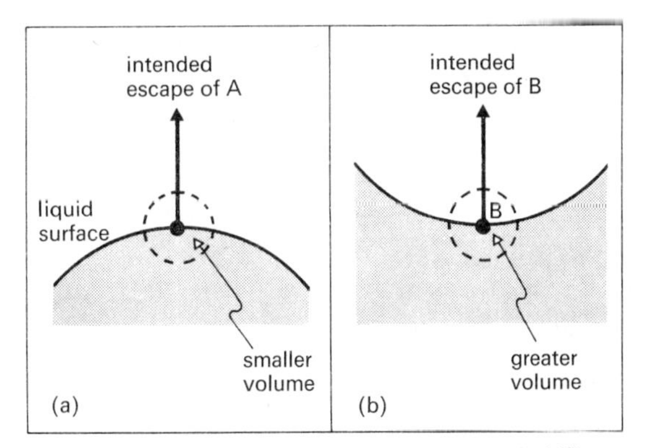

Fig. 18.11. Effect of surface curvature on probability of molecular escape: (a) convex surface, and (b) concave surface.

(*e*) The effect of surface tension encourages

(1) the *evaporation* of small droplets, and

(2) the *collapse* of small vapour bubbles.

(1) implies that a vapour may be considerably *supersaturated* before condensation occurs (unless some suitable nucleus is available). (2) implies that a liquid may be considerably *superheated* before boiling occurs, in the same way.

(1) and (2) are important in the design of the cloud and bubble chambers respectively (p. 471).

18.9 Measurement of the Angle of Contact θ

θ is controlled by the previous history of the liquid surface (fig. 18.12), and is very much affected by the cleanliness of the surface involved.

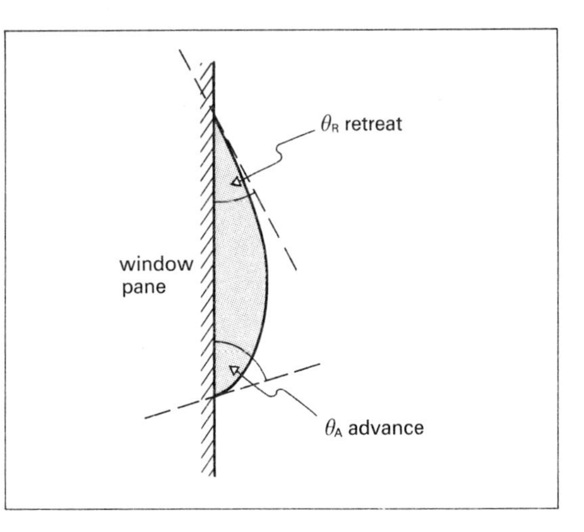

Fig. 18.12. A raindrop illustrating the advancing and retreating angles of contact.

(*a*) Round-Bottom Flask Method

The liquid is poured into a round-bottom flask (fig. 18.13) until the meniscus is horizontal.

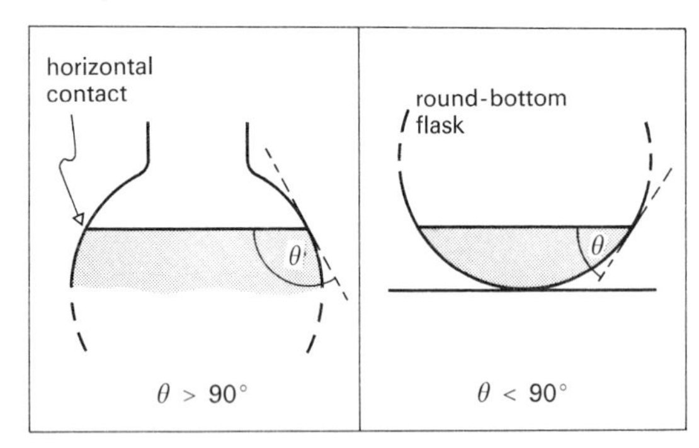

Fig. 18.13. The round-bottom flask method for θ.

(*b*) Rotating Plate Method

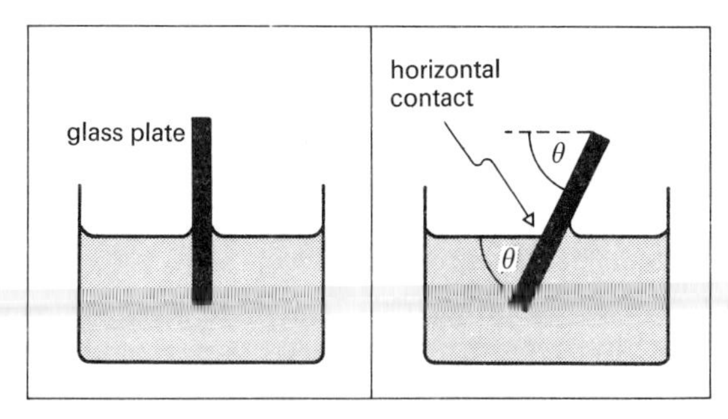

Fig. 18.14. The rotating plate method for θ.

18.10 Measurement of Surface Tension γ

No method will produce reliable results unless *absolute cleanliness* is observed throughout.

(*a*) Chemical Balance Method

(1) *Liquid which forms a stable film* (e.g. soap solution).

Balance the beam with the soap film in position. Puncture the film, and remove a mass m from the other pan to restore balance.

Weight removed = downward pull of film (two surfaces)

$$mg = 2\gamma l$$

(2) *For a liquid* (which may not form a stable film) one may pull a cleaned microscope slide or a wire ring from the surface. The method is useful for *comparing* γ for different liquids.

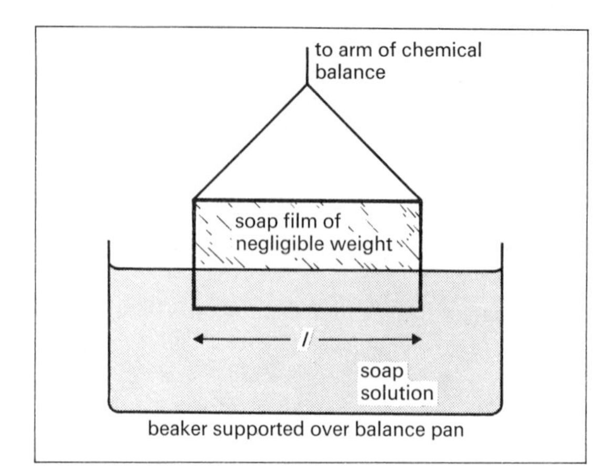

Fig. 18.15. Measurement of γ by direct weighting.

In practice a *torsion* balance is often used in place of a beam balance.

(*b*) Capillary Rise Method

Use
$$\gamma = \frac{rg\rho h}{2\cos\theta} \qquad \text{(p. 140)}$$

This is the international standard method.

Notes.

(*i*) The tube must be wetted so that θ is known.

(*ii*) θ must be measured separately.

(*iii*) h can be measured by travelling microscope (as can r).

(*iv*) r is measured (after breaking the tube) at the *meniscus level.*

Because mercury shows a depression and is not transparent the method of fig. 18.16 must be adopted.

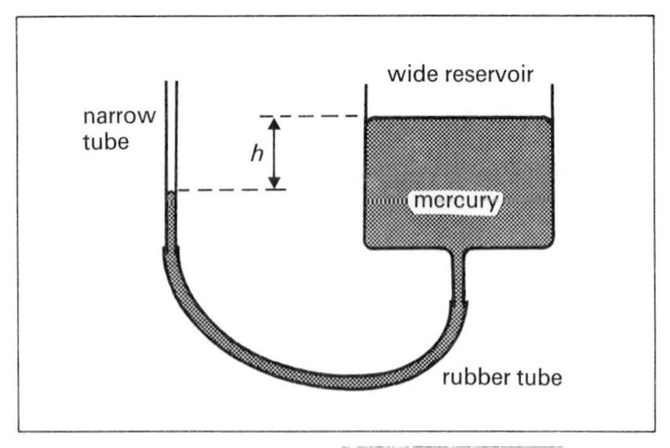

Fig. 18.16. Measurement of capillary depression for mercury.

(c) Jaeger's Bubble Method

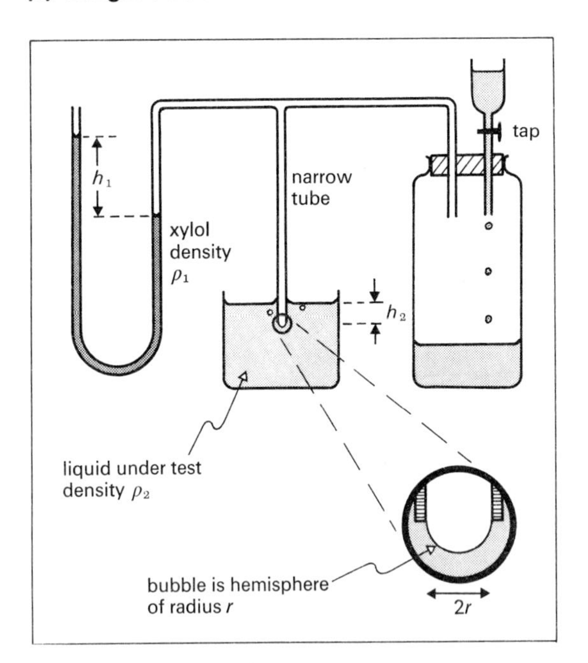

Fig. 18.17. Apparatus for Jaeger's method.

Fig. 18.17 shows the apparatus for Jaeger's method, which has the following advantages:

(*i*) It is easily adapted to investigate how γ varies with temperature.

(*ii*) Contamination is reduced because a fresh liquid surface is formed continually.

(*iii*) A knowledge of the angle of contact is not needed.

h_1 is read just as the bubble breaks away from the narrow tube—this is its maximum value, and corresponds to maximum bubble excess pressure, and so minimum r.

Since $$A + h_2\rho_2 g + \frac{2\gamma}{r} = A + h_1\rho_1 g$$

(where A is the atmospheric pressure),

$$\frac{2\gamma}{r} = (h_1\rho_1 - h_2\rho_2)g$$

The proper value of r is near to the radius of the narrow tube. It can be found by calibration (using a liquid whose γ has been found by the capillary rise method).

Other Methods

These include

(*d*) measurements on large sessile drops,

(*e*) measurement of wavelength of ripples (e.g. on a molten metal),

(*f*) the drop weight method.

19 Viscosity

19.1 The Nature of Fluid Flow

(*a*) Viscous and Non-Viscous Flow

(i) Non-Viscous Flow

When the flow is *laminar*, then the flow pattern can be described by streamlines and tubes of flow. If, in addition, no ordered energy is converted to random internal energy by work being done against viscous forces, then the *Bernoulli Equation* applies (p. 71). It represents the law of conservation of energy adapted for this particular situation.

(ii) Viscous Flow

Although a fluid at rest cannot *permanently* resist the attempt of a shear stress to change its shape, viscous forces can be brought into play which oppose the *relative motion* between different layers of the fluid. **Viscosity** is thus an internal friction between different layers of a fluid moving with different velocities.

For a viscous fluid, steady external forces produce a steady rate of flow, with no acceleration of any part of the fluid.

(*b*) Microscopic (Molecular) Description

(i) Liquids

A liquid differs from a gas in that while neither can permanently resist a shear stress, a liquid does resist both compressive and tensile stresses.

In fig. 19.1, the molecular approach is accompanied by a decrease of intermolecular p.e. and an increase in molecular k.c. If the molecular k.e. becomes disordered, a temporary bond is formed,

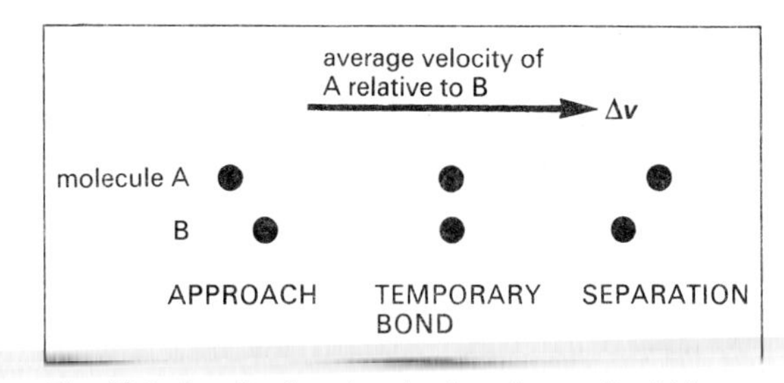

Fig. 19.1. A pair of molecules in adjacent liquid layers with different velocities.

which means that an external agent must do work if the molecules are later to be separated. The work done by the external agent equals the increase in random (internal) energy.

(1) A *temperature increase* means that the molecules have a greater thermal speed, which in turn allows a smaller time in which the gain of molecular k.e. can be disordered (the temporary bond is less strong). *The viscosity of most* (*but not all*) *liquids decreases with temperature.*

(2) A *pressure increase* brings the molecules marginally closer, which results in an *increase* in viscosity.

(ii) Gases

It is important not to confuse the *drift* momentum of a gas molecule with the momentum associated with its random *thermal* motion.

In fig. 19.2 the *slow* molecule moving up across the boundary CD acquires a larger drift velocity, i.e. gains drift momentum, and so has experienced a force to the right (*Newton II*). This means it has exerted a force to the left (*Newton* III) which attempts to retard the faster layer.

Similarly the *faster* molecule moving down across CD exerts a force to the right on the slower layer into which it moves.

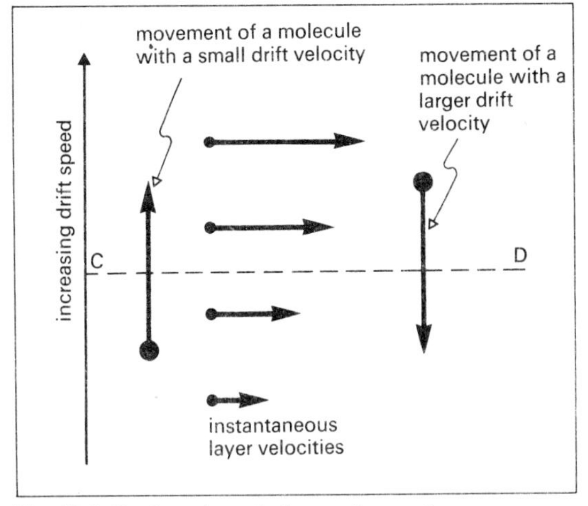

Fig. 19.2. Explanation of viscous forces in a gas.

(1) A *temperature increase* means that the molecules have a greater thermal speed, which increases the rate at which they cross the boundary CD: *the viscosity of a gas increases with temperature.*

(2) Advanced analysis predicts that *the viscosity of a gas should be independent of pressure*, provided the pressure is not too small. This has been verified experimentally, and is strong support for the kinetic theory.

A quantitative treatment is given in chapter **24**.

* 19.2 Definitions

Tangential Stress τ

This is sometimes called **shear**, or **shearing stress**.

Refer to fig. 19.3.

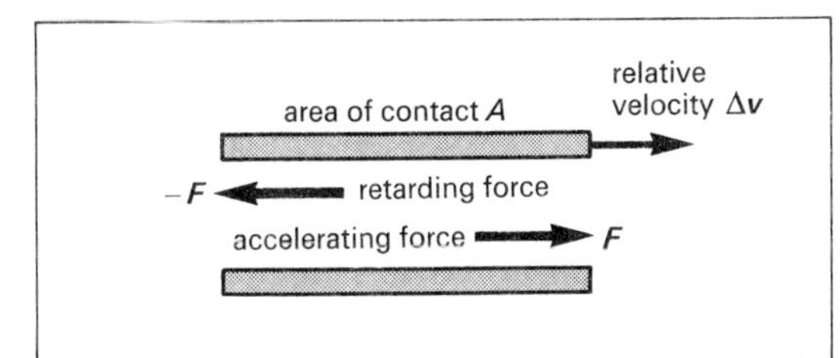

*Fig. 19.3. Adjacent layers of a fluid undergoing viscous flow. According to Newton's third law, the two forces **F** have the same magnitude.*

The tangential stress τ *is defined by the equation*

$$\boxed{\tau = \frac{F}{A}}$$

$$\tau[\mathrm{Pa}] = \frac{F[\mathrm{N}]}{A[\mathrm{m}^2]}$$

Velocity Gradient

In fig. 19.4 the average velocity gradient between the two upper layers is $\delta v/\delta y$. As $\delta y \to 0$, the particles shown are in adjacent layers.

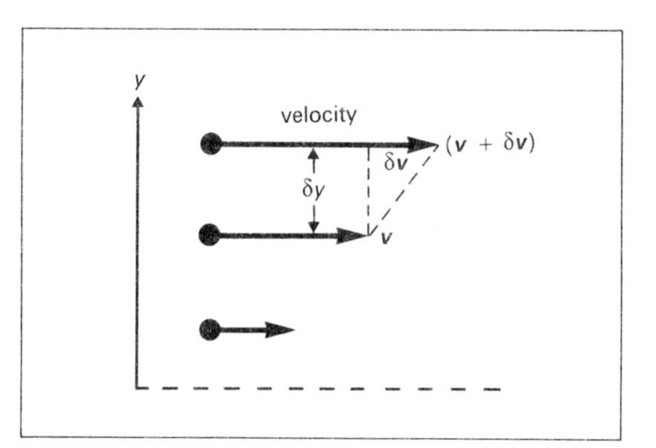

Fig. 19.4. To define velocity gradient.

The velocity gradient *is defined by the equation*

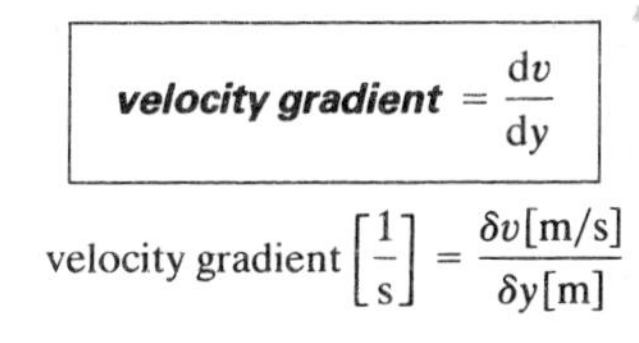

$$\boxed{\textit{velocity gradient} = \frac{\mathrm{d}v}{\mathrm{d}y}}$$

$$\text{velocity gradient}\left[\frac{1}{\mathrm{s}}\right] = \frac{\delta v[\mathrm{m/s}]}{\delta y[\mathrm{m}]}$$

Notes.

(*a*) It has dimensions $[\mathrm{T}^{-1}]$.

(*b*) Here v increases as y increases, and this makes $\mathrm{d}v/\mathrm{d}y$ positive.

(*c*) The velocity gradient is sometimes referred to as the *rate of shear.*

Coefficient of Viscosity η

Fig. 19.5 (overleaf) shows the graph obtained when one plots tangential stress against velocity gradient.

For an ideal *gas* one can *predict* a straight line.

Experimentally many pure *liquids* also give a straight line—they are called **Newtonian** liquids.

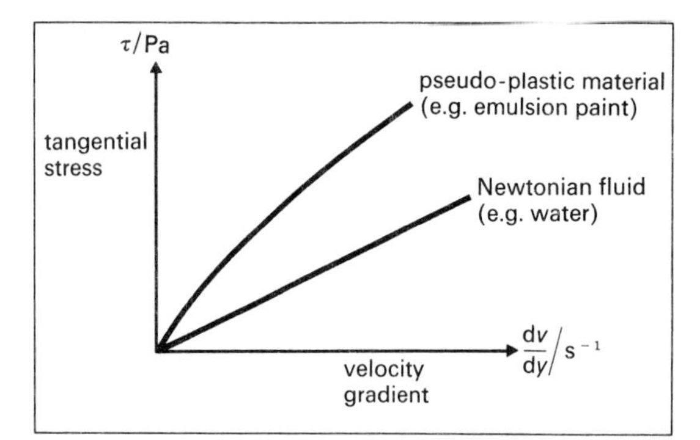

Fig. 19.5. Newtonian and non-Newtonian fluids.

We define the coefficient of viscosity η by the equation

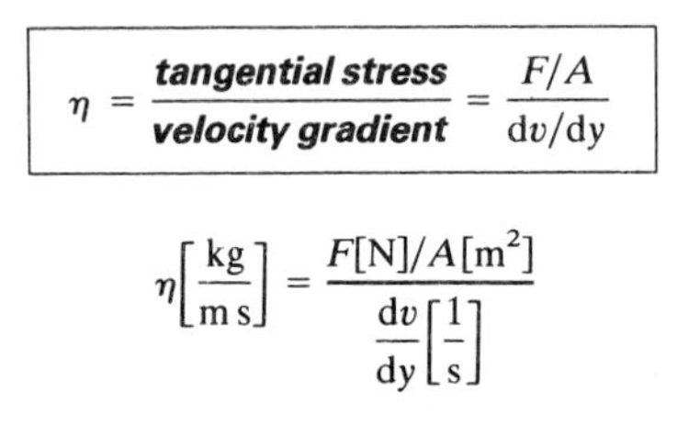

$$\eta = \frac{\textbf{tangential stress}}{\textbf{velocity gradient}} = \frac{F/A}{\mathrm{d}v/\mathrm{d}y}$$

$$\eta\left[\frac{\mathrm{kg}}{\mathrm{m\,s}}\right] = \frac{F[\mathrm{N}]/A[\mathrm{m}^2]}{\dfrac{\mathrm{d}v}{\mathrm{d}y}\left[\dfrac{1}{\mathrm{s}}\right]}$$

Notes.

(*a*) The unit of η is $1\ \mathrm{N\,s\,m^{-2}} = 1\ \mathrm{kg\,m^{-1}\,s^{-1}} = 1\ \mathrm{Pa\,s}$.

(*b*) The dimensions of η are $[\mathrm{M^1\,L^{-1}\,T^{-1}}]$.

(*c*) The coefficient of viscosity of a fluid is analogous to the shear modulus G of a solid:

$$G = \frac{\text{stress}}{\text{strain}} = \frac{\tau}{\gamma}$$

$$\eta = \frac{\text{stress}}{\text{rate of strain}} = \frac{\tau}{\dot{\gamma}}$$

The velocity gradient $\mathrm{d}v/\mathrm{d}y$ is analogous to the rate of change $\dot{\gamma}$ of the angle of shear (p. 130).

(*d*) Orders of magnitude (at 293 K)

Substance	η/Pa s
air	1.8×10^{-5}
water	1.0×10^{-3}
glycerine	8.3×10^{-1}
golden syrup	1.0×10^{2}

(*e*) η is very sensitive to changes of temperature. For pitch

$$\eta = 5 \times 10^{10}\ \mathrm{Pa\,s\ at\ 273\ K,\ and}$$

$$\eta = 1 \times 10^{1}\ \mathrm{Pa\,s\ at\ 373\ K.}$$

(*f*) It is seen from fig. 19.5.

(*i*) that for *Newtonian* fluids η has a unique value (at a particular temperature),

(*ii*) that pseudo-plastic materials have an *apparent* coefficient of viscosity whose value decreases as the tangential stress is increased. They are very important in practice, e.g. in the paint industry.

* 19.3 Forces Acting on a Liquid Layer

In fig. 19.6 the liquid element experiences three forces:

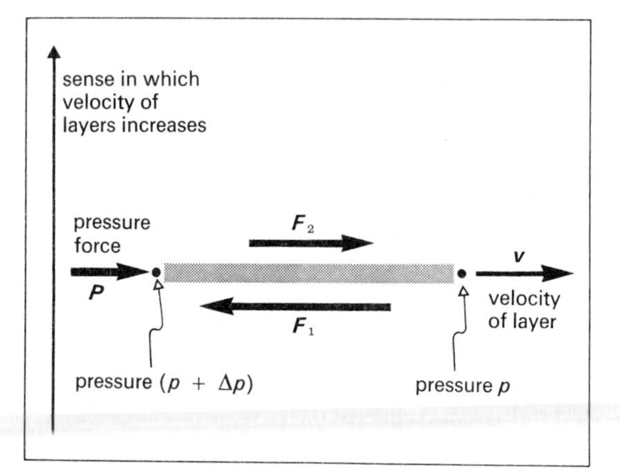

Fig. 19.6. The forces acting on a liquid layer.

(*a*) $\boldsymbol{F}_1$ is the drag force exerted by a slower layer underneath,

(*b*) $\boldsymbol{F}_2$ is the force in the flow direction exerted by a faster layer above,

(*c*) $\boldsymbol{P}$ is the force that results from the pressure difference Δp across the ends of the fluid element.

For steady (non-accelerated) flow

$$\boldsymbol{P} + \boldsymbol{F}_2 + \boldsymbol{F}_1 = 0$$

Note that $F_1 > F_2$, showing that the velocity gradient is not constant.

* 19.4 Turbulence and Reynolds's Number (*Re*)

Streamline flow occurs only when the rate of flow is small. The **critical speed** is the speed of bulk flow at which the flow becomes turbulent (fig. 19.7).

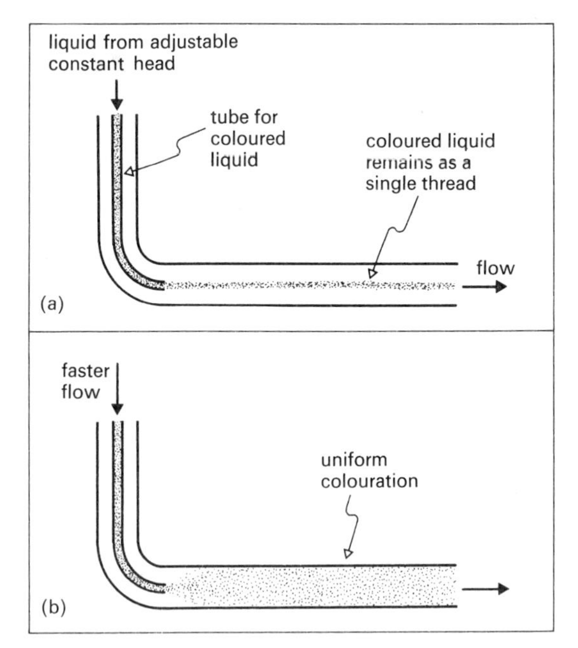

Fig. 19.7. Demonstration of turbulence: (a) streamline flow, and (b) turbulent flow.

Suppose we put $2r$ = tube diameter, ρ = liquid density and v = speed of bulk flow, defined as rate of volume flow divided by cross-sectional area.

Then we define **Reynolds's Number** (Re) by

$$(Re) = \frac{2vr\rho}{\eta}$$

It is a *dimensionless* combination of v, $2r$, ρ and η and (obviously) has no unique value. It is found empirically that for circular tubes, when

(*a*) $(Re) < 2200$, the flow is laminar,
(*b*) $(Re) \sim 2200$, the flow is unstable,
(*c*) $(Re) > 2200$, the flow is usually turbulent.

(Re) is thus a convenient parameter for measuring the stability of flow. The exact value of (Re) for instability depends on the shape of the entrance to the tube.

* 19.5 Dimensional Methods

Examples (*a*), (*b*) and (*c*) refer to fig. 19.8.

(*a*) Maximum Speed of Flow v_{max}

Suppose the speed at the centre is given by

$$v_{max} = k(\eta)^x\left(\frac{\Delta p}{l}\right)^y(r)^z$$

in which

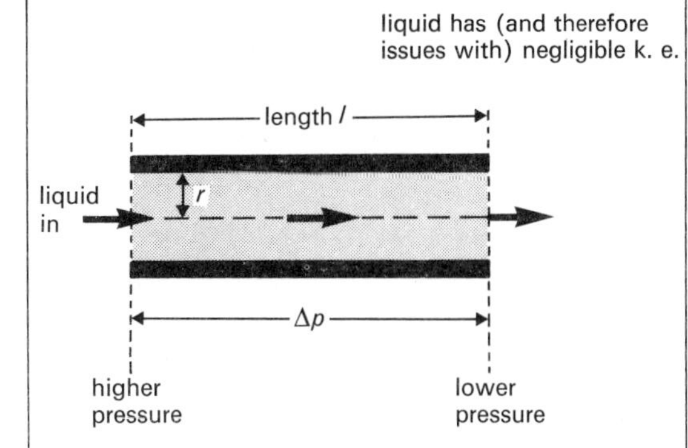

Fig. 19.8. Streamline liquid flow through a horizontal cylindrical capillary tube.

(*i*) k is a dimensionless constant,
(*ii*) $\Delta p/l$ is called the **pressure gradient**,
(*iii*) we have *not* inserted ρ since the liquid is not accelerated (its mass is not relevant).

Equating dimensions

$$[LT^{-1}] = [ML^{-1}T^{-1}]^x\left[\frac{MLT^{-2}}{L^2L}\right]^y[L]^z$$

from which

$$x = -1, \quad y = +1, \quad z = +2 \quad \text{so} \quad v_{max} = k\frac{r^2}{\eta}\left(\frac{\Delta p}{l}\right)$$

Full analysis shows that

$$v_{max} = \frac{1}{4}\frac{r^2}{\eta}\left(\frac{\Delta p}{l}\right)$$

(*b*) Rate of Volume Flow V/t

Suppose $\qquad V/t = k(\eta)^x\left(\frac{\Delta p}{l}\right)^y(r)^z$

Equating dimensions as before, we find

$$x = -1, \quad y = +1, \quad z = +4 \quad \text{so} \quad \frac{V}{t} = k\frac{r^4}{\eta}\left(\frac{\Delta p}{l}\right)$$

Note that the rate of volume flow is proportional to

(*i*) pressure gradient,
(*ii*) r^4: this seems reasonable because both the cross-sectional area and the average liquid speed are $\propto r^2$,
(*iii*) η^{-1}: this means that a small η gives a large rate of flow. η^{-1} is called the **fluidity** ϕ of the liquid. Full analysis shows that $k = \pi/8$, so

$$\boxed{\frac{V}{t} = \frac{\pi}{8}\frac{r^4}{\eta}\left(\frac{\Delta p}{l}\right)}$$

which is called **Poiseuille's Equation**. It can be used for the measurement of η.

(c) Critical Speed v_c

Turbulence starts to set in when the speed of bulk flow $= v_c$. Suppose

$$v_c = k(\rho)^x(\eta)^y(2r)^z$$

in which

(i) ρ is included because turbulence involves a change of momentum,

(ii) η is included because a very viscous liquid is less likely to form eddies,

(iii) $2r$ is a linear dimension of the channel (e.g. the diameter of a tube).

Equating dimensions shows

$$x = -1$$
$$y = +1 \quad \text{so} \quad v_c = k\left(\frac{\eta}{2\rho r}\right)$$
$$z = -1$$

It follows that

$$k = \frac{2v_c r\rho}{\eta}, \quad \text{but} \quad (Re) = \frac{2vr\rho}{\eta}$$

so $k = (Re)_c$: the constant is the value that *Reynolds's Number* has when the flow is unstable (turbulence about to occur).

$$v_c = (Re)_c\left(\frac{\eta}{2\rho r}\right)$$

(d) Stokes's Law

The dimensional part of this law can be deduced by the same method—see the next paragraph.

19.6 Stokes's Law and Terminal Velocity

(a) Derivation of Stokes's Law

In fig. 19.9 a viscous drag exists because there exists a velocity gradient between the boundary layer and more distant points in the viscous medium.

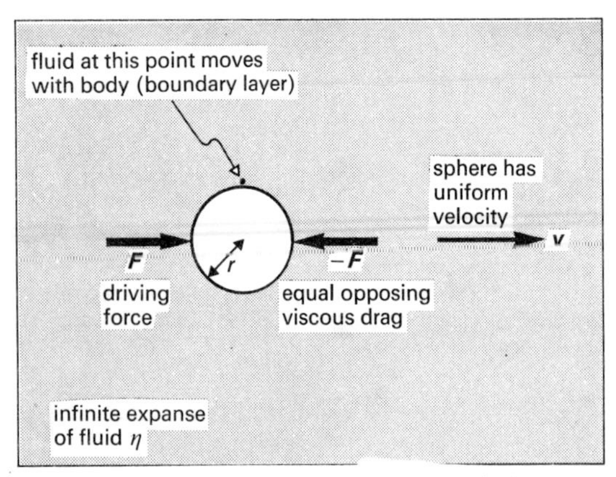

Fig. 19.9. A sphere moving at constant velocity through an infinite viscous fluid.

Suppose

$$\boldsymbol{F} = k(r)^x(\eta)^y(\boldsymbol{v})^z$$

in which we have not considered ρ because we are not concerned with net changes in the liquid's momentum, even on a local scale, if the flow is streamline.

Equating dimensions, we find

$$x = y = z = +1$$

So

$$\boldsymbol{F} = k\eta r\boldsymbol{v}$$

Full analysis shows that

$$\boxed{\boldsymbol{F} = 6\pi\eta r\boldsymbol{v}}$$

an equation which represents **Stokes's Law**.

The ***resistive*** force experienced by the body is more correctly given by $-6\pi\eta r\boldsymbol{v}$.

Notes.

(i) The equation only applies *at small speeds*, such that

(1) the conditions are steamline,

(2) there is no slipping between the liquid and the sphere (the effect of adhesive forces).

(ii) Corrections are necessary for a (real) medium of limited extent.

(iii) The equation must be modified when the size of the particles of the medium becomes significant relative to the size of the sphere. (It cannot be applied without correction to *Millikan's* experiment p. 440.)

(b) Terminal Velocity

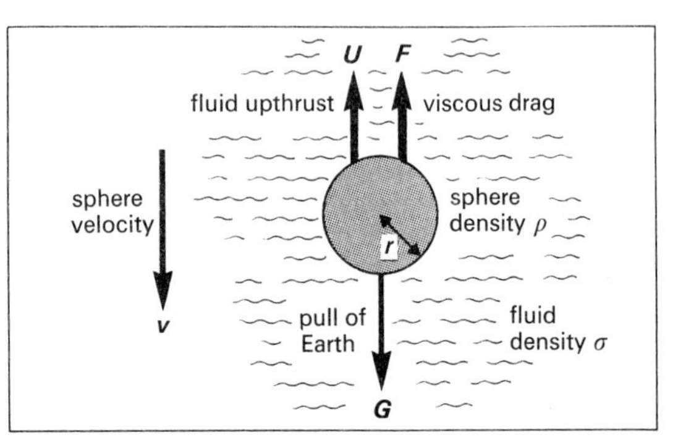

Fig. 19.10. A falling sphere,

When a sphere has reached a terminal velocity of magnitude v_t,

$$G = U + F$$
$$G - U = F$$
$$\tfrac{4}{3}\pi r^3(\rho - \sigma)g = 6\pi\eta r v_t$$

So

$$v_t = \frac{2r^2(\rho - \sigma)g}{9\eta}$$

Note that $v_t \propto r^2$.

This equation is applicable (with corrections) to

(*i*) the oil drops in *Millikan's* experiment,

(*ii*) small water droplets in clouds. (It cannot be used for large rain drops, for which $v_t > v_c$, and so turbulence sets in.)

*19.7 Resistance Offered by Fluids

(*a*) Low Speeds

When the relative velocity is less than the critical velocity, the flow is *streamline*, and *viscous forces* are responsible for the resistance (fig. 19.11). This is sometimes called **skin friction**.

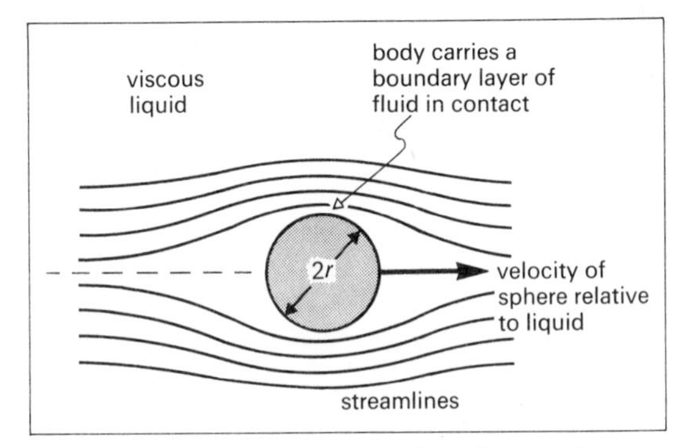

Fig. 19.11. Relative movement of a sphere and a viscous liquid (streamline conditions).

The force
$$F \propto r$$
$$\propto \eta$$
$$\propto v$$

but $\boldsymbol{F}$ is independent of ρ (*Stokes's Law*).

Note that $\boldsymbol{F}$ would be zero in a non-viscous fluid (if it were not for the turbulence that must then ensue).

(*b*) High Speeds

When the relative velocity exceeds the critical velocity, then streamlines no longer exist and the flow becomes turbulent. The ordered energy (of the fluid flow or body) is converted to the random kinetic energy associated with eddy formation (fig. 19.12).

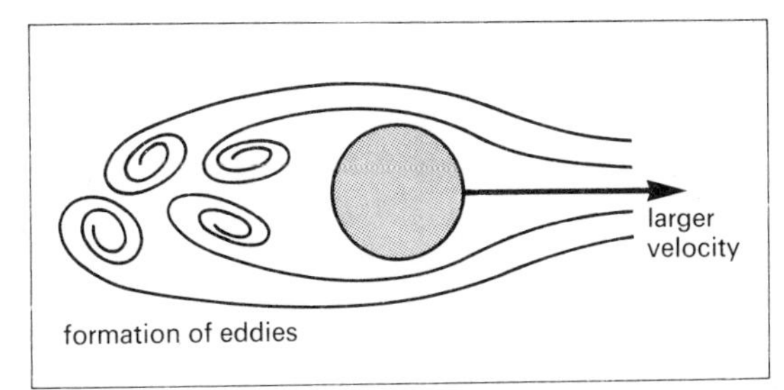

Fig. 19.12. Turbulent conditions.

The force
$$F \propto r^2$$
$$\propto \rho$$
$$\propto v^2$$

but $\boldsymbol{F}$ is independent of η.

If we assume any one of these relationships we can use the method of dimensions to show how $\boldsymbol{F}$ is related to the other three quantities.

We can also say

$F \propto$ (rate of change of liquid momentum)

$$\propto \begin{pmatrix}\text{mass of liquid whose velocity is}\\ \text{changed in unit time}\end{pmatrix} . \begin{pmatrix}\text{velocity}\\ \text{change}\end{pmatrix}$$
$$\propto (\rho r^2 v)v$$
$$\propto r^2 \rho v^2$$

The critical value of *Reynolds's Number*, and hence the critical velocity, are determined by the *shape* of the body. The purpose of *streamlining* is to reduce the onset of turbulence at the trailing edge.

(*c*) Intermediate Speeds

In (*a*) and (*b*) we take the effects of ρ and η respectively to be negligible. In practice $\boldsymbol{F}$ will depend (to some extent) on both ρ and η.

* 19.8 Experimental Determination of η

(*a*) Poiseuille's Tube Method

Use
$$\frac{V}{t} = \frac{\pi}{8}\left(\frac{\Delta p}{l}\right)\frac{r^4}{\eta}$$

(*i*) A preliminary experiment indicates the onset of turbulence (fig. 19.13 overleaf).

For example when $l \sim 0.2$ m and $r \sim 1$ mm, the flow is unstable if Δp is about 9×10^2 Pa (equivalent to a head of water of 90 mm).

(*ii*) A constant head apparatus is needed.

(*iii*) The k.e. of emerging liquid must be near zero.

(*iv*) r^4 is needed for the calculation, so great care must be taken in measuring r. The mercury weight method does give an average value for r^2. (Cf. p. 142.)

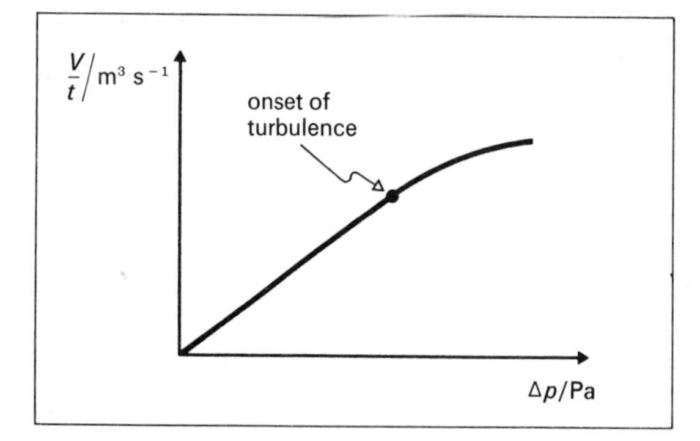

Fig. 19.13. Ensuring streamline conditions when η is being measured.

(*v*) The liquid temperature must be constant. η varies very rapidly with changes of T.

(*b*) Stokes's Law Method

Use $$\boldsymbol{F} = 6\pi\eta r\boldsymbol{v}$$

(*i*) The method is useful for viscous liquids like glycerine, for which the flow is still streamline at the terminal velocity $\boldsymbol{v}_t$.

(*ii*) r must be very much less than the diameter of the container, otherwise the walls will affect the shape of the streamlines. (For the average measuring cylinder we want $r \lesssim 2$ mm.)

(*iii*) Temperature control is important.

(*iv*) The falling sphere must be freed from clinging air bubbles.

20 Gravitation

20.1 Historical Background

The history of the study of gravitation is an important example of how ideas in Physics sometimes develop.

(1) *Copernicus* (1543) suggested that the Sun (and not the Earth) was the centre of the Solar System.

(2) *Tycho Brahe* (1590's) made accurate observations on the planetary motions.

(3) *Kepler* (1618) summarized, in **Kepler's Laws**, the kinematic facts accumulated by *Brahe*. These *laws* indicate briefly important regularities in the planetary orbits.

(4) *Newton* (1686) suggested an *hypothesis* to explain why *Kepler's* laws were obeyed—this was the inspired step of genius. The hypothesis was used to forecast the result of possible measurements, such as

(*a*) the relationship between the acceleration of free-fall g_0 at the Earth's surface, and that of the Moon in its orbit,

(*b*) the timing of eclipses, and

(*c*) the positions of undiscovered planets.

(5) When *Newton's* hypothesis had been thus confirmed by experiment, it was accepted as a *theory*. It is commonly expressed by the equation called **Newton's Law of Gravitation**.

(6) Further refinement of instrumental technique led to observations, such as those on the precession of Mercury's orbit, which could not be explained in detail by *Newton's* theory.

We now have a *modified theory*, **Einstein's General Theory of Relativity** (1905), of which *Newton's Law* is a special limiting case.

Suppose we consider a small body moving under a gravitational force near a much more massive body.

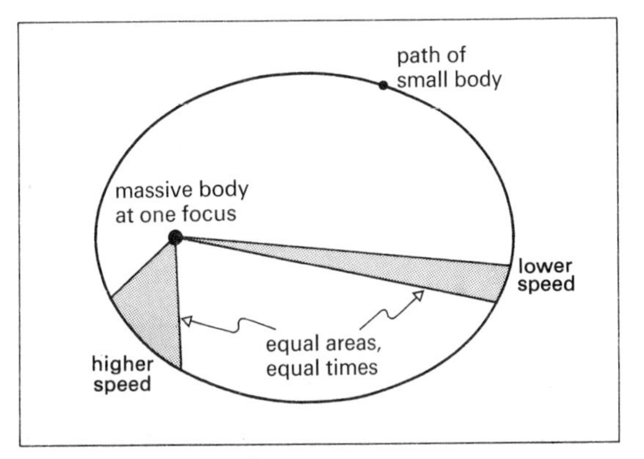

Fig. 20.1. The law of areas.

Then for this situation **Kepler's Laws** can be written:

Law I **The Law of Orbits**. *The path of the small body is a conic section having the large body as one focus. The particular section is determined by the speed of the small body at a given instant.*

Law II **The Law of Areas**. *The area swept out in a given time by the radius vector joining the large body to the small body is always the same* (fig. 20.1).

Law III The Law of Periods. *For closed orbits (ellipses) the ratio*

$$\frac{(\text{orbital period})^2}{(\text{semimajor axis})^3}$$

is always the same.

Kepler's Laws were originally stated for the solar system, which exemplifies them. They follow from the application of

(*a*) *Newton's Laws* of motion and
(*b*) *Newton's Law* of gravitation.

Law II is a consequence of the gravitational force being a *central* force.

20.2 Newton's Law of Universal Gravitation

Newton's Law of gravitation (1686) ***states that***

$$F \propto \frac{m_1 m_2}{r^2}$$

where the symbols are defined in fig. 20.2.

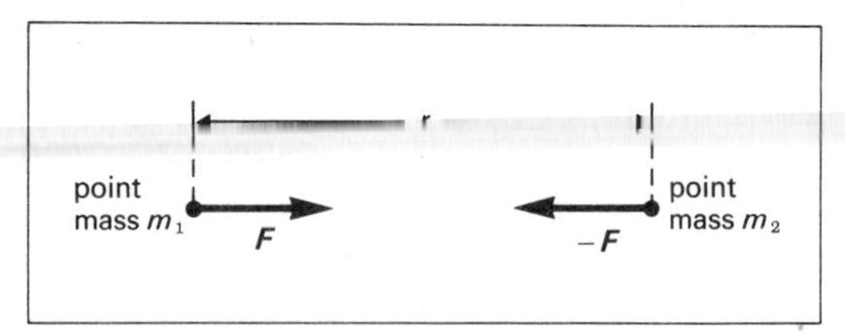

Fig. 20.2. Newton's Law of gravitation.

Notes on the law.

(*a*) Note carefully that the law deals with *point masses*. It can be extended to deal with finite bodies by integration, or by application of *Gauss's Law*.

(*b*) *m* represents the **gravitational mass** of the particle. *Inertial* and gravitational masses are identical (or at least proportional), and we will not distinguish between them.

(*c*) We already have units for **F**[N], *m*[kg] and *r*[m]. We must therefore write

$$\boxed{F = G \cdot \frac{m_1 m_2}{r^2}}$$

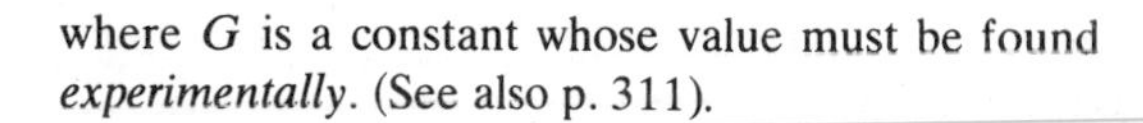

where *G* is a constant whose value must be found *experimentally*. (See also p. 311).

(*d*) The gravitational force is a **central** force, because it acts along the line joining the points. The gravitational forces that we consider are all *attractive* (p. 310).

The Universal Constant of Gravitation

$$G = \frac{Fr^2}{m_1 m_2}$$

(*a*) *G* has a magnitude 6.7×10^{-11} N m^2 kg^{-2}. (p. 159).

(*b*) *G* has dimensions

$$[G] = \frac{[\text{MLT}^{-2}][\text{L}^2]}{[\text{M}^2]} = [\text{M}^{-1}\text{L}^3\text{T}^{-2}]$$

This shows that *G* could be expressed in m^3 kg^{-1} s^{-2}. This is a less useful form than N m^2 kg^{-2}.

(*c*) *G* is a *universal constant*. Its physical magnitude does not depend on:

(1) the type of substance considered,
(2) the nature of intervening (screening) materials,
(3) temperature, etc.

(*d*) Although *G* is so small (gravitation being the *weakest interaction*), it becomes an important factor in problems dealing with massive bodies. (Refer to the discussion on the ranges of different types of force (p. 120).)

The small value of *G* illustrates why gravitation is *not* responsible for interatomic and intermolecular forces (p. 119).

20.3 The Gravitational Field

A **gravitational field** is said to exist at a point if a force of gravitational origin is exerted on a test mass placed at that point.

We consider the *interaction* from two viewpoints, as shown in fig. 20.3.

As with electric field (p. 313) and the magnetic field (p. 372) the problem is twofold:

(*a*) What is the field established by a given mass distribution?

(*b*) What force is experienced by a particular mass distribution placed in the given field?

In practice there are two factors which simplify the problem enormously:

(*i*) gravitational forces are mostly negligible unless at least one very massive body is involved, and

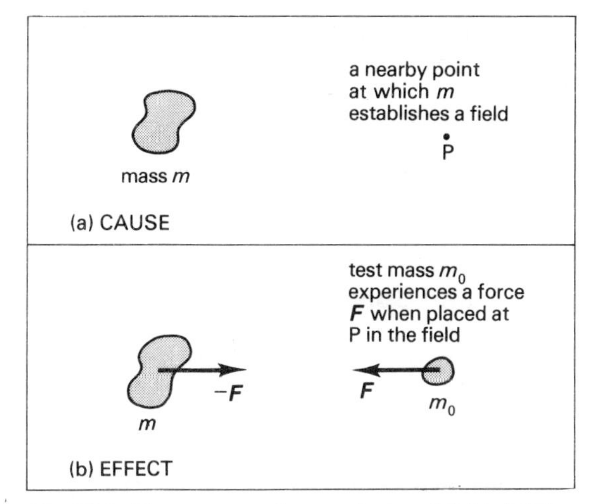

Fig. 20.3. The concept of a gravitational field: (a) the course of the field, and (b) the effect of the field.

(*ii*) one of the masses concerned is usually very large and possesses spherical symmetry to a high degree, while the other is either similar or so small as to approximate to a particle.

Gravitational Field Strength

Suppose a test mass m_0 placed at a point in a gravitational field experiences a force $\boldsymbol{F}$.

The gravitational field strength g at that point is defined by the equation

$$\boldsymbol{g} = \frac{\boldsymbol{F}}{m_0}$$

$$\boldsymbol{g}\left[\frac{\text{N}}{\text{kg}}\right] = \frac{\boldsymbol{F}[\text{N}]}{m_0[\text{kg}]}$$

Notes.

(*a*) $\boldsymbol{g}$ represents a vector quantity whose directions is that of the force exerted on the test mass.

(*b*) We see from *Newton's* second law ($\boldsymbol{F} = m\boldsymbol{a}$) that if the gravitational force is the resultant force on the test mass, then an acceleration $\boldsymbol{g}$ is produced. $\boldsymbol{g}$ has the alternative unit m s^{-2}, and in many circumstances is referred to as the **acceleration due to gravity**.

(*c*) The gravitational force exerted by the Earth on a small mass is sometimes referred to as its **weight**, but the term is not defined uniquely.

Gravitational Field Lines

A **gravitational field line** is a convenient fictional concept which can be developed to aid the visualization of a gravitational field. A line is drawn such that the *tangent* to it at any point gives the direction of $\boldsymbol{g}$ at that point.

Gravitational field lines have the general properties ascribed to electric field lines (p. 314). They are not used widely because most important gravitational fields have the same symmetry (with inward-pointing radial field lines) and are easy to visualize.

A body projected in the Earth's gravitational field follows a trajectory which is very nearly parabolic: this demonstrates that a field line does *not*, in general, display the path taken by a mass in a gravitational field.

* 20.4 Gauss's Law

Gravitational Field Flux

Suppose a gravitational field $\boldsymbol{g}$ makes an angle θ with the outward normal to a small area $\Delta\boldsymbol{A}$, as shown in fig. 20.4.

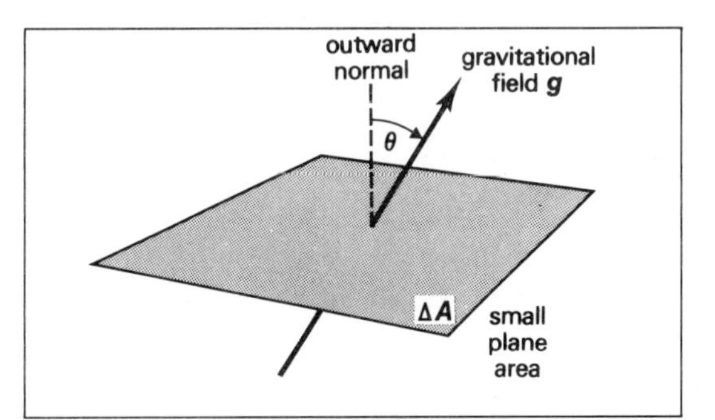

Fig. 20.4. Definition of gravitational field flux ψ_g.

Since $\Delta\boldsymbol{A}$ is small, $\boldsymbol{g}$ has effectively the same size and direction at each point on the surface.

The gravitational field flux ψ_g through the area ΔA is defined by the equation

$$\psi_g = g \cos\theta \cdot \Delta A$$

$$\psi_g\left[\frac{\text{N m}^2}{\text{kg}}\right] = g\cos\theta\left[\frac{\text{N}}{\text{kg}}\right]\cdot \Delta A[\text{m}^2]$$

ψ_g has the dimensions $[\text{L}^3\text{T}^{-2}]$.

The calculation of ψ_g through an arbitrary surface can become a complex matter (p. 317), but for the situations which we will need to consider $\boldsymbol{g}$ will have the same value at each point on the surface, and θ will usually equal π rad.

Gauss's law

***Gauss's law** in gravitation states that for any closed surface*

$$\psi_{\boldsymbol{g}} = -4\pi G\Sigma m$$

where Σm represents the total mass enclosed by the surface, and $\psi_{\boldsymbol{g}}$ is the total gravitational field flux through that surface.

A more detailed discussion of *Gauss*'s law will be found in chapter **44**. For the moment we merely note that it is a statement, alternative to *Newton*'s law, for expressing the fact that gravitational force obeys an inverse square law. Its experimental verification is largely indirect, relying on confirmation that the predictions of the law are observed in practice.

The negative sign in the statement of the law is required because we always deal with gravitational forces which are attractive. Field lines point inwards to large isolated masses, causing θ to equal π rad, and so $\cos\theta$ to be -1. (See below.)

Newton's Law deduced from Gauss's Law

In fig. 20.5 we want to find the force $\boldsymbol{F}$ acting on a point test mass m_2 placed a distance r from a point mass m_1. For convenience the *Gaussian* surface selected is spherical.

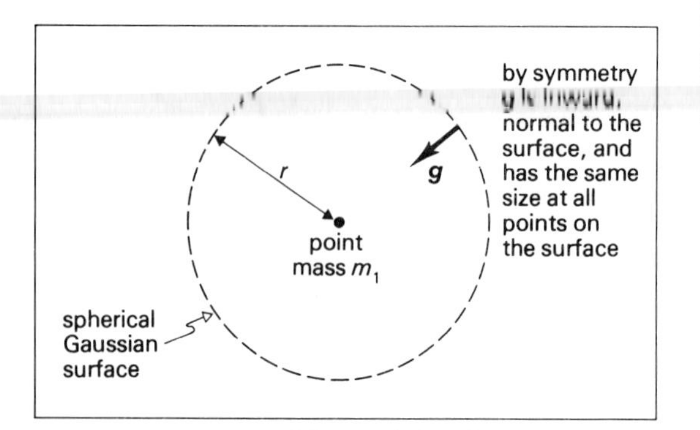

Fig. 20.5. Deduction of Newton's law from Gauss's law.

Gauss's law is

$$\psi_{\boldsymbol{g}} = -4\pi G\Sigma m$$

in which

$$\begin{aligned}\psi_{\boldsymbol{g}} &= \Sigma g\cos\theta\cdot\Delta A\\ &= (g\cos\pi)(4\pi r^2)\\ &= -4\pi r^2 g\end{aligned}$$

and

$$\Sigma m = m_1$$

Substituting

$$-4\pi r^2 g = -4\pi G m_1$$

$$g = \frac{Gm_1}{r^2}$$

As, by definition of $\boldsymbol{g}$,

$$\boldsymbol{F} = m_2\boldsymbol{g}$$

this gives

$$F = G\frac{m_1 m_2}{r^2}$$

which is *Newton*'s law of gravitation.

* 20.5 The Fields of Spherical Shells and Spheres

The two results of this paragraph follow closely from *Gauss*'s law. They may also be proved by the integration of *Newton*'s law applied to point masses.

The Field Established by a Spherical Shell

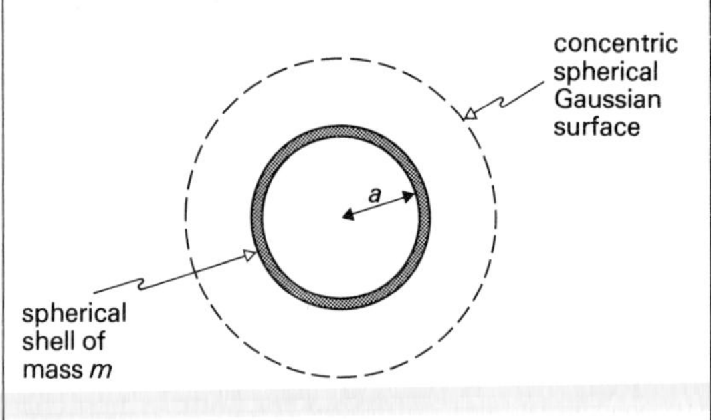

Fig. 20.6. A Gaussian surface of symmetry appropriate to a spherical shell.

The thin uniform spherical shell of fig. 20.6 has spherical symmetry. This means that the field it produces

(*a*) has the same size at each point on the *Gaussian* surface (the surface being chosen to ensure this), and

(*b*) cuts the *Gaussian* surface everywhere at right angles.

We can follow the argument of paragraph **20.4** to the point where

$$g = \frac{Gm}{r^2} \quad (\text{for } r > a)$$

The gravitational field established outside itself by a thin uniform spherical shell is the same as it would be if the mass of the shell were concentrated at its centre.

When $r < a$ our *Gaussian* surface would appear inside the shell. The argument would then run:

$$\Sigma m = 0$$

$$\therefore\ \psi_{\boldsymbol{g}} = 0 \qquad (\textit{Gauss}\text{'s law})$$

The symmetry of the situation would enable us to conclude that

The gravitational field established inside itself by a thin uniform spherical shell is zero everywhere.

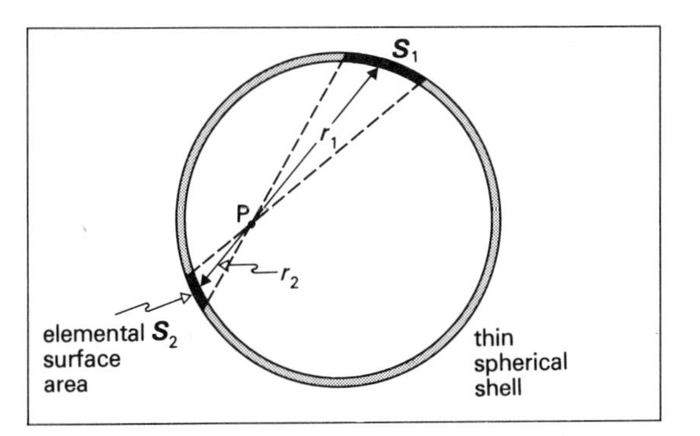

Fig. 20.7. A different viewpoint to investigate the field within a thin spherical shell.

Fig. 20.7 shows this result to be in direct agreement with *Newton*'s statement of the law of gravitation, when we remember that geometry requires that

$$\frac{S_1}{r_1^2} = \frac{S_2}{r_2^2}$$

Note that there is *no shielding effect* in gravitation as there is in electrostatics and in magnetism. A gravitational field, established by some second body, *can* exist within a uniform spherical shell.

The Fields Established by Spherical Bodies

The conclusions reached for thin spherical shells can be extended simply not only to uniform spheres, but also to other spherical bodies whose density at a given volume element depends only upon the distance of that element from the centre of the sphere (and not upon direction). This is a condition which is satisfied, to a good approximation, by most planets and stars.

The gravitational field established outside itself by a spherically symmetric body is the same as it would be if the mass of the body were concentrated at its centre.

For points inside such a body there is no contribution from that part of its mass which lies at a greater distance from the centre than the point considered, and we apply the result above using the remaining mass.

20.6 The Earth's Gravitational Field

The general principles established in this paragraph apply to any massive body which possesses the necessary spherical symmetry.

(a) Relation between G and g_0

The sizes of bodies on the Earth's surface are so much less than the Earth's radius R that we may treat them as particles: we will use *Newton*'s law.

Consider a body of mass m on the surface of the Earth of mass M. We can equate the value of the gravitational force calculated from *Newton*'s law to that found from the definition of $\boldsymbol{g}$:

$$G\frac{mM}{R^2} = mg_0$$

On the assumption that we cannot distinguish inertial mass from gravitational mass, we have

$$g_0 = \frac{GM}{R^2}$$

A knowledge of G or M enables the other to be found: that is why the measurement of G is known colloquially as 'weighing the Earth'. It also enables us to find a value for the Earth's mean density.

(b) The Field outside the Earth ($r > R$)

(*i*) ***Large distances***. The argument in (*a*) above leads to

$$g = \frac{GM}{r^2}$$

where g is the field strength a distance r from the centre.

*(*ii*) ***Small distances***. When $r > R$ we may apply

$$g = \frac{GM}{r^2}$$

$$\therefore \ln g = \ln (GM) - \ln (r^2)$$
$$= \ln (GM) - 2\ln r$$

Differentiating

$$\frac{\delta g}{g} = -2\frac{\delta r}{r}$$

If we move a *small* distance $h(\ll R)$ away from the earth's surface

$$\delta g \approx -2\left(\frac{h}{R}\right)g_0$$

or

$$g \approx g_0\left(1 - \frac{2h}{R}\right)$$

(c) The Field inside the Earth ($r < R$)

Although the Earth has approximately spherical symmetry, its density is non-uniform, varying from about $2.8 \times 10^3\ \text{kg m}^{-3}$ in the crust to $9.7 \times 10^3\ \text{kg m}^{-3}$ at the surface of the core. The average density is $5.5 \times 10^3\ \text{kg m}^{-3}$. This increasing density causes the value of g to continue to *increase* with depth for some distance below the surface.

Suppose, for discussion, that the Earth has uniform density ρ. At a distance r from the centre ($r < R$)

$$\begin{aligned} g &= G(\tfrac{4}{3}\pi r^3\rho)/r^2 \\ &= (\tfrac{4}{3}G\pi\rho)r \end{aligned}$$

Thus $g \propto r$. At a depth d below the surface

$$g = g_0\left(1 - \frac{d}{R}\right)$$

The results of this paragraph are summarized in fig. 20.8.

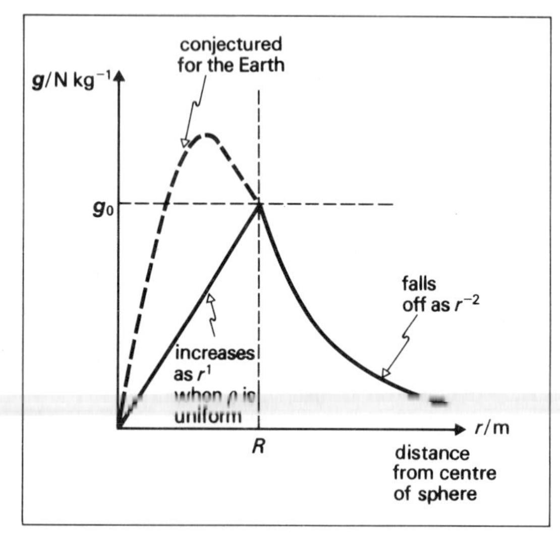

*Fig. 20.8. Variation of **g** established by bodies of spherical symmetry. (To be compared with fig. 45.7.)*

20.7 Gravitational Potential

The detailed discussion of electric potential (chapter **45**) applies broadly to gravitational potential. A concise summary only of some of the essential ideas is given here.

(*a*) The **gravitational potential energy** of a system is the work done by an external agent in assembling a particular configuration of masses. It depends upon the sizes of the masses and their relative separations. It is *always negative*, since gravitational forces are attractive.

(*b*) Suppose an external agent does work W in bringing a test mass m_0 from infinity to a particular point in a gravitational field. (W will of course be negative.)

We define the gravitational potential U at that point by the equation

$$U = \frac{W}{m_0}$$

$$U\left[\frac{\text{J}}{\text{kg}}\right] = \frac{W[\text{J}]}{m_0[\text{kg}]}$$

(*c*) **Potential differences** are calculated from

$$U_{ab} = U_b - U_a = W_{ab}/m_0$$

(*d*) The **theoretical zero of potential** is that of a point at infinity, but it is sometimes convenient to treat the surface of the Earth as the *practical* zero. This is valid when (as nearly always) we are dealing with *differences* of potential.

(*e*) Since the gravitational field is *conservative* (p. 46) gravitational potentials are unique—there is only one possible value for the potential at a given point. This is one reason why gravitational potential is such an important and useful concept.

(*f*) An **equipotential surface** is one on which the potential is the same at all points—no work is done if a test mass is moved from one point to another on the same surface.

(*g*) Field strength and potential lend themselves to alternative (but equivalent) ways of describing the same field. We can

(*i*) give values of g *or* U at various points,
(*ii*) calculate the forces on, *or* potential energies of, given masses at a point, and
(*iii*) draw field lines *or* equipotential surfaces.

(*h*) In a *uniform* field we can relate potential changes and field by

$$\Delta U = g \cdot \Delta x$$

where Δx is a displacement parallel to the field lines. The sign of the potential change can be inserted by inspection.

(*i*) For a non-uniform field

$$g = -\frac{dU}{dx}$$

The gravitational field strength is numerically equal to the potential gradient (a vector quantity). The negative sign is necessary because the vector g points in the direction along which U decreases.

*20.8 The Potential Due to a Spherical Mass

Refer to fig. 20.9. We want to find the work W' done by an external agent that brings a test mass m_0 from infinity to a point distance r from a spherical mass M.

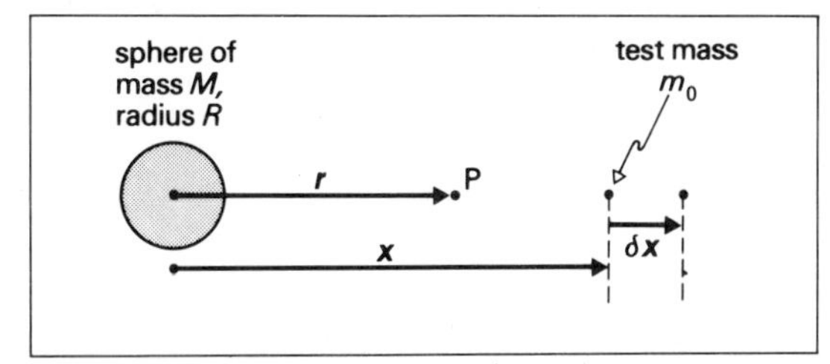

Fig. 20.9. Calculation of the gravitational potential established by a sphere.

The work δW done by the external agent in causing the displacement δx *away* from the sphere is

$$\delta W = F \cdot \delta x$$
$$= \left(G\frac{Mm_0}{x^2}\right)\delta x$$

To move from P to infinity would require work

$$W = \int_r^{\infty} G\frac{Mm_0}{x^2} \cdot \mathrm{d}x$$
$$= \frac{GMm_0}{r}$$

The work W' done by an external agent in bringing the test mass *from* infinity to P is negative, since the agent would exert a restraining force.

Using $W' = -W$, and the definition $U = W'/m_0$, we have

$$U_P = -\frac{GM}{r}$$

The gravitational potential established inside itself by a spherical shell of radius R has the same value at every point, and is equal to the value at the surface, $-GM/R$.

Application to Escape Speed

At the surface of a sphere

$$U = -GM/R$$

The p.e. E_p of a mass m at the surface is given by

$$E_p = Um = -GMm/R$$

The p.e. of the mass an infinite distance away is zero. To escape completely from the sphere the mass must gain p.e. GMm/R. If the mass leaves the sphere at speed v_e and just escapes

$$\tfrac{1}{2}mv_e^2 = \frac{GMm}{R}$$

$$v_e = \sqrt{2GM/R}$$

This value of v_e is referred to as the **escape speed**. Since the mass may be moving in any direction above the horizontal the size of v_e has an important bearing on the composition of a planet's atmosphere.

20.9 Orbital Motion

(*a*) Verification of Kepler's Third Law

Consider a light body of mass m moving at angular speed ω in a circular orbit of radius r around a massive planet of mass M.

Using $$F = ma$$

$$\frac{GmM}{r^2} = m\omega^2 r$$

So $$GM = \omega^2 r^3 = 4\pi^2 \cdot \frac{r^3}{T^2} \qquad \left(T = \frac{2\pi}{\omega}\right)$$

Thus the ratio T^2/r^3 is the same for all circular orbits of such light bodies around the parent planet, since G is a constant.

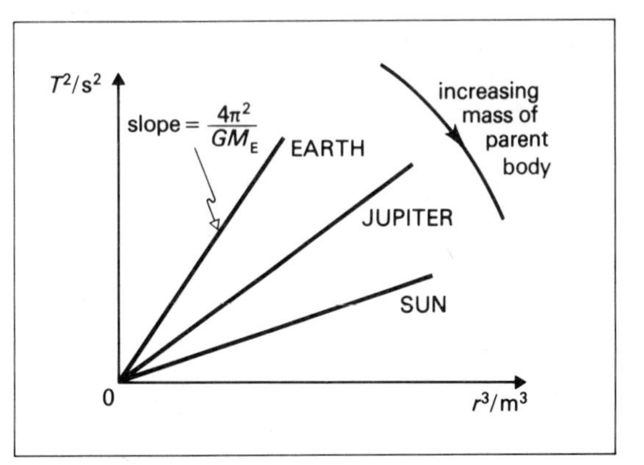

Fig. 20.10. Graphs for three planetary systems which illustrate Kepler's third law. (Not to scale.)

Kepler's third law applies universally, and is illustrated (fig. 20.10) by

(1) The Sun and its planets (since in fact it applies for all elliptical orbits),

(2) Jupiter and its moons,

(3) the Earth and its system of satellites and the Moon.

(*b*) Satellites

For a satellite in a circular orbit moving at speed v_0

$$\frac{GmM}{r^2} = \frac{mv_0^2}{r}$$

$$v_0^2 = \frac{GM}{r}$$

For a given radius of orbit, there is a particular orbital speed, and vice versa.

(1) When $r = R$, the radius of the planet,

$$v_0 = \sqrt{\frac{GM}{R}}$$

Thus $$v_e = \sqrt{2}\, v_0$$

(2) If the satellite is given a different speed, its orbit could be (fig. 20.11)

(*i*) a *hyperbola* at high speeds,
(*ii*) a *parabola* at the escape speed, or
(*iii*) an *ellipse* at lower speeds. (A circle is possible: it is an ellipse of zero eccentricity.)

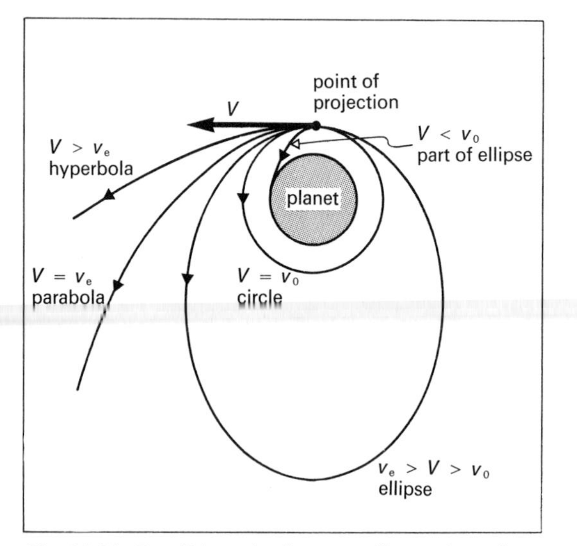

Fig. 20.11. Possible paths for a satellite projected close to the surface of a planet at speed V.

Orders of Magnitude for the Earth

$v_0 = 7.9 \times 10^3\ \mathrm{m\,s^{-1}}$ (when $r = R_E$)
$v_e = 11.2 \times 10^3\ \mathrm{m\,s^{-1}}$

If a satellite has a nearly circular path, but encounters air resistance, there will be a conversion of mechanical energy to internal energy. This reduction of total energy causes a spiral path in which the gravitational attraction of the Earth has a small component in the direction of motion. The result is an increase in the kinetic energy (see right) as the satellite spirals down to the Earth's surface.

*(c) Energy and Satellite Motion

Imagine a satellite of mass m a distance r from the centre of the parent planet of mass M ($M \gg m$).

Since $$\frac{GmM}{r^2} = m\left(\frac{v^2}{r}\right) \qquad (\boldsymbol{F} = m\boldsymbol{a})$$

the kinetic energy $$E_k = \tfrac{1}{2}mv^2 = \frac{GmM}{2r},$$

which shows that E_k is always positive.

But the p.e. $$E_p = -\frac{GmM}{r},$$

which shows that E_p is always negative.

The total energy

$$E_{tot} = E_k + E_p = -\frac{GmM}{2r}$$

Our choice of the zero of p.e. makes E_{tot} negative. As r decreases, so E_p decreases, E_k increases (but less rapidly) and so E_{tot} decreases.

These ideas are illustrated by Fig. 20.12.

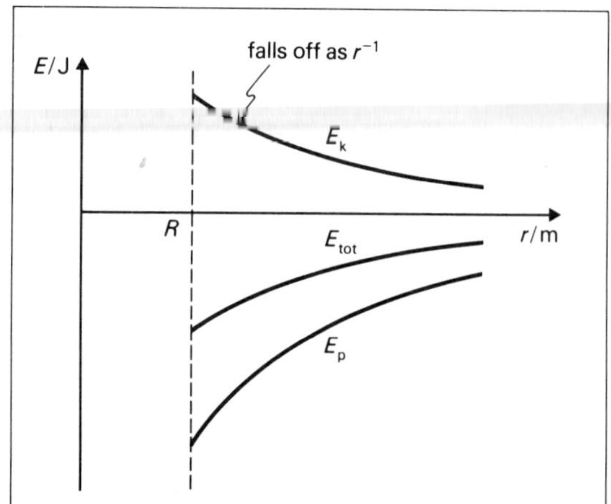

Fig. 20.12. The energies of satellite motion.

20.10 Tunnels and Tides

*(a) Tunnels through the Earth

Fig. 20.8 shows that a body moving along a tunnel through the Earth (along either a diameter or a chord) would describe s.h.m. It can be shown that the time period would be 84 minutes, the same as that of a satellite in a circular orbit close to the surface of the Earth. (This assumes the Earth's density is uniform.)

(*b*) The Tides

The two tides per day are caused by the *unequal attractions* of the Moon (and Sun) for masses at different sides of the Earth, by the rotation of the Earth about its axis, and by the rotation of the Earth and Moon about their common centre of mass (which lies within the Earth).

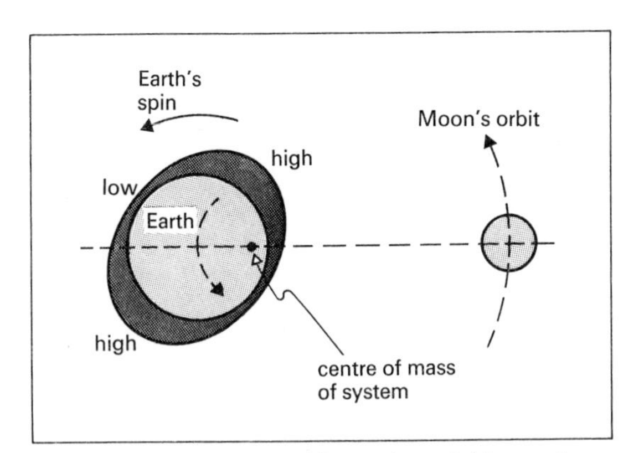

Fig. 20.13. Explanation of formation of tides on the Earth (not to scale).

The Sun causes two tides per day, and the Moon two tides every 25 hours. These tides exhibit *beats* (p. 109): **spring** tides occur when they are in phase, and **neap** tides when they are out of phase.

A high tide is experienced *after* the Moon has passed overhead.

20.11 Measurement of G

Several methods have been used.

(*a*) *Large-scale methods*, using natural masses, include:

(*i*) *Bouguer's* method (1740), using a mountain as an attracting mass, and

(*ii*) in principle, *Airy's* method (1854). This would involve measuring g_0 and g at the bottom of a mine. The experiment is complicated by the fact that the Earth does not have a uniform density.

(*b*) *Laboratory methods* include:

(*i*) *Cavendish's* torsion balance method (1798) and its improvement by *Boys* (1895).

(*ii*) *Poynting's* chemical balance method (1893), and

(*iii*) *Heyl's* vibrating mass method (1930).

Cavendish's Determination of G

The whole apparatus was enclosed to reduce the effects of air currents. Suppose the suspended system

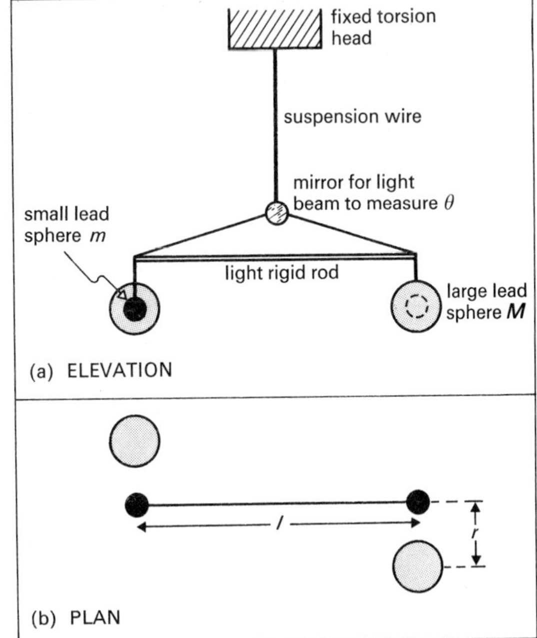

Fig. 20.14. Cavendish's torsion balance.

is deflected by an angle θ [rad] by the torque of the gravitational forces. At equilibrium

(gravitational torque) = (elastic restoring torque)

$$\frac{GmM}{r^2}l = c\theta \qquad \text{(p. 134)}$$

The suspension constant c [N m rad^{-1}] was calculated from $T = 2\pi\sqrt{I/c}$ by a separate torsional oscillation experiment (p. 134).

The experiment is clumsy and needs many corrections.

Boys improved *Cavendish's* apparatus by the introduction of fine vitreous silica (quartz) suspension fibres. These have:

(*i*) great tensile strength, but

(*ii*) a very small radius r, and a low shear modulus (therefore small c, since $c \propto r^4$).

They enabled him to reduce the size of the apparatus *without diminishing the sensitivity*. For *Cavendish* l was ~2 m, but for *Boys* l was ~20 mm.

No experiment has yet detected a substance which plays the same screening role in gravitation as that played by dielectrics in electrostatics, and by ferromagnetic materials in electromagnetism.

V Thermal Properties of Matter

21 Temperature

21.1 The Zeroth Law of Thermodynamics

Temperature (like mass, length and time) is a *fundamental quantity*, and as such, cannot be defined in terms of simpler quantities (as velocity can be, for example). We can therefore make an *arbitrary* choice as to the *unit* in which to measure temperature, but the procedure is involved.

Suppose that bodies A and B in fig. 21.1 are left for a long period of time, so that they reach **thermal equilibrium** (defined on p. 185). The property that becomes common to the two bodies A and B is **temperature**.

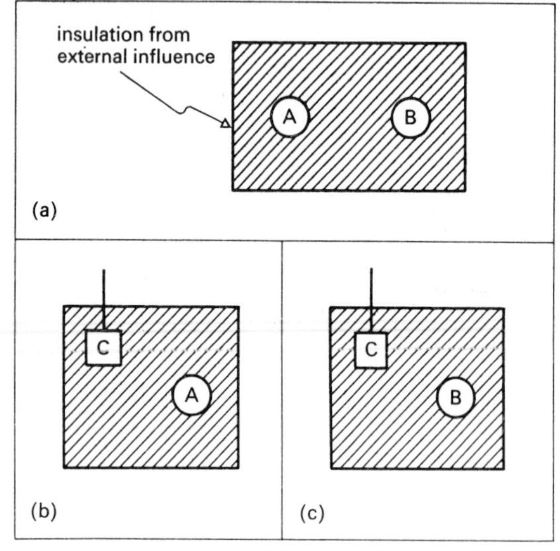

Fig. 21.1. Thermal equilibrium and the zeroth law of thermodynamics.

> **The zeroth law of thermodynamics** *states that if two bodies A and B are separately in thermal equilibrium with a third body C, then A and B are in thermal equilibrium with each other.*

C could be a *thermometer*. If C reads the same in both containers, then A and B (of fig. 21.1*b* and *c*) are at the same temperature.

A **thermometer** is a device that measures temperature (using the procedure described in this chapter).

Temperature

(*a*) is that *scalar* quantity, sometimes called degree of hotness, that is common to two systems in thermodynamic equilibrium.

(*b*) is described by a number (on an arbitrarily chosen scale) such that when two bodies are put into contact, heat flows from the body of higher temperature to that of lower temperature.

21.2 Defining a Temperature Scale

To describe a given degree of hotness by a value (number and unit) which has meaning to different people (i.e. which is *communicable*) we adopt the following procedure:

(1) We choose a given body of a given substance.

(2) We select a **thermometric property** of that body—some property whose value varies *continuously* with degree of hotness.

(3) We select two **fixed points** and measure the values of the thermometric property at the lower and upper points, and at the degree of hotness which we wish to measure. Let the measured values of X be X_l, X_u and X_m respectively.

(4) We now allot a value t_m to the degree of hotness, where

$$t_m = \frac{X_m - X_l}{X_u - X_l} \cdot N + t_l$$

in which N is the number of divisions into which we choose to divide the interval between the upper and lower fixed points, and t_l is the temperature of the lower fixed point. The interval is called the **fundamental interval**.

The equation defines the temperature t_m *on this particular scale*. Note that we have *postulated* a linear relationship between the thermometric property and the temperature measured *on that scale* (fig. 21.2). This is justified because the procedure *defines* temperature.

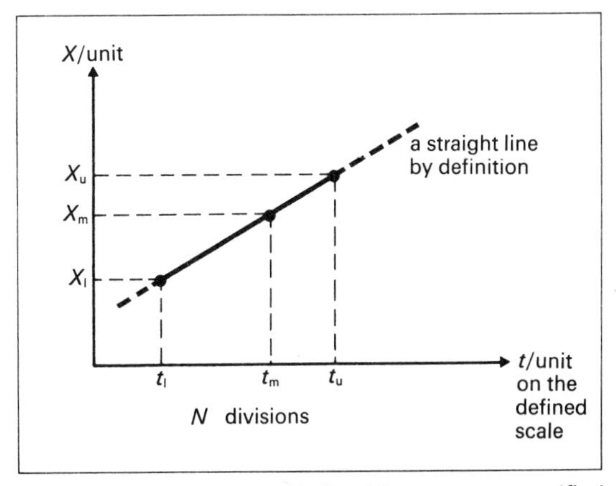

Fig. 21.2. Temperature t defined for some unspecified thermometric property X.

Thermometric Properties

The following measurable physical quantities are some of those which prove to be suitable thermometric properties:

(*a*) the length of a liquid column in a glass capillary tube,

(*b*) the electrical resistance of a platinum wire wound into a coil,

(*c*) the pressure of a gas whose volume is kept constant,

(*d*) the s.v.p. of a liquid,

(*e*) the e.m.f. of a thermocouple,

(*f*) the quality ('colour') of electromagnetic radiation emitted by a very hot source,

(*g*) the speed of sound through a gas.

Notes.

(*i*) We are entitled to extrapolate our measured values of t_m outside the fixed points, provided we remember that the scale has real meaning only over the range in which the thermometric property is measurable.

(*ii*) Temperature scales defined according to this procedure do *not* necessarily give the same number for the same degree of hotness (see p. 166 for an example). The numbers will coincide at the **calibration points**.

21.3 Realization of a Useful Temperature Scale

It is highly inconvenient to have an indefinite number of arbitrary temperature scales, on which a given degree of hotness is expressed by different values. In practice we agree to use a scale at which we arrive in the following way:

(1) The Kelvin Scale

There exists an absolute **thermodynamic scale** proposed by *Kelvin*, which is based on the theoretical efficiency of a perfectly reversible heat engine (**26.12**). Unfortunately, although this is not dependent on the properties of any particular substance, it is *theoretical* and cannot be put directly to use.

(2) The Ideal-Gas Scale

For this scale the working substance is a fixed mass of ideal gas, whose volume is to be kept at a fixed value. The thermometric property is the pressure p of this gas. This scale is made to coincide with the thermodynamic scale by appropriate choice of fixed points.

(*a*) The *lower fixed point* is **absolute zero**, to which a value 0 K is allotted. This is the temperature at which the energy of molecular motion becomes a minimum, as a molecule possesses only **zero-point energy**—its *thermal* energy is zero.

(*b*) The *upper fixed point* is the temperature of the **triple point** of pure air-free water, to which (for historical reasons) a value of 273.16 K is allotted. (This is that unique temperature at which water exists in the solid, liquid and gaseous phases in equilibrium.)

The choice of these two fixed points defines our unit of temperature, the **kelvin** (K) (p. 10).

For our defining equation

$$t_m = \frac{X_m - X_l}{X_u - X_l} \cdot N + t_l$$

We write $t_m = T$ $\quad X_m = p_T$

$t_l = 0\text{ K}$ $\quad X_u = p_{tr}$

$N = 273.16\text{ K}$ $\quad X_l = p_0 = 0\text{ Pa}$

We now define the* ideal-gas temperature *T (equivalent to the thermodynamic temperature) by

$$T = \frac{p_T}{p_{tr}} \times 273.16\text{ K}$$

This defining equation is consistent with our definition of an ideal gas (on a macroscopic scale) as one which satisfies the equation pV = constant, at constant temperature, for all values of p. It follows that

$$pV \propto T$$

which is illustrated in fig. 21.3.

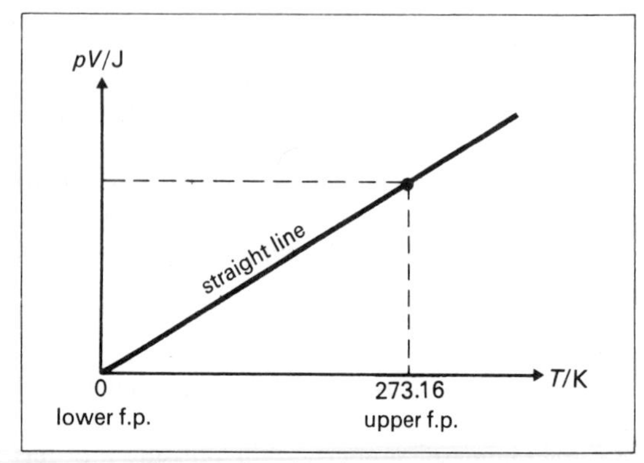

Fig. 21.3. The absolute ideal-gas scale.

(3) Real-Gas Scales

Although the ideal gas does not exist, it is found in practice that different **real-gas scales** give particularly close agreement over a wide range of temperatures when the gas pressure is low (i.e. when the molecules are well separated). When we extrapolate to *zero pressure*, the behaviour of a real gas becomes ideal, and the agreement is exact (p. 178).

Real-gas temperature is calculated from

$$T = T_{tr} \lim_{p \to 0} \frac{(pV)_T}{(pV)_{tr}}.$$

in which $T_{tr} = 273.16\text{ K}$.

In this way a *real constant volume gas thermometer* is used as a **standard** against which other thermometers can be calibrated. The thermometric substance is a 'permanent' gas (such as hydrogen, helium or nitrogen), whose pressure is the thermometric property. The values that we measure in this way are corrected by a small fraction of a kelvin to obtain the temperature that would be recorded by an ideal-gas thermometer, and so to the thermodynamic scale.

(4) The International Practical Temperature Scale (I.P.T.S.–1968)

In practice the real-gas thermometer is inconvenient (p. 167), and so to reduce experimental difficulties there is general agreement to adopt the following procedure.

(*a*) Using the standard thermometer, temperatures have been assigned to a number of reproducible equilibrium states—these are called the **defining fixed points**, and are available for *calibration* purposes.

(*b*) A particular instrument is specified for measuring temperatures within a particular range, and the temperature is calculated from a given formula. The diagram on the right shows the general scheme.

(*c*) Extrapolation is necessary beyond 1338 K:

(*i*) Optical pyrometers (p. 227) give values based on the *Planck* law of radiation.

(*ii*) Spectroscopic methods are used at very high temperatures, but the number expressing T depends upon the method.

(*d*) At very low temperatures the speed of sound through helium gas and the vapour pressure of ^{4}He and ^{3}He are used as thermometric properties.

21.4 The Celsius Temperature

Experiment shows

(*a*) that the **ice point** $T_{ice} = 273.15\text{ K}$. This is the thermodynamic temperature at which air-saturated

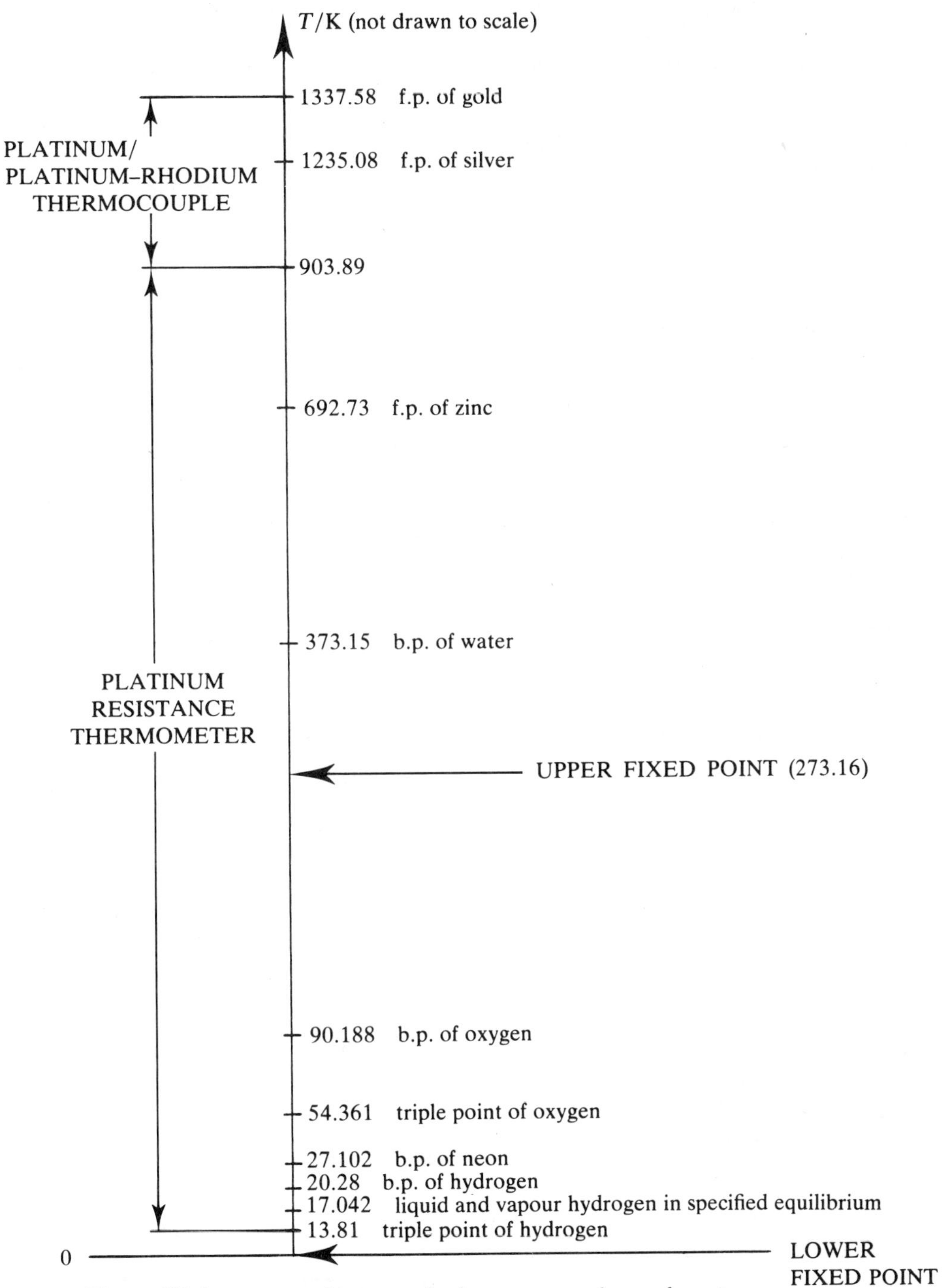

The equilibrium states refer to standard pressure p_0 where relevant.

water is in equilibrium with ice at standard pressure (101 325 Pa).

(*b*) that the **steam point** $T_{st} = 373.15$ K. This is the temperature at which water vapour is in equilibrium with pure boiling water at standard pressure.

There exists a Celsius thermodynamic temperature interval θ_C, or Celsius temperature defined by

$$\theta_C/°C = T/K - 273.15$$

Notes on the Celsius Scale

(*a*) The unit on this scale is the **degree Celsius** (°C), which represents an identical temperature *interval* to the kelvin (K).

(*b*) At the ice point $\theta_C = 0$ °C, and at the steam point $\theta_C = 100$ °C.

Many mercury thermometers in everyday use are calibrated at these temperatures, and are conveniently marked in degrees Celsius, rather than in kelvins. They have a fundamental interval of 100 degrees Celsius.

(*c*) Celsius temperatures are called customary or **common temperatures** (to distinguish them from the *thermodynamic* temperatures). [Sometimes they are incorrectly called 'centigrade temperatures'.]

(*d*) Although the degree Celsius is not an SI unit, in this book we adopt the following policy:

(1) All temperature *intervals* are in *kelvins* (K).

(2) Temperature *values* are in degrees Celsius (°C) where they give information a student is likely to obtain himself with a mercury thermometer (but see 3)). We use the symbol θ.

(3) Temperature *values* are given in kelvins in tables and in situations where thermodynamic temperatures *must* be used (as for the treatment of gases and in radiation theory). We use the symbol T.

21.5 Agreement between Temperatures Measured by Different Thermometers

The temperature recorded by one thermometer will *not* (if calculated from the equation defining temperature t on p. 163) *necessarily* agree with a second thermometer, except at the calibration points, where they *must* agree. One would obtain the same temperature from all different thermometers only if all thermometric properties varied with degree of hotness *in the same way*, which they do not.

Example. In fig. 21.4*b* the water vapour pressure thermometer was calibrated at the ice and steam points. When water vapour exerts a pressure of 92 mmHg, then in

(*a*) we see that the thermodynamic temperature is 323 K, but in

(*b*) the water vapour pressure scale temperature is close to 285 K.

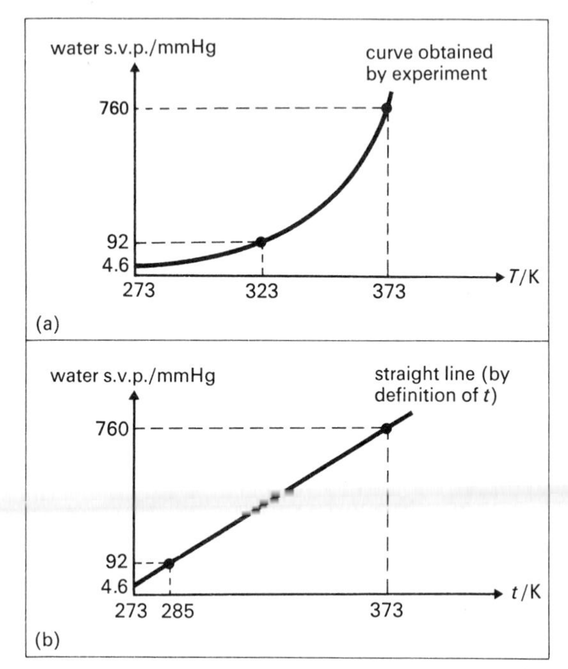

Fig. 21.4. In (a) T is measured on the thermodynamic scale, whereas in (b) t is the temperature value given by a water vapour pressure thermometer. (Not to scale.)

285 K is *not wrong*, but the fact that the water vapour pressure scale was used must be quoted explicitly if the value is to be meaningful.

In school laboratories thermometric substances and properties are chosen in such a way that for normal temperature ranges they represent a given degree of hotness by nearly equal numbers, and approximate to the thermodynamic scale.

21.6 Thermometers in Practice

These descriptions should be read together with the table of paragraph **21.7** (page 168).

(*a*) The Constant-Volume Gas Thermometer

In fig. 21.5 two adjustments are required:

(1) The height of the reservoir is adjusted until the mercury just touches the lower ivory pointer. (This ensures constant volume.)

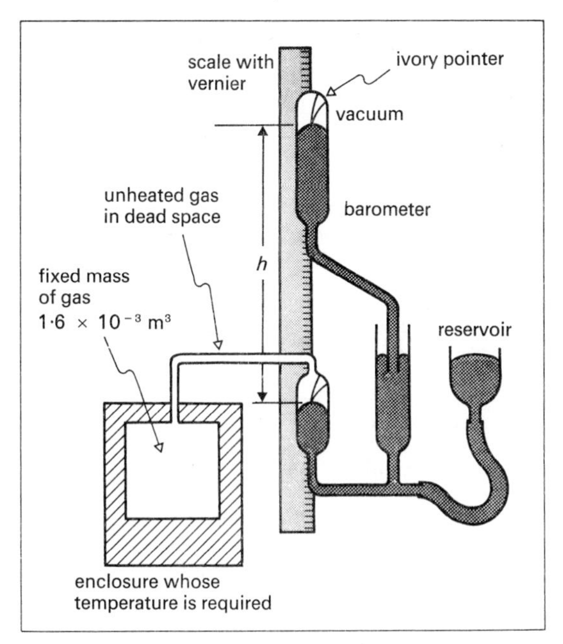

Fig. 21.5. An accurate constant-volume gas thermometer (Harker–Chappuis).

(2) The tube containing the upper ivory pointer is adjusted until the pointer just touches the barometer mercury.

The thermometric property is the gas pressure p, given by $p = h\rho g$.

Corrections are made for

(*i*) thermal expansion of bulb,
(*ii*) the gas in the dead-space not being at the required temperature,
(*iii*) capillary effects at glass–mercury contact,
(*iv*) the fact that a real gas is used—corrections are made for non-ideal behaviour.

(*b*) The Platinum Resistance Thermometer

In fig. 21.6 AB is a uniform resistance wire of resistance per unit length r.

Procedure

(*i*) Select P accurately equal to Q.
(*ii*) Position the jockey J at an appropriate point C

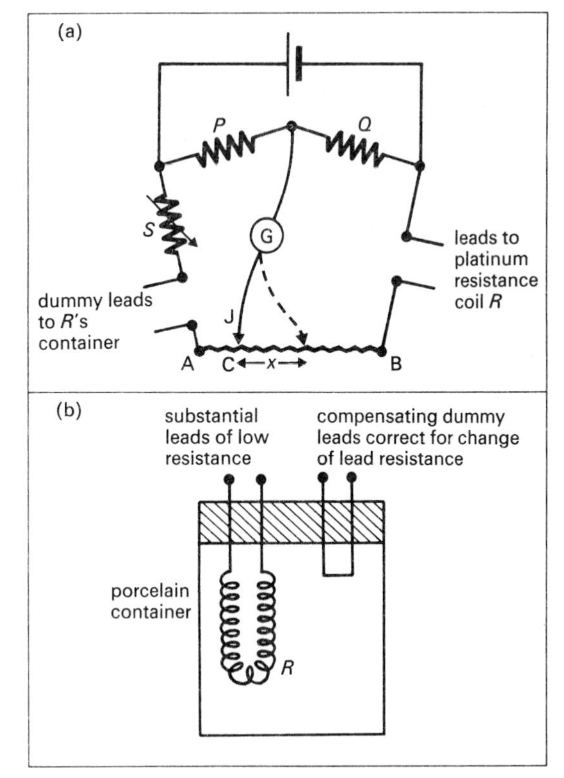

Fig. 21.6. The platinum resistance thermometer:
(a) the bridge circuit, and
(b) the platinum coil.

on AB, and alter S so as to balance the bridge. Suppose AC = l_1 and BC = l_2.

(*iii*) Put the porcelain container into the enclosure, the temperature of which is required: suppose J has to be moved a distance x to the right to retain balance.

Then because $\quad P = Q$

$$S + (l_1 + x)r = R_T + (l_2 - x)r$$

where R_T is the new coil resistance.

If one assumes a *parabolic* relation between R_T and temperature measured on the I.P.T.S., then calibration at *three* fixed points enables that temperature to be calculated.

(*c*) The Thermocouple Thermometer

The *Seebeck effect* is discussed on p. 363.

The resulting thermoelectric e.m.f. can be measured accurately by the potentiometer method described on p. 369.

21.7 Thermometers Compared

Name of thermometer	*Thermometric property*	*Maximum temperature range*/K	*Advantages*	*Disadvantages*	*Other comment*
(*a*) Mercury-in-glass	length of mercury column in capillary tube	234 to 630	(*i*) portable (*ii*) direct reading	(*i*) relatively small (*ii*) not very accurate	(*i*) the upper range can be extended beyond the normal b.p. (*ii*) calibrated for accurate work by reference to (*b*)
(*b*) Constant-volume gas	pressure of a fixed mass of low-density gas maintained at constant volume	3 to 1750	(*i*) very wide range (*ii*) very accurate (*iii*) very sensitive	(*i*) cumbersome (the bulb may have a capacity of about 10^{-3} m^3 (*ii*) very inconvenient and slow to use	(*i*) the *standard* by which others are ultimately calibrated (*ii*) He, H_2 and N_2 are used for different temperature ranges (*iii*) can be corrected to the *ideal*-gas scale
(*c*) Platinum resistance	electrical resistance of a platinum coil	13.8 to 1400	(*i*) wide range (*ii*) best for small steady differences of temperature (*iii*) the most accurate thermometer in the range 13.8 K to 904 K	not suitable for varying temperatures because of (*i*) large heat capacity, and (*ii*) use of materials of low thermal conductivity in sheath	using an agreed formula, it is used to define the temperature from 13.8 K to 904 K on the international practical temperature scale (I.P.T.S.)
(*d*) Thermo-couple	e.m.f. of two wires of dissimilar metals	25 to 1750	(*i*) wide range (*ii*) very small, so useful for both local and rapidly varying temperatures (*iii*) the most accurate thermometer in the range 904 K to 1338 K	not as accurate as (*c*) in the range 13.8 K to 904 K	(*i*) using an agreed formula, it is used to define the temperature from 904 K to 1338 K on the I.P.T.S. (*ii*) different pairs of metals are chosen for different temperature ranges
(*e*) Radiation pyrometer (see p. 226)	quality of electro-magnetic radiations emitted by hot body	beyond about 1250	the only instrument capable of measuring temperatures above about 1750 K	(*i*) cumbersome (*ii*) not direct reading	(*i*) defines I.P.T.S. temperatures above 1338 K (*ii*) readings above 1750 K are calculated from *Planck's* radiation laws

In considering their relative merits, we should remember that

(*i*) a thermometer absorbs heat from the substance whose temperature it is measuring,

(*ii*) a thermometer records its own temperature,

(*iii*) there may be a time-lag before thermal equilibrium is reached between a thermometer and its surroundings.

22 Expansion of Solids and Liquids

22.1 Molecular Behaviour

As the common temperature θ increases, so does the energy and amplitude of the molecular vibration. The asymmetry of the curve of fig. 22.1 (departure from true parabolic shape) indicates

(a) that the molecule does not describe s.h.m., and
(b) that r increases as θ increases.

This microscopic behaviour is magnified on a macroscopic scale, and appears as an increase in separation of *any two points* in an isotropic homogeneous solid.

Substances that shrink with temperature increase do not have a $E_p - r$ curve of this shape.

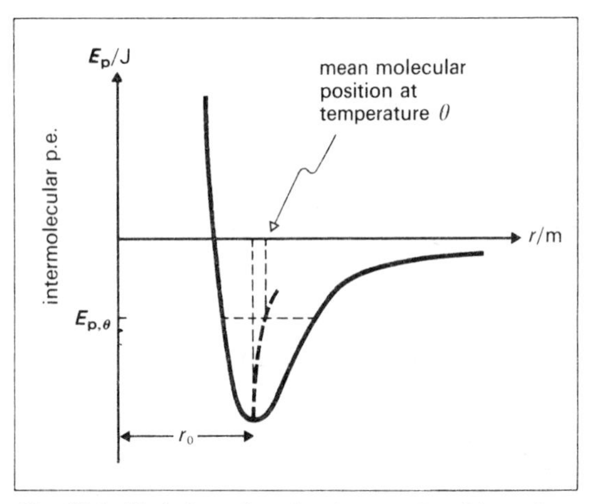

Fig. 22.1. Variation of molecular p.e. E_p with nuclear separation r.

22.2 Linear Expansion

For a change of common temperature $\Delta\theta$, experiment shows (fig. 22.2) that

(a) $\dfrac{\Delta l}{l_0} \propto \Delta\theta$ approximately, and

(b) $\dfrac{\Delta l}{l_0\,\Delta\theta}$ depends on the substance under test,

where Δl is the change of length resulting from $\Delta\theta$.

The linear expansivity α of a substance is defined by the equation

$$\boxed{\alpha = \frac{\Delta l}{l_0\,\Delta\theta}}$$

$$\alpha\left[\frac{1}{\text{K}}\right] = \frac{\Delta l[\text{m}]}{l_0[\text{m}]\,\Delta\theta[\text{K}]}$$

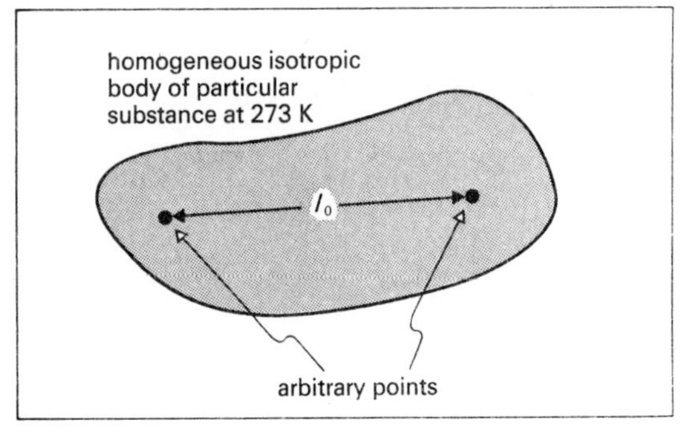

Fig. 22.2. Thermal expansion.

It follows that

$$l_\theta = l_0(1 + \alpha\theta)$$

where θ is the body's common temperature.

Although α is defined relative to l_0 measured at 273 K, it varies only a little with temperature, and we can usually justify the approximation

$$l_2 = l_1\{1 + \alpha(\theta_2 - \theta_1)\}$$

when $(\theta_2 - \theta_1)$ is the order of 100 K. (But note from fig. 22.1 that $\alpha \to 0$ at low temperatures, since the energy of vibration is then small.)

α for a metal is about $10^{-5}\,\mathrm{K}^{-1}$. To measure α we use

$$\alpha = \frac{\Delta l}{l_0\,\Delta\theta}$$

l_0 and $\Delta\theta$ are easily measured. Δl is very small, so we either

(*a*) magnify it by a device such as an optical lever, or

(*b*) use a very sensitive measuring instrument (as in the *comparator* method).

The change ΔA of a body's area can be calculated from

$$\Delta A = A_0\beta\,\Delta\theta$$

where β, the material's **superficial expansivity**, is approximately 2α. (See below.)

22.3 Volume Expansion

The cubic expansivity** γ **of a substance is defined by

$$\boxed{\gamma = \frac{\Delta V}{V_0\,\Delta\theta}}$$

$$\gamma\left[\frac{1}{\mathrm{K}}\right] = \frac{\Delta V[\mathrm{m}^3]}{V_0[\mathrm{m}^3]\,\Delta\theta[\mathrm{K}]}$$

By considering the linear expansion of the sides of a cube, it is easily shown that $\gamma \approx 3\alpha$ (since $\alpha^2 \ll \alpha$).

For a typical metal α and $\gamma \sim 10^{-5}\,\mathrm{K}^{-1}$, and so $\alpha^2 \sim 10^{-10}\,\mathrm{K}^{-2}$.

Comments on Volume Expansion

(*a*) The equation $V_\theta = V_0(1 + \gamma\theta)$ is useful.

(*b*) Hollow bodies of a given material (such as a relative density bottle) expand without distortion as though they were solid. Their internal and external dimensions change in the same proportion.

(*c*) The term *linear expansivity* has no meaning for a liquid or gas, since a fluid takes the shape of its container.

(*d*) When the temperature of a liquid changes, the internal volume of its container will also change, and one must distinguish between the **real** and **apparent cubic expansivity**.

Variation with Temperature of Density

For a fixed mass m of a substance

$$m = V_1\rho_1 = V_2\rho_2$$

$$\therefore \quad \frac{\rho_1}{\rho_2} = \frac{V_2}{V_1} = 1 + \gamma(\theta_2 - \theta_1)$$

Pressure changes in liquids are given by

$$\Delta p = \Delta h\rho g \qquad \text{(p. 68)}$$

A comparison of pressures enables densities to be compared and so γ can be evaluated. This is the basis of the *balancing column* method.

(Remember that water is anomalous in that *its density increases* when it is heated from its melting point 273 K to 277 K and then decreases on further heating.)

23 Heat Capacity

23.1 Heat and Work

> ***Work*** *W is energy that is transmitted from one system to another when each exerts on the other macroscopic forces whose points of application are moved through finite distances.*

When two parts of a body are at different temperatures the molecules in different regions will have different average vibrational energies, and this will enable the more energetic molecules to do work (but on a microscopic scale) on the less energetic.

> ***Heat*** *Q is that energy that moves from one point to another because there exists between them a temperature difference.*

Conduction and (less obviously) radiation are both methods of energy transfer that involve work being done on a microscopic scale. Work and heat thus both represent different forms of *energy in transit*, and both are measured in joules.

Once energy has been absorbed by a system we have no way of telling whether it was transferred as work or heat. Within the system it is referred to simply as *internal energy*.

> *The* ***internal energy*** *U of a system is the sum total of the kinetic and mutual potential energies of the particles that make up the system.*

The relationship between W, Q and U is discussed on p. 188 under the *first law of thermodynamics*.

23.2 Definitions

Suppose heat ΔQ added to a *body* causes an increase $\Delta\theta$ in its common temperature θ.

> ***The heat capacity C of the body is defined by***

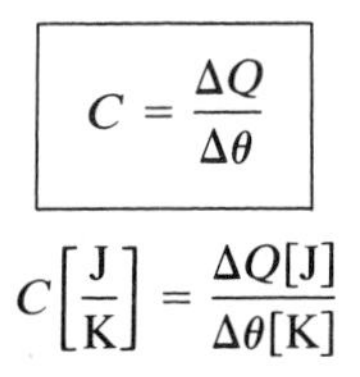

$$C = \frac{\Delta Q}{\Delta\theta}$$

$$C\left[\frac{\text{J}}{\text{K}}\right] = \frac{\Delta Q[\text{J}]}{\Delta\theta[\text{K}]}$$

The value of C may depend on the temperature θ in the vicinity of which $\Delta\theta$ is measured: strictly we measure only an *average* value.

(The word **specific** used in front of a physical quantity should always mean 'per unit mass' (e.g. kg^{-1}). An example of this use follows.)

Specific heat capacity is the heat capacity of unit mass of a *substance* making up a body. Suppose we add heat ΔQ to a mass m of a substance, and that this causes an increase $\Delta\theta$ in its common temperature.

> ***The specific heat capacity c of the substance is defined by the equation***

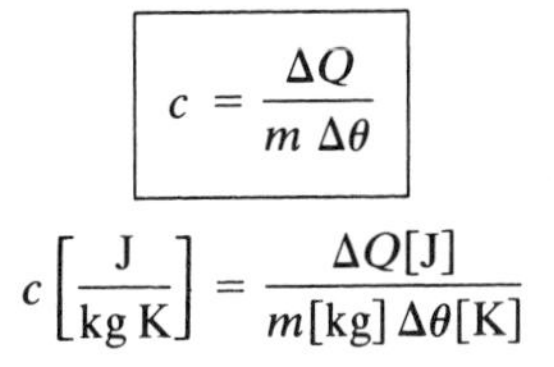

$$c = \frac{\Delta Q}{m\,\Delta\theta}$$

$$c\left[\frac{\text{J}}{\text{kg K}}\right] = \frac{\Delta Q[\text{J}]}{m[\text{kg}]\,\Delta\theta[\text{K}]}$$

Notes.

(*a*) The value of c may depend on the temperature θ in the vicinity of which $\Delta\theta$ is measured. If $\Delta\theta = (\theta_2 - \theta_1)$, then we measure only the *average* value over that range. As $\Delta\theta \to 0$, then we have measured c *at* the temperature $(\theta_1 + \theta_2)/2$.

(*b*) The equation

$$\boxed{\Delta Q = mc\,\Delta\theta}$$

is a useful relationship that follows from the definition. When the temperature is close to 300 K, or above, we usually assume that c remains constant over the range $\Delta\theta$.

(*c*) $C = \Delta Q/\Delta\theta = mc$ is a useful relationship for calculating the heat capacity C.

(*d*) For gases the **molar heat capacity** C_m (i.e. heat capacity *per mole*) is a more useful quantity than *specific* heat capacity. It is defined on p. 188.

(*e*) The dimensions of c are $[L^2T^{-2}\Theta^{-1}]$. Those of C_m are $[ML^2T^{-2}\Theta^{-1}N^{-1}]$.

Orders of Magnitude

Substance	*Temperature of measurement*	$c/\mathrm{J\,kg^{-1}\,K^{-1}}$	$C_m/\mathrm{J\,mol^{-1}\,K^{-1}}$
helium (at constant pressure)	10–3000 K	5.2×10^3	21
copper	23 K	1.4×10^1	0.9
copper	293 K	3.9×10^2	24
lead	293–373 K	1.3×10^2	26
water	288 K	4.2×10^3	75

23.3 General Notes on Calorimetry

Summary of Available Methods

Method	*Can be used to find (specific) heat capacity of*		
	Solid	*Liquid*	*Gas*
(*a*) electrical	yes	yes	$C_{V,m}$
(*b*) continuous-flow (usually electrical)	no	yes	$C_{p,m}$
(*c*) cooling	no	yes	no
change of phase	yes	yes	$C_{V,m}$
mixtures	yes	yes	$C_{p,m}$

Methods (*a*), (*b*) and (*c*) are described in this chapter.

Minimizing the Effects of Heat Loss

Several techniques are available, which include

(*a*) Polishing to reduce radiation loss; an outer container to reduce convection and conduction loss, and insulating supports to reduce conduction.

(*b*) Arranging (by preliminary experiment) for the initial temperature to be as much below room temperature as the final one will be above.

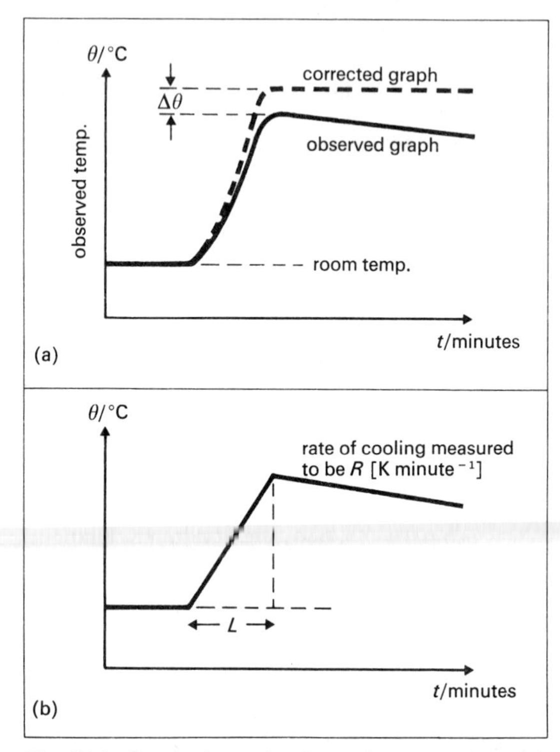

Fig. 23.1. Graphs for a simple cooling correction; (a) a real $\theta - t$ curve, and (b) an idealized simplification.

(*c*) Surrounding the calorimeter with an electrically heated jacket at calorimeter temperature. A thermocouple can be used to operate the heater circuit switch.

(*d*) *A simple cooling correction*:
Record θ throughout the experiment, and plot a $\theta - t$ graph (fig. 23.1).

Suppose the rate of cooling at the maximum observed temperature is R. Then the average rate of cooling over the heating period $L \approx R/2$.

Thus the *correction* to the observed maximum is given by

$$\Delta\theta\,[\mathrm{K}] = \tfrac{1}{2}R\left[\frac{\mathrm{K}}{\mathrm{min}}\right]L[\mathrm{min}]$$

For this correction we assume that the rate of loss of heat is proportional to the excess temperature. This procedure is justified for heat loss by

(*a*) *conduction* (see p. 216).
(*b*) *convection*, provided either
(*i*) the convection is forced, or
(*ii*) the temperature excess is small (p. 175),
and
(*c*) *radiation*, provided the temperature excess is small (p. 225).

Methods of Measuring *c*

23.4 (a) Electrical Methods

Introduction

The electrical energy converted to thermal energy in the resistor is given by

$$W[\mathrm{J}] = V[\mathrm{V}]I[\mathrm{A}]t[\mathrm{s}]$$

W can be found since V, I and t can all be measured both accurately and conveniently.

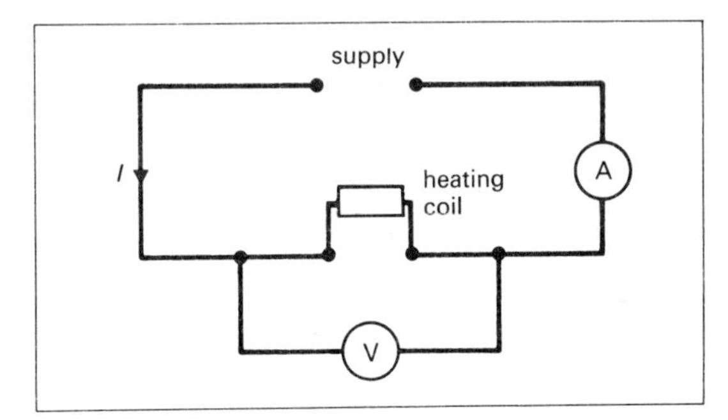

Fig. 23.2. The basic circuit.

(1) Nernst's Method for Solids

Can be used for solids that are good conductors (fig. 23.3).

Special Features

(*i*) the solid needs no calorimeter,
(*ii*) the heating coil can be used as a resistance thermometer.

In the equation

$$\Delta Q = mc\,\Delta\theta = VIt$$

we measure m, $\Delta\theta$, V, I and t for a small range $\Delta\theta$ (which can be as small as 10^{-3} K) near a particular θ.

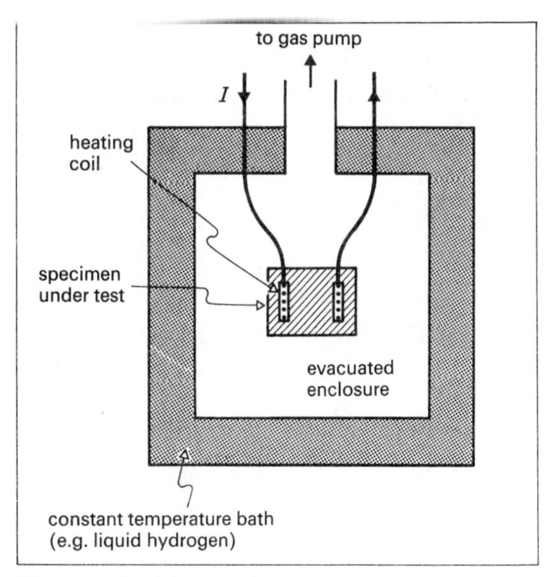

Fig. 23.3. The Nernst calorimeter.

Hydrogen (a good gaseous conductor) can then be admitted to the evacuated enclosure to change the temperature, the enclosure again evacuated, and the process repeated.

(2) A Vacuum Flask Method for Liquids

(*i*) Fill the flask (fig. 23.4) to a known depth with liquid (such as water) of known specific heat capacity c_1.

$$V_1I_1t = m_1c_1\,\Delta\theta + Q$$

where Q is the sum of the energy taken by the flask, and that lost by radiation and conduction.

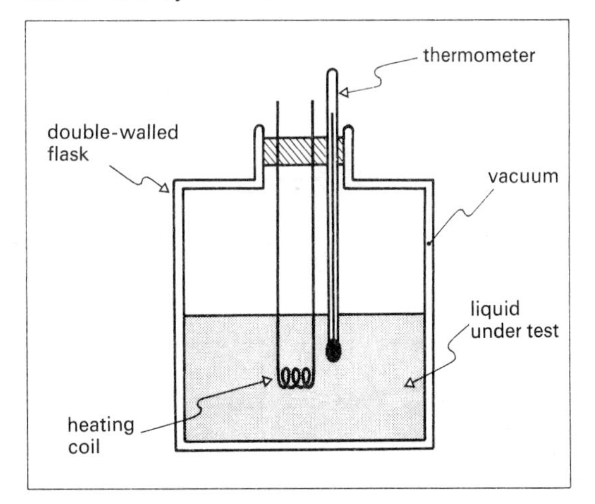

Fig. 23.4. Using a vacuum flask to measure the specific heat capacity of a liquid.

(*ii*) Fill *to the same depth* with liquid 2 whose specific heat capacity c_2 is to be found.

Adjust V and I so as to obtain the same $\Delta\theta$ in the same t.

$$V_2 I_2 t = m_2 c_2 \, \Delta\theta + Q$$

Since Q is the same as in (*i*), it can be eliminated and c_2 evaluated.

23.5 (b) Continuous-Flow Method

Procedure

(1) Pass the liquid through the calorimeter of fig. 23.5 until steady conditions hold.

Measure V_1, I_1, θ_1, θ_2, and the rate of liquid flow m_1[kg] in time t [s].

Then

$$V_1 I_1 t = m_1 c(\theta_2 - \theta_1) + Q$$

where c[J kg^{-1} K^{-1}] is the liquid specific heat capacity, and Q is the energy loss by radiation in time t *under these conditions.*

(2) Repeat, using adjusted values V_2, I_2 and m_2 so that θ_1 and θ_2 are *unaltered.*

Then for the same t

$$V_2 I_2 t = m_2 c(\theta_2 - \theta_1) + Q$$

where Q, the energy loss by radiation, is unchanged under identical conditions for θ_1 and θ_2.

(3) Subtracting

$$(V_1 I_1 - V_2 I_2)t = (m_1 - m_2)c(\theta_2 - \theta_1)$$

from which c can be calculated.

Notes.

(*a*) The values of $(V_1 I_1 - V_2 I_2)$ and $(m_1 - m_2)$ should be as large as possible to reduce the percentage error, though V_2, I_2 and m_2 should not be small.

(*b*) The method is convenient for finding the variation of the specific heat capacity of a liquid with temperature. (See also p. 172.)

Advantages

(*a*) Radiation loss is quantitatively controlled.

(*b*) Convection losses are eliminated.

(*c*) Conduction losses are negligible.

(*d*) Under steady temperature conditions no thermal energy is absorbed by the calorimeter.

(*e*) The temperatures to be recorded are static.

(*f*) $(\theta_2 - \theta_1)$ can be measured accurately with a differential resistance thermometer.

(*g*) V and I are accurate readings made by potentiometer methods.

Experimental Detail

(*a*) A large quantity of liquid at constant temperature must be available.

(*b*) A constant head apparatus is essential to ensure steady flow and steady temperature conditions.

23.6 (c) A Cooling Method

(Cooling methods are not in themselves important, but are mentioned here as an example of how to tackle problems on cooling.)

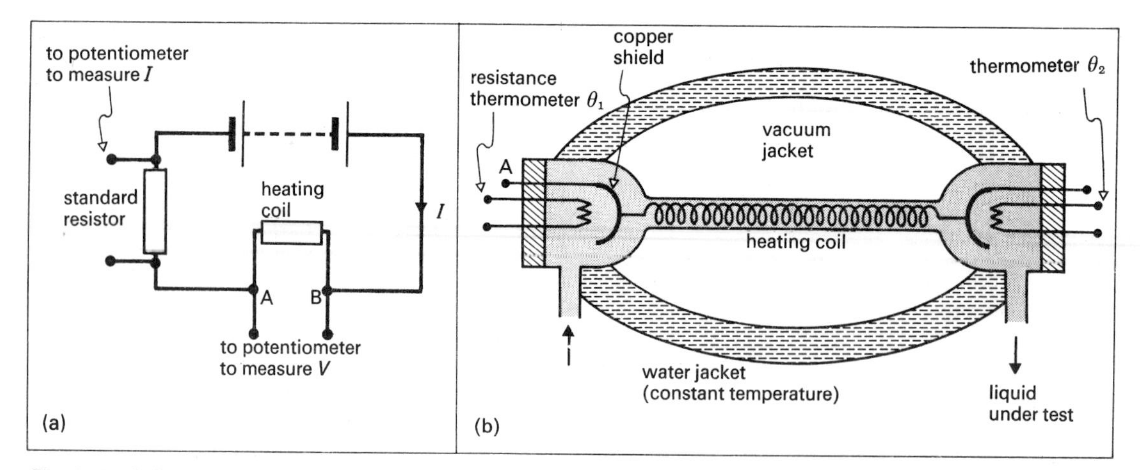

Fig. 23.5. Callendar and Barnes's continuous-flow method: (a) the circuit, and (b) details of the calorimeter.

Some Cooling Laws

(i) The Five-Fourths Power Law

For cooling by natural *convection (in a still atmosphere)*

$$(\textit{rate of heat loss}) \propto (\theta - \theta_s)^{5/4}$$

for a body at temperature θ in surroundings at temperature θ_s.

For small values of $(\theta - \theta_s)$, the five-fourths law approximates to a linear law, and for *cooling corrections* we usually assume the linear relationship (p. 173).

(ii) Newton's Law of Cooling

For cooling under conditions of forced convection (i.e. in a steady draught)

$$(\textit{rate of heat loss}) \propto (\theta - \theta_s)$$

for all methods by which the body loses heat.

This is an empirical law which holds for quite large temperature differences.

**The Use of Newton's Law*

Newton's Law states

$$\frac{dQ}{dt} \propto -(\theta - \theta_s)$$

where the minus sign indicates that the body is cooling ($d\theta/dt$ is negative).

Since $$\frac{dQ}{dt} = mc\frac{d\theta}{dt} = C\frac{d\theta}{dt}$$

we can write

$$C\frac{d\theta}{dt} \propto -(\theta - \theta_s)$$

or $$\frac{d\theta}{(\theta - \theta_s)} = -k\,dt$$

The value of k (which is usually taken to be constant) takes account of

(*a*) the heat capacity C,
(*b*) the body's surface area,
(*c*) the nature of its surface (e.g. its roughness), and
(*d*) the shape of the surface (convex, flat or concave, etc.) and its orientation (vertical, horizontal, etc.).

Suppose a body cools from a temperature excess $(\theta - \theta_s)$ of $\Delta\theta_0$ to one of $\Delta\theta$ in a time interval Δt. Integrating we have

$$\ln\left(\frac{\Delta\theta}{\Delta\theta_0}\right) = -k\,\Delta t$$

or $$\Delta\theta = \Delta\theta_0\,e^{-k\Delta t}$$

This result can be used for calculating changes of temperature under conditions to which *Newton's Law* applies.

Comparison of Liquid Specific Heat Capacities

Principle

If two bodies are to outward appearance identical (same size, shape and surface finish), then under the samc cooling conditions (same $(\theta - \theta_s)$ and degree of convection) they will lose heat *at the same rate*.

This does *not* depend on any law controlling the rate of cooling.

Method

(1) Take equal *volumes* of two liquids.
(2) Place them successively in the same calorimeter.
(3) Heat them to a conveniently high temperature.
(4) Plot a $\theta - t$ curve for their cooling under identical conditions (fig. 23.6).

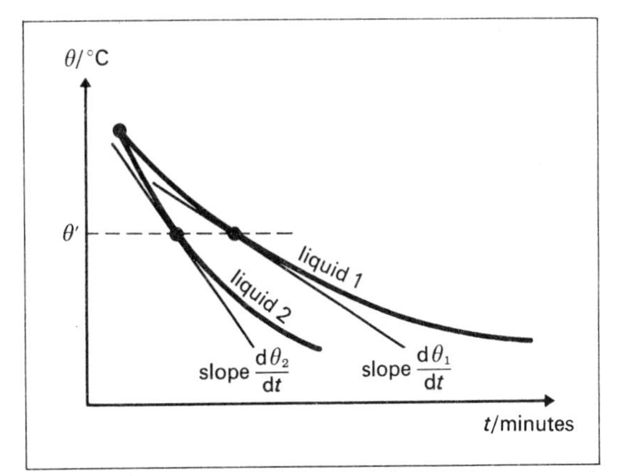

Fig. 23.6. Cooling curves being used to compare liquid specific heat capacities.

Because $Q = mc\theta$,

$$\frac{dQ}{dt} = mc\frac{d\theta}{dt} = C\frac{d\theta}{dt}$$

At the selected common temperature θ' shown on the graph, the rates of loss of heat of both liquids in their calorimeters are the same. (The nature of the liquids inside the calorimeter cannot affect dQ/dt, provided they are stirred.)

$$(C_c + m_1c_1)\frac{d\theta_1}{dt} = (C_c + m_2c_2)\frac{d\theta_2}{dt}$$

where C_c is the heat capacity of the calorimeter. If c_1 is known, c_2 can be found.

Note that the liquid of smaller c cools the more rapidly.

23.7 The Effect of Pressure on Specific Heat Capacities

The *first law of thermodynamics* (p. 188) can be written

$$\begin{pmatrix}\text{heat given}\\\text{to body}\end{pmatrix} = \begin{pmatrix}\text{increase}\\\text{of}\\\text{internal}\\\text{energy}\end{pmatrix} + \begin{pmatrix}\text{work done}\\\text{by body}\end{pmatrix}$$

The change of internal energy controls the observed change of temperature $\Delta\theta$. When we add a given amount of heat ΔQ, the observed value of $\Delta\theta$ depends upon how much work the body does in pushing against the force exerted by its surroundings.

For convenience we distinguish these two **principal specific heat capacities**:

(a) The specific heat capacity at **constant volume** c_V is the value it takes if the volume change is zero (so that the work done by the body is zero).

(b) The specific heat capacity at **constant pressure** c_p is its value when the work done by the body (if it expands) is in opposition to a nearly constant force.

Solids and *liquids* have very large bulk moduli, which means that we cannot control their volumes. In practice, therefore, we only work with c_p. Comparison with theory, which gives c_V, requires the difference $(c_p - c_V)$ to be taken into account.

The effect of the volume change for *gases* is discussed on p. 189.

23.8 Molar Heat Capacity of Elements

The **molar heat capacity** C_m of an element (sometimes called its atomic heat) is the heat capacity of one mole of its atoms. If we put M_m = molar *mass* (*not* relative molecular mass), then we calculate C_m from

$$\boxed{C_m = M_m c}$$

$$C_m\left[\frac{\text{J}}{\text{mol K}}\right] = M_m\left[\frac{\text{kg}}{\text{mol}}\right] c\left[\frac{\text{J}}{\text{kg K}}\right]$$

Dulong and Petit (1819) found empirically that for many solid elements C_m is about 25 J mol^{-1} K^{-1} at ordinary temperatures.

Hard elements such as diamond, are important exceptions to this rule, having molar heat capacities that are lower at room temperature, but which approach 25 J mol^{-1} K^{-1} at higher temperatures (fig. 23.7).

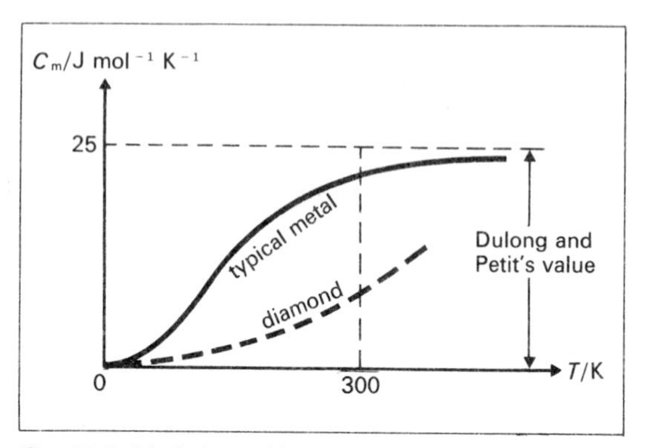

Fig. 23.7. Variation with temperature T of molar heat capacity C_m.

If we ignore the free electrons in a metal, the high temperature values can be explained by classical kinetic theory (below). The shape of the curve at low temperatures needs quantum theory treatment.

* Kinetic Theory Prediction of Molar Heat Capacity

Each ion in a metallic lattice has 6 degrees of freedom (p. 191), made up of two degrees for each of the possible vibrations along three mutually perpendicular directions.

The average energy for each ion is $6(\frac{1}{2}kT)$ according to the *principle of equipartition of energy* (p. 192).

The internal energy U for an amount of substance μ [mol] is given by

$$\begin{aligned} U &= \mu N_A 3kT \\ &= 3\mu RT \quad \text{since} \quad R = N_A k \end{aligned}$$

At constant volume the molar heat capacity $C_{V,m}$ is given by

$$C_{V,m} = \frac{\Delta Q}{\mu\,\Delta T} = \frac{\Delta U}{\mu\,\Delta T} = 3R$$

Since R = 8.31 J mol^{-1} K^{-1}, we predict that the value of $C_{V,m}$ should be 24.9 J mol^{-1} K^{-1}.

24 Ideal Gases: Kinetic Theory

24.1 Introduction

Definitions

The **Avogadro constant** N_A *is numerically equal to the number of atoms in exactly* 12×10^{-3} kg *of the nuclide carbon*-12.

By experiment

$$N_A = 6.022\,169 \times 10^{23}\ \text{mol}^{-1}$$

N_A has the dimensions of [amount of substance]$^{-1}$.

The* relative molecular mass** M_r (commonly miscalled molecular weight) ***is defined by

$$M_r = \frac{\textit{mass of a molecule}}{\textit{mass of } ^{12}_{6}\text{C } \textit{atom}} \times 12$$

Because M_r is a *dimensionless* quantity it has no unit.

Examples

Substance	M_r	M_m/kg mol^{-1}
hydrogen	2	2×10^{-3}
helium	4	4×10^{-3}
carbon-12	12 (exactly)	12×10^{-3}
nitrogen	28	28×10^{-3}
oxygen	32	32×10^{-3}

The **mole** was defined on page 10. The *mass* of a mole is called the **molar mass** M_m, and is measured in kg mol^{-1} (see the above table).

Take great care not to confuse the *relative molecular mass* M_r and the *molar mass* M_m: they have different dimensions, and in the SI their numerical values differ by a factor of 10^{-3}.

Thus $$M_m = \frac{M_r}{1000}\ \text{kg mol}^{-1}$$

Symbols

Note the following carefully. Complete familiarity with them now will prevent confusion later.

(*a*) For a *sample of gas* in a particular enclosure

N = total number of gas molecules

n = number of molecules in unit volume (the **number density** of molecules), which has dimensions [L^{-3}]

(N_A = the *Avogadro constant*)

†μ = the amount of gas (which will be expressed in moles)

M_g = the mass of the gas sample.

† (The recommended symbol for an amount of substance is n; our use of n for number density forces us to choose a different symbol.)

(*b*) For a *particular molecular species*

m = mass of a molecule

M_r = **r**elative molecular mass

M_m = **m**olar mass.

These relationships follow

(*i*) $\mu\,[\text{mol}] = \dfrac{N\,[\text{number of molecules}]}{N_A[\text{number of molecules mol}^{-1}]}$

$$\mu = \frac{N}{N_A}$$

(*ii*) $\mu\,[\text{mol}] = \dfrac{M_g[\text{kg}]}{M_m[\text{kg mol}^{-1}]} = \dfrac{Nm}{M_m}$

$$\mu = \frac{M_g}{M_m} = \frac{Nm}{M_m}$$

24.2 The Universal Molar Gas Constant *R*

Macroscopic Observations

Early experiments on a fixed amount or a fixed mass of gas showed that many common gases conformed fairly well to the behaviour described by the so-called *gas laws*.

(1) **Boyle's Law**

If T is constant, pV = constant.

(2) **Charles's Law**

If p is constant, V/T = constant.

(3) **Pressure Law**

If V is constant, p/T = constant

The temperature T was found by adding 273 K to the *Celsius* reading of a mercury thermometer. This empirical behaviour is summarized by the gas equation

$$\boxed{\frac{p_1V_1}{T_1} = \frac{p_2V_2}{T_2}}$$

A Universal Constant

Consider a particular gas, and take samples containing different amounts of gas μ [mol]. Suppose we vary the pressure p, the volume V and the thermodynamic temperature T (p. 163) of these samples. Plot the graph shown in fig. 24.1. We find

(*a*) a series of curves, *all* of which cross the $pV/\mu T$ axis *at the same point* when extrapolated,

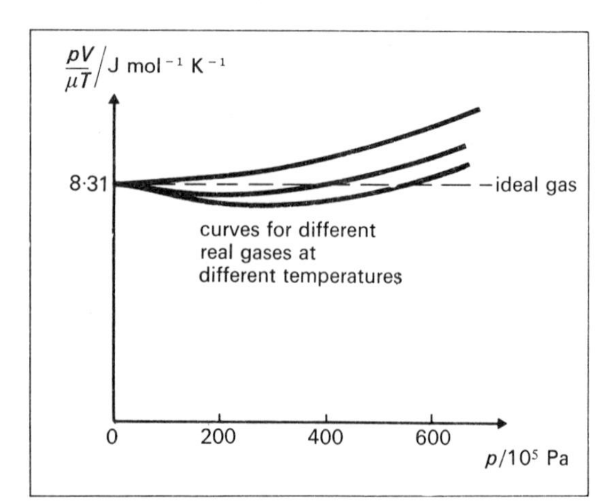

Fig. 24.1. Origin of the universal molar gas constant R.

(*b*) that other gases subjected to the same treatment give curves that cross the $pV/\mu T$ axis *at the same point.*

We deduce that the quantity $pV/\mu T$ *is the same for all gases* as the pressure tends to zero.

The Ideal Gas

On a macroscopic scale we define an ideal gas to be one which obeys *Boyle's* law exactly, i.e.

$$pV = \text{constant}$$

for a fixed μ and T, and for all values of p.

We define *ideal-gas temperature* by

$$pV \propto T. \qquad \text{(p. 164)}$$

By combining these two definitions we arrive at the **equation of state** for an ideal gas

$$\boxed{pV = \mu RT}$$

$$p\left[\frac{\text{N}}{\text{m}^2}\right]V[\text{m}^3] = \mu[\text{mol}]R\left[\frac{\text{J}}{\text{mol K}}\right]T[\text{K}]$$

The incorporation of μ into the equation has enabled us to write

$$pV/\mu T = R = 8.31\ \text{J mol}^{-1}\,\text{K}^{-1},$$

where R is a *universal* constant, called the **universal molar ideal gas constant**.

The Status of the Gas Laws

It is now seen that an ideal gas will automatically obey the gas laws, since for such a gas they are no more than definitions.

Real gases which do obey the laws are showing near-ideal behaviour.

Other forms of the Equation of State

We can write the equation in the following forms:

(*a*) For 1 mole of gas,

$$pV = (1\ \text{mol})RT$$

or

$$pV_m = RT$$

where V_m is the **molar volume** and has the unit $m^3\ mol^{-1}$.

(*b*) For a mass M_g of gas

$$pV = \left(\frac{M_g}{M_m}\right)RT = \left(\frac{\text{Nm}}{\text{M}_m}\right)RT$$

(*c*) For N molecules

$$pV = \frac{N}{N_A}RT$$

(A form using the *Boltzmann constant* k is given on p. 192.)

The Molar Volume V_m

Consider 1 mole of any ideal gas at s.t.p.

$$V = \frac{\mu RT}{p}$$

$$V_m = \frac{V}{\mu} = \frac{\left(8.31 \dfrac{\text{J}}{\text{mol K}}\right) \times (273\ \text{K})}{\left(1.01 \times 10^5 \dfrac{\text{N}}{\text{m}^2}\right)}$$

$$= 2.24 \times 10^{-2}\ \text{m}^3\ \text{mol}^{-1}$$

One mole of any ideal gas at s.t.p. has a volume of $22.4 \times 10^{-3}\ m^3$. This information is equivalent to being told that $R = 8.31\ J\ mol^{-1}\ K^{-1}$.
(A volume of $1 \times 10^{-3}\ m^3$ used to be called a *litre*.)

24.3 Assumptions of the Kinetic Theory

General Aim

(*a*) To make somc assumptions about the nature and properties of molecules, and

(*b*) To apply the laws of mechanics to gas molecules, in an attempt to relate their *microscopic* properties (speeds, masses and number per unit volume) to the *macroscopic* behaviour of an ideal gas.

Assumptions of the Kinetic Theory

(1) A gas consists of particles called *molecules.*

(2) The molecules are in constant *random motion*, but because we are dealing with perhaps 10^{23} molecules there are as many molecules travelling in one direction as any other, and so the centre of mass of the gas is at rest.

(3) The *range of intermolecular forces* is small compared to the average separation of the gas molecules, so

(*i*) the intermolecular forces (both attractive and repulsive) are negligible except during a collision,

(*ii*) the volume *of* the gas molecules is negligible compared to the volume *occupied by* the gas,

(*iii*) the duration of a collision is negligible compared to the time spent in free motion.

(4) A molecule moves with uniform velocity between collisions—we ignore gravitational effects.

(5) On average we can consider the collisions of the molecules with one another and with the walls to be *perfectly elastic* (the effects of inelastic and superelastic collisions cancel out).

(6) We can apply *Newtonian mechanics* to molecular collisions.

These assumptions define, on a *microscopic* scale, what we mean by an **ideal gas**. (See also p. 178.)

24.4 The Speeds of Gas Molecules

The graph of fig 24.2 is called the **Maxwellian distribution**. Its shape can be calculated using statistical mechanics.

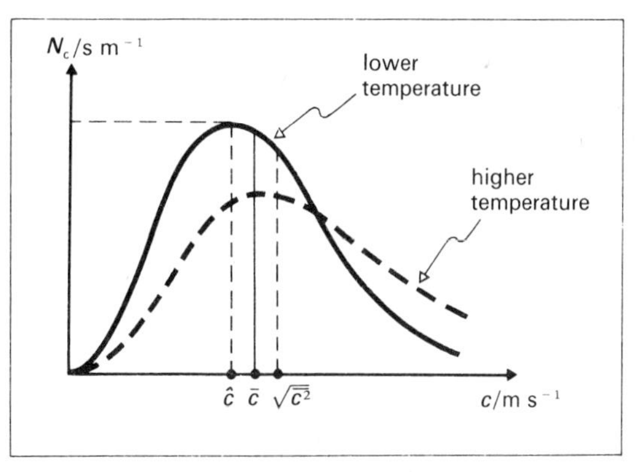

Fig. 24.2. The Maxwellian distribution. N is defined so that $N_c\delta c$ is the number of molecules with speeds between c and $(c+\delta c)$ at a particular temperature.

Suppose a gas has N molecules. We define

(a) $\hat{c}$, the *most probable speed*, as that possessed by the greatest fraction of molecules.

(b) $\bar{c}$, the *average* or *mean speed*, by

$$\bar{c} = \frac{1}{N}(c_1 + c_2 + \cdots + c_N)$$

(c) $\overline{c^2}$, the *mean square speed*, by

$$\overline{c^2} = \frac{1}{N}(c_1^2 + c_2^2 + \cdots + c_N^2).$$

It can be shown that, for a large value of N

(i) $\hat{c} \approx 0.8\sqrt{(\overline{c^2})}$, and

(ii) $\bar{c} \approx 0.9\sqrt{(\overline{c^2})}$.

24.5 The Pressure Exerted by an Ideal Gas

(The bold **A** at the side indicates the use of an assumption.)

Suppose there are N molecules in a rectangular box of dimensions $a \times b \times l$ (fig. 24.3).

Suppose that the molecule shown has a velocity $\mathbf{c}$ of components c_x, c_y and c_z. When the molecule hits the shaded face, the change of momentum is

$$-mc_x - (mc_x) = -2mc_x$$

to the right. We assume an elastic collision. **A**

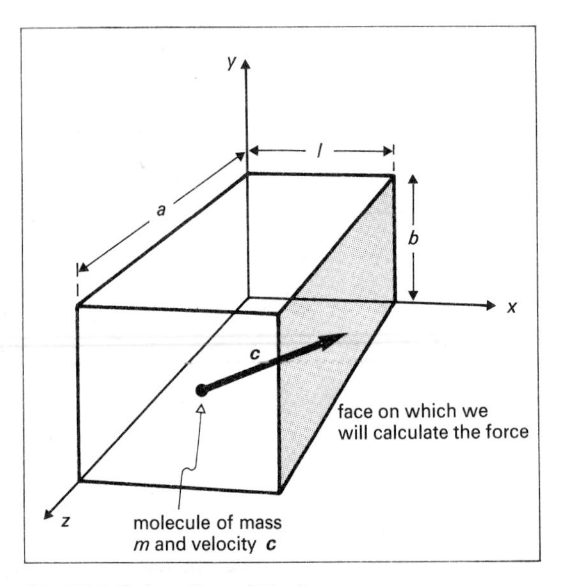

Fig. 24.3. Calculation of ideal gas pressure.

The impulse imparted to the shaded face is

$$+2mc_x \quad \text{(to the right)}$$

The time interval before the same molecule makes a second collision at the face is

$$\frac{2l}{c_x} \quad \text{(if molecular diameter} \ll l) \qquad \mathbf{A}$$

Hence the number of collisions on the shaded face of this molecule in unit time is $c_x/2l$.

Force exerted by this molecule on the face is

$$\begin{pmatrix}\text{rate at which molecule transfers} \\ \text{momentum to the face—\textit{Newton} II}\end{pmatrix}$$

$$= 2mc_x \frac{c_x}{2l} = \frac{mc_x^2}{l}$$

Pressure at shaded face is

$$\frac{\text{total force exerted by } N \text{ molecules}}{\text{area of shaded face}}$$

$$p = \frac{(m/l)[(c_x^2)_1 + (c_x^2)_2 + \cdots + (c_x^2)_N]}{ab}$$

Substitute $n = N/abl$, and gas density $\rho = nm$, so

$$p = \rho\overline{c_x^2}$$

For any molecule $c^2 = c_x^2 + c_y^2 + c_z^2$.

Because (i) N is large, and (ii) the molecules have random motion, **AA**

$$\overline{c_x^2} = \overline{c_y^2} = \overline{c_z^2} = \tfrac{1}{3}\overline{c^2}$$

This implies that they do not move parallel to any axis preferentially, or indeed parallel to any of these three axes.

We conclude

$$\boxed{p = \tfrac{1}{3}\rho\overline{c^2}}$$

Notes.

(a) The result is valid for all shapes of container.

(b) According to *Pascal's principle*, the pressure on the other faces will be the same if we neglect the weight of the gas—assumption (4). **A**

(c) Assumption (3) implies that we can neglect intermolecular collisions. Any such collisions cannot alter the net rate of transfer of momentum in any direction, and so the result is valid. **A**

24.6 Deductions from $p = \frac{1}{3}\rho\overline{c^2}$

(a) Kinetic Interpretation of Temperature

From $p = \frac{1}{3}\rho\overline{c^2}$, it follows that

$$pV = \tfrac{1}{3}Nm\overline{c^2}$$
$$= \tfrac{2}{3}N(\tfrac{1}{2}m\overline{c^2})$$

Compare

(*i*) $pV \propto \frac{1}{2}m\overline{c^2}$, the *consequence* of our kinetic theory assumptions, with

(*ii*) $pV \propto T$, our *definition of temperature* on the ideal-gas scale (p. 164), which corresponds to the thermodynamic scale.

We conclude that

The thermodynamic temperature of an ideal gas is proportional to the mean translational k.e. of its molecules (and to the total translational kinetic energy).

This interpretation is consistent with the *ideal-gas equation*, and so (p. 178) with the gas laws.

The idea is developed further on page 192.

Because $\frac{1}{2}m\overline{c^2} \propto T$

(1) for a particular gas (m constant),

$$\overline{c^2} \propto T, \quad \text{and}$$

(2) for different gases at the same T

$$\overline{c^2} \propto \frac{1}{m}$$

(b) Avogadro's Law

Consider two ideal gases

$$p_1 V_1 = \tfrac{1}{3}N_1 m_1 \overline{c_1^2}$$
$$p_2 V_2 = \tfrac{1}{3}N_2 m_2 \overline{c_2^2}$$

If they have the same pressure, volume and temperature, then

$$p_1 = p_2$$
$$V_1 = V_2$$
$$\tfrac{1}{2}m_1\overline{c_1^2} = \tfrac{1}{2}m_2\overline{c_2^2}$$

which shows

$$N_1 = N_2$$

Equal volumes of all ideal gases under the same conditions of temperature and pressure contain the same number of molecules.

(c) Gaseous Diffusion and Graham's Law

Since $\quad pV = \frac{1}{3}M_g\overline{c^2}$

and $\quad pV = (M_g/M_m)RT$

it follows that

$$\overline{c^2} = 3RT/M_m$$

or

$$\boxed{c_{\text{r.m.s.}} = \sqrt{\frac{3RT}{M_m}} = \sqrt{\frac{3p}{\rho}}}$$

Gaseous diffusion is treated as a transport phenomenon in **24.8**, and as a random process in **26.5**. For a simple treatment we will assume that the number of particles passing through a given small hole in unit time $\propto \bar{c} \propto c_{\text{r.m.s.}}$ Then examination of the boxed line above shows

The number of molecules escaping in unit time

(*a*) *at a fixed temperature is inversely proportional to the square root of the relative molecular mass*, and

(*b*) *at a fixed pressure is inversely proportional to the square root of the density.*

To calculate the rate at which the mass escapes we would need to take into account the mass of the molecules.

(d) Dalton's Law of Partial Pressures

Consider a mixture of ideal gases in volume V.

Then $\quad p_1 V = \frac{1}{3}N_1 m_1\overline{c_1^2}$

and $\quad p_2 V = \frac{1}{3}N_2 m_2\overline{c_2^2}$

When the gases have acquired the same temperature

$$m_1\overline{c_1^2} = m_2\overline{c_2^2} = m\overline{c^2} \quad \text{(say)}$$

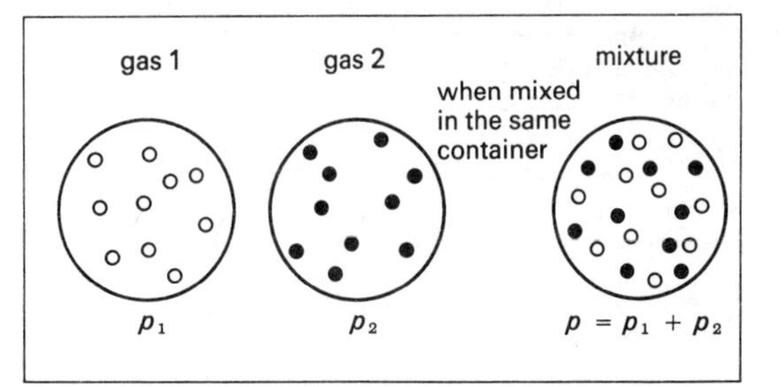

Fig. 24.4. Symbolic representation of Dalton's Law.

Adding

$$(p_1 + p_2 + \cdots)V = \tfrac{1}{3}m\overline{c^2}(N_1 + N_2 + \cdots)$$

Hence total pressure $p = \sum p_1$.

When two or more gases which do not react chemically are present in the same container, the total pressure is the sum of the partial pressures (the pressure which each gas would exert if isolated in the container).

Results (a) to (d) should be read with the comments on p. 127.

*24.7 Mean Free Path λ

The **mean free path** λ of a gas molecule is the average distance it travels between collisions. An approximate value for λ can be found from the following simplified argument.

Let r_0 (fig. 24.5) be the effective molecular *diameter.*

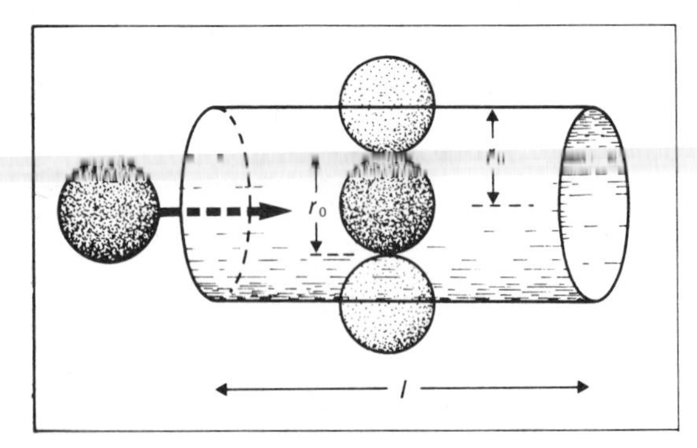

Fig. 24.5. Calculation of mean free path λ.

Then if the molecule travels a distance l in time Δt the volume swept out will be $\pi r_0{}^2 l$. We assume that the molecule will collide with any other molecule within this volume. The volume contains $\pi r_0{}^2 ln$ molecules (where $n[\mathrm{m}^{-3}]$ is their number density), and this is the number of collisions during the interval Δt.

$$\text{Mean free path } \lambda = \frac{\text{distance travelled}}{\text{number of collisions}}$$

$$= \frac{l}{\pi r_0^2 ln} = \frac{1}{\pi r_0^2 n}$$

$$\lambda[\mathrm{m}] = \frac{1}{\pi r_0^2[\mathrm{m}^2]n[\mathrm{m}^{-3}]}$$

Orders of Magnitude

Gas	n/m^{-3}	λ/m	*Pressure* p/Pa
electron gas in copper	10^{29}	10^{-8}	—
atmosphere			
—sea level	10^{25}	10^{-7}	10^5
—at 100 km	10^{17}	1	10^{-1}
—at 300 km	10^{14}	10^4	10^{-4}

*24.8 Transport Phenomena: Molecular Diffusion

Introduction

The continuous random motion of gas particles gives rise, by the statistical processes described in chapter **26** to the net macroscopic transport of

(a) matter—molecular *diffusion*
(b) energy—*thermal conduction*
(c) momentum—*viscosity.*

These transfers, together with the net movement of electric charge (*electrical conduction*) constitute a group of processes with common features called the **transport phenomena.**

In this chapter only gaseous diffusion is analysed in detail, but mention is made of gaseous conduction and viscosity, since their study involves similar techniques.

Molecular Diffusion

When the number density n of molecules varies from place to place, diffusion (the net transfer of molecules) occurs in the direction in which n decreases, i.e. from a region of high to one of low concentration, thereby equalizing the distribution of diffusing molecules. Diffusion is a statistical process caused by random motion (see **26.5**). We will consider molecules of one type only **(self-diffusion)**.

Fick's Law

Suppose that gas molecules are diffusing in the $\boldsymbol{x}$-direction. Let N molecules cross an area $\boldsymbol{A}$ placed normal to that direction in time Δt. Then the **particle current density** $\boldsymbol{j_N}$ is given by

$$j_N = \frac{N}{A\,\Delta t}$$

$$j_N\left[\frac{1}{\mathrm{m}^2\,\mathrm{s}}\right] = \frac{N}{A[\mathrm{m}^2]\,\Delta t[\mathrm{s}]}$$

Experiment shows that, so long as statistical results are applicable, the particle current density is proportional to the rate at which the concentration changes in the x-direction.

$$j_N \propto \left(\frac{\mathrm{d}n}{\mathrm{d}x}\right)$$

$$\boxed{j_N = -D\left(\frac{\mathrm{d}n}{\mathrm{d}x}\right)}$$

$$j_N\left[\frac{1}{\mathrm{m}^2\,\mathrm{s}}\right] = -D\left[\frac{\mathrm{m}^2}{\mathrm{s}}\right]\frac{\mathrm{d}n}{\mathrm{d}x}\left[\frac{1}{\mathrm{m}^4}\right]$$

Notes.

(*a*) The boxed equation is called **Fick's law**.

(*b*) The constant of proportionality D is called the **diffusion coefficient.** It has the unit $\mathrm{m}^2\,\mathrm{s}^{-1}$.

(*c*) The negative sign indicates that the molecules move in the direction in which n decreases.

Calculation of the Diffusion Coefficient

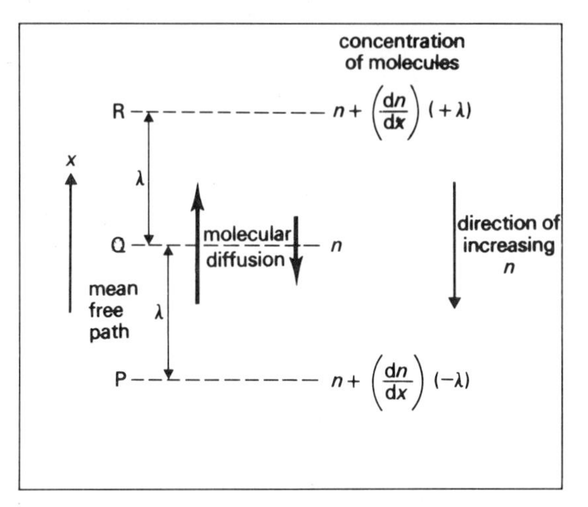

Fig. 24.6. Calculation of the diffusion coefficient. P, Q and R are planes drawn normal to the x-direction.

In fig. 24.6 molecules are diffusing in the positive x-direction. This means that $\mathrm{d}n/\mathrm{d}x$ is negative, but we do not anticipate this by writing $-(\mathrm{d}n/\mathrm{d}x)$.

(*a*) Since a molecule travels an average distance λ between collisions, the concentrations at planes R and P are of the form

$$n + \left(\frac{\mathrm{d}n}{\mathrm{d}x}\right)\Delta x = n \pm \left(\frac{\mathrm{d}n}{\mathrm{d}x}\right)\lambda,$$

as shown in the diagram. (For plane R $\Delta x = +\lambda$, and for plane P $\Delta x = -\lambda$. Both are measured relative to Q.)

(*b*) If the molecular motion is *random*, at any instant we may visualize one sixth of the molecules to be travelling in the positive x-direction, and one sixth in the negative. Thus in time Δt the number of molecules that cross an area A in the central plane Q is

(*i*) $\frac{1}{6}\left(n - \lambda\frac{\mathrm{d}n}{\mathrm{d}x}\right)A\bar{c}\Delta t$, travelling upwards, and

(*ii*) $\frac{1}{6}\left(n + \lambda\frac{\mathrm{d}n}{\mathrm{d}x}\right)A\bar{c}\Delta t$, travelling downwards.

(*c*) The *net* upward movement

$$N = -\tfrac{1}{3}\lambda\left(\frac{\mathrm{d}n}{\mathrm{d}x}\right)A\bar{c}\,\Delta t,$$

and so $$j_N = \frac{N}{A\,\Delta t} = -\tfrac{1}{3}\lambda\bar{c}\left(\frac{\mathrm{d}n}{\mathrm{d}x}\right)$$

(*d*) Comparison with the definition of D yields

$$\boxed{D = \tfrac{1}{3}\lambda\bar{c}}$$

Correlation with Experiment

At fixed pressure

$$\lambda \propto 1/\rho \propto T,$$

and $$\bar{c} \propto \sqrt{T},$$

since $$\tfrac{1}{2}m\overline{c^2} = \tfrac{3}{2}kT$$

We predict that, at fixed pressure

$$D \propto T^{3/2}$$

Experiment yields a slightly higher power.

*24.9 Thermal Conduction

Thermal conduction occurs when the temperature of a substance varies from place to place, the transport of energy being in that direction in which the temperature decreases. Following our treatment of diffusion we define an **energy current density** j_Q by

$$j_Q = \frac{Q}{A\,\Delta t},$$

leading to the experimental result

$$\boxed{j_Q = -k\frac{\mathrm{d}T}{\mathrm{d}x}}$$

This equation states **Fourier's law**, and defines for us the *thermal conductivity* k. The equation is more familiar in the form

$$\frac{\mathrm{d}Q}{\mathrm{d}t} = -\lambda A\frac{\mathrm{d}\theta}{\mathrm{d}x} \qquad \text{(p. 216)}$$

in which λ is used instead of k. (In this chapter λ represents the molecular mean free path.)

The kinetic theory calculation of k for an ideal gas is similar to that for D, with the temperature T (and by implication molecular *energy*) taking the rôle of the concentration n. The energy transfer by molecules migrating at equal rates in opposite directions is related to the mass transfer (and hence the density ρ), and to the specific heat capacity at constant volume c_V. The result is

$$\boxed{k = \tfrac{1}{3}\bar{c}\rho c_V\lambda}$$

Now the product $\lambda\rho$ is independent of pressure, as are $\bar{c}$ and c_V. We predict k to be independent of pressure, and experiment agrees over a wide range of pressures. This is excellent support for the kinetic theory.

*24.10 Viscosity

Viscosity occurs when the drift or convection velocity $\boldsymbol{v}$ within a fluid varies from place to place, the transport of momentum $\boldsymbol{p}$ being in that direction in which the velocity decreases (fig. 19.2). Following our earlier treatments, we define a **momentum current density j_p** by

$$j_p = \frac{p}{A\,\Delta t},$$

leading to the result

$$\boxed{j_p = -\eta\frac{dv}{dx}}$$

This equation states the law of viscous flow, and defines for us the *coefficient of viscosity* η. The equation is more familiar in the form

$$F = -\eta A\frac{dv}{dx}, \qquad \text{(p. 146)}$$

with the negative sign being omitted in a non-vector approach. For *Newtonian* fluids (which include gases) η is found by experiment to be a constant at a fixed temperature.

The kinetic theory calculation of η for an ideal gas is similar to that for D, with the systematic drift velocity $\boldsymbol{v}$ (and by implication the drift momentum $\boldsymbol{p}$) taking the rôle of the concentration n. The result is

$$\boxed{\eta = \tfrac{1}{3}\rho\bar{c}\lambda}$$

Notes.

(*a*) Since the product $\lambda\rho$ and $\bar{c}$ are independent of pressure, we predict that η will be too. *Maxwell* confirmed experimentally that this is true over a wide range of pressures. This is further excellent support for the kinetic theory.

(*b*) The equation suggests

$$\eta \propto \bar{c} \propto \sqrt{T},$$

since $\frac{1}{2}m\overline{c^2} = \frac{3}{2}kT$, and $\lambda\rho$ is independent of temperature. Experiment indicates that η increases with temperature, but that the power exceeds $\frac{1}{2}$.

*24.11 Correlation of Transport Phenomena

The table summarizes the discussion so far.

Name of process	*Physical property being transported*	*Definition of coefficient*	*Kinetic theory prediction of coefficient*
diffusion	molecules N	$j_N = -D\dfrac{dn}{dx}$	$D = \frac{1}{3}\lambda\bar{c}$
thermal conduction	energy Q	$j_Q = -k\dfrac{dT}{dx}$	$k = \frac{1}{3}\bar{c}\rho c_V\lambda$
viscosity	momentum $\boldsymbol{p}$	$j_p = -\eta\dfrac{dv}{dx}$	$\eta = \frac{1}{3}\rho\bar{c}\lambda$

(*a*) If we combine the results for k and η, we predict

$$k/\eta c_V = 1$$

Experiments using a wide range of gases give values between 1.4 and 2.5, an order-of-magnitude agreement which supports the validity of our elementary theory.

(*b*) The simple argument of **24.7** showed

$$\lambda = 1/\pi r_0^2 n$$

This means that a measurement of D, k and/or η for a gas enables independent estimates to be made of the molecular diameter r_0. There is good correlation between the results of the three different methods.

25 Ideal Gases: Thermal Behaviour

25.1 Important Terms

A **dissipative process** is one in which an energy conversion (e.g. by the performance of work) usually shows itself as a temperature increase.

Examples

(*i*) Work done against friction,
(*ii*) The action of an electrical resistor.

A **non-dissipative** process is one in which the energy converted can be recovered by reversing the direction of the displacement that caused the energy conversion.

Examples

(*i*) The very slow compression of a gas,
(*ii*) The charging of an electrical capacitor, if we neglect *Joule* heating.

Thermodynamic Equilibrium

A system is in thermodynamic equilibrium if

(*a*) it is in chemical equilibrium, and

(*b*) there are no temperature and pressure gradients (which might cause its *state* (q.v.) to change as time passes).

The **state of a substance** is specified by quoting its pressure p and temperature T. For p and T to have unique meaning, the substance must be in *thermodynamic equilibrium* (q.v.). The **equation of state** is the mathematical relationship between p, V and T, which are the *macroscopic* properties of the system. (See p. 178.) (Be careful to distinguish between this use of *state*, and the word *phase*.)

Reversibility

A process is reversible if the state of a system is changed in such a way that the system is at all times effectively in thermodynamic equilibrium.

The implications are

(*a*) that the process would take an infinite time (be *quasi-static*),

(*b*) that at every stage it would be possible to make the direction of the process reverse by an infinitesimal opposite change in the external conditions, (*c*) that the departure from equilibrium is negligible.

(A reversible change does *not* mean that a *cycle* is necessarily considered—we do not imply that the state of the substance will be restored to its original value.)

25.2 External Work and the Indicator Diagram

External Work

To calculate the external work done by any substance which expands *reversibly*:

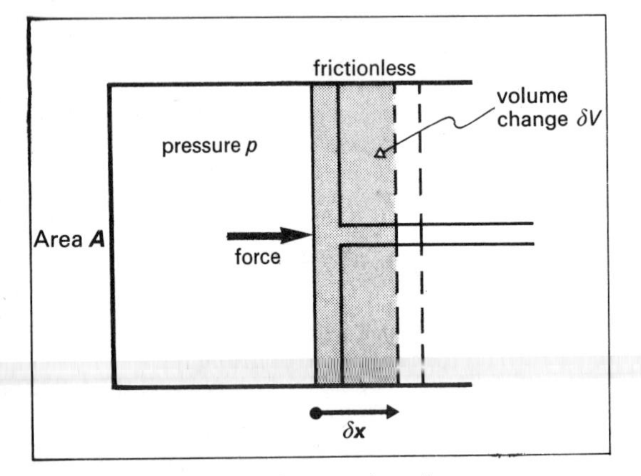

Fig. 25.1. Calculation of external work.

$$\text{work done by substance} = (\text{force}) \times (\text{displacement})$$
$$\delta W = (pA)\,\delta x$$
$$= p\,\delta V$$

For a reversible expansion which changes the volume from V_1 to V_2

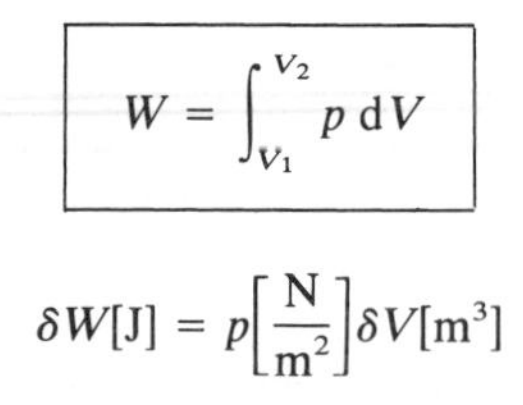

$$W = \int_{V_1}^{V_2} p\,\mathrm{d}V$$

$$\delta W[\mathrm{J}] = p\left[\frac{\mathrm{N}}{\mathrm{m}^2}\right]\delta V[\mathrm{m}^3]$$

Notes.

(*a*) For a *non-reversible* expansion the value of *p* is not properly defined.

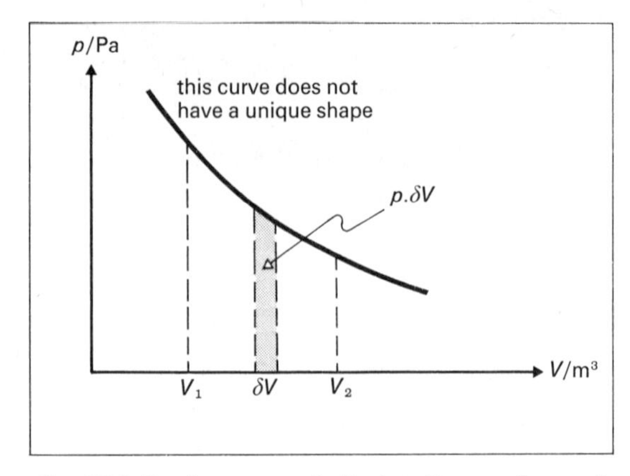

Fig. 25.2. To show on an indicator diagram the work done during a reversible expansion.

(*b*) The work done *by* the substance is positive when its pressure–force and the displacement δx are in the same direction.

(*c*) To evaluate the integral we must be able to write *p* in terms of *V*, as we can when

(*i*) *p* is constant, so that $W = p(V_2 - V_1)$,
(*ii*) the change is isothermal (p. 189),
(*iii*) the change is adiabatic (p. 190).

The Indicator Diagram

A graph of *p* against *V* is called an *indicator diagram.* The work done during a reversible change depends on the shape of the $p - V$ curve. It does *not* have a unique value for a given V_1 and V_2, as is illustrated by the examples opposite (fig. 25.3).

A *cycle* is any process during which the state of a substance (particularly a gas) undergoes changes which ultimately restore its original value. It is represented by a *closed loop* on the indicator diagram.

Example of a Cycle in which a Gas does Work

In the cycle of fig. 25.3 the state of the gas has returned to its original value (p_1, V_1, T_1) and yet the gas has done work. We explain this by saying that energy has been transferred not just by doing work, but also by *thermal energy transfer* (heat).

During the cycle there must have been given to the gas an amount of heat equal to the work done by the gas.

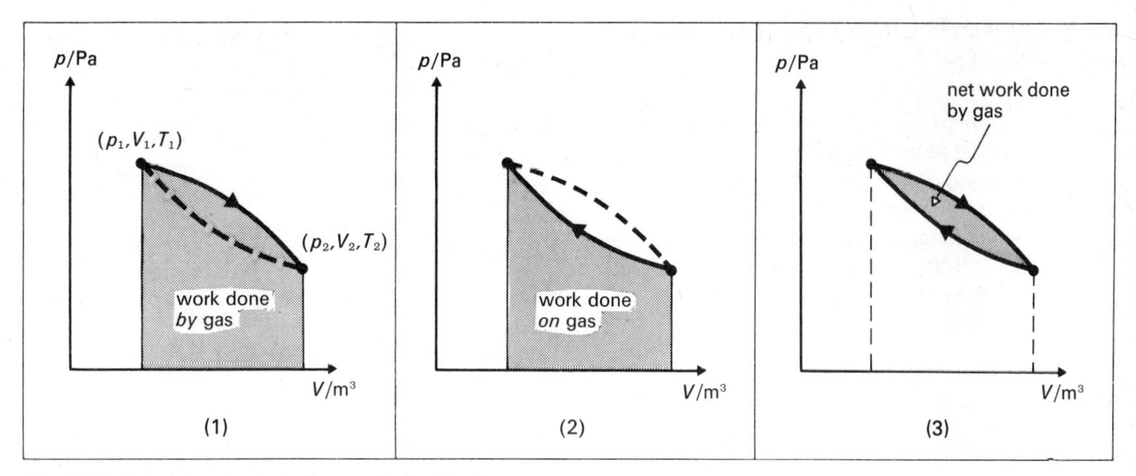

Fig. 25.3. A cycle represented on an indicator diagram.

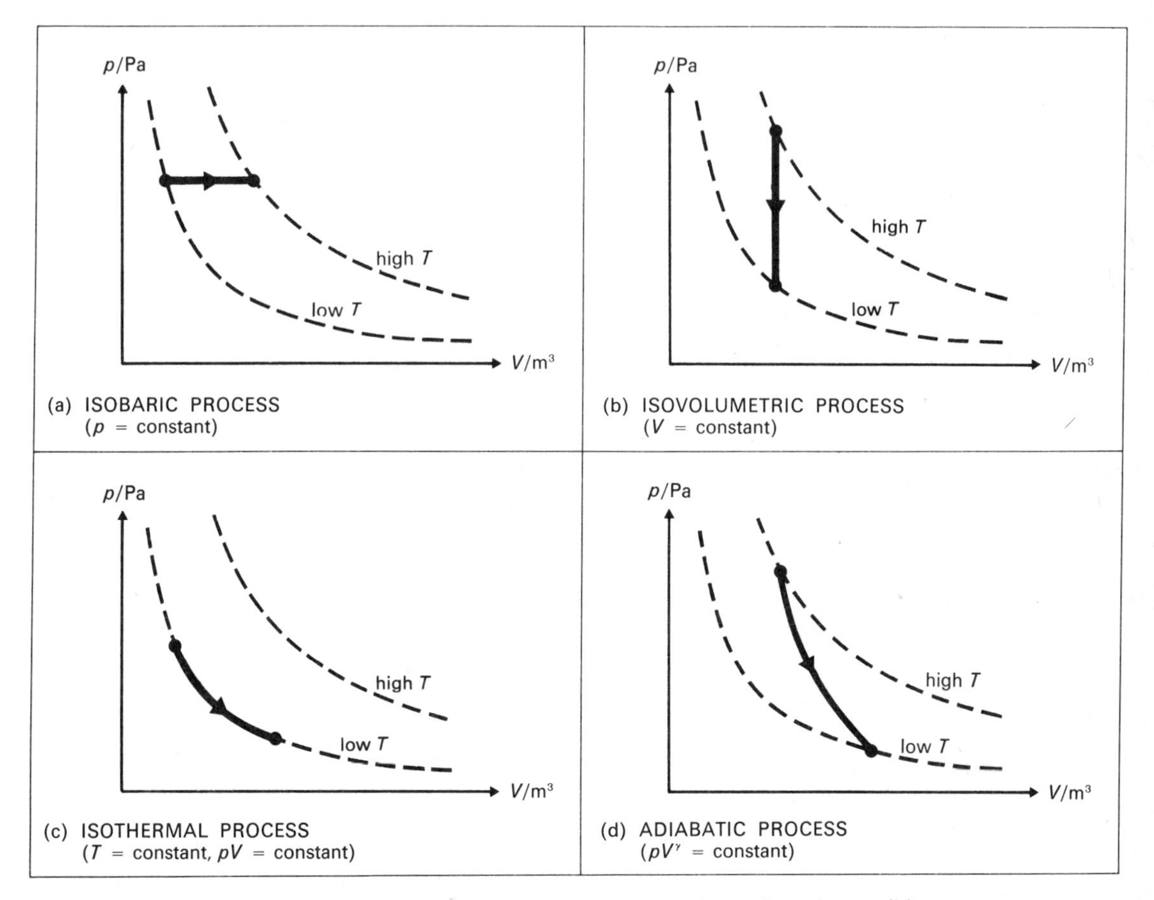

Fig. 25.4. Some ways of changing the state of a gas. Each process is conducted reversibly.

25.3 Methods of Changing the State of a Gas

The detailed shape of the p–V curve depends on the way in which the state of the gas is changed. Study fig. 25.4 carefully, which shows four *special cases.* Note the behaviour of T during each change.

The *isothermal* and *adiabatic* processes are discussed on p. 189. Part (c) shows a pair of rectangular hyperbolae (*Boyle's Law* curves).

25.4 The First Law of Thermodynamics

Suppose we supply heat ΔQ to a gas, and that this produces two effects:

(1) it raises the *internal energy* of the gas (below) from U to $(U + \Delta U)$, and

(2) it enables the gas to do *external work* ΔW.

The first law of thermodynamics *expresses the principle of conservation of energy for this situation by*

$$\boxed{\Delta Q = \Delta U + \Delta W}$$

The law recognizes two different processes by which we can transfer energy

(*i*) by doing work on a macroscopic scale—ordered energy, or

(*ii*) by doing work on a microscopic scale—heat, i.e. the transfer of disordered energy.

What happens when Heat is given to a Gas

Suppose we give heat to a gas in a situation where a volume change is allowed to occur. The energy may be used

(*a*) to do **external work** in pushing against the forces exerted by the environment ($\int p \, dV$ if the change is reversible), *and/or*

(*b*) to increase the **internal energy** U of the system.

For example it may increase

(*i*) the translational k.e. of the gas molecules,

(*ii*) the rotational energy of polyatomic molecules (p. 191),

(*iii*) the vibrational energy of polyatomic molecules (p. 192),

(*iv*) the intermolecular potential energy if work has to be done to increase the molecular separation. (This would be zero for an ideal gas.)

These five terms could be reclassified as

(1) those that depend *only on temperature*—the translational, rotational and vibrational energies, and

(2) those that depend *on volume*—the external work and intermolecular p.e.

25.5 The Molar Heat Capacities of Gases

Suppose we add heat ΔQ to an amount of gas μ [mol] under specified conditions (see box), and that this causes a temperature increase ΔT.

Then we define the* principal molar heat capacities $C_{V,m}$ *and* $C_{p,m}$ *by the equations

$C_{V,m} = \dfrac{\Delta Q}{\mu \, \Delta T}$	$C_{p,m} = \dfrac{\Delta Q}{\mu \, \Delta T}$
constant volume	***constant pressure***

$$C_m \left[\frac{\text{J}}{\text{mol K}}\right] = \frac{\Delta Q \, [\text{J}]}{\mu \, [\text{mol}] \Delta T [\text{K}]}$$

A gas has an infinite number of molar heat capacities because it can be made to do a variable amount of external work. The value of C_m may vary with temperature (p. 194).

The *internal* energy of an *ideal* gas depends only on temperature. This means that for any process which changes T to $(T + \Delta T)$, the internal energy will always change from U to $(U + \Delta U)$, and that the *same* ΔU *will always produce the same* ΔT.

For the process a–b of fig. 25.5

$$\Delta Q = \Delta U + \Delta W$$

becomes

$$\mu C_{V,m} \, \Delta T = \Delta U + 0,$$

since $\Delta V = 0$ and no external work is done.

The relation

$$\boxed{\Delta U = \mu C_{V,m} \, \Delta T}$$

holds for all methods of changing U for an ideal gas, even if the change is not isovolumetric.

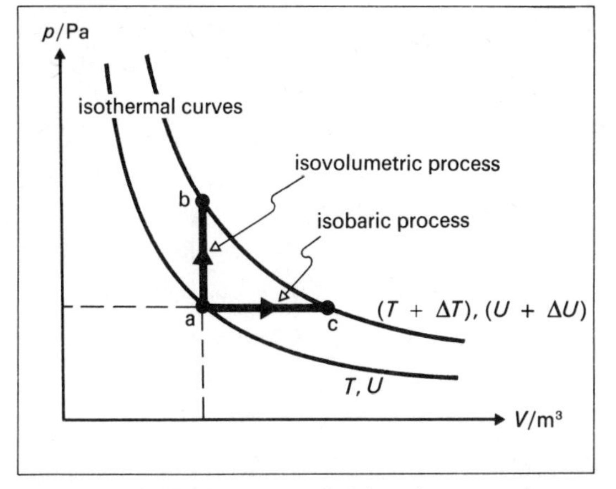

Fig. 25.5. An ideal gas experiencing the same change of internal energy ΔU while undergoing isobaric (a–c) and isovolumetric (a–b) processes.

Measurement of Gas Molar Heat Capacities

(*a*) $C_{V,\mathrm{m}}$ can be measured by *Joly's* differential steam calorimeter.

(*b*) $C_{p,\mathrm{m}}$ can be measured by
  (1) the method of mixtures (*Regnault*), or
  (2) the continuous-flow calorimeter (*Scheel* and *Heuse*).

25.6 Mayer's Equation (1842) $C_{p,\mathrm{m}} - C_{V,\mathrm{m}} = R$

Refer to fig 25.5. Suppose we take an amount μ [mol] of an *ideal* gas through the isobaric process a–c *reversibly*.

Then in $\Delta Q = \Delta U + \Delta W$, we have

(*a*) $\Delta Q = \mu C_{p,\mathrm{m}}\,\Delta T$ (by definition of $C_{p,\mathrm{m}}$)
(*b*) $\Delta U = \mu C_{V,\mathrm{m}}\Delta T$ (previous paragraph), and
(*c*) $\Delta W = p\Delta V$ (because the pressure is constant and the change reversible).

For an ideal gas

$$pV = \mu RT$$

and if only V and T are varied (as they are here),

$$p\,\Delta V = \mu R\,\Delta T$$

Then $\Delta Q = \Delta U + \Delta W$

becomes $\mu C_{p,\mathrm{m}}\,\Delta T = \mu C_{V,\mathrm{m}}\,\Delta T + \mu R\,\Delta T$

or $C_{p,\mathrm{m}} = C_{V,\mathrm{m}} + R$

$$\boxed{C_{p,\mathrm{m}} - C_{V,\mathrm{m}} = R}$$ ***Mayer's equation***

This is true for *any ideal gas* (one for which there is no intermolecular p.e.).

25.7 Introduction to Isothermal and Adiabatic Processes

It is convenient to relate the first law of thermodynamics

$$\Delta Q = \Delta U + \Delta W$$

to the four methods of changing the state of a gas discussed on p. 187.

(*a*) In the **isobaric** reversible process

$$\Delta W = p\,\Delta V$$

(*b*) In the **isovolumetric** process

$$\Delta V = 0 \qquad \Delta W = 0$$

and this led to $\Delta U = \mu C_{V,\mathrm{m}}\,\Delta T$

(*c*) In an **isothermal** process

$$\Delta T = 0 \qquad \Delta U = 0$$

and this leads to $\Delta Q = \Delta W$. There is no change of internal energy for an ideal gas.

(*d*) We define an **adiabatic** proccss to be one for which $\Delta Q = 0$, so

$$0 = \Delta U + \Delta W \quad \therefore \ -\Delta U = \Delta W$$

If a gas is made to work, there is a decrease of internal energy which becomes apparent as a temperature decrease.

Examples of processes which are approximately adiabatic are those

(*i*) in insulated containers,
(*ii*) which take place very rapidly,
(*iii*) experienced by a very large mass of a substance.

25.8 Isothermal Processes

(*a*) Equation

$$pV = \mu RT$$

is the equation of state for an ideal gas, and

$$\Delta T = 0$$

describes an isothermal process.

So

$$pV = \textit{constant}$$

is the equation which relates p and V during such a process.

(*b*) Work Done During a Reversible Process

The process of fig. 25.1 must be very slow (quasi-static) to ensure isothermal reversible conditions.

If this is so, the work done by the gas

$$W = \int_{V_1}^{V_2} p\,\mathrm{d}V$$

$$= \int_{V_1}^{V_2} \frac{\mu RT}{V}\,\mathrm{d}V$$

$$= \mu RT \ln\left(\frac{V_2}{V_1}\right)$$

$$W[\mathrm{J}] = \mu[\mathrm{mol}]R\left[\frac{\mathrm{J}}{\mathrm{mol\ K}}\right]T[\mathrm{K}]\ln\left(\frac{V_2}{V_1}\right).$$

The (natural) logarithm has no unit.

*25.9 Adiabatic Processes

These processes can be

(*a*) *non-reversible*, as in the *Joule* and the *Joule–Kelvin* experiments (p. 211), or

(*b*) *approximately reversible*, when there is only a small departure from the equilibrium situation. For example

(*i*) the *Clément and Désormes* method for measuring γ (p. 193), and

(*ii*) the alternating pressure changes associated with the passage of a sound wave (p. 302).

(*a*) Equation

Consider a *reversible* adiabatic change experienced by an amount μ[mol] of an *ideal* gas.

In $\Delta Q = \Delta U + \Delta W$, we have

(*a*) $\Delta Q = 0$ (adiabatic change)

(*b*) $\Delta U = \mu C_{V,\mathrm{m}}\,\Delta T$ (paragraph **25.5**) since the gas is ideal, and

(*c*) $\Delta W = \int p\,\mathrm{d}V$ since the change is reversible.

For an infinitesimal process

$$-\Delta U = \Delta W$$

becomes

$$-\mu C_{V,\mathrm{m}}\,\delta T = p\,\delta V \qquad \ldots(1)$$

Since the gas is ideal

$$pV = \mu RT$$

$$p\,\delta V + V\,\delta p = \mu R\,\delta T$$

$$\delta T = \frac{1}{\mu R}(p\,\delta V + V\,\delta p) \qquad \ldots(2)$$

Eliminating δT from (1) and (2)

$$-C_{V,\mathrm{m}}(p\,\delta V + V\,\delta p) = Rp\,\delta V$$

$$-C_{V,\mathrm{m}}V\,\delta p = (R + C_{V,\mathrm{m}})p\,\delta V$$

Using $C_{p,\mathrm{m}} - C_{V,\mathrm{m}} = R$ we have

$$-V\,\delta p = \left(\frac{C_{p,\mathrm{m}}}{C_{V,\mathrm{m}}}\right)p\,\delta V$$

and putting

$$\frac{C_{p,\mathrm{m}}}{C_{V,\mathrm{m}}} = \gamma$$

$$-\frac{\delta p}{p} = \gamma\frac{\delta V}{V}$$

$$\therefore\quad -\int\frac{\mathrm{d}p}{p} = \gamma\int\frac{\mathrm{d}V}{V}$$

If we can assume $C_{p,\mathrm{m}}$ and $C_{V,\mathrm{m}}$ (and hence γ) do not vary with T for this process,

then

$$-\ln p = \gamma \ln V + \text{constant}$$

Hence

$$pV^{\gamma} = \textit{constant}$$

and this equation relates p and V for a *reversible adiabatic process* for an *ideal* gas for which γ *remains constant*.

(*b*) Change of Temperature

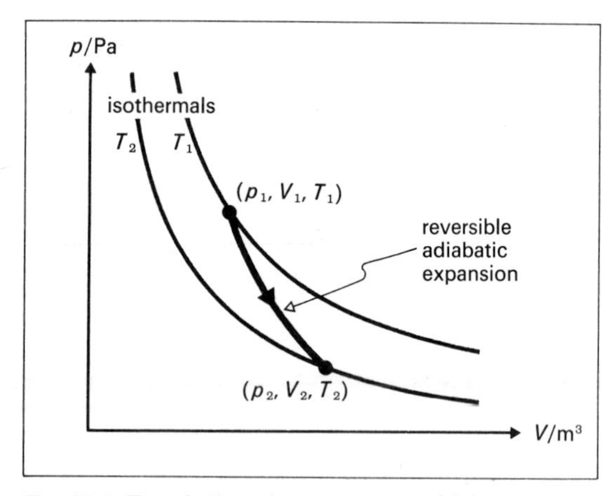

Fig. 25.6. To calculate the temperature change accompanying a reversible adiabatic expansion.

Fig. 25.6 shows a reversible adiabatic expansion during which a fall of temperature occurs ($T_1 > T_2$).

From pV = constant and pV^{γ} = constant

it follows that $\gamma = \dfrac{\text{slope of adiabatic}}{\text{slope of isothermal}}$

at a particular point.

For an ideal gas

$$p_1 V_1 = \mu R T_1 \quad \text{and} \quad p_2 V_2 = \mu R T_2$$

so

$$\frac{T_1}{T_2} = \frac{p_1 V_1}{p_2 V_2} \qquad \ldots (1)$$

For a reversible adiabatic change

$$p_1 V_1^{\gamma} = p_2 V_2^{\gamma}$$

so

$$\left(\frac{p_1}{p_2}\right) = \left(\frac{V_2}{V_1}\right)^{\gamma} \qquad \ldots (2)$$

and

$$\left(\frac{V_1}{V_2}\right) = \left(\frac{p_1}{p_2}\right)^{-1/\gamma} \qquad \ldots (3)$$

From (1) and (2)

$$\frac{T_1}{T_2} = \left(\frac{V_2}{V_1}\right)^{\gamma-1} \quad \text{or} \quad \boxed{TV^{\gamma-1} = \text{constant}}$$

From (1) and (3)

$$\frac{T_1}{T_2} = \left(\frac{p_1}{p_2}\right)^{(\gamma-1)/\gamma} \quad \boxed{T^{\gamma} p^{1-\gamma} = \text{constant}}$$

subject to the same conditions as paragraph **25.9(*a*)**.

(*c*) Work Done During a Reversible Process

Suppose the process is described validly by

$$pV^{\gamma} = \text{constant}, k$$

Then the work done

$$W = \int_{V_1}^{V_2} p \, dV$$

$$= k \int_{V_1}^{V_2} \frac{dV}{V^{\gamma}}$$

$$= k \left[\frac{V^{1-\gamma}}{1-\gamma}\right]_{V_1}^{V_2}$$

$$= \frac{k}{1-\gamma} \cdot (V_2^{1-\gamma} - V_1^{1-\gamma})$$

$$= \frac{1}{\gamma - 1} \cdot (p_1 V_1 - p_2 V_2)$$

since $p_1 V_1^{\gamma} = p_2 V_2^{\gamma} = k$.

For an *expansion*

$$T_1 > T_2 \quad \text{and} \quad p_1 V_1 > p_2 V_2$$

and since $\gamma > 1$ always, $W > 0$: the gas always does work.

Notes.

(1) A gas does more work (for a given volume increase) in a *reversible* expansion than in a *non-reversible* expansion. For example, when an ideal gas expands irreversibly into a vacuum it does no work.

(2) A gas does more work in a reversible *isothermal* expansion than in a reversible *adiabatic* expansion.

*25.10 The Second Law of Thermodynamics

The *first* law of thermodynamics (p. 188) would not be violated if either

(*i*) a heat engine were to convert internal energy entirely into mechanical energy, or

(*ii*) heat were to be transferred from a cold body to a hot one, without any loss of energy.

Nevertheless practical experience tells us that neither of the above processes is ever achieved, and we express this observation formally in the **second law of thermodynamics**. There are several different statements of the law, each of which conveys the same information. For example

(*a*) **Clausius** (1850) stated it by
No heat pump, reversible or irreversible, can transfer internal energy from a low temperature reservoir to a high temperature reservoir without work being done on it by some external agent.

(*b*) **Kelvin** (1851)
No heat engine, reversible or irreversible, operating in a cycle, can take in heat from its surroundings, and totally convert it into work.

Thus the second law determines the direction of heat flow, and implies that mechanical energy (work) cannot be produced by extracting heat from a single reservoir without returning heat to some other reservoir at a lower temperature. The law is discussed further in chapter **26**.

*25.11 Degrees of Freedom

Each method by which a system can absorb energy is called a **degree of freedom**.

Examples

(*a*) *Translational energy*
A gas molecule of mass m has translational k.e.

$$\tfrac{1}{2}mc^2 = \tfrac{1}{2}mc_x^2 + \tfrac{1}{2}mc_y^2 + \tfrac{1}{2}mc_z^2.$$

Its translational motion gives it three degrees of freedom.

(*b*) *Rotational energy*
A polyatomic gas molecule has rotational k.e. $\tfrac{1}{2}I\omega^2$ made up of terms like $\tfrac{1}{2}I_x\omega_x^2$.

Each term of this type represents a degree of freedom.

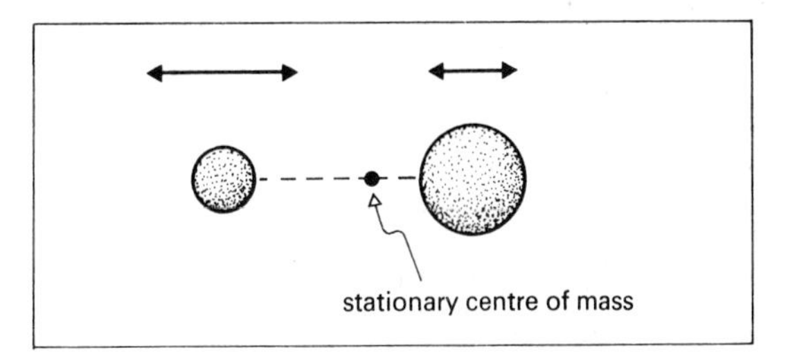

Fig. 25.7. Vibration of a diatomic molecule.

Since a gas molecule has three possible axes about which it *may* rotate, it may have up to three such degrees of freedom.

(*c*) *Vibrational energy*

The energy of vibration in fig. 25.7

$$E = E_k + E_p$$

which may be written

$$= \tfrac{1}{2}\mu v^2 + \tfrac{1}{2}kx^2$$

where μ is called the *reduced mass.*

This vibration has two degrees of freedom.

Each degree of freedom has energy written in the form

$$(\text{positive constant})(\text{variable})^2$$

Some examples are given in paragraph **25.13**.

*25.12 Equipartition of Energy

Analysis by statistical mechanics gives us the **Maxwell–Boltzmann principle**:

When we are dealing with a large number of particles in thermal equilibrium to which we can apply Newtonian mechanics, the energy associated with each degree of freedom has the same average value, and this average value depends only on the temperature.

For an example, see the *Brownian motion* (p. 125).

Application to a monatomic molecule

For any ideal gas $pV = \tfrac{2}{3}N(\tfrac{1}{2}m\overline{c^2})$ (p. 181)

and $pV = \mu RT = \dfrac{N}{N_A}RT$ (p. 178)

from which $\tfrac{2}{3}N(\tfrac{1}{2}m\overline{c^2}) = \dfrac{N}{N_A}RT$

so

$$\boxed{\tfrac{1}{2}m\overline{c^2} = \tfrac{3}{2}kT}$$

where

$$\boxed{k = \frac{R}{N_A}}$$

k is called the **Boltzmann constant**, and has the value

$$k = 1.380 \times 10^{-23}\ \mathrm{J\,K^{-1}}$$

It represents the gas constant *per molecule* for an ideal gas and has dimensions $[\mathrm{ML^2T^{-2}\Theta^{-1}}]$.

The equation of state can be written

$$\boxed{pV = NkT}$$

for N molecules of an ideal gas.

The equation

$$\tfrac{1}{2}m\overline{c^2} = \tfrac{3}{2}kT$$

tells us that energy $\tfrac{3}{2}kT$ is associated with the translational motion, which has three degrees of freedom. We deduce that for this special case, each degree of freedom has an average energy $\tfrac{1}{2}kT$.

More generally, at temperature T

$$\boxed{\begin{pmatrix}\textit{energy of a}\\ \textit{molecule}\end{pmatrix} = \begin{pmatrix}\textit{number of}\\ \textit{degrees of}\\ \textit{freedom}\end{pmatrix}\tfrac{1}{2}kT}$$

*25.13 The Kinetic Theory Prediction of Gas Molar Heat Capacities

(a) Monatomic Gases

Average energy per molecule $= \tfrac{3}{2}kT$.

Energy U for an amount of gas μ[mol] is given by

$$U = \tfrac{3}{2}\mu N_A kT$$

$$\therefore U = \tfrac{3}{2}\mu RT$$

Then

$$C_{V,m} = \frac{\Delta Q}{\mu\,\Delta T} = \frac{\Delta U}{\mu\,\Delta T}$$

since $\Delta Q = \Delta U$ when V is fixed.

$$\therefore C_{V,m} = \frac{1}{\mu}\frac{dU}{dT} = \tfrac{3}{2}R = 12.48\ \mathrm{J\,mol^{-1}\,K^{-1}}$$

and $C_{p,m} = C_{V,m} + R = 20.80\ \mathrm{J\,mol^{-1}\,K^{-1}}$.

$$\gamma = \frac{C_{p,m}}{C_{V,m}} = \tfrac{5}{3} = 1.67$$

for an ideal gas.

(b) Diatomic Gases

There are 5 degrees of freedom, 3 translational and 2 rotational. No rotation occurs about the dumb-bell axis, and no additional vibration is excited near room temperature for strongly-bound molecules such as O_2 and N_2 (see p. 194).

Hence average energy per molecule $= \tfrac{5}{2}kT$ leading to

$$C_{V,m} = \tfrac{5}{2}R = 20.8\ \mathrm{J\,mol^{-1}\,K^{-1}}$$

and

$$C_{p,m} = \tfrac{7}{2}R = 29.1\ \mathrm{J\,mol^{-1}\,K^{-1}}$$

$$\gamma = \tfrac{7}{5} = 1.40.$$

for an ideal gas.

(c) Polyatomic Gases

These have 6 degrees of freedom, 3 translational and 3 rotational (if there is no vibration). This gives

$$C_{V,m} = 3R,\quad C_{p,m} = 4R \quad \text{and} \quad \gamma = \tfrac{4}{3} = 1.33$$

For n degrees of freedom

$$\gamma = \frac{n+2}{n}$$

Compare these predictions with the experimental values for real gases near room temperature (below).

Gas	$C_{p,m}$/J mol^{-1} K^{-1}	$C_{V,m}$/J mol^{-1} K^{-1}	γ
He	20.8	12.5	1.67
Ar	20.8	12.5	1.67
H_2	28.8	20.4	1.41
O_2	29.4	21.0	1.40

The behaviour of real polyatomic gases is very complicated but some do have 6 degrees of freedom.

*25.14 The Measurement of γ

(*a*) Using the Speed of Sound

The pressure changes associated with acoustic wave propagation at audible frequencies are effectively adiabatic and reversible.

$$c = \sqrt{\left(\frac{\text{bulk modulus}}{\text{density}}\right)}$$

$$= \sqrt{\frac{\gamma p}{\rho}} \qquad \text{(see p. 302)}$$

The speeds of sound in different gases can be compared by the dust tube method (p. 306), from which γ can be deduced.

(*b*) Clément and Désormes's Method

Procedure

(1) Pump in air, and allow it to reach room temperature. Record manometer reading h_1.

(2) Remove bung for about 1 second. The gas expands roughly adiabatically and reversibly.

(3) Allow air to regain room temperature, and record the new reading $h_2(< h_1)$.

Calculation

Consider the fixed mass of air that does not escape. Suppose its volume increases from V_1 to V_2.

When the tap is opened

$$p_1 V_1^\gamma = p V_2^\gamma \qquad \ldots (1)$$

where $p_1 > p (p = \text{atmospheric pressure})$.

After warming again to room temperature

$$p_1 V_1 = p_2 V_2 \qquad \ldots (2)$$

Boyle's Law applied at room temperature.

From (1) $$\left(\frac{V_1}{V_2}\right)^\gamma = \frac{p}{p_1}$$

and (2) $$\left(\frac{V_1}{V_2}\right)^\gamma = \left(\frac{p_2}{p_1}\right)^\gamma$$

Equating $$\left(\frac{p_2}{p_1}\right)^\gamma = \frac{p}{p_1}$$

so $$\gamma = \frac{\log p - \log p_1}{\log p_2 - \log p_1}$$

in which $$p_1 = p + h_1 \rho g$$
$$p_2 = p + h_2 \rho g$$

For $h_1 \ll p$, it can be shown that

$$\gamma \approx \frac{h_1}{h_1 - h_2}$$

Note. The bung size should be adjusted to optimum diameter

(*a*) to prevent excessive air oscillation (which requires it to be small), and

(*b*) to ensure a rapid and adiabatic expansion (for which it should be large).

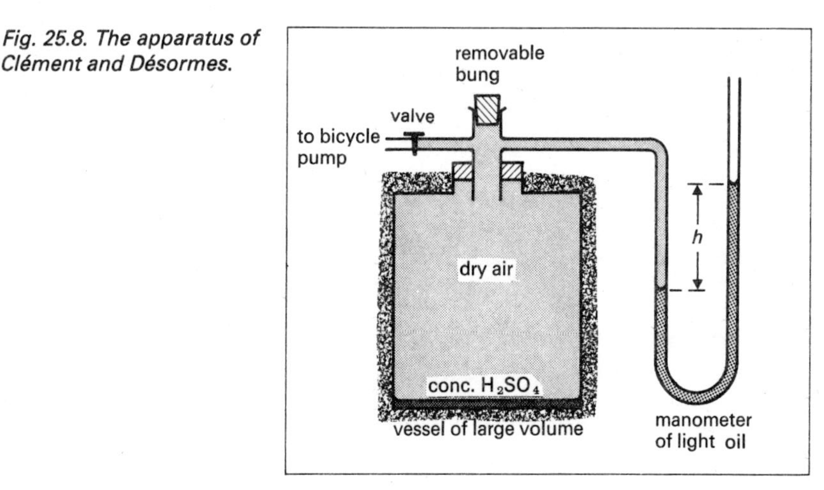

Fig. 25.8. The apparatus of Clément and Désormes.

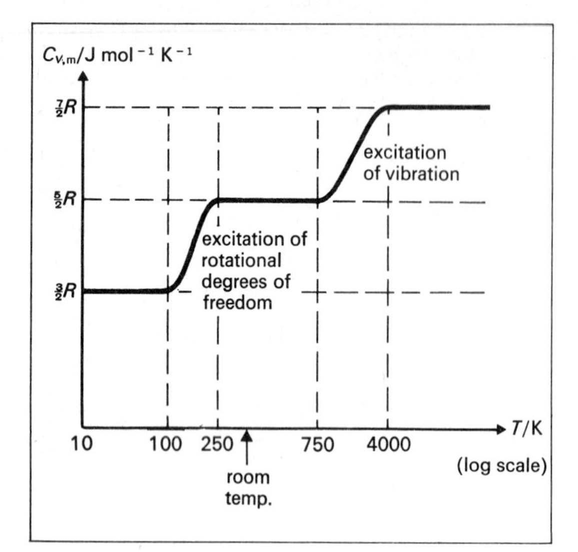

Fig. 25.9. Variation with temperature of $C_{V,m}$ for diatomic hydrogen.

*25.15 Variation of $C_{V,m}$ with Temperature

On p. 192 we showed why

$$C_{V,m} = \frac{n}{2} \cdot R$$

where n is the number of degrees of freedom of a molecule. For small (molecular) dimensions, quantum theory rather than *Newtonian* mechanics should be applied, and so the principle of equipartition of energy does not hold.

Fig. 25.9 demonstrates that the energies of rotation and vibration are *quantized.* A particular molecule may not vibrate or rotate until a sufficiently high temperature has been reached. It will then do so (fully) with its energy $\frac{1}{2}kT$ for each degree of freedom.

Of course not every molecule will start vibrating or rotating at the same temperature. For example above about 4000 K hydrogen molecules cease to behave as rigid dumb-bells, but show 7 degrees of freedom—3 translational, and 2 each rotational and vibrational. Below this temperature any vibrational energy will be that of the zero-point energy.

26 Chance, Disorder and Entropy

A proper understanding of the concepts mentioned in this chapter can be achieved only after a very full explanation. The function of this book is to provide a convenient summary of the more important ideas involved. Detailed reading can be found in, for example,

(*a*) *The Second Law*, by Bent (Oxford University Press), 1965

(*b*) *PSSC Advanced Physics Supplement* (Heath), 1966

(*c*) *Patterns in Physics*, by Bolton (McGraw-Hill), 1974

(*d*) *Heat Engines*, by Sandford (Heinemann), 1962.

*26.1 Examples of Irreversible Processes

The following illustrate processes that seem to happen in only one direction of time. The reverse is never observed to occur, and so they are said to be **irreversible**. (See also chapter **25**.)

(*a*) ***Inelastic collisions.*** The reduction of ordered (mechanical) k.e. by the gain of random molecular vibrational (internal) energy leads to an increase in temperature.

(*b*) ***Heat flow.*** The exchange of internal energy, through the agency of heat, by two bodies at different initial temperatures placed in thermal contact leads to temperature equalization.

(*c*) ***Gaseous expansion*** into a vacuum. This is discussed in more detail in **26.5.**

(*d*) The breaking of an egg.

(*e*) The growth of a plant.

(*f*) The mixing of fluids of different colours.

Although the reverse of any of these processes would not violate any of the known conservation laws, in practice they are never observed.

Such a reverse process would require a high degree of coincidence in the random motion of a number of particles. A real macroscopic physical system may contain 10^{23} atoms or more, a number so large that the chance of a reverse process occurring is virtually nil.

*26.2 Some Useful Probability Concepts

The **chance** or **probability** P of a particular event happening is defined by

$$P = \frac{\text{number of favourable observations}}{\text{total number of observations}}$$

where the total number of observations or trials is very

large, and where experience suggests that the event is equally likely to happen in each trial.

(*a*) ***Mutually exclusive events.*** Let the probability that event X will happen be P_X, and the probability that Y will happen be P_Y.

Suppose that if event X happens, then event Y cannot happen. (X and Y are then said to be **mutually exclusive**.)

Then the probability that X *or* Y will happen is

$$P_X + P_Y$$

(*b*) ***Sum of probabilities.*** In any particular trial (such as the tossing of a coin) at least one of a set of mutually exclusive events must happen. This means that the sum of their probabilities is 1.

For example if events X and Y are the *only* two possibilities, then

$$P_X + P_Y = 1$$

(*c*) ***Simultaneous events.*** In any particular trial the probability that two *independent* events will *both* happen is the *product* of the probabilities of the two single events. For example the chance that both X and Y will happen is

$$(P_X)(P_Y)$$

*26.3 Statistical Description of a Physical System

State

(*a*) ***Macroscopic.*** A description of a physical system in terms of directly observable and measurable properties, such as temperature, pressure and volume.

(*b*) ***Microscopic.*** A description which requires a detailed knowledge of the behaviour (e.g. position and energy) of every particle of the system.

Distribution

Suppose that a given system has a total number of N particles, of which

n_1 particles have energy E_1,

n_2 particles have energy E_2, etc.

The allowed values of E_1, E_2 etc. will be determined by the nature of the system. We will assume that N remains constant, where

$$N = n_1 + n_2 + n_3 + \cdots$$

The total energy of the system is given by

$$U = n_1E_1 + n_2E_2 + n_3E_3 + \cdots$$

and will be constant if the system is isolated.

The particular numbers n_1, n_2 etc. at a given instant constitute a **distribution**, or a **partition**.

Statistical Equilibrium

(*a*) For a given macroscopic state of the system there is one distribution which is more probable than any other. When that distribution exists the system is in **statistical equilibrium**.

(*b*) There will be statistical fluctuations from that equilibrium which give rise to departures from the most probable distribution.

(*c*) The average energy of the particles is given by

$$\bar{E} = \frac{U}{N} = \frac{n_1E_1 + n_2E_2 + n_3E_3 + \cdots}{n_1 + n_2 + n_3 + \cdots}$$

When N is large $\bar{E}$ has a value which determines a well-defined temperature. *Statistical* equilibrium and *thermal* equilibrium are equivalent terms.

(*d*) When an isolated system is not in equilibrium, the probability of its existing state is less than that of the equilibrium (most probable) state. Chance interactions between its component particles will cause the system to evolve until it reaches the most probable state: the system will have attained statistical equilibrium by moving to a distribution of *greater probability*.

The Aims of Statistical Mechanics

Given the composition of an isolated system, statistical mechanics attempts

(*a*) to calculate the most probable distribution corresponding to statistical equilibrium, and hence

(*b*) to evaluate the macroscopic (measurable) properties of the system when in thermal equilibrium.

*26.4 Probability and Disorder

A simple Coin Experiment

Suppose two coins are spun simultaneously. The outcome could be

HH, HT, TH or TT

There are four equally likely possible outcomes, of which two are not distinguished. The most probable outcome is thus a head and a tail, since this combination can be realized in twice as many ways as the other possibilities.

The experiment illustrate a general conclusion:

The more ways there are for realizing an event, the more likely is it to occur.

Disorder

Order and disorder are associated with the simplicity of description of a state. Two heads or two tails imply order, whereas a head and a tail imply relative disorder. The particles of a crystal are in a state of order, whereas the molecules of a gas are in a state of disorder (random energies and positions).

There are many fewer ways of arranging particles in a state of order than there are of arranging the same particles in a state of disorder. This means that, other things being equal, the state of maximum disorder is the most probable, since there are many more ways in which it can be achieved.

When large numbers are involved the most probable becomes virtually certain.

*26.5 Gaseous Expansion into a Vacuum

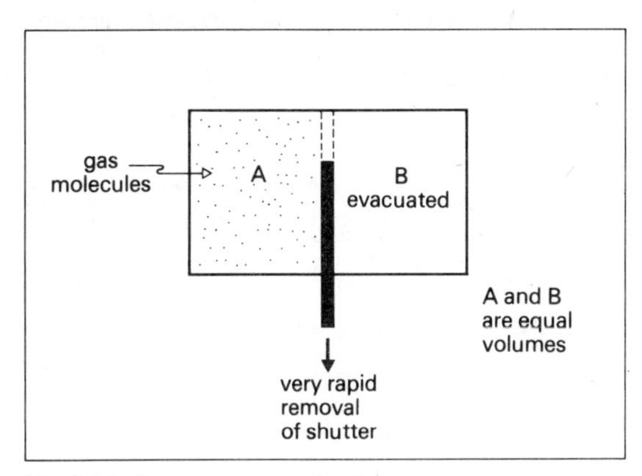

Fig. 26.1. Gaseous expansion into a vacuum.

Qualitative

The molecules of fig. 26.1 are in constant random motion, and *chance* alone can move a molecule from A to B or B to A. A particular molecule has the same chance of moving either way.

(*a*) Since all the molecules start in A, the rate of motion from A to B is initially the greater.

(*b*) When equilibrium is reached the number of molecules per unit volume in A equals that in B.

(*c*) Random motion is then equally likely to cause a molecule in A to move to B, as to cause one in B to move to A.

(*d*) The chance that random motion alone will reverse the sequence of events leading to the establishing of thermal equilibrium is completely negligible.

In general, the distribution tends to evolve (over a time period characteristic of the system) from a less probable one to a more probable one.

Quantitative

Question: What is the probability that all the molecules will, at some later time, be again found in A?

We assume that, after some suitable small time interval, a molecule is equally likely to be found in A as in B.

(*a*) *One molecule.* The probability that it is in A is $\frac{1}{2}$.

(*b*) *Two molecules.*

(*i*) The probability that one is in A is $\frac{1}{2}$.

(*ii*) The probability that the other is (independently) in A is also $\frac{1}{2}$.

(*iii*) The probability that both are in A *together is* $\frac{1}{2} \times \frac{1}{2} = \frac{1}{4}$.

(*c*) *N molecules.* The probability that all will be found in A simultaneously is

$$\tfrac{1}{2} \times \tfrac{1}{2} \times \cdots = (\tfrac{1}{2})^N$$

For one mole $N = (1\ \text{mol}) \times N_A = 6 \times 10^{23}$ molecules. $(\frac{1}{2})^{6\times 10^{23}}$ is completely negligible: *chance* alone dictates that the process will not be observed.

*26.6 Some Statistical Processes in Gases

The smaller the number of particles in a system the greater the chance of finding significant departures from the macroscopic averages associated with statistical equilibrium.

If N particles are involved, the variation is typically of the order of $\sqrt{N}$. The proportional fluctuation $\sqrt{N}/N$ becomes smaller as N increases, being negligible when $N \approx 10^{23}$.

(*a*) ***Brownian motion.*** The size of a pollen grain or a smoke particle is sufficiently small for the bombardment by fluid particles on one side to be significantly different from that on the other. (In $F = pA$, the pressure p is a statistical concept which is valid only for large numbers of molecules.)

(*b*) ***The blue sky.*** If electromagnetic radiation is scattered by an array of oscillators which are continuous and uniformly spaced, then the scattered radiation is coherent and causes no net transfer of energy perpendicular to the original direction of travel.

Consider a box of side l. When l approaches the wavelength of blue light, the number N of air molecules it contains is small enough for the proportional fluctuation $\sqrt{N}/N$ to become significant. The accompanying density fluctuation causes scattering which is responsible (in part) for the blue colour of the sky.

(*c*) ***Transport phenomena*** in gases (chapter **24**) are the consequence of random statistical processes.

*26.7 Thermal Conduction

Distributing Quanta amongst Two Atoms

Fig. 26.2 shows the number of ways in which different numbers of quanta can be distributed between two atoms.

number of quanta	distribution of quanta between atoms				number of ways
0	(0) (0)				1
1	(1) (0)	(0) (1)			2
2	(1) (1)	(2) (0)	(0) (2)		3
3	(1) (2)	(2) (1)	(3) (0)	(0) (3)	4

Fig. 26.2. Distribution of quanta between atoms.

Notes.

(*a*) The more quanta there are to be shared between the two atoms, then the greater the number of possible ways of sharing them.

(*b*) The more ways there are for distributing the quanta amongst the atoms, the more likely it is that that arrangement will happen.

Probability and Heat Flow

In fig. 26.3 we imagine two solids A and B about to be put into thermal contact. A is initially at the higher temperature.

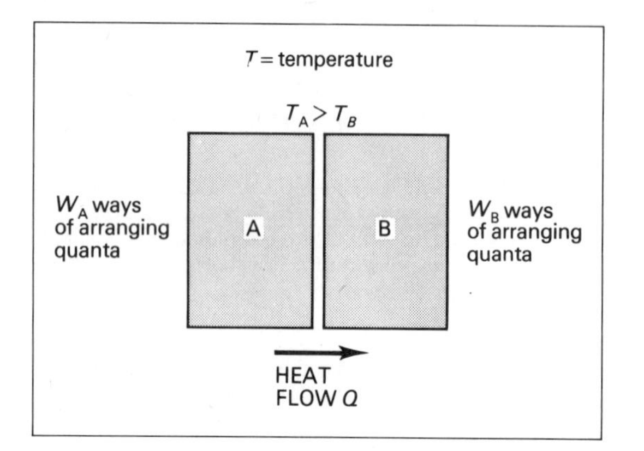

Fig. 26.3. Heat flow explained by probability.

(*a*) Suppose that there are W_A ways of arranging quanta amongst A's particles, and W_B ways for B. Then the total number of ways of arranging these quanta is $W_A \times W_B$.

(*b*) The two bodies are now put into thermal contact, and reach thermal equilibrium. To achieve this, quanta move from A to B: in consequence W_A falls to w_A, and W_B rises to w_B. The new total number of ways of arranging quanta is $w_A \times w_B$.

(*c*) Calculation shows that if $T_A > T_B$,

$$(w_A \times w_B) > (W_A \times W_B)$$

There are more ways of arranging the quanta between the two bodies if some move from A to B than if they move from B to A (or stay as they are). The decrease in the number of ways for body A is more than compensated by the increase for B.

(*d*) The more ways in which an arrangement can be made, the more likely is it to happen. We deduce that quanta are more likely to flow from A to B than from B to A. (The numbers involved ensure that the move is virtually certain.)

(*e*) The physical process of *heat flow* is identified as the random spreading of quanta of energy seeking more ways in which they can be arranged.

*26.8 Statistical Interpretation of Temperature

The situation of paragraph **26.7** leads to the following argument.

(*a*) When two objects with the *same* average number of quanta per atom are put into thermal contact, there may be *more* ways of arranging those quanta if some move from one body to the other than there would be without such movement.

(*b*) We deduce that temperature is *not* determined on a microscopic scale by the average number of quanta per atom. For the two bodies to stay at the same temperature, the effect of adding one quantum of energy to each must be to change the number of ways of arranging the quanta by the same factor.

(*c*) The addition of one quantum to a body at low temperature has a greater effect on the number of ways than the addition of one quantum has on the number for a body at a high temperature. That is, the lower the temperature, the larger the factor by which the number of ways changes.

(*d*) ***Quantitative development.*** At temperature T let there be W ways of arranging N quanta in a body, and W' ways for $(N + 1)$ quanta. Then

$$\begin{pmatrix}\text{factor by which the number} \\ \text{of ways is changed}\end{pmatrix} = \left(\frac{W'}{W}\right)$$

The numerical value of T can be made to range from 0 to ∞, as does temperature defined on the thermodynamic or *Kelvin* scale, by writing

$$\ln (W'/W) \propto 1/T.$$

Our statistical temperature T defined in this way can be made identical to thermodynamic temperature by appropriate choice of the constant of proportionality.

*26.9 Entropy and Macroscopic Ideas

Suppose that during an infinitesimal *reversible* process a system at temperature T absorbs an amount of heat δQ_{rev}.

The change in entropy δS of the system is defined by the equation

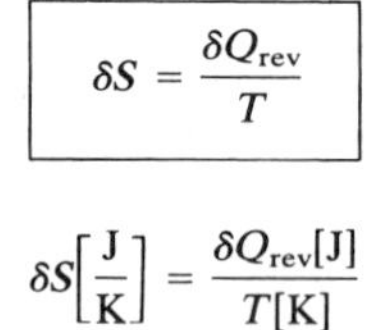

$$\delta S = \frac{\delta Q_{rev}}{T}$$

$$\delta S\left[\frac{J}{K}\right] = \frac{\delta Q_{rev}[J]}{T[K]}$$

Thus for a reversible transformation from state 1 to state 2, the change in entropy

$$\Delta S = S_2 - S_1 = \int_1^2 \frac{dQ_{rev}}{T}$$

Notes.

(*a*) ΔS depends only upon the initial and final states, and not upon the process followed.

(*b*) Since T is positive, ΔS is positive when heat is absorbed, and negative when it is rejected.

(*c*) For a *reversible* adiabatic transformation

$$\delta Q_{rev} = 0,$$

so $$\Delta S = 0$$

so $$S = \text{constant}$$

(*d*) From the defining equation, for a reversible change

$$Q_{rev} = \int_1^2 T \cdot dS$$

(*e*) When a body is taken through a complete cycle it is returned to its original state (fig. 26.4).

$$\Delta S_{cycle} = 0$$

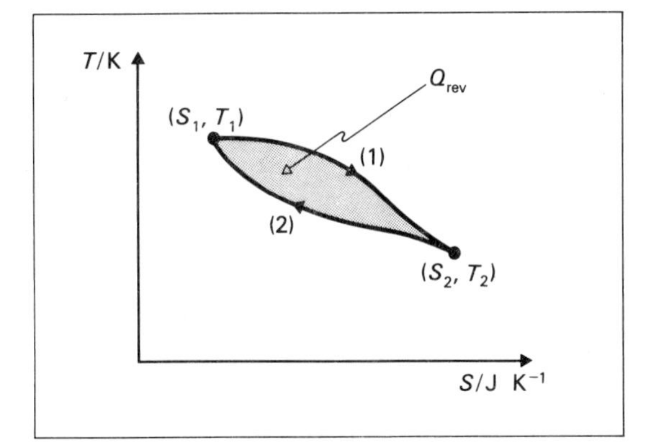

Fig. 26.4. The heat absorbed by a system undergoing a reversible cycle.

A transformation in which the entropy of the system does not change is called **isentropic**. (Thus a reversible adiabatic process is isentropic.)

$$Q_{rev} = \oint T \cdot dS$$
$$= \begin{pmatrix}\text{heat absorbed by system}\\ \text{during one cycle}\end{pmatrix}$$
$$= \begin{pmatrix}\text{work done by system}\\ \text{during that cycle}\end{pmatrix}$$

As an example the *Carnot cycle* is discussed in **26.12**.

(*f*) At any temperature a single-phase substance will have unique values of U and S. These are macroscopic variables, and they can be used to describe a process in the same way as can p, V and T.

Some Properties of Entropy

Experiment indicates the following properties to be possessed by entropy.

(*a*) *Entropy tends towards a maximum* in an isolated system. (Isolation implies lack of ability to transfer mass or energy across the boundary of the system.)

(*b*) *Entropy is extensive.* It depends, for example, on both the volume and the number of particles within the system. (Thus the mass of a system is extensive, whereas the density in **intensive** — it is unchanged by an imaginary partition placed within the system.)

(*c*) *Entropy is additive.* Suppose a system has two parts of entropies S_1 and S_2. Then the entropy of the whole is given by

$$S = S_1 + S_2$$

Calculation of Entropy Changes

(*a*) For any *reversible* process the change in entropy between states 1 and 2 is found from

$$\Delta S = S_2 - S_1 = \int_1^2 dQ_{rev}/T$$

The value of ΔS depends upon the initial and final states only.

(*b*) For an *irreversible* process we cannot evaluate $\int_1^2 dQ/T$ along the path concerned. Instead we try to find a *reversible* path which connects the same initial and final states, and evaluate $\int_1^2 dQ/T$ for that path.

Example: Free Expansion

Consider the process of fig. 26.1, in which the volume changes *irreversibly* from V to $2V$, and suppose the gas to be ideal. Now an isothermal expansion, controlled so as to occur reversibly, would connect the same initial and final states. For either process

$$\begin{aligned}\Delta S &= S_2 - S_1\\ &= \int_1^2 \frac{dQ_{rev}}{T}\\ &= \mu R \ln (2V/V)\\ &= \mu R \ln 2,\end{aligned}$$

which is *positive*. (We have used the fact that, since T = constant,

$$\Delta U = 0 \qquad \text{(ideal gas)}$$

giving $$\Delta W = +\Delta Q \qquad \text{(first law)}$$

ΔW is calculated from $\int_1^2 p \cdot dV$.)

For this irreversible adiabatic process the entropy increases. This discussion should be compared with the probability argument in **26.5.**

*26.10 Entropy and Microscopic Ideas

Although we can measure entropy changes in practice from

$$\delta S = \frac{\delta Q_{rev}}{T}$$

for reversible processes, we want to know the fundamental significance of entropy changes in terms of particle behaviour, that is to relate entropy to statistical quantities.

Now $\ln W$, the logarithm of the number of ways in which quanta of energy can be arranged in a system, shows the three experimental properties of entropy listed in **26.9.**

(*a*) In an isolated system W and $\ln W$ tend to a *maximum.*

(*b*) W and $\ln W$ depend upon the number of particles in the system, and in this sense are *extensive.*

(*c*) Whereas entropy is additive, numbers of ways are multiplicative, following the rule

$$W = W_A \times W_B$$

On the other hand $\ln W$ has the necessary *additive* characteristic, since

$$\ln W = \ln W_A + \ln W_B$$

We write

$$S \propto \ln W,$$

or

$$\boxed{S = k \ln W}$$

$\ln W$ is dimensionless, and the constant k is in fact the *Boltzmann constant* (p. 192). $k = 1.38 \times 10^{-23}\ \mathrm{J\,K^{-1}}$, and ensures that the unit of S is consistent with the earlier definition.

In a completely ordered arrangement, such as an idealized cystalline solid of a pure substance at absolute zero, we would have

$$W = 1$$

and

$$S = 0$$

*26.11 Entropy and the Second Law

The zeroth law of thermodynamics develops our concept of temperature. The first law deals with that of internal energy. The second law is bound up with the concept of entropy.

The Second Law

We have so far established that

(*a*) the more ways there are in which an event can be realized, the more likely it is to occur,

(*b*) the distribution which occurs in statistical equilibrium is that which is most probable, and

(*c*) $$S = k \ln W$$

These statements tell us that the entropy of an isolated system in statistical equilibrium has the maximum value compatible with the physical conditions of the system. More formally:

> *The* ***second law of thermodynamics*** *states that the processes most likely to occur in an isolated system are those in which the entropy either increases or remains constant.*

The usefulness of the law is that it tells us which processes in nature are likely to happen.

Notes.

(*a*) For a *reversible* process $\Delta S = 0$: such a process can go equally well in either direction.

(*b*) For a system initially *not in equilibrium*

$$\Delta S > 0$$

Not until statistical equilibrium has been achieved will further changes yield

$$\Delta S = 0$$

(*c*) For an *irreversible* process $\Delta S > 0$, and these are the spontaneous processes which have a *natural direction* in nature. Transport processes are good examples. These include molecular diffusion and thermal conduction (below).

(*d*) The second law points to a *probable* course of events, not the only possible course. There may locally be *small-scale* exceptions to the law where decreases of entropy occur spontaneously.

(*e*) The graph of fig. 26.5 shows the typical approach to equilibrium of a characteristic one-way process.

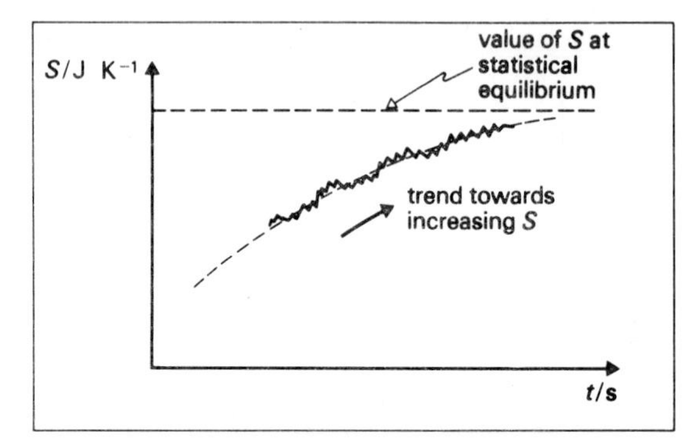

Fig. 26.5. Graph showing fluctuating value of the entropy of a system as it approaches statistical equilibrium.

A System with Components

Suppose a system has two separate components which are put into thermal contact. Initially the total entropy

$$S = S_1 + S_2$$

Finally

$$S' = S_1' + S_2'$$

The net change of entropy

$$\Delta S = S' - S$$
$$= \Delta S_1 + \Delta S_2$$
$$\geqslant 0$$

If one component loses entropy (e.g. if ΔS_2 is negative), then the other component gains entropy (ΔS_1 is positive) in such a way that the net change of entropy for the whole system is either zero or positive.

The Entropy Changes in Heat Conduction

Refer to fig. 26.3. Suppose that within an adiabatic enclosure the two bodies eventually reach a common temperature T.

Principles

(*a*) The fact that $(T_A - T_B)$ is not infinitesimally small at all stages of the operation means that this process is *irreversible.*

(*b*) To calculate the energy change for the system we must devise a *reversible* change which connects the same initial and final states, and apply

$$\Delta S = \int_1^2 \mathrm{d}Q_{\mathrm{rev}}/T$$

This can be done by means of an intermediate heat reservoir whose temperature can be adjusted to any required value.

Approximate Calculation

Let T'_A and T'_B be the mean temperatures of the bodies when heat Q is rejected and absorbed, respectively.

Then $\Delta S_\mathrm{A} \approx -\dfrac{Q}{T'_\mathrm{A}}, \quad \Delta S_\mathrm{B} \approx +\dfrac{Q}{T'_\mathrm{B}}$

$$\Delta S = \Delta S_\mathrm{A} + \Delta S_\mathrm{B}$$

$$\approx -\frac{Q}{T'_\mathrm{A}} + \frac{Q}{T'_\mathrm{B}}$$

Since $T'_\mathrm{B} < T'_\mathrm{A}$, ΔS is *positive*, as is necessary in an *irreversible* process.

*26.12 The Carnot Cycle

The *Carnot* cycle is discussed here

(*a*) as an illustration of a cyclic process,
(*b*) to exemplify earlier work on entropy, and
(*c*) because of its fundamental importance in the development of the thermodynamic scale of temperature.

The Indicator Diagram

Imagine a working substance, such as an ideal gas, contained in a cylinder by a piston. We subject it to the four processes shown in the indicator diagram of fig. 26.6.

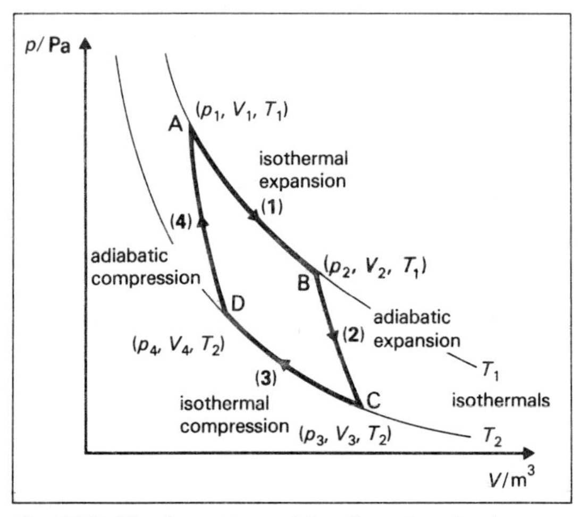

Fig. 26.6. The four steps of the Carnot cycle shown on an indicator diagram.

(1) The gas undergoes a slow reversible *isothermal expansion* by absorbing heat Q_1 at temperature T_1. It does work.

(2) The gas is allowed to *expand adiabatically* and reversibly. It does work, and its temperature falls to T_2.

(3) The gas is made to undergo a slow reversible *isothermal compression*, by having work done on it at temperature T_2. It rejects heat Q_2.

(4) The gas is *compressed adiabatically* and reversibly. Work is done on it, and its temperature rises to T_1.

The gas has been returned to its original state (p_1, V_1, T_1), and so there is no change in its internal energy. This means that the work W *done by* the system during the cycle is equal to the net heat Q *supplied to* the system per cycle, where

$$Q = Q_1 - Q_2$$

The system is a **heat engine.**

Entropy Changes during the Carnot Cycle

Refer to the T–S diagram of fig. 26.7, which represents the same *Carnot* cycle.

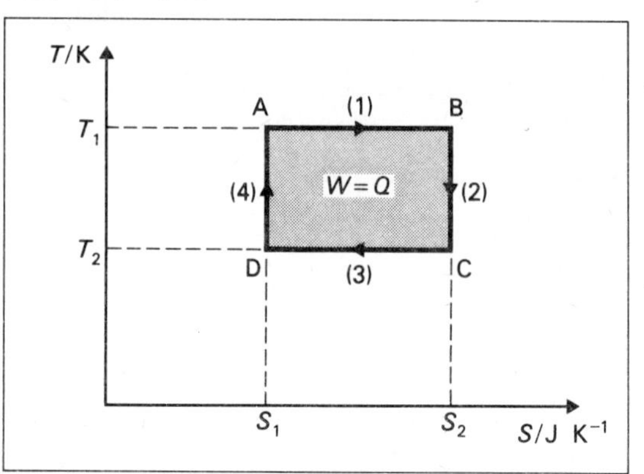

Fig. 26.7. The Carnot cycle represented on a T–S diagram.

The table summarizes the entropy changes for each process.

process	ΔW	ΔT	ΔQ	ΔS
(1) isothermal expansion	+ve	0	Q_1 (+ve)	Q_1/T_1
(2) adiabatic expansion	+ve	−ve	0	0
(3) isothermal compression	−ve	0	$-Q_2$ (−ve)	$-Q_2/T_2$
(4) adiabatic compression	−ve	+ve	0	0

Since $\Delta S_{\mathrm{cycle}} = 0$

$$\frac{Q_1}{T_1} - \frac{Q_2}{T_2} = 0$$

$$\boxed{\frac{Q_1}{T_1} = \frac{Q_2}{T_2}}$$

The work done by the system per cycle is given by the shaded rectangle:

$$\begin{aligned} W &= Q_1 - Q_2 \\ &= (T_1 - T_2)(S_2 - S_1) \end{aligned}$$

Thermal Efficiency

Although we have imagined the working substance to be an ideal gas, the result above would apply to any working substance.

In general, during a reversible isothermal process

$$\begin{aligned} S_2 - S_1 &= \int_1^2 \frac{dQ_{rev}}{T} \\ &= \frac{Q}{T} \qquad \text{(since } T = \text{constant)} \end{aligned}$$

Thus for our *Carnot* cycle

$$Q_1 = T_1(S_2 - S_1)$$

The **efficiency** of a heat engine is found from

$$(\text{efficiency}) = \frac{\left(\begin{array}{c}\text{net work done by engine during} \\ \text{one cycle}\end{array}\right)}{\left(\begin{array}{c}\text{heat taken from the high temperature} \\ \text{source during that cycle}\end{array}\right)}$$

Here efficiency

$$\begin{aligned} E &= W/Q_1 \\ &= \frac{(T_1 - T_2)(S_2 - S_1)}{T_1(S_2 - S_1)} \end{aligned}$$

$$\boxed{E = \frac{T_1 - T_2}{T_1}}$$

Notes.

(*a*) An amount of heat Q_2 is delivered to the lower temperature reservoir at a temperature T_2.

(*b*) If the cycle is performed in reverse, heat Q_2 is *removed* from the lower temperature reservoir, and the system constitutes a **refrigerator**. Work is done (or energy provided) by an external agent.

(*c*) The last boxed equation is used as the basis of the definition of temperature on the thermodynamic scale. This is designed to be identical to the ideal-gas scale.

(*d*) This equation also summarizes **Carnot's theorem**:

The efficiency of a thermal engine operating in a Carnot cycle depends only upon the two operating temperatures, and is independent of the working substance.

*26.13 Degradation of Energy

Internal energy is unique amongst energy forms in that it may be converted to other forms only partially and temporarily.

In the *Carnot* cycle of **26.12** the engine returned an amount of heat Q_2 from the working substance to the low temperature reservoir at temperature T_2. Even if no frictional or other losses are involved, i.e. if the practical design of the *machine is ideal,* only an amount $(Q_1 - Q_2)$ of the heat drawn from the high temperature reservoir can be converted into useful work.

We can imagine a fossil fuel being burned to provide steam, the steam being made to do work, and the surplus energy being disposed of via a cooling tower. *Energy* is conserved (first law), but what is not conserved is the *availability* of some of that energy (second law). The efficiency of the engine is limited by the requirement that the entropy of an isolated system must increase or stay the same.

Such processes happen continually in nature, causing *Kelvin* to recognize the **principle of degradation of energy**. The first law controls the *quantity* of energy in a system, while the second law deals with its *quality*, which is degraded with time.

In summary

(*a*) *irreversible processes continually lead to an increase in the disorder of energy,*
(*b*) *the entropy of the Universe increases inexorably towards a maximum,* and
(*c*) *when that maximum is reached no further energy will be available for the performance of work.*

27 Change of Phase

27.1 Change of Phase and Latent Heat L

Molecular behaviour and Energy changes

(a) Fusion

In solids molecules vibrate about fixed positions, and the pattern is *ordered.* In liquids molecules vibrate about variable positions, and the pattern is *disordered.*

To convert solid to liquid we provide energy (the **latent heat** L) which

(*i*) increases the p.e. on changing from the ordered to the disordered phase (while keeping roughly the same average molecular k.e.; i.e. the same temperature), and

(*ii*) does the external work (which is relatively small). If the substance contracts on melting, the external work is negative.

(b) Vaporization

The molecular separation in the gaseous phase is several times that in the liquid phase.

To convert liquid to vapour we provide energy (the latent heat L) which

(*i*) increases the p.e. of a molecule from about -0.9ε in the liquid phase (p. 120) to nearly zero in the gaseous phase (this energy enables work to be done against intermolecular forces), and

(*ii*) does the external work, which is usually much larger than in (*a*).

(c) Sublimation

This is the process by which a solid is converted directly into a vapour.

In the first law of thermodynamics (p. 188),

$$L \equiv \Delta Q = \Delta U + \Delta W,$$

we provide heat ΔQ, which enables the substance to do work ΔW, and which increases the internal energy by ΔU (but without a significant change of average molecular k.e.).

During a phase change ΔU represents basically a change of p.e.

Definition of Specific Latent Heat of Transformation l

Suppose heat L added to a mass m changes its phase without causing a temperature change.

The specific latent heat of transformation l is defined by the equation

$$\boxed{l = \frac{L}{m}}$$

$$l\left[\frac{\text{J}}{\text{kg}}\right] = \frac{L\,[\text{J}]}{m\,[\text{kg}]}$$

l would be called the specific latent heat of *fusion*, *vaporization* or *sublimation*. $[l] = [\text{L}^2\,\text{T}^{-2}]$.

The value of l depends on the pressure at which it is measured. It is usually quoted for *standard pressure* (and so for normal melting and boiling point).

Orders of Magnitude

Substance	$T_{\text{m.p.}}/\text{K}$	*Fusion* $l/\text{J kg}^{-1}$	$T_{\text{b.p.}}/\text{K}$	*Vaporization* $l/\text{J kg}^{-1}$
water	273	33×10^4	373	23×10^5
oxygen	55	1.4×10^4	90	2.1×10^5
copper	1356	18×10^4	2573	73×10^5
lead	600	2.5×10^4	1893	7.3×10^5

Summary

Note that either (*a*) the molecules *gain k.e.* (and at the same time p.e.) which shows as a *temperature increase* for a particular phase, or

(*b*) there is a gain of *p.e.* (with hardly any change of k.e.), which shows as a *phase change* at constant temperature.

In (*a*) the slope of the graph is inversely proportional to the substance's specific heat capacity, and in

(*b*) the length of a horizontal portion is proportional to a specific latent heat of transformation.

At a particular temperature molecules of any ideal gas (whatever its chemical nature) would have the same average translational k.e. (p. 181).

27.2 Measurement of Specific Latent Heat l

In $\Delta Q = ml$, we measure m and ΔQ.

(*a*) *Fusion*
- (*i*) Method of mixtures
- (*ii*) *Bunsen's* ice calorimeter

(*b*) *Vaporization*
- (*i*) Method of mixtures
- (*ii*) Continuous-flow, as in

Henning's Method

This adaptation includes the principle of the *self-jacketing vaporizer.*

Allow the apparatus to warm up throughout by prolonged continuous boiling, so that the vapour jacket is at the temperature of the boiling liquid, and a *steady state* is maintained.

Measure V_1, I_1 and the mass m_1 [kg] of liquid that evaporates and condenses in a time t [s]. Then

$$V_1 I_1 t = m_1 l + Q$$

where Q is the (small) heat loss through the jacket.

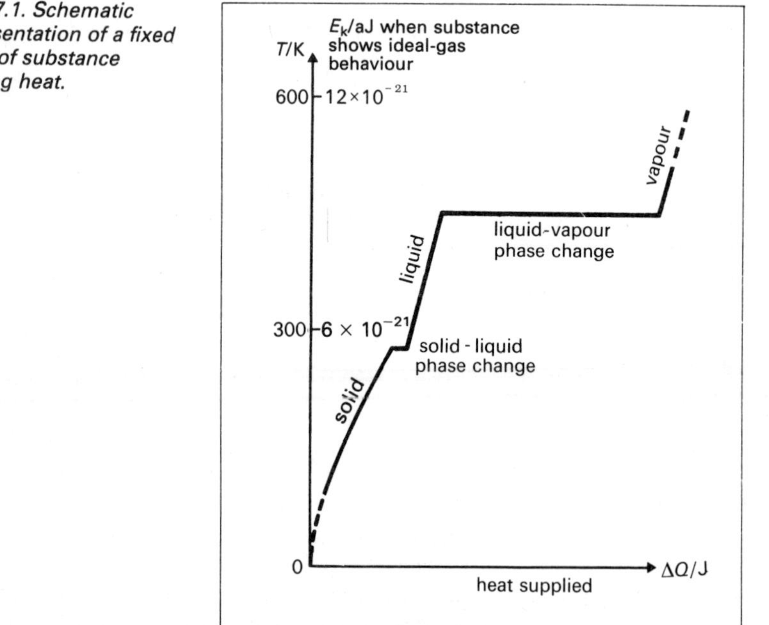

Fig. 27.1. Schematic representation of a fixed mass of substance gaining heat.

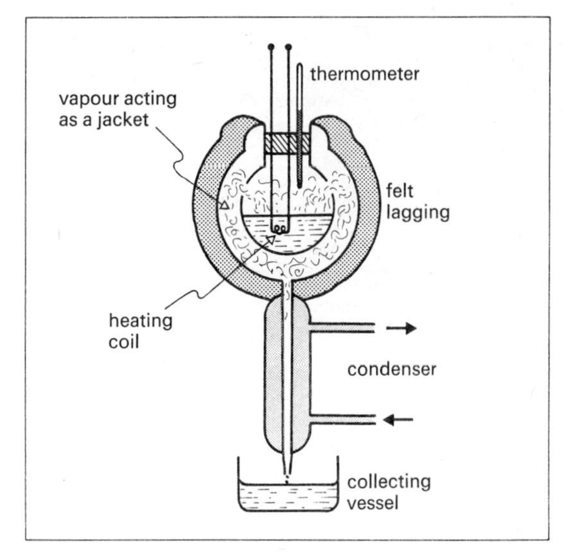

Fig. 27.2. The self-jacketing vaporizer.

Repeat the experiment with a significantly different power input V_2I_2, and collect a mass m_2 in the *same time t.*

$$V_2I_2t = m_2l + Q$$

Q is the same because in an identical environment we have the same power loss, and this loss occurs for the same time. Subtracting to eliminate Q

$$l\left[\frac{\text{J}}{\text{kg}}\right] = \left(\frac{V_1I_1 - V_2I_2}{m_1 - m_2}\right)\cdot t$$

The method has the usual advantages of continuous-flow calorimetry (p. 174).

27.3 Evaporation

Molecular Behaviour

Evaporation is the escape of molecules from the *surface* of a liquid. Upward-moving molecules near the surface may have enough energy

(*a*) to escape completely, or

(*b*) to escape to a position from which they are repelled back into the liquid by other molecules above the surface, or

(*c*) to move only a little way before other liquid molecules pull them back.

The *rate* of evaporation for a given liquid can be increased by

(1) increasing the surface *area*,

(2) having faster (more energetic) molecules: this is achieved by a *temperature* increase.

(3) having fewer molecules above the surface (by reducing the *pressure*), and/or

(4) removing escaped molecules in a *draught.*

Evaporation occurs *at all temperatures*, but because of (2) is faster at high temperatures.

Vapour Pressure

When the space above a liquid is closed, it quickly becomes **saturated** with vapour, and a *dynamic equilibrium* is established. The vapour molecules exert a **saturation vapour pressure** (s.v.p.) whose value depends only on the temperature; it is independent of any external pressure. If the volume of the space is reduced, some of the vapour liquefies, but there is no change in the pressure.

A *saturated* vapour does *not* obey the gas laws. An unsaturated vapour obeys them reasonably well.

Mixtures of Gases and Vapours

A vapour can be liquefied by pressure alone, whereas a gas cannot be unless it is first cooled (p. 210). To solve problems involving a mixture of the two:

(*a*) apply *Dalton's law* (p. 181) to separate the

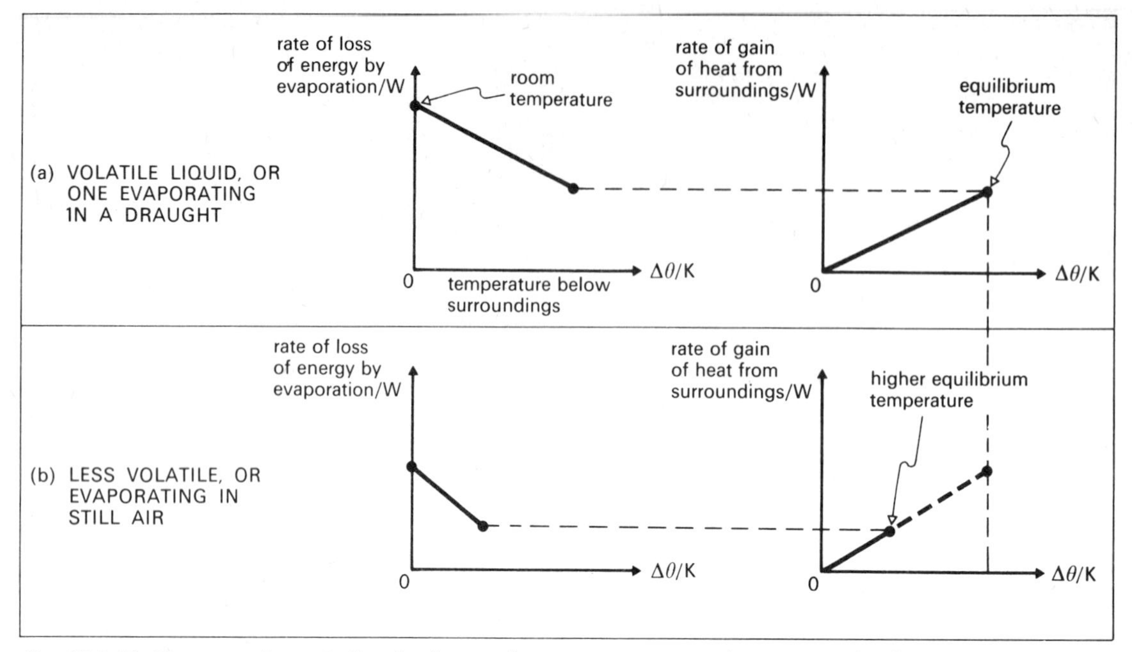

Fig. 27.3. To illustrate schematically why the steady-state temperature of an evaporating liquid is below that of its surroundings.

total pressure into the partial pressures of the gas and vapour respectively.

(*b*) apply the gas equation (p. 178) to evaluate what happens to the gas and any *unsaturated* vapour,

(*c*) calculate the pressure in any space *saturated* by vapour by reference to its s.v.p. at the temperature concerned, and

(*d*) calculate the final pressure by further application of *Dalton's law*.

Cooling due to Evaporation

If it were equally likely for *any* molecule to escape, the liquid would lose energy and mass in the same proportion, and its temperature would not change. Because the *faster* molecules escape, the average molecular k.e. (and hence temperature) decreases.

The fall of temperature brought about by evaporation depends on the rate of evaporation and thus on the nature of the liquid, and on external factors. As soon as the liquid temperature falls below that of the surroundings, they give heat to the liquid.

A steady state is reached when

$$\begin{pmatrix}\text{rate of gain of heat}\\ \text{from surroundings}\end{pmatrix} = \begin{pmatrix}\text{rate of loss of energy}\\ \text{by evaporation}\end{pmatrix}$$

Then the temperature remains steady at a value below that of the surroundings, while the liquid mass continues to decrease. This is shown by fig. 27.3.

The lowering of temperature will be different at different temperatures, and can be increased by

(*i*) using a volatile liquid, and
(*ii*) a strong draught.

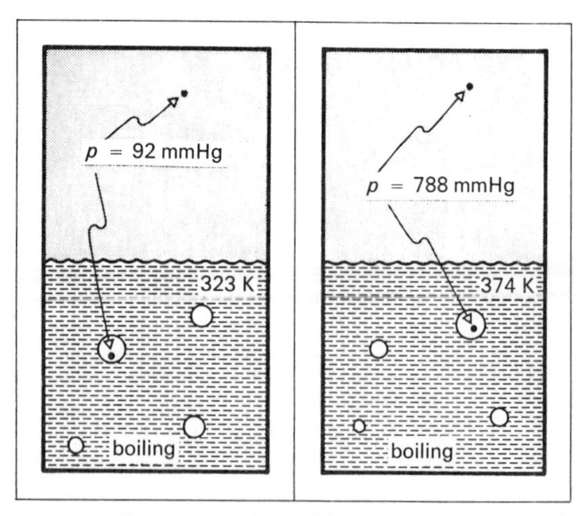

Fig. 27.4. Water boiling at different temperatures when the external pressures are different.

27.4 Boiling

Boiling occurs when molecules escape in the form of bubbles of vapour from the *body* of the liquid. If we ignore surface tension effects (p. 141), this occurs when the pressure inside such a bubble is equal to the pressure in the liquid at that depth. The pressure inside a bubble will be the s.v.p. at that temperature, so

A liquid boils at the temperature at which its s.v.p. is equal to the external pressure.

Thus boiling occurs

(*a*) at a *temperature* controlled by the external pressure, but

(*b*) at a *rate* controlled by the power of the heat source.

Note that a graph which plots s.v.p. against temperature, also represents external pressure against boiling point.

Standard boiling point is the temperature at which the s.v.p. of a liquid equals standard pressure (101 325 Pa, which is equivalent to that exerted by a 760 mm column of mercury).

The standard boiling point of pure water is 373 K: this is in effect (for elementary *practical* purposes) a *definition* and not an experimental result, since it is a fixed point on the International Practical Temperature Scale.

27.5 Measurement of s.v.p.

(*a*) Static (Barometer) Method

Usually used for s.v.p. in the range 50–400 mmHg. The sources of error include

(1) change with temperature of the density of mercury,
(2) surface tension effects from the required surplus of liquid,
(3) the pressure exerted by this surplus, and
(4) the difficulty of measuring accurately the height difference involved.

The method *cannot* measure an s.v.p. greater than atmospheric pressure.

(*b*) Dynamic (Boiling) Method

Usually used for s.v.p. from 50 mmHg upwards.

The apparatus shown in fig. 27.5 is a means of boiling liquid under a variable pressure. We measure

(*i*) the boiling point, and

(*ii*) the corresponding external pressure (which equals the s.v.p.): this is calculated from h and the atmospheric pressure.

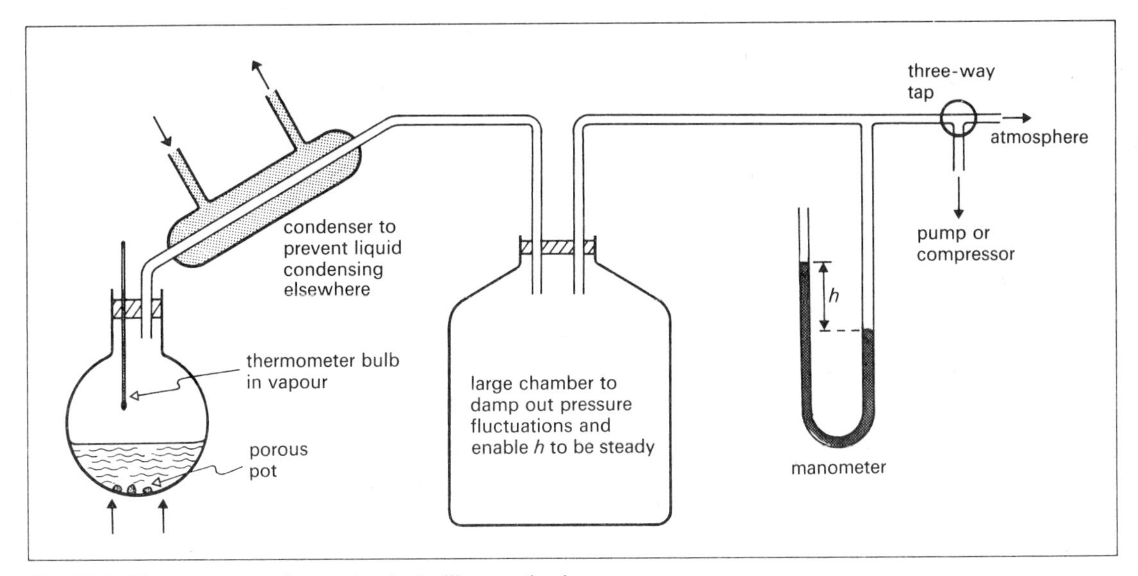

Fig. 27.5. Measurement of s.v.p. by the boiling method.

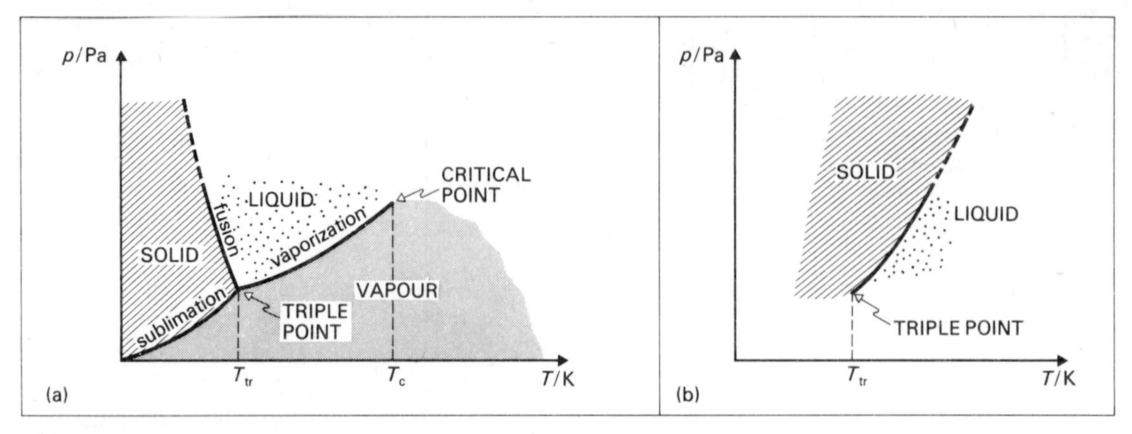

Fig. 27.6. (a) A p–T graph for a substance (such as water) which contracts on melting. (b) The fusion curve for a substance which expands on melting. (Neither graph to scale).

27.6 *p–T* Graphs and The Triple Point

Fig. 27.6 shows the graphs obtained when experimental values of pressure are plotted against temperature (as in simple boiling point experiments).

We can identify three curves.

(*a*) The **sublimation curve** connects points at which vapour and solid exist in equilibrium.

(*b*) The **vaporization curve** shows vapour and liquid existing in equilibrium, and is more widely known as the graph plotting s.v.p. against temperature. The curve begins at the triple point, and ends at the critical point (p. 210).

(*c*) The **fusion curve** shows liquid and solid existing in equilibrium. It has no definite upper end-point.

The Triple Point

(*a*) The sublimation, vaporization and fusion curves meet at a single point called the **triple point**.

(*b*) That *unique* temperature at which water exists in solid, liquid and gaseous phases simultaneously in equilibrium (i.e. its triple point) is chosen as the upper fixed point of the thermodynamic temperature scale.

For historical reasons it is *allotted* the value 273.16 K.

(*c*) On a p–V diagram such as fig. 28.2 the triple 'point' would appear as a *line* rather than as a point.

(*d*) The triple point occurs at a unique pressure (611 Pa in the case of water), unlike m.p. and b.p.

28 Real Gases

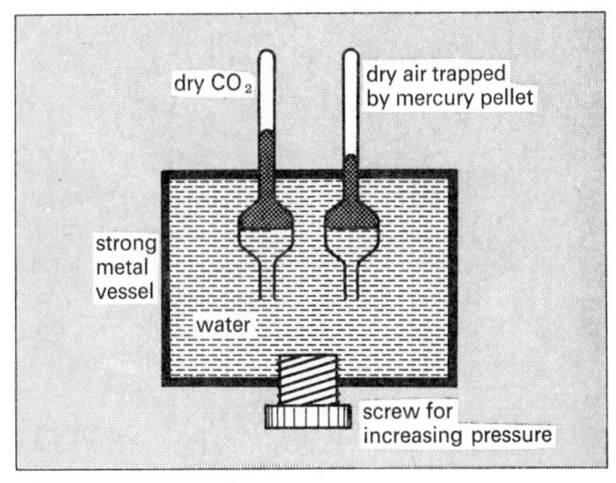

Fig. 28.1. Andrews's apparatus.

28.1 Andrews's Experiments (1863)

Using the apparatus of fig. 28.1 *Andrews* measured the volume of the trapped carbon dioxide for different values of the applied pressure. The pressure was calculated by assuming that air behaved as an ideal gas.

The experiment was repeated over a wide range of temperatures, and the results represented as a series of isothermals on an indicator diagram.

In fig. 28.2 the thick lines (full or dotted) separate the areas where it is possible for the carbon dioxide to exist as gas, vapour, liquid, etc.

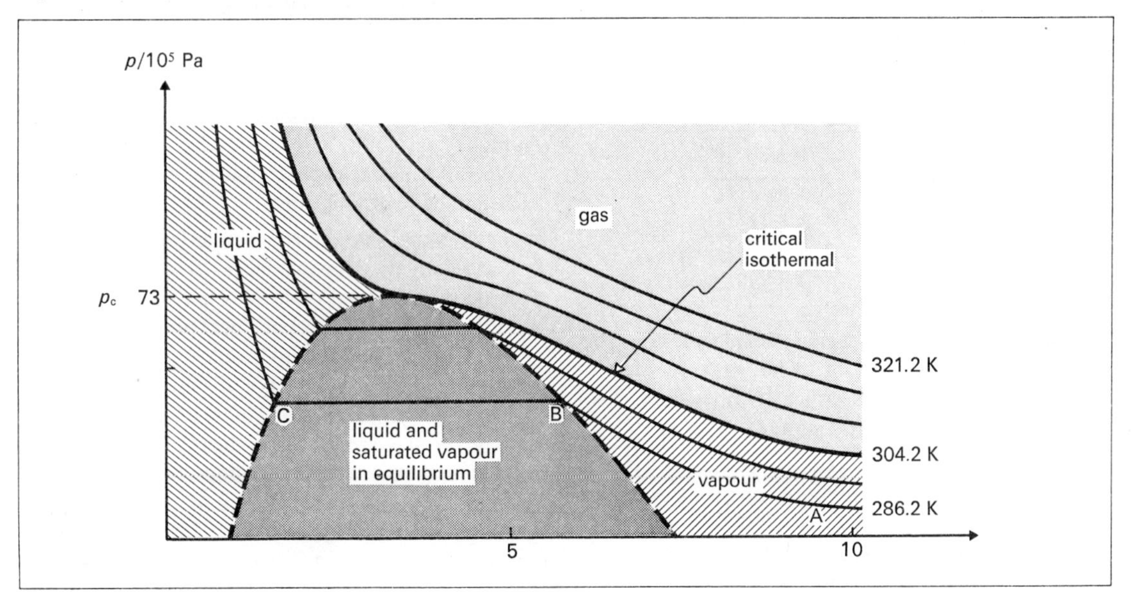

Fig. 28.2. Andrews's isothermals for 1 kg of carbon dioxide.

(a) The 321.2 K isothermal
The curve is approximately hyperbolic, as would be the curve for an ideal gas.

(b) The 304.2 K isothermal
This is called the **critical isothermal**, because it represents the dividing line between those temperatures at which CO_2 can be liquefied, and those at which it cannot. (*Andrews* measured the **critical temperature** to be 304.0 K.)

(c) The 286.2 K isothermal
At A the vapour is about to be compressed. The volume decreases as the pressure is increased (similar to the ideal-gas behaviour). At B the s.v.p. for that temperature has been reached: this means that further pressure increases are not possible while saturated vapour is present. As the volume is reduced, so the vapour continues to condense, with the pressure remaining constant at the s.v.p. until the point C, where all is liquid. Liquids are not easily compressed, so further great increases of pressure cause very little decrease in volume.

Some Conclusions

(1) CO_2 cannot be liquefied above a certain temperature: it is reasonable to suppose that every gas has its own **critical temperature** T_c, above which it cannot be liquefied by increasing the pressure.

A fluid in the gaseous phase is named a **gas** above its critical temperature, and a **vapour** below.

The **critical pressure** p_c is the pressure necessary to produce liquefaction at the critical temperature.

	He	H_2	O_2	CO_2	NH_3	H_2O
critical temperature T_c/K	5	33	155	304	406	647
critical pressure $p_c/10^5$ Pa	2.2	13	50	73	111	218

One can also define a **specific critical volume** v_c for a gas, which is the volume of 1 kg of gas subjected to its critical pressure at its critical temperature. The value for CO_2 is 2.2×10^{-3} m^3 kg^{-1}.

(2) The isothermals emphasize that carbon dioxide does not show the behaviour of an ideal gas unless either

(*a*) the temperature is high, and/or

(*b*) the pressure is very low.

An ideal gas would never liquefy.

28.2 Deviations from the Gas Laws

For all *real* gases

(*a*) experiments (such as those of *Andrews* and *Amagat*) show that $pV/\mu T$ (p. 178) does vary with temperature and pressure, and

(*b*) the assumptions of the theory that led to $p = \frac{1}{3}\rho c^2$ (p. 179) do not hold.

Deviations from $pV/\mu T$ = constant are small for the so-called *permanent* gases at normal pressures because normal temperatures are far above their critical temperatures.

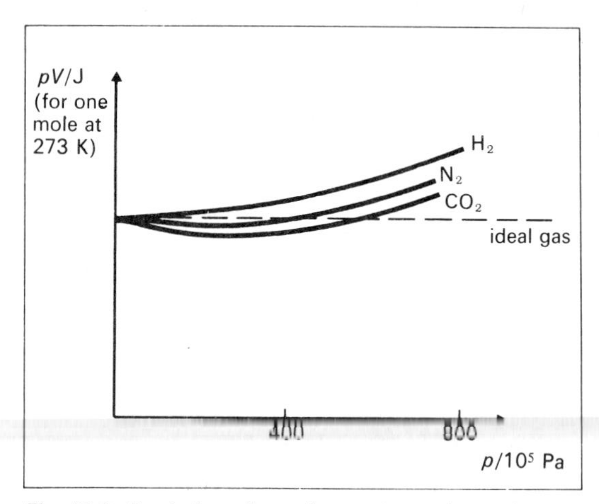

Fig. 28.3. Deviations from the gas laws shown by three different gases. (Not to scale.)

Fig. 28.3 illustrates how, for all real gases, pV increases with p, provided the pressure is large enough.

We will consider two possible breakdowns for our kinetic theory assumptions:

(*i*) The *repulsive* intermolecular forces are operative at relatively large distances. (In other words the volume of the actual molecules may be comparable with the volume of the space occupied by the gas.)

(*ii*) The *attractive* intermolecular forces are operative at relatively large distances. They may cause the formation of **molecular complexes** (p. 212), and the result is a reduction in the pressure exerted by the gas on the faces of its container. (The effect of the container's walls is considered in detail on p. 212.)

(*i*) is more important at small volumes (and so high pressures), when effect (*ii*) may be masked. For some gases (depending on the temperature) the two effects may counterbalance near a particular value of *p*.

The Boyle Temperature

The **Boyle temperature** is the temperature at which pV is most nearly constant over a wide range of pressures, because the repulsive and attractive intermolecular forces produce equal but opposing effects.

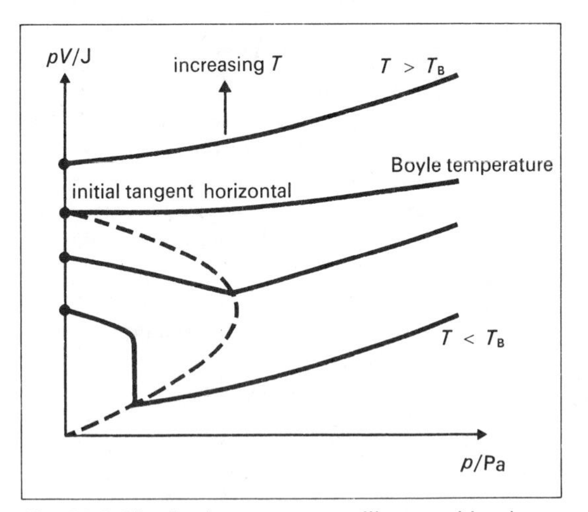

Fig. 28.4. The Boyle temperature illustrated by the variation of pV with p for a fixed mass of one gas.

Suppose that in fig. 28.4 we can write

$$pV = A + Bp + Cp^2 + \cdots \quad \text{for a real gas.}$$

Then $A = \mu RT$, and B and C etc. are progressively smaller factors, all of which are zero for an *ideal* gas. They depend on temperature, but not pressure.

For a *real* gas $B = 0$ at the *Boyle temperature*, and the value of C is negligible.

28.3 Internal Energy of Real Gases

(*a*) Joule's experiment (1845)

Suppose a mass of real gas expands adiabatically (and irreversibly) into a vacuum. Then

(*i*) the gas does zero *external* work, but

(*ii*) the energy required to separate the gas molecules (the *internal* work) is taken from the random molecular translational k.e. This process is accompanied by a temperature decrease.

Joule tried an experiment of this kind, but the high heat capacity of the apparatus made it too insensitive for the very small temperature change to be detected.

(*b*) The Joule–Kelvin experiment (1852)

Suppose a mass of gas at volume V_1 and (high) pressure p_1 is forced through a *porous-plug* to a region where its volume becomes V_2 and the (low) pressure is p_2 (fig. 28.5).

Two effects are superimposed:

(*i*) since $V_2 > V_1$, there will be an increase in intermolecular p.e., which will cause cooling, and

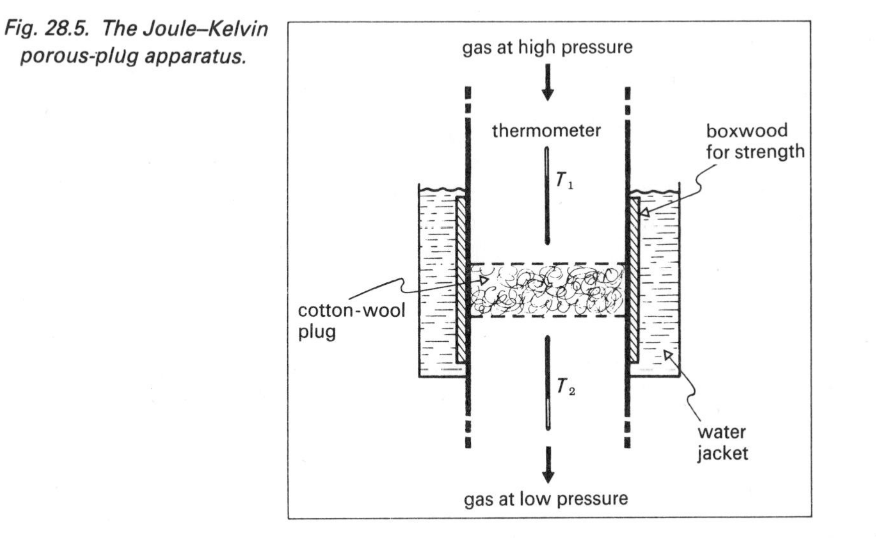

Fig. 28.5. The Joule–Kelvin porous-plug apparatus.

(*ii*) the *net* work done *by* the gas is

$$(p_2V_2 - p_1V_1)$$

Thus there may be either a cooling or a heating effect, depending on whether the real gas is below or above its *Boyle* temperature. (For an *ideal* gas $p_1V_1 = p_2V_2$.)

The observed temperature change is the *net* result of effects (*i*) and (*ii*).

Experimentally *Joule* and *Kelvin* found that hydrogen showed a heating effect (because of effect (*ii*) above), whereas *most gases showed a cooling effect* which was greater than could be accounted for by effect (*ii*). This is *evidence for attractive intermolecular forces.*

The **inversion temperature** is that temperature at which a gas shows no change of temperature. *Above* its inversion temperature a gas will show a *net heating* effect, and *below* its inversion temperature a *net cooling* effect. (The inversion temperature obviously exceeds the *Boyle* temperature.)

28.4 Van der Waals's Equation

$$p_i\left(\frac{V_i}{\mu}\right) = RT$$

is the **equation of state** for an amount μ of an *ideal* gas. *Van der Waals's* equation is the result of one attempt to find such an equation for a *real gas.*

(a) Effect of Repulsive Intermolecular Forces

Consider 1 mol of a real gas, and let it occupy a volume V_m. The actual volume of the gas molecules will reduce by b (the molar **covolume**), the free volume in which random molecular motion can take place.

Then $(V_m - b) = V_i/\mu$ is the molar volume which can be altered by changes of temperature and pressure, and which corresponds to the molar volume of an ideal gas.

(b) Effect of Attractive Forces

Attractive forces between non-ideal gas molecules may cause two or more molecules to become bound into loose complexes. Thus if (for example) *two* molecules are bound together, a **dimer** is formed. Since the total number of separate particles present is thereby reduced, so also is the number of moles of separate particles. We could indicate this *decrease* in μ by writing

$$\mu_{real} < \mu_{ideal}$$

but it is more common to *add* a correction factor to the pressure.

(Note that because a dimer and a monomer are in thermal equilibrium, on average they have equal translational k.e.: a dimer would *not* have twice the average momentum of a monomer.)

Suppose we write

(*i*) p_i = the pressure that *would* have been measured at the walls of the container if the real gas had been ideal,

(*ii*) p = the pressure as normally understood (i.e. that *actually* measured for the real gas),

(*iii*) p_a = the correction that must be added to p to take account of the formation of complexes.

Then $$p_i = (p + p_a)$$

A consideration of the chemical equilibrium between monomers and dimers (beyond the scope of this book) indicates that we should write

$$p_a = a/V_m^2$$

where the 'constant' a depends on temperature.

We conclude

$$p_i = \left(p + \frac{a}{V_m^2}\right)$$

(c) The Effect of the Container Walls

A molecule approaching very close to the wall of the container will experience an attractive force which will increase its momentum towards the wall. During this process the wall molecules will gain an equal but oppositely directed momentum, so that the overall effect of the collision at the wall is to impart the *same net momentum to the wall* as that which would have been imparted in the absence of the attractive force.

The fact that the range of the forces exerted by the wall is very small compared to the dimensions of gas container thus causes the *walls* to have zero effect on the pressure.

Summary

Experiment shows that the 'constants' a and b vary from the gas to gas, as we should expect. We now rewrite the ideal-gas equation of state

$$p_i\left(\frac{V_i}{\mu}\right) = RT$$

in the following form for a real gas

$$\left(p + \frac{a}{V_m^2}\right)(V_m - b) = RT$$

This is called **van der Waals's equation of state**.

Orders of Magnitude

Substance	a/N m^4 mol^{-2}	b/m^3 mol^{-1}
hydrogen	2.5×10^{-2}	2.7×10^{-5}
nitrogen	1.4×10^{-1}	3.9×10^{-5}
carbon dioxide	3.6×10^{-1}	4.3×10^{-5}

Note carefully the units that we allot to a and b. Their dimensions are [$ML^5T^{-2}N^{-2}$] and [L^3N^{-1}] respectively.

The Predictions of van der Waals's Equation

Fig. 28.6 shows the isothermals that are predicted by *van der Waals's* equation, which is a cubic in V_m.

The predicted curve ABC cannot be realized experimentally, but otherwise the isothermals resemble those plotted by *Andrews* for CO_2.

At large values of V_m the equation reduces to the ideal-gas equation, so the isothermals are hyperbolic.

28.5 Principles of Gas Liquefaction

A gas may be cooled by

(*a*) Mixing it with a cool fluid.

(*b*) Making it do external work in an adiabatic expansion.

(*c*) Making it (while below its inversion temperature) expand through a porous-plug from a region of high pressure to one of low pressure (the *Joule–Kelvin effect*).

A gas may be liquefied by

(*a*) Reducing the temperature of a vapour to below its boiling point at normal pressures.

(*b*) Using an evaporating liquid to cool a gas below its critical temperature, and then increasing the pressure, as in the **cascade process** (*Pictet*).

(*c*) Cooling the gas by *Joule–Kelvin* expansion, and using the cool gas to reduce the temperature of gas which is about to expand. This is called **regenerative cooling**.

(*d*) making the gas do external work in a nominally reversible adiabatic expansion, as in **Claude's process**.

(*e*) Combining a *Joule–Kelvin* expansion with *Claude's* process, as in the **Collins** process for liquefying helium.

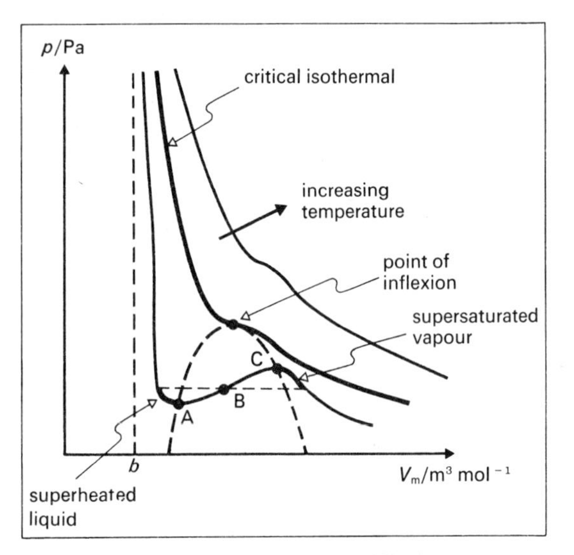

Fig. 28.6. Isothermals for a van der Waals gas.

The table shows some temperatures of importance in gas liquefaction, together with some characteristic values for comparison.

name of temperature	*origin of definition*	*page*	*value, in kelvins, for*		
			N_2	H_2	He
standard melting point	liquid and solid in equilibrium at standard pressure	203	63	14	3.5
triple point	solid, liquid and vapour in equilibrium	208	63	13.8	2.2
standard boiling point	liquid and vapour in equilibrium at standard pressure	203	77	20.3	4.27
critical temperature	coexisting liquid and vapour phases have the same density—no pressure can cause liquefaction at higher temperatures	210	121	33	5.2
Boyle temperature	initial tangent of pV–p curve is horizontal, and pV is most constant over a wide range of pressures	211	420	114	16
inversion temperature	gas shows no change of temperature in a *Joule–Kelvin* throttling process	212	650	190	25

29 Thermal Conduction

29.1 The Mechanisms of Conduction

Thermal conduction is a transport phenomenon (p. 183) in which a temperature difference causes the transfer of thermal energy from one region of a hot body to another region of the same body which is at a lower temperature. It takes place in the direction in which the temperature decreases: it therefore tends to equalize the temperature *within the body.*

(a) In gases

Conduction is the result of collisions between fast and slow mobile molecules. Such a collision transfers k.e. from the fast to the slow molecule.

(b) In liquids and solids

There are two processes:

(*i*) In *metals* (which possess **conduction** electrons) the positive ions of a lattice are in thermal equilibrium with the electron gas.

The electrons have random speeds as high as 10^6 m s^{-1}, but otherwise behave rather like gas molecules, and transfer thermal energy from the hot region of the lattice to the cold relatively quickly. (They will also do this in a liquid metal.)

(*ii*) The molecules of liquids and solids vibrate about fixed positions with a vibrational energy that increases with temperature. *Coupling* between neighbouring molecules allows transfer of part of this vibrational energy from energetic (hot) to less energetic (cold) regions.

Conduction by the electron gas suggests a close relationship with *electrical conductivity*, which is discussed on p. 217.

29.2 Temperature Gradient

Suppose **steady state** conditions hold, so that the common temperature θ at any point does not vary with time (fig. 29.1).

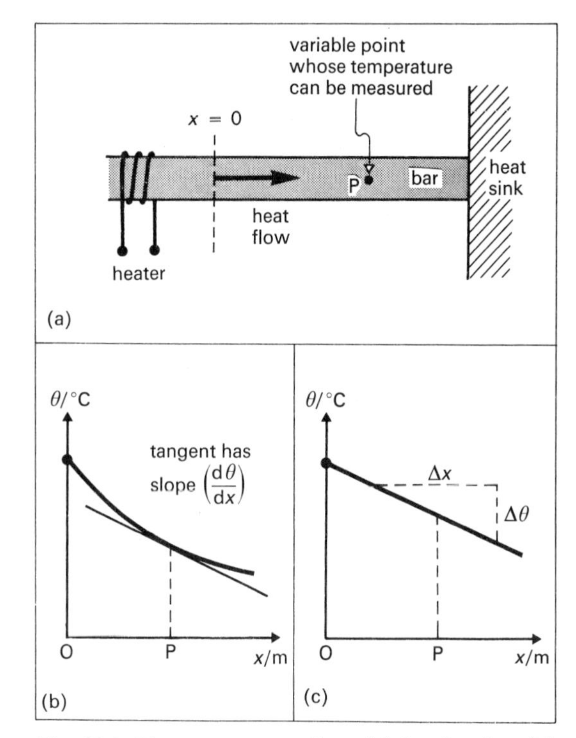

Fig. 29.1. Temperature gradient; (a) the situation, (b) and (c) graphs drawn for unlagged and perfectly insulated bars respectively.

The **temperature gradient** is defined to be $\mathrm{d}\theta/\mathrm{d}x$, which is

(*a*) for the unlagged bar, the slope of the graph *at a given point*, and

(*b*) for the perfectly insulated bar, $\Delta\theta/\Delta x$, a *constant*.

In each case it is *negative* because the heat flows in the direction in which x increases, and θ decreases in this direction. The lines of heat flow are

(*a*) *divergent* for the unlagged bar, but

(*b*) *parallel* for the perfectly insulated bar, since there is zero *lateral* heat loss.

29.3 Definition of Thermal Conductivity λ

We first define λ for the general situation (lines of heat flow divergent), and then simplify the equation for parallel heat flow.

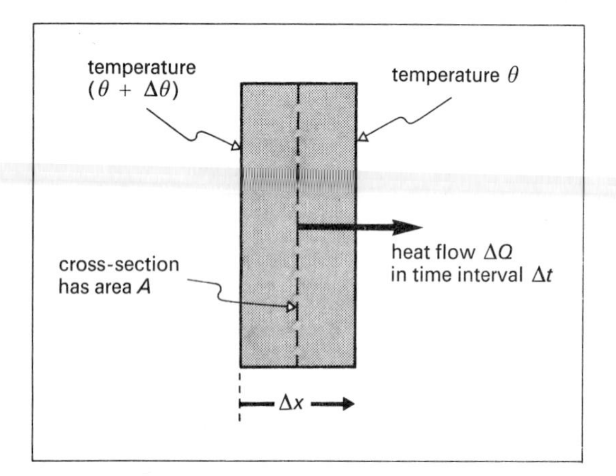

Fig. 29.2. Situation used to define λ.

Experiment shows that for small values of $\Delta\theta$ and Δx in fig. 29.2

(*a*) $\Delta Q \propto \Delta t$, and $\Delta Q \propto A$ when $\Delta\theta$ is fixed.

(*b*) $\Delta Q \propto \Delta\theta/\Delta x$, the *average* temperature gradient, when Δt and A are fixed.

(*c*) The direction of heat flow (here that of x) is that in which θ *decreases*, so is *opposite* to the temperature gradient.

Combining

$$\frac{\Delta Q}{\Delta t} \approx -\lambda A \frac{\Delta\theta}{\Delta x}$$

where λ is introduced below.

Suppose $\Delta x \to 0$, so that we have a limiting situation in which we consider a *cross-section*.

The thermal conductivity λ of a substance is defined by the equation

$$\boxed{\frac{\mathrm{d}Q}{\mathrm{d}t} = -\lambda A \frac{\mathrm{d}\theta}{\mathrm{d}x}}$$

in which the symbols are defined in fig. 29.2.

$$\frac{\mathrm{d}Q}{\mathrm{d}t}[\mathrm{W}] = -\lambda\left[\frac{\mathrm{W}}{\mathrm{m\,K}}\right] A[\mathrm{m}^2]\frac{\mathrm{d}\theta}{\mathrm{d}x}\left[\frac{\mathrm{K}}{\mathrm{m}}\right]$$

The insertion of the negative sign ensures that the constant of proportionality λ is positive.

Notes.

(*a*) λ is numerically equal to the rate of heat flow through a cross-section of unit area at which there is unit temperature gradient. It has the unit $\mathrm{W\,m^{-1}\,K^{-1}}$, and the dimensions $[\mathrm{MLT^{-3}\,\Theta^{-1}}]$.

(*b*) *Typical values*

Substance	$\lambda/\mathrm{W\,m^{-1}\,K^{-1}}$	*Measured at T/K*
copper	3.9×10^{2}	293
asbestos	8×10^{-2}	293
air	2.4×10^{-2}	273
hydrogen	14×10^{-2}	273

Although these vary by a factor of about 10^4, the variation of electric conductivity between good conductors and good insulators is greater by many orders of magnitude (p. 346).

(*c*) When the *lines of heat flow are parallel*, the temperature gradients at all cross-sections are equal. For the situation of fig. 29.3

$$-\frac{\mathrm{d}\theta}{\mathrm{d}x} = \frac{\theta_2 - \theta_1}{x}$$

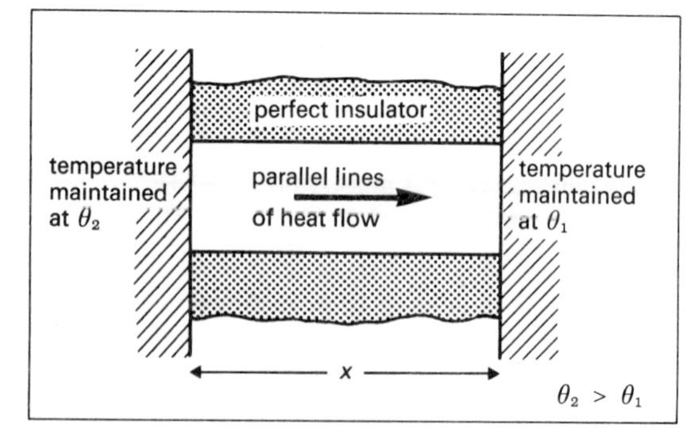

Fig. 29.3. Situation for a special case of the defining equation for λ.

and the defining equation for λ can be written

$$\frac{Q}{t} = \lambda A\left(\frac{\theta_2 - \theta_1}{x}\right)$$

a form which is applicable to most elementary problems.

Because the internal energy of the rod remains constant with time, the heat Q crossing *any* cross-section of the rod in time t is the same.

(*d*) λ varies slightly with temperature, but for calculation purposes this variation is usually negligible.

(*e*) The calculation of λ for an ideal gas is given on p. 184.

29.4 Comparison with Electrical Conductivity

In this table we compare some analogous quantities:

Heat conduction		*Electrical conduction*	
quantity	*symbol*	*symbol*	*quantity*
rate of heat flow	$\frac{dQ}{dt}$	$I = \frac{dQ}{dt}$	rate of charge flow
temperature gradient	$\frac{d\theta}{dx}$	$\frac{dV}{dx}$	potential gradient (numerically equal to field strength)
thermal conductivity	λ	σ	electrical conductivity

In heat transfer

$$\frac{dQ}{dt} = -\lambda A \frac{d\theta}{dx}$$

under conditions of zero lateral heat loss can be written

$$\frac{Q}{t} = \lambda A\left(\frac{\theta_2 - \theta_1}{x}\right)$$

In electricity we have

$$R = \frac{l}{\sigma A} \qquad \text{(p. 346)}$$

and

$$R = \frac{V_2 - V_1}{I} \qquad \text{(p. 346)}$$

whence

$$\frac{V_2 - V_1}{I} = \frac{l}{\sigma A}$$

$$I = \sigma A\left(\frac{V_2 - V_1}{l}\right)$$

The close analogy between the defining equations for λ and σ, and the fact that electrons are largely responsible for transport of thermal energy and electric charge in a metal, both underline the close relationship that exists between these two processes.

Wiedemann and Franz (1853) pointed out that, at a given temperature, the ratio (λ/σ) is nearly the same for all pure metals.

Lorentz (1872) showed in addition that, except at low temperatures,

$$\left(\frac{\lambda}{\sigma}\right) \propto T$$

the thermodynamic temperature, for most pure metals.

These experimental results can be explained by the theory of the behaviour of electrons in metals, which, though important, is beyond the scope of this book.

29.5 The Compound Slab

Suppose that in fig. 29.4

(*a*) a steady state has been achieved (so that θ does not depend on t, and specific heat capacities are not involved), and

(*b*) that the lines of heat flow are parallel.

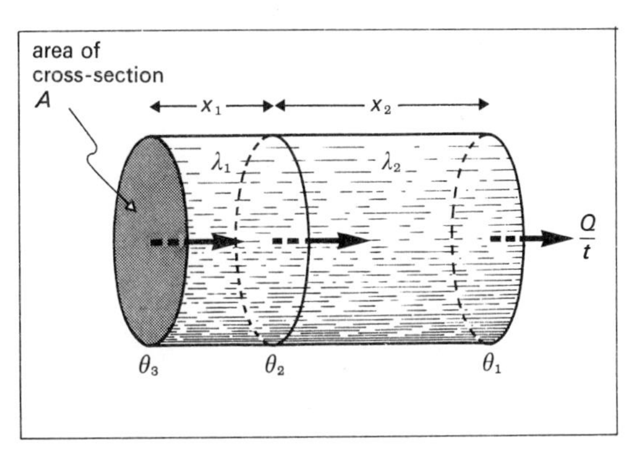

Fig. 29.4. Steady heat flow through a compound slab.

It follows that the rate of heat flow through all cross-sections is the same.

Using

$$\frac{dQ}{dt} = -\lambda A \frac{d\theta}{dx}$$

we have

$$\frac{Q}{t} = \lambda_1 A\left(\frac{\theta_3 - \theta_2}{x_1}\right) = \lambda_2 A\left(\frac{\theta_2 - \theta_1}{x_2}\right)$$

from which θ_2 (for example) could be calculated.

Notes.

(*a*) Suppose $\lambda_1 > \lambda_2$, so it follows that

$$\left(\frac{\theta_3 - \theta_2}{x_1}\right) < \left(\frac{\theta_2 - \theta_1}{x_2}\right)$$

The temperature gradient is *greater* within the *poorer* conductor.

(*b*) The situation is of practical importance when a poor conductor (such as an air-film) controls the rate of heat flow to a good conductor (such as the copper base of a saucepan). It may then be convenient to replace the compound slab for calculation purposes by its equivalent thickness of one material.

*29.6 Use of the Defining Equation

It is emphasized that the equation

$$\frac{Q}{t} = \lambda A\left(\frac{\theta_2 - \theta_1}{x}\right)$$

is a simplification that is not applicable in most real situations. The situation at a given *cross-section* is described by

$$dQ/dt = -\lambda A(d\theta/dx)$$

and in general this equation must be integrated to find the rate of heat flow. As illustrations of possible complications, note that

(*a*) at low temperatures λ may vary significantly with θ,

(*b*) the value of $\boldsymbol{A}$ may depend upon $\boldsymbol{x}$ (as in the example below), and

(*c*) lines of heat flow are generally not parallel, which means that $d\theta/dx$ is usually *not a constant.*

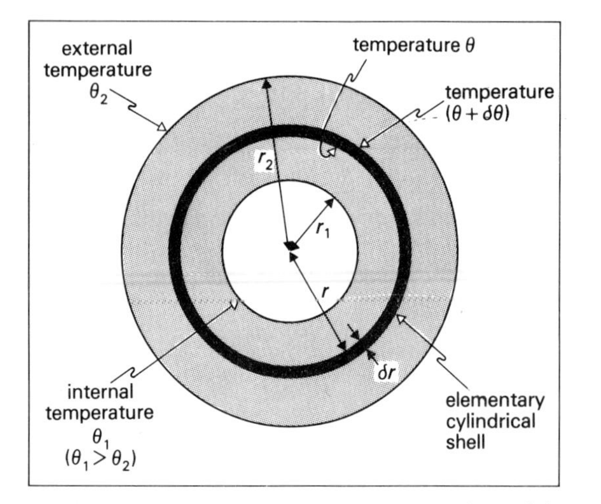

Fig. 29.5. Radial heat flow between a pair of coaxial cylinders.

Radial Lines of Heat Flow

Refer to fig. 29.5. The aim is to find the rate at which heat is conducted from the inner cylinder to the outer cylinder under steady-state conditions. Both cylinders have length l. We assume that λ is independent of θ.

To apply

$$dQ/dt = -\lambda A(d\theta/dx)$$

to the elementary cylindrical shell shown, we note that

(*a*) $A = 2\pi rl$,

(*b*) the lines of heat flow are radial, so that $\delta\theta/\delta r$ is dependent upon r, and

(*c*) $\delta\theta$ is negative.

Substituting $$\frac{dQ}{dt} = -\lambda(2\pi rl)\frac{d\theta}{dr}$$

$$\frac{dQ}{dt}\int_{r_1}^{r_2}\frac{dr}{r} = -2\pi\lambda l\int_{\theta_1}^{\theta_2} d\theta$$

$$\frac{dQ}{dt}\ln\left(\frac{r_2}{r_1}\right) = 2\pi\lambda l(\theta_1 - \theta_2)$$

which gives $$\left(\frac{dQ}{dt}\right) = \frac{2\pi\lambda l(\theta_1 - \theta_2)}{\ln(r_2/r_1)}$$

The result is relevant to both the electrical and thermal insulation of a cable dissipating electrical energy.

29.7 Measurement of Conductivity

We aim to use

$$\frac{Q}{t} = \lambda A\left(\frac{\theta_2 - \theta_1}{x}\right)$$

which implies:

(*a*) We must prevent lateral heat flow so that the lines of heat flow are parallel.

(*b*) We must use a well insulated *rod* for a *good conductor.* This means that the large x and small $\boldsymbol{A}$ make $(\theta_2 - \theta_1)$ sufficiently big to measure.

(*c*) We must use a *disc* for a *bad conductor.* This reduces lateral conduction, and the large $\boldsymbol{A}$ and small x make Q/t sufficiently big to measure.

For *fluids* we adapt the disc method for a bad conductor, but it is difficult to obtain an accurate value for λ, because

(*i*) there will nearly always be convection, and

(*ii*) thermal energy transferred by convection and radiation, for which it is difficult to account, becomes large relative to that transferred by conduction (since λ is small). One can correct for radiation through a gas by a control experiment in a vacuum.

29.8 Conductivity of a Good Conductor

Using the apparatus shown in fig. 29.6, we measure the following:

The mass m of cooling water that flows in a time t
Temperature increase of water $= \Delta\theta$
Average temperature gradient between P and Q $= \left(\dfrac{\theta_2 - \theta_1}{x}\right)$

Cross-sectional area $= \mathbf{A}$

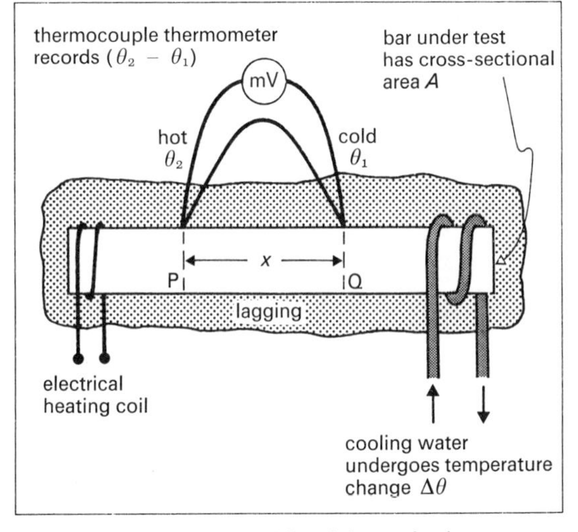

Fig. 29.6. The principle of Searle's method.

Then using $$\frac{Q}{t} = \lambda A\left(\frac{\theta_2 - \theta_1}{x}\right)$$

$$\left(\frac{m}{t}\right)c\,\Delta\theta = \lambda A\left(\frac{\theta_2 - \theta_1}{x}\right)$$

in which c is the specific heat capacity of water. λ is then the only unknown.

The difference between the electrical rate of energy supply, and the rate of energy absorption by the cooling water provides a useful check on power losses by lateral conduction. (We have assumed these to be zero.)

Improvement

A better approximation to parallel heat flow is obtained by using a *guard-ring*, shown in fig. 29.7.

The idea behind the guard-ring should be compared to that used in the parallel-plate capacitor (p. 332). In this example the outer layers of the block constitute the guard-ring for the central cylinder.

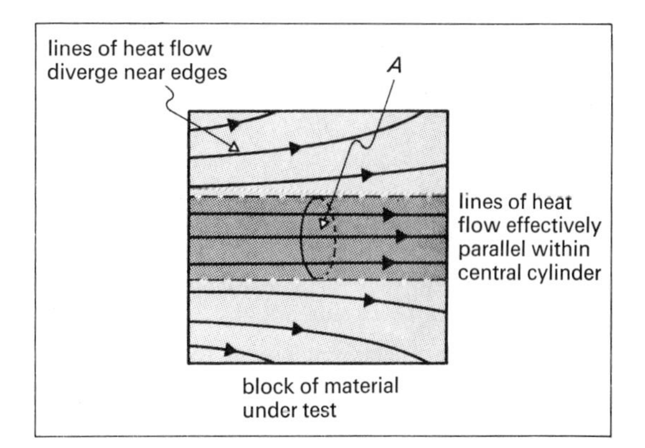

Fig. 29.7. The principle of the guard-ring.

29.9 Conductivity of a Poor Conductor

These symbols refer to fig. 29.8.

θ_0 = room temperature,
$\mathbf{A}$ = *cross-sectional* area of sample and identical copper discs,
A_1, A_2, A_3 and A_s = *emitting* surface areas of discs and sample, all with identical enamelled surfaces,
P = power emitted from unit area of these surfaces, for unit temperature difference between surface and surroundings. (P would be measured in W m^{-2} K^{-1}.)

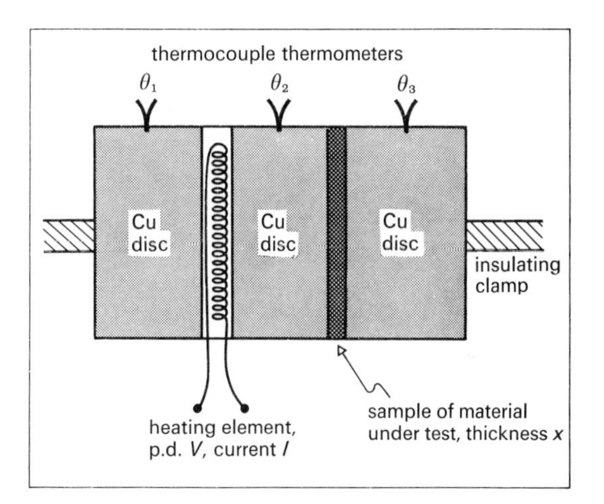

Fig. 29.8. One version of Lees's disc method. (Not to scale.)

Allow a *steady state* to be attained.

(a) Power supplied electrically equals rate of heat loss from surfaces:

$$VI = P\Big[A_1(\theta_1 - \theta_0) + A_2(\theta_2 - \theta_0) + A_s\Big(\frac{\theta_2 + \theta_3}{2} - \theta_0\Big) + A_3(\theta_3 - \theta_0)\Big]$$

in which P, the only unknown, can be calculated.

(b) Effective rate of heat flow across sample equals average of heat entering and heat leaving in each second:

$$\lambda A\Big(\frac{\theta_2 - \theta_3}{x}\Big) = \frac{1}{2}P\Bigg\{\underbrace{\Big[A_s\Big(\frac{\theta_2 + \theta_3}{2} - \theta_0\Big) + A_3(\theta_3 - \theta_0)\Big]}_{\text{ENTERING}} + \underbrace{\Big[A_3(\theta_3 - \theta_0)\Big]}_{\text{LEAVING}}\Bigg\}$$

Since P has been found in (a), λ is the only unknown.

Notes.

(i) Good thermal contact between the sample and copper is aided by using glycerine.

(ii) To reduce convection currents in liquids, one

(a) uses the apparatus in a horizontal position, and

(b) heats from above.

30 Thermal Radiation

30.1 The Nature of Radiation

Thermal radiation is energy that travels from one place to another by means of an electromagnetic wave motion (p. 260). When absorbed by matter it may increase the vibrational or translational k.e. of atoms or molecules: this increase of internal energy will usually become apparent as a temperature increase. *All* frequencies of the electromagnetic spectrum produce such a heating effect when waves are absorbed.

The vibrational frequencies of atoms at room temperature are about 10^{14} Hz (p. 298). A wave of frequency about 10^{14} Hz would cause *resonance* to occur, and in general would thus be efficient at transferring the electromagnetic wave energy to the electrically charged particles of which matter is composed.

Waves whose frequencies are close to 10^{14} Hz are called **infra-red** waves. They are both radiated and absorbed by bodies at normal temperatures.

For convenience the infra-red part of the spectrum can be classified thus:

	Near	*Intermediate*	*Far*
$\lambda/10^{-6}$ m	0.8–3	3–10	10 – (1000)
ν/Hz	4×10^{14}	$10^{14} - 3 \times 10^{13}$	3×10^{11}

Infra-red radiation is emitted when thermal agitation causes changes in the vibrational and rotational energy states of molecules. A study of this radiation helps us to elucidate molecular structure.

It has all the usual *properties* of an electromagnetic wave motion (p. 265).

A **diathermanous** body, such as a calcium fluoride prism, is one that absorbs little of the radiation passing through it (cf. a body *translucent* to visible light).

An **adiathermanous** body, such as a mass of water is one that absorbs strongly (cf. a body *opaque* to visible light).

For example glass is diathermanous when

$$0.4 \times 10^{-6}\,\text{m} < \lambda < 2.5 \times 10^{-6}\,\text{m}$$

but adiathermanous for longer wavelengths such as 10^{-5} m. This fact can be used to explain the action of a greenhouse.

30.2 Detection of Radiant Energy

Photodetectors depend upon the photoelectric effect, and photoconductivity. They respond if a **photon** of incident radiation has a certain minimum energy (and so maximum wavelength (p. 457)).

Thermal detectors are *thermometers*, that is they respond to the temperature change that accompanies the absorption of *all* frequencies.

The human skin is one example; others include

(*a*) The sensitive differential air thermometer.

(*b*) *Boys's radiomicrometer* (1889), in which a thermocouple is incorporated into a moving-coil galvanometer suspended by a quartz fibre.

(*c*) *The thermopile*, which consists of about 25 or more thermocouples joined in series. Radiation falls on the blackened hot junction, while the cold junction remains shielded.

(*d*) *Langley's bolometer* (1881)

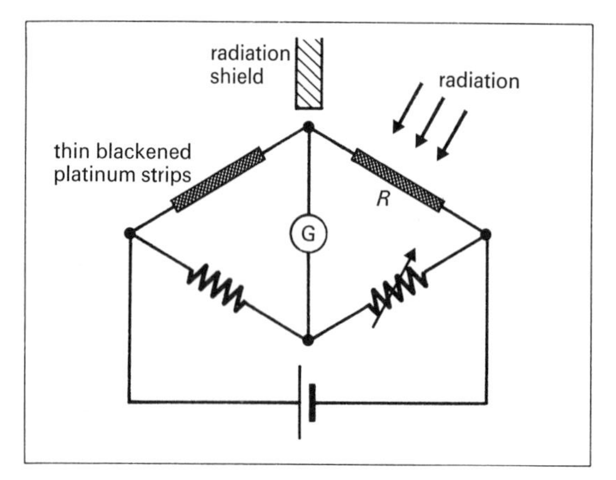

Fig. 30.1. The principle of Langley's bolometer.

This is a differential resistance thermometer (fig. 30.1).

Procedure

(1) Balance the bridge.

(2) Allow the incident radiation to change R to $(R + \Delta R)$ using one arm.

(3) Rebalance the bridge, and calculate ΔR and hence the temperature change ΔT.

(4) Repeat, using the arm that was previously shielded.

The bolometer is made more sensitive if the platinum is replaced by a semiconductor of high negative temperature coefficient of resistance (p. 354).

*30.3 The Black Body

Fig. 30.2 shows what may happen when radiation is incident on a body.

One must distinguish carefully between

(*i*) absorbed and re-radiated energy,
(*ii*) transmitted energy, and
(*iii*) scattered and reflected energy.

The Concept of a Black Body

By analogy with visible optics, a **black body** is one that absorbs *all* the radiation (of all wavelengths) falling on it. It

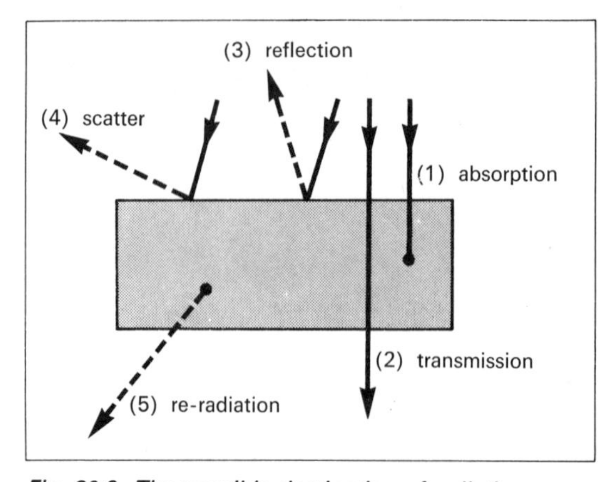

Fig. 30.2. The possible destination of radiation incident on matter.

is (like the ideal gas of kinetic theory) an *idealized concept* that can nearly be realized in practice, and which is invaluable for the formulation of the laws of radiation.

All the energy incident on the hole in the cavity of fig. 30.3 is absorbed.

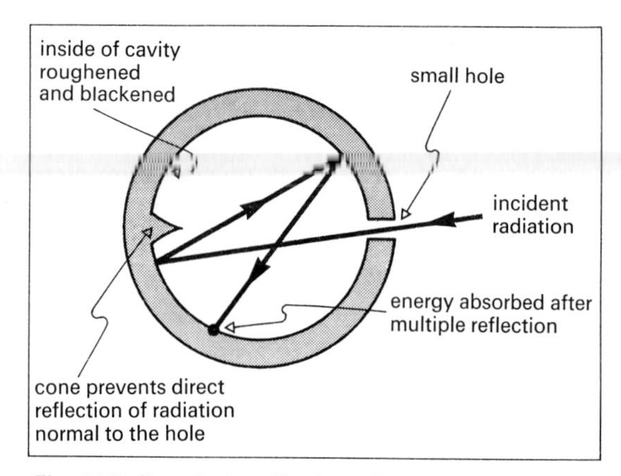

Fig. 30.3. Practical realization of a black body (absorber).

Black-body Radiation

The radiation inside an **equal temperature enclosure** depends only upon the *temperature*: it is *not* controlled by the *nature* of the walls.

A *small* hole in the side of the **cavity** of fig. 30.4 then approximates to a *perfect* radiator.

The radiation emitted is called variously **black-body radiation**, *full* radiation, *cavity* radiation or *temperature* radiation, and has the characteristic feature that the intensity of each frequency has some well-defined value which is determined by the temperature (p. 226).

If the cavity is at (say) 2000 K, the small hole (the 'black' body) emits visible radiation whose colour is characteristic of that temperature.

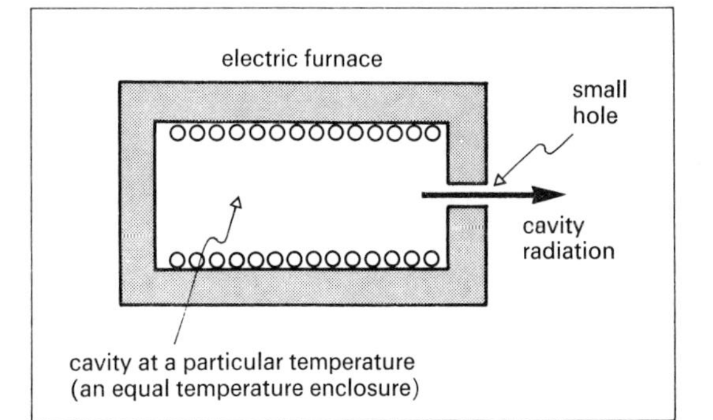

Fig. 30.4. Practical realization of a black body (radiator).

Black-body temperature is the temperature of a black body which emits radiation of the same description as the body under consideration.

*30.4 Important Terms

(a) Reflectance (*reflexion factor*) ρ, and Transmittance (*transmission factor*) τ

Suppose radiant power Φ_0 is incident on a surface, and power Φ_r is reflected.

Then $\rho = \Phi_r/\Phi_0$, and τ is defined in a similar way.

Neither has a unit, and they are not discussed further in this book.

(b) Spectral Radiant Exitance M_λ

Suppose a surface of area A radiates power Φ_λ in the waveband λ to $(\lambda + \delta\lambda)$.

The spectral radiant exitance M_λ of the surface for this waveband is defined by the equation

$$M_\lambda \delta\lambda = \frac{\Phi_\lambda}{A}$$

$$M_\lambda\left[\frac{\text{W}}{\text{m}^3}\right]\delta\lambda[\text{m}] = \frac{\Phi_\lambda\,[\text{W}]}{A\,[\text{m}^2]}$$

Spectral radiant exitance has the unit W m^{-3}.

(c) Total Radiant Exitance M

Suppose a surface of area A radiates power Φ over *all* wavelengths (so that Φ is the *total* power radiated).

The total radiant exitance M of the surface is defined by the equation

$$M = \frac{\Phi}{A}$$

$$M\left[\frac{\text{W}}{\text{m}^2}\right] = \frac{\Phi\,[\text{W}]}{A\,[\text{m}^2]}$$

Notes.

(*i*) M_λ and M have different dimensions.

(*ii*) It follows that

$$M = \int_0^\infty M_\lambda \, d\lambda$$

(*iii*) M_λ and M for a *black body* will be written in this book as $M_{\lambda,B}$ and M_B respectively.

(d) Spectral Emissivity ε_λ

The spectral emissivity ε_λ of a surface for the waveband λ to $(\lambda + \delta\lambda)$ is defined by

$$\varepsilon_\lambda = \frac{M_\lambda}{M_{\lambda,B}}$$

Notes.

(*i*) ε_λ is dimensionless, and has no unit.

(*ii*) $\varepsilon_\lambda = 0$ for a perfect transmitter or a perfect reflector.

(*iii*) $\varepsilon_\lambda = 1$ for a perfect emitter (such as a hole in a cavity).

(*iv*) $\varepsilon_\lambda \leq 1$ for all non-luminescent bodies.

(*v*) $M_\lambda = \varepsilon_\lambda M_{\lambda,B}$

$$\begin{pmatrix}\text{spectral}\\ \text{radiant}\\ \text{exitance of}\\ \text{a surface}\end{pmatrix} = \begin{pmatrix}\text{spectral}\\ \text{emissivity}\end{pmatrix}\begin{pmatrix}\text{spectral radiant}\\ \text{exitance of a}\\ \text{black body at the}\\ \text{same temperature}\end{pmatrix}$$

(e) Spectral Absorptance (*absorption factor*) α_λ

Suppose radiant power $\Phi_{\lambda,0}$ in the waveband λ to $(\lambda + \delta\lambda)$ is incident on a surface, and that the surface absorbs power $\Phi_{\lambda,a}$.

The spectral absorptance α_λ of the surface is defined for this waveband by

$$\alpha_\lambda = \frac{\Phi_{\lambda,a}}{\Phi_{\lambda,0}}$$

Notes.

(*i*) α_λ is dimensionless, and has no unit.

(*ii*) $\alpha_\lambda = 0$ for a perfect transmitter or a perfect reflector.

(*iii*) $\alpha_\lambda = 1$ for a perfect absorber (i.e. a black body).

Compare these notes with the corresponding notes (*i*) to (*iii*) for ε_λ.

(f) The Solar Constant

Suppose Φ_0 is the total power radiated from the Sun to an area A placed normal to the radiation at the edge of the Earth's atmosphere, while the Earth is at its mean distance from the Sun. (Φ_0 is called the **radiant flux.**)

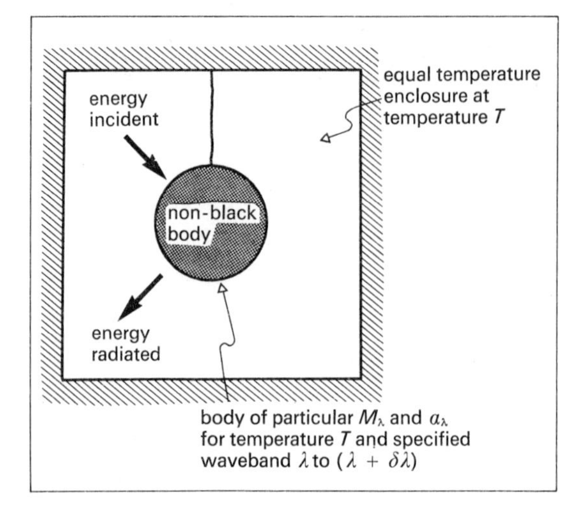

Fig. 30.5. Derivation of Kirchhoff's Law.

The quantity Φ_0/A is called the **solar constant**, and is the **irradiance** E of this particular surface. Its value is found by experiment to be 1.35 kW m^{-2}, and can be used (together with *Stefan's Law*) to give an estimate of the surface black-body temperature of the Sun.

30.5 Prévost's Theory (1792)

The theory states:

Every body radiates energy at a rate controlled by its thermodynamic temperature T.

We deduce

(*a*) When $T_{\text{body}} > T_{\text{surroundings}}$, there is a *net loss* of radiant energy.

(*b*) When $T_b < T_s$, there is a *net gain* of radiant energy, but

(*c*) When $T_b = T_s$, equilibrium is attained, but the energy exchange continues—*the equilibrium is dynamic.*

Notes.

(*a*) If the temperature of a *good absorber* is not to rise spontaneously above that of its surroundings, it must also be a *good emitter* of radiation. (See *Kirchhoff's Law.*)

(*b*) A thermometer immersed in a diathermanous medium (such as air) will record a temperature influenced by the environment from which it receives radiation, as well as by the medium.

*30.6 Kirchhoff's Law (1858)

The law which will be deduced applies for a *specified waveband* λ to $(\lambda + \delta\lambda)$, and a *specified temperature T.*

Suppose $E_{\lambda,0}$ is the radiant power in the specified waveband incident *on unit area* of the body of area A (fig. 30.5).

The rate of absorption of this energy by the body is

$$E_{\lambda,a}A = \alpha_\lambda(E_{\lambda,0}A) \qquad \text{(def. of } \alpha_\lambda\text{)}$$

The rate of emission of radiant energy is

$$A(M_\lambda \delta\lambda) \qquad \text{(def. of } M_\lambda\text{)}$$

Since the body is in an equal temperature enclosure, these are equal, so

$$\alpha_\lambda E_{\lambda,0}A = AM_\lambda\, \delta\lambda$$

$$\frac{M_\lambda}{\alpha_\lambda} = \left(\frac{E_{\lambda,0}}{\delta\lambda}\right)$$

Now $E_{\lambda,0}$ and $\delta\lambda$ are controlled by the waveband $\delta\lambda$ and the temperature T, but *not* by the nature of the body.

It follows that M_λ/α_λ *has the same value* (*for a given waveband and temperature*) *for all bodies.* This is **Kirchhoff's Law**.

For a *black body* the constant of *Kirchhoff's Law* is given by

$$\frac{M_\lambda}{\alpha_\lambda} = \frac{M_{\lambda,B}}{1}$$

We deduce

(*a*) this constant is the spectral radiant exitance $M_{\lambda,B}$ of a black body, and

(h)
$$\alpha_\lambda = M_\lambda / M_{\lambda,B} \quad \text{(from } \textit{Kirchhoff's Law}\text{)}$$
$$= \varepsilon_\lambda \quad \text{(def. of } \varepsilon_\lambda\text{)}$$

For a specified waveband and temperature, the spectral emissivity of a body equals its spectral absorptance.

Demonstration (See fig. 30.6)

(i) Emission

Excited sodium atoms in a bunsen flame emit characteristic yellow light (p. 250).

(ii) Absorption

The observed D-lines are less bright than the rest of the spectrum, as Na vapour (at the *same temperature* as in (*i*)) has absorbed the energy of *this wavelength*, and reduced its intensity by re-radiation *in all directions.*

*30.7 Stefan's Law (1879)

This law is also called the *Stefan–Boltzmann Law*; it was derived by *Boltzmann* from the thermodynamic reasoning (1884).

The law states that the power radiated over all wavelengths from a black body is proportional to the fourth power of the thermodynamic temperature T.

Thus *total* radiant exitance $M_B \propto T^4$,

$$M_B = \sigma T^4$$

but $M = \Phi/A$ so

$$\boxed{\Phi = A\sigma T^4}$$

$$\Phi[\text{W}] = A\,[\text{m}^2]\,\sigma\left[\frac{\text{W}}{\text{m}^2\text{K}^4}\right] T^4\,[\text{K}^4]$$

σ is called the **Stefan–Boltzmann constant**, and has the value $5.67 \times 10^{-8}\ \text{W m}^{-2}\ \text{K}^{-4}$.

$$[\sigma] = [\text{MT}^{-3}\Theta^{-4}]$$

For a non-black body

$$\Phi = \varepsilon A\sigma T^4$$

where ε is the *total* emissivity.

Net Power Loss or Gain

For a black body at temperature T in an environment at T_0

$$\Phi_{\text{net}} = \underset{\text{(loss)}}{A\sigma T^4} - \underset{\text{(gain)}}{A\sigma T_0^4}$$
$$= A\sigma(T^4 - T_0^4)$$

For a non-black body, net power loss is

$$\varepsilon A\sigma T^4 - \alpha A\sigma T_0^4 - \varepsilon A\sigma(T^4 - T_0^4)$$

where we have assumed $\varepsilon = \alpha$.

Special Cases

(*a*) If $T \gg T_0$, then T_0^4 is negligible, so for a black body the loss P_{net} is given by

$$P_{\text{net}} \approx A\sigma T^4$$

(*b*) If $(T - T_0)$ is small, for a black body

$$\Phi_{\text{net}} \approx A\sigma(T^2 + T_0^2)(T^2 - T_0^2)$$
$$\approx A\sigma(2T_0^2)(2T_0)(T - T_0)$$
$$\approx (4A\sigma T_0^3)(T - T_0)$$

The rate of loss of energy by radiation is approximately $\propto (T - T_0)$ when $T \approx T_0$. One sometimes assumes this result when making a simple cooling correction (p. 173).

*30.8 Energy in the Black-body Spectrum

The problem is to measure the way in which the energy radiated from a black body is distributed between the different wavelengths, and to explain the experimental results.

The measurements were taken by *Lummer* and *Pringsheim* (1893), but can be explained only by using the quantum theory (*Planck* 1900).

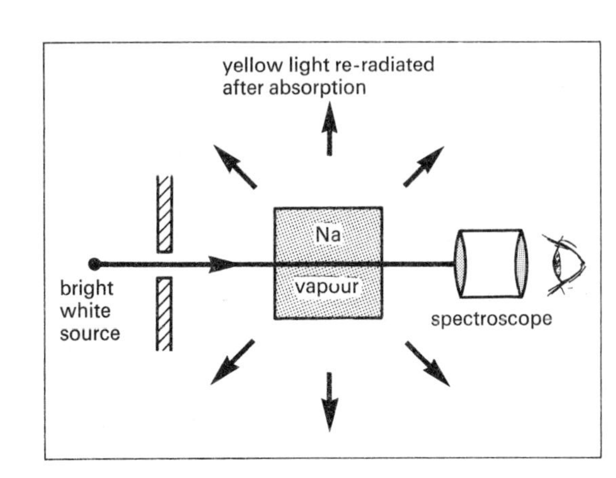

Fig. 30.6. Observation of Fraunhöfer lines.

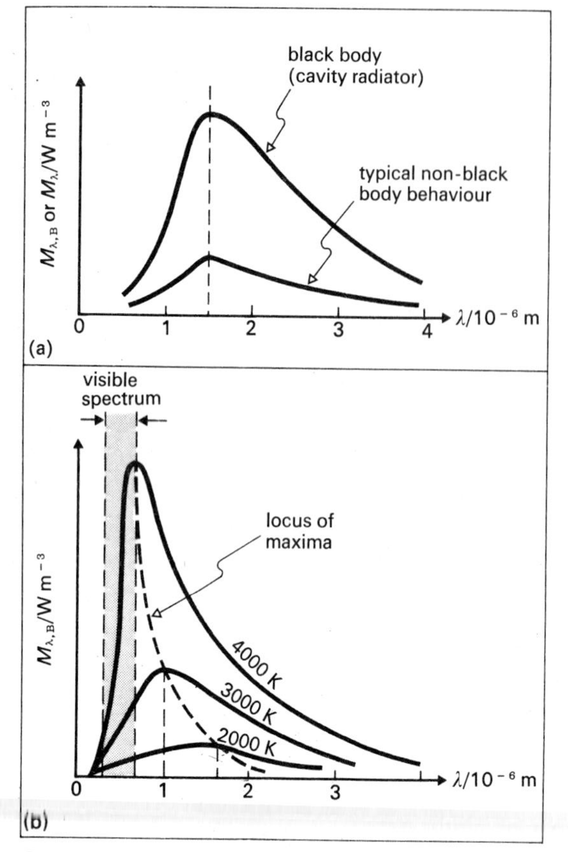

Fig. 30.7. Distribution of radiant energy for various spectra; (a) variation between bodies at a given temperature (about 2000 K), and (b) variation between different temperatures for a given (black) body. $M_{\lambda,B}$ is proportional to the intensity of black-body radiation at wavelength λ.

Distinguish carefully between the situations represented by fig. 30.7*a* and *b*.

Notes on fig. 30.7*b*.

(*a*) At each temperature T there is a wavelength λ_{max} for which the intensity is a maximum.

These are related by **Wien's displacement law**

$$\lambda_{max} T = \text{constant}$$

The law was based on thermodynamic reasoning. The constant, found by experiment, is 2.9×10^{-3} m K for a black-body radiator.

The law illustrates why the colour of the visible light *radiated* by a very hot body changes from red to yellow as the temperature increases.

(*b*) The total area under each curve

$$M_B = \int_0^{\infty} M_{\lambda,B}\, d\lambda$$

This represents the total radiant exitance for a black body *at that particular temperature.*

Stefan's Law states $M_B = \sigma T^4$. We find experimentally that (the area under the curve for each temperature) $\propto T^4$, which agrees with the law.

Calculation of the Sun's Temperature

For the Sun $\lambda_{max} \sim 5 \times 10^{-7}$ m

so using $\lambda_{max} T_S = 2.9 \times 10^{-3}$ m K

we deduce that the *black-body temperature* of the Sun is about 6000 K. The Sun's temperature varies, being $>10^7$ K at the centre. 6000 K is the temperature of the radiating surface.

30.9 Radiation Pyrometers

These are generally used to measure $T > 1250$ K.

(*a*) Total Radiation Pyrometer (Féry)

This responds to both visible and invisible radiation.

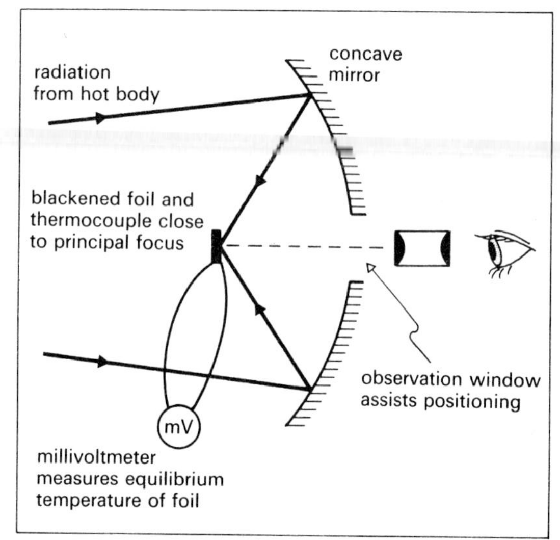

Fig. 30.8. A total radiation pyrometer.

Notes.

(*a*) It can be calibrated by using gold, freezing at 1338 K.

(*b*) The recording of temperature can be made automatic.

(*c*) Black-body temperatures can be calculated by using *Stefan's Law*. This will give reliable temperatures when the radiation is from a hole in the side of a furnace.

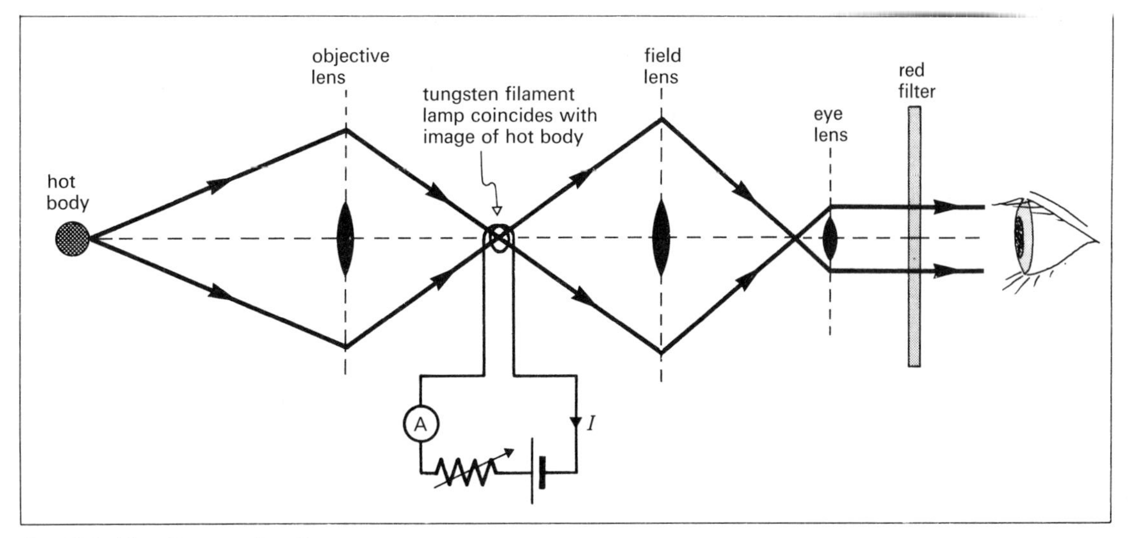

Fig. 30.9. The disappearing filament pyrometer.

(*b*) Optical Pyrometer (disappearing filament)

This responds to visible radiation only.

The current I is adjusted until the image of the filament cannot be distinguished.

The instrument is simpler and more accurate than the total radiation instrument. It is used to establish the international temperature scale from the gold point to about 2300 K.

VI Geometrical Optics

31 Principles of Geometrical Optics

31.1 Introduction and the Ray Concept

By *light* we mean the part of the electromagnetic spectrum that has the correct frequency range to stimulate the nerve endings of the human retina. To study the production and absorption of these waves one must use the *quantum theory*. To study the interaction of the waves with each other one must use *wave theory*, and this is done in section VII.

The wavelength of all visible light is about 10^{-7} m.

When we consider objects and apertures of dimensions greater than about 10^{-5} m, diffraction effects are often considered as relatively unimportant. This will enable us to treat the behaviour of light using *rays* rather than waves.

A **ray** is the path taken by light energy in travelling from a source to a receiver: in an isotropic medium the path is a straight line.

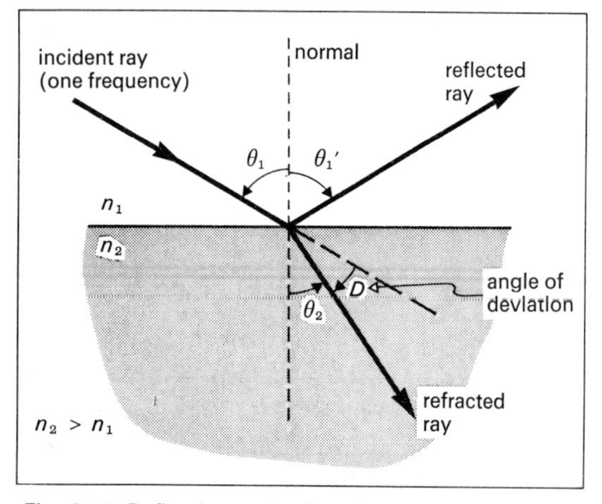

Fig. 31.1. Reflection and refraction at a plane interface.

31.2 The Laws of Reflection and Refraction

I *The incident ray and the normal define a plane which contains both the reflected ray and the refracted ray.*

II *For reflection*:
The angle of incidence equals the angle of reflection.

$$\theta_1 = \theta_1'$$

III For *refraction*:
For a wave of given frequency, and a given pair of media

$$\frac{\sin \theta_1}{\sin \theta_2} = \textbf{\textit{a constant}}$$

This is **Snell's Law**. Wave theory shows that the constant is the (relative) **refractive index** for light passing from medium 1 to medium 2 (p. 95).

Note that $\theta_1 = \theta_1'$ for both *regular* (specular) and diffuse reflection, but that when *diffuse* reflection occurs the normals at many different points on the surface are not parallel.

31.3 Refractive Index and Snell's Law

Absolute refractive index n of a medium for a given frequency is defined by

$$n = \frac{\textit{speed of light in a vacuum}}{\textit{speed of light in the medium}}$$

$$\boxed{n = \frac{c}{c_n}}$$

n varies with the frequency of the light wave.

When a ray is refracted at the interface between two media, the *relative* refractive index determines the deviation. We can avoid the use of relative refractive index as follows (fig. 31.2):

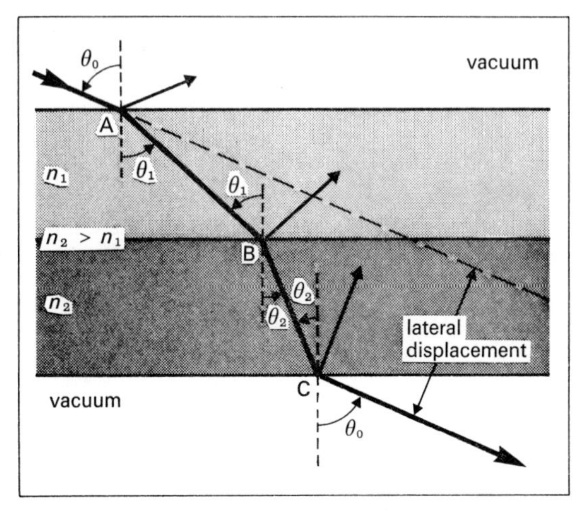

Fig. 31.2. The symmetrical form of Snell's Law.

At A $$n_1 = \frac{c}{c_1} = \frac{\sin\theta_0}{\sin\theta_1}$$

at C $$n_2 = \frac{c}{c_2} = \frac{\sin\theta_0}{\sin\theta_2}$$

Dividing, $$\frac{n_2}{n_1} = \frac{c_1}{c_2} = \frac{\sin\theta_1}{\sin\theta_2} = {}_1n_2 \quad \text{(refraction at B)}$$

The relative refractive index ${}_1n_2 = n_2/n_1$, and *Snell's Law* can be written in the symmetrical form

$$\boxed{n_1 \sin\theta_1 = n_2 \sin\theta_2}$$

where θ_1 and θ_2 are the angles between the ray and the normal in the media n_1 and n_2.

By definition the (absolute) refractive index of a vacuum is 1 exactly. That of air at 101 325 Pa pressure and 293 K is 1.000 3, which is usually taken as 1.

31.4 Approximations in Optics

Many equations used in geometrical optics are derived on the assumption that

$$\sin\theta \approx \tan\theta \approx \theta\,[\text{rad}]$$

where θ is measured in *radians*. The diagrams show some examples.

SITUATION	EXACT STATEMENT	BUT WE CAN SAY
(i) θ, a, b	$\tan\theta = \frac{a}{b}$	$\theta = \frac{a}{b}$
(ii) θ, r, h	$\sin\theta = \frac{h}{r}$	$h = r\theta$
(iii) $\frac{1}{2}\theta$, $\frac{1}{2}\theta$, d, h	$\tan\frac{\theta}{2} = \frac{h/2}{d}$	$h = d\theta$

Fig. 31.3. Approximations.

This approximation is valid, for most purposes, provided θ is small. When $\theta = 10° \approx 0.17$ rad, then the error is about 1%.

In the situation of fig. 31.3 (*iii*) we might have a lens of 50 mm diameter (h) placed 200 mm (d) from a light-box. Then $\frac{1}{2}\theta \approx 0.12$ rad $\approx 7°$, and the difference between the exact and approximate value for θ is likely to be less than the experimental error.

Diagrams are usually drawn with an exaggerated vertical scale in order to show clearly the related lengths. For the same reason we usually exaggerate small angles.

31.5 Real and Apparent Depth

In fig. 31.4 the image position I is displaced relative to the object position O. We will find the position I for paraxial rays (rays for which θ_a and θ_n are *small*).

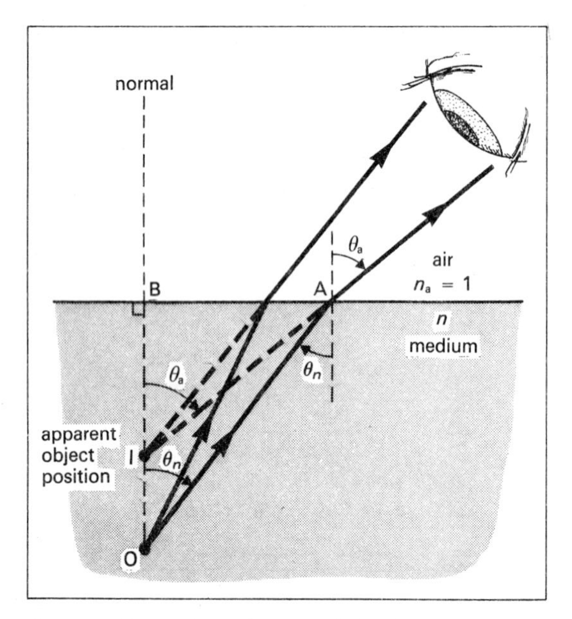

Fig. 31.4. Calculation of apparent depth.

For refraction at A

$$n_a \sin \theta_a = n \sin \theta_n$$

$$n = \frac{\sin \theta_a}{\sin \theta_n} \qquad (n_a = 1)$$

$$\approx \frac{\tan \theta_a}{\tan \theta_n}$$

$$= \frac{AB/IB}{AB/OB}$$

$$= \frac{OB}{IB}$$

$$IB = \frac{OB}{n} = \text{constant}$$

so that the position of I is the same for *all* rays making a small angle with the normal. This shows that I is the *image* of O. (For the concept of an image, see p. 234.)

The result is usually stated by

$$n = \frac{\text{real depth of object}}{\text{apparent depth of object}}$$

Notes.

(*a*) If the eye position is changed so that θ_a becomes larger, the point I describes a *caustic curve*, whose *cusp* is at the limiting position.

(*b*) The distance OI is called the *displacement*.

(*c*) When an object is viewed through several media, the *net* displacement is the sum of the displacements that each medium would have produced in the absence of the others.

31.6 Total Internal Reflection and Critical Angle

When light travels from a medium 1 to an optically denser medium 2, the maximum angle of refraction occurs when $\theta_1 = 90°$. Then

$$n_1 \sin \theta_1 = n_2 \sin \theta_2$$

$$n_1 \sin 90° = n_2 \sin \theta_{max}$$

This maximum angle is called the **critical angle**, c, given by

$$\sin c = \frac{n_1}{n_2} \qquad (n_1 < n_2)$$

The exact value of c is controlled by the frequency of the wave, and hence by the colour of the light.

Suppose (fig. 31.5) light travels from medium 2 toward the interface with medium 1.

When $\theta_2 \leqslant c$, we see *partial* reflection and partial transmission (refraction).

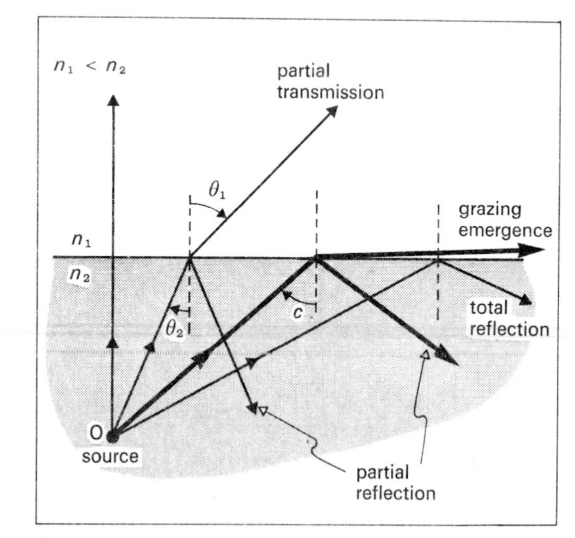

Fig. 31.5. Total internal reflection and the critical angle.

When $\theta_2 > c$, refraction *cannot* occur, and the ray is *totally* reflected back inside medium 2.

Note the *conditions* for total internal reflection

(1) $n_2 > n_1$,
(2) $\theta_2 > c$, where $\sin c = n_1/n_2$.

Total internal reflection is a phenomenon associated with all types of wave motion. In light it finds application in the light pipe, in prisms for optical instruments, and in the measurement of critical angle.

31.7 Methods of Measuring Refractive Index

For Solids

(*a*) Measure θ_1 and θ_2, and use

$$n_1 \sin \theta_1 = n_2 \sin \theta_2$$

(*b*) Measure real and apparent depth using a travelling microscope.

(*c*) Use a prism of the substance on the table of a spectrometer (p. 249). This is a very accurate method.

For Liquids

(*a*) Measure the critical angle, e.g. by
  (*i*) the air-cell method (see below), or
  (*ii*) the **Pulfrich refractometer**, useful when very little liquid is available.

(*b*) As method (*b*) for solids.

(*c*) As method (*c*) for solids: the liquid is confined to a hollow prism whose sides are made accurately parallel.

The Air-Cell Method

The cell of fig. 31.6 is rotated about a vertical axis until the field of view is half bright and half dark. For refraction at A and B

$$n_l \sin \theta_l = n_g \sin \theta_g = n_a \sin \theta_a = 1$$

so

$$n_l = \frac{1}{\sin \theta_l}$$

Notes.

(*a*) The glass surfaces are parallel and a knowledge of n_g is not needed.

(*b*) $2\theta_l$ is measured in practice, by a second rotation to cut-off in the opposite sense. This simplifies the procedure, and reduces the percentage error in θ_l.

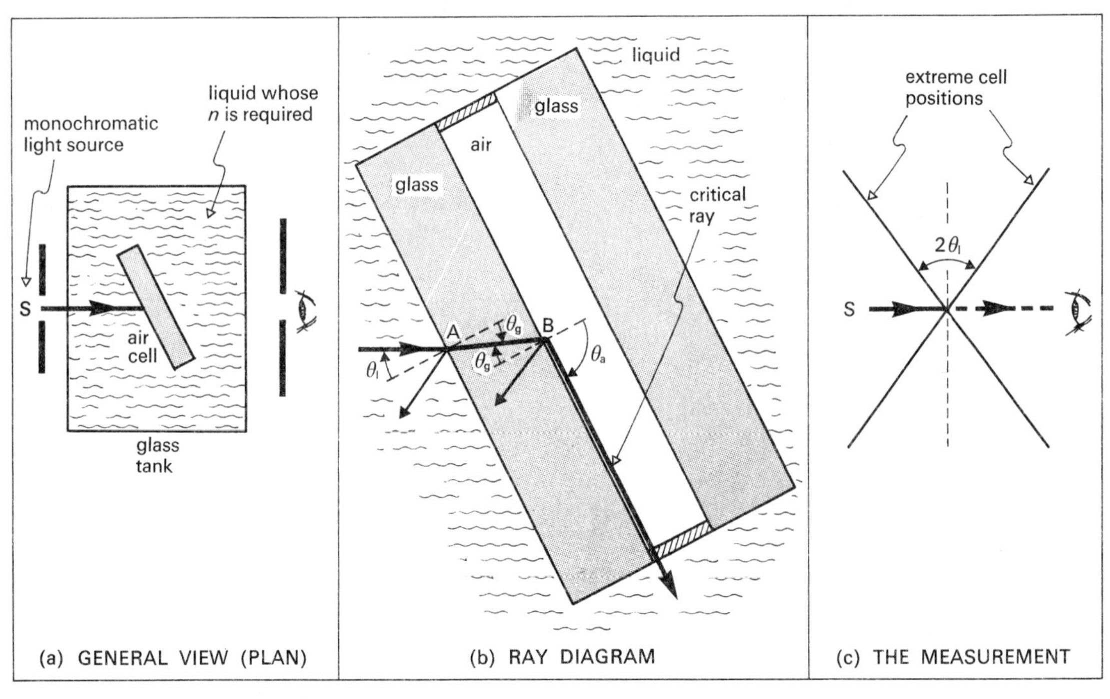

Fig. 31.6. The air-cell method for n_l.

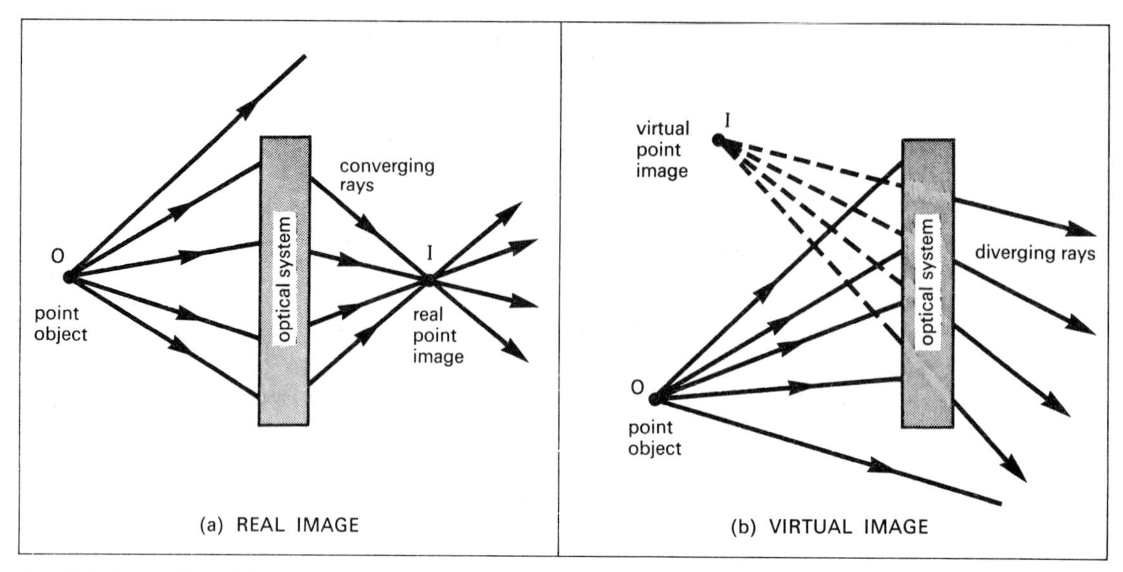

Fig. 31.7. Formation of images.

31.8 The Principles of Image Formation

Refer to fig. 31.7.

A **point image** I of the point object O is formed if *all* the rays intercepted by the system either

(*a*) pass through a single point I, after emergence—this is called a **real image**—or

(*b*) *seem* to have come from a single point I—this is called a **virtual image**.

Both types of image can be seen and photographed.

A *real* image can be photographed by simply placing a photographic emulsion at the site of the image.

Virtual images can only be photographed with the help of a converging optical system which serves to collect the divergent light energy, and concentrate it upon a single point (i.e. to form a real image by using the original virtual image as a virtual object (p. 239)).

The image of an *extended object* is built up from the point images of all the point objects which constitute the object. If the image is to be undistorted the point images must maintain the same relative spacings as their respective point objects.

An **aberration** is the phenomenon of an object not giving rise to an image as defined above. (See p. 244.)

32 The Prism and Thin Lens

32.1 Refraction through a Prism

These relationships follow from the *geometry* of fig. 32.1:

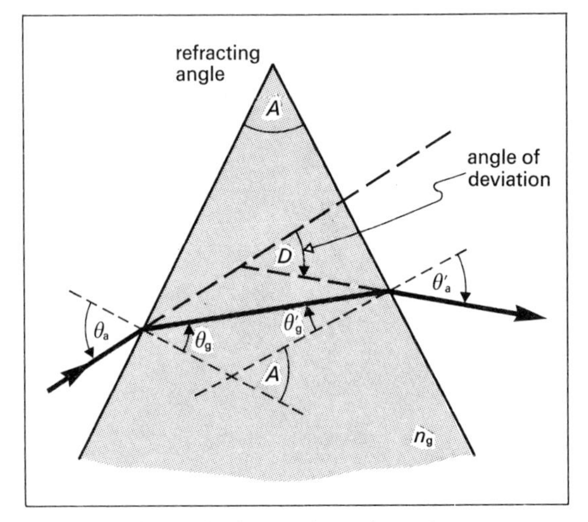

Fig. 32.1. The path of a ray through a prism.

$$\text{Deviation } D = (\theta_a - \theta_g) + (\theta'_a - \theta'_g)$$

$$A = (\theta_g + \theta'_g)$$

According to the **principle of reversibility**, the light energy can travel *in either direction* along the path shown.

32.2 Minimum Deviation

Suppose D is measured for various angles of incidence θ_a.

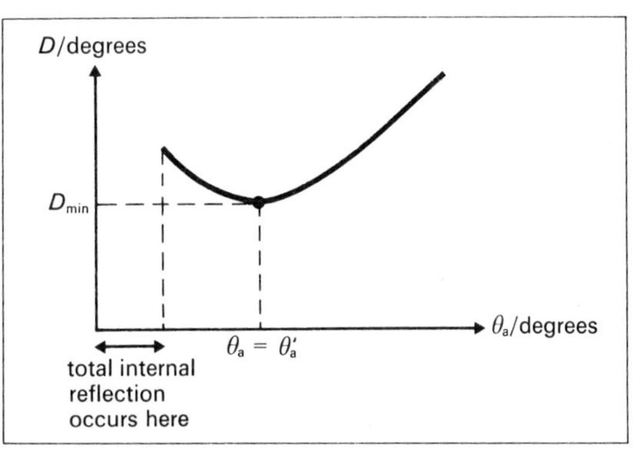

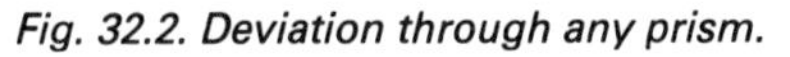

Fig. 32.2. Deviation through any prism.

It can be proved theoretically, and demonstrated in practice, that the deviation is a *minimum* when the ray passes *symmetrically* through the prism. Then

$$\theta_a = \theta'_a \quad \text{and} \quad \theta_g = \theta'_g$$

From the previous paragraph

$$\theta_a = \frac{A + D_{min}}{2} \quad \text{and} \quad \theta_g = \frac{A}{2}$$

But $$n_a \sin\theta_a = n_g \sin\theta_g$$

so $$n_g = \frac{\sin \theta_a}{\sin \theta_g} \qquad (n_a = 1)$$

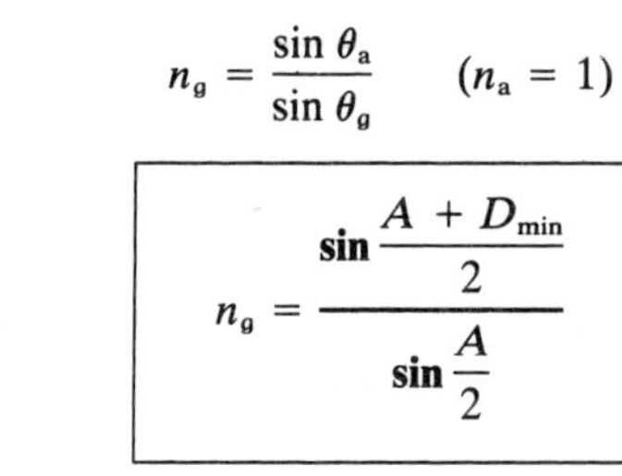

$$n_g = \frac{\sin \dfrac{A + D_{min}}{2}}{\sin \dfrac{A}{2}}$$

Note carefully that this equation is valid for a prism of *any* angle A less than twice the critical angle, but *only when the deviation is a minimum.*

The result is also proved, using wave ideas, on p. 102.

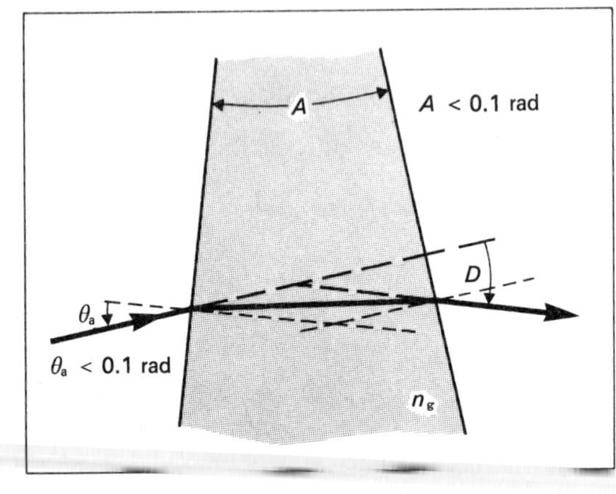

Fig. 32.3. Deviation by a small-angled prism.

32.3 Small-angled Prisms

The result proved here is fundamental for the method used in this book to develop lens theory.

Suppose that (a) $\theta_a < 0.1$ rad, and
(b) $A < 0.1$ rad (fig. 32.3).

Then it follows that each of θ_g, θ_g' and θ_a' is less than 0.1 rad, and is therefore approximately equal to its sine.

So $$n_g = \frac{\sin \theta_a}{\sin \theta_g} \approx \frac{\theta_a}{\theta_g}$$

and $$n_g \approx \frac{\theta_a'}{\theta_g'}.$$

Then
$$\begin{aligned} D &= (\theta_a - \theta_g) + (\theta_a' - \theta_g') \\ &= n\theta_g + n\theta_g' - (\theta_g + \theta_g') \\ &= (n - 1)A \qquad \text{using} \quad A = \theta_g + \theta_g' \end{aligned}$$

$$\boxed{D = (n - 1)A}$$

Note carefully that this equation is valid

(a) for angles $A < 0.1$ rad (approx.), and then only

(b) for all values of $\theta_a < 0.1$ rad.

It shows that, for a prism of given angle A, *all rays* making small angles of incidence undergo the *same deviation.*

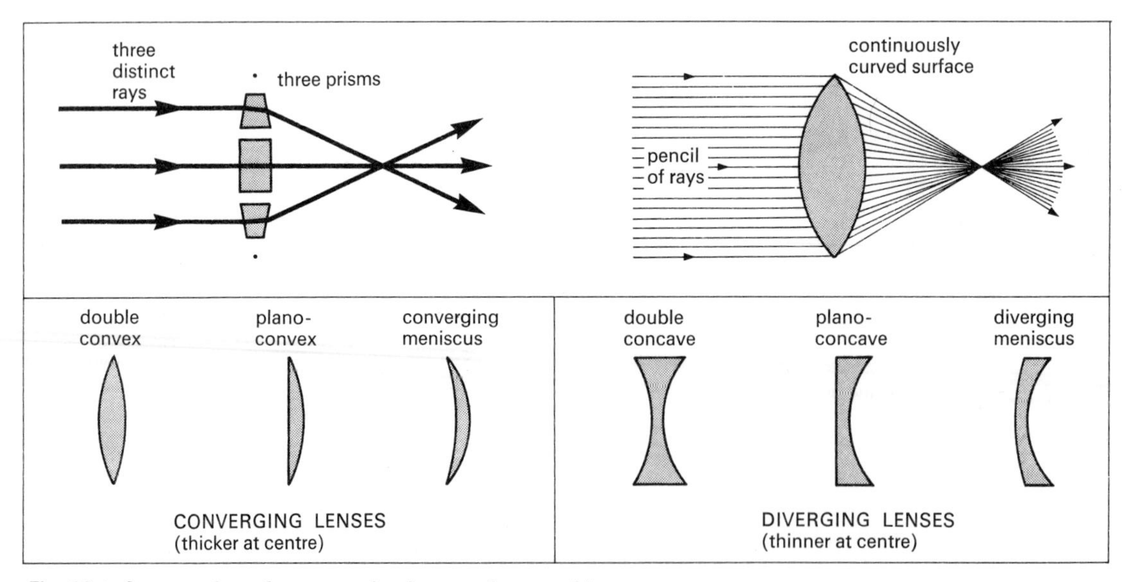

Fig. 32.4. Construction of a converging lens, and types of lenses.

32.4 Lenses and their Construction

Fig. 32.4 illustrates

(*a*) that a series of correctly-angled prisms can be built into a system that causes several *distinct* rays to cross at a point, and

(*b*) that a lens of continuously curved surfaces can be made to focus a *pencil* of rays at a point.

In the next paragraph we prove that an optical system made of a medium bounded by two *spherical* surfaces will cause such a pencil of rays to cross the axis at approximately the same point. This means that a spherical lens has the ability to form *images* (see p. 234).

Fig. 32.5 illustrates terms used for lenses. f is the image distance for an infinite object distance.

The **principal axis** joins the centres of curvature of the two spherical surfaces.

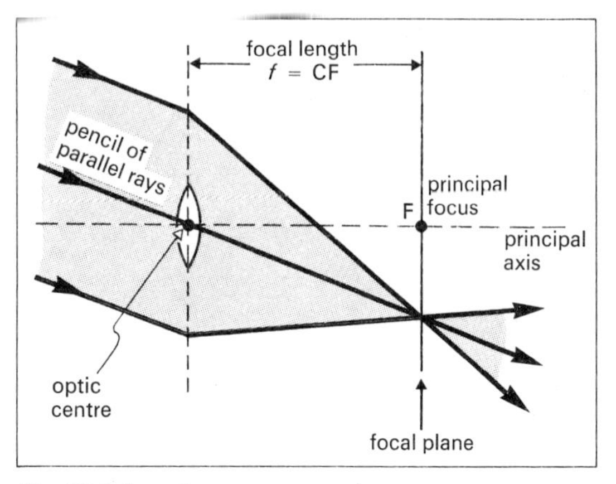

Fig. 32.5. Lens terms.

A ray passing through the **optic centre** suffers *zero deviation*, and, for a thin lens, negligible lateral *displacement*.

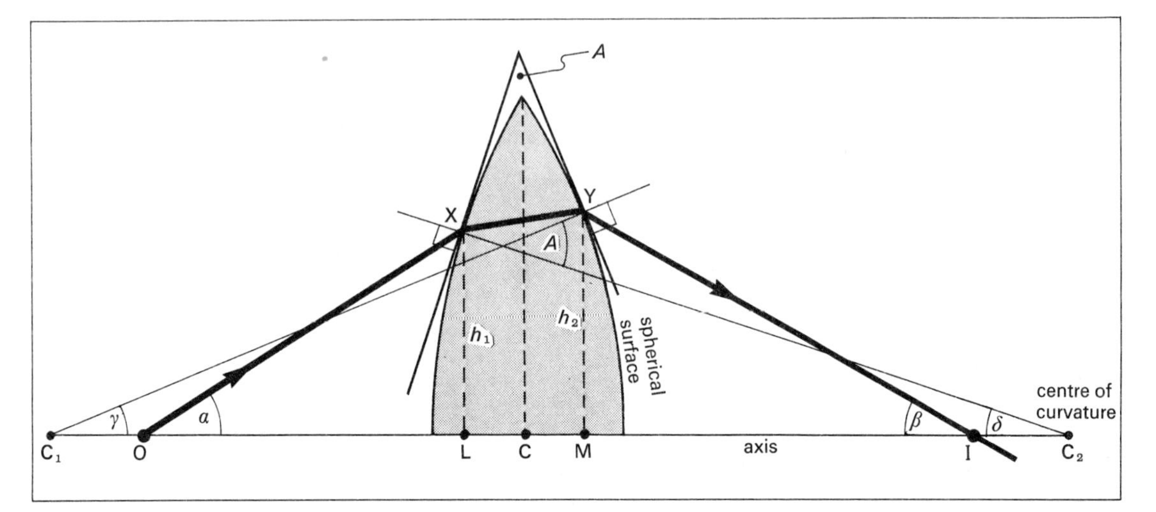

Fig. 32.6. Details of a ray passing through the top half of a thin lens (all features are considerably distorted for clarity).

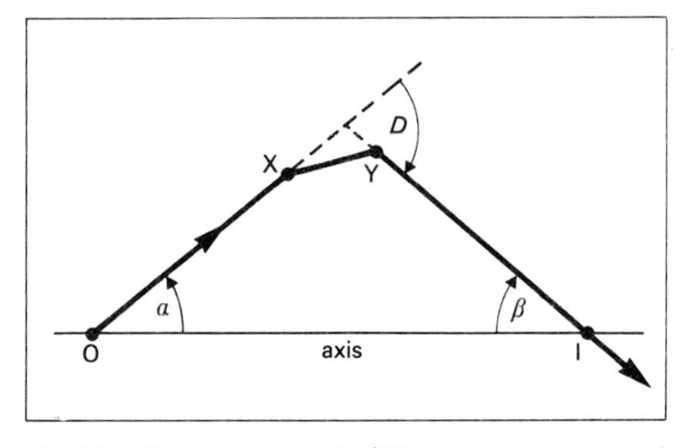

Fig. 32.7. To show $\alpha + \beta = D$. (All angles exaggerated.)

32.5 A Lens Formula

Refer to figs. 32.6 and 32.7. The aim is to find a numerical relationship between OC, IC, CC_1 and CC_2 for this situation, in which the rays are paraxial.

We assume

(*a*) that the lens can be considered as a system of small-angled prisms, and

(*b*) that angles of incidence are small.

In the diagram $D = \alpha + \beta$

and $A = \gamma + \delta$

Using $$D = (n - 1)A$$

for the small-angled prism,

$$\alpha + \beta = (n - 1)(\gamma + \delta)$$

Because α, β, γ and δ are small, we replace them by their tangents. So

$$\frac{h}{\mathrm{OC}} + \frac{h}{\mathrm{IC}} = (n - 1)\left(\frac{h}{\mathrm{CC}_1} + \frac{h}{\mathrm{CC}_2}\right)$$

in which we have put $h_1 = h_2 = h$, because the lens is to be considered thin, so that L, M and C approximately coincide.

Dividing by h

$$\frac{1}{\mathrm{OC}} + \frac{1}{\mathrm{IC}} = (n - 1)\left(\frac{1}{\mathrm{CC}_1} + \frac{1}{\mathrm{CC}_2}\right)$$

Notes.

(*a*) For this particular lens n, CC_1 and CC_2 are constant. The equation shows that for a fixed value OC, IC is *not* dependent on the angle α. Therefore *all* rays from O that are intercepted by the lens pass through the same point I. Thus I is the *image* of O (see paragraph **31.8**).

(*b*) The equation is a *numerical* relationship between the lengths OC, IC, CC_1 and CC_2 for this particular type of lens. We will now adopt a *sign convention* for the distances involved; this will enable us to derive an *algebraic* relationship which will serve for all types of lens used in all possible ways.

32.6 Sign Convention

We adopt the following **symbols**:

u = object distance, f = focal length,

v = image distance, R = radius of curvature.

A symbol represents three things: a sign, a number and a unit. These symbols should not be given any sign until they are replaced by a numerical value. Several sign conventions are used. In this book we adopt the **Real is Positive** convention.

The Convention

(*a*) All distances are measured to the optic centre.

(*b*) Distances are taken to be *positive* if actually traversed by the light ray (distances to *real* objects and images).

(*c*) Distances are taken to be *negative* if only apparently traversed by the light ray (distances to *virtual* objects and images).

(*d*) More generally the bracket $(n - 1)$ should read

$$\left(\frac{n_{\mathrm{lens}}}{n_{\mathrm{medium}}} - 1\right)$$

It will be taken to be positive, so that it represents the magnitude but not the sense of the deviation.

(*e*) The sense of the deviation will be shown by the sign allotted to R (fig. 32.8). This means that R is positive for surfaces which cause a deviation towards the axis.

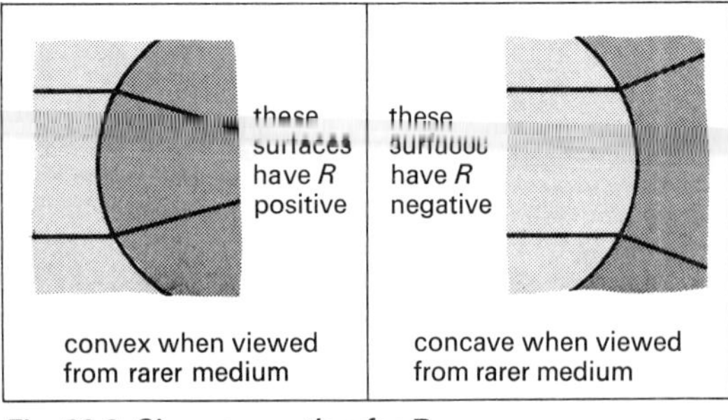

Fig. 32.8. Sign convention for R.

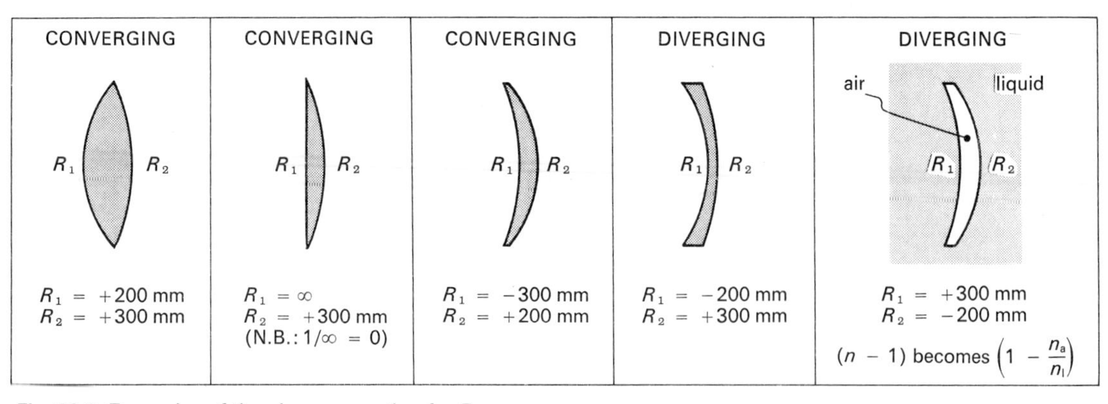

Fig. 32.9. Examples of the sign convention for R.

32.7 The Lens-Maker's Formula

Using the sign convention we can write down an equation which can be used for any lens.

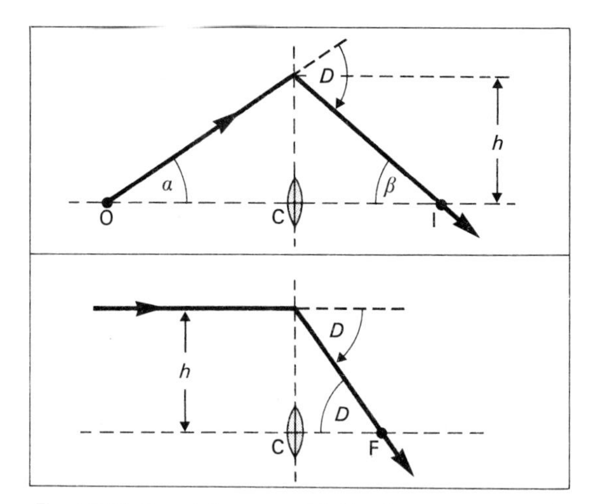

Fig. 32.10. To relate u, v and f. (Angles exaggerated.)

In radian measure

$$D = \alpha + \beta$$

$$\approx \frac{h}{\mathrm{OC}} + \frac{h}{\mathrm{IC}}$$

and because D is the same for the same value of h

$$D \approx \frac{h}{\mathrm{CF}}$$

so

$$\frac{1}{\mathrm{OC}} + \frac{1}{\mathrm{IC}} = \frac{1}{\mathrm{CF}}$$

We combine this equation and paragraphs **32.5** and **32.6** to give

$$\frac{1}{u} + \frac{1}{v} = (n - 1)\left(\frac{1}{R_1} + \frac{1}{R_2}\right) = \frac{1}{f}$$

which is commonly called the **lens-maker's formula.** f will be positive for a converging lens, and negative for a diverging lens. Its value is often deduced from the graph of fig. 32.11.

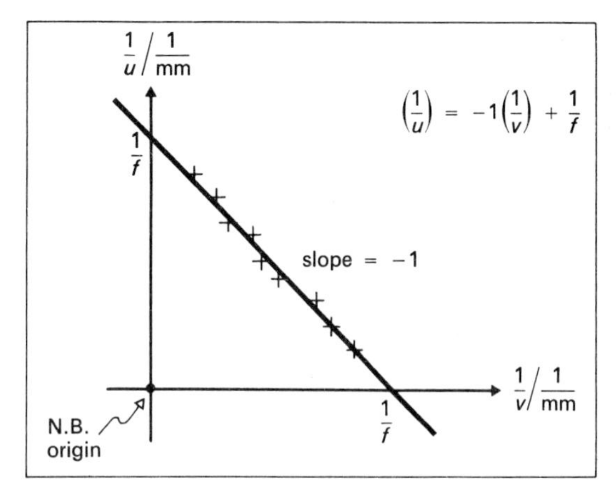

Fig. 32.11. A graph of $1/u$ plotted against $1/v$ for a converging lens.

32.8 The Virtual Object

When rays converging to a point are intercepted by an optical system, that point is the **virtual object** for any subsequent image (fig. 32.12). An example is discussed in paragraph **32.10.**

Fig. 32.12. Example of a virtual object.

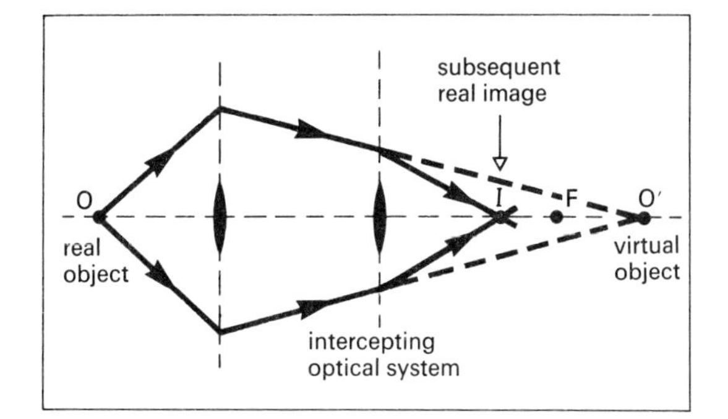

32.9 Magnification

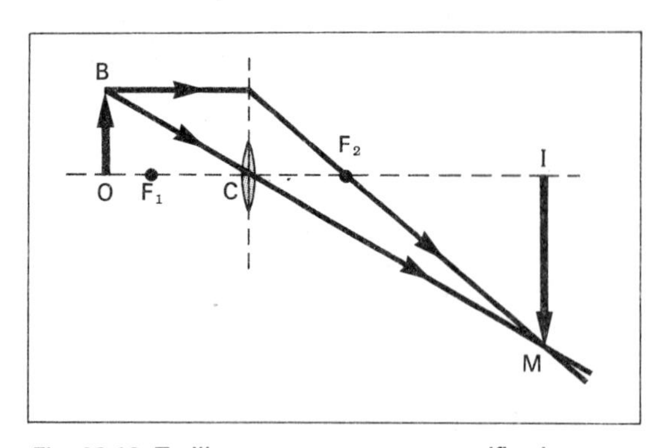

Fig. 32.13. To illustrate transverse magnification.

The transverse magnification m is defined by

$$m = \frac{\textbf{\textit{height of transverse image}}}{\textbf{\textit{height of transverse object}}}$$

$$\boxed{m = \frac{\mathrm{IM}}{\mathrm{OB}}}$$

In magnitude $m = \dfrac{v}{u} = \dfrac{v}{f} - 1$

If m is negative, this means that either O or I is virtual. The negative sign has no other physical significance.

* *The* **longitudinal magnification** *is defined as*

$$\frac{\textit{length of image}}{\textit{length of object}}$$

where each lies parallel to the axis.

Its value cannot be calculated simply in terms of u and v, which are no longer uniquely defined.

32.10 Thin Lenses in Contact

In fig. 32.14 a, b and c represent lengths.
For lens 1,

$$\frac{1}{+a} + \frac{1}{+b} = \frac{1}{f_1} \qquad (1)$$

For lens 2,

$$\frac{1}{-b} + \frac{1}{+c} = \frac{1}{f_2} \qquad (2)$$

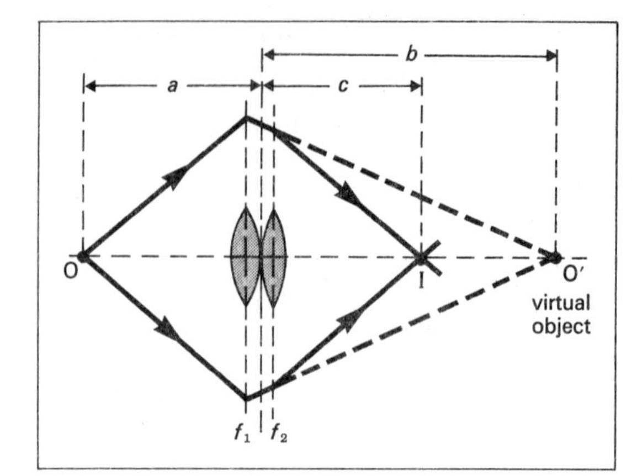

Fig. 32.14. Two thin lenses in contact.

For the composite lens

$$\frac{1}{+a} + \frac{1}{+c} = \frac{1}{f}$$

Adding (1) and (2), we see

$$\frac{1}{f_1} + \frac{1}{f_2} = \frac{1}{f}$$

where f would be the focal length of the single lens equivalent to the other two *in contact*.

32.11 The Power of a Lens

The power F of a lens is defined by the equation

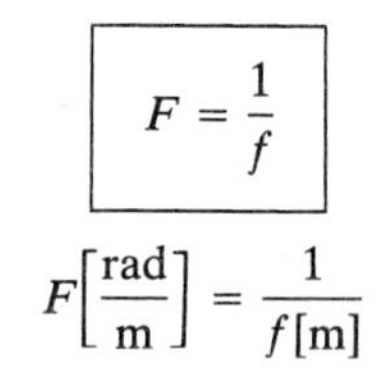

$$\boxed{F = \frac{1}{f}}$$

$$F\left[\frac{\mathrm{rad}}{\mathrm{m}}\right] = \frac{1}{f[\mathrm{m}]}$$

The equation CF $\approx h/D$ (from paragraph **32.7**) indicates why the rad m^{-1} is an appropriate unit for the power of a lens.

From the defining equation it follows that

(*a*) *diverging* lenses have a power which is *negative*, and

(*b*) for a *small* number of *thin* lenses in contact

$$F = F_1 + F_2 + F_3 + \cdots$$

Note that f must be expressed in *metres*. If it is quoted in mm, then F can be calculated from

$$F\Big/\frac{\text{rad}}{\text{m}} = \frac{1000}{f/\text{mm}}$$

32.12 Newton's Formula

This is an alternative way of representing the lens formula. In fig. 32.13 suppose

$$OF_1 = x \quad \text{so} \quad u = x + f$$
$$IF_2 = y \quad \text{so} \quad v = y + f$$

Then
$$\frac{1}{x+f} + \frac{1}{y+f} = \frac{1}{f}$$

which simplifies to

$$\boxed{xy = f^2} \qquad \textit{Newton's formula}$$

The result is helpful for finding the focal length of an inaccessible lens.

32.13 Object-Image Distances

Experimentally it is found that if D is big enough (fig. 32.15) *two* lens positions give an image on the screen.

Rewriting the lens formula in terms of D and a,

$$D = 2b + a \quad \text{so} \quad b = \frac{D-a}{2}$$

$$D = b + c \quad \text{so} \quad c = \frac{D+a}{2}$$

Then
$$\frac{2}{D-a} + \frac{2}{D+a} = \frac{1}{f}$$

giving
$$D^2 - 4Df - a^2 = 0 \qquad \ldots (1)$$

Solving the quadratic for D, we find

$$D = \frac{4f \pm \sqrt{16f^2 + 4a^2}}{2}$$

For a real image and object, D must be positive and greater than zero.

$$D_{\text{min}} = \frac{4f + 4f}{2} = 4f$$

Thus the *minimum distance* between a real object and its image is $4f$. This is important in the design of the terrestial telescope which uses a converging erecting lens.

($D = 0$ corresponds to a real object—virtual image.)

Equation (1) can be written as

$$f = \frac{D^2 - a^2}{4D}$$

This gives an accurate method of measuring f when the exact location of the *optic centre* is not accessible.

32.14 Measurement of Lens Radius of Curvature

(a) Concave Surface

Treat as a concave spherical mirror (p. 243).

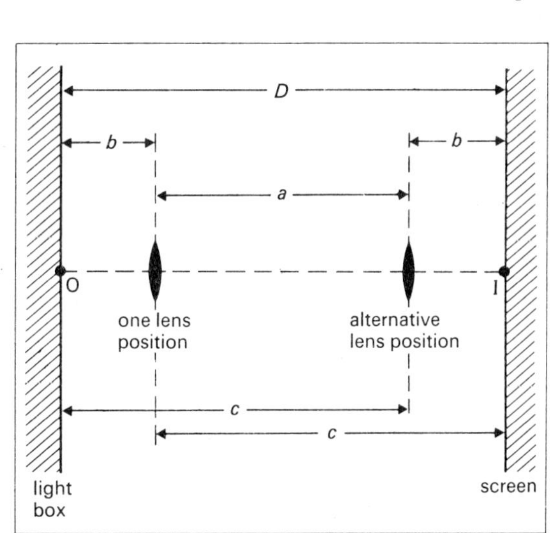

Fig. 32.15. Object-image distances.

(b) Convex Surface
Use the **Boys image** of fig. 32.16.

When rays strike the second surface at right angles
(*i*) there is a virtual image at C, and
(*ii*) reflection (followed by a further refraction) forms a real ghost image beside the object.

For the lens formula

$$u = +a$$
$$v = -b$$

If we know f, and measure a, we can find b.

For this lens $R = +b$ for the right-hand surface. If f, R_1 and R_2 are measured, then n can be calculated from the lens-maker's formula.

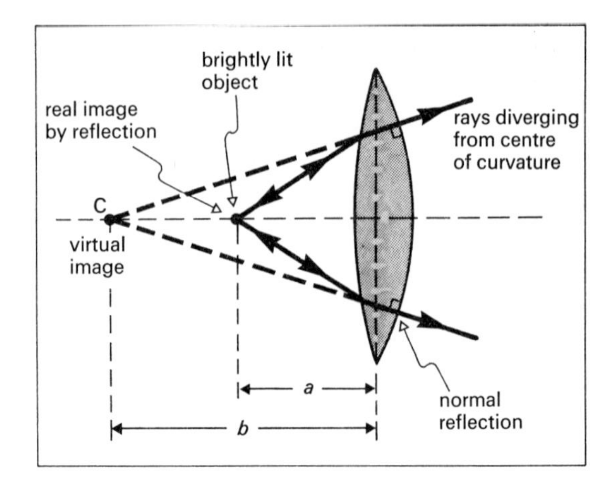

Fig. 32.16. To illustrate the Boys image.

33 Spherical Mirrors

33.1 The Mirror Formula

This derivation follows the argument of paragraphs **32.5** and **32.7** (the lens formula).

In outline,

$$\beta = \alpha + \theta$$

$$\gamma = \alpha + 2\theta$$

So $$\alpha + \gamma = 2\beta$$

Thus when the rays are *paraxial* (when AP $\lesssim$ 1/10 of the smallest of OP, CP and IP)

$$\frac{AP}{OP} + \frac{AP}{IP} \approx \frac{2AP}{CP}$$

which leads to the *numerical relationship*

$$\frac{1}{OP} + \frac{1}{IP} = \frac{2}{CP} \qquad \ldots (1)$$

for this mirror used in this way.

We adopt a *sign convention* which includes (a), (b) and (c) of paragraph **32.6**. In addition we specify that

(d) surfaces which reflect on the *concave* side (fig. 33.1) have a *positive* radius of curvature, and

(e) surfaces that reflect on the *convex* side have a *negative* radius of curvature.

This now leads to the *algebraic* relationship

$$\frac{1}{u} + \frac{1}{v} = \frac{2}{r}$$

in which the symbols are treated as before.

Finally we note that the *focal length f* is the image distance corresponding to infinite object distance, so

$$\frac{1}{f} = \frac{2}{r}$$

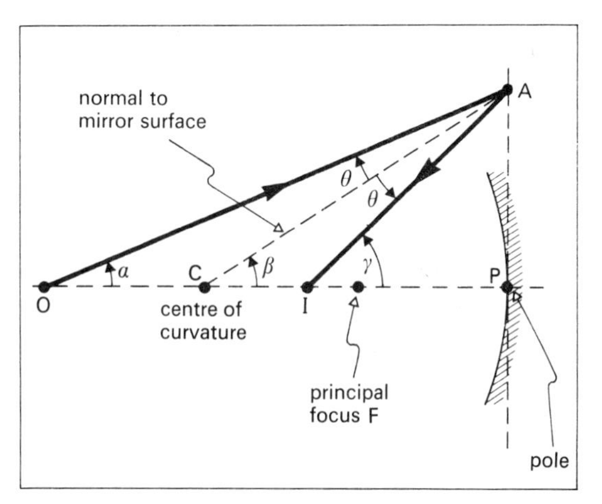

Fig. 33.1. Derivation of the mirror formula.

f is allotted the same sign as the corresponding r.

Then

$$\boxed{\frac{1}{u} + \frac{1}{v} = \frac{2}{r} = \frac{1}{f}}$$

is referred to as the **mirror equation.**

Notes.

(a) Equation (1) above shows that the location of I is independent of α if α is small. Thus a spherical mirror will form a point *image* of a point object by reflection of paraxial rays.

(b) Note the following terms used in connection with spherical mirrors: *pole, principal axis, focal plane.*

(c) *Magnification* is defined in the same way as for lenses (p. 240).

34 Lens and Mirror Aberrations

34.1 Introduction

An **aberration** is the phenomenon of a point object not giving rise to a point image. It may cause the image of an extended object to differ from the object in colour, shape and sharpness of definition.

Because light is a wave motion, it exhibits **diffraction** effects which *cannot be eliminated.* The extent of the diffraction determines the blurring of what would otherwise be a point image from a point object. It is proportional to

$$\frac{\text{wavelength of light}}{\text{diameter of optical system}}$$

and so can be *reduced* (but never eliminated) by

(a) increasing the diameter of an optical system such as a lens, and/or

(b) using light of a smaller wavelength.

It follows that we do not necessarily improve the quality of an image by stopping down a lens (p. 286).

34.2 Spherical Aberration

The mirror formula and lens-maker's formula are only approximately true. Perfectly *spherical* surfaces *cannot* produce point images of point objects simply because approximations as to the sizes of angles were made in deriving the formulae.

(*a*) Lenses

Fig. 34.1 shows a point object giving rise to a diffused circle as image. Note the terms **longitudinal spherical aberration** and **circle of least confusion**.

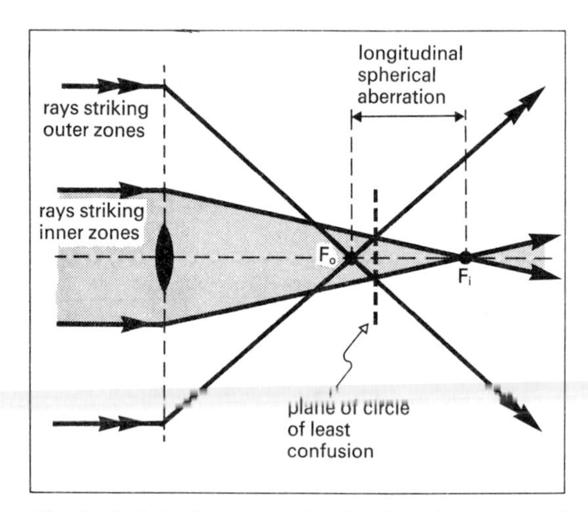

Fig. 34.1. Spherical aberration in a lens (exaggerated considerably).

(1) The effect can be *corrected* (for a given object distance) only by grinding the lens surfaces to make them suitably **aspherical**.

(2) The effect can be *minimized* by arranging for the deviation produced by the lens to be divided equally between the surfaces. This is achieved in the plano-convex lens when the curved surface receives or emits those rays which are most nearly parallel to the axis. Thus in a microscope, the plane surface faces the incident light.

(3) The effect can be *reduced* for a given lens by stopping down the lens aperture, but this makes the image less bright and introduces greater diffraction effects. For large relative aperture (e.g. the camera), aberration is usually more serious than diffraction, whereas for small relative aperture (e.g. the telescope), the reverse is usually true.

Spherical aberration is corrected within the human eye by a decrease in refractive index away from the axis.

(*b*) Mirrors

(1) The effect can be *eliminated* for point objects *on the axis* by using a **paraboloidal** surface. Then all rays parallel to the axis are reflected through the principal focus, whatever the mirror aperture (fig. 34.2).

(2) The effect can be *reduced* for a given mirror by using a smaller aperture.

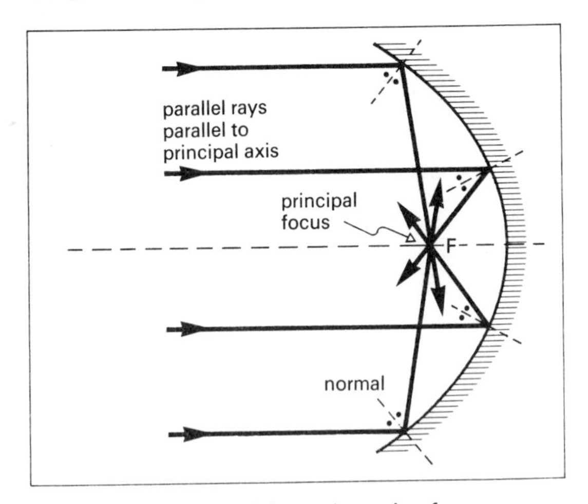

Fig. 34.2. Reflection without aberration from a paraboloidal surface.

34.3 Chromatic Aberration

The equation

$$\frac{1}{f} = (n - 1)\left(\frac{1}{R_1} + \frac{1}{R_2}\right)$$

shows that, for a given lens, f depends on n. n in turn is controlled by the frequency of the light.

$$n_{\text{violet}} > n_{\text{red}}$$

so $$f_{\text{red}} > f_{\text{violet}}$$

Note the term **longitudinal chromatic aberration** in fig. 34.3.

(1) For two lenses of the same material we can make f independent of λ by putting their separation equal to the mean of their focal lengths. This system is satisfactory for viewing a virtual image.

(2) The effect can be eliminated for *two* colours (and improved for all) by using an **achromatic doublet**, fig. 34.4.

The crown and flint glass lenses have different powers, but produce equal and opposite dispersion (p. 247).

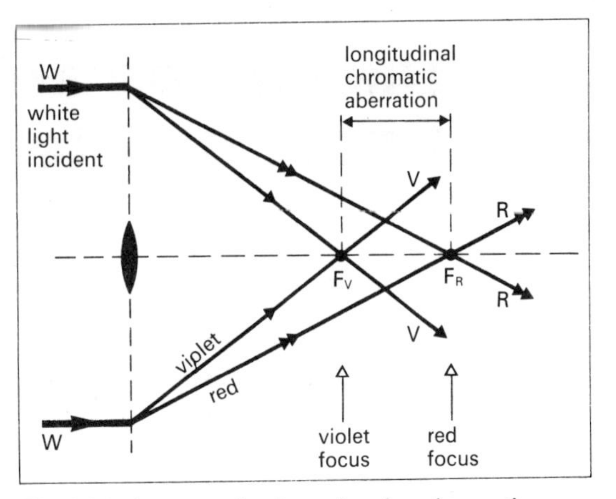

Fig. 34.3. A converging lens showing chromatic aberration (exaggerated considerably).

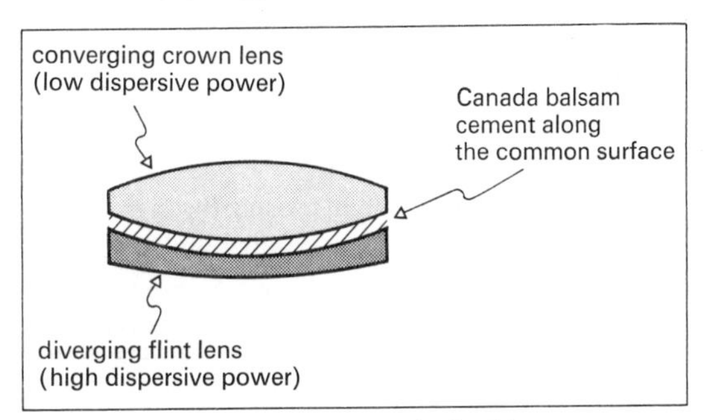

Fig. 34.4. A design of achromatic doublet.

34.4 Practical Considerations

Large mirrors are constructed by silvering or aluminizing paraboloidal glass surfaces. If the reflecting surface is damaged or tarnished it can then be repaired more easily than if the whole mirror were metal.

Telescopes are made to have a *large aperture*

(*a*) to reduce diffraction effects, and

(*b*) to make the image of low power sources bright enough.

They are *reflecting* because

(1) there is *no* chromatic aberration

(2) it is relatively easy to make a paraboloidal reflecting surface

(3) it is difficult to maintain uniform refractive index throughout a large volume of glass

(4) the suspension of a large lens is more difficult to achieve than that of a large mirror.

Note that the paraboloidal mirror does *not* form a point image of a point object *off* the axis.

35 Dispersion and the Spectrometer

35.1 Dispersion and Dispersive Power

The deviation of a ray of light passing through a prism depends on refractive index n, which in turn is controlled by the light frequency. Light waves of different frequency (and therefore different wavelength in a vacuum) are deviated by the prism through different angles and are said to be **dispersed** (fig. 35.1).

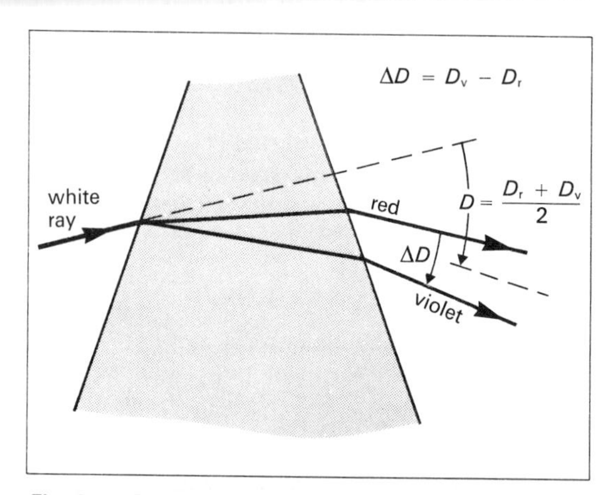

Fig. 35.1. Exaggerated dispersion of a ray of white light.

Suppose n is measured for different wavelengths, and the graph of n against λ plotted for different glasses (fig. 35.2).

Different glasses disperse the light through different angles ΔD for a fixed mean deviation D.

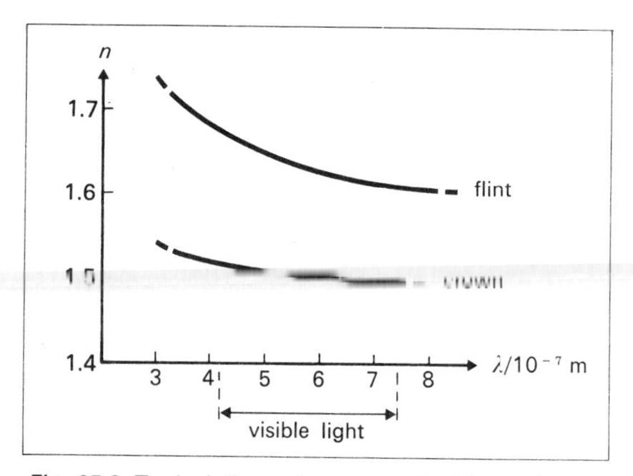

Fig. 35.2. Typical dispersion curves for flint and crown glasses.

The dispersive power ω ***of a material is defined by***

$$\omega = \frac{n_b - n_r}{n_y - 1}$$

and determines the size of ΔD. n_r, n_y and n_b are the refractive indices of the material for light of particular frequencies defined by the C, D and F *Fraunhöfer* lines (p. 250). Thus we might find for crown glass $\omega \sim 0.016$, and for flint glass $\omega \sim 0.029$.

Example. Consider a prism of small angle A working at small angles of incidence, so that

$$D = (n - 1)A$$

It follows that

$$\Delta D = \Delta n A$$
$$(D_b - D_r) = (n_b - n_r)A$$

and that the mean deviation

$$D_y = (n_y - 1)A$$

So for this *special case*

$$\omega = \frac{D_b - D_r}{D_y}$$

Special Prism Combinations

(*a*) *The achromatic prism* is a means of producing resultant deviation with zero resultant dispersion. Using two prisms of different glasses, rays of only two particular wavelengths can be made parallel.

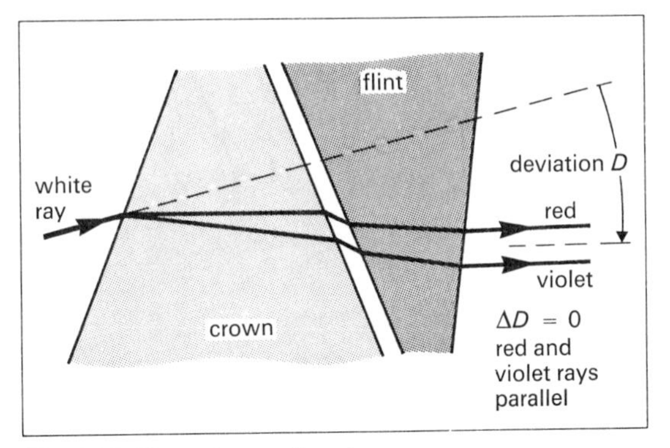

Fig. 35.3. The achromatic prism.

In fig. 35.3 if a *beam* of white light is used, its top edge is coloured red, and the bottom violet. This is the basis of the achromatic lens (p. 245).

(*b*) *The direct vision prism* is a means of producing resultant dispersion without deviation for a particular wavelength near the middle of the spectrum.

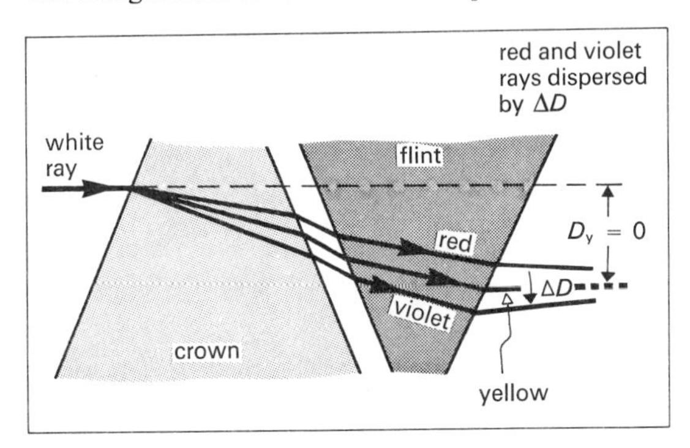

Fig. 35.4. The principle of the direct vision prism.

The direct vision spectroscope uses a compound prism of this type for the rapid observation of spectra.

35.2 The Spectrometer

This is an instrument suitable for

(*a*) producing, viewing and taking measurements on a *pure* spectrum (using either a prism or a diffraction grating), and

(*b*) measuring accurately the refractive index of a material in the form of a prism.

Its Construction

The essential parts are (fig. 35.5):

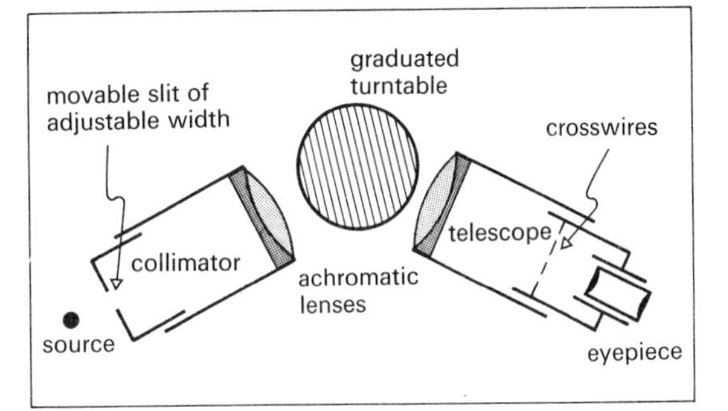

Fig. 35.5. Parts of the spectrometer.

(*a*) The **collimator** (which is fixed to the base of the instrument), consisting of a **slit** of variable width, and an achromatic lens.

(*b*) The **table**, which can be rotated, and to which a prism or grating can be attached. The circular edge of the table has a scale graduated in degrees. (Sometimes we may need to convert from degrees to radians.)

(*c*) The **telescope**, which can also be rotated. A **vernier scale** is fitted to the telescope where it adjoins the table, enabling their relative orientation to be measured to 0.1°, or less.

Adjustments

These should be made in the following order:

(1) The **eyepiece** is focussed on the crosswires.

(2) The objective lens of the *telescope* is focussed so that the crosswires are in its focal plane.

(3) Using a slit of width appropriate to the source brightness, the *collimator* lens is moved so that the slit is in its focal plane.

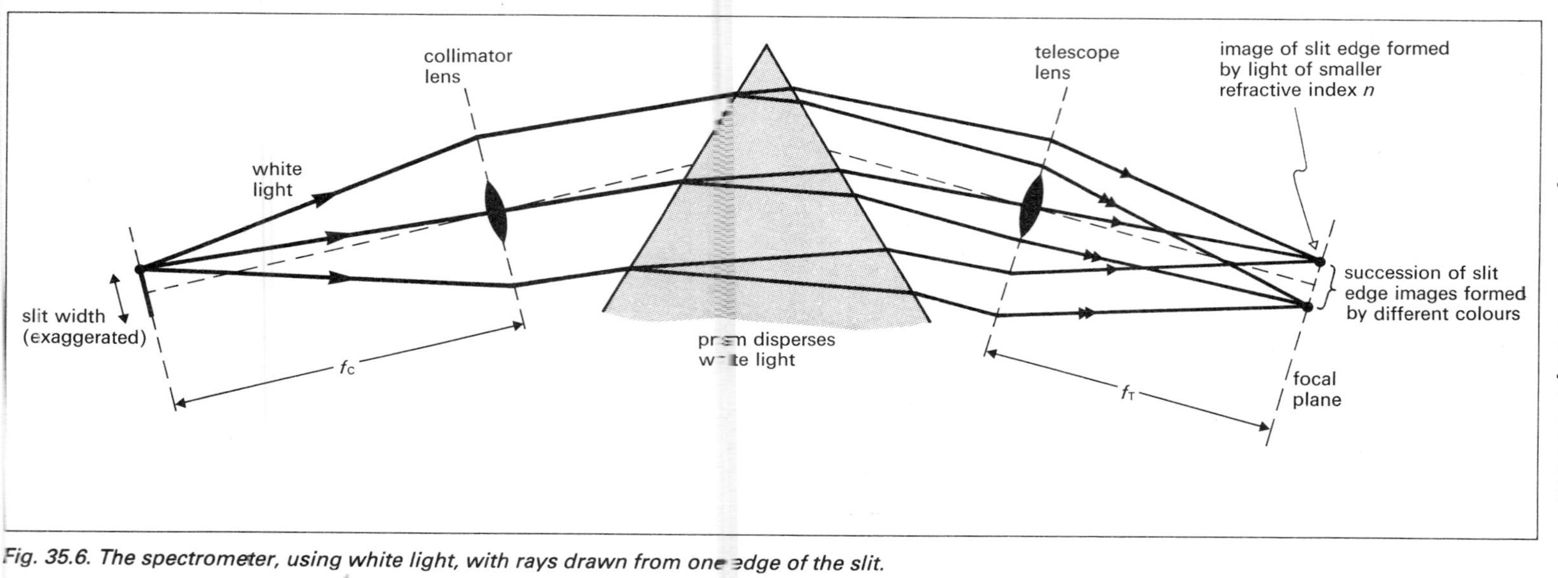

Fig. 35.6. The spectrometer, using white light, with rays drawn from one edge of the slit.

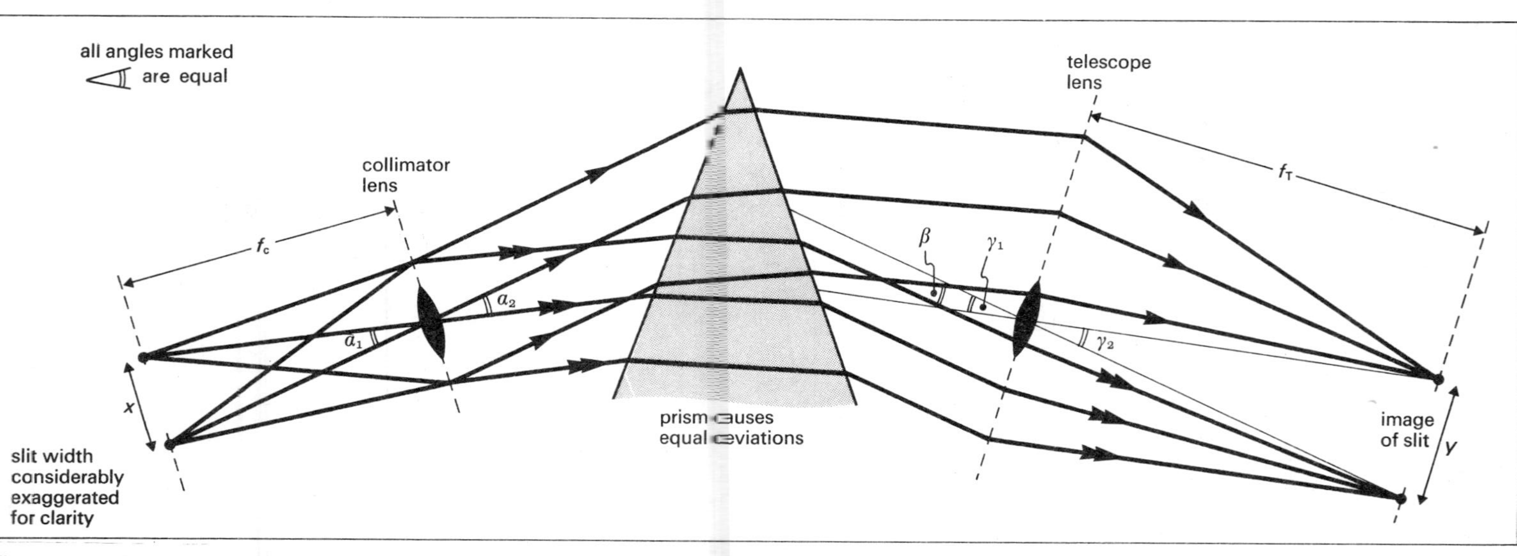

Fig. 35.7. The spectrometer, using monochromatic light, with rays drawn from both edges of the slit.

(4) Using the table levelling screws, the axis of the *table* is made perpendicular to the plane containing the principal axes of the collimator and telescope lenses.

35.3 Optical System of the Spectrometer

Figs. 35.6 and 35.7 and their captions need most careful study.

In fig. 35.7, the two angles marked α are equal since they are vertically opposite.

$\alpha = \beta$, because parallel monochromatic rays are deviated through approximately equal angles by the prism when the rays from the centre of the slit undergo minimum deviation.

$\beta = \gamma$, alternate angles between parallel lines. The diagram shows construction lines drawn through the centre of the lens.

The two angles marked γ are equal since they are vertically opposite.

Because $\alpha_1 = \gamma_2$

$$\frac{x}{f_C} = \frac{y}{f_T}$$

This means that the image width y can be calculated.

35.4 Measurement of Refractive Index

A knowledge of A and D_{min} enables n to be found from

$$n = \frac{\sin[(A + D_{min})/2]}{\sin(A/2)}$$

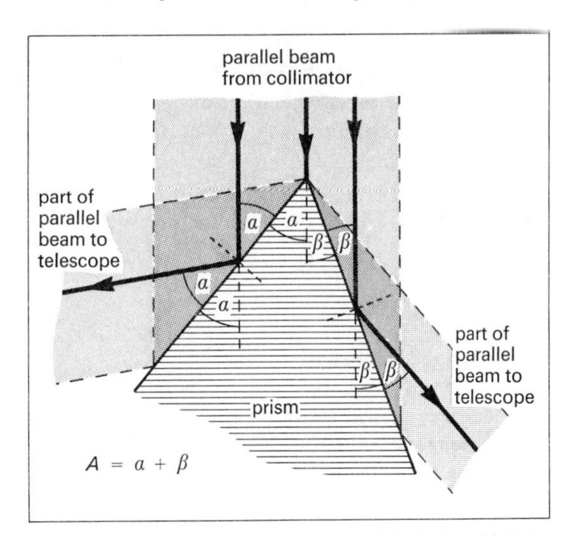

Fig. 35.8. Three rays from one beam of light which is split into two beams by reflection.

(*a*) Optical Measurement of A

(1) Place prism on table centre.

(2) Record two telescope positions where reflected beams are observed (fig. 35.8).

(3) Angle between rays shown $= 2(\alpha + \beta)$

$$\text{Angle of prism} = A = (\alpha + \beta)$$

Any kind of light can be used for this part of the experiment.

(*b*) Measurement of D_{min}

This part of the experiment requires a source of the wavelength at which n is to be measured.

(1) Set the telescope to accept rays that pass approximately symmetrically through the prism.

(2) Clamp the telescope in such a position that one

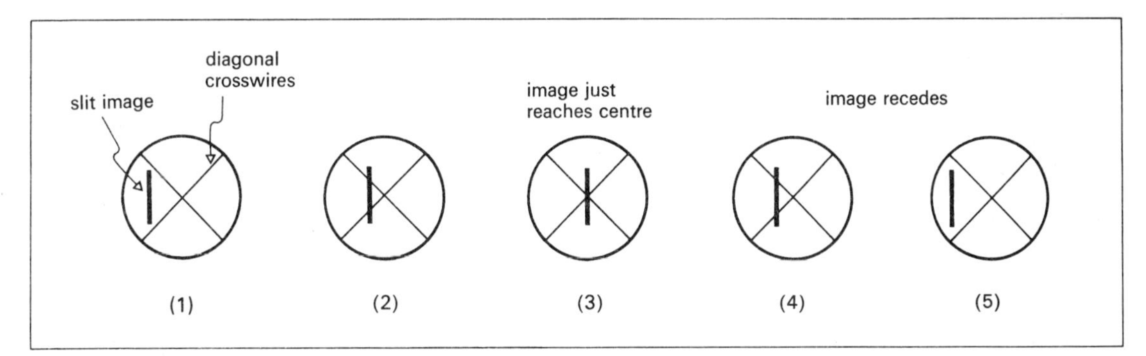

Fig. 35.9. Successive views seen as the table and prism are rotated together through the position of minimum deviation.

sees this succession of views (fig. 35.9) as the table is rotated consistently in the same direction.

(3) Clamp the table when the centre view is seen. Note the scale reading.

(4) Remove the prism, and record the scale reading when the telescope receives light from the collimator directly. D_{min} is the difference between these two readings.

35.5 Spectroscopy

Spectroscopy is the study of emitted or absorbed electromagnetic radiation. It involves techniques of very high precision, and very high sensitivity (only a small mass of source material need be used).

A **spectrograph** is an apparatus producing dispersion, in which the spectrum is usually photographed.

A **spectroscope** lends itself for viewing by eye directly.

A **spectrometer** is designed for measurement. Sources can be

(*a*) **Incandescent** when a body is at a high temperature (>1100 K or so), or

(*b*) **Luminescent.**

(*i*) **Fluorescence** occurs when a substance emits visible light when exposed to u.v. radiation.

(*ii*) **Phosphorescence** occurs when the emission of visible light takes place even though the u.v. source is no longer present.

35.6 Classification of Spectra

Spectra are of two fundamental types:

(*a*) An **emission spectrum** is what is seen when energy travels from the source to the observer uninterrupted.

(*b*) An **absorption spectrum** is what is seen when, before analysis, part of the emitted energy is absorbed by some intervening medium.

Both types are shown as

(1) a line spectrum, and

(2) a band (fluted) spectrum, while the emission spectrum can also be

(3) a continuous spectrum.

A continuous absorption spectrum may reduce the intensity of frequencies in one *range* of the spectrum, while another part of the spectrum is not absorbed.

(1) Line Spectra

(a) Emission

This is the result of exciting (giving extra energy to) *substantially independent atoms.* They are therefore emitted by luminous gases and vapours at low density.

(b) Absorption

If light of all frequencies is passed through a gas, then the gas absorbs light of the same frequency as it would emit at the same temperature (p. 225). This is an example of resonance. The light is then reradiated in *all* directions. This causes a *reduction* of intensity in the direction of the observer.

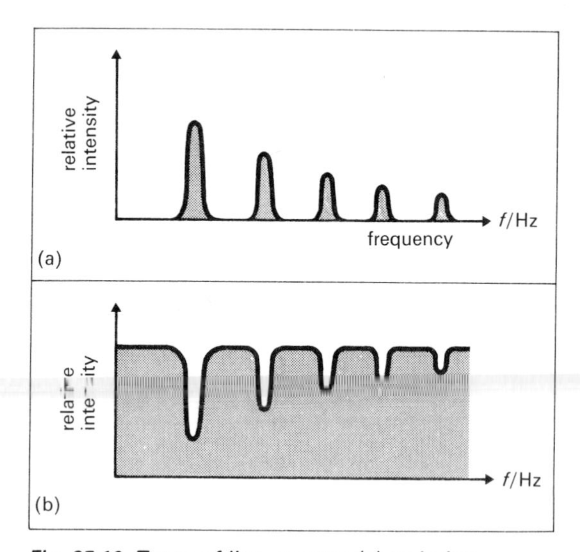

Fig. 35.10. Types of line spectra: (a) emission spectrum, (b) absorption spectrum for same material under identical conditions.

In practice only a few of the lines in an emission spectrum can be matched with absorption lines.

Fraunhöfer lines are *dark lines* crossing the otherwise continuous spectrum of the *Sun.* While the **photosphere** emits white light, energy of some frequencies is absorbed in the cooler gases of the **chromosphere**, resulting in a reduced intensity of these frequencies at the *Earth's* surface.

Fraunhöfer labelled the most important lines A to H. They correspond to *bright lines* of emission spectra of atoms on *Earth* e.g.

(*i*) the D line(s) correspond to the yellow light emitted by sodium, and

(*ii*) the C and F lines are the red and blue lines emitted by hydrogen.

These frequencies were used to define dispersive power (p. 246), but the sodium D line is now being replaced by the helium d line.

(2) Band Spectra

(a) Emission

This is the result of exciting *substantially independent polyatomic molecules* (e.g. molecules in a low pressure gas). This could be done by electrical discharge. The emitted frequency corresponds to changes in

(*i*) *electron energy*, giving visible light,
(*ii*) *molecular vibrational energy*, giving frequencies in the near infra-red, and
(*iii*) *molecular rotational energy*, giving frequencies in the far infra-red.

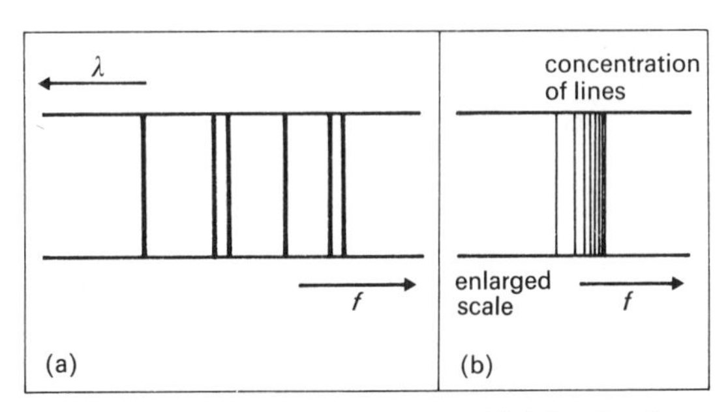

Fig. 35.11. (a) A band spectrum, and (b) details of one band.

A **band** (fig. 35.11) is a large number of close lines which could be produced, for example, by superimposing rotational changes on a vibrational change.

(b) Absorption

This is what one sees when observing a continuous spectrum after it has passed through a region which contains a band-emitting substance. It is useful for analysing molecular structure.

(3) Continuous Spectra

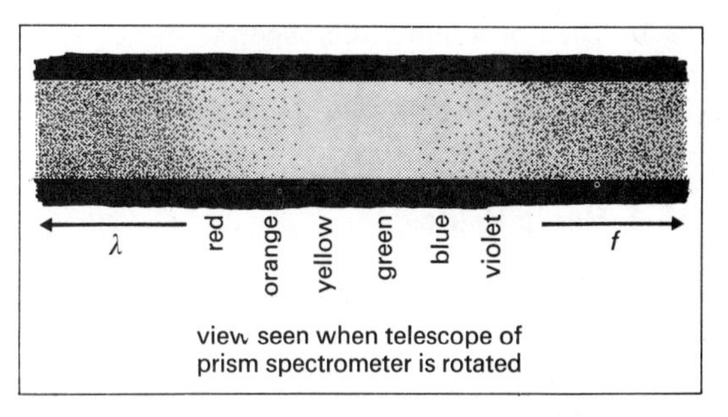

Fig. 35.12. Continuous spectrum from a carbon arc.

When molecular collisions are very frequent (as in a compressed gas such as the Sun), or there is other strong molecular interaction (as in the liquid and solid phases), *all* frequencies are emitted. Thus a *continuous* spectrum is *not* characteristic of the nature of the source.

36 Optical Instruments

36.1 Basic Principles

The **near point** is the position of the nearest object that can be focussed both clearly and without strain by the unaided eye.

The **far point** is the position of the farthest object that can be focussed by the unaided eye.

For the **normal eye**

(*a*) the far point is at infinity, and

(*b*) the near point is taken to be a distance 250 mm from the eye. This is called the **least distance of distinct vision** D.

The *apparent size of an object* (and hence the discernible detail) is determined by the size of the image it forms on the retina, and so by the magnitude of the **visual angle** θ (fig. 36.1).

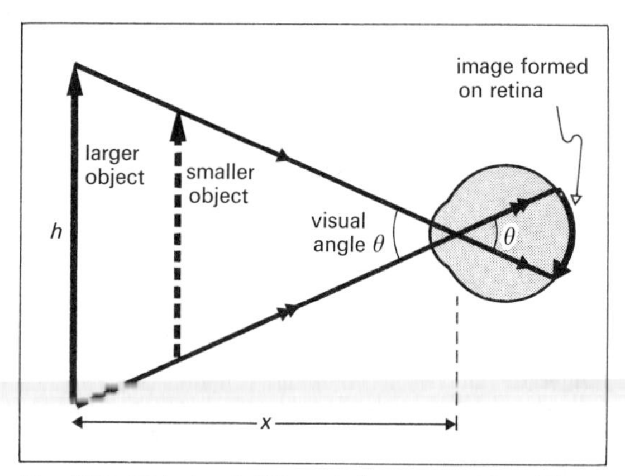

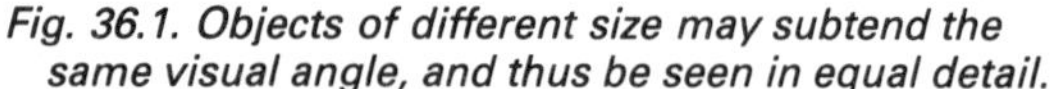

Fig. 36.1. Objects of different size may subtend the same visual angle, and thus be seen in equal detail.

Angular magnification M ***(magnifying power) is defined by the equation***

$$M = \frac{\theta_I}{\theta_O}$$

where θ_I and θ_O are defined in fig. 36.2. In general M is *not* the same as the linear magnification.

When calculating θ_I, we imagine the eye to be placed close to the eye-lens of an instrument.

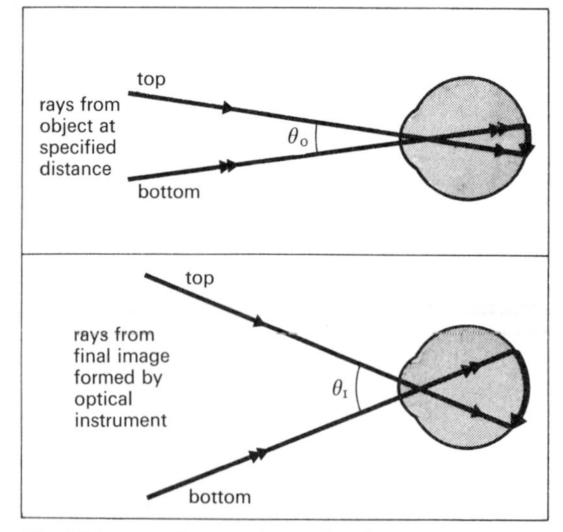

Fig. 36.2. θ_O and θ_I are the angles that determine magnifying power.

Diagrams

When considering optical instruments, it is often convenient to draw a diagram from which one can make calculations simply. Fig. 36.3*a* and *b* should be compared carefully.

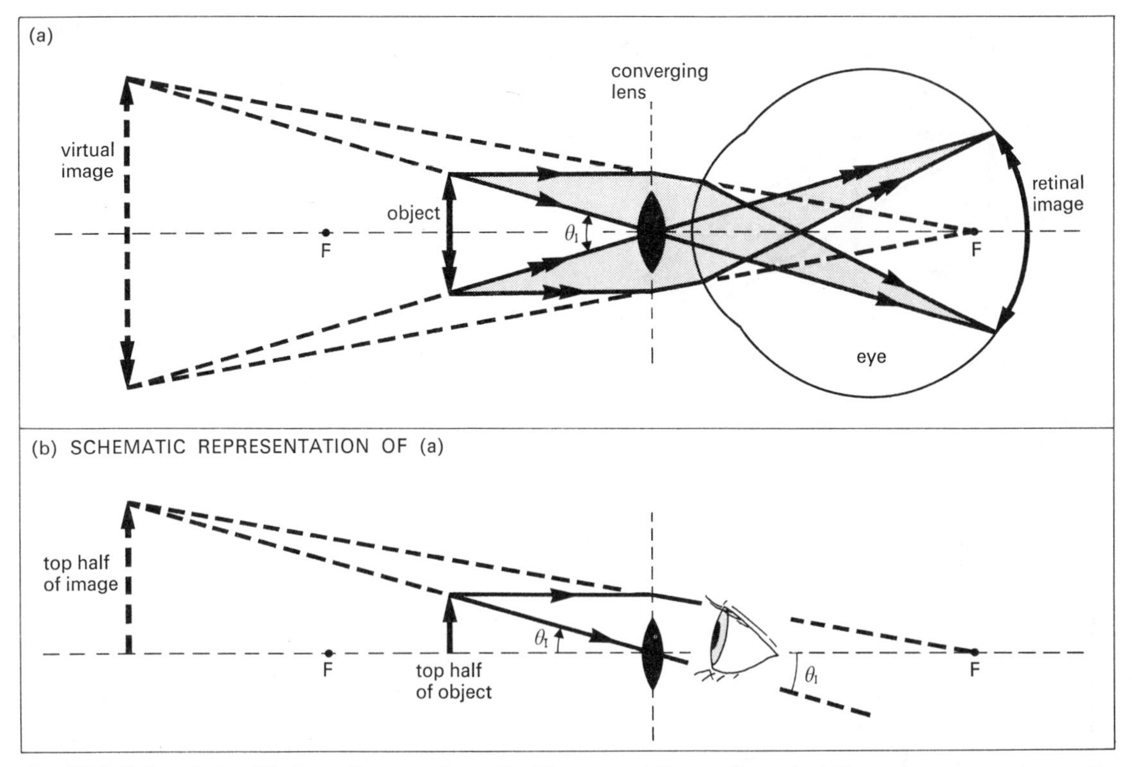

Fig. 36.3. Full and simplified ray diagrams for optical instrument theory. For calculation purposes we assume the eye to be placed very close to the eye-lens, as in (a).

Throughout this chapter diagrams similar to fig. 36.3*b* will be used, and frequently the angle to which we refer as θ is *strictly* half the visual angle. For clarity θ has been drawn grossly exaggerated.

Normal Adjustment

An optical instrument is in normal adjustment when the final image is formed in a specified position.

(*a*) For a *telescope*, the image is at infinity, so that the eye is most relaxed.

(*b*) For a *microscope*, the image is at the near point, so that maximum angular magnification is obtained.

Approximations

We assume that all angles are small, so that angles may be replaced by their tangents. Thus in fig. 36.1, we may write $\theta = h/x$.

36.2 The Simple Magnifying Glass

By convention, θ_O is the angle subtended when the object is placed at the near point (fig. 36.4).

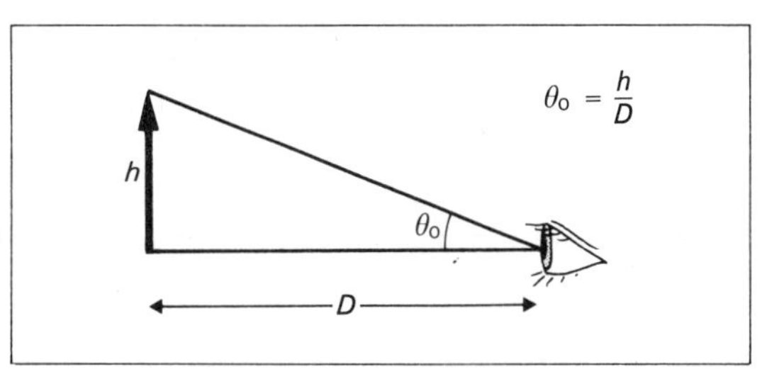

Fig. 36.4. Visual angle for object seen by unaided eye (for microscope calculations).

(*a*) Image at Infinity (fig. 36.5)

Angular magnification

$$M = \frac{\theta_I}{\theta_O} = \frac{h/f}{h/D} = \frac{D}{f}$$

where D is the *magnitude* of the least distance of distinct vision. This is the smallest value that M can have.

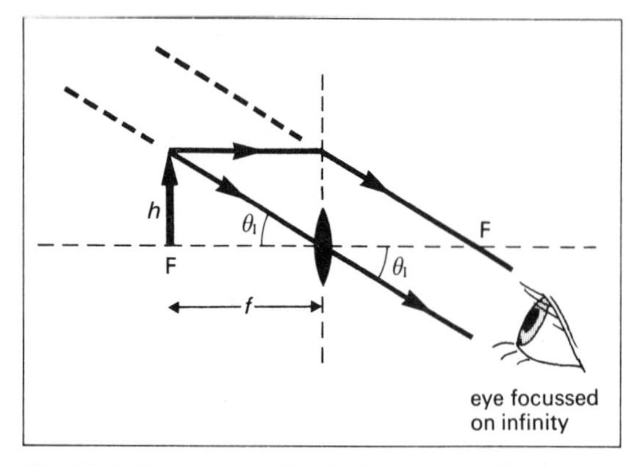

Fig. 36.5. Object placed in the focal plane of a converging lens.

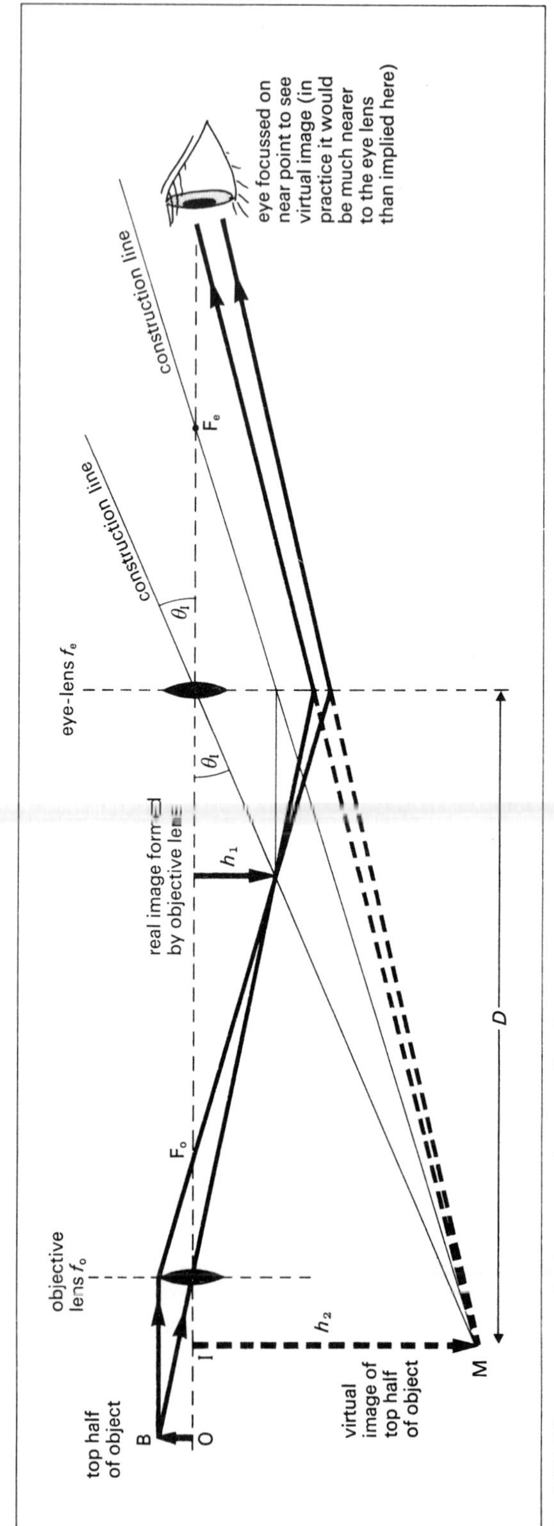

Fig. 36.7. The compound microscope in normal adjustment.

(b) Image at Near Point (fig. 36.6).

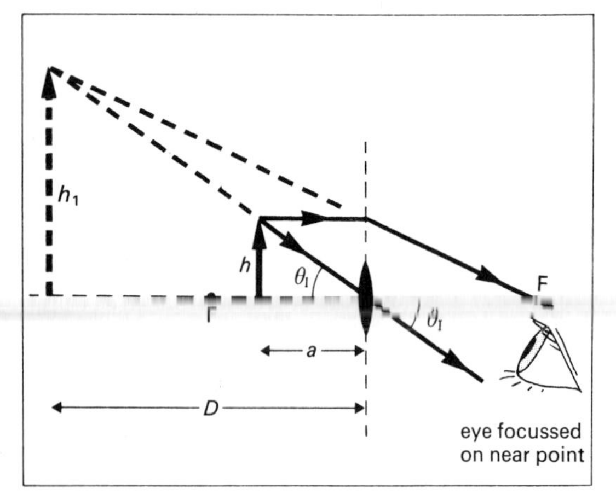

Fig. 36.6. Object placed on the lens side of the focal plane.

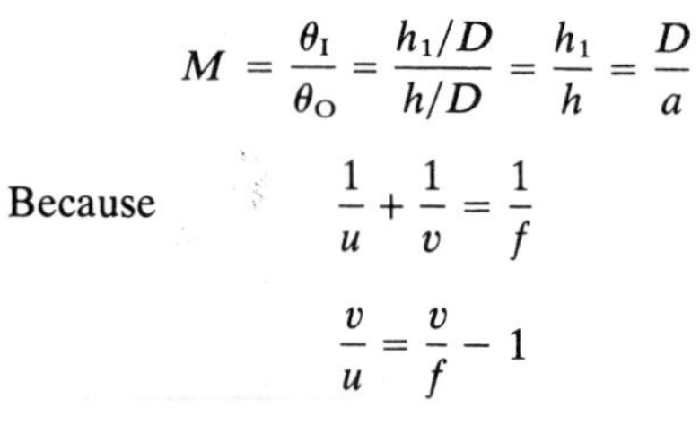

$$M = \frac{\theta_I}{\theta_O} = \frac{h_1/D}{h/D} = \frac{h_1}{h} = \frac{D}{a}$$

Because $$\frac{1}{u} + \frac{1}{v} = \frac{1}{f}$$

$$\frac{v}{u} = \frac{v}{f} - 1$$

Here $$\left|\frac{v}{u}\right| = \frac{D}{a} = \frac{D}{f} + 1 \quad \text{since} \quad v = -D$$

So the magnitude of M is $\left(1 + \frac{D}{f}\right)$. This is the *largest* value that M can have.

M is increased by using a smaller value for f, but the aberrations involved may make the image more obscure.

36.3 The Compound Microscope

The *aim* of the instrument is to produce an image on the retina larger than that obtainable by placing a small *accessible* object at the near point.

Refer to fig. 36.7. The objective lens produces an *enlarged real* image of the object. The eye uses the eye-lens as a magnifying glass to view this as a real object. In normal adjustment an *enlarged virtual* image is formed *at the near point,* 250 mm from the normal eye.

θ_O is evaluated when the object is placed at the near point (fig. 36.4). So

$$M = \frac{\theta_I}{\theta_O} = \frac{h_2/D}{h/D} = \frac{h_2}{h} = \frac{h_2}{h_1}\cdot\frac{h_1}{h}$$

$$M = \begin{pmatrix}\text{linear magnification}\\\text{produced by eyepiece}\end{pmatrix} \times \begin{pmatrix}\text{linear magnification}\\\text{produced by objective}\end{pmatrix}$$

This can be made larger by using smaller values for f_o and f_e.

Calculations on the microscope should *always* be carried out from first principles. Never attempt to remember formulae.

36.4 Astronomical Telescopes

(a) Refracting Telescope

The *aim* is to produce an enlarged retinal image of a distant *inaccessible* object.

Refer to fig. 36.8. Rays from the top of the object subtend an angle θ_o with rays from the bottom. Rays from any one point on the distant object are effectively parallel at the objective lens, so the objective lens forms a *diminished* real image in its focal plane. The eye-lens is used as a magnifying glass, and the final image is formed *at infinity* when the instrument is in normal adjustment.

$$M = \frac{\theta_I}{\theta_o} = \frac{h_1/f_e}{h_1/f_o} = \frac{f_o}{f_e}$$

Notes.

(*a*) To make M large, we want a large f_o, but a small f_e.

(*b*) The length of the telescope tube = $(f_o + f_e)$.

(*c*) The resolving power depends on the diameter and quality of the objective lens. Large lenses are difficult to make, but small lenses do not deteriorate as fast as mirror surfaces. (See also p. 245).

(*d*) The telescope drawn in fig. 36.8 can be converted to a **terrestrial** telescope, producing an upright image, by means of a third converging lens (the *erecting* lens) placed between the objective and the eyepiece. The length of the telescope is increased by a minimum of four times the focal length of this lens (**32.13**). The **Galilean** telescope produces an upright image using only two lenses. The eyepiece is a diverging lens and the separation of the lenses is made equal to the difference of their focal lengths.

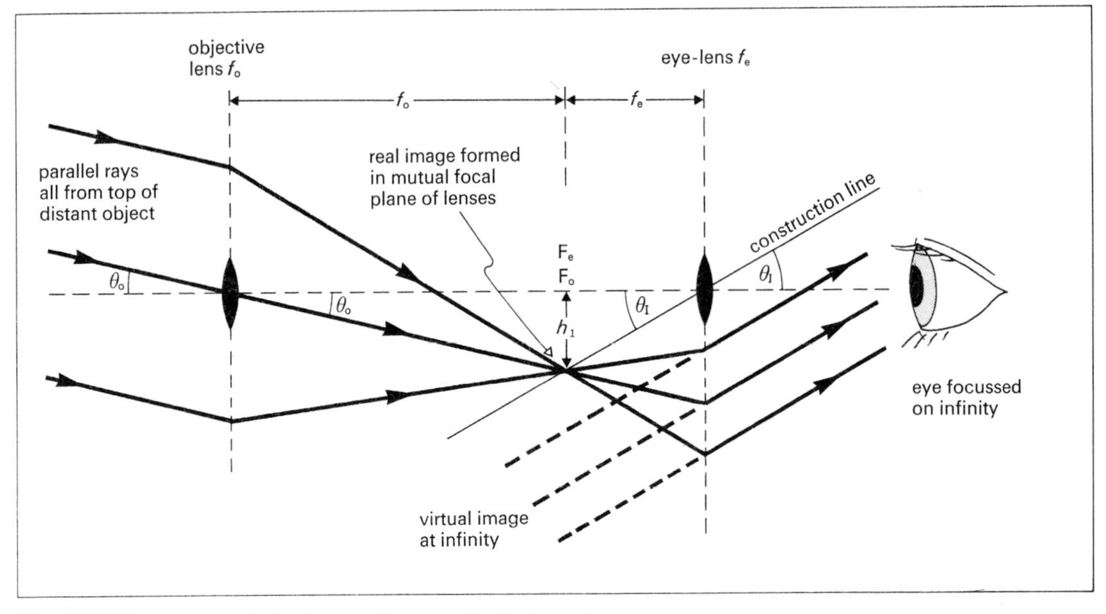

Fig. 36.8. The astronomical telescope in normal adjustment.

(b) Reflecting Telescope

For the advantages of a reflecting system, see p. 245. The maximum aperture can be as much as 5 m, as opposed to 1 m for a refracting telescope.

If the image is to be photographed, the plate is placed at I, where a *real* image is formed (fig. 36.9). (Note that in practice the smaller mirror has an area which is small in comparison with that of the concave one.)

The *Cassegrain* reflector uses a convex mirror to bring the light to a convenient focus. The *Newtonian* reflector uses a plane mirror oriented so that the eye-lens axis is perpendicular to the mirror axis. The *Coudé* reflector is a combination of the two.

Procedure for Drawing the Reflecting Telescope Diagram (fig. 36.9)

(*a*) Draw the principal axis, the concave mirror and its principal focus *F*.

(*b*) Locate ① and ②, and draw the plane mirror mid-way between them.

(*c*) Draw the eye-lens and construction line so that $\theta_I > \theta_O$.

(*d*) Draw arbitrary rays *x* and *y* reflected as to ②, then through ① and finally parallel to the construction line.

36.5 The Eye-Ring

In fig. 36.10 the circular area of diameter A′B′ is the *smallest* area through which *all* rays from the distant object, which are refracted by both the objective lens and the eye-lens are made to pass.

An observer should ideally

(*a*) place his eye pupil at A′B′, and

(*b*) have a pupil diameter equal to A′B′.

A circular stop, of diameter not greater than that of the average eye pupil, is placed at A′B′. It is called the **eye-ring** and represents the *optimum position* for the observer's eye (fig. 36.11).

We can show that

$$\frac{AB}{A'B'} = \frac{f_o}{f_e} = M$$

for a telescope in normal adjustment.

36.6 Resolving Power of a Telescope

Diffraction (chapter **39**) by the objective is the factor that limits the ability of a telescope to distinguish between two point objects that subtend a very small angle at the observer. Consider a double star system that can just be resolved according to the *Rayleigh*

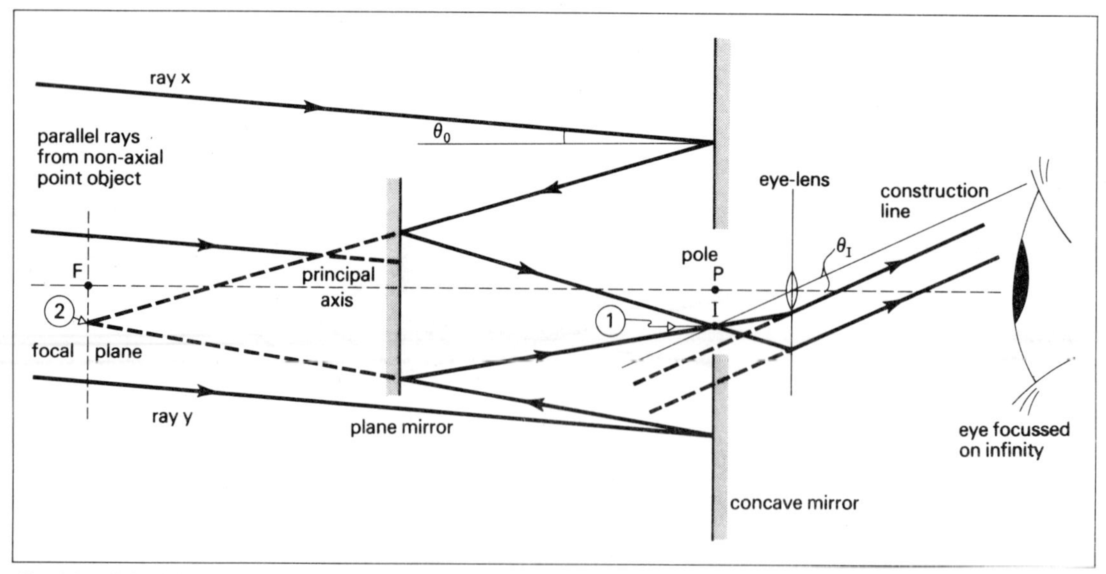

Fig. 36.9. Principles of the reflecting telescope.

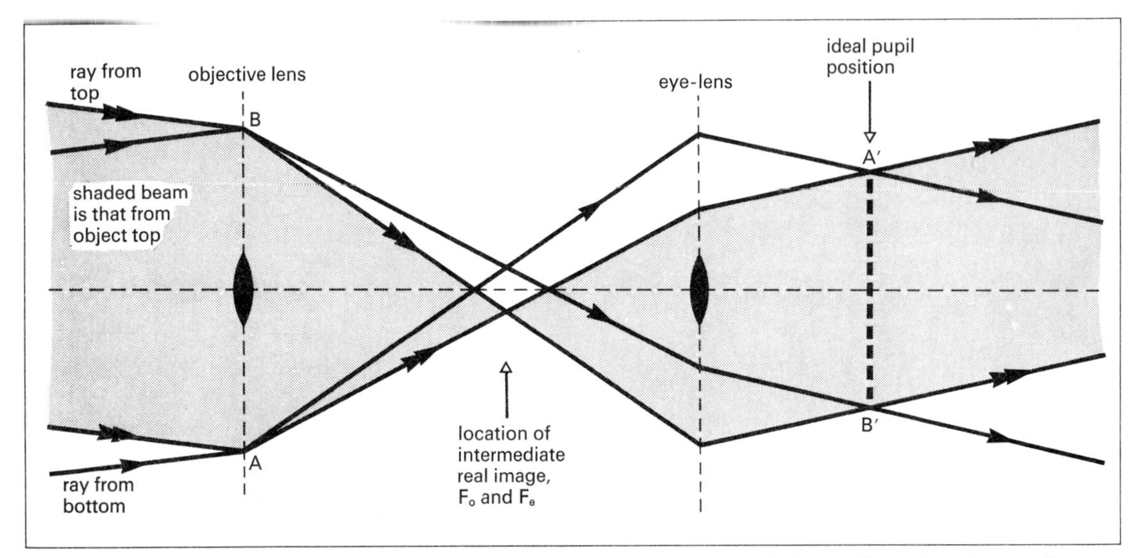

Fig. 36.10. To illustrate the behaviour of rays after they have passed through the eye-lens of a telescope in normal adjustment. (Not to scale.)

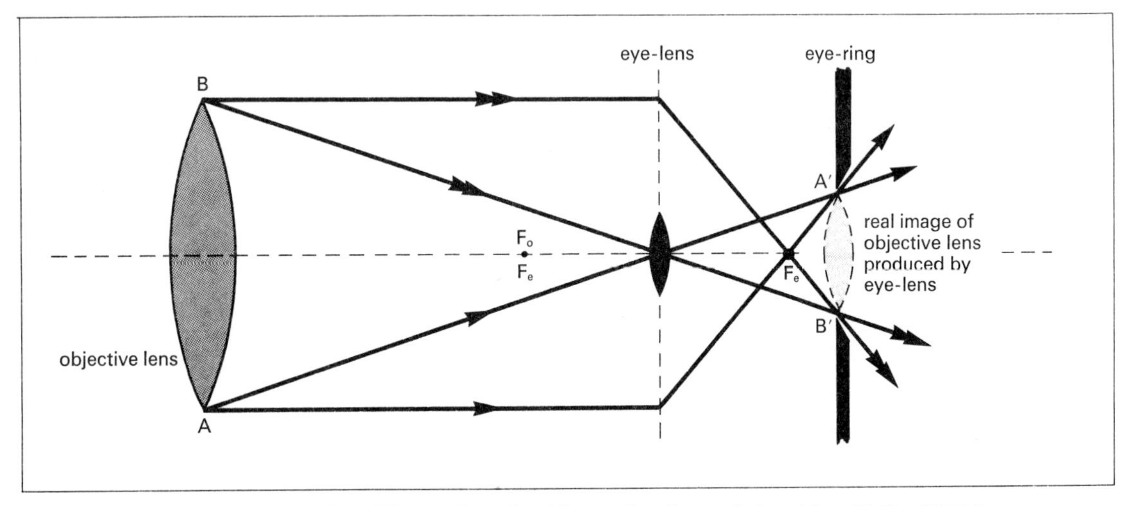

Fig. 36.11. Location of the eye-ring. (The student should note the close relationship with fig. 36.10.)

criterion (p. 286). If θ is the angular separation of the two stars, λ the wavelength being examined and D the diameter of the objective, then since $\lambda \ll D$

$$\theta = 1.22\frac{\lambda}{D}$$

The resolving power is the reciprocal of this angle which just allows resolution. The refracting telescope at the *Yerkes* Observatory has a diameter of 1.0 m. Thus for observations using $\lambda = 500$ nm, we have $\theta = 0.6$ μrad. For the human eye $\theta \approx 0.2$ mrad (200 μrad).

VII Wave Properties of Light

37 The Nature of Electromagnetic Radiation

37.1 The Wave Nature of Electromagnetic Radiation

Using the method of p. 93, we can classify electromagnetic waves as

(1) *periodic* waves

(2) *progressive* waves (usually, but see the comment on p. 116 and *Wiener's* experiment p. 266)

(3) *transverse*, in which

(4) the *disturbance* is a time-variation of the magnitude of the electric and magnetic fields $\boldsymbol{E}$ and $\boldsymbol{B}$ at a point.

Electromagnetic waves are generated by accelerated (often oscillating) electric charges, or by magnetic dipoles. Fig. 37.1 shows the nature of the disturbance.

Suppose the source of radiation is a straight vertical conductor (an antenna) that causes the radiation to travel outwards radially and symmetrically.

Fig. 37.2 is a horizontal section of such a source which shows the *wavefronts* and the associated *rays* (in this example *wave-normals*) which travel along the z-direction of fig. 37.1.

All electromagnetic waves can be propagated through a vacuum.

The relation between $\boldsymbol{E}$, $\boldsymbol{B}$ and $\boldsymbol{c}$ can be found by considering the energy density in the electromagnetic wave. It can be shown that $c = E/B$.

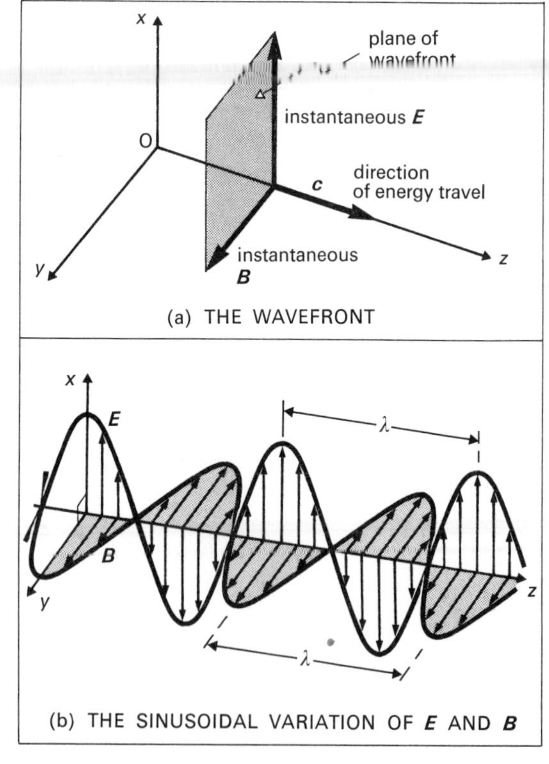

Fig. 37.1. The relative orientation of $\boldsymbol{E}$, $\boldsymbol{B}$ *and wave velocity* $\boldsymbol{c}$ *in an electromagnetic wave.*

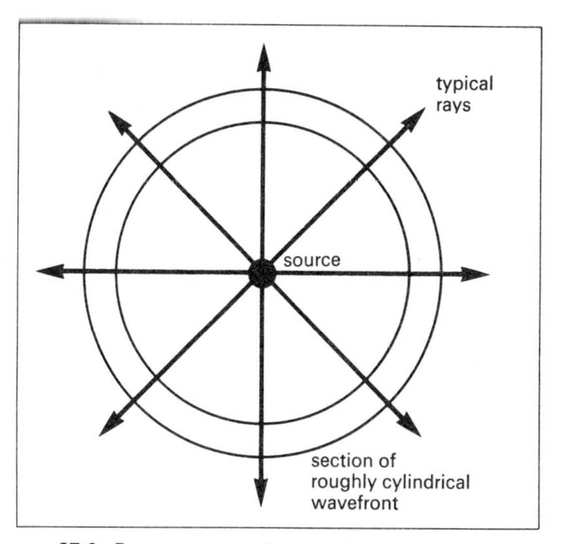

37.2. *Rays and wavefronts related.*

Some Orders of Magnitude

Suppose a light beam has an intensity (p. 99) of 100 W m^{-2}. Then it can be shown that

$$\boldsymbol{E} \approx 300 \text{ V m}^{-1}$$
$$\boldsymbol{B} \approx 10^{-6} \text{ Wb m}^{-2} = 10^{-6} \text{ T}$$

Relative to typical laboratory fields, $\boldsymbol{E}$ is much larger than $\boldsymbol{B}$. For this reason, and because the source of $\boldsymbol{E}$ (and not that of $\boldsymbol{B}$) can exert forces on *stationary* electric charges, we are primarily concerned with the *electric vector* for optical effects. (Nevertheless the energies associated with the magnetic and electric fields are equal.)

37.2 Types of Electromagnetic Radiation

(a) Summary of Parts of Electromagnetic Spectrum (see table on p. 262)

The names given to different parts of the spectrum are for convenience only: they do not imply a rapid change of properties. The ranges of γ-rays and X-rays overlap because they are classified by *origin* rather than frequency. Note that the three sets of numbers are presented on logarithmic scales. (Compare with the graph of fig. 37.4, in which the scale is linear.)

(b) Properties of Electromagnetic Radiation (See table on p. 263.)

37.3 Polarization and Coherence

Much electromagnetic radiation (e.g. that from radio and radar antennae) is **polarized** (p. 97) and relatively continuous: the plane containing the electric vector retains the same orientation. Infra-red, visible and ultra-violet waves are emitted *independently* by the individual atoms and molecules that make up the source. Since the atoms do not act **coherently**, the separate wavetrains (p. 269) that each produces are completely unrelated to those emitted by their neighbours. Their planes of vibration and instants of emission are random.

Polarization is discussed more fully in chapter **40**, and coherence in chapter **38**.

The **laser** (**l**ight **a**mplification by the **s**timulated **e**mission of **r**adiation) is a device in which one wavetrain *creates* another from a previously excited atom such that the two are *in phase*. The process multiplies itself very rapidly, and results in a *beam of coherent light*, i.e. coherence on a large scale.

37.4 The Quantum Nature of Electromagnetic Radiation

Planck suggested that radiation is *emitted* in packets or **quanta**, which for i.r., visible light and the higher frequencies are called **photons**.

The quantum nature of light is *not* inconsistent with our observation of uniform illumination (fig. 37.3).

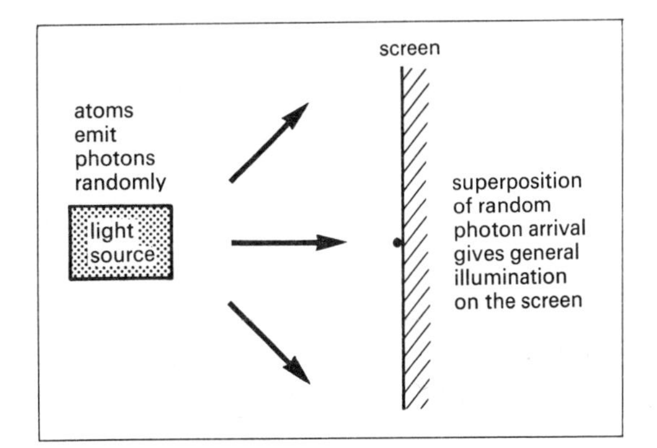

Fig. 37.3. Quantum nature of light.

SUMMARY OF PARTS OF ELECTROMAGNETIC SPECTRUM

Name and (approximate) range of radiation	†*Energy per photon* W/J	*Frequency* f/Hz	*Wavelength* λ/m	*Common length units for comparison*
	10^{-11}	10^{22}		
γ-rays	10^{-12}	10^{21}	10^{-13}	
	10^{-13}	10^{20}	10^{-12}	picometre pm
	10^{-14}	10^{19}	10^{-11}	
X-rays	10^{-15}	10^{18}	10^{-10}	
	10^{-16}	10^{17}	10^{-9}	nanometre nm
u.v.	10^{-17}	10^{16}	10^{-8}	
	10^{-18}	10^{15}	10^{-7}	
visible light	10^{-19}	10^{14}	10^{-6}	micrometre μm
i.r.	10^{-20}	10^{13}	10^{-5}	
	10^{-21}	10^{12}	10^{-4}	
	10^{-22}	10^{11}	10^{-3}	millimetre mm
microwaves	10^{-23}	10^{10}	10^{-2}	
	10^{-24}	10^{9}	10^{-1}	
short radio waves	10^{-25}	10^{8}	1	metre m
	10^{-26}	10^{7}	10^{1}	
standard broadcast	10^{-27}	10^{6}	10^{2}	
	10^{-28}	10^{5}	10^{3}	kilometre km
long radio waves	10^{-29}	10^{4}	10^{4}	
	10^{-30}	10^{3}	10^{5}	

† W can be expressed relative to the quantity *electron volt* (p. 323) by dividing these numbers by 1.6×10^{-19}.

PROPERTIES OF ELECTROMAGNETIC RADIATION

N.B. All have the wave properties associated with a transverse wave motion

<table>
<tr><th>Name</th><th>Generated by</th><th>Dispersed by</th><th>Detected by</th><th>Particular properties</th></tr>
<tr><td>γ-rays</td><td>Changes of energy levels in the nucleus</td><td></td><td rowspan="2">(a) Photography
(b) Ionization chamber
(c) Phosphorescence (e.g. ZnS)</td><td rowspan="2">(a) Penetrate matter (e.g. radiography)
(b) Ionize gases
(c) Cause fluorescence
(d) Cause photoelectric emission from metals
(e) Reflected and diffracted by crystals, enabling ionic lattice spacing and N_A (or wavelength λ) to be measured</td></tr>
<tr><td>X-rays</td><td>(a) Rapid deceleration of fast-moving electrons (e.g. by a tungsten target)
(b) Changes in energy of innermost orbital electrons</td><td></td></tr>
<tr><td>ultra-violet</td><td>Orbital electrons of atoms, as in high voltage gas discharge tubes, the arc, the Sun, and the mercury vapour lamp</td><td>quartz
fluorite (CaF_2)</td><td>(a) Photography
(b) Photoelectric cell
(c) Fluorescence</td><td>(a) Absorbed by glass
(b) Can cause many chemical reactions, e.g. the tanning of the human skin
(c) Ionize atoms in atmosphere, resulting in the ionosphere</td></tr>
<tr><td>visible light</td><td>Re-arrangement of outer orbital electrons in atoms and molecules, e.g. gas discharge tube, incandescent solids and liquids</td><td>glass</td><td>(a) Eye
(b) Photography
(c) Photocell</td><td>(a) Detected by stimulating nerve endings of human retina
(b) Can cause chemical action</td></tr>
<tr><td>infra-red</td><td>(a) Outer electrons in atoms and molecules
(b) Change of molecular rotational and vibrational energies
e.g. incandescent matter</td><td>rock salt</td><td>(a) Photography by special plate
(b) Special heating effect (e.g. radiometer, bolometer, etc.)
(c) Photoconductive cells (e.g. PbS)</td><td>(a) Useful for elucidating molecular structure
(b) Less scattered than visible light by atmospheric particles—useful for haze photography</td></tr>
<tr><td>microwaves</td><td>Special electronic devices such as klystron tube</td><td>paraffin wax</td><td>Valve circuit arranged as microwave receiver
Point-contact diodes
Thermistor bolometers</td><td>(a) Radar communication
(b) Analysis of fine details of molecular and atomic structure
(c) Since $\lambda \sim 3 \times 10^{-2}$ m useful for demonstration of all wave properties on macroscopic scale</td></tr>
<tr><td>radiowaves</td><td>Oscillating electrons in special circuits coupled to radio aerials</td><td></td><td>Tuned oscillatory electric circuit (i.e. radio receiver)</td><td>Different wavelengths find specialized uses in radio communications</td></tr>
</table>

Orders of Magnitude

The *energy* carried by a photon

$$W = hf$$

where h is the *Planck* constant (p. 456).

For u.v. photons

$$W \approx 10^{-18}\ \text{J}$$

For γ-rays

$$W \approx 10^{-14}\ \text{J}$$

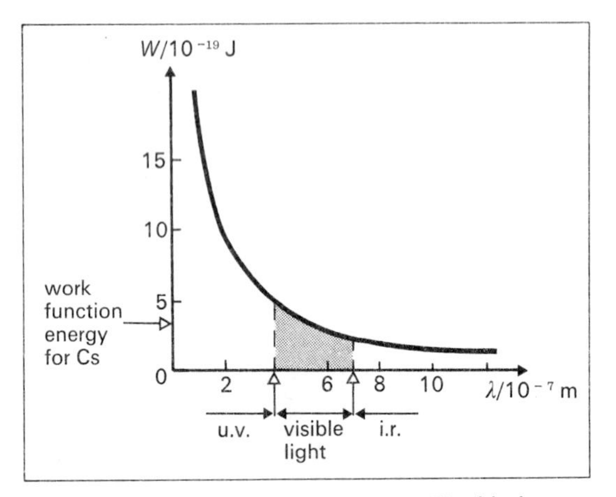

Fig. 37.4. Variation of photon energy W with the wavelength λ of electromagnetic radiation.

Thus a lamp rated at 100 W will emit about 10^{20} photons in each second. (Note that this explains fig. 37.3.) The *duration of emission* of a photon varies from atom to atom, but typically

$$\Delta t \approx 10^{-9}\ \text{s}$$

Since $c = 3 \times 10^8\ \text{m s}^{-1}$, the *length* of the wave-train associated with a typical photon is only

$$\left(3 \times 10^8 \frac{\text{m}}{\text{s}}\right) \times (10^{-9}\ \text{s}) = 3 \times 10^{-1}\ \text{m}$$

Some photons are, however, associated with much longer wave-trains.

* When electromagnetic radiation strikes an electron in a material, the $\boldsymbol{E}$-field and $\boldsymbol{B}$-field associated with the wave exert forces $\boldsymbol{F}_e$ and $\boldsymbol{F}_m$ as shown in fig. 37.5.

The **radiation force** $\boldsymbol{F}_r$ exerted on the material by the e.m. radiation is equal to the magnetic force $\boldsymbol{F}_m$, and so is given by

$$F_r = Bev. \qquad \text{(p.373)}$$

As $B = \frac{E}{c}$, $\quad F_r = \frac{(Ee)v}{c} = \frac{F_e v}{c}$

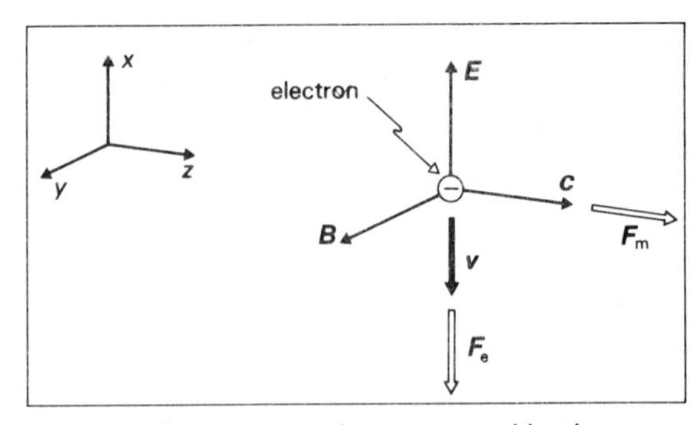

Fig. 37.5. Forces on an electron caused by the incidence of e.m. radiation.

$\boldsymbol{F}_e$ (and not $\boldsymbol{F}_m$) is the force that does work on the electron, so

$$F_r = \frac{P}{c}$$

where P is the power of the wave.

This expression is for total absorption of radiation—if there is complete reflection, the radiation force is twice as great. The **radiation pressure** $p_r = F_r/A$ and so, for complete absorption or reflection,

$$p_r = \frac{I}{c} \quad \text{or} \quad \frac{2I}{c},$$

respectively, where I is the intensity of radiation.

The radiation pressure at the Earth's surface is about 10^{-6} Pa, whereas atmospheric pressure is about 10^5 Pa.

By writing $\quad F_r = \frac{P}{c}$

in the form $\quad \frac{\mathrm{d}p}{\mathrm{d}t} = \frac{\mathrm{d}W}{\mathrm{d}t}\Big/c$

and integrating, we see that the **linear momentum** $\boldsymbol{p}$ carried by a photon becomes

$$p = \frac{W}{c}$$

in the direction of $\boldsymbol{c}$.

As $W = hf$ and $c = f\lambda$, $\boldsymbol{p}$ is also given by

$$p = \frac{h}{\lambda}$$

For a photon of blue light the linear momentum

$$p \approx 10^{-27}\ \text{N s}$$

Wave-Particle Duality

Electromagnetic radiation consists of waves emitted in packets which sometimes appear to behave like particles.

Some properties are better explained by a wave theory (refraction, diffraction, etc.), whereas others are amenable to a particle explanation (the photoelectric effect, p. 455, and the *Compton* effect, p. 463).

The particle aspects of behaviour are more pronounced where the photons carry a large amount of energy (the higher frequencies), and the wave aspects are pronounced where the photons have smaller energies so that the energy flow is more continuous (the lower frequencies).

* De Broglie Waves

In 1924 *de Broglie* suggested that material particles such as electrons might show wave properties. The *de Broglie* wavelength λ of a material particle is given by

$$\lambda = \frac{h}{p} = \frac{h}{mv}$$

where m is the relativistic mass and v the velocity of the particle.

In 1927 *Davisson* and *Germer* confirmed the prediction of the wave nature of particles with their **electron diffraction** experiments. Theirs was a *reflection* experiment, in which a beam of electrons was diffracted by a nickel target in the form of a single crystal. (The arrangement was similar to that shown in fig. 37.6.) The measured wavelength associated with the electrons agreed well with the *de Broglie* wavelength calculated from the accelerating p.d. V.

$$\text{k.e. of electrons } E_k = Ve = \tfrac{1}{2}m_e v^2,$$

and so

$$\lambda = \frac{h}{mv} = \frac{h}{\sqrt{2Vem_e}}$$

Diffraction effects have since been shown for various other particles such as hydrogen atoms, helium atoms and neutrons. Neutron diffraction is an excellent way of studying crystal structures. An electron in its orbit can be replaced by a *de Broglie* wave of wavelength $\lambda = h/(m_e v)$. If each circular orbit round the nucleus is to contain an integral number of waves, then orbits of only certain radii are possible. If r is the radius of the orbit and n is an integer, then

$$n\lambda = 2\pi r,$$

and so

$$m_e vr = n\frac{h}{2\pi}$$

As will be seen later (p. 467) this, *Bohr's* first postulate, was an assumption rather than a wave mechanics derivation.

The angular momentum carried by a photon is $h/2\pi$, being determined by a change in a quantum number of the atom emitting the photon. This ensures that the atom–photon system conserves angular momentum when the photon is emitted.

The waves described are not real, but the particles move in a manner which is better described by a wave equation (*Schrödinger*) than by *Newtonian* mechanics.

37.5 Examples of Wave Properties

Electromagnetic waves exhibit all the wave properties shown on p. 94, but the effects may be difficult to demonstrate. The most striking properties are discussed in chapters **38** to **40**.

Beats

Two electromagnetic waves beating together would have a frequency

$$f_{beat} = (f_2 - f_1)$$

Consider the two D-lines of sodium. Their respective frequencies are

$$f_2 = 5.090 \times 10^{14}\ \text{Hz}$$

and

$$f_1 = 5.085 \times 10^{14}\ \text{Hz}$$

f_{beat} is about 5×10^{11} Hz.

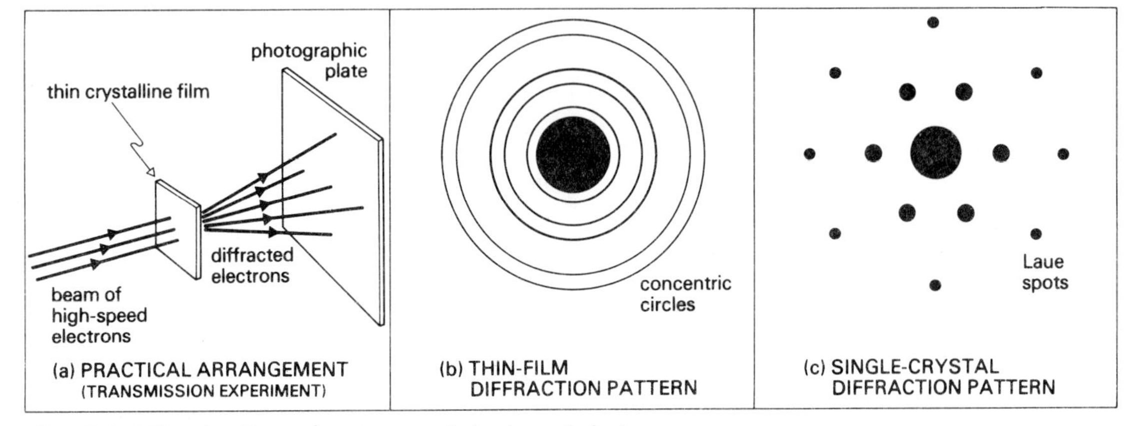

Fig. 37.6. Diffraction illustrating the wave behaviour of electrons.

This effect for visible light obviously cannot be detected using separate sources when one remembers that they will be non-coherent. Nevertheless beats have been observed from a beam superposed on its own reflection from a *moving* mirror. (The *Doppler effect* (p. 112) causes the two beams to have *slightly* different frequencies.)

It is also possible to demonstrate beats from a laser being made to oscillate in two longitudinal modes. The beat frequency of about 150 MHz is easily shown by shining the laser output onto a photodiode, and feeding the output into the aerial socket of a radio.

Stationary Waves

Fig. 37.7 shows how *Wiener* (1889) demonstrated stationary light waves. The emulsion was most exposed where it cut the *electric vector antinodes.*

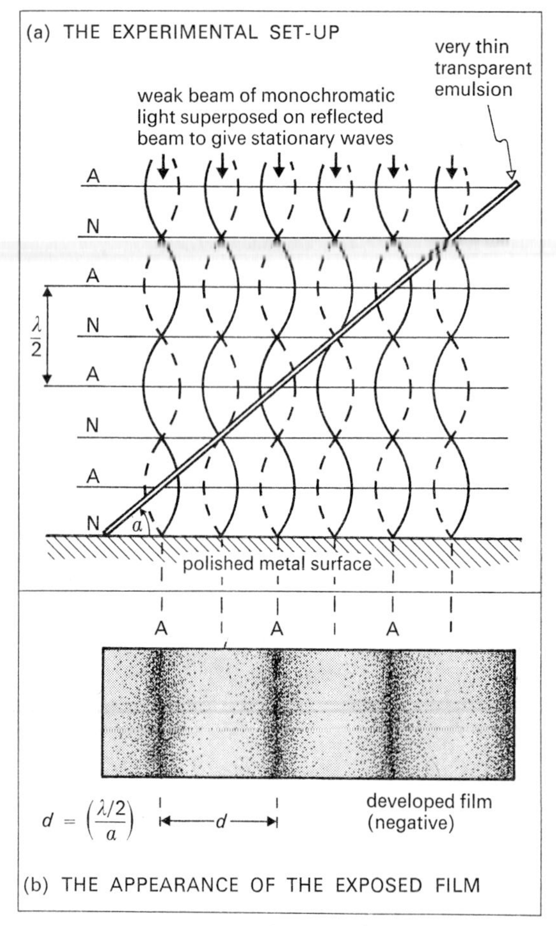

Fig. 37.7. Wiener's experiment to demonstrate stationary light waves.

Orders of Magnitude

If $\lambda \sim 5 \times 10^{-7}$ m

and $\alpha \sim 2 \times 10^{-2}$ rad

then $d \sim 1.5 \times 10^{-5}$ m.

Obviously α needs to be very small.

The thickness of the emulsion was about $\lambda/20 \sim 20$ nm.

Standing microwaves and radiowaves are relatively easy to demonstrate. Fig. 37.8 shows a typical experimental arrangement.

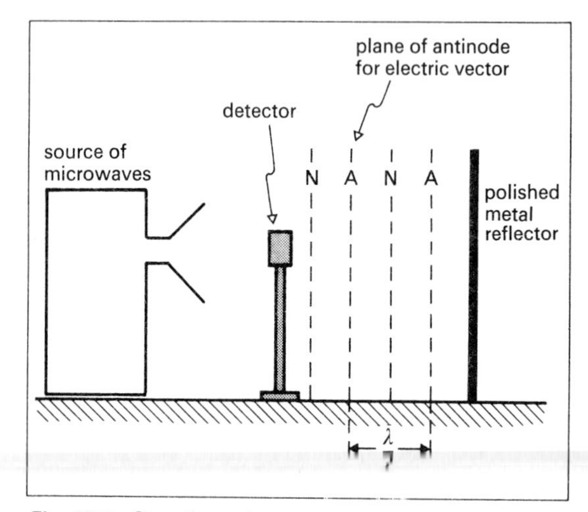

Fig. 37.8. Standing microwaves.

37.6 The Speed c of Electromagnetic Radiation

Importance of a Knowledge of c

(*a*) For confirmation of *Maxwell's* theory of electromagnetic radiation, which had predicted

$$c = \frac{1}{\sqrt{\varepsilon_0 \mu_0}}$$

in which μ_0 has a defined value and ε_0 has a value that must be found by experiment. It can be seen that this equation is dimensionally consistent since

$$[\varepsilon_0] = [\mathrm{M^{-1}L^{-3}T^4I^2}]$$

and

$$[\mu_0] = [\mathrm{MLT^{-2}I^{-2}}]$$

N.B. Nowadays c can be measured more accurately than ε_0, so one measures c, and then calculates ε_0 from the above equation. See also p. 333.

(*b*) Using c = distance/time, one can measure large distances (e.g. in astronomy) by timing radio pulses.

(*c*) The important part played by c in relativistic physics, and hence in modern atomic and nuclear physics. (For example the well-known equation $E = mc^2$.)

Historical review

(1) Roemer's method (1675)
He deduced the relatively large time taken for light to travel across the diameter of the Earth's orbit round the Sun.

(2) Bradley's method (1725)
The orbital velocity of the Earth combines vectorially with the velocity of light from a star, and may cause an aberration between the star's true and apparent direction. The speed of light can be calculated from the necessary change in the orientation of the telescope.

(3) Fizeau's method (1849)
A rotatable *toothed wheel* served two purposes:

(*i*) It chopped, or *modulated*, a continuous beam of light into pulses.

(*ii*) It measured the time interval separating adjacent pulses, and so their transit time.

An echo technique was used so that the pulses covered a known distance in a known time.

(4) Foucault's method (1860)
He used a rotating mirror to measure the speed of light through air, and through water, and discovered that the latter was smaller. This experiment was for a long time considered to be a crucial one for comparing the different explanations offered by the wave and corpuscular theories for refraction (p. 102).

(5) Michelson's method (1926)
This is discussed separately. The modulation was achieved by a rotating *octagonal mirror.*

(6) Anderson's method (1941)
Similar in principle to *Fizeau's* and *Michelson's*. The modulation was achieved by an *electronic shutter* called a *Kerr Cell.*

(7) Recent methods
Measurements made more recently have been based on the standing waves set up in a cavity by short radio waves.

The value of c recommended by the International Union of Pure and Applied Physics is

$$(2.997\,924\,580 \pm 0.000\,000\,012) \times 10^8\ \mathrm{m\ s^{-1}}$$

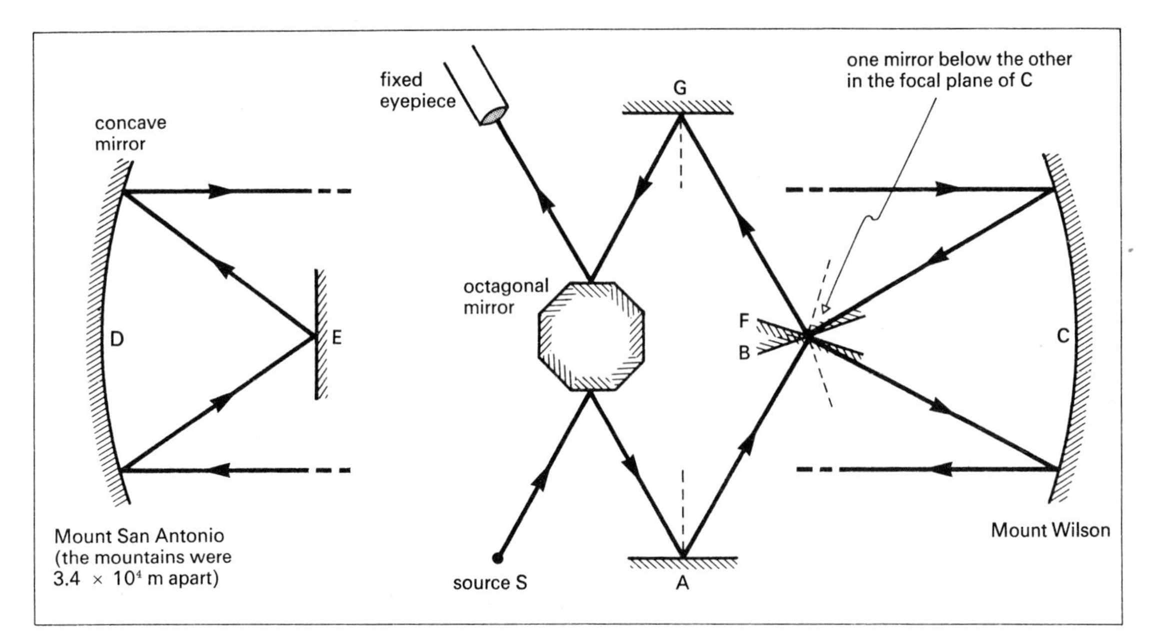

Fig. 37.9. Michelson's method for finding the speed of light.

37.7 Michelson's Method

Refer to fig. 37.9. The light from the source follows this path: S, octagonal mirror, A, B, C, D, E, D, C, F, G, octagonal mirror, eyepiece.

During this time the mirror is made to rotate exactly one eighth of a revolution, so that an image is seen in the eyepiece. For slightly higher or lower rates of rotation, no image will be seen.

Details

(*a*) The rotational frequency of the octagonal mirror (528 revolutions per second) can be measured stroboscopically.

(*b*) The separation of the mountains was measured with an uncertainty of about 1 in 10^7.

Calculation

$$\text{Speed} = \frac{\text{distance travelled}}{\text{time taken}}$$

$$= \frac{\text{twice mountain separation}}{\text{time for mirror to make } \frac{1}{8} \text{ rev.}}$$

38 Interference

38.1 Conditions for Wave Interference to be Visible

By **interference** *we mean the superposition of wave-trains from a finite number of coherent sources.* Two sources are frequently used.

For All Types of Wave Motion

The following conditions must be observed if we are to see interference effects.

(*a*) The sources must have roughly the *same amplitude*.

(*b*) The sources must be **coherent**, i.e. they must have

(*i*) a *constant phase relationship*, which implies

(*ii*) the *same frequency* (and therefore the same wavelength where they are to be superposed).

Sources which have a constant phase relationship are said to be **phase-linked**.

(*c*) Transverse waves must be either unpolarized, or have significant resolved parts in the same plane.

When the Conditions are not Observed

(*a*) *Amplitudes unequal.* Fig. 38.1 shows the effect on the intensity-location curve. Cancellation is not complete, and the *contrast* between constructive and destructive superposition may not be detected easily.

(*b*) *A time-dependent phase difference* between the sources would result in a varying phase difference between the superposed wavetrains. The fringe pattern would be constantly moving, perhaps too rapidly for detection by an eye (whose retinal image persists for at least 0.05 s).

(*c*) *Polarization*: vectors must lie along the same direction if they are to be able to cancel one another.

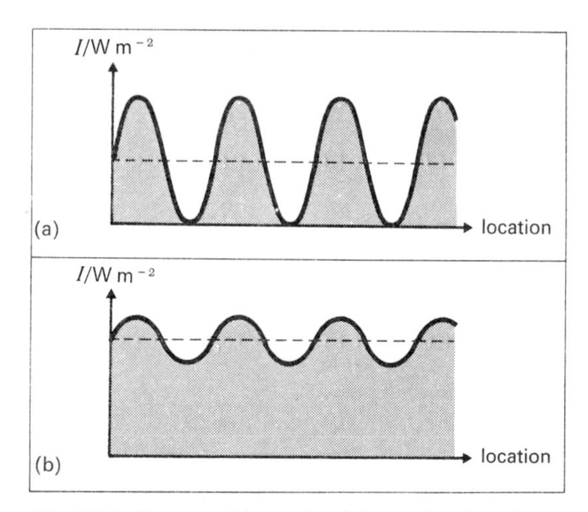

Fig. 38.1. Superposition using (a) equal and (b) unequal wave amplitudes.

Visible Light Wavetrains

Refer to the comments on the quantum nature of light (p. 261). Two coherent visible light wavetrains can be produced by deriving them either from a single slit source, or from a laser.

Fig. 38.2 shows a way of visualizing a photon emitted by a typical light source. The difference

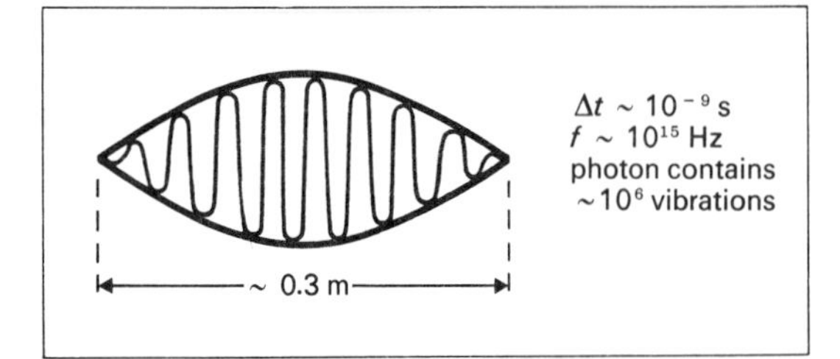

Fig. 38.2. A way of visualizing a typical photon.

between the paths travelled by two wavetrains must not exceed the length of such a photon, or else the two wavetrains will not overlap. The twin sources (derived from the single source) are usually placed close together. (If they are not, the fringe width may be too small to measure.)

Reflection

There is a *phase change* of π rad when the light wave is reflected at a rare-dense interface (such as that between air and glass). This is illustrated by fig. 38.3.

This behaviour resembles that of a transverse wave on a string (p. 94).

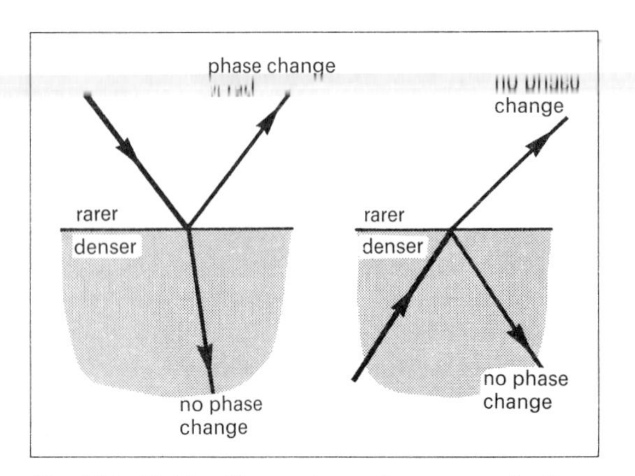

Fig. 38.3. Light suffers a phase change on reflection at a denser medium.

Brightness and the Electric Vector

Brightness seen by the eye corresponds to maximum intensity (rate of arrival of energy per unit area). For visible light we see

(*a*) *bright fringes* where the *amplitude* of the electric field vector is large (its magnitude but not its direction is important)

(*b*) *dark fringes* where the electric field vector is *permanently* smaller than at adjoining places.

(We discuss the electric vector for convenience: equal energy is associated with the electric and magnetic fields of the electromagnetic wave.)

38.2 Optical Path

When light of frequency f and free-space wavelength λ_0 crosses into a medium, f does not change, but λ_n becomes (λ_0/n) where n is the absolute refractive index of the medium.

The number of waves contained in a distance l is l/λ, where λ refers to the medium concerned (fig. 38.4).

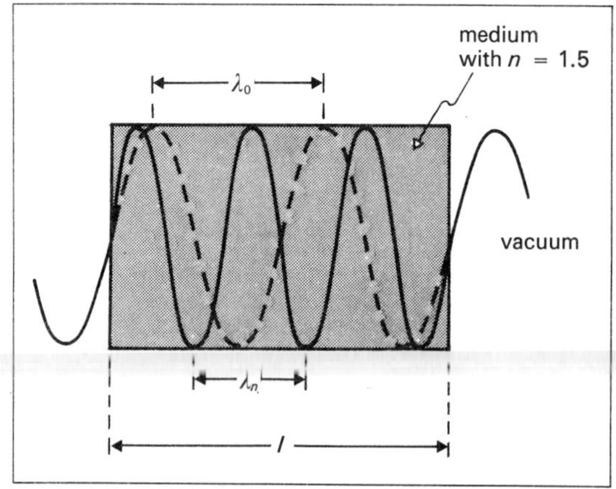

Fig. 38.4. The effect of a denser medium ($n = 1.5$) on the wavelength of light: three wavelengths fit into a space previously occupied by two.

In a length l in a vacuum, there are (lf/c) waves.
In a length l in a medium, there are

$$\frac{lf}{c_n} = \frac{lf}{(c/n)} = n\left(\frac{lf}{c}\right) \text{ waves}$$

The same number of waves fit into a length nl in a vacuum as into a length l in the medium. *nl is called the* **optical path** for that medium.

The total optical path for a wave traversing several media $= \sum nl$: it is this quantity that determines the phase of a wave at any point, since

$$(\text{phase difference}) = \left(\frac{2\pi}{\lambda}\right) \cdot \left(\begin{matrix}\text{total optical} \\ \text{path difference}\end{matrix}\right)$$

38.3 Classification of Fringe Systems

(a) Classification by Fringe Location

(i) Localized fringes

These have a position on which the eye or microscope must be focussed if they are to be seen. Fig. 38.5 gives some examples.

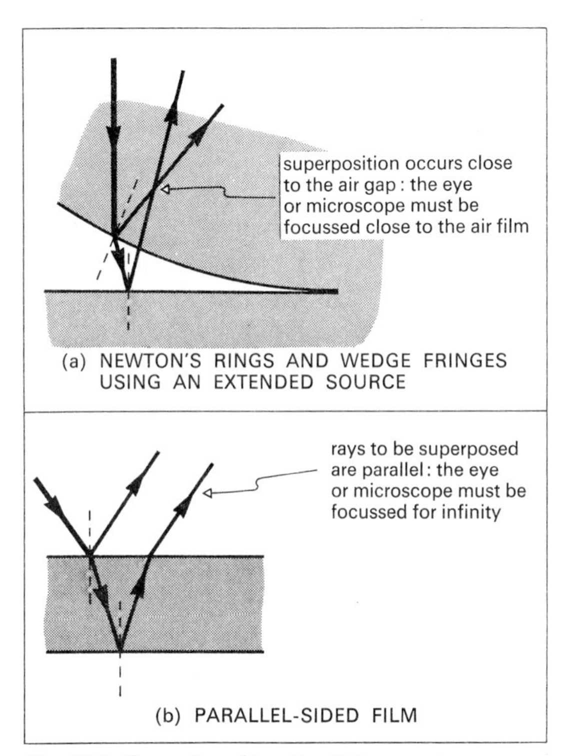

Fig. 38.5. Examples of localized fringe systems.

(ii) Non-Localized Fringes

These can be picked up even if one alters the location of the screen in a two-source experiment such as *Young's*, *Fresnel's* or *Lloyd's*. The fringes are non-localized in the sense that they can be seen at the intersection of the loci of antinodal lines (or surfaces) with an *arbitrary* screen (fig. 38.6).

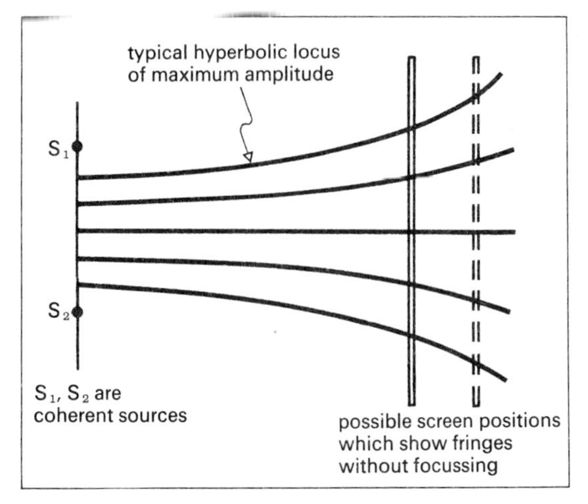

Fig. 38.6. Non-localized fringes in a two-source interference experiment.

(b) Classification by Origin of Wavefronts

(i) Division of Wavefront

In a typical two-source experiment the superposed wavetrains are obtained by dividing the wavefront given out by a single source (fig. 38.7).

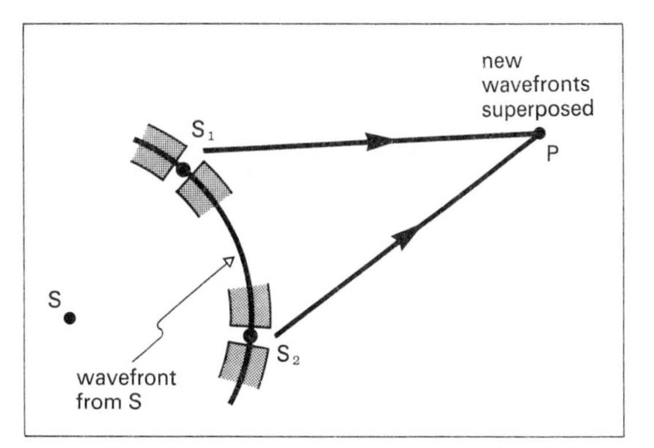

Fig. 38.7. Dividing a wavefront to obtain two coherent light sources, S_1 and S_2 (where usually $S_1S_2 \ll S_1P$ or S_2P).

(ii) Division of Amplitude

In *Newton's* rings, wedge fringes and the effects obtained from a parallel-sided film, the incident wavefront is divided into two wavetrains by partial reflection and partial transmission. The new wavefronts each take part of the amplitude (and energy) of the incident wave.

38.4 Young's Double-slit Experiment

The fringe system obtained by *Young* (1801) gave positive evidence to support the theory of the wave nature of light. This experiment (and other pioneer experiments such as the *Newton's* rings experiment) were originally performed using *white* light.

The apparatus is shown in fig. 38.8 overleaf.

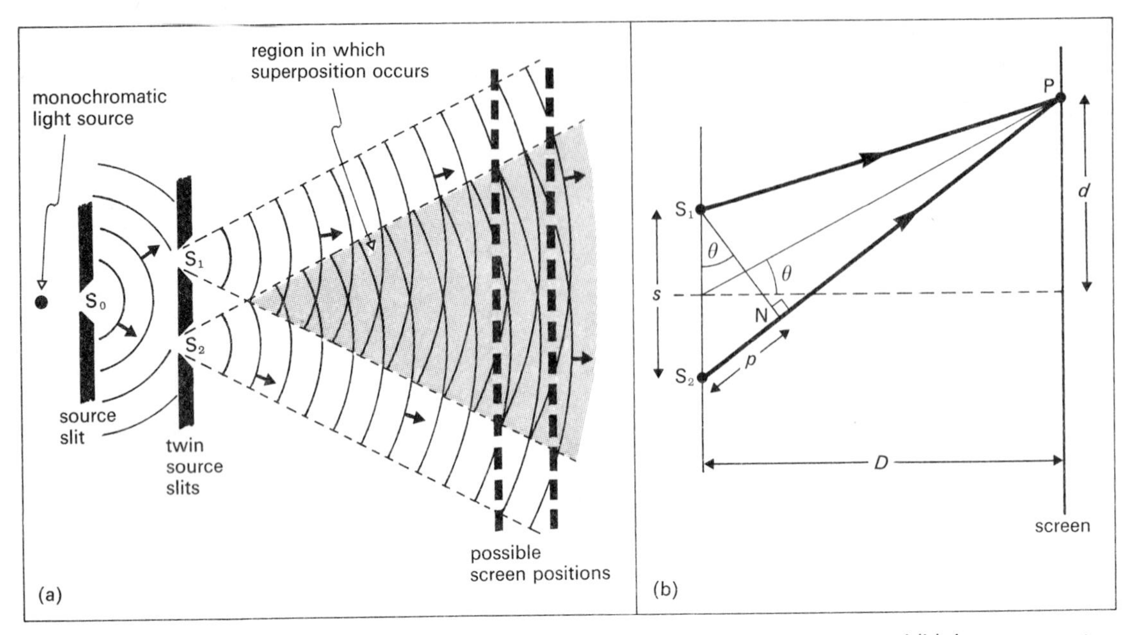

Fig. 38.8 The scheme of a Young's experiment, showing (a) why superposition occurs, and (b) the geometry to calculate the fringe separation. (In this and succeeding diagrams, the wavelength of light is grossly exaggerated. Since, in practice, $D \gg d$, S_1P can be considered parallel to S_2P.)

Light from the monochromatic source is diffracted at S_0 so as to illuminate S_1 and S_2 *coherently*. *Diffraction causes them to emit spreading coherent wave-trains* which overlap *provided S_1 and S_2 are narrow enough.*

Calculation of Path Difference

In fig. 38.8*b*

$$\sin\theta = \frac{p}{s}$$

where p is the path difference ($S_2P - S_1P$) for waves that are to be superposed at P. But

$$\tan\theta = d/D.$$

If $\theta < 1/10$ rad, then $\tan\theta \approx \sin\theta$

$$\frac{d}{D} = \frac{p}{s}$$

so

$$p = \frac{ds}{D}$$

This is a purely geometrical relationship.

Condition for Bright Fringes

When

$$p = \frac{ds}{D} = m\lambda$$

the screen will show positions of reinforcement: these will appear as fringes parallel to the slits (fig. 38.9). (m is any integer.)

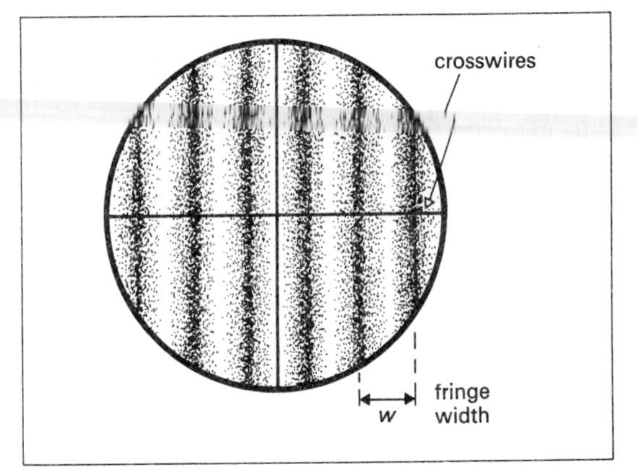

Fig. 38.9. View through an eyepiece placed at P in fig. 38.8.

Thus

$$d = \frac{m\lambda D}{s} \text{ for brightness}$$

The fringe width w is given by

$$d_{m+1} - d_m = \frac{\lambda D}{s}$$

$$\boxed{w = \frac{\lambda D}{s}}$$

Note explicitly that the fringe width is directly proportional to λ and D, but inversely proportional to s.

Condition for Dark Fringes

Destructive superposition occurs where

$$p = \frac{ds}{D} = (m + \tfrac{1}{2})\lambda,$$

so that the bright and dark fringes show the same spacing.

Measurement of Wavelength λ

These orders of magnitude show why we can justify

$$\tan\theta \approx \sin\theta$$

If

$$\lambda \sim 5 \times 10^{-7}\ \text{m},$$
$$s \sim 1 \times 10^{-3}\ \text{m} = 1\ \text{mm}$$
$$D \sim 1\ \text{m}$$

then the fringe width would be about

$$5 \times 10^{-4}\ \text{m} = 0.5\ \text{mm}.$$

Measurement by travelling microscope of s and the fringe width enables λ to be calculated. The fringe width is measured by traversing ten or more fringes using a backlash-free screw.

The observer sees a field-of-view like fig. 38.9. To a first approximation the fringes are equally spaced.

The illumination varies across the field of view as shown in fig. 14.6, page 108.

*38.5 Other Two-source Experiments

The geometry of methods (a) and (b) below is the same as that for *Young's* method. The result

$$\text{fringe width } w = \frac{\lambda D}{s}$$

can be applied.

(a) Lloyd's Mirror

The two sources are the slit S (parallel to the mirror) and its virtual image S′ (fig. 38.10).

Notes. (i) If the screen is moved so that the point O touches the edge of the glass plate, the *geometrical* path difference for the two wavetrains is zero. The phase change of π rad on reflection at the denser medium then causes a *dark* fringe to be formed.

(ii) This interference pattern is frequently seen in a ripple tank when one uses a wavetrain to demonstrate the law of reflection.

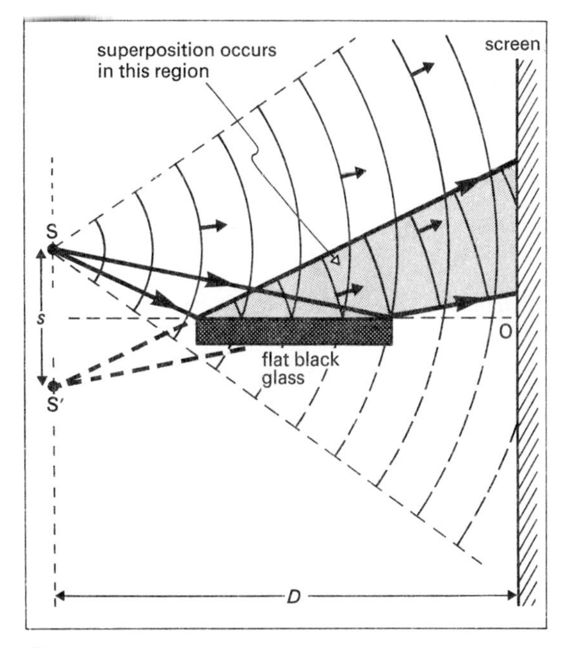

Fig. 38.10. Lloyd's mirror arrangement.

(iii) Note the *real* waves propagating as though coming from the *virtual* source S′.

(b) Fresnel's biprism

The two sources are S_1 and S_2 (both being virtual images of the source slit S_0) formed by refraction in the two halves of the double prism (fig. 38.11).

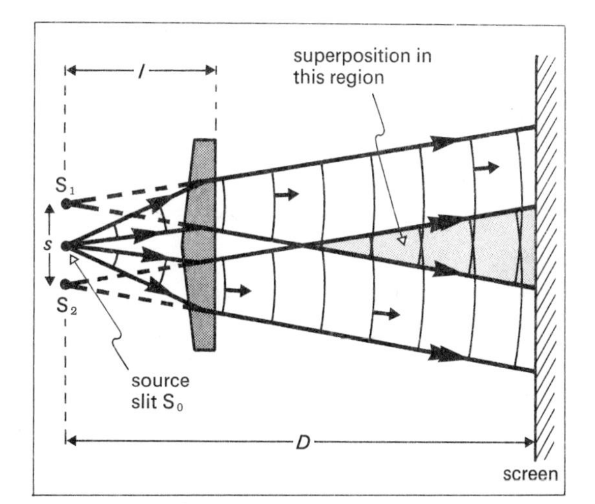

Fig. 38.11. Fresnel's biprism.

The fringes (which are brighter than those from double-slits) are observed using a travelling eyepiece in the plane of the screen.

Measurement of s

(1) Introduce a converging lens between prism and screen. Focus a real image of the two virtual sources. Note their separation s_1, and the object and image distances a and b.

(2) Repeat with the lens in the conjugate position, and note the new measurements s_2, b and a respectively.

From (1)
$$\frac{s}{s_1} = \frac{b}{a},$$

and from (2)
$$\frac{s}{s_2} = \frac{a}{b}$$

so
$$s = \sqrt{s_1 s_2}$$

Alternatively one can measure the angular separation θ of the virtual images (using parallel light from a spectrometer), and then apply $s = l\theta$ (fig. 38.11).

(c) Pohl's Interferometer

In fig. 38.12 the point P lies on a particular fringe (bright or dark) if

$$(S_1P - S_2P) = \text{a constant}$$

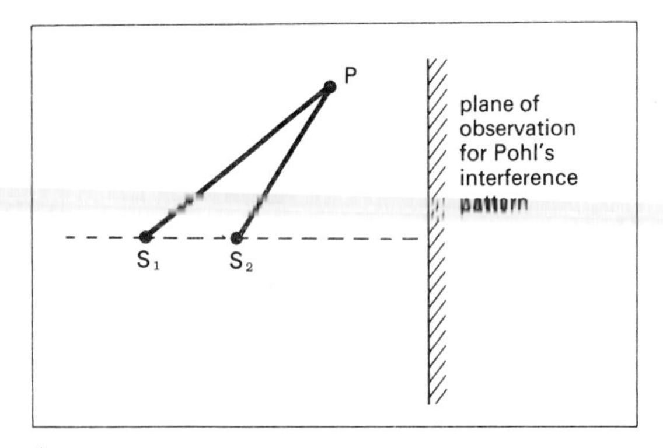

Fig. 38.12. Discussion of locus of a particular fringe.

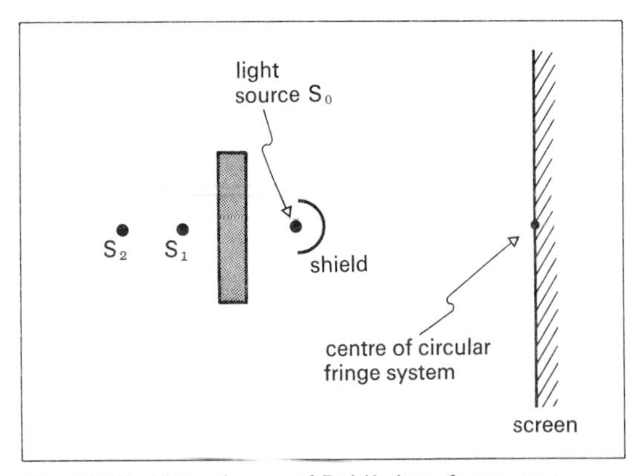

Fig. 38.13. The scheme of Pohl's interferometer.

The fringe system consists of a series of surfaces each being a hyperboloid of revolution about S_1S_2 as axis. A plane placed perpendicular to S_1S_2 intersects these surfaces in a series of *concentric circles.*

In *Pohl's* interferometer the two sources are the virtual images of a light source, formed by reflection at the two surfaces of a mica sheet about 10^{-5} m thick. A series of large concentric fringes can be produced on the screen of fig. 38.13.

38.6 White Light Interference

In a two-source experiment like *Young's*, the fringe width $w \propto \lambda$.

When white light is used, different colours produce their own fringe systems which overlap (fig. 38.14). At P the eye sees the overlapped fringes, which give general illumination, but analysis by a spectroscope would reveal that some colours had suffered destructive superposition.

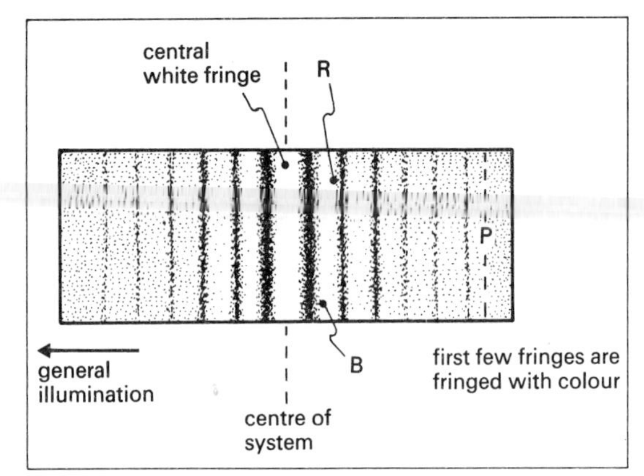

Fig. 38.14. Appearance at the centre of a Young's fringe system using white light.

The central white fringe is the only location at which all colours show *zero optical path difference*: it is invaluable for locating the centre of a fringe system, and so for measuring shifts of the whole system.

Shifting a System of Fringes

The central fringe of a ripple tank pattern is moved by making one source show a constant phase shift (say π rad) relative to the other. The same effect can be seen in a *Young's* two-slit experiment by introducing a different *optical* path length for one wavetrain while

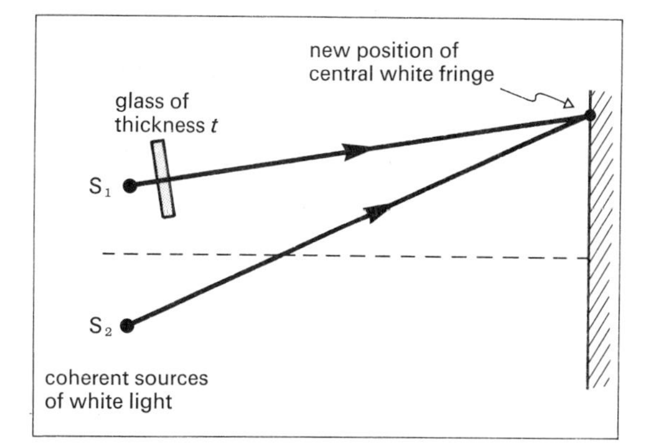

Fig. 38.15. Shifting a fringe system.

leaving the geometrical arrangement unchanged (fig. 38.15).

The optical path through the glass $= nt$

The *extra* optical path introduced $= nt - t$

$= (n - 1)t.$

This will cause the zero of the system to shift by

$$(n - 1)t\frac{D}{s}$$

since

$$d = \frac{pD}{s} \qquad \text{(p. 272)}$$

Measurement of this shift enables either n or t to be calculated.

*38.7 Parallel-Sided Thin Films

Fig. 38.16 shows an arbitrary ray of light incident on a film, and undergoing partial reflections and transmissions. We want to calculate the optical path difference for rays that travel

(*i*) along the path AE in air, and
(*ii*) along the path ABCD through the film.

The parallel wavefronts formed by partial reflection will be coherent, and so will give interference fringes when brought together in the focal plane of the eye or some other converging lens.

The phase change of π rad at A is equivalent to an effective path difference $\lambda/2$.

The net optical path difference is then

$$n(\text{AB} + \text{BC}) + \frac{\lambda}{2} \qquad \text{since} \quad \text{AE} = n(\text{CD})$$

By geometry

$$(\text{AB} + \text{BC}) = \text{A'C} = 2t\cos\theta_n$$

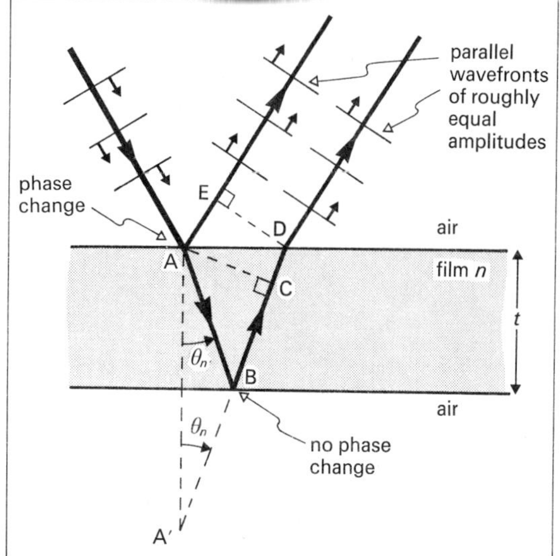

Fig. 38.16. Calculation of path difference for thin film interference using reflected light. (N.B. Light transmitted through the film has not been shown.)

The net optical path difference is

$$2nt\cos\theta_n + \frac{\lambda}{2}$$

Constructive superposition occurs where

$$2nt\cos\theta_n + \frac{\lambda}{2} = m\lambda$$

and destructive superposition where

$$2nt\cos\theta_n + \frac{\lambda}{2} = (m + \tfrac{1}{2})\lambda$$

$$2nt\cos\theta_n = m\lambda$$

Observation of Fringe Systems

(a) Using a Point Source

In fig. 38.17 the eye is focused to receive *parallel* rays reflected from the film surfaces. The image of the point source formed on the retina may be dark or bright, depending on θ_a (which controls θ_n).

Any interference pattern is only revealed *when the eye moves* so as to sample (in turn) a range of values for θ_a.

(b) Fringes using an Extended Source

If the source is extended, then an eye (or lens) at a given point receives rays leaving the film over a range of values θ_a (fig. 38.18). Because the film has parallel sides, $t =$ *constant*. A particular dark fringe is seen along a direction where

$$2nt\cos\theta_n = m\lambda$$

so

$$\cos\theta_n = \text{constant}$$

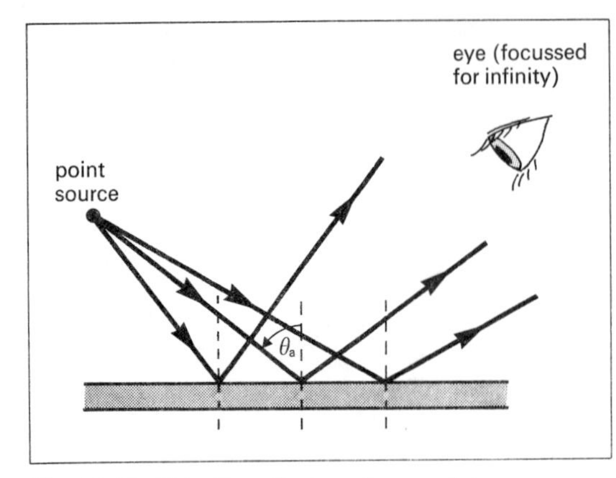

Fig. 38.17. Thin film effects using a point monochromatic source.

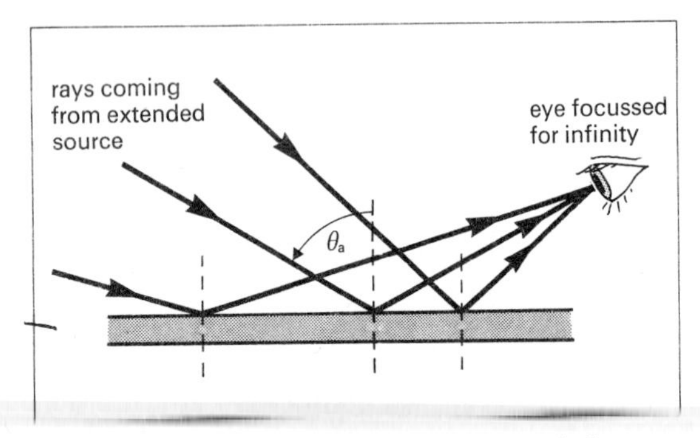

Fig. 38.18. Thin film effects using an extended monochromatic source.

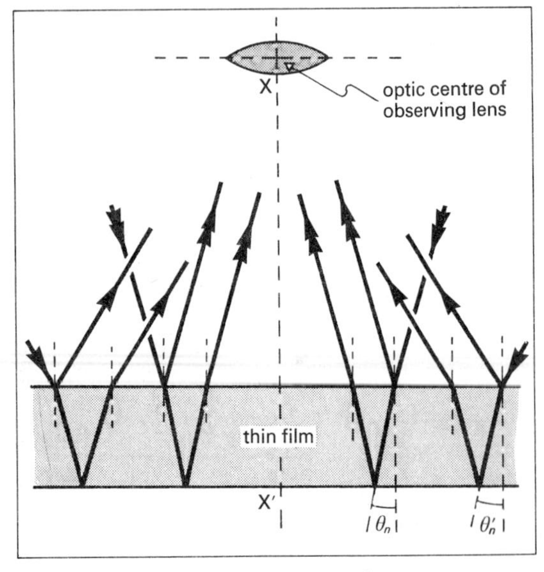

Fig. 38.19. Fringes of equal inclination appear as circles about XX' as axis.

The fringes are called **fringes of equal inclination**: fig. 38.19 shows that they will appear as concentric circles about XX' as axis in the focal plane of the lens used to superpose the parallel rays. The fringes alter position as this lens is moved.

Miscellaneous Effects

(a) Colours of Thin Films

Suppose that t is small, and that the incident light is white. Reinforcement occurs along directions satisfying

$$2nt \cos \theta_n + \frac{\lambda}{2} = m\lambda$$

When t is small one or two colours may reinforce along a direction in which others cancel. The film will then display brilliant colours, as does oil on water.

(b) Thick Films

When t is large, several values of λ satisfy the equation for a given value of θ_n. The film will appear to be generally illuminated.

(c) Blooming

Refer to fig. 38.20, which shows a coated lens.

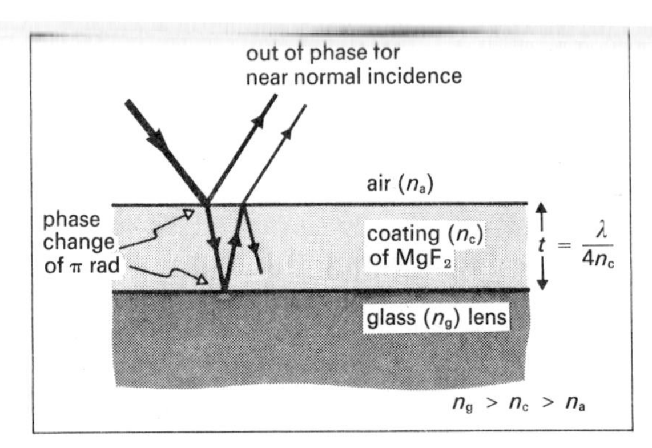

Fig. 38.20. Coating a lens can reduce wasted energy.

Suppose $n_c t = \lambda/4$ where λ is the free-space wavelength for green light (chosen because it lies in the middle of the visible spectrum).

The reflected light undergoes destructive superposition because the net optical path difference is $\lambda/2$ for near normal incidence. The wasted reflected energy is reduced if the refractive index of the coating material is correctly chosen.

The lens appears purple, because the reflected light is deficient in green.

*38.8 Wedge Fringes

In this paragraph we again apply

$$2nt\cos\theta_n + \frac{\lambda}{2} = m\lambda$$

as the condition for brightness, but we consider *near-normal incidence*, so that

$$\theta_a = \theta_n = 0$$

A given fringe must then satisfy

$$t = \text{constant.}$$

The fringes are called **contour fringes**, or **fringes of equal thickness**.

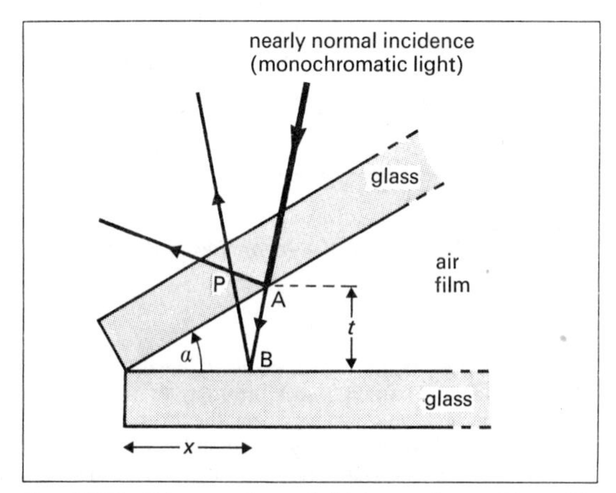

Fig. 38.21. Fringes of equal thickness formed in a thin air wedge ($\alpha \approx 10^{-4}$ rad, and is exaggerated).

In fig. 38.21, α is so small that refraction has been ignored. A phase change occurs at B, but not at A. The net optical path difference between rays superposed at P is

$$2t + \frac{\lambda}{2}$$

For bright fringes at P

$$2t + \frac{\lambda}{2} = m\lambda$$

in which $$t = \alpha x$$

so $$x = (m - \tfrac{1}{2})\frac{\lambda}{2\alpha}$$

The fringe separation is given by

$$x_{m+1} - x_m = \frac{\lambda}{2\alpha}$$

A Draining Soap Film

If the film drains uniformly, it will show horizontal bands in monochromatic light. Where $t \ll \lambda$ (near the top of the film), destructive superposition occurs for reflected light (because of the phase change). The film appears dark. By transmitted light it would appear bright.

Location of the Fringes

(a) Using a point source

The two surfaces each produce reflected wavefronts: their superposition could occur anywhere above the wedge, so the fringes would be *non-localized*.

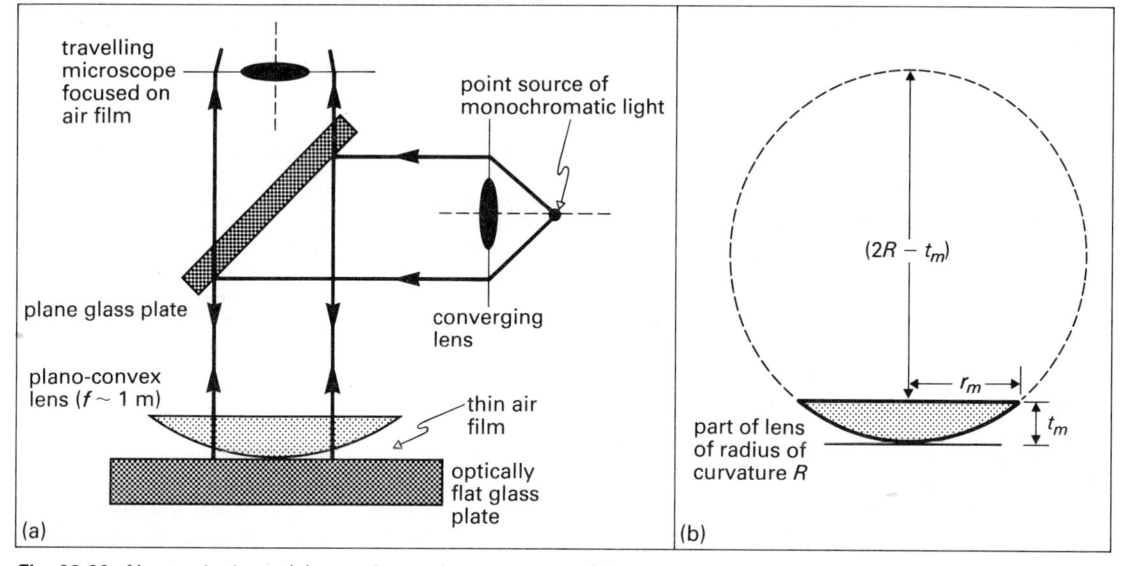

Fig. 38.22. Newton's rings: (a) experimental arrangement, (b) geometry.

(b) Using an extended source

Each point of the source produces its own non-localized fringe system. The only place where the systems coincide for all such point sources is close to the plane of the wedge: *it is here that the fringes become localized.*

Use of Wedge Fringes

Using an experimental arrangement similar to that of fig. 38.21, one sees a series of parallel equally spaced straight fringes.

(*a*) Measurement of their separation allows α and t to be calculated. This method is adapted in *Fizeau's apparatus* for measuring distance changes of about 10^{-7} m.

(*b*) The regularity of the fringe system is a useful guide for the manufacture of an **optically flat** surface.

*38.9 Newton's Rings

Newton's rings represent a system of contour fringes with radial symmetry (fig. 38.22).

Following paragraph **38.7**, the path difference for an air film thickness t_m is

$$2t_m + \frac{\lambda}{2}$$

For the mth dark fringe

$$2t_m + \frac{\lambda}{2} = (m + \tfrac{1}{2})\lambda$$

or

$$2t_m = m\lambda$$

From the intersecting chords of fig. 38.22*b*,

$$r_m^2 = (2R - t_m)t_m$$

so

$$r_m^2 \approx 2Rt_m \qquad \text{since} \quad t_m \ll R$$

Then dark fringes occur where

$$\frac{r_m^2}{R} = m\lambda$$

When m is constant, r_m will be constant: the fringe system consists of unequally spaced concentric circles.

Since

$$2r_m \frac{dr_m}{dm} = \lambda R$$

$$\frac{dr_m}{dm} = \frac{\lambda R}{2r_m}$$

As r_m increases, the circles are closer together.

Notes.

(*a*) For fringes viewed by *reflected* light, the *centre* of the fringe system is *dark* provided there is true contact between lens and plate. (A dust speck of thickness $\lambda/4$ could cause a central bright fringe.)

(*b*) Fringes formed by *transmitted light* are complementary (bright and dark exchange positions): the central fringe is bright. There is less contrast than in (*a*).

Measurement of λ

Suppose the mth ring has *diameter* d_m. Then

$$\frac{r_m^2}{R} = m\lambda$$

can be written

$$d_m^2 = (4R\lambda)m$$

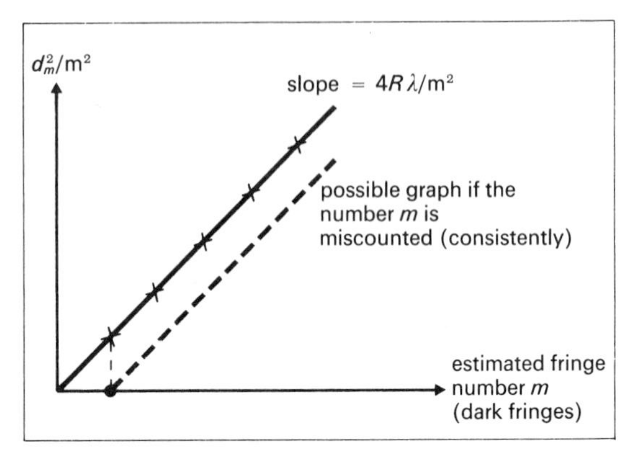

Fig. 38.23. Calculation of λ from Newton's rings measurements.

A graph (fig. 38.23) which plots d_m^2 against m has a gradient ($4R\lambda$). The graph goes through the origin only if there is perfect contact between the lens and plate. Nevertheless the *slope* of the graph enables λ to be found, and the graph *will* be a straight line if the fringe number m is counted *consistently*.

d_m is measured by travelling microscope.

R is measured by a separate experiment.

*38.10 Summary of Thin Film Effects

$2nt \cos \theta_n + (\lambda/2) = m\lambda$ is the condition for brightness.

Films of constant thickness	*Films of varying thickness*
Fringes are formed where θ_n = const., and are called fringes of equal inclination.	Fringes are formed where t = const., and are called contour fringes.
θ_n varies from fringe to fringe.	In this book we assume near-normal incidence, i.e. $\theta_n = 0$.
Examples Colour of oil films Blooming of lenses	*Examples* Wedge fringes (e.g. a draining soap film). *Newton's rings*

39 Diffraction

39.1 The Phenomenon of Diffraction

If, when using *Huygens's* construction (p.100) we assume the secondary wavelets to be effective only over the surface of tangency, then *we predict rectilinear propagation.* By introducing a factor (1 + cos θ), where θ is the angle between the direction of wave travel and some other direction, we recognize that the amplitude is finite along other directions. We are thereby led to the possibility (within one medium) of a *change in direction* of the wave energy.

> **Diffraction** *is the result of superposing (light) waves from coherent sources on the same wavefront after the wavefront has been distorted by some obstacle.*

By implication we superpose an *infinite* number of *infinitesimal* wavelets.

Typical diffracting obstacles include

(*a*) lens edges and mirrors,
(*b*) stops in optical instruments,
(*c*) the source slit of a spectrometer.

These objects will be assumed to be *passive.* (An object which absorbs and re-radiates incident energy is said to **scatter** the light.)

The effects of diffraction become more obvious when the linear dimensions of the obstacle approach the wavelength of the wave motion. Most diffraction effects can only be observed with *small sources.* Each point on a large source produces its own diffraction pattern, but when these patterns overlap, only general illumination can be seen.

The Principle of Diffraction Calculations

In fig. 39.1 the amplitude at P is found by superposing wavelets originating in the gap. It differs from

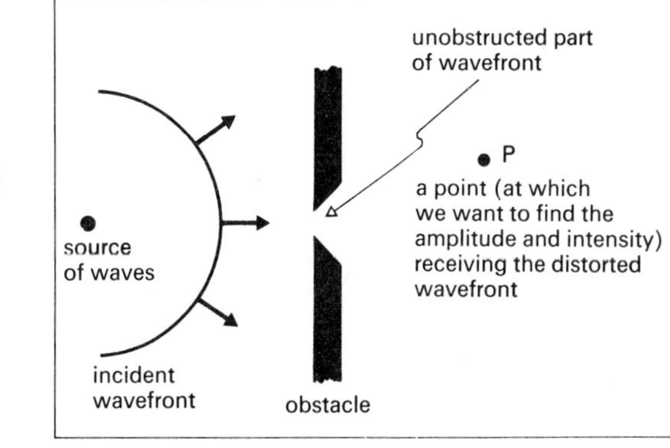

Fig. 39.1. Scheme for calculating the effects of a diffracting obstacle.

what would be observed in the absence of the obstacle

(*a*) because the obstacle removes some secondary wavelets from the wavefront, and

(*b*) because the secondary wavelets originate at differing distances from P: this

(*i*) affects their relative amplitudes, and
(*ii*) controls their phase relationships.

In practice this calculation may become complex mathematically.

Classes of Diffraction

(*a*) **Fresnel diffraction** (fig. 39.2*a*) concerns *non-plane wavefronts*, so that the source S and/or the point P are to be a finite distance from the diffracting obstacle. Experimentally this is very simple, but analysis proves to be very complex.

(*b*) **Fraunhöfer diffraction** (fig. 39.2*b*) deals with *wavefronts* that are *plane* on arrival, and an effective viewing distance of *infinity*. Experimentally this is achieved by using lenses (fig. 39.2c). Mathematically the problem is simplified because we handle parallel rays.

Notes.

(*i*) It follows that *Fraunhöfer* diffraction is an important special case of *Fresnel* diffraction.

(*ii*) When doing quantitative work on *Young's experiment* (p. 272), we assume the screen to be relatively distant, i.e. that we have *Fraunhöfer* conditions.

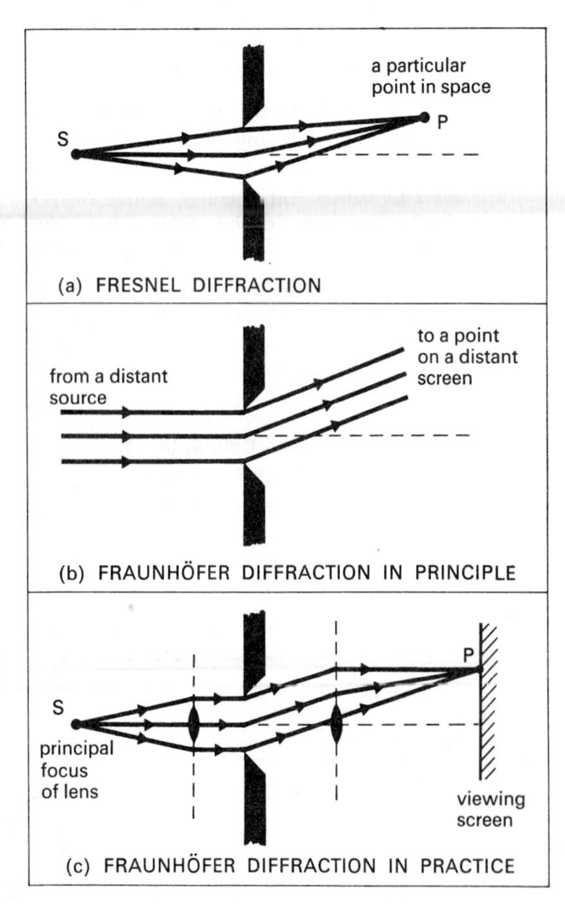

Fig. 39.2. A way of classifying diffraction.

39.2 The Diffraction Grating

An optical transmission diffraction grating consists of a large number of fine equidistant parallel lines ruled onto a transparent plate (such as glass) so that light can pass between the lines, but not through them.

If each gap is sufficiently narrow (relative to λ), *diffraction* causes it to act as a centre of secondary disturbance. Nearly semicircular wavefronts (in section) are superposed on the far side of the grating (fig. 39.3).

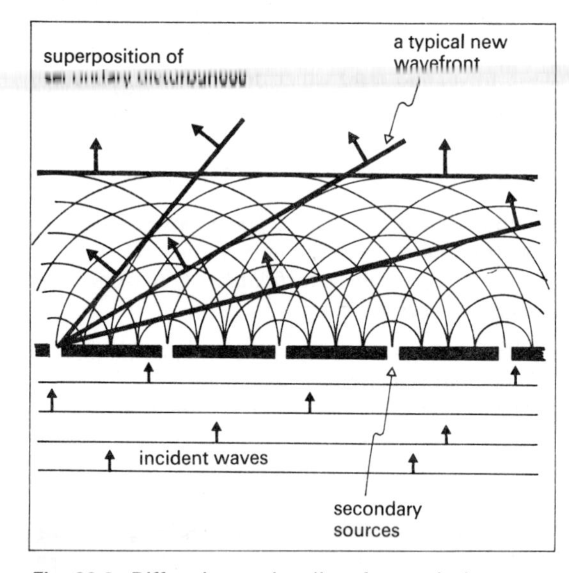

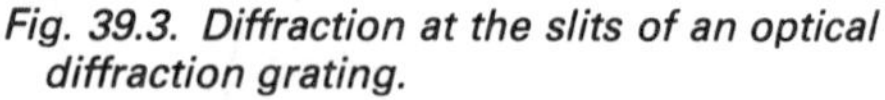
Fig. 39.3. Diffraction at the slits of an optical diffraction grating.

Along a few specified *directions* (rather than at some points) these secondary wavelets combine to produce a continuous wavefront. Along others their superposition is destructive, and no wavefront exists.

Fig. 39.4 shows rays of path difference p being brought to a point P by a lens. (All rays have the same

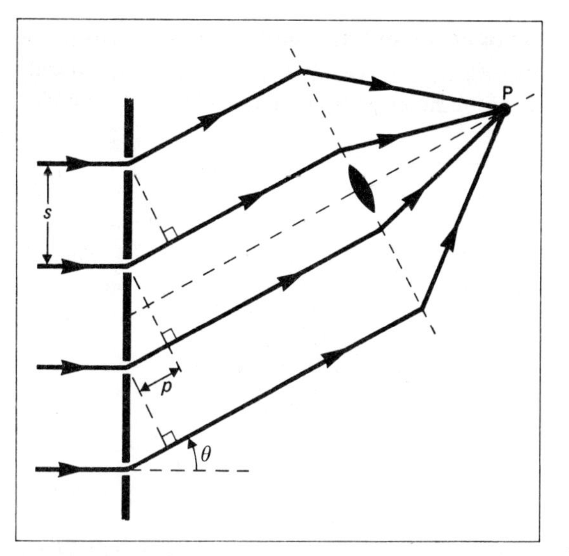

Fig. 39.4. Superposition of secondary wavelets brought together in the focal plane of a lens.

optical path through the lens, so their phase relationships are unaltered.)

The superposition will be *constructive* (and result in brightness) when

$$p = m\lambda$$

where m is an integer, i.e. when

$$\boxed{s \sin \theta_m = m\lambda}$$

where s, the separation of the centres of neighbouring gaps, is called the **grating spacing**.

Example

Suppose a grating has 5×10^5 lines per metre, so that $s = 2 \times 10^{-6}$ m, and that monochromatic light of wavelength 6×10^{-7} m is used. Then when

$$m = 0, \quad \theta_0 = 0°$$
$$m = 1, \quad \theta_1 = 17°$$
$$m = 2, \quad \theta_2 = 37°$$
$$m = 3, \quad \theta_3 = 64°$$

but $\quad m = 4$ predicts $\sin \theta_4 = 1.2$

so there is no value θ_4 or above.

This grating would show 7 *directions* of brightness (fig. 39.5). Along all other directions the secondary wavelets undergo destructive superposition, and effectively there would be darkness.

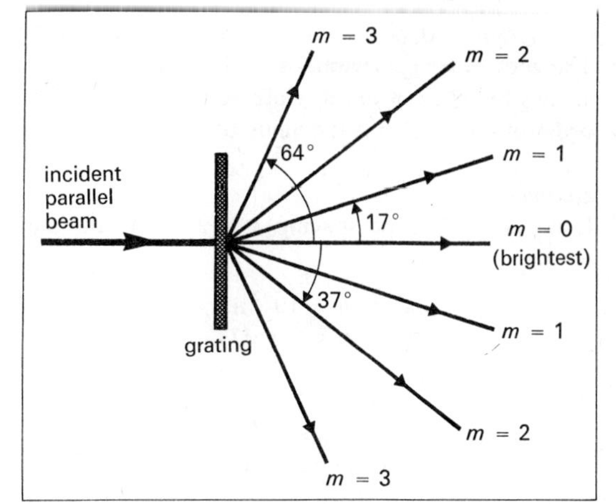

Fig. 39.5. Directions of reinforcement for a grating of $s = 2 \times 10^{-6}$ m using light of $\lambda = 6 \times 10^{-7}$ m.

39.3 Grating Spectra

Since

$$\sin \theta_m = \frac{m\lambda}{s}$$

$$\sin \theta \propto \lambda$$

for a given value of m.

If white light is used, different *wavelengths* reinforce along different directions (fig. 39.6), and so a *spectrum* is formed in the focal plane of the focusing lens.

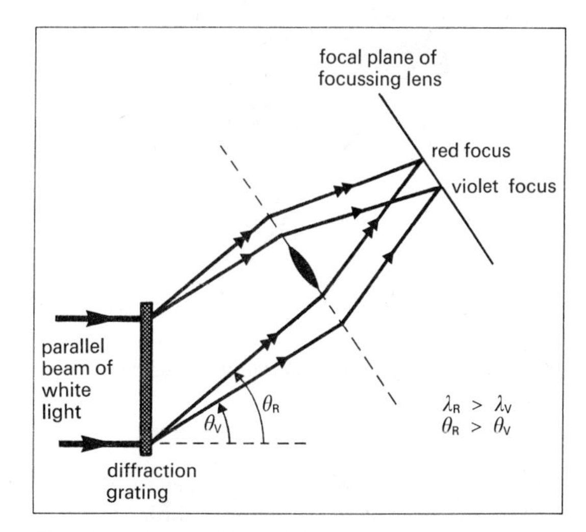

Fig. 39.6. A diffraction grating separates colours according to wavelength.

When $m = 0$, all colours reinforce along $\theta_0 = 0°$. The **zero-order spectrum**, or central maximum, will always be white when a white source is used. The **order** of a spectrum is the value of m.

Example
Using a grating of spacing $s = 2 \times 10^{-6}$ m, and taking

$$\lambda_V = 4 \times 10^{-7}\text{ m}$$
$$\lambda_R = 7.5 \times 10^{-7}\text{ m},$$

we have

$$(\theta_1)_V = 12°$$
$$(\theta_1)_R = 22°$$

The angular dispersion of the first-order spectrum is

$$(22° - 12°) = 10°$$

The second-order spectrum ranges from 24° to 49° (a dispersion of 25°). The angular dispersion increases with the order m (see fig. 39.7).

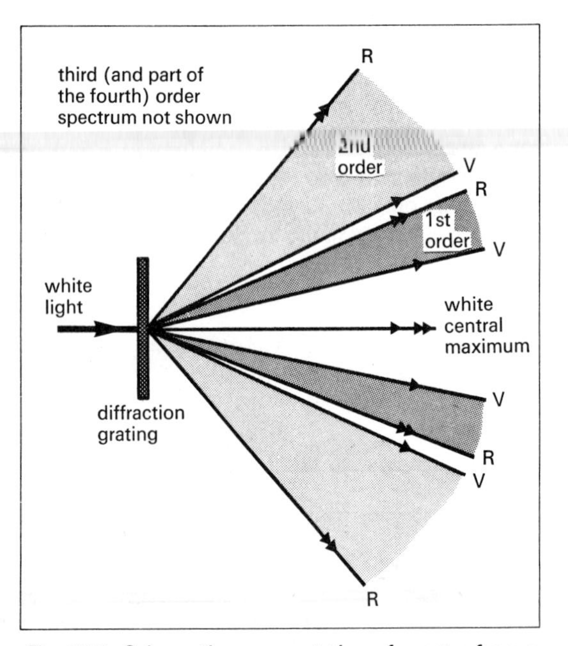

Fig. 39.7. Schematic representation of spectra from a grating.

Spectra of different orders can *overlap.* For this grating $(\theta_3)_V = 37°$, and so the third-order overlaps the second by 12°. The part of the fourth order that is present overlaps the third.

Comparison of Prism and Grating Spectra

Spectra are observed when the grating is mounted on a spectrometer table, as in fig. 39.8.

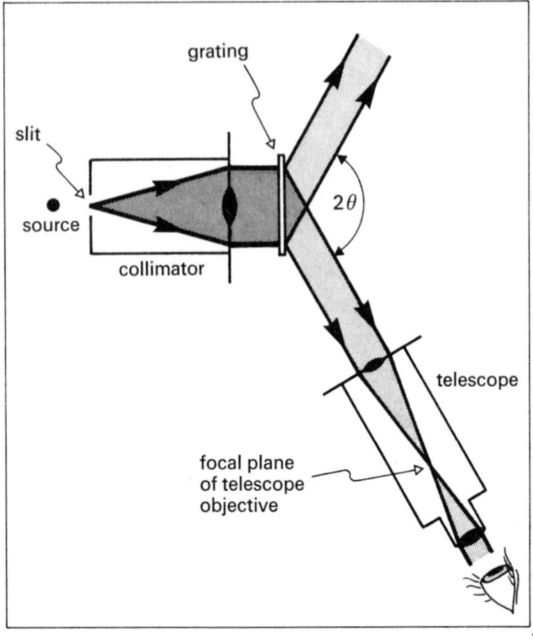

Fig. 39.8. A grating spectrometer.

(*a*) Since the energy from the slit is distributed over several spectra, their intensity is less than that of a prism spectrum. (Nevertheless it is possible to shape a groove profile so as to concentrate the light in a particular order of spectrum.)

(*b*) The wavelengths are distributed fairly evenly through the grating spectrum, whereas in the prism spectrum wavelengths are compressed at the red end, and extended at the violet.

(*c*) The **resolving power** of a spectrometer is its ability to distinguish very close wavelengths. It is generally greater for a grating than for a prism. It does not depend separately upon the size and spacing of the grating ruling, but it increases with the total number of rulings (i.e. *cetera paribus* it increases with the overall width of the grating). It also increases with the order of the spectrum.

(*d*) Gratings can be made *reflecting* by ruling them on metal. If the metal is shaped as a concave mirror, lenses are not required, and it is possible to investigate wavelengths otherwise absorbed by glass.

*39.4 Fresnel Diffraction: the Nature of Shadows

These *Fresnel* patterns are fringed images of the obstructing obstacles. The angular relationships involved are *dependent on wavelength,* so that if white light is used, a detailed analysis shows that the shadows have coloured edges.

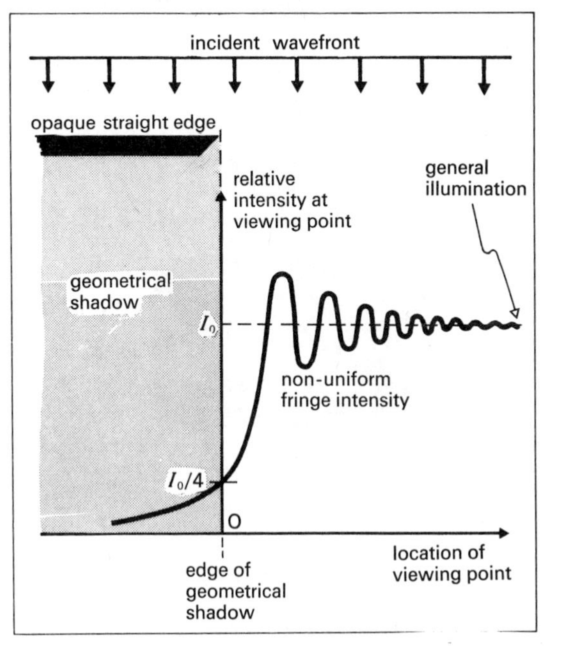

Fig. 39.9. Fresnel diffraction at a straight edge, showing the arrangement and the result.

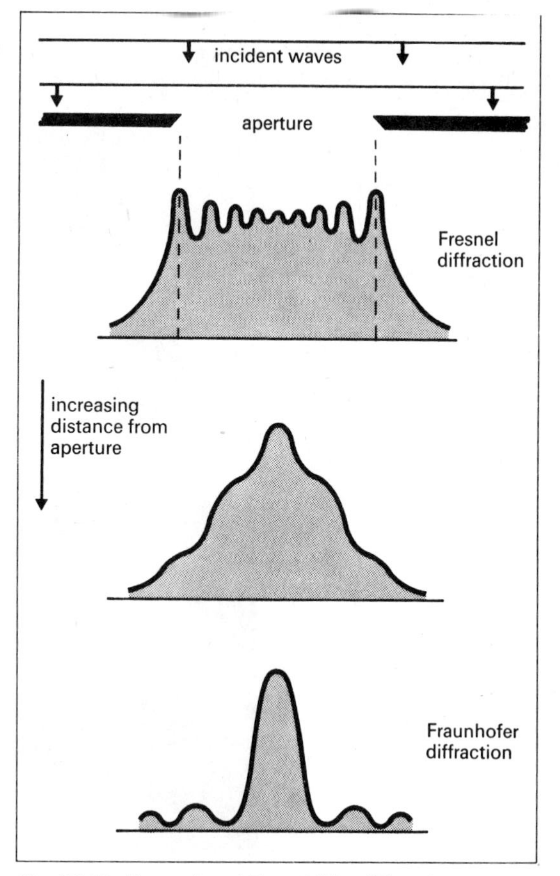

Fig. 39.10. Fresnel and Fraunhöfer diffraction patterns for a circular aperture.

(*a*) The Straight Edge (fig. 39.9)

The viewing screen shows a diffraction fringe system when viewed through a magnifying glass (fig. 39.11*a*). The intensity falls swiftly within the geometrical shadow. The actual location of a fringe is determined by an angular relationship, i.e. by *relative* distances.

(*b*) An Opaque Disc (fig. 39.11*b*)

The exact pattern depends on the separation of the viewing screen from the disc, but there is always a *bright spot at the centre of the geometrical shadow.* The disc's shadow is surrounded by a complex of circular fringes of varying brightness.

(*c*) The Circular Aperture (fig. 39.10)

The general pattern (fig. 39.11*c*) is similar to (b), but the central spot is sometimes dark (depending on the viewing distance). At *large* distances we approach *Fraunhöfer* patterns, and the central spot will be *bright* (*fig.* 39.14).

(*d*) The Long Rectangular Slit (fig. 39.11*d, e, f*)

The diffraction effects are emphasized as the width a approaches λ. When $a \gg \lambda$, we have approximate rectilinear propagation. The *shadow* patterns of fig. 39.11 should be compared to the diffracted *wave* patterns shown on p. 111, in which details of fringes are not shown.

(*e*) The Rectangular Aperture (fig. 39.11*g*)

Note that the pattern is extended along the direction of smaller aperture width ($a < b$).

*39.5 Fraunhöfer Diffraction at a Single Slit

In fig. 39.12 we want to investigate the diffraction pattern on the screen. The point O will show that there is always a central maximum.

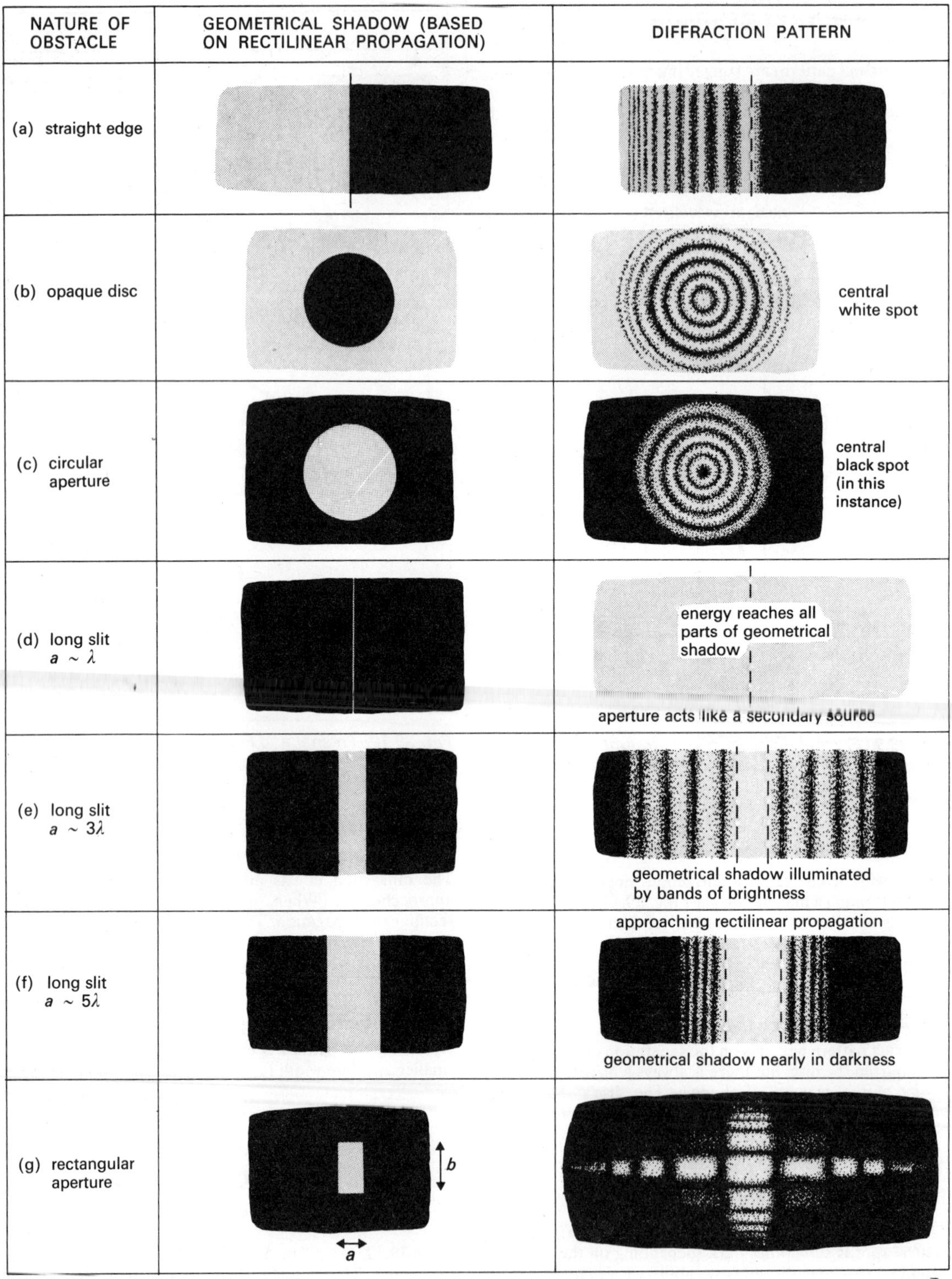

NATURE OF OBSTACLE	GEOMETRICAL SHADOW (BASED ON RECTILINEAR PROPAGATION)	DIFFRACTION PATTERN
(a) straight edge		
(b) opaque disc		central white spot
(c) circular aperture		central black spot (in this instance)
(d) long slit $a \sim \lambda$		energy reaches all parts of geometrical shadow; aperture acts like a secondary source
(e) long slit $a \sim 3\lambda$		geometrical shadow illuminated by bands of brightness
(f) long slit $a \sim 5\lambda$		approaching rectilinear propagation; geometrical shadow nearly in darkness
(g) rectangular aperture	b; a	

Fig. 39.11. Diffraction patterns compared with geometrical shadows.

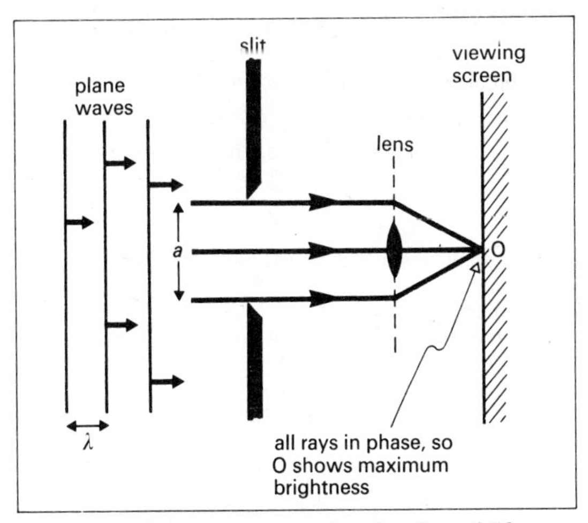

Fig. 39.12. Arrangement for showing Fraunhöfer diffraction at a single slit.

When locating the *direction* in which we expect to find the first minimum (fig. 39.13) we use the fact that the lens does not change the relative phases of the different rays that it brings to a point on the screen.

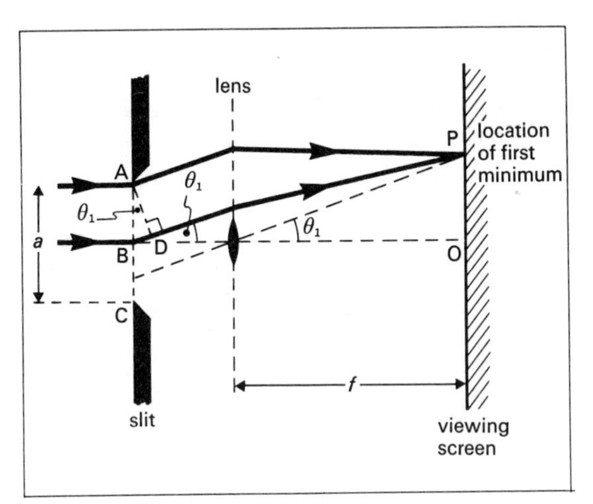

Fig. 39.13. Path differences calculated for pairs of rays diffracted through the slits.

If BD = $\lambda/2$, then the two rays shown are out of phase, and their superposition at P gives *zero* amplitude.

Every ray that originates between A and B has a corresponding ray originating a distance $a/2$ below between B and C, so that the *resultant* amplitude (and intensity) at P is zero.

θ_1 can be found from

$$\frac{\mathrm{BD}}{\mathrm{AB}} = \sin\theta_1$$

$$\frac{\lambda/2}{a/2} = \sin\theta_1$$

$$a\sin\theta_1 = \lambda$$

for the first *minimum*.

Note that this is an *angular* relationship. The *distance* OP can be found from

$$\frac{\mathrm{OP}}{f} = \tan\theta_1$$

This argument can be extended to show that the angular displacements θ_m of the *minima* are given by

$$\boxed{a\sin\theta_m = m\lambda}$$

where $m = \pm 1, 2, 3$, etc. ($m = 0$ gives the central *maximum*.)

Maxima occur roughly half way between the minima. Full analysis enables the relative intensities of the different maxima to be calculated. The result is given in fig. 39.14. For example, when

(*i*) $a/\lambda \sim 5$, then $\theta_1 = 12°$,
(*ii*) $a/\lambda \sim 10$, then $\theta_1 = 6°$.

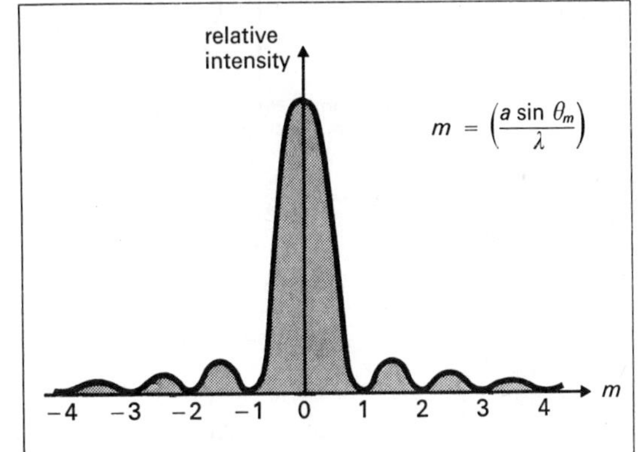

Fig. 39.14. Relative intensities of the different maxima. (Scale approximate only.)

Notes.

(*a*) As a increases, so θ_1 decreases. Then the conditions approach more nearly those assumed in geometrical optics.

(*b*) As a decreases, so θ_1 increases.
When $a = \lambda$, $\theta_1 = 90°$; this means that the central maximum spreads fully into the geometrical shadow. We generally assume this to hold

(*i*) for the source-slit and twin-slits of a *Young's* experiment, and
(*ii*) for the diffraction grating.

(*c*) If θ_1 is small, we may approximate and write $\theta_1 \approx \lambda/a$ for the first minimum.

(*d*) The central bright fringe has several times the intensity and twice the width of any other.

The relation of this pattern to the *Fresnel* pattern is shown on page 283.

*39.6 Resolving Power

The **angular resolving power** of an obstacle (which term includes a slit or an aperture) is the reciprocal of the minimum angle subtended by the directions of travel of incident waves coming from distant point sources such that we can distinguish their diffraction patterns. (See fig. 39.15.)

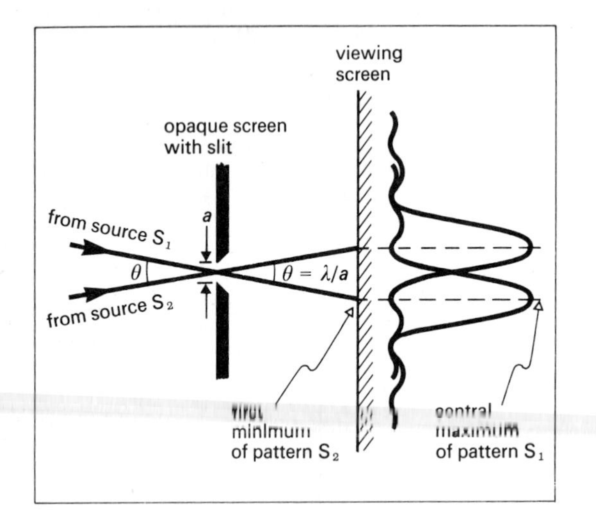

Fig. 39.15. Diffraction patterns distinguished according to Rayleigh's criterion

Rayleigh's criterion *suggests that we agree to call the patterns distinguishable when the first minimum of one falls on the central maximum of the other (or better).*

Using this criterion, we find the following.

(*a*) The resolving power of a *slit* is found from

$$\theta = \frac{\lambda}{a}$$

We cannot distinguish the images of S_1 and S_2 if they subtend an angle less than λ/a. θ can be regarded as the minimum angle of resolution.

(*b*) The resolving power of a *circular aperture* can be calculated from

$$\theta \approx \sin\theta = 1.22\frac{\lambda}{D}$$

where D is the aperture diameter. The result is difficult to prove mathematically, but it is important when we consider optimum lens and mirror diameters for optical instruments such as telescopes (p. 256).

(*c*) The resolving power of the *eye* (when the pupil is small) is such that θ is about 10^{-4} rad of arc. This controls the minimum separation of points that are to be distinguishable: we can just resolve points 0.1 mm apart when they are 250 mm from the eye. (This resolving power is matched by the structure of the retina, which is made of discrete rods and cones.)

Fig. 39.16 shows different degrees of resolution of diffraction images from two slit sources—the intensities from the separate patterns combine to produce the resultant pattern in each case.

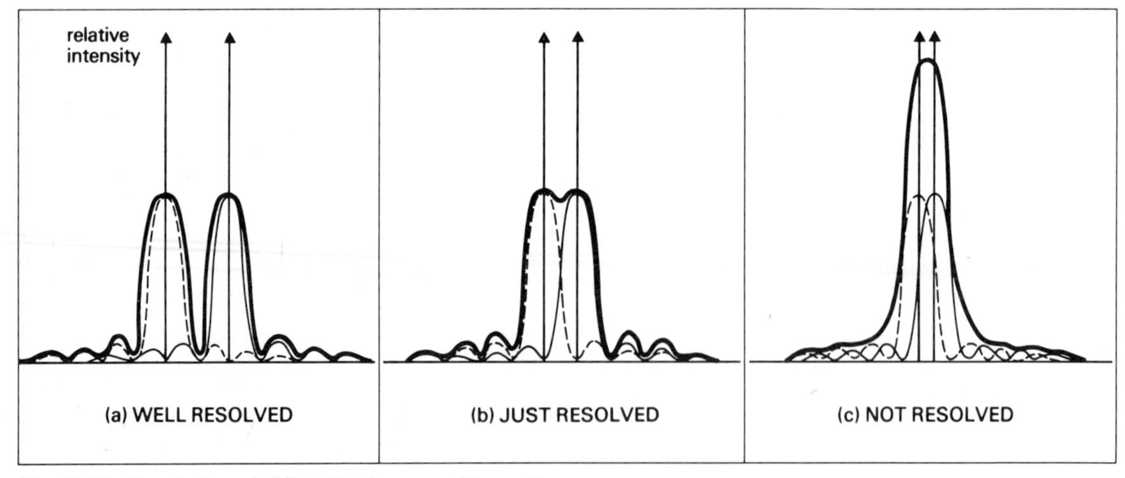

Fig. 39.16. Resolution of diffraction images of two slit sources.

*39.7 X-Ray Diffraction

See p. 459 for the production and properties of X-rays.

Von Laue's Experiment (1912)

The practical arrangement is described in fig. 39.17. The experiment established the wave nature of X-rays. At this time it confirmed that they could not be matter (earlier than *de Broglie*) and must be electromagnetic radiation. The very small grating spacing demanded by the very small wavelength of the radiation could not then be ruled mechanically. The crystal's periodic three-dimensional spacing of ions acts as a 3-D transmission diffraction grating. Thus the many diffracting centres produce constructive superposition in certain sharply-defined directions. The angles of diffraction will be large if the wavelength is comparable with the repetition length r_0.

Interpretation of the transmission pattern is not simple. Positions and intensities of *Laue* spots enable atomic arrangements within the crystal to be deduced. The fundamental repetitive diffracting unit is called the *unit cell*—for the NaCl crystal it has four sodium ions and four chloride ions associated with it.

(*a*) *Directions* of diffracted X-ray beams indicate basic *symmetry of unit cells* in the crystal.

(*b*) *Intensities* show the *distribution of electrons* within the unit cell—it is the electrons rather than the nuclei that diffract the X-rays.

Unit Cells and Reflecting Planes

Fig. 39.18 shows a *two-dimensional* representation of a NaCl crystal, its diffraction centres and a possible family of planes passing through these centres.

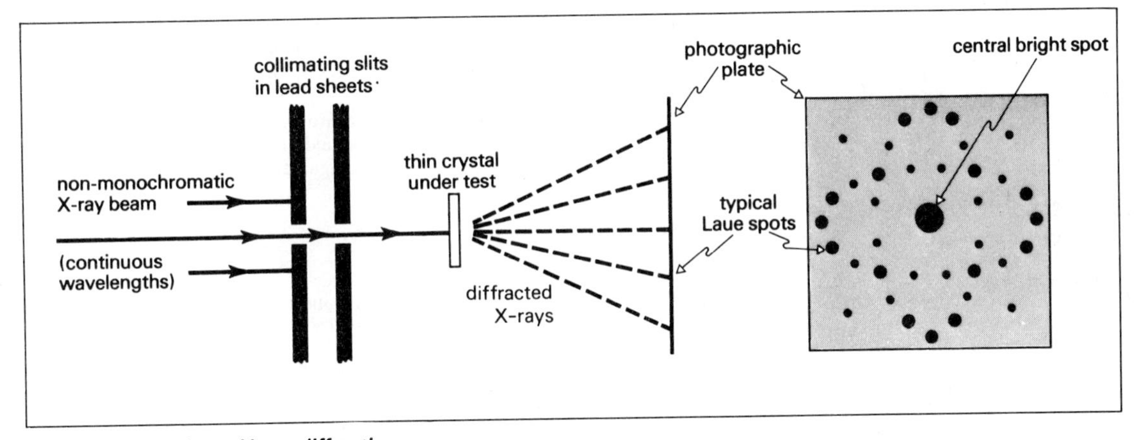

Fig. 39.17. Von Laue X-ray diffraction.

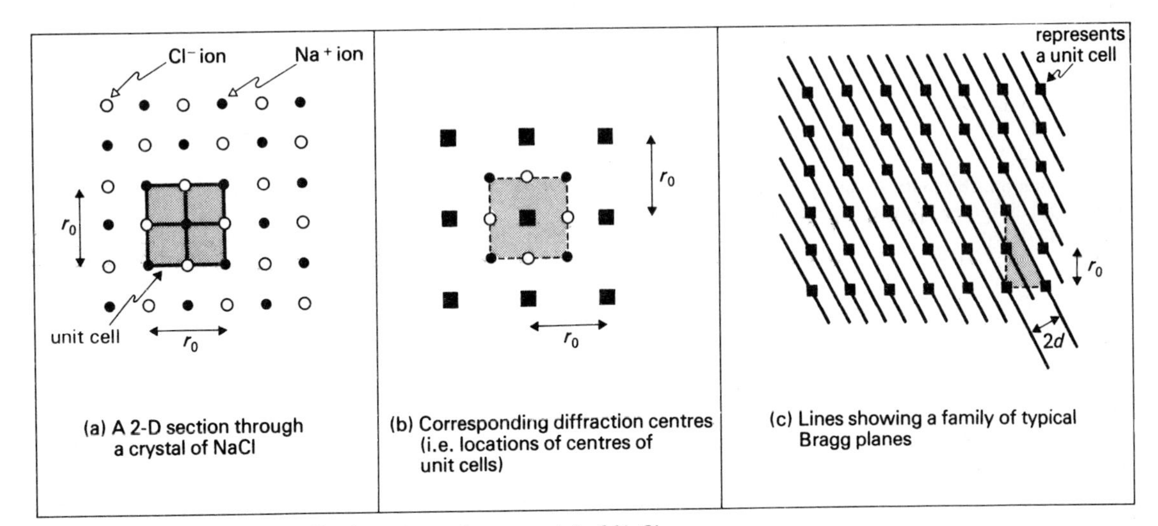

Fig. 39.18. Unit cells and reflecting planes for a crystal of NaCl.

When X-radiation falls on a crystal, each unit cell acts as a diffracting centre. Superposition of the scattered secondary wavelets from those unit cells lying in a particular family of planes produces an apparently reflected beam which can be interpreted quite easily. The perpendicular spacing of the planes is closely related to the unit cell dimension—in fig. 39.18(*c*) the spacing $d = r_0/\sqrt{5}$.

The Bragg Equation

(*a*) A particular surface plane always acts like a mirror, giving a weak reflected beam for any angle of incidence.

(*b*) An entire family of planes will produce a reflected beam by constructive superposition only when the path difference from adjacent planes obeys the proper condition.

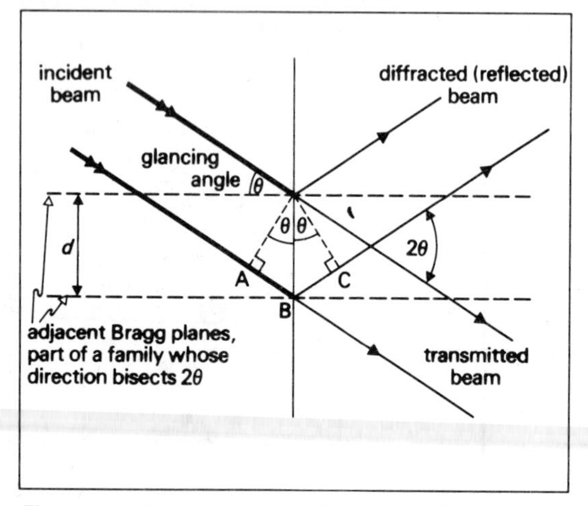

Fig. 39.19. Derivation of the Bragg equation.

In fig. 39.19, the path difference for adjacent planes is AB + BC = $2d \sin\theta$. For constructive superposition this must equal $m\lambda$.

$2d \sin\theta = m\lambda$	**W. L. Bragg's Law**

m = 1, 2, 3, etc., the order of diffraction.

(*a*) For monochromatic X-rays incident at arbitrary θ, no (reflected) diffracted beam appears, and the beam is totally transmitted, but

(*b*) for beams of a mixture of X-ray wavelengths, diffracted beams appear when

$$\lambda = 2d \sin\theta/m$$

is satisfied.

Bragg's law can be applied to the expression for the *de Broglie* wavelength discussed in electron diffraction (p. 265).

$$2d \sin\theta = m\lambda = \frac{mh}{\sqrt{2Vem_e}}$$

This equation can be used to confirm *de Broglie's* theory.

The X-Ray Spectrometer

The practical arrangement is shown in fig. 39.20. The wavelength can be measured using a crystal of known unit cell dimensions. The value of λ can then be used to measure d and r_0, and so to analyse the nature of the atomic pattern in crystals, and the structure of complex substances. The ionization chamber and crystal are rotated to keep the angle of incidence equal to the angle of 'reflection'. The intensity of the diffracted X-rays is deduced from the ionization current.

In the school laboratory analogous X-ray diffraction experiments can be carried out using 30 mm electromagnetic waves and an arrangement of polystyrene spheres or tiles.

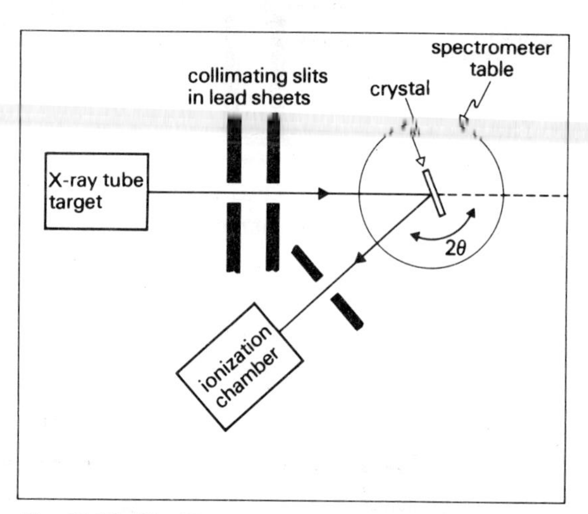

Fig. 39.20. The X-ray spectrometer.

40 Polarization

40.1 Light as a Transverse Wave Motion

The term **transverse** was defined on p. 92. The ideas of wave polarization were discussed on p. 97, and polarization in optics was mentioned on p. 261.

Fig. 37.1 showed a propagating electromagnetic wave in which the plane containing the electric vector is fixed: it is therefore **plane-polarized**. For historical reasons (p. 291) the **plane of polarization** was chosen to be that *perpendicular* to the electric vector.

The **plane of vibration** is that which contains the *electric* vector and the direction of propagation. We define it relative to the electric field, since it is $\boldsymbol{E}$ that enables a force to be exerted on a *stationary* charge, and thus which is primarily responsible for the interaction of electromagnetic waves and matter.

Light is said to be **unpolarized** if the plane of vibration and the magnitude of $\boldsymbol{E}$ continually and randomly change.

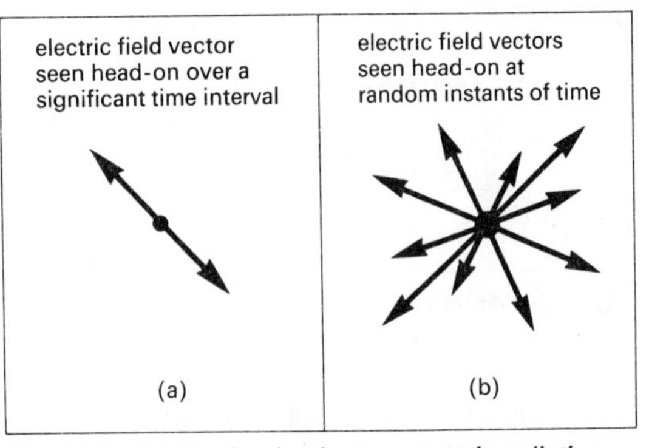

Fig. 40.1. Polarization in electromagnetic radiation: (a) polarized radiation, (b) unpolarized radiation.

Examples

(*a*) Many radio waves, radar waves and the microwaves we use for laboratory demonstrations *are* plane-polarized as a result of the nature of the source (fig. 40.1*a*).

(*b*) Most visible light, being emitted *randomly* by atoms and molecules which are only a small part of a light source, is unpolarized (fig. 40.1*b*).

Fig. 40.1*a* could be demonstrated by rotating a detector of microwaves through an angle of 90°. If it previously registered a maximum signal, it would now show zero (and vice-versa).

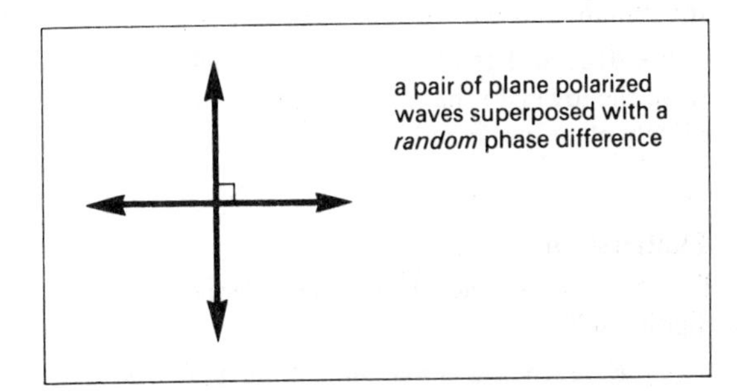

Fig. 40.2. An equivalent representation of fig. 40.1b.

Fig. 40.2 shows an alternative way of representing fig. 40.1*b*. At any given instant of time, the particular resultant of fig. 40.1*b* could be resolved into its resolved parts along the directions shown.

40.2 Malus's Experiment and Brewster's Law

Malus's Experiment (1809)

In fig. 40.3 as the observer rotates the plate AB about an axis CD he views the parallel beam reflected from the plate.

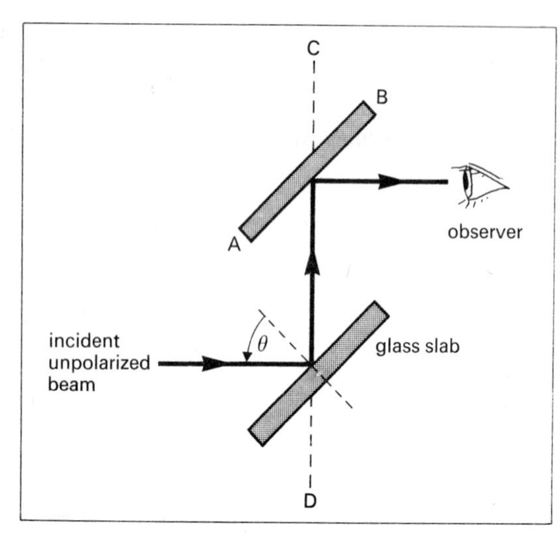

Fig. 40.3. Malus's experiment.

(*a*) When the beam is reflected to left or right (in the plane of the paper) the viewed intensity is a maximum.

(*b*) When the beam is reflected into or out of the paper normally, the intensity is a minimum.

(*c*) When θ takes a value θ_p, the **polarizing angle**, the minimum intensity becomes *zero*.

We deduce that when $\theta = \theta_p$, the light travelling from the lower to the upper plate has become completely plane-polarized by reflection.

Explanation

Fig. 40.4 shows the electric field radiated from a dipole oscillator.

(*a*) $\boldsymbol{E}$ reaches a maximum along the *transverse* dipole axis.

(*b*) $\boldsymbol{E}$ is zero along the longitudinal dipole axis.

We conclude that since intensity $\propto E^2$, no energy is radiated along the direction of oscillation.

When a linearly-polarized ray (of correct orientation) is incident at the polarizing angle (fig. 40.5) *there is no reflected ray*: the oscillators set into motion by the incident ray cannot send energy along the direction of the usual reflected ray if it would be perpendicular to the refracted ray.

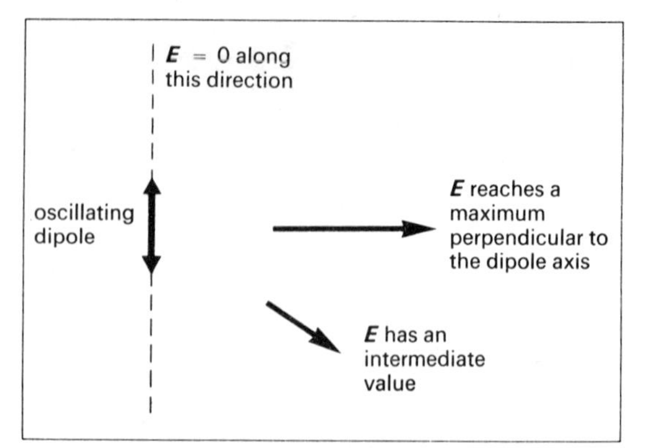

Fig. 40.4. Two-dimensional diagram of radiation from an oscillating electric dipole.

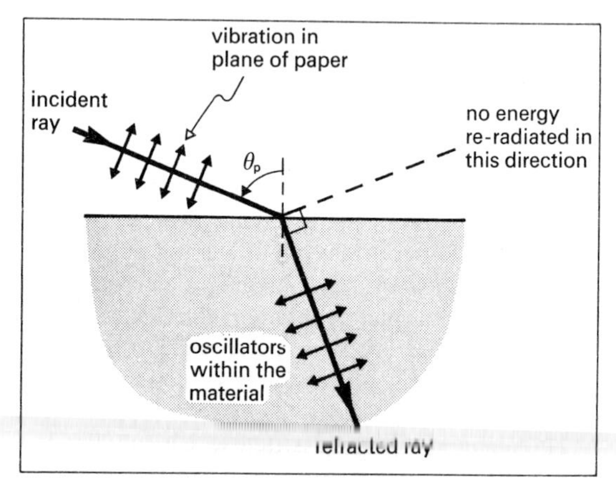

Fig. 40.5. When the angle of incidence equals θ_p the refracted ray is perpendicular to the usual direction of the reflected ray.

Brewster's Law (1812)

Fig. 40.6 shows unpolarized light incident at the polarizing angle. Since the reflected ray is completely plane polarized

$$\theta_1 + \theta_2 = 90°$$

By *Snell's Law*

$$n_1 \sin \theta_1 = n_2 \sin \theta_2$$
$$= n_2 \cos \theta_1$$

where $\theta_1 = \theta_p$.

$$\therefore \quad \frac{n_2}{n_1} = \tan \theta_p$$

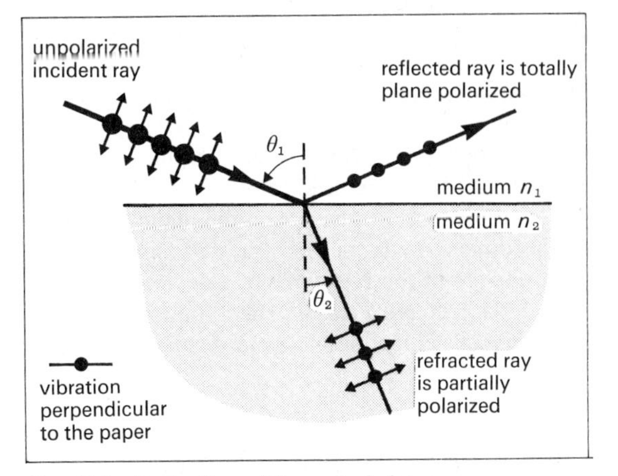

Fig. 40.6. Derivation of Brewster's Law.

If medium 1 is air or vacuum

$$n_2 = \tan \theta_p \qquad \textit{Brewster's Law}$$

Notes.

(*a*) *Brewster* discovered the law experimentally.

(*b*) The reflected beam is *always partially* plane-polarized ($\theta_1 \neq 0$).

(*c*) For glass of $n = 1.5$, $\theta_p = 57°$.
Since θ_p depends upon n, we need to use a parallel *monochromatic* beam for accurate measurements.

(*d*) The plane containing the incident and reflected (plane-polarized) rays of fig. 40.6 was chosen *historically* to be called the **plane of polarization**.

40.3 Production of Plane-Polarized Light

Fig. 40.7 illustrates the function of a **polarizer**.

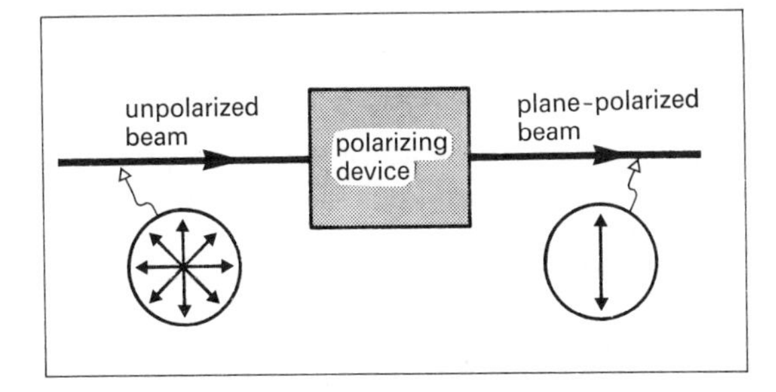

(*a*) Using Reflection

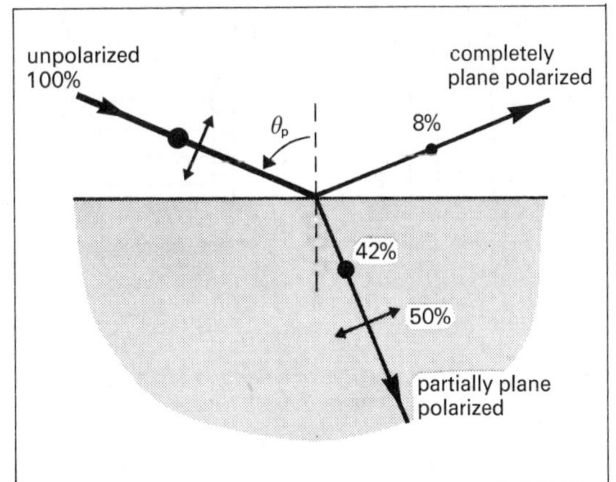

Fig. 40.8. Proportion of energy taken by reflected and refracted rays.

Fig. 40.8 shows the possible distribution of energy from the incident beam.

(*b*) Using Normal Refraction

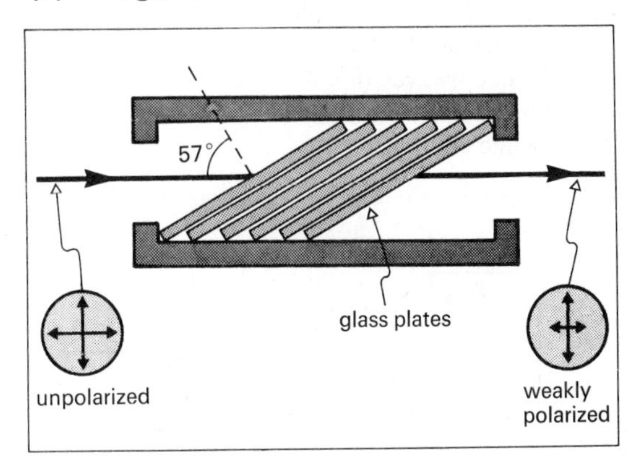

Fig. 40.9. An attempt to produce plane-polarized light by successive refractions.

Fig. 40.8 illustrates how a transmitted ray is *partially* plane-polarized. Fig. 40.9 shows how *successive* refractions can produce a transmitted ray that is *nearly* plane-polarized.

(*c*) Using Selective Absorption

(*i*) *The microwave analogy*

Fig. 40.10 shows how completely polarized microwaves are absorbed by conducting wires *parallel* to their plane of vibration, but transmitted by wires

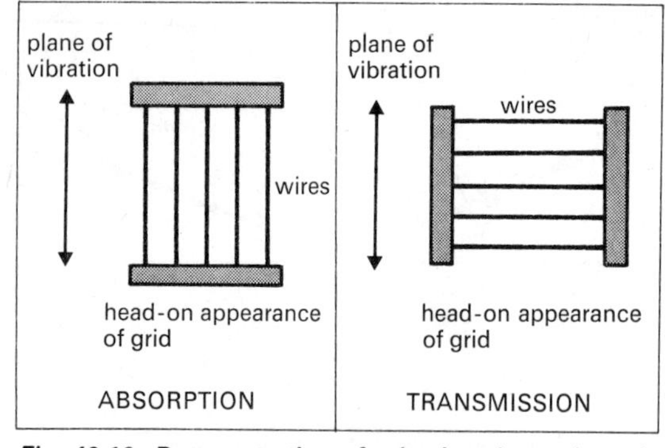

Fig. 40.10. Demonstration of selective absorption of polarized microwaves by parallel wires.

perpendicular to that plane. During absorption the microwave energy is used to set the wire's electric charge into oscillation along the length of the wire.

(ii) Dichroism

Materials which transmit one component of polarization but attenuate the mutually perpendicular component exhibit **dichroism**. *Polaroid* is a commercial material which contains long molecules whose axes have been aligned during manufacture. They have a similar effect on light waves to that of conducting wires on microwaves (fig. 40.10 and 40.11). *Tourmaline* is a crystal which occurs naturally, and which has the same effect.

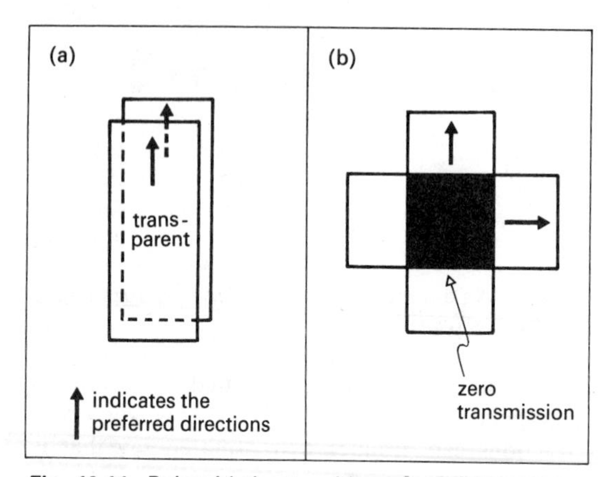

Fig. 40.11. Polaroid sheets with preferred directions (a) parallel, allowing transmission, and (b) crossed at 90° causing extinction of unpolarized light.

(*d*) Using Double Refraction

Many crystals show **double refraction** or **birefringence**: an unpolarized ray which enters the crystal is split into an **ordinary ray**, and an **extraordinary ray** (which does not obey *Snell's Law*). Each ray is plane-polarized, their planes of vibration being perpendicular.

A calcite crystal (which is cut correctly and then reconstructed) can be used to separate the ordinary ray from the extraordinary, and the plane-polarized extraordinary ray is then transmitted alone. This is the principle of the **Nicol prism**, a device which has the same effect as a *Polaroid* sheet (quinine iodosulphate crystals embedded in a plastic material).

*40.4 Scattering and Polarization

Electromagnetic radiation energy is absorbed by an atom or molecule when the $\boldsymbol{E}$-field of the wave acts on the electrons causing them to become forced oscillating electric dipoles with their axes along the direction of the *field*. The re-radiation of the electromagnetic wave in all directions is known as **scattering**, and results in a reduction of intensity from the initial radiation direction (p. 224). Fig. 40.4 shows that, if the incident radiation is linearly *polarized*, the intensity of the scattered wave is a maximum along the transverse dipole axis, and that it is zero along the longitudinal axis where the scattering angle is 90°. If the incident radiation is *unpolarized*, the dipole will oscillate in numerous different directions. The re-radiation is

(*a*) completely polarized when scattered through 90°,
(*b*) completely unpolarized in the direction of the incident wave, and
(*c*) partially polarized in all other directions.

The natural oscillation frequencies within a typical atom or molecule are in the u.v. region of the electromagnetic spectrum. The greater the frequency of waves in the visible region the closer they become to the u.v. resonant frequency, and hence the larger the amplitude of the forced oscillations. Blue light is therefore scattered more effectively than red light. This contributes to the blue colour of the sky and to the redness of the sun at sunrise and sunset. Smoke or dust particles and water droplets also cause scattering, as does a colloidal suspension of particles in a liquid (**Tyndall effect**).

40.5 Detection of Plane-Polarized Light

An **analyser** is a device which can detect polarized light. It has the same effect on the light as a polarizer and is constructed in the same way.

Consider (fig. 40.12) plane-polarized light incident on an analyser. The component of $\boldsymbol{E}_0$ that is transmitted is $E_0 \cos\theta$. Since intensity

$$I \propto E^2 \qquad \text{(p. 99)}$$

the transmitted intensity

$$I \propto E_0^2 \cos^2 \theta$$

or

$I = I_0 \cos^2 \theta$	*Malus's Law* (1809)

This equation should be related to *Malus's* experiment on p. 290.

If an analyser is rotated in a plane-polarized beam, the observer sees two maxima and two minima (zero) for each rotation.

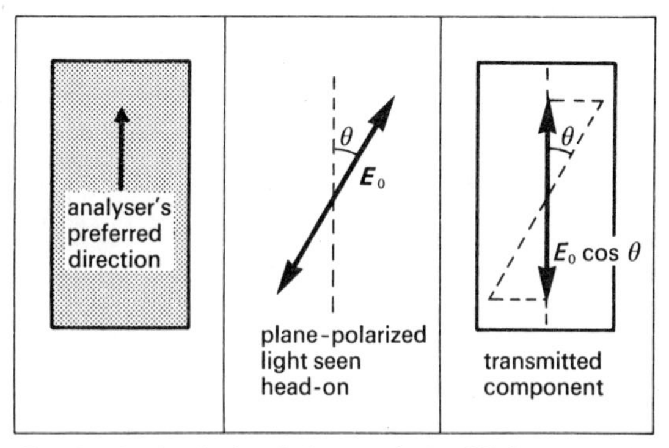

Fig. 40.12. Analysis of plane-polarized light.

40.6 Uses of Polarization

(a) Academic

(*i*) Polarization is the only wave property of those on p. 94 which can be used to distinguish *transverse* from longitudinal waves. *Arago* and *Fresnel* (1815) had shown that the two beams emerging from a doubly-refracting crystal, although otherwise coherent, did not give interference effects. *Young* (1817) suggested that this resulted from a lack of symmetry: this is only possible if the quantity associated with the periodic fluctuations is a *transverse vector.*

(*ii*) Polarized light is very useful for general experiments on the *coherency* of different light beams. For example two beams of plane-polarized light which were originally derived from perpendicular components of unpolarized light, and which are later rotated into the same plane, *cannot* be made to exhibit typical interference fringes.

(*iii*) The phenomenon can be used to show that the *electric vector* is largely responsible for the interaction of electromagnetic waves and matter.

(b) Applications

(i) Reduction of Glare

Sunglasses and camera filters are made of polarizing materials to reduce the glare of light polarized by reflection from shiny surfaces (such as water).

(ii) Solar Compass

The blue light scattered from gas molecules in the sky is partly plane-polarized. In polar regions, where a magnetic compass becomes useless, one uses a solar compass, which analyses the polarization of scattered sunlight.

Bees have eyes which are polarization-sensitive: these enable them to orientate themselves.

(iii) Optical Activity

Many crystals (such as quartz) and some liquids (especially sugar solutions) have the ability to cause rotation of the plane of vibration of plane-polarized light.

The angle of rotation α for solutions depends on

(1) the concentration, and

(2) the length of solution through which light passes.

A **polarimeter** is a device which measures α, from which the concentration of the solution can be found. **Optical activity** is the name given to the phenomenon.

(iv) Photoelasticity

Glass and some plastic materials become doubly refracting only while stressed. If polarized white light is then passed through them, and analysed, bright coloured lines show the existence of strains. In engineering work plastic models of structures are constructed, and weaknesses are exposed in this way.

(v) Kerr Cell

This is a light shutter with no mechanical parts, whose working is based on the **Kerr electro-optic effect** (1875). (A material which is usually **isotropic** may become anisotropic when subjected to a strong electric field.)

(vi) Comparison of Light Intensities

This can be done by using

$$I = I_0 \cos^2 \theta \quad (\textit{Malus's Law})$$

if one measures θ.

VIII Sound Waves

41 Sound Waves

41.1 Classification of Sound Waves

Refer to the wave classification on p. 93. Sound waves

(1) (*a*) can be *periodic*, e.g. those emitted by a tuning fork, *or*

(*b*) *non-periodic*, e.g. the shock wave emitted by an aircraft which breaks the sound barrier

(2) (*a*) can be *progressive*, e.g. tuning fork waves, *or*

(*b*) *stationary*, e.g. the sound waves *inside* an organ pipe

(3) are *longitudinal*, and

(4) the *disturbance* is a variation of the pressure and density of the medium through which the wave passes. The disturbance is thus a *scalar* quantity.

(4) above indicates that a *medium* is necessary for the transmission of the waves, which (unlike electromagnetic waves) cannot travel through a vacuum. This medium can be solid, liquid or gas.

The *source* of a sound wave is an object which vibrates so as to send out waves through the medium (frequently air) in which it is immersed.

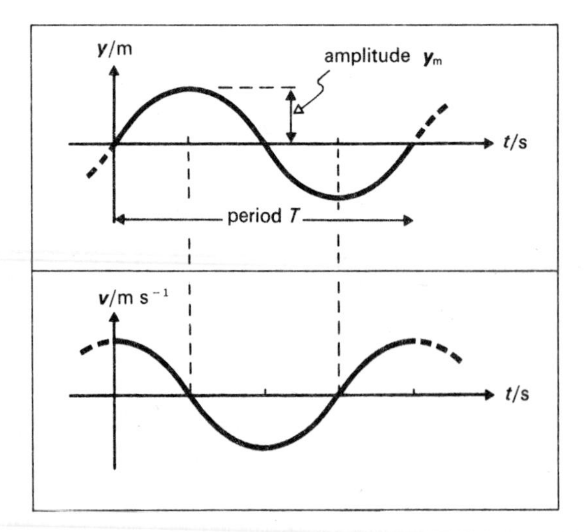

Fig. 41.1. Displacement–time and velocity–time curves for a particle disturbed by a sinusoidal wave.

41.2 Displacement, Velocity, Excess Pressure and Density

When a longitudinal wave disturbs a particle of the medium through which it passes, an oscillation is superimposed on the particle's previous movement. Gas molecules are already in random motion, but we are only concerned here with the oscillation.

If the wave is sinusoidal, the variation *with time* of the displacement (y) and velocity (v) of a typical particle are given by fig. 41.1.

Notes.

(a) These variations differ in phase by $\pi/2$ rad.

(b) Particles at other locations show the same variations, but with systematic phase shifts relative to their near neighbours.

The displacement of a particle from its usual position causes a variation in the *number of molecules per unit volume*. Recalling (p. 180) that

$$p = \tfrac{1}{3}nm\overline{c^2}$$
$$= \tfrac{1}{3}\rho\overline{c^2}$$

we deduce that there are also variations in

(1) *pressure*, and

(2) *density*, relative to their usual values.

The disturbance consists of a train of **compressions** or **condensations** (of higher pressure and density than usual) and **rarefactions** (lower pressure and density). Fig. 41.2 relates particle displacement and excess pressure (or excess medium density) to usual particle position.

	At a compression	*When excess pressure and density are zero*	*At a rare-faction*
then displacement y is	zero	$\pm y_m$	zero
then particle velocity $\dot{y}$ is	maximum and positive	zero	maximum and negative

Notes.

(a) Graphs such as those in fig. 41.2 necessarily represent a *longitudinal* displacement *as though* it were transverse. They cannot be a pictorial representation of reality.

(b) A **waveform** is a graph showing either

(i) pressure variation plotted against time at a given location (as one would display on an oscilloscope), or

(ii) pressure variation plotted against position at an instant of time.

(c) It can be shown that the size of the pressure variation is proportional to the size of the particle displacement: therefore fig. 41.2 is also a waveform.

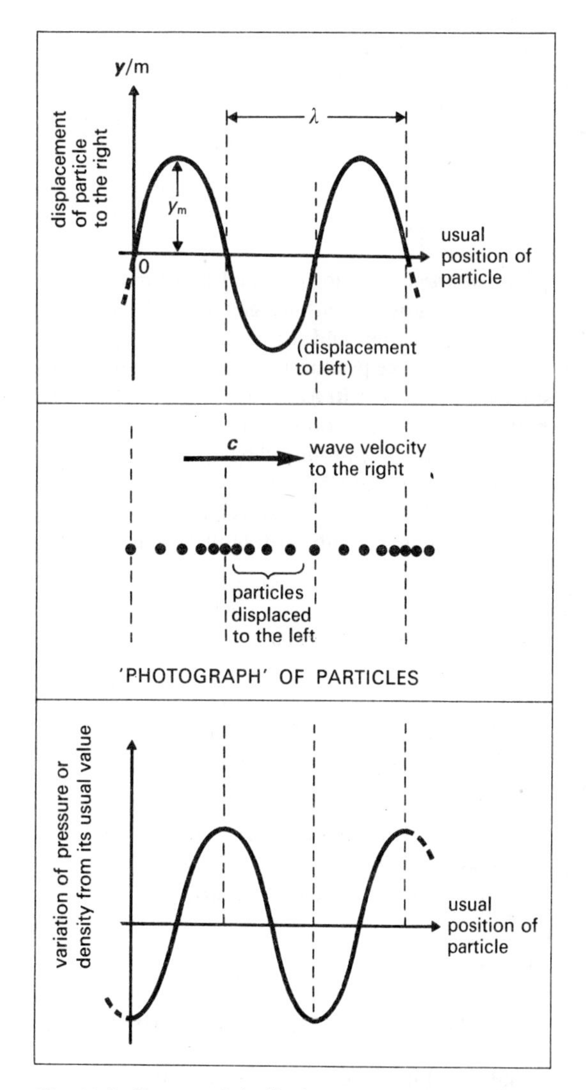

Fig. 41.2. Zero particle displacement corresponds to maximum variation in pressure and density. The graphs correspond to one instant of time.

41.3 Facts about Sound Waves

Spectrum

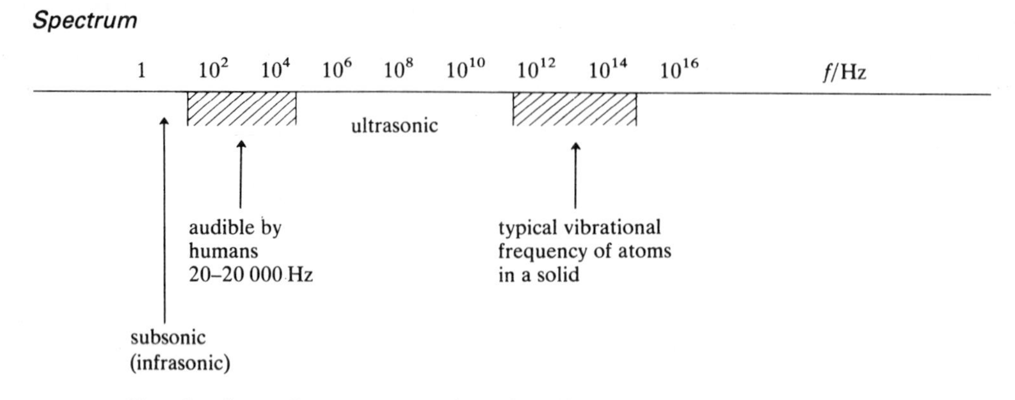

Note that the numbers are presented on a logarithmic scale.

Sound waves are longitudinal mechanical waves audible to humans, and are generated by vibrating strings, air columns and surfaces.

An earthquake produces examples of **subsonic** (or **infrasonic**) waves. **Ultrasonic** waves of frequencies as high as 6×10^8 Hz (with a wavelength in air of about 5×10^{-7} m) can be produced by using the piezoelectric effect in a quartz crystal.

Bats produce waves of wavelength about 20 to 30 mm, a range beyond adult human hearing.

Some Uses of Ultrasonics

(*a*) Low intensity
 (*i*) echo sounding,
 (*ii*) finding flaws in materials,
 (*iii*) locating tumours in the body.

(*b*) High intensity
 (*i*) destruction of living cells suspended in liquids,
 (*ii*) removal of grease and dirt.

Speeds

Substance	*Speed* $c/10^2$ m s^{-1}
granite (at 293 K)	60
fresh water (at 298 K)	15
air (at 273 K)	3.3

Amplitudes (for $f = 10^3$ Hz)

	Air molecule displacement amplitude y_m/m	*Variation in air pressure* Δp/Pa
for faintest sound that can be heard	10^{-11}	2×10^{-5}
for loudest sound that can be tolerated	10^{-5}	30

Intensity

The intensity I of a wave is defined by

$$I = \frac{P}{A} = \frac{W}{At} \qquad \text{(p. 99)}$$

$$I\left[\frac{\mathrm{W}}{\mathrm{m}^2}\right] = \frac{P[\mathrm{W}]}{A[\mathrm{m}^2]}$$

Examples. ($f \approx 10^3$ Hz)

Loudness of sound	I/W m^{-2}
threshold of hearing	10^{-12}
a person talking	10^{-6}
threshold of feeling	$10^0 = 1$

The intensity decreases away from the source

(*a*) because the energy is distributed through a larger volume: (for a point source and isotropic medium the intensity will obey an *inverse square law*), and

(*b*) because of **attenuation**: the wave energy is gradually converted by an imperfectly elastic medium into random energy of the medium's molecules (its internal or thermal energy).

The *power* of a source is the rate at which it converts other forms of energy into sound wave energy. The power of a full orchestra is $\sim$ 70 W.

The Human Ear

The assessment of **loudness** is controlled by both

(*a*) the *wave intensity*, which depends on

(*i*) (amplitude)^2

(*ii*) (frequency)^2

(*iii*) speed, and

(*b*) the *sensitivity* of the hearer to the particular frequency being sounded.

The ear responds to pressure variations, rather than air molecule displacement.

It can be shown that

$$I = \frac{(\text{pressure amplitude})^2}{2(\text{air density})(\text{wave speed})}$$

$$I = \frac{\Delta p_{\max}^2}{2\rho c}$$

The frequency does not appear in this expression, so there will often be an advantage in measuring *pressure* rather than *displacement* amplitude.

Fig. 41.3 indicates in a general way the variation of sensitivity of the human ear.

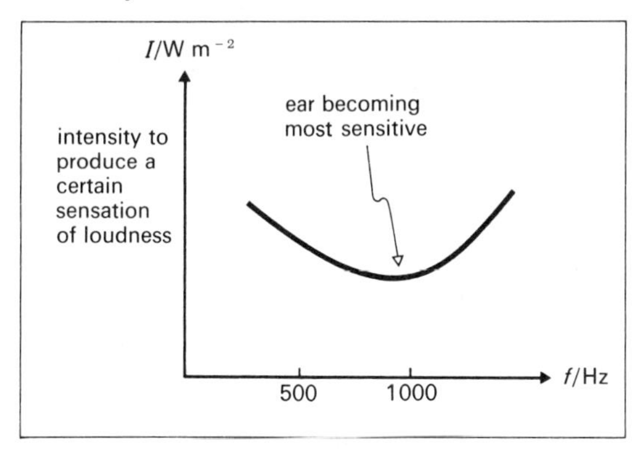

Fig. 41.3. The ear's sensitivity depends on frequency.

Fig. 41.4. Waveform (a) has a sinusoidal shape and sounds characterless beside the sound from (b), whose source is not vibrating with s.h.m.

41.4 Noise and Musical Notes

A **noise** is an audible sound which is not harmonious to the ear: it is heard when the ear-drum is disturbed by a wave whose waveform is *non-periodic*.

A **musical note** is produced when the sound is pleasant: it is heard when the waveform is *periodic*. The air molecules which transmit the energy undergo a periodic motion, which is the result of superposing a relatively small number of simple harmonic motions.

Musical notes are described conveniently by three properties or characteristics:

(1) **Loudness**, which was discussed in **41.3.**

(2) **Pitch**. Pitch is a subjective property of a musical note that is determined principally by the *frequency* of vibration of the air molecules, which in turn is fixed by the frequency of vibration of the source. It determines the position of a particular note in the musical scale, being analogous to colour in the visible spectrum.

(3) **Quality**. The quality of a note is controlled by the detailed behaviour of the air molecules, which is determined by the exact mode of vibration of the source.

Two notes of the same amplitude and frequency from different instruments will show *different waveforms* on an oscilloscope, and are said to have different quality or **timbre** (fig. 41.4).

(See also *overtones* and *harmonics* on p. 304.)

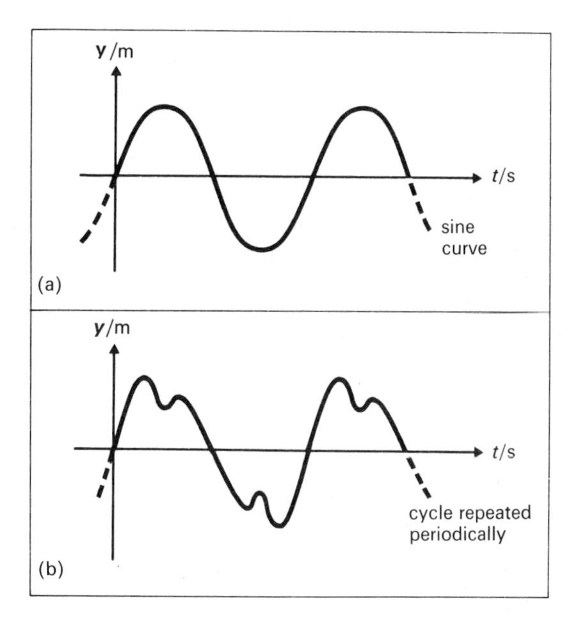

41.5 General Properties of Sound Waves

Being a longitudinal wave motion, sound has all the properties of waves discussed on p. 94 except polarization.

(a) Reflection

Reflection occurs such that the law of reflection is obeyed: this enables images (**echoes**) to be formed. *Periodic* reflection of a noise from a set of railings results in a musical note.

(b) Refraction

When sound energy is incident on a boundary at which the wave speed changes, we observe

(1) part of the energy carried back into the first medium by *partial reflection*,

(2) part of the energy *transmitted* into the second medium (unless there is total reflection).

The transmitted energy shows.

(*i*) a change of direction of travel, and

(*ii*) a change of wavelength λ, since the frequency (being determined by the source) does not alter.

For a given pair of media we can define a **relative refractive index** by

$$_1n_2 = \frac{c_1}{c_2}$$

and according to *Snell's Law* (p. 95)

$$_1n_2 = \frac{c_1}{c_2} = \frac{\sin\theta_1}{\sin\theta_2} = \text{constant}$$

for waves of a given frequency.

Ripple tank analogies exist. A balloon filled with a relatively dense gas would become a *converging* agent (p. 96); one filled with a gas of low density would be a diverging agent. Simple experiments on focusing are complicated by diffraction effects (since when $f \sim 300$ Hz (say), then $\lambda \sim 1$ m).

The effect of temperature change on wave speed is discussed on p. 302.

(c) Total Internal Reflection

This is a general wave property, and is not confined to light waves.

If (1) medium 2 is rarer than medium 1, *and*

(2) θ_1 exceeds the critical angle θ_c

then (apart from local effects) no energy is transmitted into medium 2. (We can find θ_c from

$$\theta_c = \arcsin {_1n_2} = \operatorname{arccosec} {_2n_1}.)$$

Examples.

(*i*) The Whispering Gallery of St. Paul's Cathedral.

(*ii*) Ships' speaking tubes (fig. 41.5).

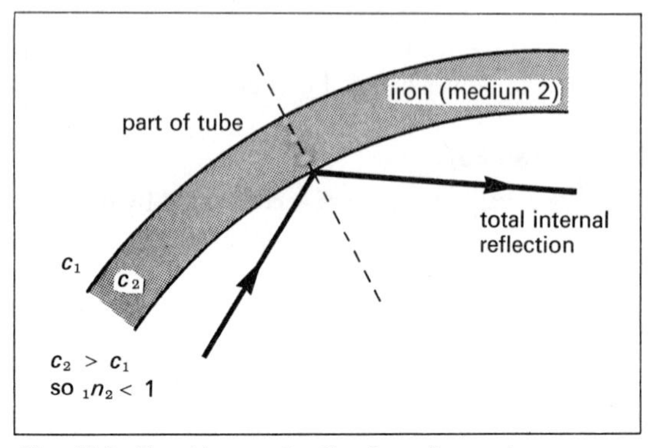

Fig. 41.5. Total internal reflection of sound energy.

(d) Dispersion

For waves whose frequency lies in the *audible range*, the wave speed in air is independent of wavelength, i.e. dispersion does *not occur*. (The different harmonics of a cymbal clash are not separated in time for an observer 100 m distant.) Dispersion does occur at ultrasonic frequencies.

(e) Speed

The speed c of sound in air is about 330 m s^{-1}. Sources of sound often have speeds v such that (v/c) is relatively large. Hence the *Doppler effect* (p. 112) is frequently observed.

Reverberation

The energy given out by a sound source in a large hall reaches the listener

(*i*) directly,

(*ii*) by regular reflection (echoes), and

(*iii*) by diffuse reflection.

When each route produces sounds of appreciable intensity, a noise emitted at a given instant by the source takes some time to die away at the location of the listener. The phenomenon is called **reverberation**, and is of importance in the acoustics of buildings.

41.6 Properties Based on Superposition

(a) Interference

A sounding tuning fork shows an interference pattern. If the fork is held with the shaft vertical, and rotated once about a vertical axis close to the ear, the loudness of the sound shows four maxima, and four minima.

Interference experiments are often complicated by unwanted reflections, but a *Young's* experiment can be carried out using two identical speakers activated by the same oscillator. Unless a pure note is used, the fringe pattern will be analogous to that produced by white light (p. 374). *Quincke's method* (p. 306) gives a simple demonstration of superposition in sound.

(b) Beats

(Refer to the explanations on p. 109.)

When two notes of comparable frequency are sounded together, the *intensity* or loudness at a given location varies in time with a frequency

$$f_{beat} = (f_1 \sim f_2)$$

Suppose f_1 is known, and f_{beat} is measured, then f_2 can be calculated from

$$f_2 = f_1 \pm f_{beat}$$

To discover whether

$$f_1 < f_2 \quad \text{or} \quad f_1 > f_2$$

one varies either in a *known* direction (e.g. loading a tuning fork decreases f), and observes whether f_{beat} increases or decreases. This method is used for *tuning*.

Beats will also be heard when a wave is superposed on its own reflection from a *moving* reflector. According to the *Doppler effect* the reflected wave has a changed frequency (p.112).

(c) Stationary Waves

(Refer to the full account on p. 104.)

Superposition between waves emitted by identical loudspeakers (energized by the same alternating current) set up facing each other would result in a stationary wave pattern. There would be no net energy transfer along the line joining them.

In practice it is easier to superpose a wave on its reflection from a *still* reflector. It is most important to distinguish

(1) *displacement nodes* and *pressure antinodes* which occur together, from

(2) *displacement antinodes* and *pressure nodes*, which also occur together.

The adjacent node-antinode separation for each is always $\lambda/4$, but it is safest to specify which type is considered. (See p. 303.)

(d) Diffraction

Diffraction is discussed fully in the section on physical optics (chapter **39**). Results obtained there apply fully to sound waves, but of course the scale of apparatus is modified to take into account the greater wavelength involved.

Diffraction effects are important in the design of foghorns and directional loudspeakers.

41.7 Speed of Sound by a Dimensional Argument

Suppose the wave speed c through a gas is governed by

(a) an elasticity factor, its bulk modulus K,
(b) an inertia factor, its undisturbed density ρ, and
(c) the wavelength λ of the waves.

Provided no relevant factor has been omitted (of which we cannot be sure), then

$$c = (\text{dimensionless constant})K^x\rho^y\lambda^z$$

Equating dimensions

$$[\mathrm{LT^{-1}}] = [\mathrm{MLT^{-2} \cdot L^{-2}}]^x[\mathrm{ML^{-3}}]^y[\mathrm{L}]^z$$

Equating separately the dimensions of

T, we have $\quad -1 = -2x$
M, $\quad 0 = x + y$
L, $\quad 1 = -x - 3y + z$

from which

$$x = \tfrac{1}{2}, \quad y = -\tfrac{1}{2} \quad \text{and} \quad z = 0$$

Thus

$$c \propto \sqrt{\frac{K}{\rho}}$$

The dimensionless constant is shown by full analysis to be 1.

$$\boxed{c = \sqrt{\frac{K}{\rho}}}$$

According to this argument the speed does not depend on the wavelength, unless this controls the value of K.

If we assume that conditions are

(*a*) *isothermal* (p. 189), as *Newton* did, we can show

$$K_{\text{iso}} = p$$

(*b*) *adiabatic* (p. 190), as *Laplace* did, we can show

$$K_{\text{ad}} = \gamma p,$$

where

$$\gamma = \frac{C_{p,\text{m}}}{C_{V,\text{m}}} \qquad \text{(p. 190)}$$

Experiment indicates that $K = \gamma p$, i.e. that sound waves in the audible range are propagated *under adiabatic conditions*.

The explanation is complex, but we suppose that at audible frequencies there is insufficient time for the wave energy to flow as heat from a compression to a relatively distant rarefaction, because the temperature gradient is too small. We use

$$\boxed{c = \sqrt{\frac{\gamma p}{\rho}}}$$

Speed of Sound through Bars

An expression for the speed of any mechanical wave through an elastic medium will be of the form

$$c = \sqrt{\frac{\text{elasticity factor}}{\text{inertia factor}}}$$

For example, the speed of longitudinal waves in a solid rod, such as that used in the dust tube experiment, is given by $c = \sqrt{E/\rho}$, where E is the material's *Young* modulus and ρ its density.

41.8 Variations in the Speed of Sound

For an amount μ[mol] of an ideal gas

$$pV = \mu RT$$

If we put M_{m} = molar mass, and M_{g} = mass of a particular gas sample, then

$$\rho = \frac{M_{\text{g}}}{V} = \frac{\mu M_{\text{m}}}{V} = \frac{pM_{\text{m}}}{RT}$$

Substituting in

$$c = \sqrt{\frac{\gamma p}{\rho}}$$

we have

$$c = \sqrt{\frac{\gamma RT}{M_{\text{m}}}}$$

(a) Pressure

Changes in pressure are proportional to changes in density, and do not affect γ, R or M_{m}. At a fixed temperature, the speed of sound is *independent of pressure*.

(b) Temperature

Provided the temperature change does not alter either γ or M_{m}, we have shown

$$c \propto \sqrt{T}, \quad \text{i.e.} \quad \frac{c_1}{c_2} = \sqrt{\frac{T_1}{T_2}}$$

This is an important factor in the tuning of wind instruments.

(c) Nature of Gas

If the nature of the transmitting gas changes, then so may γ and M_{m}. Thus

(*i*) γ decreases as the gas atomicity increases (p. 192): this decreases c, but

(*ii*) a low value of M_{m} corresponds to a low gas density, and a higher value for c.

Variations in Atmospheric Conditions

(a) Effect of Humidity

An increase in the proportion of (triatomic) water vapour molecules decreases γ slightly and M_{m} considerably. Since the latter effect predominates, the wave speed in humid air slightly exceeds that in dry air.

(b) Effect of Wind

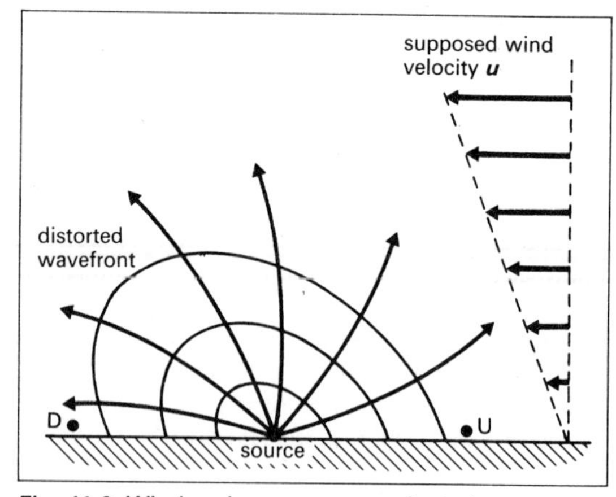

Fig. 41.6. Wind makes a source easier to hear downwind (at D), but more difficult upwind (at U).

The wind velocity $\boldsymbol{u}$ is added vectorially to that of the sound: since $\boldsymbol{u}$ usually increases with height, the wave-front is distorted according to fig. 41.6.

(c) Audibility and Temperature Gradient

Fig. 41.7 is explained by

(*i*) the speed of sound in warm air exceeding that in cold air, and

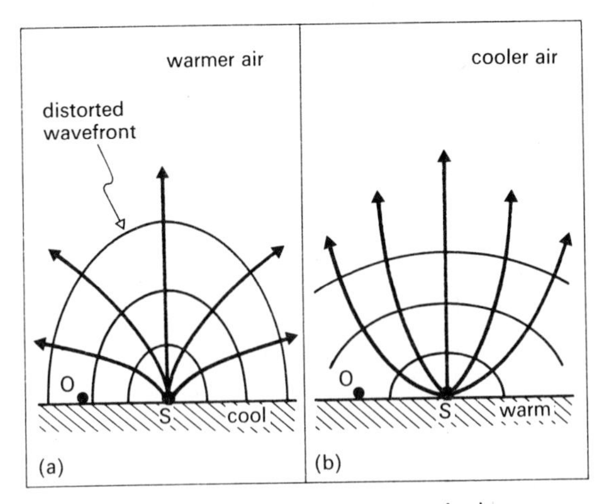

Fig. 41.7. In (a) (perhaps a summer evening) refraction causes distant sounds to be heard distinctly, whereas in (b) (a summer morning) refraction has taken the sound away from the Earth.

(*ii*) air temperature being largely controlled by that of the ground, or an open expanse of water, rather than by direct radiation from the Sun.

41.9. Vibration of Air Columns

Principles

(*a*) If the air inside a pipe is disturbed by a source of sound such as a vibrating reed, waves travel in both directions away from the source.

(*b*) The conditions at the ends of the pipe (the *boundary* conditions) will determine the phase of any reflected waves.

(*c*) Reflected waves superimposed on incident waves will then establish a *stationary wave pattern* according to paragraph **14.2** (p. 104).

(*d*) *Resonance* will occur only for those waves whose wavelengths are correctly matched to the length of the pipe. On reflection, such waves cause stimulated emission of sound from the source, thereby increasing their intensity.

Boundary Conditions

(*a*) *At a closed end,*

(*i*) the displacement amplitude must be zero

(*ii*) the pressure amplitude is therefore a maximum

(*iii*) the reflected pressure wave is in phase with the incident wave.

(*b*) *At an open end* (supposing the tube to be narrow relative to the wavelength),

(*i*) the pressure amplitude must be zero

(*ii*) the displacement amplitude is therefore a maximum

(*iii*) the reflected wave is effectively π rad out of phase with the incident wave.

A small length of air outside a tube is also set into vibration, so the effective length of a vibrating air

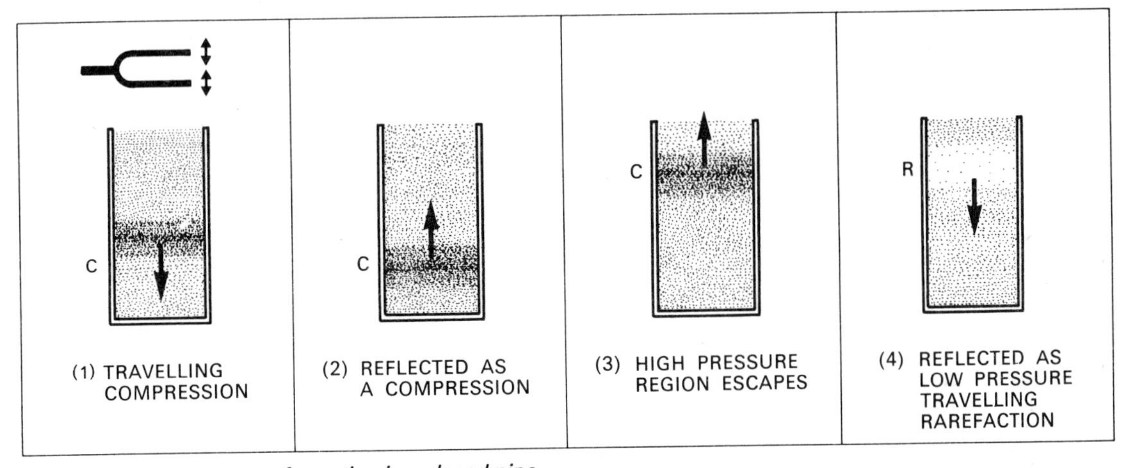

Fig. 41.8. Stationary wave formation in a closed pipe.

column exceeds the length of the pipe by an **end-correction**. The correction (about 0.6 × (pipe radius) for each open end) is illustrated on p. 307.

Example of Stationary Wave Formation

In fig. 41.8 stationary waves result from superposition of reflected waves on the incident progressive waves.

Resonance would occur if the reflected rarefaction in (4) entered the pipe at the same instant as a rarefaction from the tuning fork (constructive superposition).

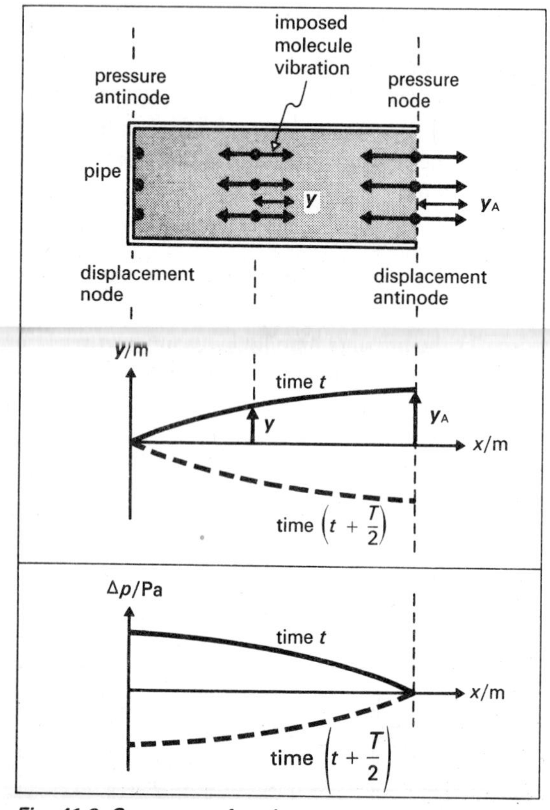

Fig. 41.9. Summary of stationary wave conditions inside a pipe.

Fig. 41.9 shows how longitudinal molecule displacement y and excess pressure Δp are related to molecular location x.

Modes of Vibration

Definitions

(*a*) The **fundamental** (mode of vibration) is that whose frequency f_0 is the lowest obtainable from the pipe.

(*b*) An **harmonic** is a note whose frequency is a whole number multiple of f_0.

(*c*) An **overtone** is a note whose frequency is actually obtainable from a particular pipe.

For the diagrams that follow, N and A represent the locations of *displacement* nodes and antinodes. The distance AN = $\lambda/4$.

(a) Closed (Stopped) Pipes

Fig. 41.10 demonstrates that the first overtone is

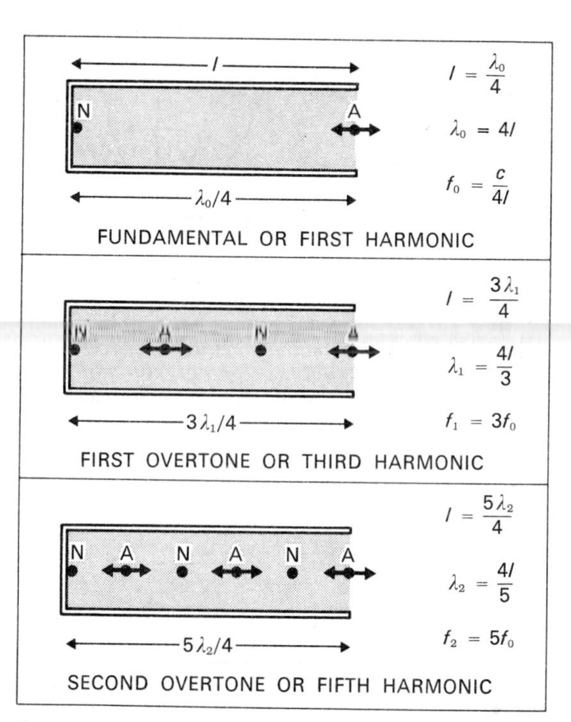

Fig. 41.10. First three modes of vibration of a closed pipe (a pipe stopped at one end).

the third harmonic, the second overtone is the fifth, etc.

A closed pipe gives odd harmonics only

The overtone frequencies are

$$f_m = (2m + 1)f_0$$

where m is the number of the overtone.

(b) Open Pipes

Fig. 41.11 demonstrates that *open* pipes give both *even and odd* harmonics, the *m*th overtone having a frequency

$$f_m = (m + 1)f_0$$

Notes.

(*a*) For a given length of pipe, the fundamental of the closed pipe has half the frequency of that of the open.

(*b*) The presence of extra overtones in the note from an open pipe gives it a richer tone than that from a closed pipe. The higher overtones are promoted by hard blowing in a musical instrument.

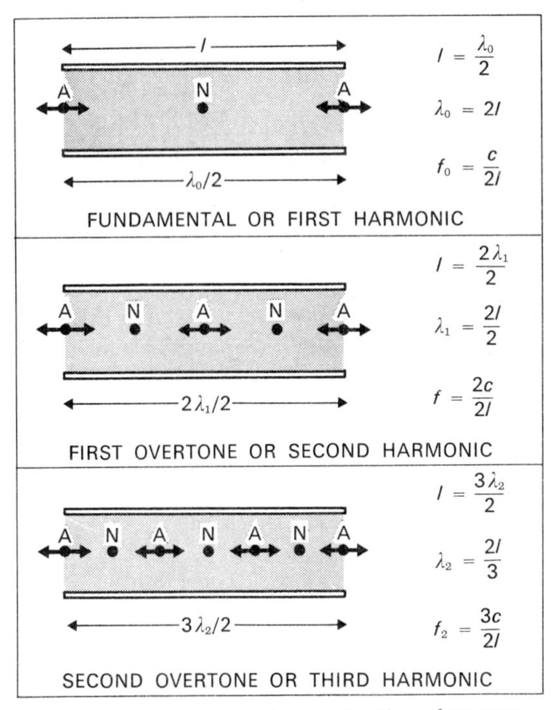

Fig. 41.11. First three modes of vibration of an open pipe.

The location of nodes and antinodes can be demonstrated by

(*a*) *jumping sand* on taut silk: the sand is buffeted by moving air molecules at *displacement antinodes*

(*b*) *manometric capsule*: a rubber membrane responds actively to the pressure variations at *pressure antinodes*, and can cause a flame to flicker.

41.10 Vibration of Strings

Principles and Boundary Conditions

(*a*) A transverse wave travels at speed c along a string mass per unit length μ and tension T, where

$$c = \sqrt{\frac{T}{\mu}}$$

Remembering that $[\mu] = [\mathrm{ML}^{-1}]$, the reader should verify this equation using dimensional analysis.

(*b*) A stretched string is necessarily fixed at each end: the ends are therefore *displacement nodes.*

(*c*) When the string is set vibrating at frequency f, a *stationary wave* pattern of wavelength λ is established, where

$$\lambda = \frac{1}{f}\sqrt{\frac{T}{\mu}}$$

(*d*) *Resonance* occurs for those waves whose wavelengths are correctly matched to the length of the string. This is illustrated by *Melde's experiment* (fig. 41.12).

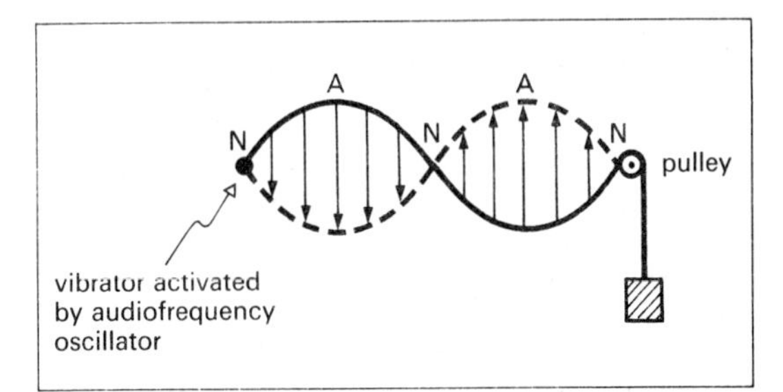

Fig. 41.12. Melde's experiment: the string vibrates with a large amplitude only when the frequency is correctly adjusted.

Notes.

(*i*) The point to which the vibrator is attached is (surprisingly) nearly a displacement *node.*

(*ii*) The relative phases of the adjacent string segments are clearly revealed if it is illuminated by a flashing xenon lamp (a **stroboscopic** device).

Modes of Vibration

Fig. 41.13 illustrates that a stretched string produces both odd and even harmonics.

The overtone frequencies are given by

$$f_m = (m + 1)f_0 = \frac{m + 1}{2l}\sqrt{\frac{T}{\mu}}$$

where m is the number of the overtone.

A string can vibrate in several modes at the same time. Their frequencies and relative intensities are determined by the original disturbance, and control the *quality* of the note emitted (as a travelling wave) from the string.

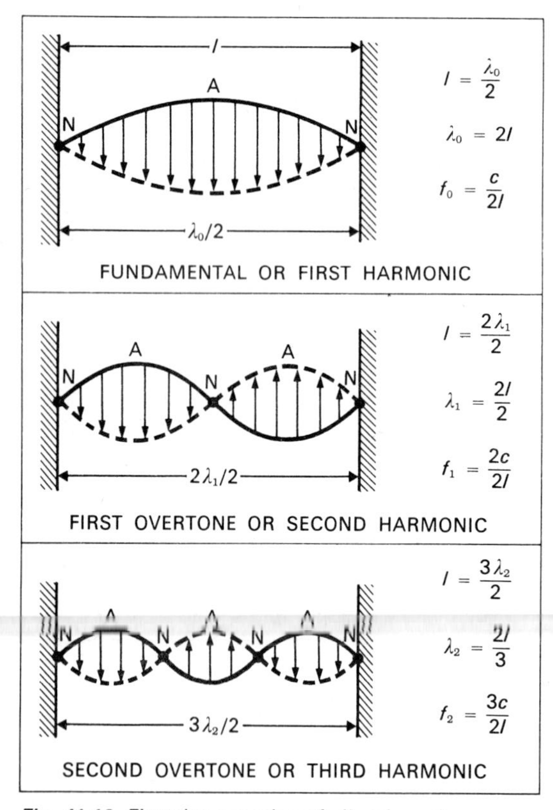

Fig. 41.13. First three modes of vibration of a stretched string.

41.11 Measurement of c, f and λ

Since $c = f\lambda$, if we measure (or assume) any two, the value of the third can be calculated. The reader should refer to a textbook on sound for details.

(a) Speed c

(*i*) Time sound over a known distance.

(*ii*) *Quincke's tube method* (fig. 41.14).

The sound from S has two possible paths (A and B) to the detector D. If the path difference is an odd number of half wavelengths, compressions from A meet rarefactions from B at D. The sliding tube is moved to find two consecutive positions of minimum intensity. The distance between these positions is $\lambda/2$.

(*iii*) *Hebb's method* is very similar, and gives a more precise result.

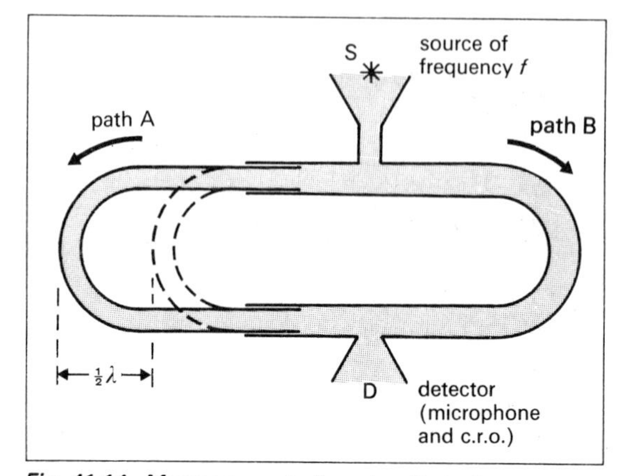

Fig. 41.14. Measurement of c using Quincke's tube.

(b) Frequency f

(*i*) The frequency of a tuning fork can be measured by the *rotating disc method*, in which the prong causes a sinusoidal mark on a disc rotating at known speed.

(*ii*) The frequency of a vibrating object can be found by viewing it through an adjustable *stroboscope*.

(*iii*) Two frequencies which are close can be related by their *beat frequency*.

The known frequency could come from

(1) a standard fork,
(2) a variable audio-frequency oscillator and loudspeaker,
(3) a siren, or
(4) a stretched wire (the **sonometer**).

(c) Wavelength λ

(*i*) *Kundt's dust tube*: illustrated by fig. 41.15. Standing waves in a gas are revealed by the periodic pattern made by dust in an enclosed tube. The dust may be agitated either by rubbing a brass rod which is clamped at its mid-point with one end projecting into the dust tube, or by means of a loudspeaker fitted over one end of the tube. The experiment is useful for comparing the speeds of sound in different gases.

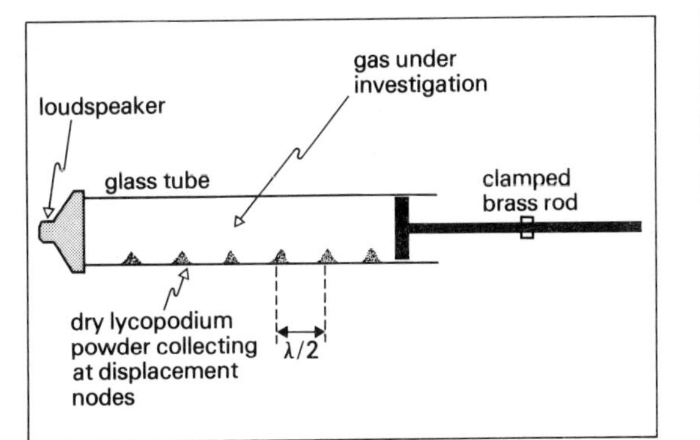

Fig. 41.15. Kundt's dust tube. (The loudspeaker and brass rod are alternatives.)

(*ii*) *The resonance tube*: illustrated by fig. 41.16. If l_0 and l_1 are the lengths of the tube for the first two positions of resonance and δ is the end-correction, then

$$\frac{\lambda}{4} = l_0 + \delta \quad \text{and} \quad \frac{3\lambda}{4} = l_1 + \delta$$

$$\therefore (l_1 - l_0) = x = \frac{\lambda}{2}$$

Measurement of x therefore eliminates the end-correction. If the tuning fork frequency f is known, the value of c can be calculated.

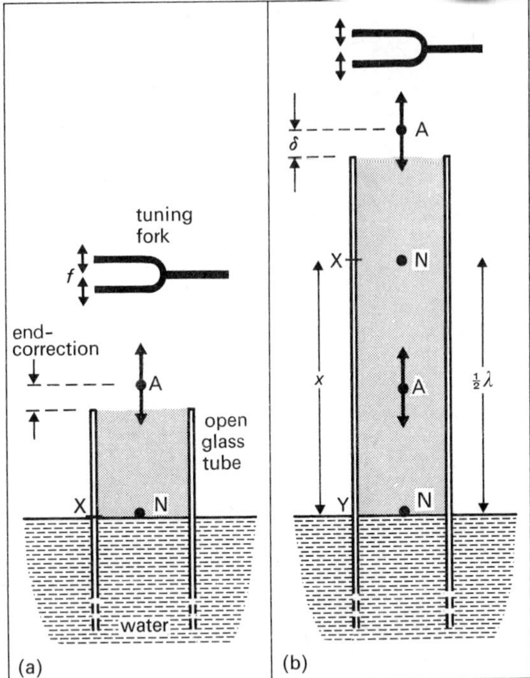

Fig. 41.16. Finding λ by resonance tube. (a) First resonance (fundamental), (b) second resonance (first overtone of the longer tube).

IX Electrostatics

42 Charge and Coulomb's Law

42.1 Conductors and Insulators

Elementary experiments on rubbed bodies indicate that

(*a*) there are two, and only two, kinds of electric charge,
(*b*) unlike charges attract each other, but
(*c*) like charges repel.

The charge associated with rubbed *ebonite* is *negative* by choice. This choice determines that the charge of an electron is also *negative*.

A charge placed on an **insulator** (or **dielectric** material) is confined to the region in which it was placed. An insulator has no *charge-carriers* that are free to migrate within the boundaries of the body.

A charge placed on a **conductor** may be allowed to spread over the whole surface of the body, because the body has charge-carriers which are free to migrate. The **Hall Effect** (p. 375) enables us to find the *sign* of the charge-carriers.

(*a*) In *metals*, which are good conductors, the charges are carried by **free** or **conduction electrons** (negative).

(*b*) In *electrolytes*, which are intermediate conductors, the charge-carriers are ions and have both signs, whereas

(*c*) in *semiconductors* (p. 351) the charge carriers may appear to be positive, or negative, or both.

42.2 Coulomb's Law (1785)

The law was formulated from experiments using the *torsion balance* of fig. 42.1.

Coulomb twisted the torsion head to maintain equilibrium: the angle of twist θ is a measure of the (electric) force $\boldsymbol{F}$ which acts on *each* (*Newton's Law III*).

He varied

(*a*) the charge separation r, and showed

$$F \propto \frac{1}{r^2}$$

(*b*) the charge magnitudes Q_1 and Q_2 to show

$$F \propto Q_1 Q_2$$

He concluded that for *point charges*

$$F \propto \frac{Q_1 Q_2}{r^2}$$

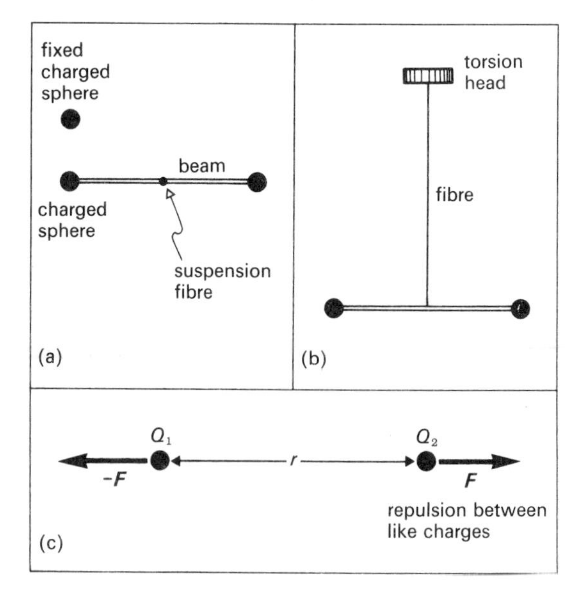

Fig. 42.1. Coulomb's torsion balance experiment: (a) plan view, (b) elevation, (c) symbols.

The electric force between static electric charges is sometimes referred to as the **Coulomb force**.

This force

(*a*) obeys the *superposition principle*—a third charge will not affect the forces exerted by two charges on each other

(*b*) is a *conservative* force (p. 46), and

(*c*) is a **central** force—it acts along the line joining the two point charges.

Verification of Coulomb's Law

(1) The *torsion balance* is accurate to a few per cent.

(2) The field inside a hollow charged conductor can only be zero for a strict inverse square law. *Cavendish* showed (to about 2%) that this was so for a sphere.

(3) *Plimpton* and *Lawton* repeated *Cavendish's* experiment to show an inverse nth law,

where $$n = 2 \pm 2 \times 10^{-9}$$

(4) *Rutherford's* scattering experiments with α-particles and the nucleus (a very close approach to a point charge) show that the law holds down to at least 10^{-15} m (p. 464).

42.3 The Permittivity Constant

Coulomb's Law can be written for a vacuum as

$$F = k \cdot \frac{Q_1 Q_2}{r^2}$$

in which k is a constant. We already have units for

(*a*) F [N] where $1\text{ N} = 1\text{ kg m s}^{-2}$

(*b*) r [m], and

(*c*) Q [coulomb, symbol C] (p. 344)

so k must be *found by experiment.* (Cf. the value of G on p. 152.)

$$k = \frac{Fr^2}{Q_1 Q_2} \approx 9.0 \times 10^9 \frac{\text{N m}^2}{\text{C}^2}$$

In the **rationalized system** of units it is convenient to write

$$k = \frac{1}{4\pi\varepsilon_0}$$

where ε_0 is called the **permittivity constant,** the **permittivity of free space**, or sometimes simply the **electric constant**. It follows that

$$\varepsilon_0 = 8.9 \times 10^{-12} \frac{\text{C}^2}{\text{N m}^2}$$

which is more usually written

$$8.9 \times 10^{-12}\text{ F m}^{-1} \qquad \text{(p. 330)}$$

ε_0 has the dimensions $[\text{M}^{-1}\text{L}^{-3}\text{T}^4\text{I}^2]$.

Thus *for a vacuum* we write **Coulomb's Law** as

$$\boxed{F = \left(\frac{1}{4\pi\varepsilon_0}\right) \cdot \frac{Q_1 Q_2}{r^2}}$$

(See p. 321 for a comment on rationalization.)

In chapters **42** to **46**, we assume charges to be placed *in air*, in which case k has about the same value as it does in a vacuum.

42.4 Charge and Matter

Matter can be considered as made up of electrons, protons and neutrons (p. 118). Consider a hydrogen atom made up of an electron encircling a proton. Using experimental values for G and ε_0 we calculate the forces they exert on each other as follows:

(*a*) $$F_{\text{elec}} = \frac{1}{4\pi\varepsilon_0} \cdot \frac{Q_1 Q_2}{r^2} \approx 8 \times 10^{-8}\text{ N, and}$$

(*b*) $$F_{\text{grav}} = G \cdot \frac{m_1 m_2}{r^2} \approx 4 \times 10^{-47}\text{ N.}$$

Thus $$\frac{F_{\text{elec}}}{F_{\text{grav}}} \sim 10^{39}$$

This demonstrates that *Coulomb forces* (*not* gravitational forces) are responsible for binding

(*i*) electrons to nuclei to form atoms,

(*ii*) atoms to atoms to form molecules and

(*iii*) molecules to molecules to form solids and liquids.

(The nucleus has a net positive charge—it is *not* the *Coulomb forces* that bind it together.)

42.5 Charge Quantization and Conservation

(*a*) Quantization

The charge of an electron is $-e$, where

$$e = 1.6 \times 10^{-19}\text{ coulomb}$$

No particle has been observed to carry a fraction of this charge.

Any charge Q that we observe in the laboratory is given by

$$Q = Ne$$

where N is a positive or negative integer, and e is the *fundamental electric charge* (or *quantum* of charge).

Because e is so small and N consequently so large, in the usual laboratory experiments Q behaves as though it were continuous—the 'graininess' is not apparent.

(*b*) Conservation

There are no known exceptions to the law that *charge is conserved.*

Examples

(*i*) When silk rubs glass, electrons are transferred from glass to silk, but the *pair* of them remain neutral. They constitute an *isolated* system.

(*ii*) During an annihilation process. (p. 463)

$$\underset{(-e)}{\text{electron}} + \underset{(+e)}{\text{positron}} \rightarrow \underset{\text{(zero charge)}}{2\ \gamma\text{-rays}}$$

(*iii*) During a radioactive transformation, such as

$$^{238}_{92}\text{U} \rightarrow {}^{234}_{90}\text{Th} + {}^{4}_{2}\alpha$$

$$\text{the initial charge} = 92e, \text{ and}$$
$$\text{final charge} = (90 + 2)e.$$

In an isolated system, the net charge is constant.

42.6 Measurement of Charge

Electrostatic charges are conveniently measured in the laboratory using a *d.c. amplifier* (described in **46.9**). A capacitor is connected across the input, and the small charge (typically a few nanocoulombs) is put onto a conductor connected to the upper input terminal (fig. 46.14). This causes a small current to flow in the input. It is amplified by a factor of about 10^{10}, enabling the output current to operate a milli-ammeter.

43 The Electric Field

43.1 Concept of the Electric Field

An **electric field** is said to exist at a point if a force of electrical origin is exerted on a test charge placed at that point.

It is convenient to consider the *interaction* in two parts, as shown in fig. 43.1.

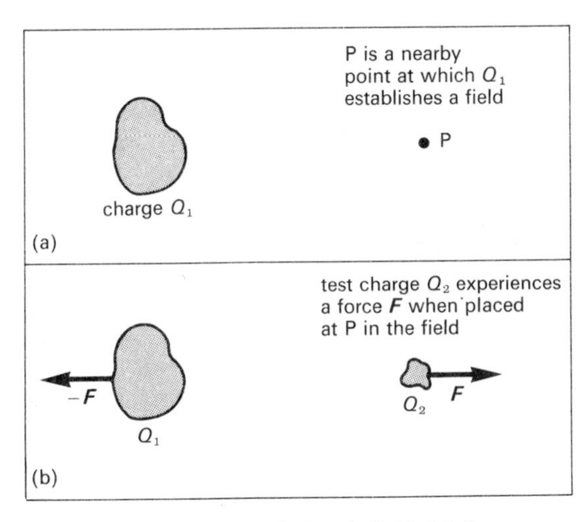

Fig. 43.1. The concept of electric field: (a) the cause of the field, and (b) the effect of the field.

The problems are

(*a*) calculating the field set up by a certain distribution of charges, and

(*b*) calculating the force experienced by a charge when placed in the field.

Analogous fields include

(*i*) the gravitational field (vector $\boldsymbol{g}$),

(*ii*) the magnetic field (vector $\boldsymbol{B}$),

(*iii*) fluid flow, which can be described by a vector $\boldsymbol{v}$, the velocity of flow at a particular point (p. 70).

The corresponding vector here is given the symbol $\boldsymbol{E}$.

43.2 Electric Field Strength E

$\boldsymbol{E}$ is called **electric field strength**, the electric *field intensity*, the electric *vector* or simply the electric field.

Suppose we place a test charge Q_0 at a point in an electric field where it experiences a force $\boldsymbol{F}$.

The electric field strength at that point is defined by the equation

$$\boldsymbol{E} = \frac{\boldsymbol{F}}{Q_0}$$

$$\boldsymbol{E}\left[\frac{\mathrm{N}}{\mathrm{C}}\right] = \frac{\boldsymbol{F}\,[\mathrm{N}]}{Q_0\,[\mathrm{C}]}$$

Notes.

(*a*) $\boldsymbol{E}$ represents a *vector* quantity whose direction is that of the force that would be experienced by a *positive* test charge.

(*b*) The unit for $\boldsymbol{E}$ following from the definition is $\mathrm{N\,C^{-1}}$, but later it will be seen that a *volt metre*$^{-1}$ is an equivalent unit which is frequently used (p. 324).

(*c*) The magnitude of Q_0 must be small enough not to affect the distribution of the charges that are responsible for $\boldsymbol{E}$ – that is Q_0 *must be very small* if it is not to change the value of $\boldsymbol{E}$ that it is measuring.

(*d*) $\boldsymbol{E}$ is numerically equal to the force acting on unit charge placed at a point—but note (*c*).

(*e*) The dimensions of $\boldsymbol{E}$ are $[\mathrm{MLT^{-3}I^{-1}}]$.

43.3 Electric Field Lines

An **electric field line** is a convenient fictional concept developed to aid the visualization of an electric field. The field *line* is imaginary, but it represents a real *field*.

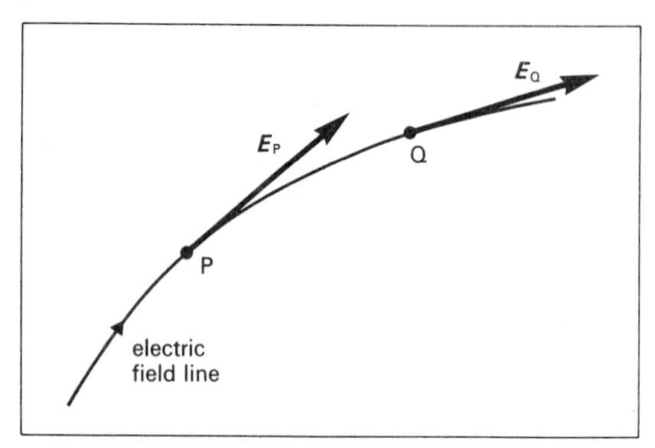

Fig. 43.2. An electric field line.

Field lines (fig. 43.2) are drawn such that

(*a*) the *tangent* to a field line at a point gives the *direction* of $\boldsymbol{E}$ at that point, and

(*b*) the *number* of field lines drawn per unit cross-sectional area is proportional to the *magnitude E*.

Field lines can be demonstrated (in strong fields) by grass seeds, which behave as iron filings do in a magnetic field.

Properties of Electric Field lines

(*a*) They begin from positive charges, and end on equal negative charges. (Note that lines of $\boldsymbol{B}$ (p. 372) do not end, but form closed loops.)

(*b*) They cannot cross (since the direction of $\boldsymbol{E}$ at a point is unique).

(*c*) Where the lines are

(*i*) close together, the field is strong,
(*ii*) far apart, the field is weak, and
(*iii*) parallel and equally-spaced, the field is uniform.

(*d*) As one travels *along* a field line in the direction of the arrow, the electric potential (p. 322) is *decreasing*.

(*e*) They cut equipotential surfaces (such as those of conductors) at right angles (p. 323).

Although a field line shows the *initial* direction of movement of a resting positive charge, it does *not* show the path of a charged particle in an electric field. (Think of the trajectory of a projectile in the Earth's gravitational field.)

The reader should be able to sketch the field lines that represent the fields of the following charge distributions:

(*a*) a negatively charged sphere,
(*b*) an infinite sheet of positive charge,
(*c*) two equal like charges, and
(*d*) two equal unlike charges.

43.4 Calculation of E at a Point

(a) Due to a Point Charge

Consider a point distance r from a point charge Q (and strictly in a vacuum).

By *Coulomb's Law*

$$F = \left(\frac{1}{4\pi\varepsilon_0}\right)\cdot\frac{QQ_0}{r^2}$$

gives the force $\boldsymbol{F}$ on a test charge Q_0.

By the definition of $\boldsymbol{E}$

$$E = \frac{F}{Q_0} = \left(\frac{1}{4\pi\varepsilon_0}\right)\cdot\frac{Q}{r^2}$$

N.B. Radial field lines ONLY

gives the magnitude E. Its direction is radial.

Note that although this is a useful result, it is for the *special case* of a field established by a *point charge*, and it takes no account of any charges that may be induced on nearby conductors.

(b) Due to Several Point Charges

Suppose we have several point charges Q_1, etc. Then we can

(*i*) evaluate $\boldsymbol{E}_1$, etc., as above, and
(*ii*) find $\boldsymbol{E} = \sum \boldsymbol{E}_1$

by using the principle of superposition and *vectorial* addition. There is an example in the next paragraph.

If the charge distribution is effectively continuous (e.g. throughout a volume) calculus may be necessary.

43.5 The Electric Dipole

A pair of charges $+Q$ and $-Q$ separated by a distance $2a$ constitute an **electric dipole** of electric dipole moment $\boldsymbol{p} = 2a\boldsymbol{Q}$. (See also p. 316.)

$$\boldsymbol{E}_P = \boldsymbol{E}_1 + \boldsymbol{E}_2$$

which is to be found by vector addition.

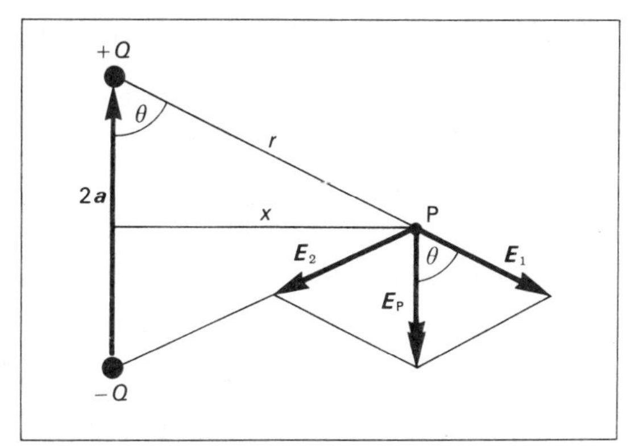

Fig. 43.3. The field set up by an electric dipole.

In magnitude

$$E_1 = E_2 = \left(\frac{1}{4\pi\varepsilon_0}\right)\cdot\frac{Q}{(a^2 + x^2)}$$

$$E_P = 2E_1 \cos\theta$$

$$= \frac{2}{4\pi\varepsilon_0}\cdot\frac{Q}{(a^2 + x^2)}\cdot\frac{a}{(a^2 + x^2)^{1/2}}$$

$$= \frac{1}{4\pi\varepsilon_0}\cdot\frac{2aQ}{(a^2 + x^2)^{3/2}}$$

Suppose $x \gg a$, so that $r \approx x$, then putting $p = 2aQ$

$$E_P = \left(\frac{1}{4\pi\varepsilon_0}\right)\cdot\frac{p}{r^3}$$

where P is a point along the equatorial axis of the dipole distance r away. The direction of $\boldsymbol{E}_P$ is given in fig. 43.3.

43.6 A Point Charge in an Electric Field

(a) Charge Released from Rest in a Field

The particle in fig. 43.4 does not experience forces resulting from its own field (unless its charge changes the charge distribution of the plates). The mass m carrying the charge Q is accelerated as is a mass in a gravitational field.

$$\boldsymbol{a} = \frac{\boldsymbol{F}}{m} = \frac{EQ}{m} \quad \text{and is uniform.}$$

The k.e. acquired by the particle in moving distance y

$= $ work done by electric force

$= QEy$

(Compare this with p.e. $= mgh$ for a gravitational field.)

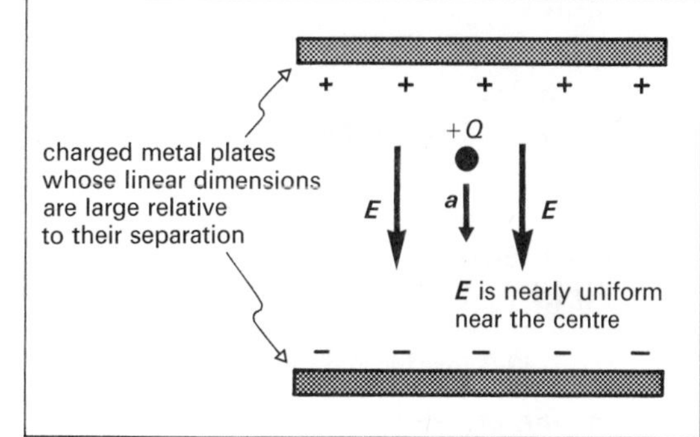

Fig. 43.4. A charged point mass in an electric field.

Application: *Millikan's* experiment (p. 440), in which an oil drop acquires a terminal velocity.

(b) Charge Moving Through a Field

Suppose an electron (fig. 43.5) in the beam has an initial horizontal velocity $\boldsymbol{u}$. Its subsequent upward acceleration while between the plates has a size (Ee/m_e), and is uniform if the electric field lines are parallel.

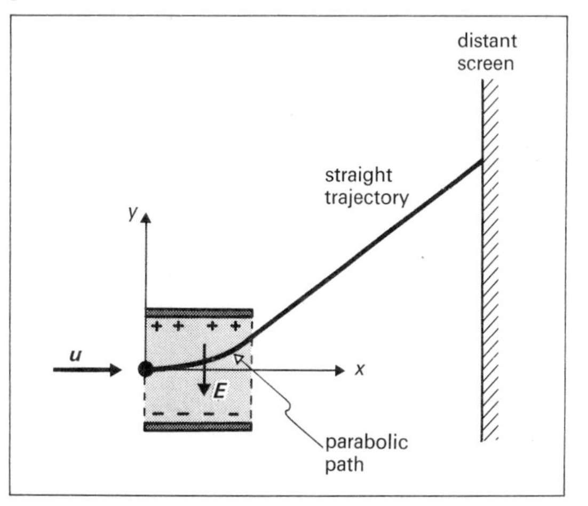

Fig. 43.5. An electron beam being deflected by an electric field.

Thus using $s = ut + \frac{1}{2}at^2$

horizontally $x = ut$

vertically $y = \dfrac{\frac{1}{2}(Ee)t^2}{m_e}$

Eliminating t

$$y = \left(\frac{eE}{2m_e u^2}\right)x^2$$

which shows that the path between the plates is parabolic (since it is of the form $y = kx^2$ (p. 25)).

Applications.

(*i*) Deflection experiments to measure e/m_e (p. 438)

(*ii*) The cathode ray oscilloscope (p. 441).

43.7 An Electric Dipole in an Electric Field

Refer to fig. 43.3 for the symbols $\boldsymbol{a}$ and Q.

We define the* electric dipole moment *by the equation

$$\boxed{\boldsymbol{p} = 2\boldsymbol{a}Q}$$

$$\boldsymbol{p}[(\mathrm{C\,m}] = 2\boldsymbol{a}[\mathrm{m}]Q[\mathrm{C}]$$

The concept of an electric dipole is an important one which is discussed further on p. 338.

(*a*) Torque Acting on a Dipole

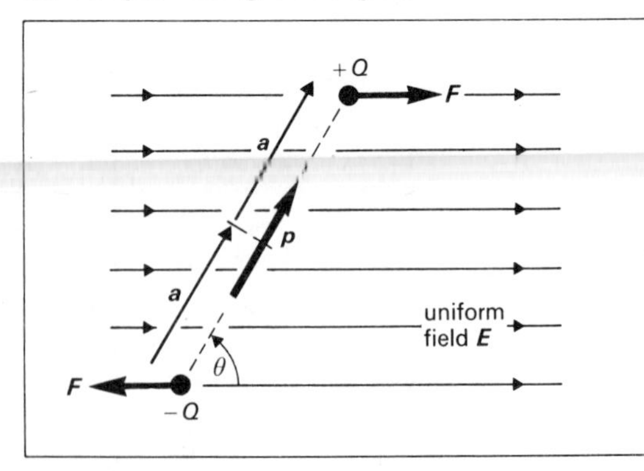

Fig. 43.6. An electric dipole experiences a torque in a uniform field but zero resultant force.

Refer to fig. 43.6.

Using $\boldsymbol{E} = \dfrac{\boldsymbol{F}}{Q_0}$ we have $\boldsymbol{F} = \boldsymbol{E}Q$

The *net force* on the dipole is *zero*.

The torque of the couple

$$\begin{aligned} T &= F(\text{arm of couple}) \\ &= EQ2a \sin\theta \\ &= (2aQ)E \sin\theta \\ &= pE\sin\theta \end{aligned}$$

Notes.

(*a*) The torque tends to align $\boldsymbol{p}$ and $\boldsymbol{E}$.

(*b*) $\boldsymbol{p}$ is a vector whose direction is shown in fig. 43.6.

(*c*) $T = pE\sin\theta$ *could* have been used to *define* electric dipole moment. (See p. 379.)

*(*b*) Energy and Dipole Orientation

Suppose an external agent exerts a torque of magnitude T while θ increases to $(\theta + \delta\theta)$ in fig. 43.7.

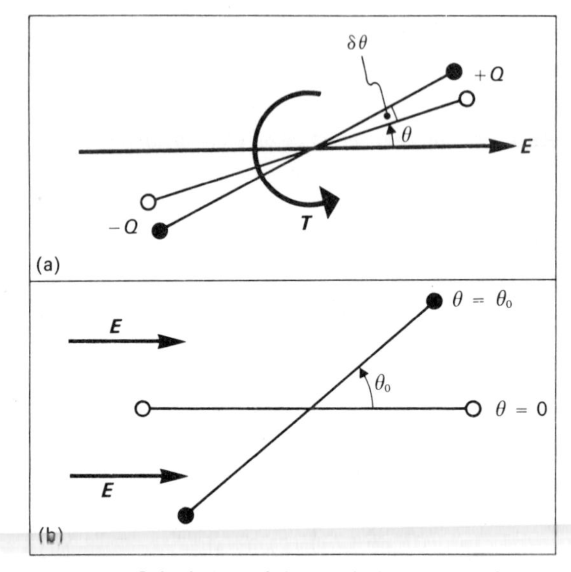

Fig. 43.7. Calculation of the work done in rotating a dipole: (a) the increment, and (b) the net change.

We assume the axis of $\boldsymbol{T}$ to be parallel to that of $\delta\boldsymbol{\theta}$. The work done by the external agent

$$\begin{aligned} \delta W &= T\delta\theta \qquad \text{(p. 57)} \\ &= (pE\sin\theta)\delta\theta \end{aligned}$$

The work done in rotating the dipole from $\theta = 0$ to $\theta = \theta_0$

$$\begin{aligned} W &= \int \delta W \\ &= \int_0^{\theta_0} pE\sin\theta\, \mathrm{d}\theta \\ &= pE(1 - \cos\theta_0) \end{aligned}$$

Notes.

(*a*) The maximum amount of work has been done when $\theta_0 = \pi$ rad: the dipole and $\boldsymbol{E}$ are then *anti-parallel*.

(*b*) The work done is stored as potential energy in the electric field set up by the system (the dipole and charges responsible for the external field).

(*c*) If we arbitrarily choose a zero of potential energy E_p at $\theta = \pi/2$ rad, we would have

$$E_p = -pE\cos\theta$$

Compare this result with that for the magnetic dipole (p. 380).

44 Gauss's Law

44.1 Electric Field Flux ψ_E

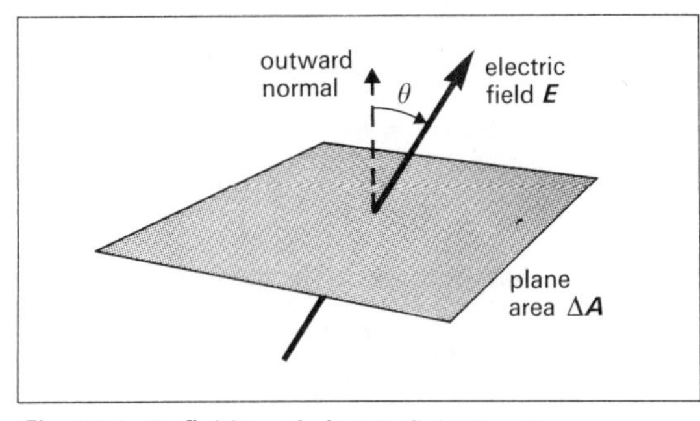

Fig. 44.1. Definition of electric field flux ψ_E.

Suppose an electric field $\boldsymbol{E}$ makes an angle θ with the outward normal to a small area $\Delta\boldsymbol{A}$, as shown in fig. 44.1. Since $\Delta\boldsymbol{A}$ is small, then $\boldsymbol{E}$ has effectively the same value at each point on the surface.

The electric field flux ψ_E through the area ΔA is defined by the equation

$$\boxed{\psi_E = E\cos\theta\,\Delta A}$$

$$\psi_E\left[\frac{\text{N m}^2}{\text{C}}\right] = E\left[\frac{\text{N}}{\text{C}}\right]\cdot\Delta\boldsymbol{A}\,[\text{m}^2]\cdot\cos\theta$$

Notes.

(*a*) We can visualize ψ_E as being proportional to the number of electric field lines that are drawn through the area $\Delta\boldsymbol{A}$.

(*b*) When $\boldsymbol{E}$ is uniform, we can calculate ψ_E from

$$\psi_E = \begin{pmatrix}\text{resolved part of } \boldsymbol{E}\\ \text{parallel to the}\\ \text{normal to an area}\end{pmatrix} \times (\text{the area})$$

(*c*) A closed surface can frequently be divided conveniently into several elementary areas such as $\Delta\boldsymbol{A}$. The total flux ψ_E through the closed surface can be found by the scalar addition of the flux through the elementary areas. (See the examples of paragraphs **44.9** and **44.10**.) ψ_E represents a *scalar* quantity.

(*d*) For a closed surface the value of ψ_E over a specified area is taken to be *positive* if the electric field lines point *outwards* ($\theta < 90°$), and *negative* if they point *inwards.*

*(*e*) If $\boldsymbol{E}$ is non-uniform, or $\boldsymbol{A}$ non-planar, then

$$\psi_E = \int E\cos\theta\,\mathrm{d}A$$

*(*f*) For a closed surface, the flux emerging is written

$$\psi_E = \oint E\cos\theta\,\mathrm{d}A$$

This means that the integral is to be taken over the whole surface.

(*g*) $[\psi_E] = [\text{ML}^3\text{T}^{-3}\text{I}^{-1}]$

44.2 Gauss's Law

Gauss's Law *states that for any closed surface*

$$\varepsilon_0 \psi_E = \sum Q$$

where $\sum Q$ represents the algebraic sum of the charges enclosed by the surface.

(This means that if a surface encloses a pair of equal but opposite charges, the *net* charge is zero.)

The actual *location* of a charge *inside* a surface is not relevant. A charge *outside* the surface makes no net contribution to ψ_E.

In the next paragraph we show that *Gauss's Law* is an alternative statement of *Coulomb's Law*: both are fundamental laws in electromagnetic theory. *Gauss's Law* can be *proved* by assuming an *inverse square law*, and is often simpler to apply than *Coulomb's Law*.

Experimental Verification of Gauss's Law

It is shown below that *Gauss's Law* predicts that the charge on an insulated conductor moves to its outer surface. Several workers have devised methods to show that this is in fact so (p. 311). Their work is an experimental verification of the law.

The Usefulness of Gauss's Law

(*a*) If we have a situation in which the charge distribution has sufficient symmetry for us to be able to calculate ψ_E, then we may be able to *evaluate* $\boldsymbol{E}$ at various points on a closed surface.

(*b*) If we know $\boldsymbol{E}$ at points on a closed surface, then calculation of ψ_E will enable us to find $\sum Q$. If in any situation we find that ψ_E is positive, this means that there is a net positive charge within that surface.

The Technique of Using the Law

(1) Draw the situation.

(2) Imagine a closed surface to be constructed which is appropriate to the symmetry. (This imaginary surface is frequently called a **Gaussian surface**.)

(3) Apply the law.

In paragraphs **44.3** to **44.10** this is done

(*a*) to deduce *Coulomb's Law*.

(*b*) to investigate the charges of insulated conductors, and

(*c*) to investigate the field established by various charge distributions.

44.3 Coulomb's Law deduced from Gauss's Law

We want to find the force $\boldsymbol{F}$ acting on a point charge Q_2 placed in a vacuum a distance r from a point charge Q_1 (fig. 44.2).

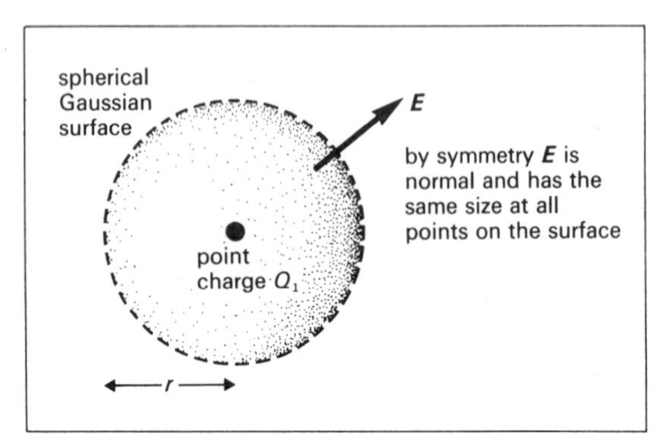

Fig. 44.2. Deduction of Coulomb's Law.

By *Gauss's Law*

$$\varepsilon_0 \psi_E = Q_1$$

in which

$$\psi_E = EA\cos\theta = E4\pi r^2$$

so

$$\varepsilon_0 E 4\pi r^2 = Q_1$$

$$E = \left(\frac{1}{4\pi\varepsilon_0}\right)\cdot\frac{Q_1}{r^2}$$

For the test charge Q_2 placed at the surface

$$F = EQ_2 = \left(\frac{1}{4\pi\varepsilon_0}\right)\cdot\frac{Q_1 Q_2}{r^2}$$

which is Coulomb's Law.

44.4 The Net Charge Migrates to the Surface of an Insulated Charged Conductor

Any charge distributes itself over a conductor very quickly until electrostatic equilibrium has been reached. Then there is no *systematic* drift of the mobile charge-carriers. Since they are not accelerated, the average resultant force acting on them is zero: the field *within the material* of the conductor must be zero at every point.

At every point on the *Gaussian* surface of fig. 44.3

$$\boldsymbol{E} = 0, \quad \text{and so} \quad \psi_E = 0$$

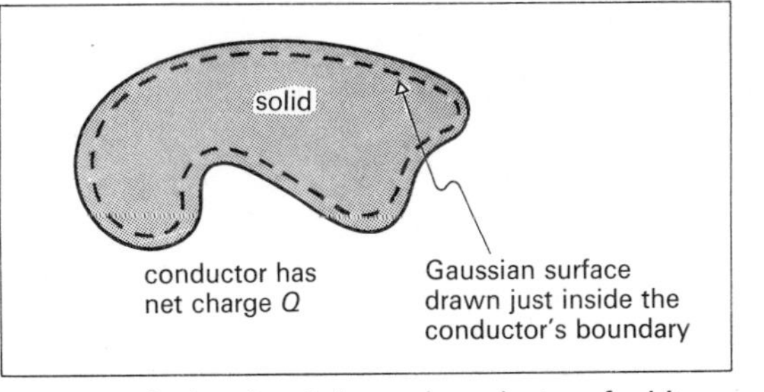

Fig. 44.3. An insulated charged conductor of arbitrary shape.

By *Gauss's Law*

$$\varepsilon_0 \psi_{\boldsymbol{E}} = \sum Q$$

in which $\psi_{\boldsymbol{E}} = 0$

so $\sum Q = 0$

There is no net charge inside the *Gaussian* surface.

Any net charge placed on an insulated conductor resides entirely on its outer surface.

(See *Faraday's* ice-pail experiment, below.)

44.5 *E* inside an Insulated Hollow Charged Conductor

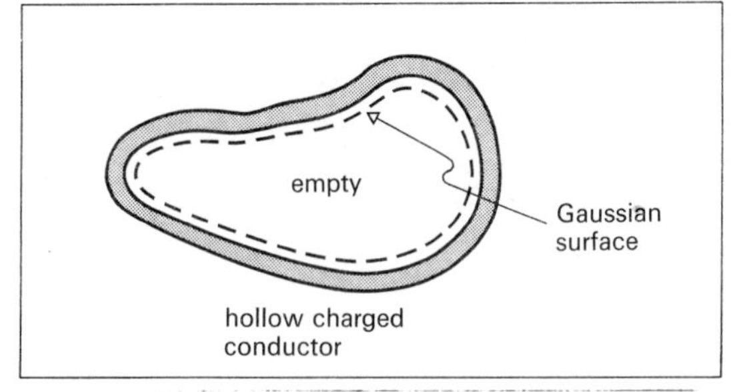

Fig. 44.4. An empty hollow charged conductor.

Since the hollow conductor of fig. 44.4 is empty, the charge Q inside the *Gaussian* surface is zero. Using

$$\varepsilon_0 \psi_{\boldsymbol{E}} = \sum Q$$

$$\psi_{\boldsymbol{E}} = 0$$

If there were a field at *any* point on the *Gaussian* surface, it would have the same direction (out of or into the surface) at *every* point. Thus $\psi_{\boldsymbol{E}} = 0$ *here* requires $\boldsymbol{E} = 0$ (but not in general).

The field must be zero at every point inside an empty hollow charged conductor.

44.6 The Magnitude of an Induced Charge

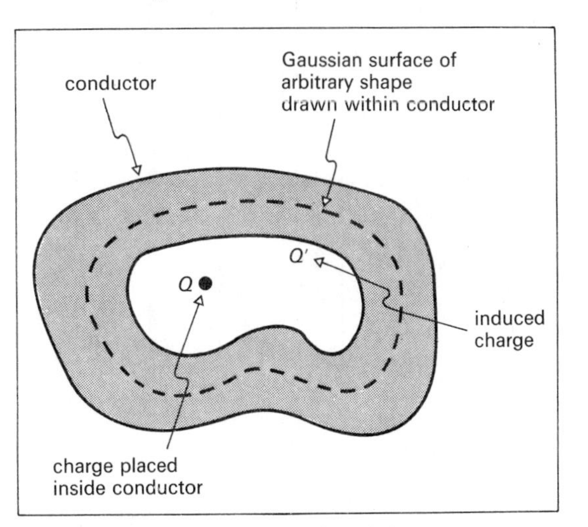

Fig. 44.5. Calculation of an induced charge.

Suppose the conductor of fig. 44.5 *totally* encloses the charge $+Q$, and that there is an induced charge Q'. At every point on the *Gaussian* surface

$$\boldsymbol{E} = 0$$

By *Gauss's Law*

$$\varepsilon_0 \psi_{\boldsymbol{E}} = \sum (\text{enclosed charge})$$

in which $\psi_{\boldsymbol{E}} = 0$

so $\sum (\text{enclosed charge}) = Q + Q' = 0$

$$\therefore Q = -Q'$$

The induced charge is equal in magnitude but opposite in sign.

(See *Faraday's* ice-pail experiment, below.)

If the conductor were originally electrically neutral, it would now carry a charge $+Q$ on its outer surface (charge conservation).

44.7. Faraday's Ice-Pail Experiment

This illustrates the ideas of paragraphs **44.4** and **44.6.** Refer to fig. 44.6 overleaf.

In (a), provided the can is deep (so as to enclose the sphere), the original charge and the positive and negative induced charges all have the same magnitude.

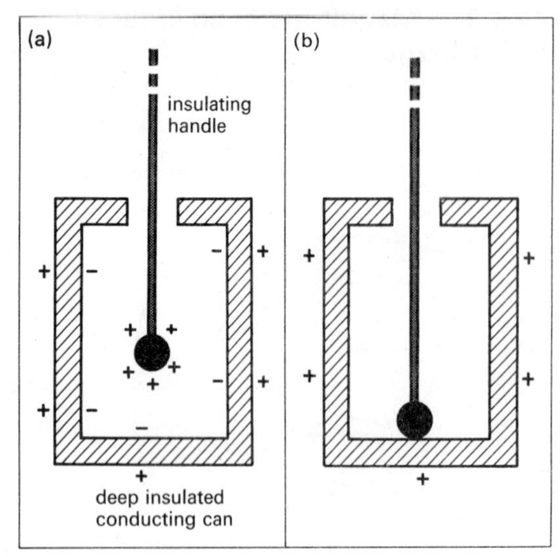

Fig. 44.6. Faraday's ice-pail experiment: (a) the induced charges, (b) charge transfer on contact.

In (*b*) neutralization has occurred: the net effect is that the ball's entire charge is transferred to the *outside* of the can.

This idea is used in the *Van de Graaff generator.*

44.8 *E* due to a Spherically Symmetric Charge Distribution

Suppose the charge density $\rho[\mathrm{C\,m^{-3}}]$ in any small volume is a function of r only (but not of direction).

By choice the *Gaussian* surface of fig. 44.7 is concentric with the spherical charge distribution.

By *Gauss's Law*

$$\varepsilon_0 \psi_E = Q$$

in which $$\psi_E = E 4\pi r^2$$

so $$\varepsilon_0 E 4\pi r^2 = Q$$

$$E = \left(\frac{1}{4\pi\varepsilon_0}\right)\cdot\frac{Q}{r^2}$$

At points outside the charge distribution, the charge behaves as though it were all placed at the centre.

A spherical shell makes no contribution to the value of *E inside* the shell. (Refer to the diagram on page 155.)

We use these results in gravitation, where frequently they are derived by integration from the inverse square law (p. 152).

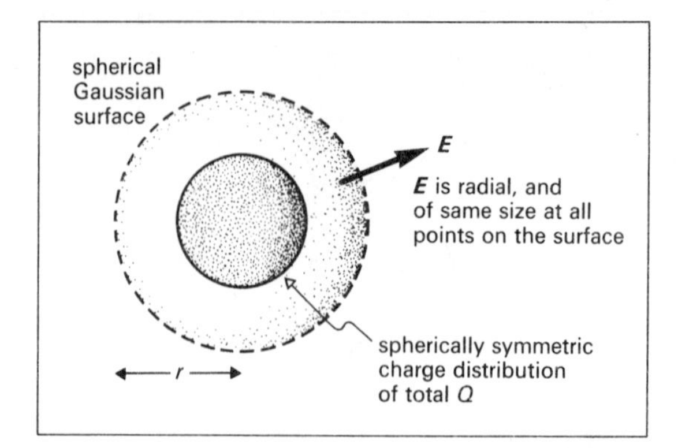

Fig. 44.7. A spherically symmetric charge distribution

44.9 *E* Due to a Line of Charge

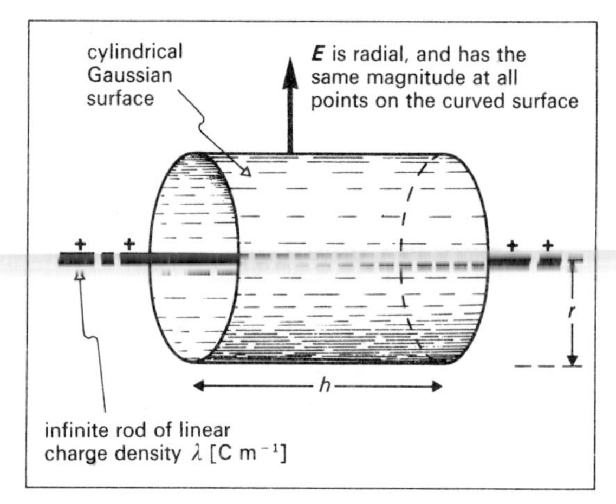

Fig. 44.8. **E** *due to a line of charge.*

In fig. 44.8, by the choice of *Gaussian* surface, the flux emerging from it is made up from

(*a*) zero at the plane ends (where **E** is parallel to the surface), and

(*b*) $E2\pi rh$ over the curved surface (where **E** is everywhere perpendicular to the surface).

$$\varepsilon_0 \psi_E = Q$$

becomes $$\varepsilon_0 E 2\pi r h = \lambda h$$

so $$E = \frac{\lambda}{2\pi\varepsilon_0 r}$$

and is radial in direction.

This result is used on p. 327.

Rationalization (mentioned on p. 311) is a convenient (but arbitrary) procedure. As a result of it situations involving

(*i*) cylindrical symmetry (such as this one) give equations with factors of 2π,

(*ii*) spherical symmetry (e.g. p. 320) show factors of 4π, and those with

(*iii*) plane symmetry (next paragraph) involve equations without π.

44.10 *E* due to an Infinite Conducting Sheet

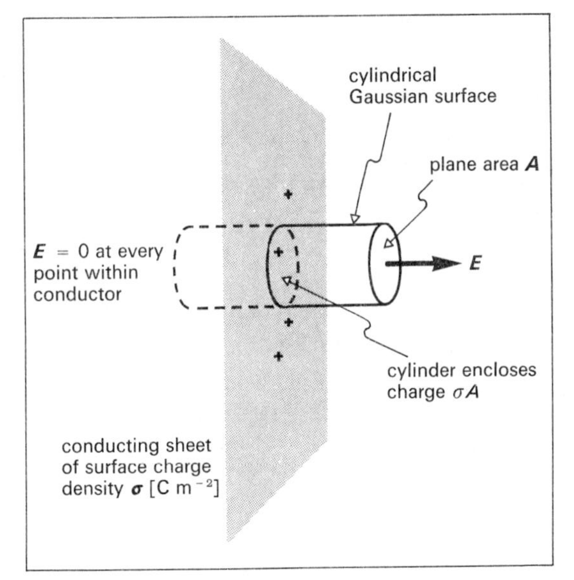

Fig. 44.9. An infinite conducting sheet.

In fig. 44.9, by the choice of *Gaussian* surface, the flux emerging from the closed surface is

(*a*) zero over the left plane end (since $\boldsymbol{E} = 0$ at every point),

(*b*) zero over the curved surface, and

(*c*) EA over the right plane end.

$$\varepsilon_0 \psi_{\boldsymbol{E}} = Q$$

becomes

$$\varepsilon_0 EA = \sigma A$$

so

$$\boldsymbol{E} = \frac{\boldsymbol{\sigma}}{\varepsilon_0}$$

and is normal to the infinite sheet.

This result applies

(*i*) at all distances from an *infinite* conducting sheet, and

(*ii*) to points *close* to a *finite* sheet (but not near the edges where the field lines are not parallel).

If the sheet is made of *insulating* material the charge is not *necessarily* on the surface, and this must be taken into account.

E Close to the Surface of a Charged Conductor

The previous result gives the value of $\boldsymbol{E}$ *very close* to the surface of *any charged conductor* whose surface charge density is $\boldsymbol{\sigma}$.

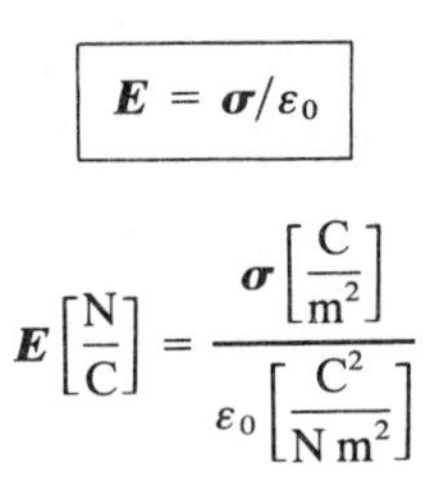

$$\boxed{\boldsymbol{E} = \boldsymbol{\sigma}/\varepsilon_0}$$

$$\boldsymbol{E}\left[\frac{\mathrm{N}}{\mathrm{C}}\right] = \frac{\boldsymbol{\sigma}\left[\frac{\mathrm{C}}{\mathrm{m}^2}\right]}{\varepsilon_0\left[\frac{\mathrm{C}^2}{\mathrm{N\,m}^2}\right]}$$

The direction of $\boldsymbol{E}$ at the surface of a *conductor* is *perpendicular to the surface*: if it were not, its resolved part along the surface would accelerate charges along the surface. This cannot happen if a conductor is to be in electrostatic equilibrium.

Field Strength Between Parallel Charged Plates

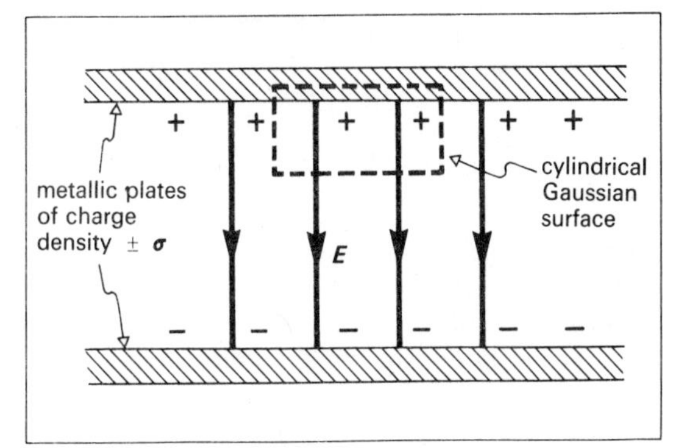

Fig. 44.10. Parallel charged plates.

The field between the plates

$$\boldsymbol{E} = \boldsymbol{\sigma}/\varepsilon_0$$

This can be verified by considering the *Gaussian* surface shown (fig. 44.10).

45 Electric Potential

45.1 Electric Potential Energy, Potential and Potential Difference

Electric Potential Energy and Potential

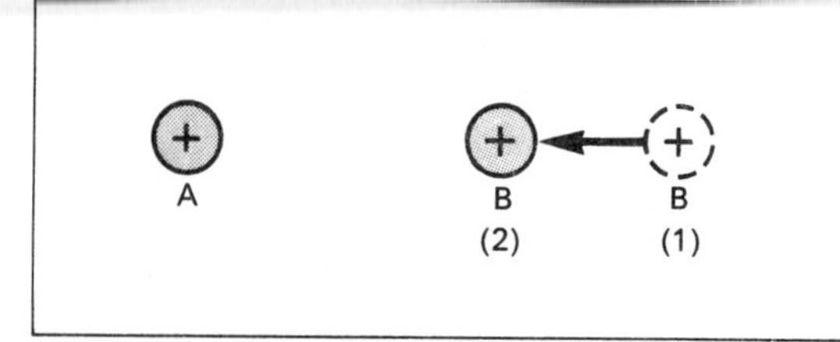

Fig. 45.1. To illustrate electric p.e.

To move charge B from position (1) to (2) in fig 45.1 requires an external agent to do work, because B is in the field established by A. The work done by the agent increases the **electric potential energy** of the system.

The increase of electric p.e. depends on

(*a*) the *magnitude of the charge*, and

(*b*) the *locations in the field* of positions (1) and (2).

We say that different points in the field have different **electric potentials**.

Compare

(*i*) *electric potential*, which is a property of a *point* in an electric field, even in the absence of a test charge, with

(*ii*) *electric potential energy*, which depends on both the position and size of a *charge*.

Electric potential V and field strength $\boldsymbol{E}$ lead to equivalent (but alternative) ways of describing a given electric field. Their relationship is discussed on p. 323.

Electric Potential V

Suppose an external agent does work W in bringing a test charge Q_0 from infinity to a particular point in an electric field.

Then we define the electric potential V at that point by the equation

$$V = \frac{W}{Q_0}$$

$$V\left[\frac{\text{J}}{\text{C}}\right] = \frac{W\,[\text{J}]}{Q_0\,[\text{C}]}$$

The unit of potential, a *joule coulomb*$^{-1}$, occurs so often it is convenient to call it a **volt** (V).

Notes.

(*a*) The potential at a point is numerically equal to the work done in bringing unit positive charge from infinity to the point (provided the field is not disturbed by the presence of such a large charge).

(*b*) $[V] = [ML^2T^{-3}I^{-1}]$.

(*c*) If the external agent does *negative work* (p. 43), then either

(*i*) it moves a negative charge to a point of positive potential, or

(*ii*) it moves a positive charge to a point of negative potential.

The agent then exerts a *restraining* force on the charge that it moves.

(*d*) Electric potential is a *scalar* quantity.

Potential Difference V_{ab}

We are nearly always concerned with the *difference* of electrical potential between two points in a field. Suppose an external agent does work W_{ab} in taking a test charge Q_0 from a point A to a point B.

Then we calculate the* potential difference V_{ab} *between the points A and B by the equation

$$V_{ab} = V_b - V_a = \frac{W_{ab}}{Q_0}$$

$$V_{ab}\,[\mathrm{V}] = \frac{W_{ab}\,[\mathrm{J}]}{Q_0\,[\mathrm{C}]}$$

$V_b > V_a$ if an external agent does *positive* work when moving a *positive* test charge.

The Zero of Potential

(*a*) The *practical* zero is that of the Earth (see p. 331).

(*b*) The *theoretical* zero, according to the definition of V, is that of a point at infinity.

Because we are nearly always concerned with *differences* of potential, the zero is not of fundamental importance.

The Uniqueness of Potential

In electrostatics the electric field is *conservative* (p. 46). In fig. 45.2 the work W_{ab} has the same value whatever path is taken in moving the test charge. So

$$V_{ab} = V_b - V_a = \frac{W_{ab}}{Q_0}$$

has the same value, and V_b and V_a are unique for the points A and B.

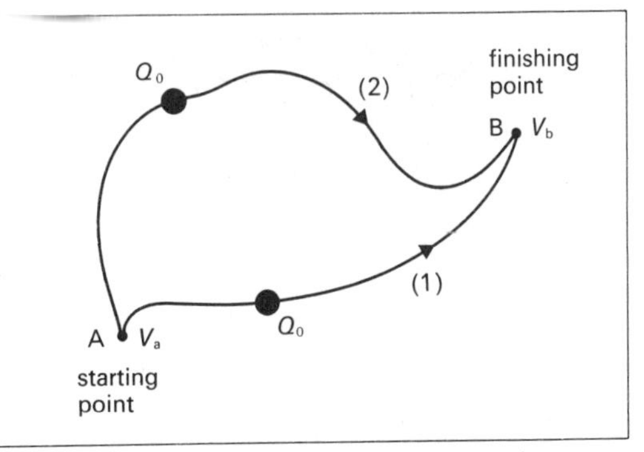

Fig. 45.2. Movement of a test charge along alternative paths.

The Electronvolt

When a charged particle is accelerated by an electric field, its increased (mechanical) k.e. is accounted for by a reduction in (electrical) p.e.

Suppose an electron ($Q = -1.6 \times 10^{-19}$ C) moves through a p.d. of 1 volt: then the change of electrical potential energy is found from

$$\begin{aligned} W &= VQ \qquad \text{(def. of } V\text{)} \\ &= (1\,\mathrm{J\,C^{-1}}) \times (-1.6 \times 10^{-19}\,\mathrm{C}) \\ &= -1.6 \times 10^{-19}\,\mathrm{J} \end{aligned}$$

This quantity of energy is convenient for much work in atomic and molecular physics: it is called an **electronvolt** (eV). (Strictly it is not a *unit*, and does not form part of the SI.)

45.2 Relationship between Potential and Field

We now have two different ways of describing the same field. We can either

(*a*) (*i*) give values of $\boldsymbol{E}$,

(*ii*) calculate the *force* on a charge at a point, and

(*iii*) draw electric field lines,

or (*b*) (*i*) give values of V,

(*ii*) calculate the potential *energy* of a charge at a point, and

(*iii*) draw equipotential lines or surfaces.

Equipotential surfaces are surfaces on which the potential is the same at all points: this means that no work has to be done in moving a test charge between any two points.

It follows that equipotential surfaces and electric field lines are perpendicular to each other at any crossing point, since there can be no component of the $\boldsymbol{E}$-field parallel to the equipotential surface. In electrostatics the surface of a conductor is an equipotential surface. (This is not so when a current flows.)

Equipotential surfaces due to a point charge are a family of concentric spheres (p. 325).

Field Strength and Potential Gradient

Refer to fig. 45.3.

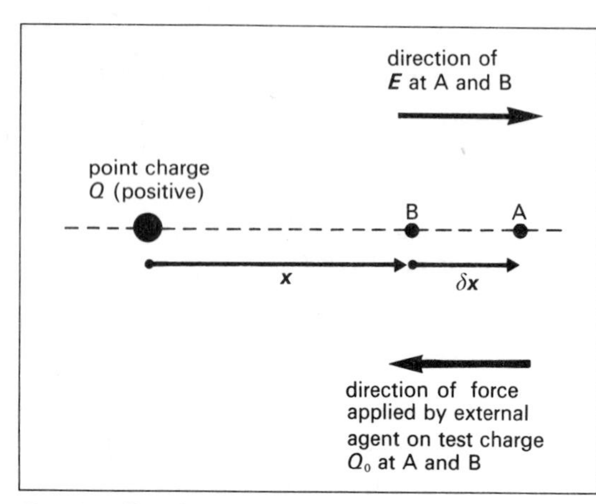

Fig. 45.3. To relate $\mathbf{E}$ *and potential gradient.*

The *electric* force on a positive test charge Q_0 at A or B is radially outward, and of magnitude EQ_0. $\boldsymbol{x}$ is measured in the same direction as $\boldsymbol{E}$. The work δW done *by an external agent* in moving the charge from A to B is positive, being given by

$$\delta W = -EQ_0\,\delta x$$

(In a move from A to B δx is negative.)

$$E = -\frac{1}{Q_0}\frac{\delta W}{\delta x}$$

$$= -\frac{\mathrm{d}V}{\mathrm{d}x} \qquad \left(V = \frac{W}{Q_0}\right)$$

Let A and B be very close, so $\delta x \to 0$. Then

$$\boxed{E = -\frac{\mathrm{d}V}{\mathrm{d}x}}$$

$$E\left[\frac{\text{volt}}{\text{metre}}\right] = -\frac{\delta V\,[\text{volt}]}{\delta x\,[\text{m}]}$$

Notes.

(*a*) The *volt metre*$^{-1}$ is alternative to $\mathrm{N\,C^{-1}}$ as a unit for $\boldsymbol{E}$.

$$1\frac{\mathrm{V}}{\mathrm{m}} = 1\frac{\mathrm{J}}{\mathrm{C}}\frac{1}{\mathrm{m}} = 1\frac{\mathrm{N}}{\mathrm{C}}$$

(*b*) $\mathrm{d}V/\mathrm{d}x$ represents the **potential gradient** (in the x-direction) at a point—this is another term for the resolved part of the electric field strength in that direction. Potential gradient is a *vector* quantity.

(*c*) The negative sign indicates that although the vectors potential gradient and electric field strength are along the same line, they are oppositely directed.

(*d*) For a *uniform* field

$$E = -\frac{\mathrm{d}V}{\mathrm{d}x}$$

can be written $\qquad \Delta V = -E\Delta x$

The sign of the potential change is usually inserted by inspection, and we often write

$$\begin{pmatrix}\text{change of}\\ \text{potential}\end{pmatrix} = \begin{pmatrix}\text{uniform}\\ \text{field}\end{pmatrix}(\text{distance})$$

(*e*) It follows from $E = -\mathrm{d}V/\mathrm{d}x$ that where $\boldsymbol{E}$ is zero, V is the same at all points. Thus all points inside an empty hollow conductor are at the same electric potential (that of the conductor).

45.3 The Gravitational Analogy

Electrostatics	*Gravitation*
Coulomb's Law $F = \left(\frac{1}{4\pi\varepsilon_0}\right)\cdot\frac{Q_1Q_2}{r^2}$	*Newton's Law* $F = \frac{Gm_1m_2}{r^2}$
Electric field vector $\boldsymbol{E} = \frac{\boldsymbol{F}}{Q_0}$	Gravitational field vector $\boldsymbol{g} = \frac{\boldsymbol{F}}{m_0}$
Electric potential $V = \frac{W}{Q_0}$	Gravitational potential $V = \frac{W}{m_0}$

Although the gravitational field vector $\boldsymbol{g}$ is usually quoted in $\mathrm{m\,s^{-2}}$, the $\mathrm{N\,kg^{-1}}$ is an equivalent unit. In fig. 45.4 we assume it takes the value

$$\boldsymbol{g} = 10\,\mathrm{N\,kg^{-1}} = 10\,\mathrm{m\,s^{-2}}$$

vertically downwards at all points.

Example of a Gravitational Field

If the Earth had a spherically symmetric mass distribution, all points on a sphere concentric with its centre would have the same gravitational potential. (The resolved part of $\boldsymbol{g}$ along the tangent to the sphere would be zero.) This is nearly true in practice.

In fig. 45.4, a mass 5 kg placed at P (10 m above the Earth's surface) has 500 J more gravitational potential energy than it would have at the surface.

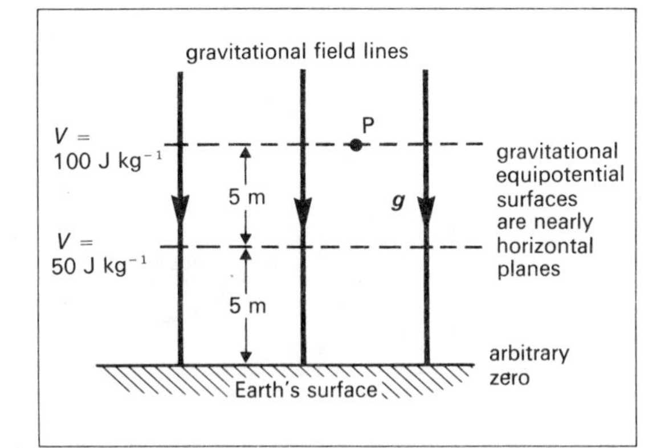

Fig. 45.4. Representation of an approximately uniform gravitational field.

The gravitational potential difference between P and the surface is

$$\frac{500\,\text{J}}{5\,\text{kg}} = 100\,\text{J kg}^{-1}$$

This could have been found from

$$\begin{aligned}\begin{pmatrix}\text{difference of}\\ \text{potential}\end{pmatrix} &= (\text{field})(\text{distance})\\ &= (10\,\text{N kg}^{-1}) \times (10\,\text{m})\\ &= 100\,\text{J kg}^{-1}\end{aligned}$$

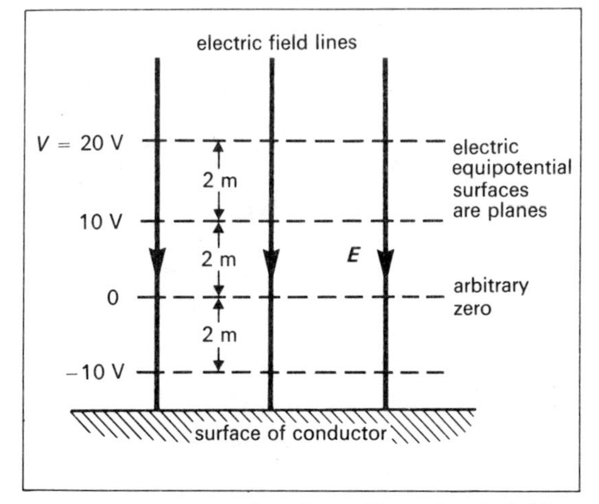

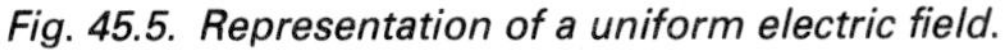
Fig. 45.5. Representation of a uniform electric field.

A Similar Electric Field

At all points in fig. 45.5

$\boldsymbol{E} = 5\,\text{N C}^{-1} = 5\,\text{V m}^{-1}$ vertically downwards.

45.4 Potential due to a Point Charge

*Suppose that in fig. 45.3 the external agent brings the test charge Q_0 from infinity through a vacuum to a point distance r from a point charge Q.

$$\delta W = -EQ_0\delta x \qquad \text{(p. 324)}$$

The work W done by the agent

$$\begin{aligned}W &= -\int_{\infty}^{r} EQ_0\,\mathrm{d}x\\ &= -\frac{QQ_0}{4\pi\varepsilon_0}\int_{\infty}^{r}\frac{\mathrm{d}x}{x^2}\\ &= \frac{QQ_0}{4\pi\varepsilon_0}\cdot\frac{1}{r}\end{aligned}$$

Since $V = W/Q_0$, the potential at the point P is

$$\boxed{V_P = \frac{1}{4\pi\varepsilon_0}\cdot\frac{Q}{r}}$$

N.B. Radial field lines ONLY

Notes. This equation

(*a*) enables us to calculate the potential due to a *point charge*, but does *not* constitute a *definition* of potential,

(*b*) demonstrates that the equipotential surfaces associated with a point charge are spheres concentric with the charge, but

(*c*) takes no account of the potential resulting from any charges induced on nearby conductors.

Potential due to Many Point Charges

Potential is a scalar. If several charges contribute to the potential at a point, the net potential can be found by algebraic addition without regard to *direction*. This addition may require the use of calculus.

45.5 A Conducting Material in an Electric Field

Fig. 45.6 shows

(*a*) a pair of oppositely charged plates which establish a uniform field between them,

(*b*) a $V - x$ graph, and (*c*) an $\boldsymbol{E} - \boldsymbol{x}$ graph, both before and after a conducting plate is inserted as shown.

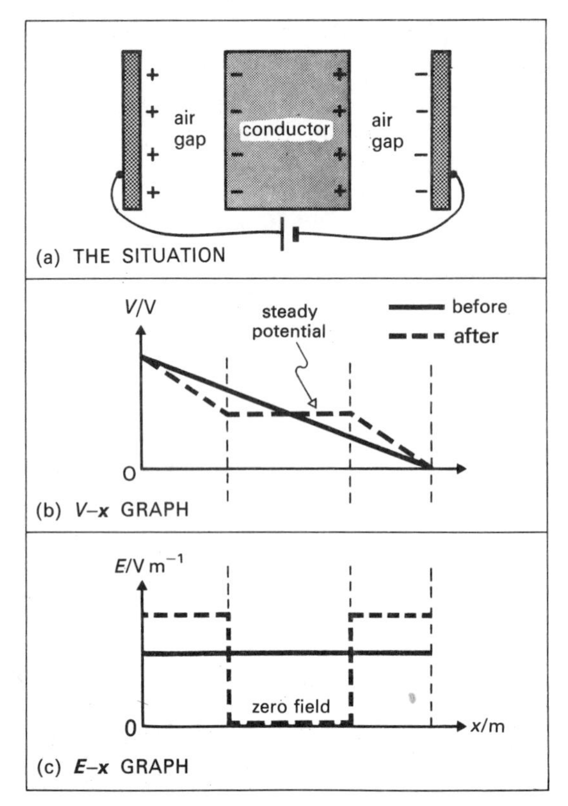

*Fig. 45.6. The effect of a conducting plate on values of **E** and V nearby.*

The insertion of the conductor is accompanied by the flow of an **induced charge**. Opposite sides of the plate acquire opposite charges so that $\boldsymbol{E}$ inside the plate is again zero at equilibrium. The plate's net charge remains zero. Note the effect of the plate on the values of V and $\boldsymbol{E}$ in the air gap.

45.6 Field and Potential due to an Insulated Conducting Charged Sphere

(*a*) Field

Suppose we have a sphere of radius a.

We have shown (p. 320 and p. 319)

$$E = \left(\frac{1}{4\pi\varepsilon_0}\right)\cdot\frac{Q}{r^2} \quad \text{for points outside}$$

and $\quad E = 0 \quad$ for points inside

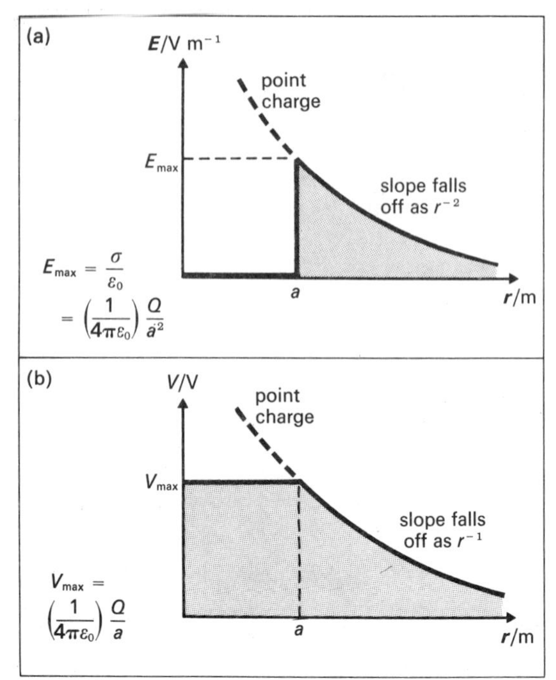

Fig. 45.7. (a) Field and (b) potential established by a charged conducting sphere. (Cf. Fig. 20.3.)

(*b*) Potential

If the sphere is *isolated*, then (p. 325)

$$V = \left(\frac{1}{4\pi\varepsilon_0}\right)\cdot\frac{Q}{r} \quad \text{for points outside}$$

and $\quad V = \left(\frac{1}{4\pi\varepsilon_0}\right)\cdot\frac{Q}{a} \quad$ for every point inside

this being the potential of the sphere.

These ideas are brought out in fig. 45.7.

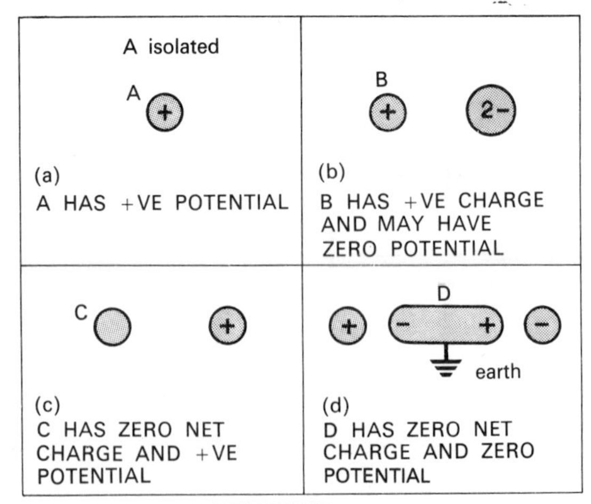

Fig. 45.8. The potential of a body is not controlled by its own charge only.

Conductor not Isolated

(*i*) The *charge* on a conductor can only be altered by electrical *contact.*

(*ii*) The *potential* of a conductor is strongly influenced by its environment. We use this fact in designing a capacitor (p. 331).

Examine fig. 45.8 carefully.

*45.7 Potential Differences Close to a Line of Charge

Gauss's Law was used in **44.9** to show that the magnitude of the radial electric field vector at a distance r from an infinitely long line of charge is given by $E = \lambda/2\pi\varepsilon_0 r$, where λ is the linear charge density. Using $E = -(\mathrm{d}V/\mathrm{d}x)$, we find the p.d. between two points distant r_1 and r_2 from the rod ($r_2 > r_1$) to be given by

$$V_2 - V_1 = -\int_{r_1}^{r_2} E \cdot \mathrm{d}r$$

If the rod is positively charged then $\boldsymbol{E}$ and $\boldsymbol{r}$ have the same direction.

$$V_2 - V_1 = -\int_{r_1}^{r_2} \frac{\lambda}{2\pi\varepsilon_0} \cdot \frac{\mathrm{d}r}{r} = -\frac{\lambda}{2\pi\varepsilon_0} \ln\left(\frac{r_2}{r_1}\right)$$

or

$$V_1 - V_2 = \left(\frac{\lambda}{2\pi\varepsilon_0}\right) \ln\left(\frac{r_2}{r_1}\right)$$

Notes.

(*a*) If V_2 is taken to be zero when $r_2 = \infty$, the above equation requires infinite potential for all finite values of r_1. This physical impossibility stems from our postulating an *infinite* line of charge, which cannot be realized in practice.

(*b*) The equation remains useful nevertheless for calculating potential *differences* between pairs of points relatively close to a *finite* line of charge. (See the coaxial capacitor on p. 333.)

45.8 Charge Distribution over a Conductor

It can be shown that the **surface density of charge** on a conductor of varying radii of curvature (fig. 45.9) is inversely proportional to the radius of curvature.

Thus $\boldsymbol{\sigma}[\mathrm{C\,m^{-2}}]$ is large where the curvature is large. But $\boldsymbol{E} = \boldsymbol{\sigma}/\varepsilon_0$, so the field is also large.

At *a sharp point* $\boldsymbol{E}$ becomes very high, and this may

(*a*) attract or repel ionized particles from the air, or

(*b*) in extreme cases *ionize* air molecules.

Movement of charge then neutralizes the charge on the conductor: the excess charge seems to stream off into the air. This is called the **action of points.** It is utilized in the lightning conductor and the field ion microscope. **Corona discharge** is the extreme process.

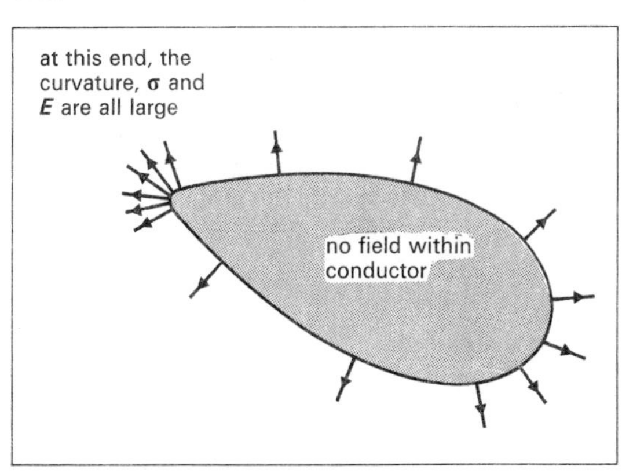

*Fig. 45.9. Variations of σ and **E** over a surface.*

45.9 The Van de Graaff Generator

In fig. 45.10 the comb at A is maintained at a positive potential, and so the moving belt acquires a positive charge. (In a small laboratory version friction between the roller and the belt serves to maintain charge on the belt.) This positive charge induces a negative charge on comb B, which streams onto the belt. The net effect is to transfer positive charge from the belt to the *outside* of the conducting sphere (p. 318).

The limit to the potential that the sphere can acquire is controlled by the air breakdown.

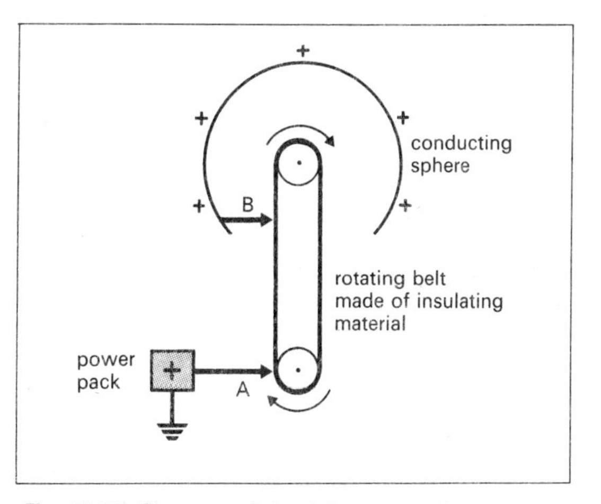

Fig. 45.10. The essentials of the Van de Graaff generator.

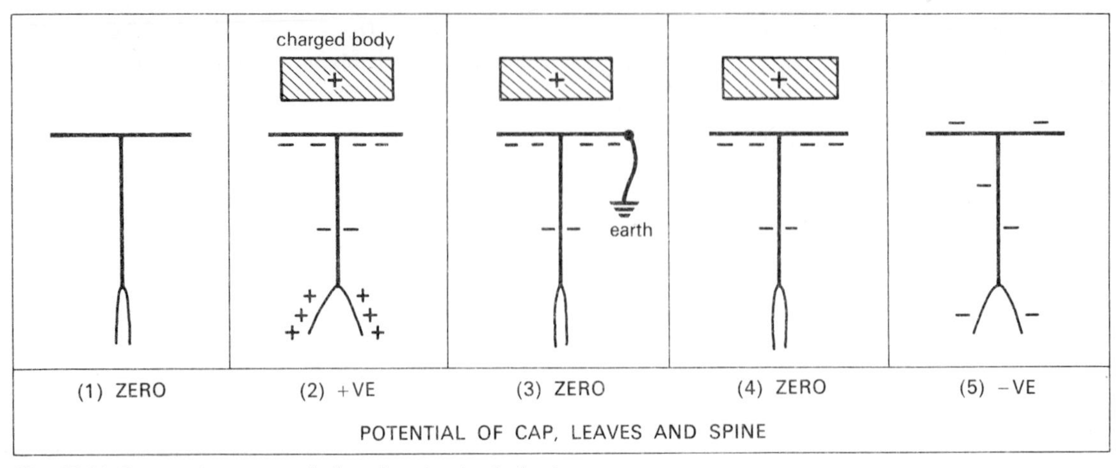

Fig. 45.11 Successive stages during charging by induction.

45.10. The Gold Leaf Electroscope

Its Construction

There are two basic components

(*a*) the *cap, spine* and *leaves*, which are connected conductors at the same potential, and

(*b*) the *case*, also a conductor, but insulated from (*a*), and usually earthed.

Why the Leaves Diverge

The leaves diverge when they experience a force because they are in the electric field established between the case (zero potential) and themselves. The presence of this field indicates a potential difference between the case and leaves: the *electroscope is a device which measures this potential difference*. The leaves diverge until they reach an equilibrium position. Suppose

(*a*) The case is earthed: then *indirectly* the electroscope measures the *charge* on the cap, since a greater charge leads to a greater p.d. between leaves and case, a greater field, and a greater deflection.

(*b*) The case is connected to the cap: a charge placed on the cap increases the potential of *both* basic components *equally*. The leaves are in a zero field, and so experience zero force.

Charging the Electroscope

(*a*) On *contact*, the cap, spine and leaves acquire charge of the same sign as the charging body.

(*b*) By *induction*, they ultimately acquire charge of opposite sign. Study fig. 45.11 carefully, in which the case is earthed.

*45.11 The Attracted Disc Electrometer

Force on Surface of a Charged Conductor

In fig. 45.12 we deduce

(*a*) net field *outside* $= \dfrac{\sigma}{2\varepsilon_0} + \dfrac{\sigma}{2\varepsilon_0} = \dfrac{\sigma}{\varepsilon_0}$,

(*b*) net field *inside* $= \dfrac{\sigma}{2\varepsilon_0} - \dfrac{\sigma}{2\varepsilon_0} = 0$.

These both satisfy the requirements of p. 321 and p. 319.

The outward force exerted on elementary area ΔA is given by

$$\boldsymbol{F} = \boldsymbol{E}'Q$$

where $\boldsymbol{E}'$ is the field set up by the *rest* of the conductor.

$$F = \frac{\sigma}{2\varepsilon_0}\cdot(\Delta A\sigma)$$

$$\boxed{F = \frac{\Delta A\sigma^2}{2\varepsilon_0}}$$

$$F[\mathrm{N}] = \frac{1}{2}\frac{\Delta A[\mathrm{m}^2]\sigma^2\left[\dfrac{\mathrm{C}^2}{\mathrm{m}^4}\right]}{\varepsilon_0\left[\dfrac{\mathrm{C}^2}{\mathrm{N\,m}^2}\right]}$$

It will act outwards whatever the sign of σ.

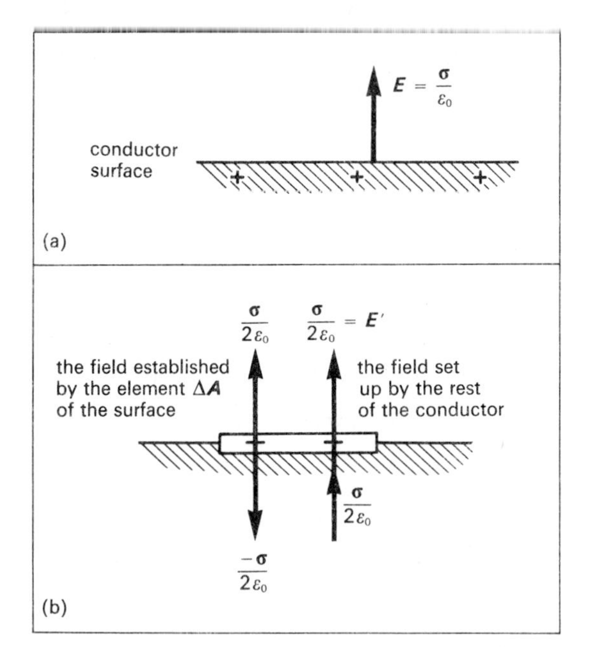

Fig. 45.12. Analysis of the field at the surface of a conductor: (a) the total field, (b) components of the field.

The Principle of the Electrometer

This is an **absolute** instrument which enables a potential difference to be measured in terms of mechanical quantities. Fig. 45.13 shows only the principle on which it works.

Procedure

(1) Arrange the upper plate to be in the plane of the guard-ring (p. 332), whose function is to create a uniform field.

(2) Apply a potential difference V between the plates.

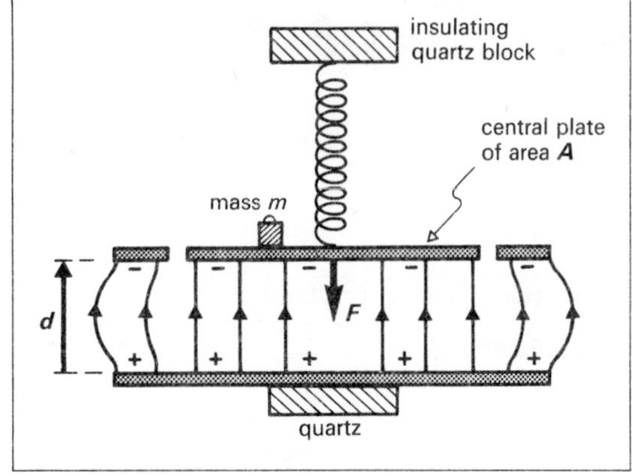

Fig. 45.13. The essentials of the attracted disc electrometer.

(3) Remove a mass m to return the central plate to its original height. This compensates for the downward electrical force. Then

$$F = mg = \frac{A\sigma^2}{2\varepsilon_0}$$

in which $\frac{\sigma}{\varepsilon_0} = \boldsymbol{E}$, and $V = Ed$

so
$$mg = \frac{\varepsilon_0 V^2 A}{2d^2}$$

V can be calculated from this equation, since m, g, d, ε_0 and $\boldsymbol{A}$ are known.

$$\varepsilon_0 = 9 \times 10^{-12}\ \mathrm{C^2\,N^{-1}\,m^{-2}}$$

so $\boldsymbol{F}$ is very small. Typically the electrometer measures potential differences of a few kilovolts.

(*Note.* ε_0 does not have a defined value. This means that the electrometer may be regarded as *absolute* only if the assumed value of ε_0 is measured by the method involving the speed of light (p. 333), since this requires only mechanical measurements.)

46 Capacitors

46.1 Capacitance for Charge

Suppose we put a charge Q on an insulated conductor, and that its potential changes by V.

> ***Then we define the* capacitance *C of the conductor by the equation***

$$\boxed{C = \frac{Q}{V}}$$

$$C\left[\frac{\text{C}}{\text{V}}\right] = \frac{Q\,[\text{C}]}{V\,[\text{V}]}$$

Notes.

(*a*) The unit of capacitance, a *coulomb volt*$^{-1}$, occurs so often that it is convenient to give it the name **farad** (F). The farad is an extremely large unit, and we use more often

(*i*) the *microfarad* $1\ \mu\text{F} = 10^{-6}\ \text{F}$, and
(*ii*) the *picofarad* $1\ \text{pF} = 10^{-12}\ \text{F}$

(*b*) The change in V brought about by a given change in Q depends upon a conductor's size, the nearness of other conductors, and the material surrounding it. It can be shown theoretically and experimentally, that, for a given environment,

$$Q \propto V$$

i.e. that *C is a constant.*

(*c*) Capacitance is numerically equal to the charge in coulombs that changes the potential of a conductor by one volt.

$$[C] = [\text{M}^{-1}\text{L}^{-2}\text{T}^4\text{I}^2]$$

(*d*) Note the analogy with *heat capacity*:
(*i*) For heat $\Delta Q = C\,\Delta\theta$ (p. 171)
(*ii*) For electric charge $Q = CV$

where V represents a *change* in potential resulting from a change Q in the total charge.

Example. For an isolated conducting sphere, of radius a and carrying a charge Q,

$$V = \left(\frac{1}{4\pi\varepsilon_0}\right)\cdot\frac{Q}{a} \qquad \text{(p. 325)}$$

so

$$C = \frac{Q}{V} = 4\pi\varepsilon_0 a$$

Capacitance always has dimensions

$$[\varepsilon_0]\,[\text{L}^1]$$

A sphere of radius 9×10^9 m (more than $10^3 \times$ that of the Earth) would have a capacitance of about one farad.

The Unit of the Permittivity Constant

Compare $\varepsilon_0 = \dfrac{C}{4\pi a}$

(from the capacitance of an isolated sphere)

with $\varepsilon_0 = \dfrac{Q^2}{4\pi F r^2}$

(from *Coulomb's Law*).

The first equation shows that a possible unit for ε_0 is the *farad metre*$^{-1}$: this unit, which is equivalent to the $C^2 N^{-1} m^{-2}$, is the one usually quoted.

$$\begin{aligned} 1\,C^2 N^{-1} m^{-2} &= 1\,C\,(C N^{-1} m^{-1})\, m^{-1} \\ &= 1\,C V^{-1} m^{-1} \\ &= 1\,F m^{-1} \end{aligned}$$

46.2 Earthing

When a small conducting body shares its excess charge by contact with a large conducting neutral body, it loses most of it. The two bodies acquire the same potential, which is less than that of the small body's original potential.

The Earth has a capacitance given by

$$\begin{aligned} C &= 4\pi\varepsilon_0 a \\ &= (\tfrac{1}{9} \times 10^{-9}\,F m^{-1}) \times (6 \times 10^6\,m) \\ &= 7 \times 10^{-4}\,F, \end{aligned}$$

since it may be regarded as a conducting body.

This capacitance is so large that we say isolated bodies that are **earthed** (i.e. share their charge with the Earth) effectively lose that charge. The body and the Earth acquire a common potential which we call the *practical zero of potential.*

46.3 Capacitors

The charge that can be put onto a conductor is limited by the large surface fields that develop, which result in electrical breakdown, sparking and consequent discharge.

A **capacitor** is a system of high capacitance which has been designed for the storage of separated positive and negative charges.

In fig. 46.1 the potential of each conductor is affected by the presence of the other – the *p.d. between them is reduced.*

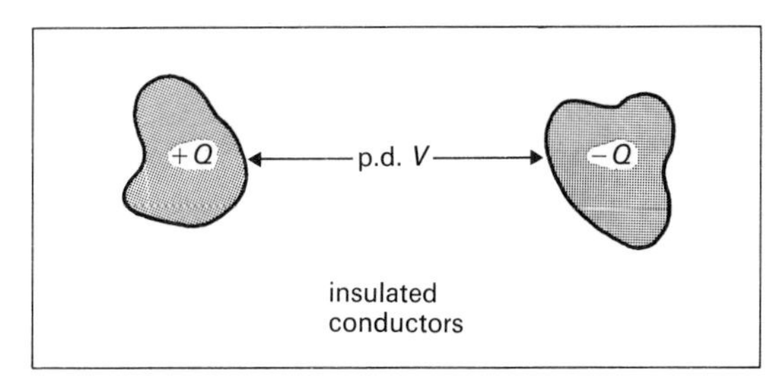

Fig. 46.1. The principle of the capacitor.

The capacitance of the system is

$$C = \frac{\text{magnitude of either charge}}{\text{p.d. between conductors}} = \frac{Q}{V}$$

Thus for a fixed Q, C is made large by lowering V: this is achieved by bringing a second conductor, which is frequently earthed, very close to the first conductor.

Demonstration of the Parallel-Plate Capacitor

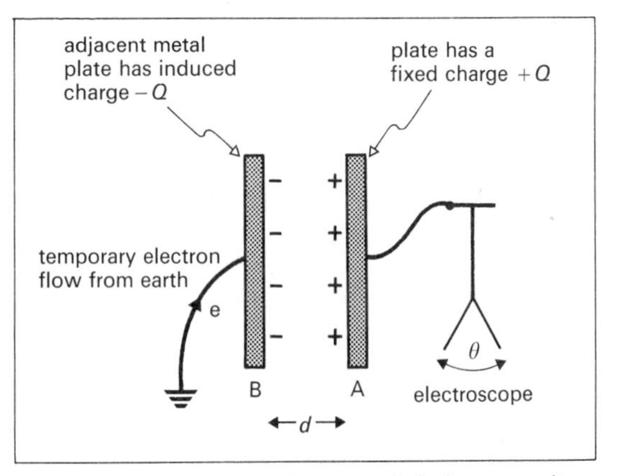

Fig. 46.2. Investigation of the parallel-plate capacitor.

The negative charge induced on plate B (fig. 46.2) attracts and tends to anchor the positive charge on plate A. Since both B and the electroscope case are earthed,

$$\theta \propto V, \quad \text{the p.d. between the plates}$$

But $C = Q/V$, and Q is fixed, so

$$C \propto \frac{1}{\theta}$$

If we

(*a*) decrease d, then θ decreases, so C increases,

(*b*) increase the common plate area A, θ decreases and so C increases.

We have shown experimentally, in a general way

$$C \propto \frac{A}{d}$$

Because the plates attract (as in the electrometer p. 329) they are kept apart by an insulator (p. 338).

The Importance of Capacitors

(*a*) They can be used to establish electric fields of required pattern, e.g. for

(*i*) *Millikan's* experiment, (p. 440)
(*ii*) the study of dielectric materials. (p. 337)

(*b*) They store not only charge, but *energy* in the electric field between the plates.

(*c*) They form an integral part of most a.c. circuits, e.g. in

(*i*) smoothing circuits, (p. 426)
(*ii*) oscillatory circuits, and (p. 453)
(*iii*) time bases. (p. 442)

Capacitors in Practice

(*a*) *Tuning capacitors* (in wireless sets) have a variable A and so a variable C.

(*b*) *Electrolytic capacitors* have a small d and so a large C.

(*c*) The *guard-ring capacitor* prevents distortion of the electric field lines.

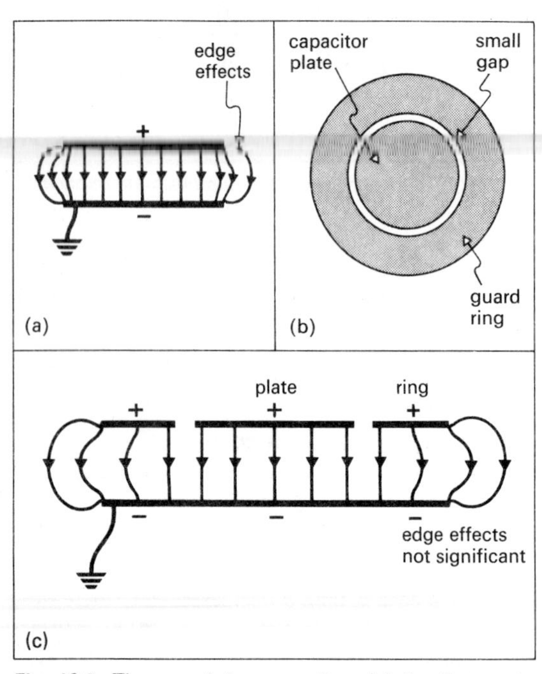

Fig. 46.3. The guard-ring capacitor: (a) the distorted field, (b) plan view, (c) side view (different scale).

In fig. 46.3 the ring and plate are maintained at the same potential, but are not joined so their charges are kept separate.

46.4 Calculation of Capacitance

Procedure

(1) Imagine a charge $(+Q)$ to be put on one conductor. Then in an *ideal capacitor* a charge $(-Q)$ is induced on the other.

(2) Calculate the change of potential V that results.

(3) Apply the defining equation $C = Q/V$

The result should have dimensions

$$[\varepsilon_0]\,[\mathrm{L}^1]$$

(2) is the difficult step, and can be done only when the geometry is simple.

(*a*) The Parallel-Plate Capacitor

(1) Refer to fig. 46.4.

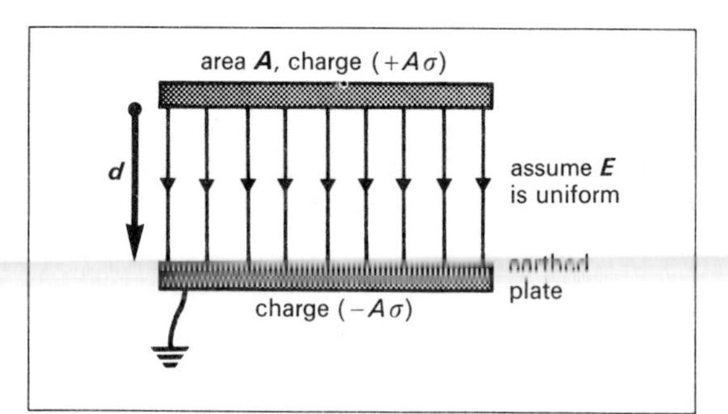

Fig. 46.4. The parallel-plate capacitor.

(2) Using $V = Ed$ (p. 324)

$$V = \frac{\sigma}{\varepsilon_0}d \quad \text{since} \quad \boldsymbol{E} = \frac{\sigma}{\varepsilon_0} \quad \text{(p. 321)}$$

(3)

$$C = \frac{Q}{V} = \frac{A\sigma}{\dfrac{\sigma}{\varepsilon_0}d}$$

$$\boxed{C = \frac{\varepsilon_0 A}{d}}$$

As required, the dimensions of C are

$$[\varepsilon_0]\,[\mathrm{L}^1]$$

*(b) The Coaxial Cylindrical Capacitor

(1) Refer to fig. 46.5.

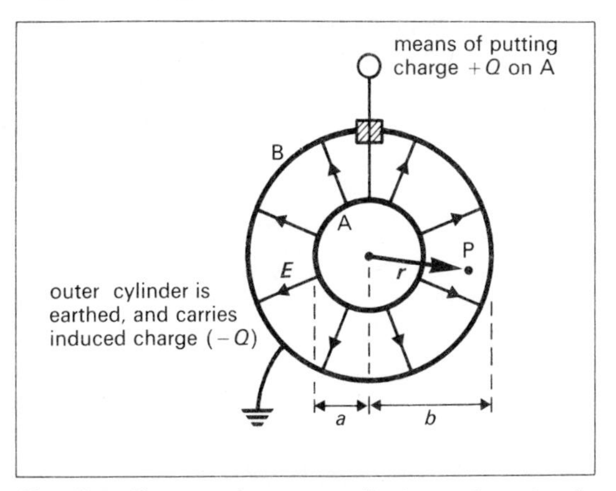

Fig. 46.5. Plane section perpendicular to the axis of a coaxial cylindrical capacitor of length l.

(2) In **45.7** the p.d. between two points close to a line of charge was shown to be

$$V_1 - V_2 = \frac{\lambda}{2\pi\varepsilon_0} \ln \frac{r_2}{r_1}$$

The p.d. between a point on cylinder A and a point on cylinder B is given by

$$V = \frac{\lambda}{2\pi\varepsilon_0} \ln \frac{b}{a} = \frac{Q}{2\pi\varepsilon_0 l} \ln \frac{b}{a}$$

(3) $$C = \frac{Q}{V} = \frac{2\pi\varepsilon_0 l}{\ln (b/a)}$$

This has the required dimensions. Except at the ends, where fringing occurs, it enables us to calculate the capacitance per unit length of coaxial cables.

(c) The Concentric Spheres Capacitor

Using the symbols of fig. 46.5 one can show

$$C = 4\pi\varepsilon_0 \left(\frac{ab}{b - a}\right)$$

46.5 Measurement of Capacitance and ε_0

The measurement of ε_0

Although ε_0 is defined from the statement of *Coulomb's Law* (p. 311), the situation cannot be used for a direct experimental determination of its size.

(*a*) The determination of C for a parallel-plate capacitor in the school laboratory (below) enables ε_0 to be calculated from $C = \varepsilon_0 A/d$, since we need measure only $\boldsymbol{A}$ and $\boldsymbol{d}$ in addition.

(*b*) *Sir J. J. Thomson* used a cylindrical capacitor with cylindrical guard rings at the ends (1890).

(*c*) *Rosa* and *Dorsey* used a concentric spherical capacitor (1907), and obtained an uncertainty of 1 in 10^4.

(*d*) Experiments of this kind have all confirmed *Maxwell's* prediction that $c = 1/\sqrt{\varepsilon_0\mu_0}$. The most accurate way of evaluating ε_0 is to assume the truth of this equation, and to *calculate* the value of ε_0 from the *defined* value of μ_0 and the *measured* value of c. $\varepsilon_0 = 8.854\ \mathrm{pF\ m^{-1}}$, with an estimated uncertainty of 7 in 10^9 (1973).

Measurement of Capacitance

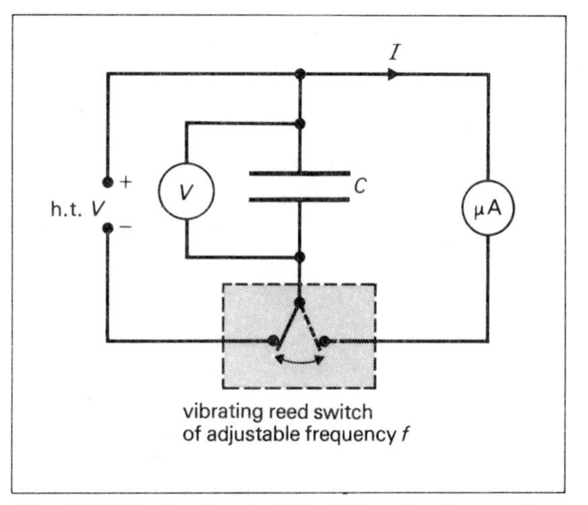

Fig. 46.6. The basic principle of the vibrating reed capacitance meter.

Refer to fig. 46.6. The capacitor is charged at a frequency f to the p.d. V across the supply, and each time discharged through the sensitive ammeter. At sufficiently high values of f the meter reading is steady.

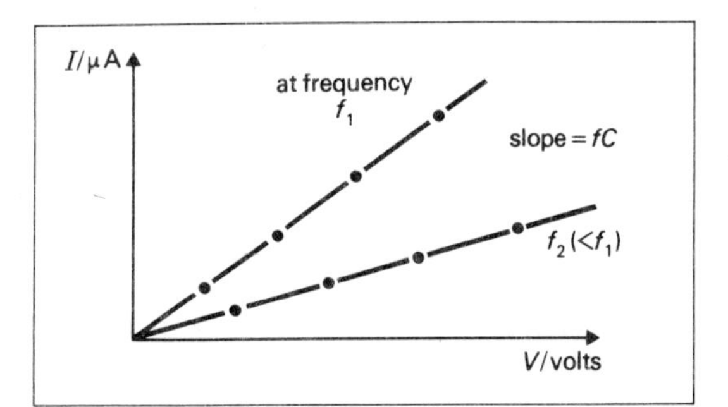

Fig. 46.7. Experimental results.

During each time interval $1/f$, a charge $Q = CV$ is passed through the meter.

$$\therefore I = \frac{Q}{(1/f)} = fCV$$

The graph of fig. 46.7 shows the experimental results, from which C is found.

46.6 Comparison of Capacitances

(a) By d.c. amplifier (p. 336).

Since $Q = CV$

$$\frac{dQ}{dt} = I = C\frac{dV}{dt}$$

$$= C\left(\frac{V}{t}\right)$$

if I is steady. When the switch is closed in fig. 46.8 the value of R is adjusted to ensure a steady current.

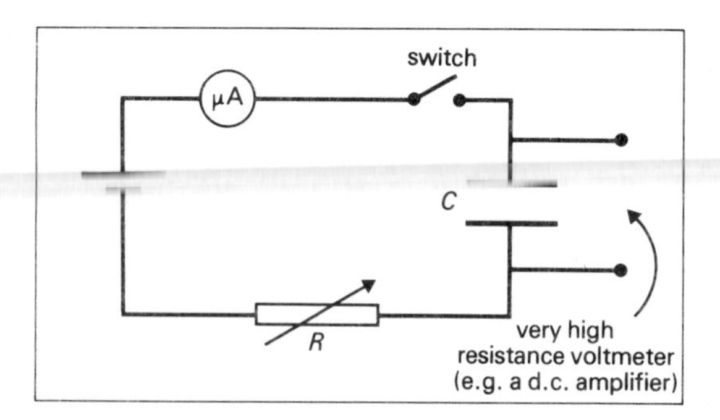

Fig. 46.8. Measurement of capacitance using a steady charging current.

Suppose that after time t the switch is opened, and for capacitor C_1 the voltmeter reading is V_1.

The same current passed for the same time into C_2 will give a reading V_2. Then

$$I = C_1\left(\frac{V_1}{t}\right) = C_2\left(\frac{V_2}{t}\right),$$

from which C_1/C_2 can be found.

(b) By ballistic galvanometer (p. 391).

(c) By vibrating reed capacitance meter (fig. 46.6).

For fixed values of V and f the value of C is proportional to I.

(d) By the De Sauty bridge (fig. 46.9).

When the sound in the headphones is adjusted to a minimum value, the bridge is balanced.

$$V_{AB} = V_{AD} \quad \text{and} \quad V_{BC} = V_{DC},$$

so $$I_1\left(\frac{1}{\omega C_1}\right) = I_2R_1 \quad \text{and} \quad I_1\left(\frac{1}{\omega C_2}\right) = I_2R_2,$$

giving $$\frac{C_1}{C_2} = \frac{R_2}{R_1}$$

Note that ω cancels from the final equation.

These methods enable the results of **46.7** to be tested experimentally.

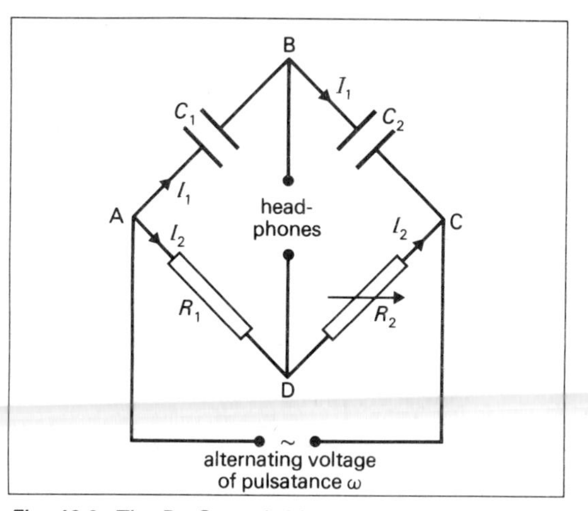

Fig. 46.9. The De Sauty bridge for comparing capacitances.

46.7 Combinations of Capacitors

(a) Capacitors in Parallel

In fig. 46.10 we want the capacitance C of the single capacitor which, connected across AB, would have the same effect as C_1 and C_2 connected as shown.

(*i*) We have the *same* p.d. V across each.

(*ii*) $Q = Q_1 + Q_2$

so $$CV = C_1V + C_2V \qquad (Q = CV)$$

$$\boxed{C = C_1 + C_2} \qquad \textit{Capacitors in parallel}$$

The result can be extended to any number of capacitors. (Compare with resistors in series p. 355.)

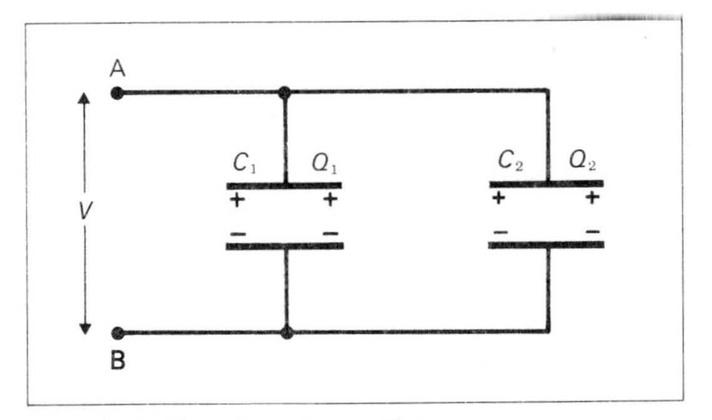

Fig. 46.10. Capacitors in parallel.

(*b*) Capacitors in Series

In fig. 46.11 we suppose the charging process has not involved any connection other than those shown, so that (*i*) below applies.

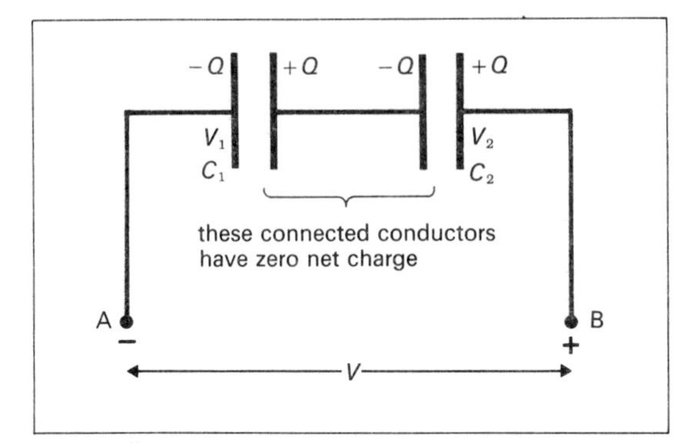

Fig. 46.11. Capacitors in series

(*i*) The magnitude Q of the *charge* induced on each capacitor plate is the *same*.

(*ii*)
$$V = V_1 + V_2$$
$$\frac{Q}{C} = \frac{Q}{C_1} + \frac{Q}{C_2} \qquad \left(V = \frac{Q}{C}\right)$$

$$\boxed{\frac{1}{C} = \frac{1}{C_1} + \frac{1}{C_2}} \qquad \textit{Capacitors in series}$$

The result can be extended to any number of capacitors. The capacitance of the combination is less than C_1 or C_2. (Compare with resistors in parallel p. 356.)

46.8 Energy of a Charged Conductor

Electric potential energy is the work done in assembling a particular configuration of charges (e.g. in putting charge onto a conductor). As a conductor is charged, its potential changes, and it becomes more difficult to add a further charge increment ΔQ.

*Suppose we bring charge δQ from infinity to a conductor already at potential V. Work done $\delta W = V\delta Q$. Total work W done in raising the charge from 0 to Q_0 is

$$W = \int \delta W = \int_0^{Q_0} V\,\mathrm{d}Q$$
$$= \int_0^{Q_0} \frac{Q}{C}\cdot \mathrm{d}Q \qquad \left(V = \frac{Q}{C}\right)$$
$$= \frac{1}{2}\frac{Q_0^2}{C}, \quad \text{since } C \text{ is constant}$$

$$\boxed{W = \tfrac{1}{2}Q_0V_0 = \tfrac{1}{2}Q_0^2/C = \tfrac{1}{2}V_0^2C}$$

$$W\,[\mathrm{J}] = \tfrac{1}{2}Q_0\,[\mathrm{C}] \times V_0\left[\frac{\mathrm{J}}{\mathrm{C}}\right]$$

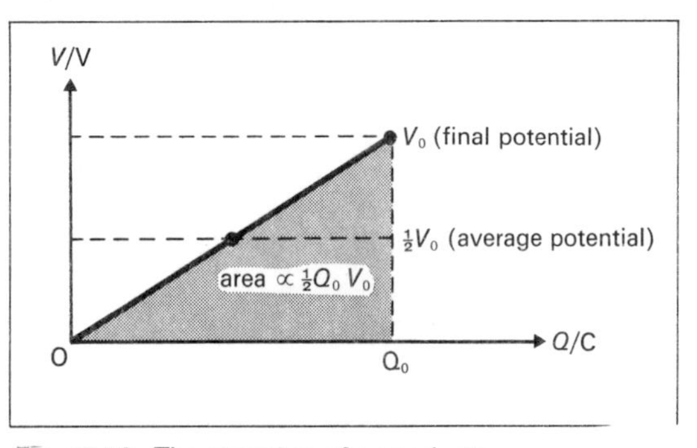

Fig. 46.12. The charging of a conductor.

Fig. 46.12 illustrates the change of potential with charge during the charging process. The *energy stored*, W, is represented by the area under the graph.

Loss of Electrical Potential Energy on Joining two Capacitors

On depressing S in fig. 46.13, there will be sparking or flow of charge (current) until the same p.d. exists across each capacitor.

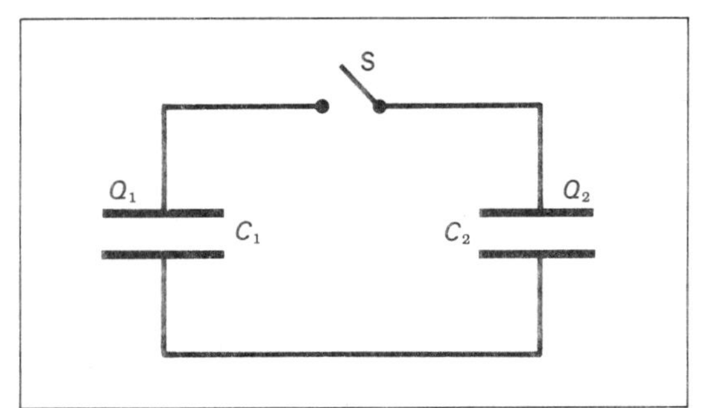

Fig. 46.13. Joining two charged capacitors.

$$\text{Initial energy} = \frac{\frac{1}{2}Q_1^2}{C_1} + \frac{\frac{1}{2}Q_2^2}{C_2}$$

$$\text{Final energy} = \frac{\frac{1}{2}Q^2}{C} = \frac{\frac{1}{2}(Q_1 + Q_2)^2}{C_1 + C_2}$$

Subtract and re-arrange terms: loss of electrical potential energy (gain of (e.g.) internal energy)

$$= \frac{\frac{1}{2}(C_2Q_1 - Q_2C_1)^2}{C_1C_2(C_1 + C_2)}$$

Notes.

(*a*) There can never be a gain of electrical p.e. (this is illustrated by the squared term).

(*b*) The electrical p.e. loss is zero when

$$C_2Q_1 = Q_2C_1$$

or

$$\frac{Q_1}{C_1} = \frac{Q_2}{C_2}$$

i.e. if

$$V_1 = V_2$$

Then there is no movement of charge.

46.9 The d.c. Amplifier

An **electrometer** is a measuring instrument that operates on a current of 1 pA or less. The type shown in fig. 46.14 uses a very high-gain d.c. amplifier. This instrument amplifies a very small input p.d. to an output p.d. which, being perhaps 10^{10} times as big, is large enough to operate a 100 μA moving-coil meter.

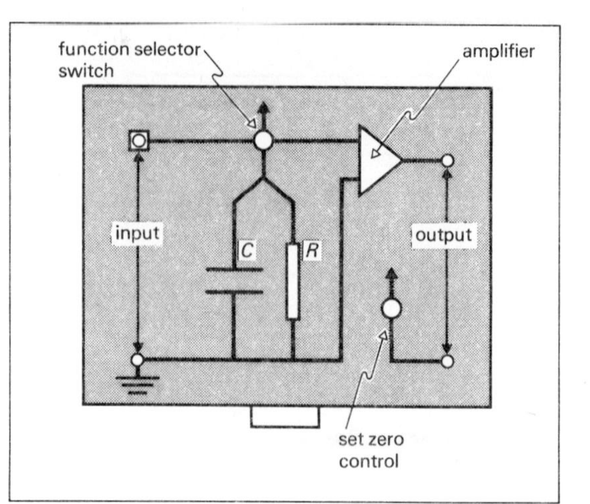

Fig. 46.14. A d.c. amplifier.

Uses

(*a*) ***To measure p.d.*** The microammeter connected across the output can be calibrated to measure the *input* p.d. An input p.d. of about 1 volt would normally give f.s.d.

(*b*) ***To measure current.*** Although the amplifier is basically a voltmeter of impedance about 10 TΩ, it can be adapted to measure small currents of about 1 pA if these are made to flow through known resistances. (For an example see p. 472.)

(*c*) ***To measure charge.*** The method was described on p. 312.

Notes.

(*a*) Types that are not mains-operated are provided with a *gain control* to compensate for the ageing of the amplifier's batteries.

(*b*) A range of capacitances and resistances can be built into the electrometer.

(*c*) A *short* button discharges the internal components between readings.

47 Dielectric Materials

47.1 Parallel-Plate Capacitor with Dielectric

(The following results can also be shown with the apparatus of fig. 46.2, in which the electroscope is used as an electrostatic voltmeter.)

(a) Capacitors with same p.d.

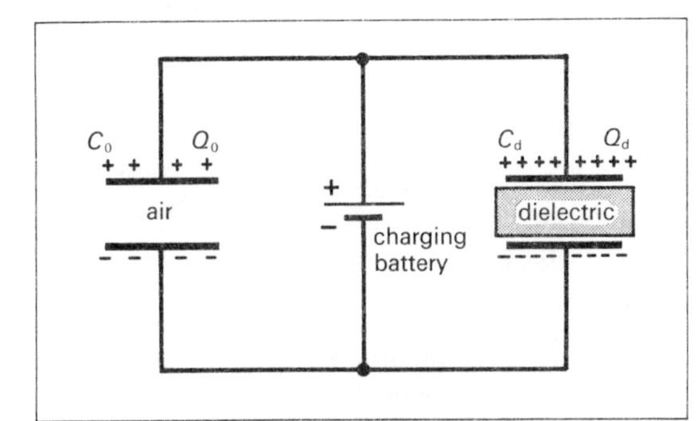

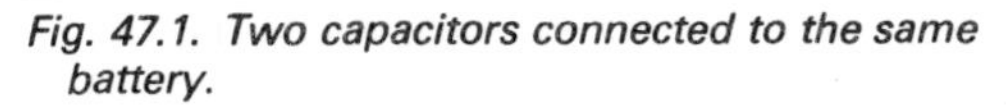

Fig. 47.1. Two capacitors connected to the same battery.

We can use a ballistic galvanometer (fig. 47.1) to show $Q_d > Q_0$, and deduce that the capacitor with the dielectric has the greater capacitance; $C_d > C_0$.

(b) Capacitors with same charge

Experimentally (fig. 47.2) we can show that $V_0 > V_d$, and deduce that $C_d > C_0$, as before.

Suppose the dielectric material *fills* the space between the plates.

Then we define the* relative permittivity ε_r *by the equation

$$\varepsilon_r = \frac{C_d}{C_0}$$

(ε_r is sometimes called the *dielectric constant.*)

Typical values of ε_r are given in the table on p. 338. ε_r is dimensionless, and has no unit.

(a) For a capacitor with V kept constant

$$\varepsilon_r = \frac{C_d}{C_0} = \frac{Q_d}{Q_0}$$

(b) For Q kept constant

$$\varepsilon_r = \frac{C_d}{C_0} = \frac{V_0}{V_d} = \frac{E_0}{E_d}$$

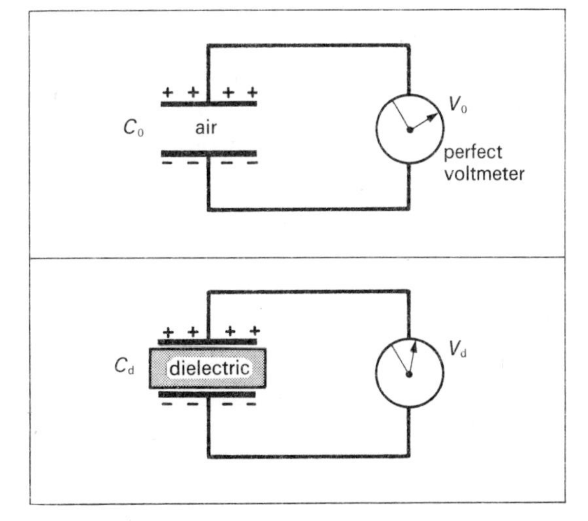

Fig. 47.2. Two capacitors with the same charge.

This idea is discussed in the next paragraph, where $\boldsymbol{E}_0$ and $\boldsymbol{E}_d$ are defined.

The electric permittivity* ε *is defined by the equation

$$\varepsilon = \varepsilon_r \varepsilon_0$$

ε is sometimes called the **absolute permittivity**. It has the same dimensions and unit as ε_0.

Capacitors and Dielectrics

Dielectric materials are used in capacitors

(*i*) to keep the plates apart,
(*ii*) to raise the capacitance, relative to air, by a factor of about ε_r, and
(*iii*) to reduce the chance of electric breakdown, and thus to enable a large V to be used.

To achieve this they require a large **dielectric strength**, which is the largest potential gradient that can exist in the material.

Material	*Relative permittivity* ε_r *measured at zero frequency*	*Dielectric strength* / 10^6 V m^{-1}
vacuum	1 exactly	—
air (at s.t.p.)	1.0006	3
water	80	(conductor)
mica	5	150
polyethene	2	50

Our capacitance equations become

(1) Parallel-plate $\quad C = \dfrac{\varepsilon A}{d}$

(2) Coaxial cylinder $\quad C = \dfrac{2\pi\varepsilon l}{\ln(b/a)}$

ε has replaced ε_0.

Measurement of ε_r

(*a*) If the dielectric material fills the space between the plates (as would a fluid), use

$$\varepsilon_r = \frac{C_d}{C_0}$$

(*b*) If the dielectric is a solid slab which does *not* fill the space, measure C_0 and C_d, and use the equation derived in the next paragraph.

47.2 Molecular Explanation of Dielectric Behaviour

(*a*) In a **polar substance** the centre of negative charge in a molecule does not coincide with the centre of positive charge. A molecule has a *permanent electric dipole moment* (p. 316).

In a uniform electric field, the molecules experience

(*i*) a torque that tends to align them (see fig. 43.6), but also
(*ii*) a tendency to random orientation because of their thermal motion.

Their partial alignment results in a field $\boldsymbol{E}'$ (opposite) in *opposition* to an external field $\boldsymbol{E}_0$.

(*b*) **Non-polar substances** acquire a temporary *induced* dipole moment in an external field (fig. 47.3).

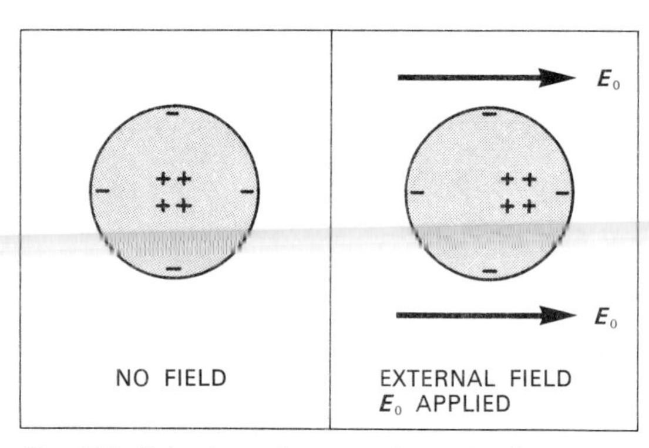

Fig. 47.3. Behaviour of a non-polar molecule.

The temporary dipoles are parallel to $\boldsymbol{E}_0$, and establish an opposing field $\boldsymbol{E}'$.

In both (*a*) and (*b*) the material is said to be **polarized** by the *microscopic* charge movement. Any small sample of material (other than at the surface) remains electrically neutral.

(*c*) **Conductors** (which are not dielectrics) have free charge-carriers. Their behaviour was discussed on p. 326. They are *completely polarized* in that they make *zero* the field within themselves.

In fig. 47.4 we note three fields

(1) $\boldsymbol{E}_0$ the external (applied) field,
(2) $\boldsymbol{E}'$ the internal field that results from the induced surface charges, and
(3) $E_d = (E_0 - E')$, the resultant internal field.

It is $\boldsymbol{E}_d$ that is related to the p.d. V across the dielectric slab (not $\boldsymbol{E}_0$ or $\boldsymbol{E}'$).

$$\varepsilon_r = \frac{E_0}{E_d}$$

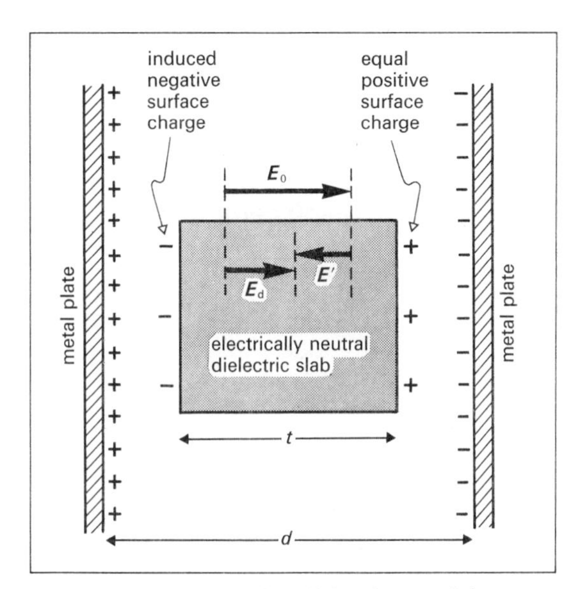

Fig. 47.4. The field inside a dielectric material.

Capacitance of a Partly Filled Capacitor

Refer again to fig. 47.4.

If no dielectric is present, the p.d. V_0 is given by

$$V_0 = E_0 d$$

When the dielectric slab is present, the p.d. V_d between the plates

$$\begin{aligned} V_d &= E_0(d - t) + E_d t \\ &= E_0(d - t) + \frac{E_0 t}{\varepsilon_r} \\ &= E_0(d - t + t/\varepsilon_r) \end{aligned}$$

Hence $$\frac{C_d}{C_0} = \frac{V_0}{V_d} = \frac{d}{d - t + t/\varepsilon_r}$$

from which ε_r can be found once C_0, C_d and d and t are measured.

Electrostatic Attraction

Fig. 47.5 shows how a charged body exerts an attractive force on a nearby neutral body which becomes polarized.

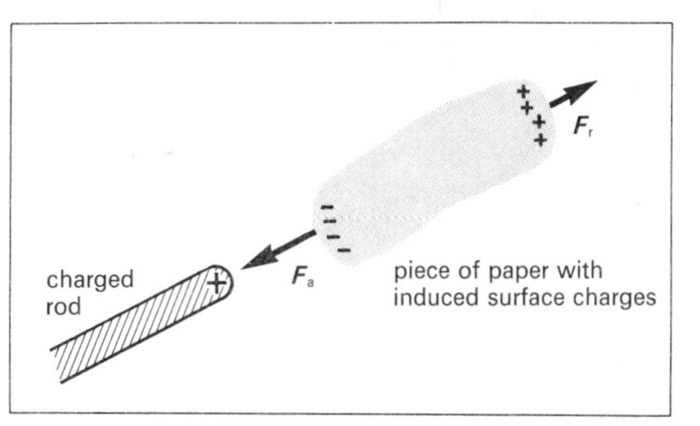

Fig. 47.5. Electrostatic attraction. $\boldsymbol{F}_a > \boldsymbol{F}_r$ since the field is non-uniform.

*47.3 Dielectrics and Coulomb's Law

For a vacuum we write

$$F = \left(\frac{1}{4\pi\varepsilon_0}\right) \cdot \frac{Q_1 Q_2}{r^2}$$

When the charges are surrounded by dielectric material, they exert the same *direct* forces on each other as in a vacuum, but the *observed* force is smaller because the dielectric material also exerts a force on each charge, but in the opposite direction.

In a dielectric we write

$$\begin{aligned} F &= \frac{1}{\varepsilon_r} \cdot \left(\frac{1}{4\pi\varepsilon_0}\right) \cdot \frac{Q_1 Q_2}{r^2} \\ &= \left(\frac{1}{4\pi\varepsilon}\right) \cdot \frac{Q_1 Q_2}{r^2} \end{aligned}$$

We modify the results on pages 314 and 325.

At a point distant r from a point charge Q

$$E = \left(\frac{1}{4\pi\varepsilon}\right) \cdot \frac{Q}{r^2} \quad \text{and} \quad V = \frac{1}{4\pi\varepsilon} \cdot \frac{Q}{r}$$

These were derived from *Coulomb's Law*, and only apply when field lines are *radial*.

*47.4 Dielectrics and Gauss's Law

(a) With no Dielectric

In fig. 47.6 *Gauss's Law* can be written

$$\varepsilon_0 E_0 A = Q$$

so $$E_0 = \frac{Q}{\varepsilon_0 A} \qquad \dots (1)$$

(b) With Dielectric

Gauss's Law can be written

$$\varepsilon_0 E_d A = Q - Q'$$

$$E_d = \frac{1}{\varepsilon_0 A}(Q - Q') \qquad \ldots(2)$$

But
$$E_d = \frac{E_0}{\varepsilon_r} \qquad \text{(p. 337)}$$

so
$$E_d = \frac{Q}{\varepsilon_r \varepsilon_0 A}, \quad \text{using (1)}$$

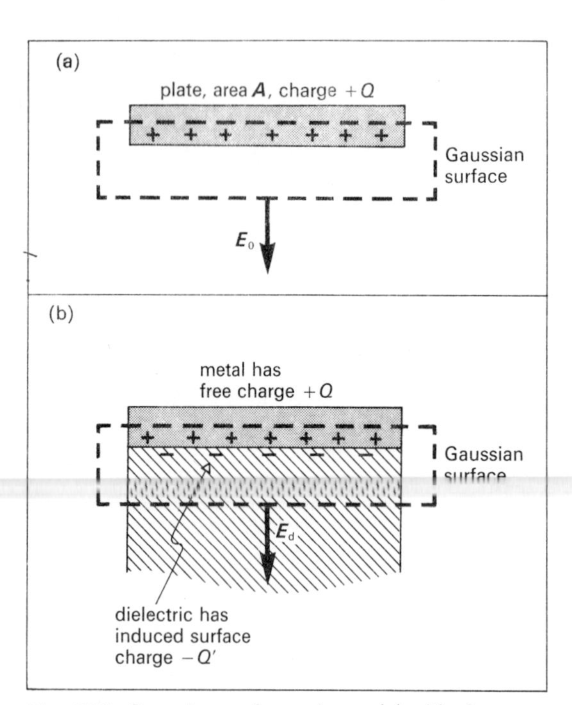

Fig. 47.6. Gaussian surfaces drawn (a) with air between the plates, and (b) with a dielectric material. (The top plate only of a parallel-plate capacitor is shown.)

Combining this with (2), we have

$$(Q - Q') = \frac{Q}{\varepsilon_r} \qquad \ldots(3)$$

This equation demonstrates that when

(*i*) $Q' = 0$, $\varepsilon_r = 1$: this corresponds to a vacuum,

(*ii*) $Q' < Q$, $\varepsilon_r > 1$: this corresponds to a dielectric material.

(Note that if we have a conducting slab, $Q' = Q$.)

Substitution for $(Q - Q')$ from (3) into (2) gives

$$\varepsilon_0 E_d A = \frac{Q}{\varepsilon_r}$$

$$\varepsilon_r \varepsilon_0 E_d A = Q$$

We can now rewrite **Gauss's Law** *as*

$$\varepsilon_r \varepsilon_0 \psi_E = \sum \begin{pmatrix}\textit{free charges inside}\\ \textit{Gaussian surface}\end{pmatrix}$$

When we insert ε_r on the left-hand side, we need only add up the *free or conduction charges* inside the *Gaussian* surface. We then omit the induced *surface* or *polarization charges*.

The law can be applied in this form to any situation met at this level.

*47.5 Electric Displacement D

The electric displacement D at a point in an electric field is defined by

$$D = \varepsilon_r \varepsilon_0 E = \varepsilon E$$

$$D\left[\frac{\text{C}}{\text{m}^2}\right] = \varepsilon_r \varepsilon_0 \left[\frac{\text{C}^2}{\text{N m}^2}\right] E\left[\frac{\text{N}}{\text{C}}\right]$$

D is a *vector* quantity which has the same unit, C m^{-2}, as surface density of charge, σ. It is convenient concept for reformulating *Gauss's Law*:

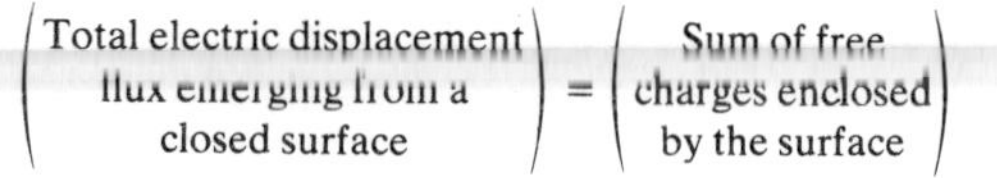

$$\begin{pmatrix}\text{Total electric displacement}\\ \text{flux emerging from a}\\ \text{closed surface}\end{pmatrix} = \begin{pmatrix}\text{Sum of free}\\ \text{charges enclosed}\\ \text{by the surface}\end{pmatrix}$$

Displacement flux ψ_D is defined in an analogous way to *field flux* (p. 317), and has the unit of DA i.e. (C m^{-2}) (m^2) = C.

47.6 Energy Changes in a Capacitor

When evaluating energy changes for a capacitor, use

$$W = \tfrac{1}{2}QV = \tfrac{1}{2}Q^2/C = \tfrac{1}{2}V^2C$$

But if

(*a*) The *battery is connected*, then V is kept constant. A change in C results in a change in Q.

(*b*) The *capacitor is insulated*, then Q is kept constant. A change in C results in a change in V.

Examples

(*a*) If the plates of a parallel-plate capacitor are pulled apart, with the battery connected, then

$$V = \text{constant}$$

so $\qquad E = \dfrac{\sigma}{\varepsilon_0}$ decreases,

so $\qquad Q$ decreases

and we conclude W decreases. (Charge has been returned to the battery.)

(*b*) If the plates are pulled apart with the charged capacitor insulated, then

$$Q = \text{constant}$$

so $\qquad E = \dfrac{\sigma}{\varepsilon_0}$ is constant

so $\qquad V = Ed$ increases

and we conclude W increases. An external agent does work against the attractive forces between the plates: the work done increases the electric potential energy stored in the field.

(*c*) Suppose a dielectric material is *slowly* inserted between the plates, as in fig. 47.7 The electric field causes work to be done *on* the dielectric slab, and so there is a decrease of electric potential energy. (Think of the gravitational analogy of a stone, held in someone's hand, being allowed to move downwards at constant speed.)

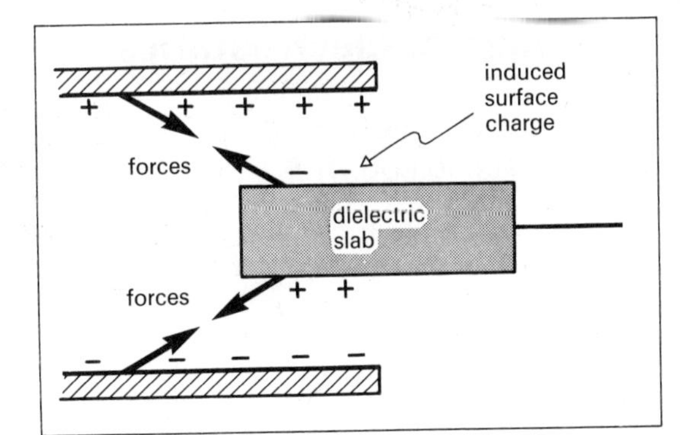

Fig. 47.7. A dielectric material attracted into a parallel-plate capacitor.

Results (*a*), (*b*) and (*c*) can alternatively be obtained by calculating the resulting change in C, and noting what stays constant.

Summary of Electrostatics

	DEFINING EQUATIONS	LAWS	IMPORTANT DERIVATIONS
Q	$Q = \int I\,\mathrm{d}t$	*Coulomb*	
ε_0	$F = \left(\frac{1}{4\pi\varepsilon_0}\right)\cdot\frac{Q_1Q_2}{r^2}$	$F \propto \frac{Q_1Q_2}{r^2}$	
$\mathbf{E}$	$\mathbf{E} = \mathbf{F}/Q_0$		
$\mathbf{p}$	$\mathbf{p} = 2\mathbf{a}Q$		$E = \left(\frac{1}{4\pi\varepsilon_0}\right)\cdot\frac{Q}{r^2}$
$\psi_{\mathbf{E}}$	$\psi_{\mathbf{E}} = EA\cos\theta$	*Gauss* $\varepsilon_0\psi_{\mathbf{E}} = \sum Q$	$\mathbf{E} = \boldsymbol{\sigma}/\varepsilon_0$
V	$V = W/Q_0$		$E = -\frac{\mathrm{d}V}{\mathrm{d}x}$
V_{ab}	$V_{ab} = W_{ab}/Q_0$		$E = \frac{V}{d}$
			$V = \left(\frac{1}{4\pi\varepsilon_0}\right)\cdot\frac{Q}{r}$
			$F = \frac{\Delta A\sigma^2}{2\varepsilon_0}$
C	$C = Q/V$		$C = \frac{\varepsilon_0 A}{d}$
			$C = C_1 + C_2$ parallel
			$\frac{1}{C} = \frac{1}{C_1} + \frac{1}{C_2}$ series
			$W = \frac{1}{2}Q_0V_0 = \frac{1}{2}\frac{Q_0^2}{C} = \frac{1}{2}V_0^2C$
ε_r	$\varepsilon_r = C_d/C_0$	*Coulomb* $F = \left(\frac{1}{4\pi\varepsilon}\right)\cdot\frac{Q_1Q_2}{r^2}$	$C = \frac{\varepsilon A}{d}$
ε	$\varepsilon = \varepsilon_r\varepsilon_0$	*Gauss* $\varepsilon_r\varepsilon_0\psi_{\mathbf{E}} = \sum$(free charges)	
$\mathbf{D}$	$\mathbf{D} = \varepsilon_r\varepsilon_0\mathbf{E}$	*Gauss* (displacement flux) = (sum of free charges)	

X Current Electricity

48 Current, Conductors and Resistance

48.1 The Nature of Current and Electrical Conductors

The Nature of Current

An **electric current** consists of the movement of electric charge. A steady current exists when there is a systematic drift of charge-carriers. The *conventional* direction of an electric current is that in which there is a *net displacement of positive charge* over a given time interval (fig. 48.1).

Current is taken to be a *primary (fundamental) quantity* in electricity in the SI. The **ampere** (A), the unit of current, is defined in terms of force and distance on page 10.

Suppose a conductor carries a current I.

Then the rate of flow of charge Q past a given cross-section is defined by the equation

$$I = \frac{dQ}{dt}$$

$$I[\text{A}] = \frac{\Delta Q\,[\text{coulombs}]}{\Delta t\,[\text{s}]}$$

Notes.

(*a*) The unit *ampere second* (A s) is called a **coulomb** (C), and is the *unit for charge.*

(*b*) The charge Q that passes a given cross-section in a given time is found from

$$Q = \int I\,dt$$

(*c*) When the current does not vary with time we can use

$$I = \frac{Q}{t} \qquad Q = It$$

The Nature of an Electrical Conductor

It was shown in **45.2** that $E = -dV/dx$. When a p.d. exists between two points an electric field is established. Any charged particles placed in that field will experience an electric force, and unless restraining forces are brought into play they will be accelerated.

A material will be a **conductor** if it allows the charged particles inside it to undergo significant *net translational motion* when a p.d. is applied. These particles are then called **mobile charge-carriers**. They may be negative or positive, and in some situations carriers of both signs will be present.

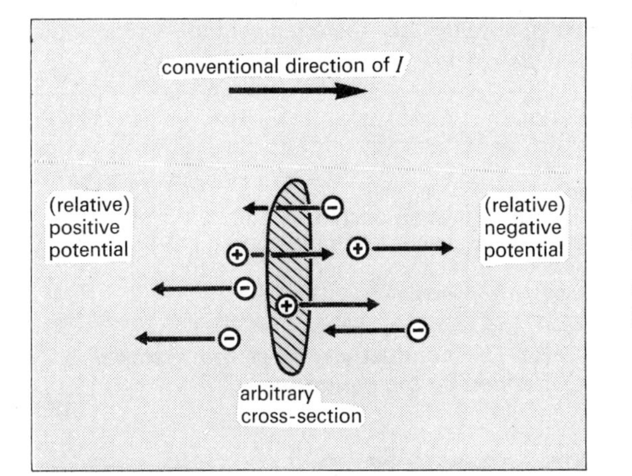

Fig. 48.1. Charge-carriers of both signs contribute to the net current I.

48.2 The Charge-carriers in Different Materials

(*a*) ***Insulators.*** The atoms' outer orbital electrons are involved in forming the covalent or ionic bonds which hold the atoms together. They are therefore not mobile, and not available for conduction.

(*b*) ***Metals.*** Each atom in a metal lattice supplies one or more *free* electrons to form the bond which holds the positive ions together. The electrons are not attached to any particular parent ion, but are mobile within the boundary of the lattice, and therefore contribute to conduction. (A fuller account is given on p. 348 and p. 349.)

(*c*) ***Semiconductors.*** The process is similar to that in metals, but the number density of charge-carriers is much smaller. The carriers can be negative (electrons), positive (holes), or of both signs. (More details are to be found on p. 349 and p. 351.)

(*d*) ***Electrolytes.*** An electrolyte is a liquid which undergoes a chemical change when it conducts charge through itself. Solid NaCl consists of rigidly-held positive sodium ions and negative chloride ions (p. 122)—the ions are not free to wander. When NaCl is dissolved in water, water's high relative permittivity enables it to weaken the electrostatic attraction, and this causes the **cations** (positive) to become dissociated from the **anions** (negative). The ions originally present in the crystal have now become mobile.

The electric current in an electrolyte consists of a steady simultaneous drift, in opposite directions, of the mobile anions and cations. Chemical changes occur at the surfaces of the electrodes.

(*e*) ***Gases.*** (*i*) High-energy photons (such as X-rays) and high-energy particles can cause a neutral gas molecule to be split into its charged constituents. These ions can conduct a steady current so long as the cause of ionization persists. See, for example, the ionization chamber (p. 470).

(*ii*) If the ions produced in (*i*) are accelerated through a sufficiently large p.d. they may gain enough energy between collisions to cause further ionization. The process is called a **Townsend discharge**.

(*iii*) If the p.d. is increased still further the rate of production of ions can equal the rate at which the electrodes collect them. The luminous effect which results is called a **glow discharge**. The *gas discharge tube* was an important subject for research at the turn of the century.

(*f*) ***Charged particles in a vacuum.*** The movement of particles is not impeded in a vacuum. Charged particles subjected to an electric field in a vacuum undergo a continuing acceleration, and quickly achieve high speeds, as electrons do in the X-ray tube. (Contrast this with their staccato drift motion through a metal p. 350.) The current involved is usually small.

48.3 Definitions

(*Word endings*: as a general rule a word ending
. . . or represents a *device*
. . . ance represents a *property* of the *device*
. . . ivity represents a property of a *material.*)

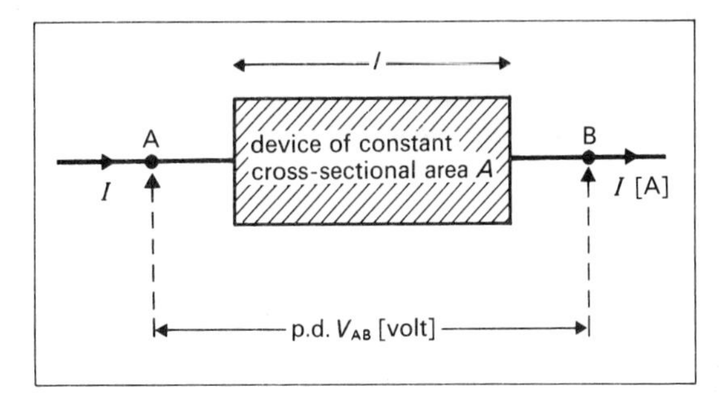

Fig. 48.2. To define V, R, G, ρ and σ.

(a) Potential difference *V* is defined by the equation

$$V_{ab} = \frac{W}{Q}$$

This is discussed on p. 323.

(b) The resistance *R* of the device is defined by the equation

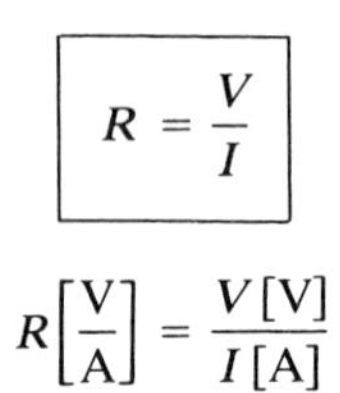

$$R = \frac{V}{I}$$

$$R\left[\frac{V}{A}\right] = \frac{V\,[V]}{I\,[A]}$$

The unit *volt ampere*$^{-1}$ is called an **ohm** (Ω). A conductor has a resistance of 1 Ω if there is a p.d. of 1 V across its terminals when a current of 1 A flows through it.

$$[R] = [ML^2T^{-3}I^{-2}]$$

(c) The conductance *G* of the device is defined by the equation

$$G = \frac{I}{V} = \frac{1}{R}$$

$$G\,[\Omega^{-1}] = \frac{I\,[A]}{V\,[V]}$$

(The unit Ω^{-1} is called **siemens** (S).)

Suppose the device shown is a *homogeneous* conductor, and that it has a cross-section of uniform area ***A*** (but any shape).

(d) The resistivity ρ of the material is defined by the equation

$$R = \frac{\rho l}{A}$$

$$R[\Omega] = \frac{\rho[\Omega\,m]\,l[m]}{A[m^2]}$$

Some values (for room temperature) are given in the table below.

Material	*Class*	ρ/Ω m	σ/S m^{-1}
copper	good conductor	1.7×10^{-8}	5.8×10^{7}
manganin	conductor	4.4×10^{-7}	2.3×10^{6}
graphite	(poor) conductor	8.0×10^{-6}	1.2×10^{5}
germanium	semiconductor	0.5	2
silicon	semiconductor	2.5×10^{3}	4×10^{-4}
quartz	insulator	5×10^{16}	2×10^{-17}

(e) The conductivity σ of the material is defined by the equation

$$\sigma = \frac{1}{\rho} = \frac{l}{RA}$$

σ is measured in Ω^{-1} m^{-1} = S m^{-1}, and has the dimensions $[M^{-1}L^{-3}T^3I^2]$.

Some values (for room temperature) are given in the table above.

Note that at room temperature σ and ρ vary from material to material by factors of up to 10^{24}. (See variations of thermal conductivity on p. 216.)

(f) The electric current density *J* at a cross-section of area *A* is defined by the equation

$$J = \frac{I}{A}$$

J is measured in A m^{-2}. The maximum safe value of ***J*** for copper is about 10^7 A m^{-2}. The concept of current density is used on p. 350 and p. 375.

(g) Suppose that a mobile charge-carrier acquires a mean drift velocity $\boldsymbol{v}_d$ when an electric field ***E*** is applied to a conducting material.

The electric mobility *u* of the charge-carriers is defined by the equation

$$u = \frac{v_d}{E}$$

$$u\left[\frac{m^2}{V\,s}\right] = \frac{v_d[m/s]}{E[V/m]}$$

The concept of mobility is used on p. 350 and p. 351.

In an electrolyte different types of ion will be present. Each may have a different mobility. This mobility will increase when the temperature rises, since the liquid viscosity is reduced.

48.4 Electrical Characteristics and Ohm's Law

Suppose we measure simultaneously the p.d. V across a two-terminal electric device and the current I through it. The graph of V (x-axis, cause) against I (y-axis, effect) is called the **electrical characteristic**. Some examples are given in fig. 48.3.

Non-linear conductors and devices are those whose characteristics are not straight lines: their graphs may, in addition, be asymmetrical relative to the origin. Such devices include, apart from those shown, semiconducting materials (such as carbon, silicon and germanium), electrolytes (such as the liquids in chemical cells), thermionic diodes and metal rectifiers.

For some circuit elements, including the inductor and capacitor, the relation between the current and the applied p.d. is *time*-dependent.

Ohm's Law

Ohm discovered in 1826 that the characteristic shown in (a) is common to a large class of conducting materials. His law was an experimental result, stated by:

Provided physical conditions (such as temperature) do not alter, then over a wide range of applied p.d.'s

$$V \propto I,$$

or $$V/I = \text{constant}$$

for many materials, particularly pure metals and alloys.

$$\boxed{R = \text{constant}} \quad \textbf{Ohm's Law}$$

(Note that $R = V/I$ *defines* R, and that it is *not* a statement of *Ohm's law*.) The physical reason for a metal having this constant resistivity is discussed in **48.6**.

Test of whether a Substance obeys Ohm's Law

(a) Ammeter–Voltmeter method

Connect the device under test into the circuit as in fig. 48.2. Use an ammeter to measure I, and a voltmeter to measure V. Plot the V–I graph, and see whether it is a straight line.

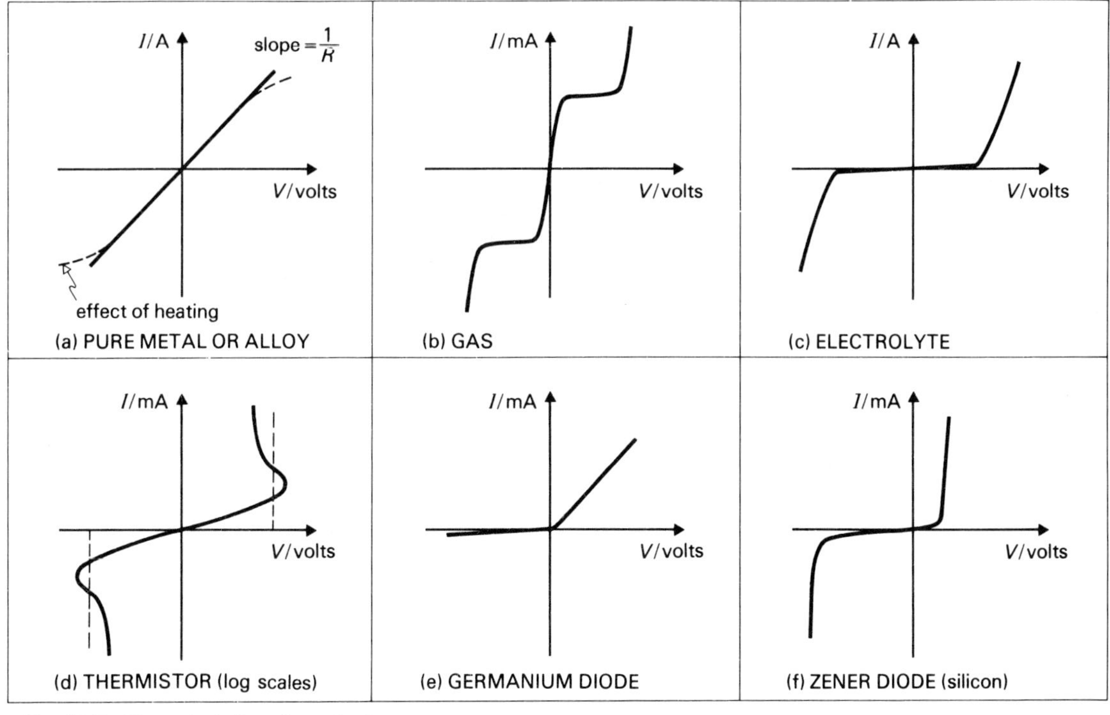

Fig. 48.3. Characteristics of conductors.

(It does not matter that the ammeter and voltmeter have been calibrated using the law, provided that *the material under test* does not form part of the instrument.)

One cannot conclude '*Ohm's Law* is not true', but only 'This material does not obey the law'.

(b) Energy method

The rate P at which a resistor converts electrical energy to internal energy is given by

$$P = I^2R$$

and can be measured by a wattmeter (p. 393), or by calorimetry. If this is done for a range of currents, the graph of P against I^2 will be a straight line only if the device has a constant resistance.

(c) Wheatstone Bridge method (p. 370)

*48.5 Band Theory of Conduction Through Solids

The Formation of Energy Bands

In *classical* theory the energy of an electron in a solid can have any value. In *quantum* theory certain discrete values only are allowed.

(*a*) ***Isolated atoms.*** The energies available to orbital electrons are confined to certain strict values. Some of those for the hydrogen atom are shown in fig. 48.4. (The theory behind these results is given in **63.11**.)

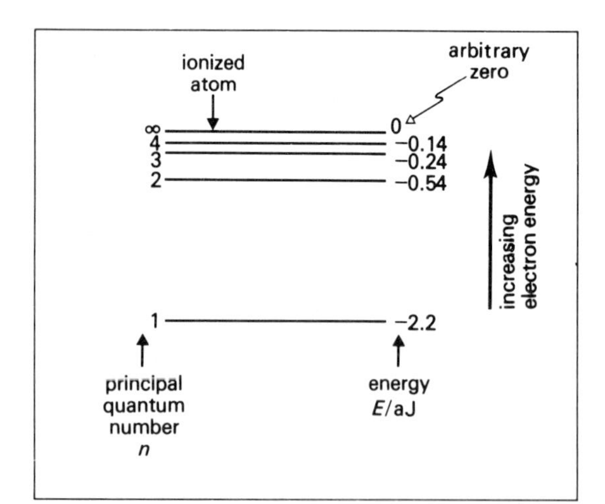

Fig. 48.4. The allowed energy levels of an isolated hydrogen atom.

(*b*) ***Interacting atoms.*** When atoms are in a position to affect each other through their electric fields, a greater number of energy levels becomes available to the electrons. The individual lines of the isolated atomic levels are split or *broadened* into a very large number of levels which are very close together (fig. 48.5.) Each group of energy levels is called a **band**.

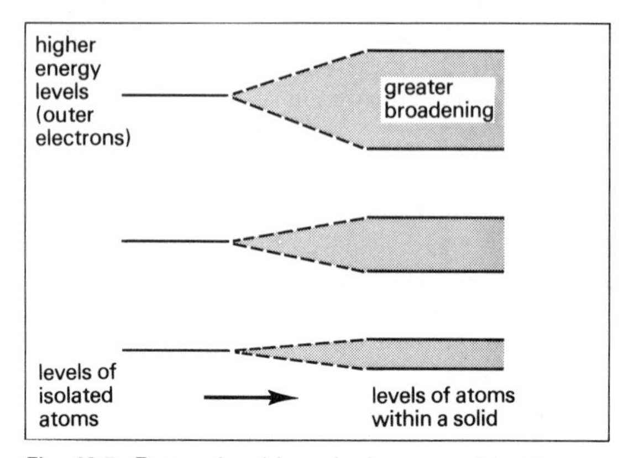

Fig. 48.5. Energy level broadening caused by the atoms of a solid affecting each other's discrete levels.

Notes.

(*a*) The outermost orbital electrons are most affected, and so their energy levels are broadened into the widest bands.

(*b*) Wider bands are produced for tighter packing, since the atomic interaction is more pronounced.

(*c*) It is possible for a higher energy band to be so broadened as to overlap a lower energy band.

How the Bands can explain Conduction

At absolute zero the electrons in a given crystal lattice occupy the lowest available energy states. (According to the *Pauli* **exclusion principle** no two electrons can be in precisely the same state at the same time.)

An increase in temperature causes the electrons to move into higher *available* energy states. They move continually between different energy states as they exchange energy.

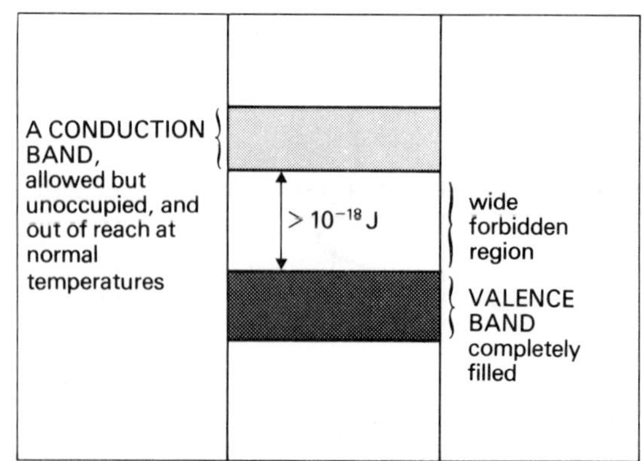

Fig. 48.6. Electron energy level band scheme for an insulator.

(*a*) ***Insulators.*** Refer to fig. 48.6. If the electrons occupy all the energy states available to them in a given band, they may be unable to change state. This would prevent them from acquiring the energy associated with the systematic drift velocity needed for the net transfer of charge. A band of this kind which is completely filled is a **valence band**.

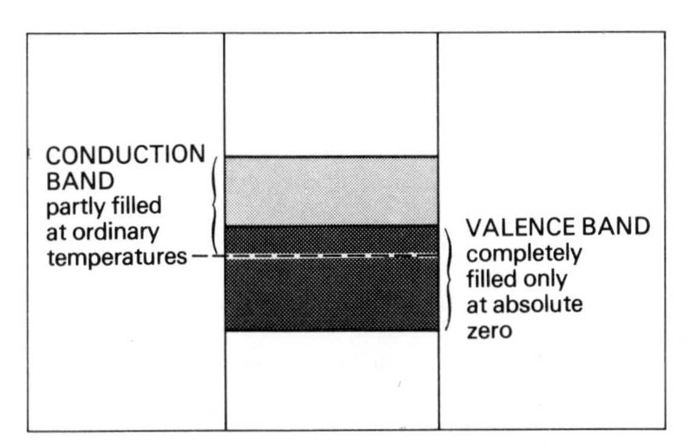

Fig. 48.7. Electron energy band scheme for a conductor in which the valence and conduction bands are overlapping.

(*b*) ***Conductors.*** An applied electric field will cause the systematic drift of electrons if those electrons in the highest occupied band do not fill it. Such a partly filled band is called a **conduction band** (fig. 48.7). The valence and conduction bands either have no gap between them, or they overlap.

(*c*) ***Insulator breakdown.*** It may be possible for electrons to be lifted across the gap from the valence band to the conduction band by the application of a sufficiently strong electric field, or by an increase in temperature. Those electrons then become mobile, since they can acquire drift k.e. by moving into different energy states in the conduction band.

(*d*) ***Intrinsic semiconductors.*** Refer to fig. 48.8. Suppose that a given substance has its highest occupied energy band completely filled at absolute zero, and that the next allowed band does not overlap it. Such a substance would be an insulator.

At ordinary temperatures it is possible for the electrons to be given sufficient energy by thermal agitation to move continually from the valence band across the **forbidden zone** into the conduction band.

Notes.

(*i*) As the temperature increases more electrons can cross the gap, and so the conductivity increases.

(*ii*) At ordinary temperatures the thermal excitation energy of electrons is about 10^{-20} J. The gap in diamond is close to 85×10^{-20} J, making it an insulator, whereas that for germanium is only 11×10^{-20} J, and so it is a semiconductor.

(*iii*) The conductivity of semiconductors is influenced considerably by the presence of impurities, as discussed in **48.8**. A *pure* semiconductor is called an **intrinsic** semiconductor. One *doped* with impurities is called an **extrinsic** semiconductor.

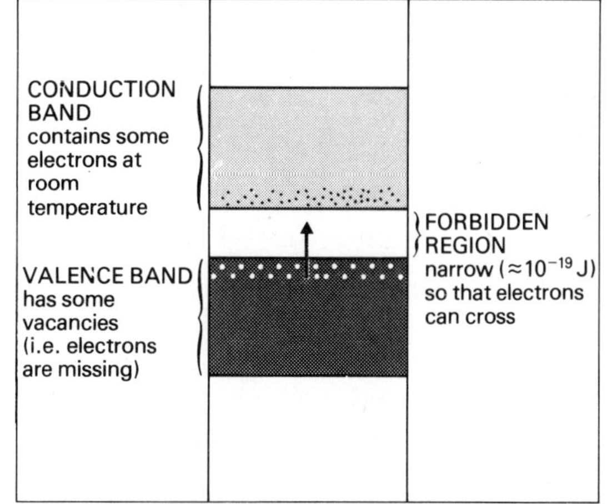

Fig. 48.8. Energy band scheme for an intrinsic semiconductor.

48.6 The Movement of Electrons Through Metals

The Current in a Metallic Conductor

The charge carried by an electron is -1.6×10^{-19} C.

$$1\,\text{C} = 1\,\text{A s}$$

so if a current of 1 A flows for 1 s past a given cross-section in a metal, 6.25×10^{18} electrons pass in the opposite direction.

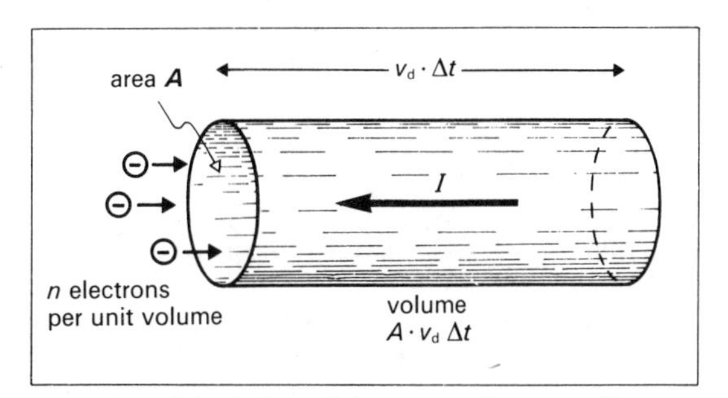

Fig. 48.9. Calculation of the current in a metallic conductor.

Suppose that the electrons carry a charge of magnitude e, move with an *average* **drift speed** v_d and that their **number density** is n. After a time interval Δt all the electrons within the volume of length $v_d \Delta t$ will have passed through the right end face.

The charge ΔQ contained is given by

$$\Delta Q = n(Av_d\,\Delta t)e$$

so

$$\frac{\Delta Q}{\Delta t} = I = nAev_d$$

$$\boxed{I = nAev_d}$$

$$\boxed{\boldsymbol{J} = ne\boldsymbol{v}_d}$$

Since $e = -1.6 \times 10^{-19}$ C, it follows that the direction of current density is opposite from that of the electron average drift velocity.

Compare

(*i*) the speed of an *electrical signal* in a conductor ($\approx 3 \times 10^8$ m s^{-1}),

(*ii*) the speeds of the electrons in their *random motion* ($\approx 10^6$ m s^{-1}), and

(*iii*) the average *drift speed* v_d of the electrons ($\approx 10^{-4}$ m s^{-1} in a typical conductor carrying a current of 5 A through a cross-section of 1 mm^2).

*The Mechanism of Current Flow

The mobile or conduction electrons in a metal are accelerated by electric forces until they collide inelastically with lattice ions: the motion is repeated very rapidly at short time intervals, as suggested in fig. 48.10 (*a*), being superimposed on the random thermal motion of the electrons.

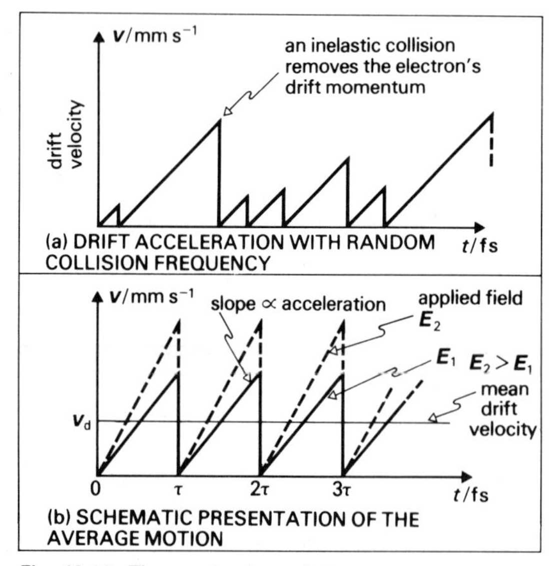

Fig. 48.10. The mechanism of electron movement through a metallic conductor carrying a current.

It can be calculated from the relationships that follow that electrons have a surprisingly large mean free path of about 100 ionic diameters.

The *macroscopic* effect is a steady current flow, and the gain of internal energy by the lattice causes an increase of temperature. *Microscopically* electric field energy is converted initially to the mechanical k.e. of the drifting electrons, and then to the k.e. and p.e. of the vibrating lattice ions. The electrons are in thermal equilibrium with these ions.

In a steady field $\boldsymbol{E}$ the electrons have a constant acceleration superimposed on their random motion. Using $\boldsymbol{a} = \boldsymbol{F}/m = \boldsymbol{E}e/m_e$, we can calculate their average drift velocity $\boldsymbol{v}_d$ from

$$\boldsymbol{v}_d = \tfrac{1}{2}(\text{maximum drift velocity})$$

$$= \tfrac{1}{2}\boldsymbol{a}t = \tfrac{1}{2}\left(\frac{\boldsymbol{E}e}{m_e}\right)\tau$$

where τ is the *characteristic time* (about 10^{-14} s) for which an electron gains speed between collisions.

But $\qquad \boldsymbol{v}_d = \boldsymbol{J}/ne \qquad$ (boxed above)

Equating

$$\frac{\boldsymbol{J}}{ne} = \frac{\boldsymbol{E}e\tau}{2m_e}$$

or

$$\boldsymbol{J} = \frac{n\boldsymbol{E}e^2\tau}{2m_e} \qquad \ldots (1)$$

Now by definition, electrical conductivity

$$\sigma = \frac{l}{RA} = \frac{lI}{VA} \qquad \text{(def. of } R\text{)}$$

$$= \frac{J}{E} \qquad \text{(def. of } \boldsymbol{J}\text{)}$$

$$\boxed{\boldsymbol{J} = \sigma\boldsymbol{E}}$$

Combining this result with equation (1) above gives

$$\sigma = \frac{ne^2\tau}{2m_e}$$

*The Physical Basis of Ohm's Law

The last equation indicates that if the values of e and m_e are fixed, the electrical conductivity of a material, and hence the resistance of a given sample, will be independent of current so long as

(*a*) the number density n of charge-carriers does not depend on I, and

(*b*) τ is independent of current. Now the *frequency of electron collision with the lattice is not changed by the current flow* because typically $v_d \approx 10^{-4}$ m s^{-1}, whereas the random speed $\approx 10^6$ m s^{-1}.

These two conditions are obeyed by metals and alloys unless the temperature changes, or unless the material undergoes a change of stress.

Electron Mobility

Since $\sigma = J/E$, and $\boldsymbol{J} = ne\boldsymbol{v}_d$,

$$\sigma = nev_d/E = neu_e$$

where u_e is the *mobility* of the electrons.

Thus one aspect of *Ohm's* law may be expressed in microscopic terms by the statement

At a fixed temperature the mobility of electrons in pure metals and some alloys is independent of current.

In addition, of course, *Ohm's* law requires n to be constant.

48.7 Intrinsic Semiconductors

The *band theory* of intrinsic (pure) semiconductors is given on p. 349. According to this theory the net transfer of charge necessary for an electric current is caused by

(*a*) the movement of *electrons*, which in effect have been shaken loose by their parent atoms by thermal agitation, together with

(*b*) the oppositely directed movement of a position from which an electron is missing. A position such as this is called a **positive hole**: it is a vacancy for an electron, and it is located in the place of a positive ion in the crystal lattice.

The electrons and positive holes in an intrinsic semiconductor *both* contribute to the conductivity, and are called **minority charge-carriers**. (Only about 1 atom in 10^{10} contributes charge-carriers in this way.)

Positive Holes

While the holes indicate *locations* in the crystal lattice, *they behave like positive particles.* Under the influence of an electric field they drift through the lattice in the same manner as electrons. The mechanism is the movement of a bound electron (in the valence band) from an atom into a neighbouring positive hole. The table shows that holes have a smaller electric mobility than conduction electrons, and so the latter contribute more to the conductivity.

Substance	*Electron mobility* $u_e / \frac{m^2}{V\,s}$	*Hole mobility* $u_h / \frac{m^2}{V\,s}$	*At* 300 K $\frac{n_e + n_h}{m^3}$
copper	0.002 8	—	1×10^{29}
germanium	0.39	0.19	5×10^{19}
silicon	0.14	0.048	3×10^{16}

Variations of Conductivity

The energy necessary to create an electron-hole pair can be provided

(*a*) by thermal agitation,
(*b*) by photons of electromagnetic radiation, or
(*c*) during avalanche breakdown (p. 446).

At equilibrium there is a balance between the regeneration and recombination of electrons and holes: this means that at a particular temperature the product of the concentrations of electrons and holes is a constant. Their sum is a minimum when the two concentrations are equal.

The product increases rapidly with rise in temperature for both silicon and germanium.

48.8 Extrinsic (Impurity) Semiconductors

Seebeck Effect

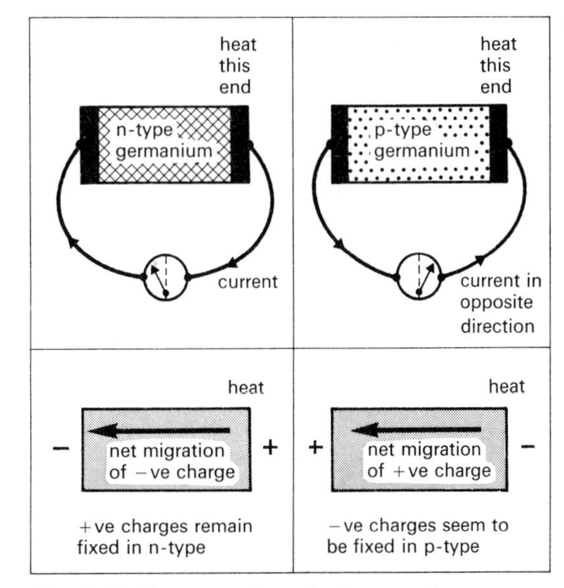

Fig. 48.11. Demonstration of existence of charge-carriers of opposite signs.

Refer to fig. 48.11. The conclusion to be drawn is that the current-flow mechanism appears to be different in the two different types of germanium. This is confirmed by the *Hall effect* (p. 375).

Their Formation

Semiconductors such as germanium can be made extremely pure, but their conductivity is considerably modified by the presence of impurities. (An impurity level of 1 in 10^7 may change the conductivity by a factor of 10^3.) These impurities can be introduced deliberately: the process is called **doping**.

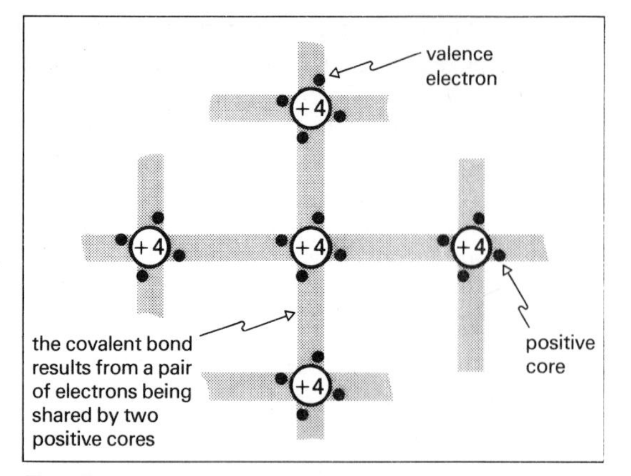

Fig. 48.12. Two-dimensional schematic representation of the three-dimensional arrangement of ions in a germanium crystal lattice.

(*a*) ***The structure of pure germanium.*** Refer to fig. 48.12. The germanium atom can be considered to consist of a positive core (the nucleus, and those orbital electrons in closed shells) surrounded by four valence electrons. Germanium is *tetravalent*, and its crystal structure is tetrahedral, like that of diamond.

Any one positive core shares its four valence electrons, one with each of the four near-neighbouring cores. No electrons are mobile at low temperatures since all the valence electrons are involved in forming these **covalent bonds**.

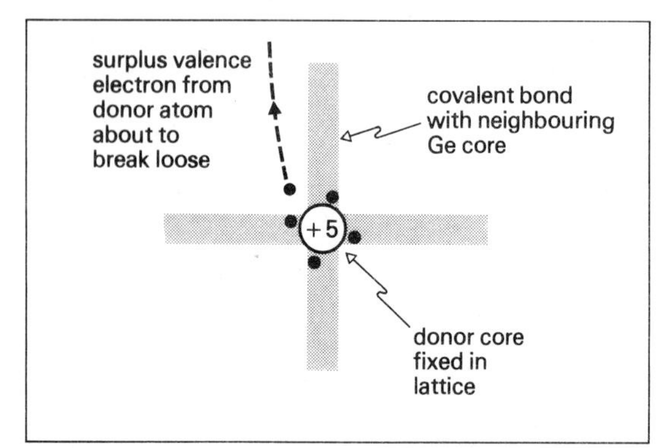

Fig. 48.13. The formation of n-type germanium.

(*b*) ***The effect of donor impurities.*** Refer to fig. 48.13. *Pentavalent* atoms, such as those of antimony, arsenic and phosphorus, have *five* valence electrons, of which only four are needed to complete the covalent bonding. The fifth electron is bound very weakly to the parent atom, and is easily liberated to become mobile. This liberation requires less energy than would be needed by an electron to escape from the Ge–Ge bond.

Notes.

(*i*) The conducting material is called **n-type**, since the main contribution to its conductivity comes from the mobile *electrons*. This is revealed experimentally by the *Hall* effect (p. 375).

(*ii*) The pentavalent material is called a **donor** impurity, since it *gives* electrons to the lattice.

(*iii*) The electrons are referred to as **majority carriers**. There will always be some hole conduction, and in an n-type material the holes are called **minority carriers**.

(*iv*) The number density of electrons considerably exceeds that of positive holes.

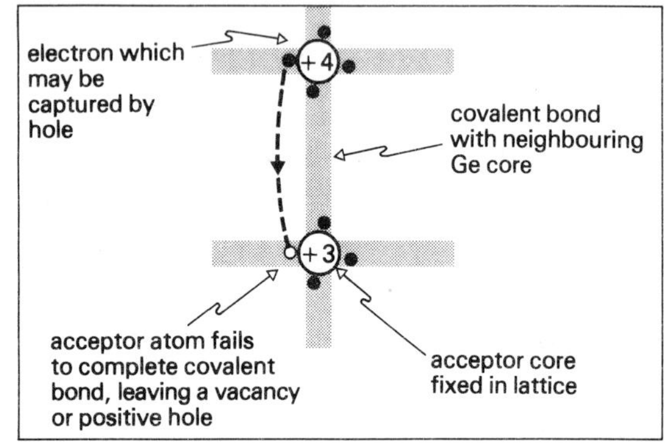

Fig. 48.14. The formation of p-type germanium.

(*c*). ***The effect of acceptor impurities.*** Refer to fig. 48.14. *Trivalent* atoms, such as those of aluminium, boron or indium, have *three* valence electrons, whereas four are needed to complete the covalent bonding. Consequently a positive hole or *vacancy* appears in the lattice, and less energy is needed for this hole to capture an electron from an adjacent germanium atom than is required for an electron to break out of the Ge–Ge bond. Such a capture causes the net translation of a location at which there is a net positive charge.

Notes.

(*i*) The conducting material is called **p-type**, since the main contribution to its conductivity comes from the mobile *positive holes*. This is revealed experimentally by the *Hall* effect (p. 375).

(*ii*) The trivalent material is called an **acceptor** impurity, since it *accepts* electrons from the lattice atoms to fill the vacancies that it creates.

(*iii*) The holes are referred to here as **majority carriers**. There will always be some electron conduction, and in a p-type material the electrons are called **minority carriers**.

(*iv*) The number density of holes considerably exceeds that of electrons.

*The Band Theory of Extrinsic Semiconductors

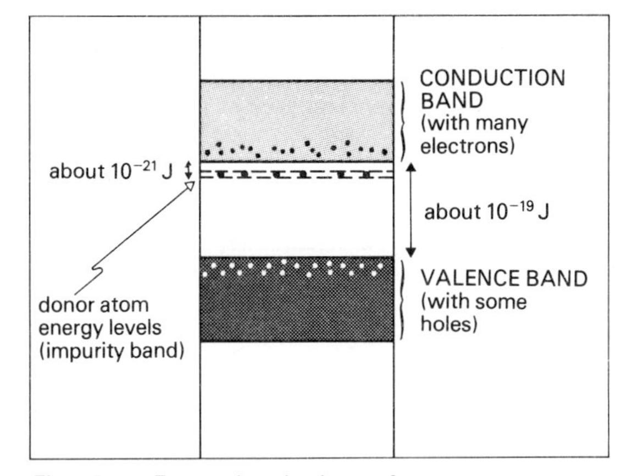

Fig. 48.15. Energy band scheme for an n-type impurity semiconductor. (Not to any scale.)

(*a*) ***n*-type.** Refer to fig. 48.15. The donor atoms contribute electrons whose available closely-spaced energy levels lie just below the conduction band. These *electrons* are moved easily by thermal excitation to available states in the conduction band, where they become *mobile*.

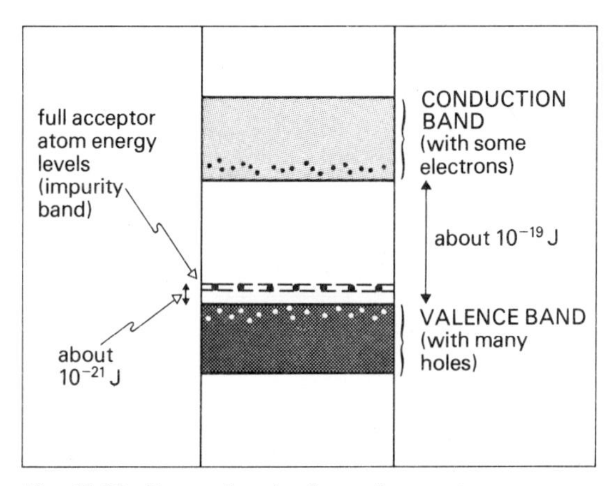

Fig. 48.16. Energy band scheme for a p-type impurity semiconductor. (Not to any scale.)

(*b*) ***p*-type.** Refer to fig. 48.16. The acceptor atoms contribute electrons whose available closely-spaced energy levels lie just above the valence band. Electrons can move easily into these levels, and so create *mobile positive holes* in the valence band.

As the acceptor energy levels are usually full, electrons in these levels cannot normally take part in conduction.

48.9 Variation of Resistance with Temperature

It was shown in **48.6** that a material's conductivity is given by

$$\sigma = ne^2\tau/2m_e$$

where τ is the characteristic time for which an electron gains speed between collisions, and n is the number density of charge-carriers.

(*a*) Metallic Conductors

(*i*) n is practically independent of temperature for most metals.

(*ii*) An increase of temperature increases the amplitude of vibration of lattice ions, which reduces τ.

Thus the conductivity falls, and the resistivity rises with increase of temperature, as shown in fig. 48.17.

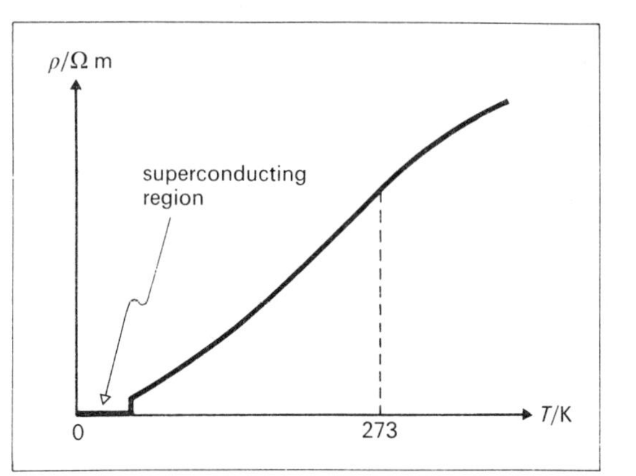

Fig. 48.17. Variation with temperature T of metal resistivity ρ. (The zero resistivity is not shown by all metals.)

We define the temperature coefficient of resistivity α by the equation

$$\alpha = \frac{1}{\rho}\frac{d\rho}{dt}$$

Near the ice-point the curve is described approximately by

$$\rho = \rho_0[1 + \alpha\theta]$$

where ρ_0 is the resistivity at the ice-point, ρ the resistivity at temperature T, and $\theta = T - 273$ K. For pure metallic elements α is positive, and has a value close to 4×10^{-3} K^{-1}. The variation of resistivity with temperature is utilized in the resistance thermometer.

As the temperature approaches absolute zero, some substances (pure metals, alloys and some compounds) become **superconductors**. Their crystal lattices are unable to take energy from the drifting electrons, and so $\rho = 0$. This effect can be explained only by the quantum theory.

Some alloys, such as constantan, have an extremely small value of α: this makes them useful for constructing standard resistors.

(b) Semiconductors

An increase in temperature

(*i*) greatly increases n (p. 349), but
(*ii*) as in metallic conductors, reduces τ.

For many semiconductors effect (*i*) is far more important than (*ii*) in certain temperature ranges.

Typically $\alpha \approx -6 \times 10^{-2}\,\mathrm{K}^{-1}$, being negative because the resistivity is reduced as the temperature increases.

48.10 Comparison of Metals and Semiconductors

	Metal	*Semiconductor*
charge-carriers	free electrons	can be mainly electrons, or positive holes, or both equally
number density at 300 K n/m^{-3}	1.1×10^{29} (Cu)	5×10^{19} (Ge)
resistivity at 300 K $\rho/\Omega\,\mathrm{m}$	10^{-5} to 10^{-8} 1.7×10^{-8} (Cu)	variable 0.5 (Ge)
Hall effect	usually negative (free electrons), but can be positive (e.g. Zn)	positive (p-type), or negative (n-type), but see below†
typical Hall coefficient $R_H/\mathrm{m}^3\,\mathrm{C}^{-1}$	-1.1×10^{-10} (Au)	depends upon doping, but for p-type Ge we might have $+10^{-2}$
typical temperature coefficient of resistance α/K^{-1}	$+4 \times 10^{-3}$	about -60×10^{-3}
effect of impurities	relatively little	impurity concentration of 1 in 10^7 can alter ρ by a factor of about 10^3 at 300 K
effect of light	very little, unless photon energies can produce an external photoelectric emission	incident photons readily liberate free charge-carriers internally

† For Ge and Si the fact that $u_e > u_h$ makes the *Hall* coefficient negative even for intrinsic materials.

Semiconductor-metal junctions

(*a*) Where a contact occurs the current is maintained by the recombination of electrons and holes, or by their regeneration, at the semiconductor surface.

(*b*) The condition of the interface at the contact will determine whether the resistance is to be ohmic or to have the ability to rectify.

48.11 The Thermistor

This is a non-linear device made of semiconducting material. The name is a contraction of *thermally sensitive resistor*. A typical characteristic is given in fig. 48.3: its shape can be explained by the large, negative and variable values taken by α at different temperatures. A given value of V might correspond to one or more possible values for I.

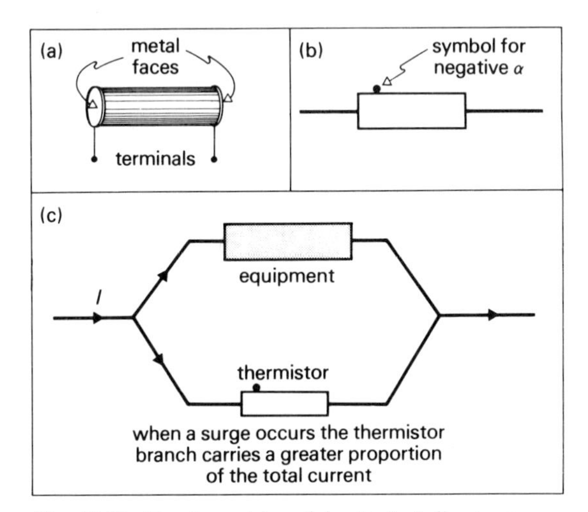

Fig. 48.18. The thermistor: (a) a typical disc-type design, (b) the circuit symbol and (c) its use as a surge suppressor.

Applications

(*a*) *Resistance thermometer*

(*i*) Small size, sensitive, and with rapid response.

(*ii*) Used in the bolometer for the measurement of energy flow in a beam of electromagnetic waves.

(*iii*) Useful range is from about 200 K to 600 K.

(*iv*) It is unstable, and needs periodic recalibration.

(*b*) It can be used to operate *relays* (in e.g. a fire alarm), since R falls markedly as T increases.

(*c*) It is used to limit surge currents through equipment when the current is switched on or off (fig. 48.18c). For certain applications (e.g. to limit the initial current through a cold filament) it is connected in series.

48.12 Combinations of Resistors

See fig. 48.19.

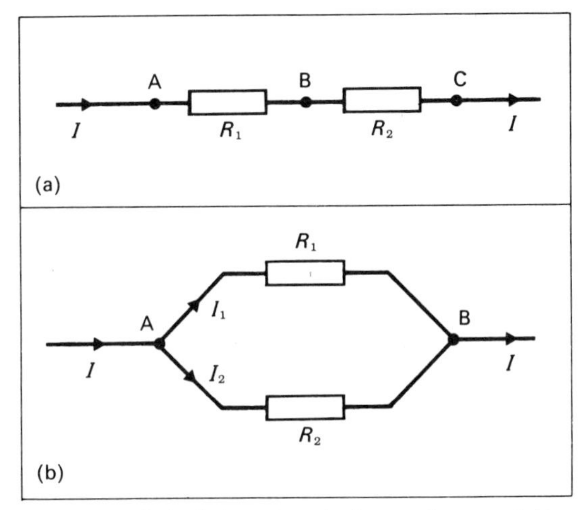

Fig. 48.19. Calculation of equivalent resistance: (a) in series, (b) in parallel.

(a) In Series

Each resistor carries the same current I. But

$$V_{AC} = V_{AB} + V_{BC}$$

(since the electric field is conservative, and energy is conserved), so

$$IR = IR_1 + IR_2$$

$$\therefore \quad \boxed{R = R_1 + R_2} \quad \textit{Series}$$

(b) In Parallel

There exists the same p.d. V across each resistor. But

$$I = I_1 + I_2$$

(law of conservation of charge), so

$$\frac{V}{R} = \frac{V}{R_1} + \frac{V}{R_2}$$

$$\therefore \quad \boxed{\frac{1}{R} = \frac{1}{R_1} + \frac{1}{R_2}} \quad \textit{Parallel}$$

The expressions can be extended to apply to any number of resistors.

48.13 The Potential Divider

A variable resistor is frequently used to control the p.d. across a device, as shown in fig. 48.20.

Since

$$\frac{V_{BC}}{V_{AC}} = \frac{R_{BC}}{R_{AC}}$$

the value of V_{BC} can be varied by moving the sliding contact B on the resistor AC.

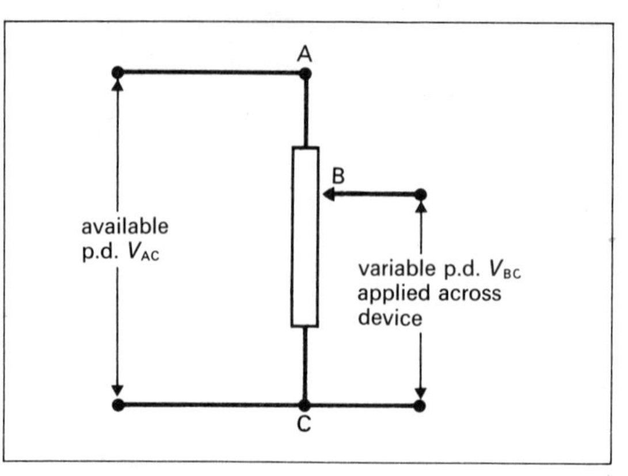

Fig. 48.20. A potential divider.

(Note that this equation applies only if the device concerned does not draw any current.)

49 Energy Transfer in the Circuit

49.1 Energy Conversion in a Load

In fig. 49.1 the *source* of electrical energy could be a battery. The *load* might be a resistor, a motor, an accumulator or a combination of devices.

By definition, whatever the nature of the load, a charge Q which passes through the load loses energy W given by

$$V_{AB} = \frac{W}{Q}$$

$$W = VQ$$

The rate at which energy is transferred from the source to the load is given by power

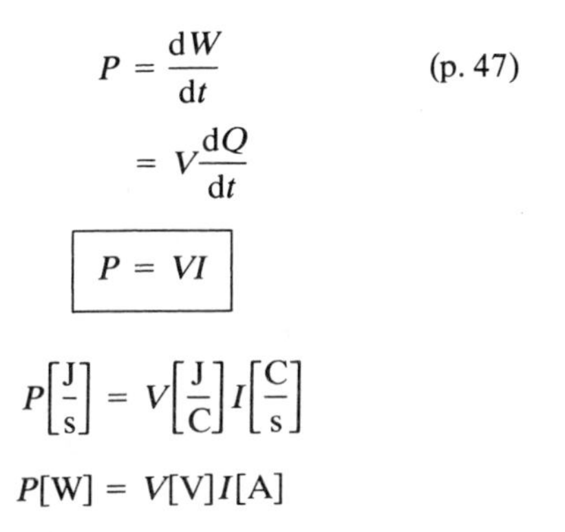

$$P = \frac{dW}{dt} \qquad \text{(p. 47)}$$

$$= V\frac{dQ}{dt}$$

$$\boxed{P = VI}$$

$$P\left[\frac{J}{s}\right] = V\left[\frac{J}{C}\right]I\left[\frac{C}{s}\right]$$

$$P[W] = V[V]I[A]$$

The equation $P = VI$ enables us to calculate the rate at which the load converts electrical energy to other forms, e.g.

(*a*) if a resistor, to internal energy (below),
(*b*) if a motor, to mechanical work done by the motor,
(*c*) if an accumulator, to stored chemical energy.

$P = VI$ *applies to any circuit element.*

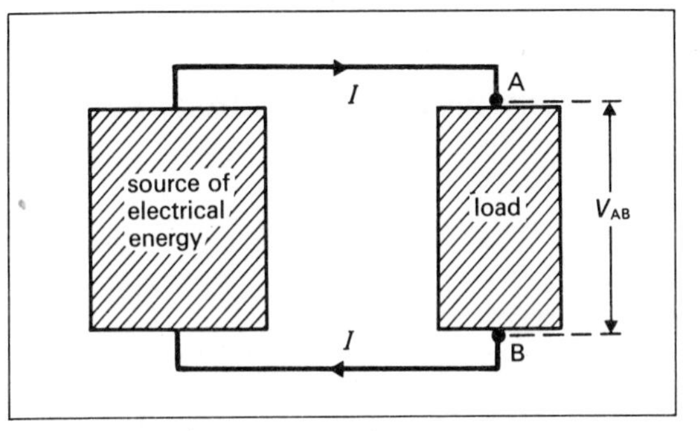

Fig. 49.1. Transfer of energy.

49.2 Joule Heating

For a resistor carrying a steady current I, we have

$$\frac{W}{t} = VI \quad \text{(from the def. of } V\text{)}$$

and

$$R = \frac{V}{I} \quad \text{(the def. of } R\text{)}$$

Combining these expressions, the electrical energy W converted to internal energy of the lattice in time t is given by

$$\boxed{W = \frac{V^2}{R}t = I^2Rt} \quad \textbf{Joule's Laws}$$

Joule's Laws represent energy conservation when a resistor converts electrical energy into internal energy. The mechanism is given on p. 350. The laws apply *only to a resistor*.

Notes.

(*a*) This energy conversion is *irreversible* (p.185).

(*b*) The Department of Trade and Industry measures electrical energy in kilowatt hours (kWh), where 1 kWh is the electrical energy converted to other forms when a power of 1 kW is used for 1 hour.

Thus
$$\begin{aligned} 1\ \text{kWh} &= (1000\ \text{W}) \times (3600\ \text{s}) \\ &= 3.6 \times 10^6\ \text{W s} \\ &= 3.6 \times 10^6\ \text{J} \end{aligned}$$

The kilowatt hour is no longer recognized by physicists as a *unit*.

49.3 Transmission of Electrical Power

In fig. 49.2

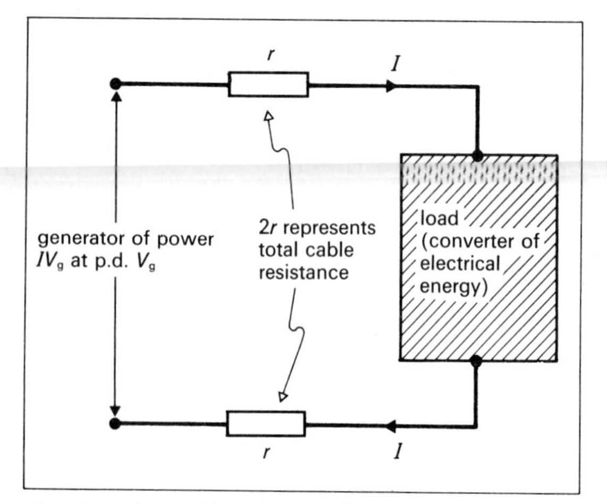

Fig. 49.2. Calculation of power losses.

power generated $= IV_g$
power dissipated in cables $= 2I^2r$
power available at load $= IV_g - 2I^2r$

The power loss is reduced by making V_g large so that I, and $2I^2r$, are small. An economic balance must be struck with insulation requirements.

Potential differences are stepped up and down using transformers, which function *only* on alternating current (p. 401). (Be sure not to confuse V_g in problems of this type with the potential drop $2Ir$ along the cables.)

*49.4 The Maximum Power Theorem

We want to calculate the value of R (fig. 49.3) for which the battery supplies maximum power to the external circuit.

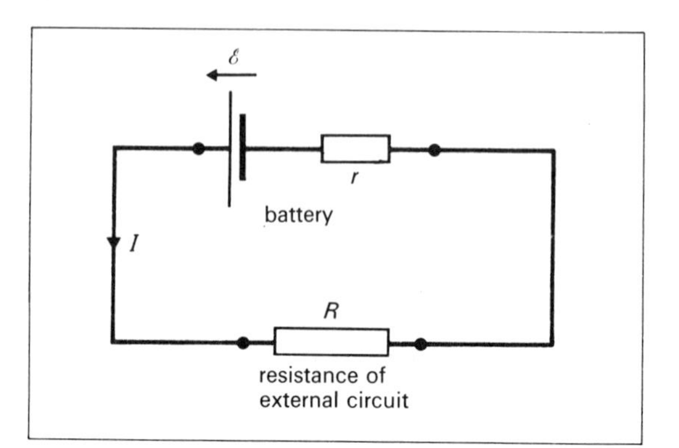

Fig. 49.3. Circuit for maximum power theorem.

External power supplied

$$P = I^2R$$
$$= \frac{\mathscr{E}^2R}{(R+r)^2} \qquad \text{since } \mathscr{E} = I(R+r)$$
$$\frac{dP}{dR} = \mathscr{E}^2\left[\frac{(R+r) - 2R}{(R+r)^3}\right]$$

For a stationary value

$$\frac{dP}{dR} = 0$$

so
$$R + r = 2R$$

Either (*a*) $R = 0$ or $\infty\,\Omega$, in which case the power supply is zero (minimum), or

(*b*) $R = r$, which corresponds to maximum rate of energy supply.

When $R = r$ half the electrical energy available is converted to internal energy in the source. This does *not* represent maximum *efficiency*.

* 49.5 How the Current Divides at a Junction

In fig. 48.7*b* we want to find how I_1 and I_2 are related if least power is to be dissipated as internal energy in the resistors

The power dissipated

$$P = I_1^2R_1 + I_2^2R_2$$
$$= I_1^2R_1 + (I - I_1)^2R_2$$

$$\frac{dP}{dI_1} = 2I_1R_1 + 2(I - I_1)(-1)R_2$$

For a stationary value

$$\frac{dP}{dI_1} = 0$$

$$\therefore I_1R_1 = (I - I_1)R_2$$
$$= I_2R_2$$

The current splits so that

(*a*) $I \propto 1/R$, and
(*b*) IR = constant. We have the same p.d. across each resistor. This result was assumed on p. 355.

(We have assumed R_1 and R_2 to be independent of current. The p.d. across each resistor remains equal even if this is not true, but the power dissipated is then not a minimum.)

50 E.M.F. and Circuits

50.1 Electromotive Force (E.M.F.)

A seat of e.m.f. is a device which can supply energy to an electric current. It does this by producing and maintaining a potential difference (and of course an electric field) between two points to which it is attached.

For example a chemical cell causes forces to be exerted on positive charge-carriers which take them from a point of low potential (negative terminal) to a point of higher potential (positive terminal), and vice-versa for negative charges. It exerts a nonconservative force on the charge-carriers to drive them against the electrostatic force in the seat of e.m.f., and in so doing achieves a *reversible transformation* (p. 185) between non-electrical and electrical energy.

Notes. (*i*) The e.m.f. exists whether the current flows or not (see the theory of the transformer with the secondary on open circuit (p. 401)).

(*ii*) An e.m.f. can exist in a vacuum—a conductor does not have to be present. (See the discussion of induced e.m.f. on p. 396.)

Examples. Each of these examples has its counterpart (indicated in brackets) in which electrical energy is converted to non-electrical energy:

(*a*) the electrochemical cell or battery (the voltameter),
(*b*) the thermocouple, using the *Seebeck effect*, p. 363 (the *Peltier effect* at a metal junction),
(*c*) the photoelectric or photovoltaic cell (the fluorescence of a phosphor),
(*d*) the electric generator or dynamo (the electric motor).

Joule heating is a non-reversible effect, so we do not call the p.d. across a resistor an e.m.f.

Definition of e.m.f.

Suppose that non-electrical energy W is converted to electrical energy when a charge Q is taken through the source (seat of e.m.f.).

Then we define the* e.m.f. $\mathscr{E}$ *of the source by the equation

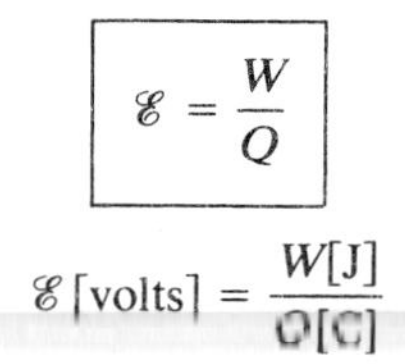

$$\mathscr{E} = \frac{W}{Q}$$

$$\mathscr{E}\,[\text{volts}] = \frac{W[\text{J}]}{Q[\text{C}]}$$

Notes.

(*a*) The unit of e.m.f. is the same as that of p.d.—*volt.*

(*b*) $\mathscr{E}$ equals the p.d. between the terminals of the source when no current flows through the source—when it is on open circuit.

(*c*) If a charge Q passes through such a seat of e.m.f. in the direction shown in fig. 50.1*a*, the energy W given to it is $\mathscr{E}Q$.

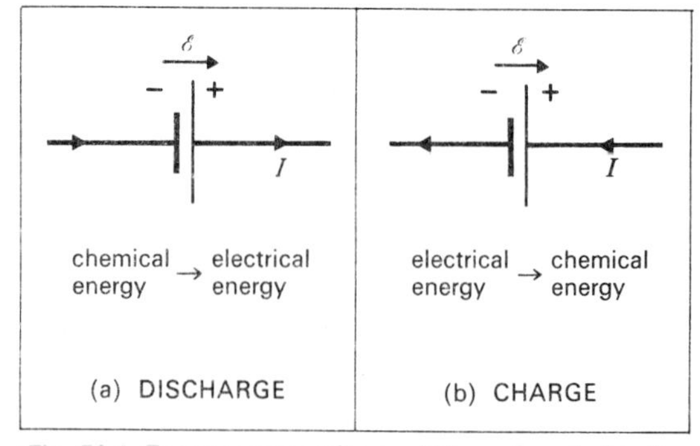

Fig. 50.1. Energy conversions within a chemical cell.

Combination of Cells

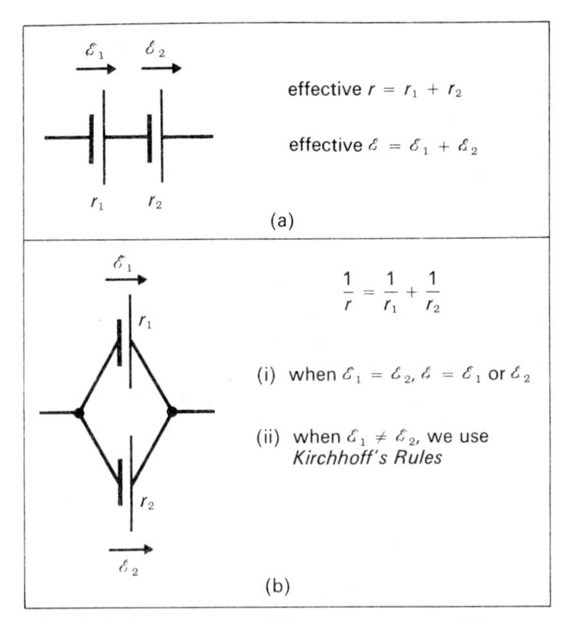

Fig. 50.2. Cells connected (a) in series, and (b) in parallel.

Fig. 50.2 shows the effective values of e.m.f. and resistance when several cells are used together.

50.2 The Circuit Equation

Rate of Energy Conversion in a Cell

From the previous paragraph

$$W = \mathscr{E}Q$$

so differentiating

$$\frac{dW}{dt} = \mathscr{E}\frac{dQ}{dt}$$

$$P = \mathscr{E}I$$

The *power* of a cell is the rate at which it converts chemical energy into electrical energy, and is given by

$$P = \mathscr{E}I$$

Application to the Circuit (Fig. 50.3)

Rate of production of electrical energy by the cell

$$= \mathscr{E}I$$

Rate of dissipation of this energy (as *Joule heating*)

(*a*) in the external resistance R, $= I^2R$,

(*b*) in the resistance r of the cell (its **internal resistance**), $= I^2r$.

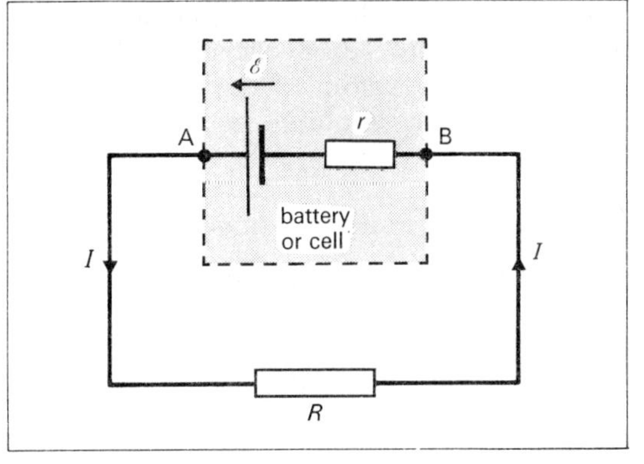

Fig. 50.3. The simple circuit.

By the law of the conservation of energy for the circuit, these are equal

$$\mathscr{E}I = I^2R + I^2r$$

$$\boxed{\mathscr{E} = I(R + r)}$$ *The Circuit Equation*

The p.d. across the cell terminals

$$= V_{AB} = IR \quad \text{(def.)}$$

This equals the p.d. across the cell

$$= (\mathscr{E} - Ir)$$

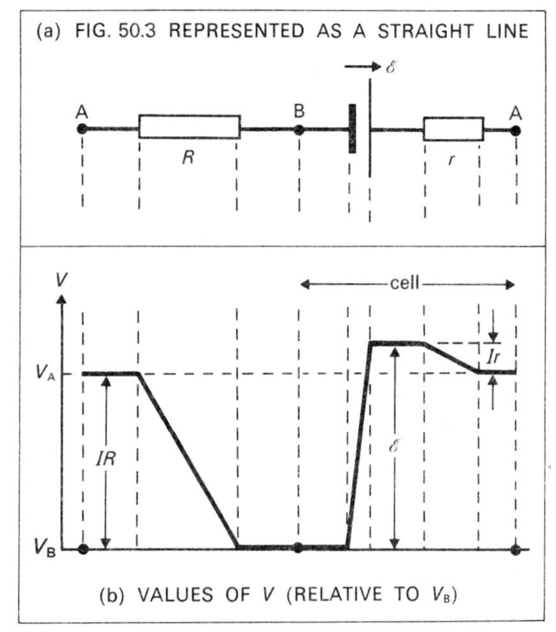

Fig. 50.4. Variation of potential from point to point in a simple circuit.

The *terminal* p.d. of the cell is less than the e.m.f. $\mathscr{E}$ by Ir (the **lost volts**), because the internal resistance of the cell dissipates electrical energy.

As mentioned in the previous paragraph,

$$V_{AB} = \mathscr{E} \quad \text{when} \quad I = 0$$

The current can be calculated from

$$I = \frac{\mathscr{E}}{R + r} = \frac{V_{AB}}{R} = \frac{\mathscr{E} - V_{AB}}{r}$$

Fig. 50.4 shows how the potential varies from point to point in the circuit.

50.3 Kirchhoff's Rules

What the Rules do

They (*a*) summarize the conditions of current and p.d. in circuits with steady currents, and

(*b*) state the *conservation laws* of

(1) electric charge, and

(2) energy

using the quantities associated with circuits.

Thus they do not introduce any new ideas.

Their Statement

Rule I *The algebraic sum of the currents at a junction is zero.*

$$\sum I = 0$$

The rule expresses the fact that charge may not accumulate when we are dealing with steady currents.

Rule II *Round any closed circuit the algebraic sum of the e.m.f.'s is equal to the algebraic sum of the products of current and resistance.*

$$\sum \mathscr{E} = \sum IR \qquad \textit{The Loop Equation}$$

The rule expresses the law of conservation of energy.

Solution of Circuit Problems

(1) *Always* draw a circuit diagram.

(2) Make sure that all the quantities given are expressed in appropriate units (e.g. A not mA).

(3) Quote equations in symbols before substituting numbers and units.

(4) The equations

$$V = IR, \qquad P = VI$$

and

$$W = VIt = VQ$$

apply to *particular devices*, since in each case V is the p.d. between *two particular points*. State which points you are considering.

(5) Use of *Kirchhoff's Rules*:

(*a*) *Rule I*: Mark on your circuit diagram unknown currents in terms of known currents using as few unknowns as possible. If the *wrong sense* is selected for a current, this will be shown by a negative sign when the equations are solved. (Use of x, y, z, etc., for current symbols leads to fewer errors than I_1, I_2, I_3, etc.)

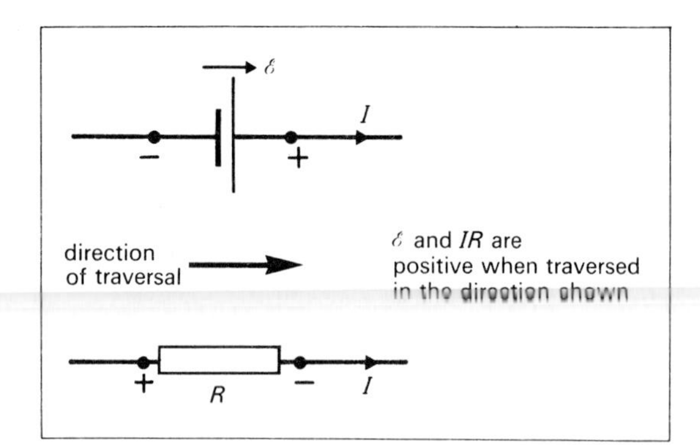

Fig. 50.5. The sign convention for Kirchhoff's Rule II.

(*b*) *Rule II*: Traverse each electrical device in a particular loop *in the same sense* (either clockwise or anticlockwise). The term *algebraic* means paying regard to *sign* (fig. 50.5).

50.4 Thermoelectricity

Contact Potential Difference

Different metals do not necessarily contain the same number density n of conduction electrons. If two metals are placed in contact, the conduction electrons diffuse across the boundary, showing a net flow in a direction determined by the work functions of the metals concerned. The resulting p.d. across the boundary, which is equal to the difference of the work function potentials, prevents further diffusion and so a *dynamic equilibrium* is reached. (See fig. 50.6*a*.)

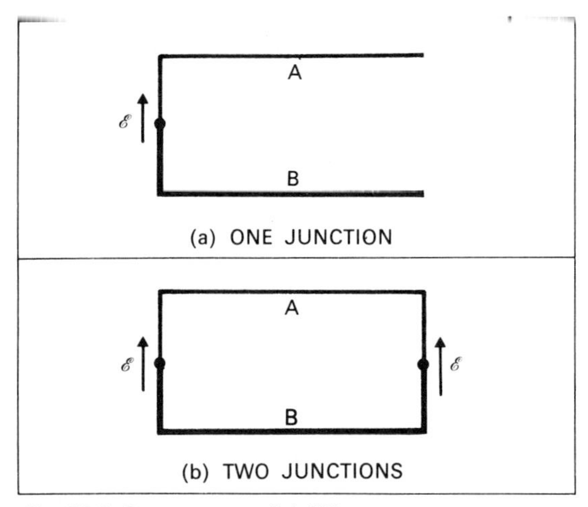

Fig. 50.6. Contact potential differences.

If the two metals are joined a second time (fig. 50.6*b*) there is no current because the second contact p.d. is equal in magnitude, but opposite in direction, to the first.

The Seebeck Effect (1821)

If the temperatures of the two junctions are different, the two p.d.'s will be different, and a current will flow. *The temperature difference causes an e.m.f.* (See fig. 50.7.) The device is called a **thermocouple**. The effect is the **Seebeck Effect**.

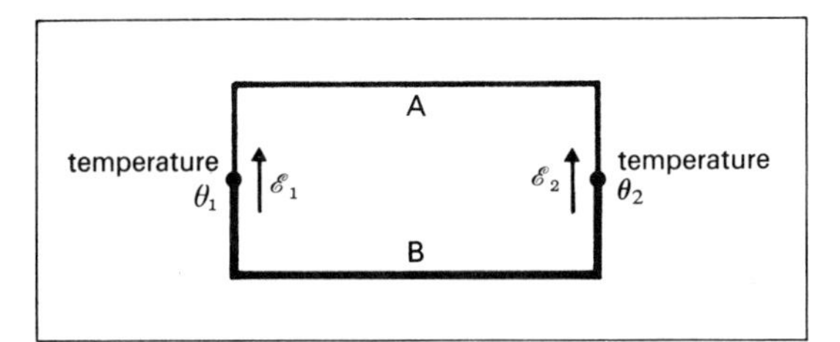

Fig. 50.7. The Seebeck effect.

*The Peltier Effect (1834)

When a current is passed through a junction between dissimilar metals, the junction is heated or cooled. *The current causes a temperature difference.* This is the **Peltier Effect.**

This effect is *reversible*, being proportional to I. (Cf. *Joule* heating which is proportional to I^2.) Thus a heating effect becomes a cooling effect if the direction of current flow is reversed, and vice-versa. (This is why we consider the *Seebeck* effect to produce an *electromotive force* as opposed to a *potential difference.*)

When the junctions are at the same temperature, the *Peltier* effect that accompanies a current in the circuit causes a temperature difference between junctions. The *Seebeck* effect that results takes the form of an e.m.f. which tries to drive a current in opposition to the original current (an example of *Le Chatelier's Principle*).

Laws of Thermoelectricity

(a) Law of intermediate metals

If a wire C of a third metal is connected as in fig. 50.8, the net e.m.f. of the circuit (and so of the thermocouple) is not changed, provided the two A–C junctions are at the same temperature. This means that a meter *could* be used in such a circuit to measure the current. To measure the e.m.f. we would choose a potentiometer (p. 369).

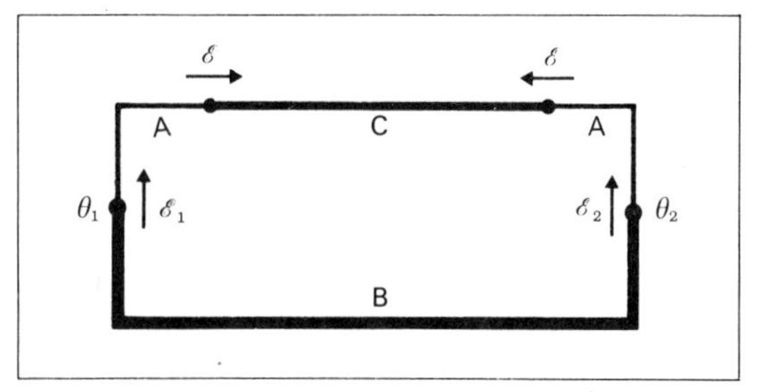

Fig. 50.8. The law of intermediate metals.

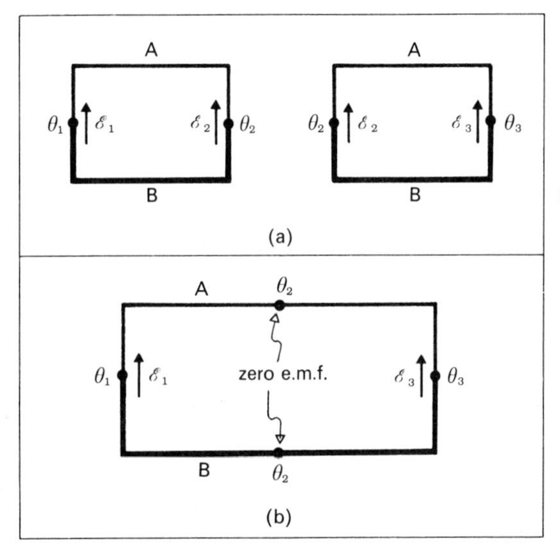

Fig. 50.9. Law of intermediate temperatures: (a) thermocouples separate, and (b) joined in series.

* (b) Law of intermediate temperatures

In fig. 50.9

(*a*) the net e.m.f.'s are ($\mathscr{E}_2 - \mathscr{E}_1$) and ($\mathscr{E}_3 - \mathscr{E}_2$) respectively,

(*b*) the net e.m.f. is ($\mathscr{E}_3 - \mathscr{E}_1$).

Thus the e.m.f. when the junctions are at θ_3 and θ_1 equals the sum of the e.m.f.'s when the junctions are kept first at θ_2 and θ_1, and then at θ_3 and θ_2.

The e.m.f. of a Thermocouple

The size of the e.m.f. (≈mV) is too small to be useful for the large scale production of electrical energy from thermal energy.

The e.m.f. varies with the temperature difference between the junctions according to fig. 50.10.

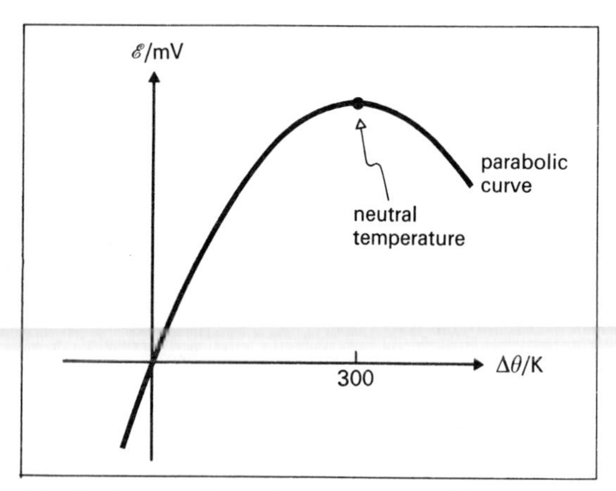

Fig. 50.10. Variation with temperature difference Δθ of a typical thermoelectric e.m.f.

The thermoelectric e.m.f. can be used as a *thermometric property* i.e. to measure temperature. The e.m.f.-temperature curve is always approximately a *parabolic curve.* To measure a temperature, one

(*a*) places one junction where its temperature is kept steady (usually at 273 K)

(*b*) chooses a pair of pure metals, alloys or semiconductors which

(*i*) give an approximately linear graph in the required temperature range, and

(*ii*) produce a big e.m.f.

(Note that fig. 50.10 would have shown *a straight line* if we had plotted $\mathscr{E}$ against temperatures measured on the *thermocouple temperature scale.*)

*50.5 The *RC* Circuit

(a) Growth of Charge on Capacitor Plates

Connect switch to terminal A in fig. 50.11. The *loop equation* (p. 362) can be quoted as

$$\sum \mathscr{E} = \sum V$$

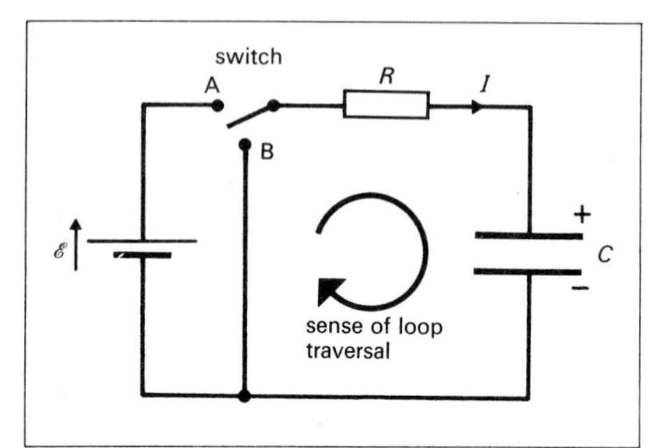

Fig. 50.11. Circuit for charge and discharge of a capacitor through a resistor.

where V represents the *decrease* of potential in traversing a device.

Using $C = Q/V$, and traversing the loop clockwise

$$\mathscr{E} = IR + \frac{Q}{C}$$

$$\therefore \mathscr{E} = R\frac{dQ}{dt} + \frac{Q}{C} \qquad \text{since} \quad I = \frac{dQ}{dt}$$

This differential equation in Q has the solution

$$Q = \mathscr{E}C(1 - e^{-t/RC})$$

and this can be checked by substitution.

It follows that the charging current is given by

$$I = \frac{dQ}{dt} = \frac{\mathscr{E}}{R}(e^{-t/RC})$$

These results are shown in fig. 50.12.

The graphs show

(*i*) when $t = 0$,

$$Q = 0 \qquad I_0 = \frac{\mathscr{E}}{R}$$

(*ii*) when $t \to \infty$

$$Q \to Q_0 = C\mathscr{E} \text{ (the final charge)}$$
$$I \to 0$$

(*iii*) when $t = RC$, then

$$Q = (1 - e^{-1})Q_0$$

i.e. Q is 0.63 of its final value Q_0.

Similarly $\quad I = \dfrac{I_0}{e} = 0.37\, I_0$

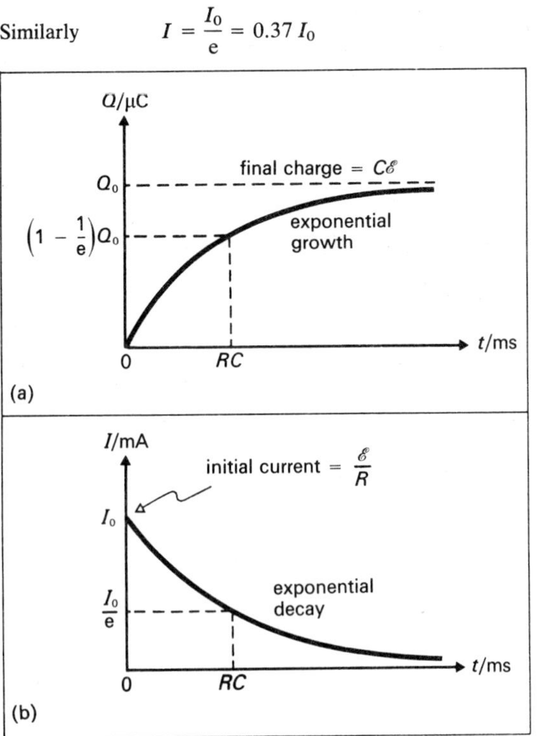

Fig. 50.12. Graphs showing variation of (a) Q and (b) I against t for a capacitor being charged.

RC is called the **capacitative time constant** (τ_c)

$$\tau_c = RC$$

$$\tau_c[\mathrm{s}] = R\left[\frac{\mathrm{V}}{\mathrm{C\,s^{-1}}}\right]C\left[\frac{\mathrm{C}}{\mathrm{V}}\right]$$

E.g. when $R = 10^3\ \Omega$, and $C = 10^{-6}$ F, then

$$\tau_c = RC = 10^{-3}\ \mathrm{s}.$$

(*b*) Discharge Through a Resistor

After a time $t(\gg RC)$ throw the switch from A to B.

The loop equation becomes

$$0 = IR + \frac{Q}{C}$$

$$\therefore\ 0 = R\frac{dQ}{dt} + \frac{Q}{C}$$

which has the solution

$$Q = Q_0\,e^{-t/RC}$$

where $Q = Q_0$ when $t = 0$.
When $t = RC$,

$$Q = \frac{Q_0}{e}$$

$$= 0.37\, Q_0$$

This is shown in fig. 50.13.

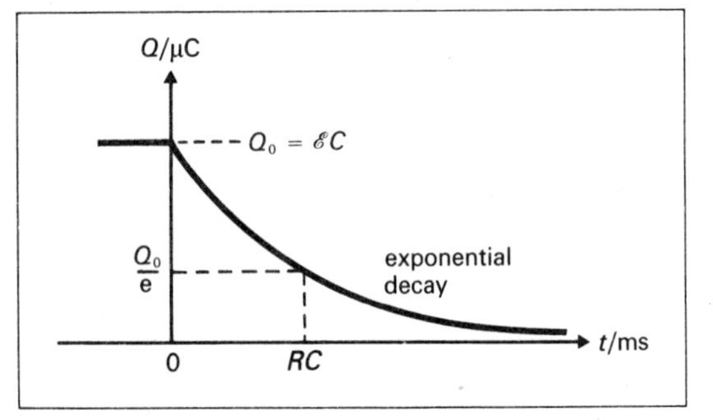

Fig. 50.13. Decay of charge on the plates of a discharging capacitor.

During discharge

$$I = \frac{dQ}{dt}$$

$$= -\frac{Q_0}{RC}e^{-t/RC}$$

$$= -\frac{\mathscr{E}}{R}e^{-t/RC}$$

The *minus sign* indicates that, in the absence of the battery, the current I flows in a direction *opposite* from that shown in fig. 50.11. (Refer to the note on *Kirchhoff's Rule* on p. 362.)

51 Principles of Electrical Measurements

51.1 Laboratory Standards

The **basic standards** are used in standardizing laboratories, where operations are carried out with an accuracy of 1 in 10^5. These standards are *based on fundamental definitions.* They include

(*a*) *The Lorenz rotating disc apparatus* (p. 402) which is used to measure the resistance of standard coils.

(*b*) *A standard mutual inductance* (p. 40[illegible]) which is used for the measurement of inductance and capacitance.

(*c*) *A current balance* (p. 391) which is used to measure

(*i*) current, and then
(*ii*) e.m.f. (by using a standard resistance coil).

In practice a laboratory will usually have a standard cell which has been calibrated by reference to the basic standards. For example the e.m.f. of one design of *Weston cadmium cell* at 293 K is 1.0186 V. Since it is a primary cell, ideally it should never be required to deliver a current of more than a few μA.

51.2 Ammeters and Voltmeters

Ammeters and voltmeters are usually adaptations of the moving-coil galvanometer (p. 388). Suppose that the galvanometer has a resistance R_G, and that full scale deflection (f.s.d.) is produced by a current of I_G *through the galvanometer.* The maximum p.d. that should be applied across the terminals is given by

$$V_G = I_G R_G$$

(*a*) Use as an Ammeter

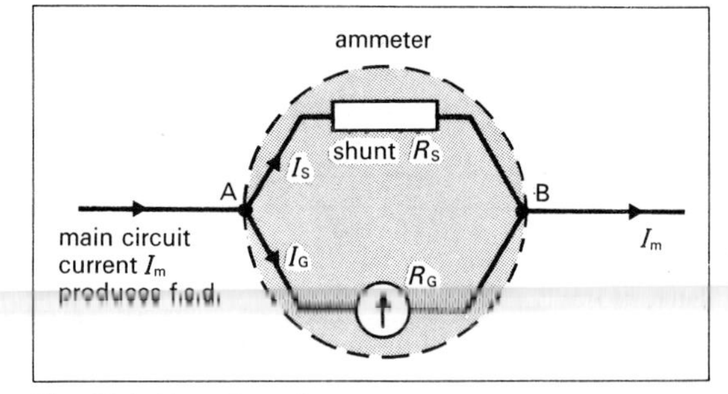

Fig. 51.1. Use of a galvanometer as an ammeter.

To measure currents up to I_m, add a **shunt** (a resistor of *small* resistance R_S) placed in *parallel*, such that

$$I_G R_G = I_S R_S \quad \text{where} \quad I_S = (I_m - I_G)$$

The ammeter as a whole

(*i*) has a very low resistance, and
(*ii*) is connected in series with the circuit (so one has to break the circuit to insert the ammeter).

(*b*) Use as a Voltmeter

To measure V_{AB} in fig. 51.2, add a **bobbin** (a resistor of *large* resistance R_B) in *series.* If V_m is the largest p.d. to be measured, then

$$V_m = I_G R_G + I_G R_B$$

The voltmeter as a whole

(*i*) has a very high resistance, and
(*ii*) is connected across the points A and B whose p.d. is to be measured (i.e. in parallel with the device concerned).

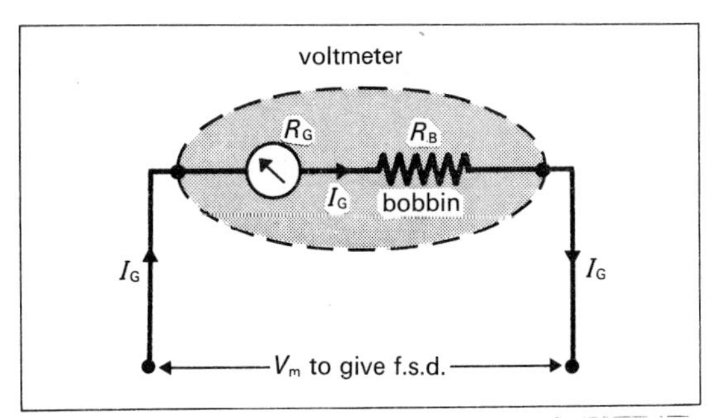

Fig. 51.2. Use of a galvanometer as a voltmeter.

A typical teaching laboratory galvanometer, for which 15 mA gives f.s.d., may have a resistance of 5.0 Ω. If it is converted into

(*a*) an ammeter of f.s.d. 1.5 A, the ammeter would have a resistance of 0.05 Ω, or

(*b*) a voltmeter of f.s.d. 1.5 V, the voltmeter would have a resistance of 100 Ω.

Ideal Instruments

(*a*) The ideal *ammeter* has *zero resistance* so that when inserted into a circuit, it does not reduce the current that was previously flowing.

(*b*) The ideal *voltmeter* has *infinite resistance* so that it takes no current. A finite resistance causes it to take current from the circuit, and to lower the p.d. between the points to which it is connected. *It registers this disturbed value.* If its resistance is relatively high, the error may not be significant.

A valve voltmeter, cathode ray oscilloscope or potentiometer can all be used as voltmeters when the resistance of an ordinary voltmeter is not large relative to the device with which it is connected in parallel. These instruments will not alter significantly the p.d. they are measuring.

51.3 The Principle of the Potentiometer

Fig. 51.3 shows a potentiometer circuit in which the wire is being calibrated by using a standard cell.

Potential differences are balanced against the p.d. between A and P when the galvanoscope gives a zero reading.

Causes for Lack of Balance

No balance point can be found if

(*a*) the high potential terminal of the p.d. under test is connected to the low potential terminal of the potentiometer wire, and/or

(*b*) the p.d. under test exceeds the p.d. between the points A and B. (In fig. 51.3, if $V_{AB} < \mathscr{E}$.)

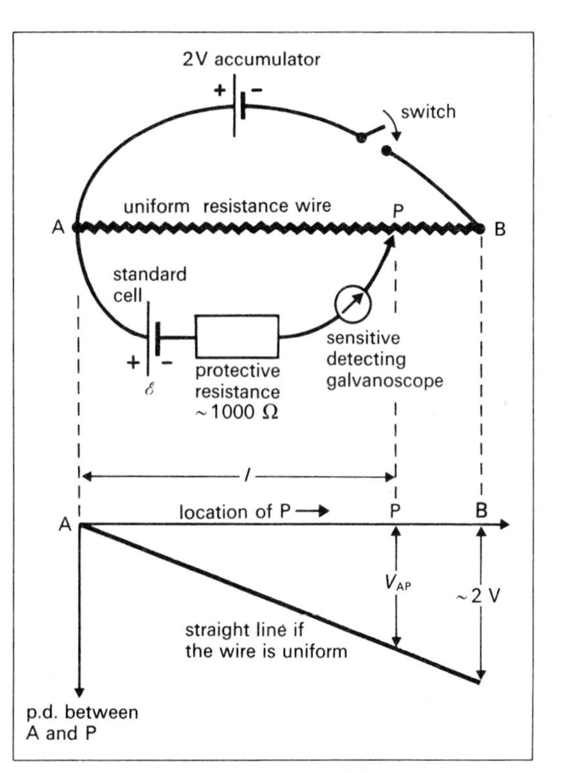

Fig. 51.3. The potentiometer principle.

Notes.

(*a*) When the jockey P is correctly positioned no current flows between the potentiometer circuit and that to which it is connected. The *potentiometer does not disturb the other circuit*, whereas a voltmeter does.

(*b*) Using a cell of known e.m.f. $\mathscr{E}$, one can calibrate the wire AB as for use with a particular accumulator. (The e.m.f. of the accumulator need not be known, but *it must be steady*.) The wire then provides a continuously variable p.d. for comparison purposes.

(*c*) Best results are obtained when *AP* is as large as possible. (Cf. the metre bridge where the jockey is near the mid-point of the uniform wire.)

51.4 Uses of the Potentiometer

The potentiometer is a device which compares (or measures) potential differences.

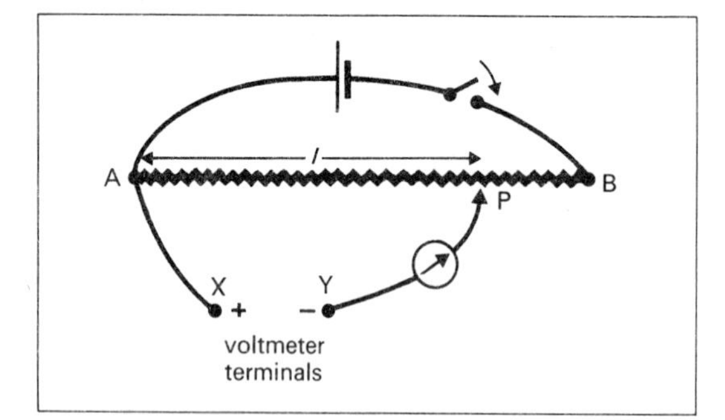

Fig. 51.4. The potentiometer as a voltmeter.

(*a*) Measurement of p.d.

The potentiometer is applied exactly as would be a voltmeter. See fig. 51.4.

The potentiometer can be used to calibrate a voltmeter.

(*b*) Comparison of e.m.f.'s

In fig. 51.3

$$\frac{\mathscr{E}_1}{\mathscr{E}_2} = \frac{l_1}{l_2}$$

One does compare genuine *e.m.f.*'s (not terminal p.d.'s) because at balance the cells are on open circuit.

(*c*) Measurement of Current

The potentiometer measures the p.d. V across a standard resistance R carrying the current.

Then

$$I = \frac{V}{R}$$

This method is often used to calibrate an ammeter.

(*d*) Comparison of Resistances

In fig. 51.5 the two resistors are carrying the same current.

$$V_1 = IR_1, \qquad V_2 = IR_2$$

So

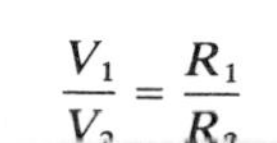
$$\frac{V_1}{V_2} = \frac{R_1}{R_2}$$

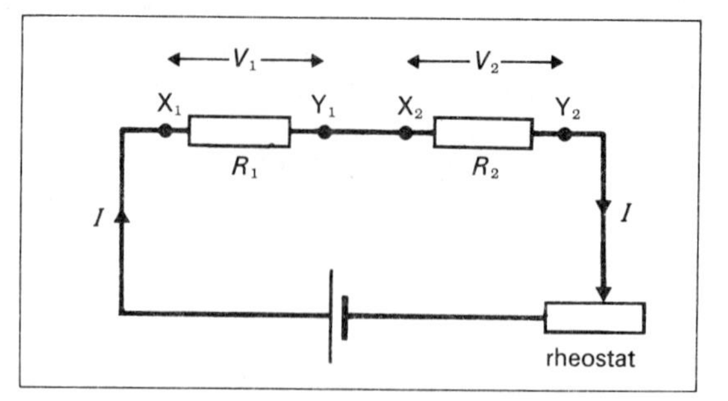

Fig. 51.5. Comparison of two resistances.

V_1 and V_2 are measured successively by connecting the terminals X and Y of fig. 51.4

(*i*) to X_1 and Y_1, and
(*ii*) to X_2 and Y_2,

Then

$$\frac{R_1}{R_2} = \frac{l_1}{l_2}$$

(*e*) Measurement of Internal Resistance

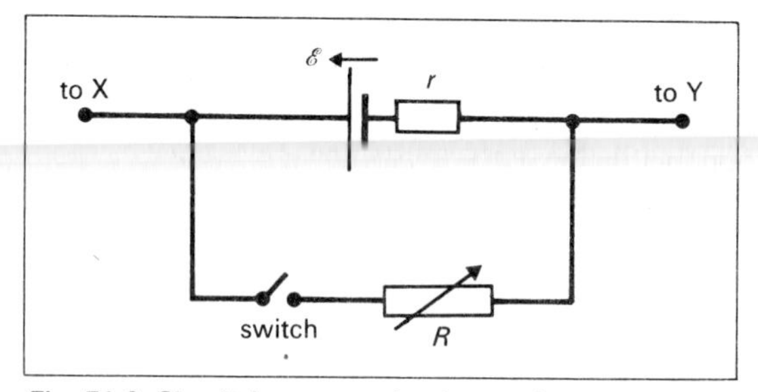

Fig. 51.6. Circuit for measuring internal resistance.

(*i*) Record the balance length l_0 when the switch is open: this corresponds to the e.m.f. $\mathscr{E}$.

(*ii*) Record the length l when the switch is closed: this corresponds to the terminal p.d. $(\mathscr{E} - Ir)$.

$$\therefore \quad \frac{\mathscr{E} - Ir}{\mathscr{E}} = \frac{l}{l_0}$$

Using $\mathscr{E} = I(R + r)$, we have

$$\frac{R}{R + r} = \frac{l}{l_0}$$

which on re-arranging gives

$$\frac{1}{l} = \left(\frac{r}{l_0}\right)\frac{1}{R} + \frac{1}{l_0}$$

Because R is a known variable resistance, we can plot a straight-line graph of $1/l$ against $1/R$. The slope is r/l_0, from which r can be found. (Note that if the graph is not a straight line, it may be that r is non-ohmic.)

(*f*) Measurement of a Very Small p.d. ($\approx$mV)

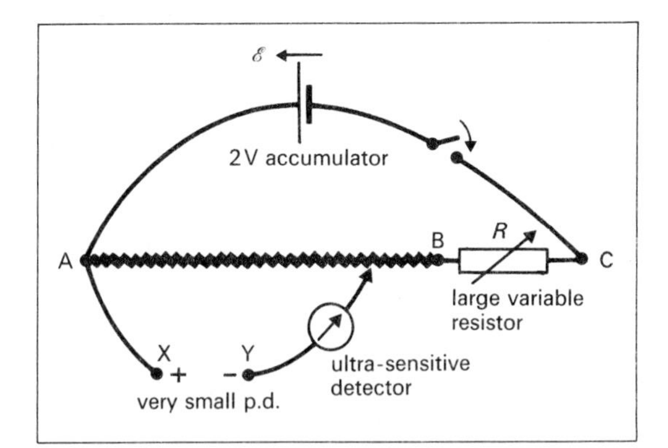

Fig. 51.7. Adapting the potentiometer to compare very small p.d.'s.

In principle one wants to extend the wire (to a length of perhaps 200 m) so that the balance length can be measured without a large experimental error.

In practice this is achieved by adding a large resistance in series with a metre wire: its value can be adjusted to suit the p.d. being measured.

In fig. 51.7,

$$V_{AB} = \left(\frac{r}{R + r}\right) V_{AC}$$

For example to measure a p.d. of up to 0.1 V we might have $V_{AC} \sim 2$ V, $r = 4\,\Omega$ and we would arrange $R = 76\,\Omega$.

If the wire were to be *calibrated* by a standard cell we would need to put about half of this resistance between A and the accumulator. We could then *measure* (rather than compare) the small potential differences.

Applications

(*i*) To measure the *e.m.f. of a thermocouple*, which may be $\approx$ millivolts.

(*ii*) To *compare small resistances*. To prevent contact resistance, connections are made as shown in fig. 51.8.

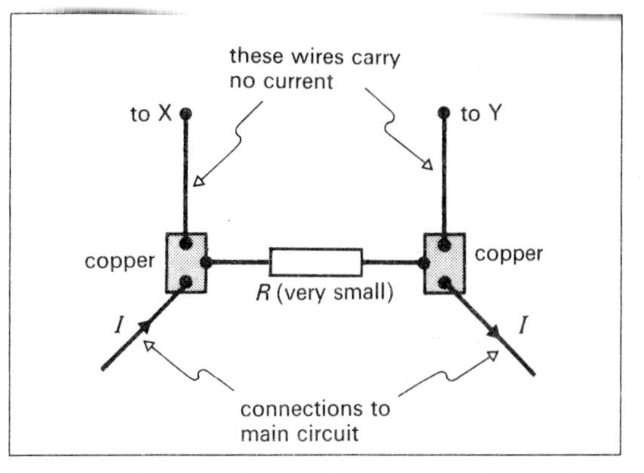

Fig. 51.8. Method of connecting a potentiometer to a very small resistance.

Because R is so small, its connecting wires must not include any resistance wire.

51.5 Bridge Circuits

(a) The Wheatstone Bridge

The resistors of fig. 51.9 are arranged in a *Wheatstone Bridge* network.

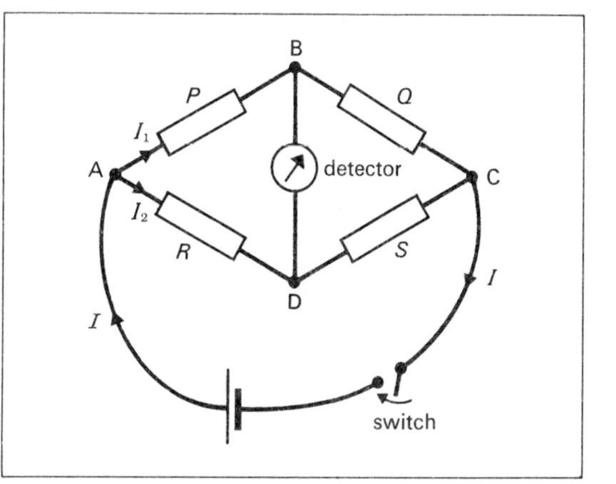

Fig. 51.9. The Wheatstone bridge network.

When G shows zero deflection, B and D are at the same potential.

$$\therefore \quad V_{AB} = V_{AD} \quad \text{and} \quad V_{BC} = V_{DC}$$

so

$$I_1 P = I_2 R \quad \text{and} \quad I_1 Q = I_2 S$$

Dividing

$$\frac{P}{Q} = \frac{R}{S}$$

If any three of P, Q, R and S are known, the fourth can be found. P and Q should be interchanged to obtain a second value for their ratio.

Applications

(*i*) The resistance thermometer (p. 167).

(*ii*) The bolometer (p. 222).

(*iii*) **To test whether a material obeys Ohm's Law.**

Make R and S of *identical* resistance coils (using any material). Make P a long thick wire, and Q a short thin wire, of the material under test.

(1) Using a small current, change the length of resistance Q so that $P = Q$, and the bridge is balanced.

(2) Change the main circuit current by a significant amount. If the bridge remains balanced (at least until heating occurs), then the material of P and Q is ohmic.

This material could now be used in conjunction with a potentiometer to calibrate a moving-coil instrument, which could then be used to test further materials as indicated on p. 347.

(*b*) The Metre Bridge

The metre bridge is the practical application of the *Wheatstone* network principle in which the *ratio* of two of the resistances is deduced from the ratio of their balance lengths.

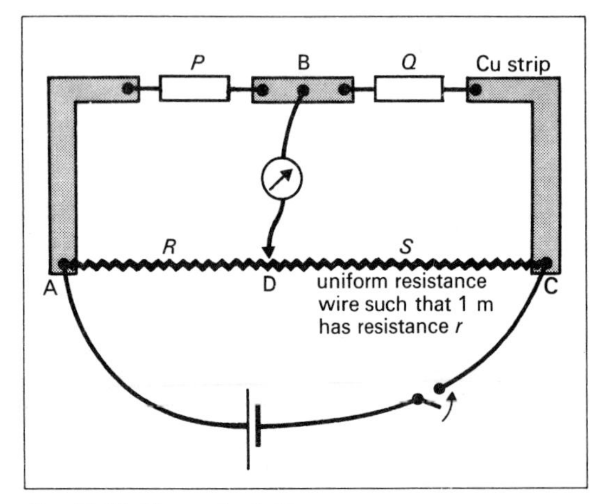

Fig. 51.10. The metre bridge.

The balance condition

$$\frac{P}{Q} = \frac{R}{S}$$

becomes

$$\frac{P}{Q} = \frac{rl_{AD}}{rl_{CD}} = \frac{l_{AD}}{l_{CD}}$$

Notes.

(*i*) For accuracy D should be near the mid-point of AC. (Cf. the potentiometer, where P should be near one end.)

(*ii*) P and Q should again be interchanged, and an average value calculated for their ratio. (This procedure also makes it possible to eliminate end-corrections.)

(*iii*) The resistance wire is heavily soldered to the copper strips so that the resistance of the join is negligible.

(*iv*) The e.m.f. of the driving cell need not be steady, and need only be large enough to make the galvanometer deflect significantly when the bridge is just not balanced.

51.6 The Measurement of Resistance

These methods are collected here for reference.

(a) Ammeter-Voltmeter Method

Using the defining equation $R = V/I$, measure I using an ammeter in series, and V with a voltmeter in parallel.

(b) Substitution method

In the circuit of fig. 51.11

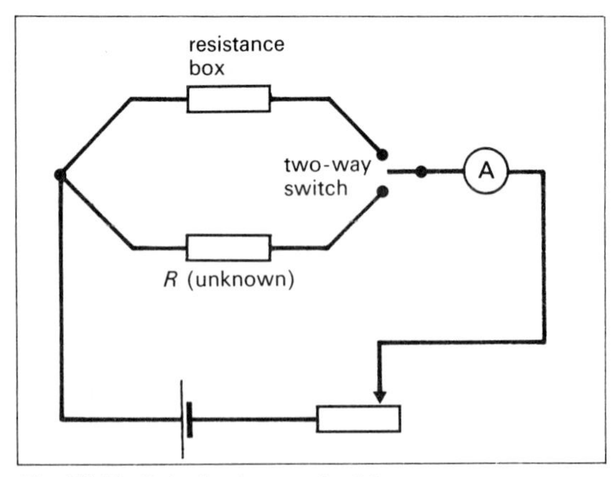

Fig. 51.11. Substitution method for R.

(1) using the unknown resistance R, adjust the rheostat until the ammeter gives a suitable reading,
(2) using the resistance box, find the (known) value that gives that same reading.

(c) Absolute Measurement

Details of the *Lorenz* method are given under electromagnetic induction (p. 402).

(d) Comparison method

Compare the unknown with a known resistance, using

(*i*) *Wheatstone* or metre bridge, or
(*ii*) the potentiometer (modified if the resistances are small).

52 The Magnetic Field Defined from its Effect

52.1 The Magnetic Force on a Moving Charge

Quite apart from the *Coulomb* forces, a *moving* electric charge exerts a force on another *moving* charge. The first moving charge sets up a **magnetic field**, and the second experiences a **magnetic force** because it *moves* in the magnetic field. We represent the magnitude, direction and sense of the magnetic field by the **magnetic flux density *B***, which is also called the magnetic *induction*, or simply the magnetic field.

A magnetic field can be represented by **magnetic field lines**, in the same way that we represent an electric field by electric field lines. The magnetic field line is not real: it is a concept developed as an aid to visualizing the effect of a field. Field lines are drawn as in fig. 52.1 such that

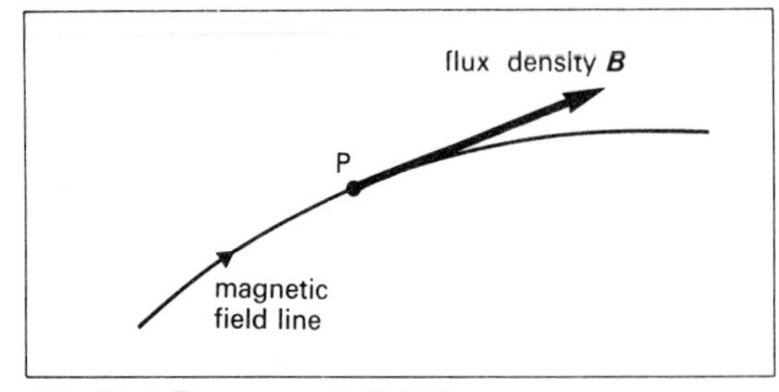

Fig. 52.1. The magnetic field line.

(*a*) the *tangent* to a field line at a point gives the *direction* of ***B*** at that point, and

(*b*) the *number* of field lines drawn per unit cross-sectional area is proportional to the magnitude *B*. If the field is uniform, the field lines are evenly spaced.

Consider a beam of charged particles moving in a region where there exists a magnetic field, as in fig. 52.2.

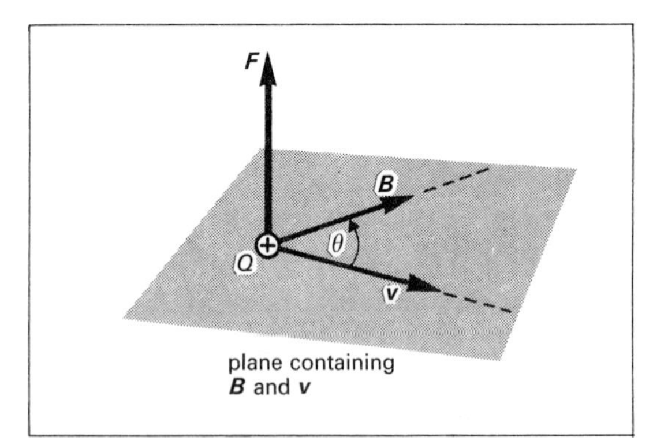

Fig. 52.2 The magnetic force on a moving charge.

By experiment, one finds

(*a*) the magnetic force ***F*** is perpendicular to the particle velocity ***v***, (*b*) $F \propto Q$, the magnitude of the charge, but that the direction of $\boldsymbol{F}$ is reversed for a charge of opposite sign,

(*c*) $F \propto$ magnitude of $\boldsymbol{v}$,

(*d*) $\boldsymbol{F} = 0$ for a particular direction of $\boldsymbol{v}$ (the direction of the magnetic field),

(*e*) $\begin{pmatrix}\text{The magnitude}\\ F\end{pmatrix} \propto \begin{pmatrix}\text{resolved part of } \boldsymbol{v}\\ \text{normal to the line of}\\ \boldsymbol{B}\end{pmatrix}$.

$$\text{i.e.} \quad F \propto v \sin\theta.$$

F is thus a maximum when θ is $\pi/2$ rad.

52.2 Definition of *B*

These experimental results are summarized by

$$F \propto Qv \sin\theta$$

We define the magnitude of* magnetic flux density *B by the equation

$$\boxed{F = BQv \sin\theta}$$

$$B = \frac{F}{Qv \sin\theta}$$

$$B\left[\frac{\text{N}}{\text{A m}}\right] = \frac{F[\text{N}]}{Q[\text{C}]v\left[\frac{\text{m}}{\text{s}}\right] \sin\theta}$$

$\boldsymbol{B}$ is a *pseudovector* quantity (p. 19) whose direction follows from **Fleming's left-hand motor rule** (fig. 52.3).

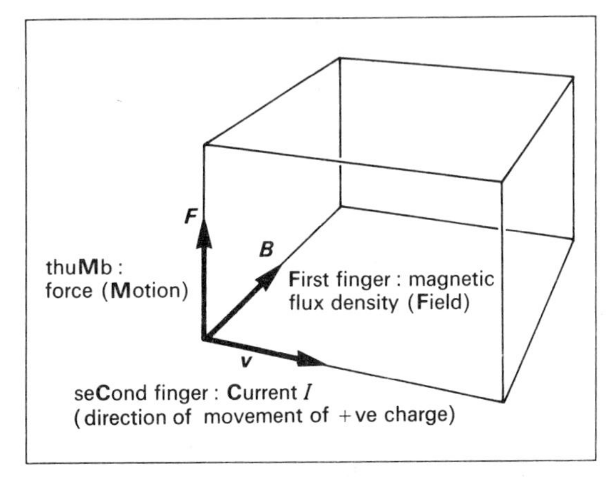

Fig. 52.3. Schematic representation of Fleming's left-hand motor rule.

Notes.

(*a*) The unit for $\boldsymbol{B}$ is the N A^{-1} m^{-1}, but for convenience we define a new unit called the **weber** (Wb), such that

$$1 \text{ Wb m}^{-2} = 1 \text{ N A}^{-1} \text{ m}^{-1}$$

or

$$1 \text{ Wb} = 1 \text{ N m A}^{-1} = 1 \text{ V s}$$

The unit Wb m^{-2} for $\boldsymbol{B}$ is consistent with the idea that $\boldsymbol{B}$ is a flux *density*.

(*b*) The SI unit for $\boldsymbol{B}$ is called a **tesla** (T), defined so that

$$1 \text{ T} = 1 \text{ Wb m}^{-2} = 1 \text{ N A}^{-1} \text{ m}^{-1}$$

(*c*) $[\boldsymbol{B}] = [\text{MT}^{-2}\text{I}^{-1}]$.

(*d*) The equation

$$B = \frac{F_m}{Qv \sin\theta}$$

which defines $\boldsymbol{B}$, should be compared with

$$\boldsymbol{E} = \frac{\boldsymbol{F}_e}{Q}$$

which defines $\boldsymbol{E}$. Note particularly the different directional relationships.

(*e*) Because $\boldsymbol{B}$ is a *pseudovector* quantity, we must add two or more flux densities by the parallelogram law.

Orders of Magnitude for B

In Earth's field $B \approx 5 \times 10^{-5}$ T

At centre of circular coil of 1000 turns, each of radius 10^{-1} m, and carrying 1.5 A $B \approx 1 \times 10^{-2}$ T

Between poles of large electromagnet $B \approx 2$ T

52.3 Magnetic Flux Φ

Refer to fig. 52.4.

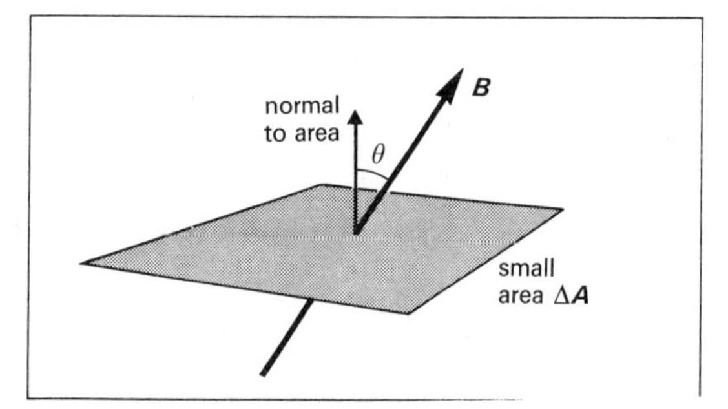

Fig. 52.4 Definition of magnetic flux Φ.

The magnetic flux $\Delta\Phi$ passing through the small area ΔA shown is defined by

$$\boxed{\Delta\Phi = B\cos\theta\,\Delta A}$$

$$\Delta\Phi[\text{Wb}] = B\cos\theta\left[\frac{\text{Wb}}{\text{m}^2}\right]\Delta A[\text{m}^2]$$

Φ has the dimensions $[\text{ML}^2\text{T}^{-2}\text{I}^{-1}]$.

Notes.

(*a*) For a *uniform* magnetic flux density cutting a plane surface of area A

$$\Phi = BA\cos\theta$$

(*b*) When $\boldsymbol{B}$ is *perpendicular* to the surface

$$\Phi = BA$$

(*c*) Because of the way in which Φ is defined (by scalar product), it is a *scalar* quantity even though $\boldsymbol{B}$ and $\boldsymbol{A}$ are vectors.

The concept of magnetic flux is fundamental to the ideas of electromagnetic induction. We can visualize Φ as being proportional to the number of magnetic field lines that cut through a hypothetical surface placed in a field.

52.4 The Motion of Charged Particles in a Magnetic Field

(*a*) If the direction of $\boldsymbol{v}$ is *parallel* or *antiparallel* to $\boldsymbol{B}$,

$$\boldsymbol{F} = 0$$

The trajectory is a *straight line.*

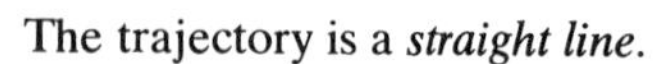

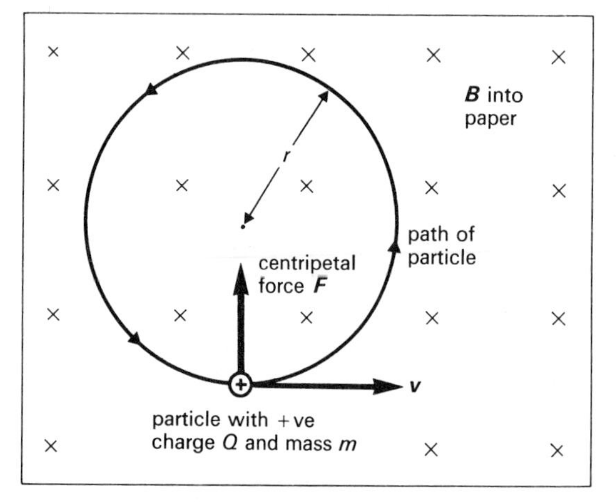

Fig. 52.5. Charged particles describing circular orbits in a uniform magnetic field.

(*b*) If $\boldsymbol{v}$ is *perpendicular* to $\boldsymbol{B}$, then $\theta = \pi/2$ rad so

$$F = BQv$$

and the direction of $\boldsymbol{F}$ is perpendicular to the plane containing $\boldsymbol{B}$ and $\boldsymbol{v}$ at the instantaneous particle position (fig. 52.5).

The particles, moving under a constant centripetal force, describe a *circle.*
Using

$$\boldsymbol{F} = m\boldsymbol{a}$$

we have

$$BQv = \frac{mv^2}{r}$$

$$BQ = \frac{mv}{r} = m\omega$$

This shows that ω, and hence the period of revolution $T\ (=2\pi/\omega)$, are *independent* of the speed. The radius r of the orbit *is* determined by the speed.

*(*c*) If the velocity direction makes an angle θ with $\boldsymbol{B}$ (fig. 52.6), then the particle moves in a *helix*, the axis of which is parallel to $\boldsymbol{B}$.

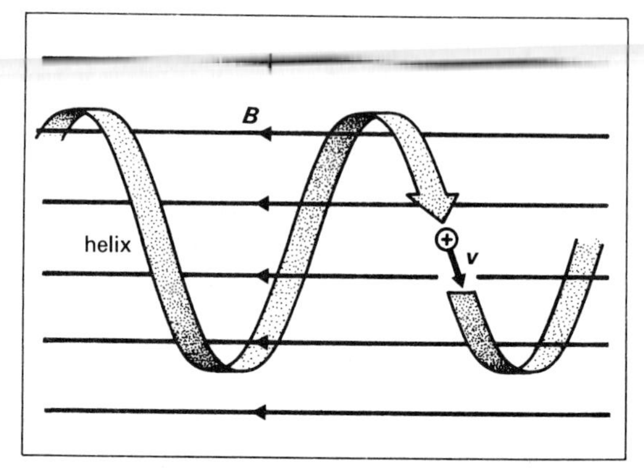

Fig. 52.6. The helical path of a charged particle moving in a magnetic field.

Its motion is the result of superposing

(*i*) a uniform circular motion in which it has speed $v\sin\theta$ in a plane perpendicular to the direction of $\boldsymbol{B}$, on

(*ii*) a steady axial speed of magnitude $v\cos\theta$ along the direction of $\boldsymbol{B}$.

The frequency f of this circular motion is given by

$$f = \frac{\omega}{2\pi} = \frac{Q}{2\pi m}B$$

and is called the **cyclotron frequency**. It does *not* depend on v, θ or r, but (for a given particle) is controlled only by $\boldsymbol{B}$.

*52.5 The Cyclotron

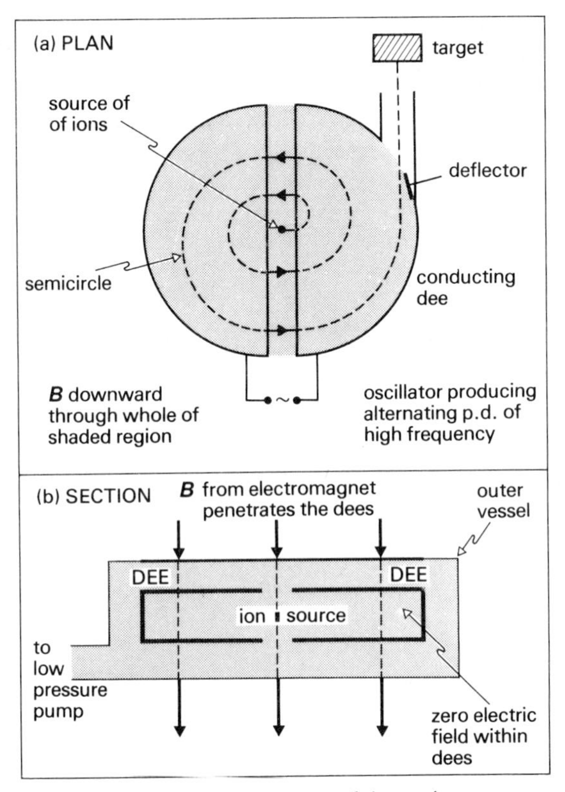

Fig. 52.7. The essential features of the cyclotron.

Refer to fig. 52.7. The purpose of the machine is to accelerate ions from the source at the centre to high energies, before causing them to be deflected from the dees so as to bombard a target.

(a) The frequency f of the alternating p.d. is adjusted to equal the cyclotron frequency $Bq/2\pi$ (above), where q is the ions' specific charge.

(b) As the k.e. increases, f remains unchanged while r increases: *synchronism* is preserved.

(c) The energy W acquired by an ion can be found from

$$W = \begin{pmatrix}\text{total number of}\\ \text{transitions between}\\ \text{the dees}\end{pmatrix} \times \begin{pmatrix}\text{p.d. between the}\\ \text{dees at the}\\ \text{relevant instant}\end{pmatrix} \times \begin{pmatrix}\text{charge}\\ \text{on}\\ \text{ion}\end{pmatrix}$$

(d) The relativistic increase of mass with gain of speed complicates the synchronism between the electric field variation and the rotation of the ions. This fact makes the cyclotron unsuitable for electrons, since they acquire very high speeds at relatively low energies.

Orders of Magnitude

This simple design of cyclotron might have a dee diameter of 4.5 m, use a B of 2 T, and produce ions of energy 12 pJ. It has now been superseded, but provided useful information about the nucleus.

52.6 The Hall Effect

Fig. 52.8 shows the consequence (for the majority carriers) of applying a transverse magnetic field to samples of *extrinsic* n-type and p-type material.

A magnetic field $\boldsymbol{B}$ *causes* the electric fields $\boldsymbol{E}_p$ and $\boldsymbol{E}_n$, which are the *effects*. $\boldsymbol{E}_p$ and $\boldsymbol{E}_n$ are oppositely directed, and so a voltmeter connected between the points X and Y would give deflections in opposite directions, corresponding to the different polarities of the p.d.'s.

Intrinsic semiconductors do show a small p.d.: this is caused by the mobility of the electrons being greater than that of the holes. The polarity of the p.d. is usually the same as that of an n-type extrinsic material.

Metals usually show the same effect as n-type material, but to a much smaller degree. Some metals (e.g. Zn) give a p.d. of the opposite polarity.

The Hall effect enables us to identify the sign of majority charge-carriers in a conductor.

52.7 The Hall Coefficient

General Theory

Suppose that the material shown in fig. 52.9 has a number density n of majority charge-carriers, each carrying a charge e.

(1) Since $\quad I = nAev_d \quad$ (p. 350)

the average drift velocity

$$v_d = \frac{I}{nAe} = \frac{J}{ne},$$

where $\boldsymbol{J}$ is the current density (p. 346).

(2) The transverse magnetic force $\boldsymbol{F}_m$ on each carrier is given by

$$F_m = Bev_d = \frac{BeJ}{ne} = BJ/n$$

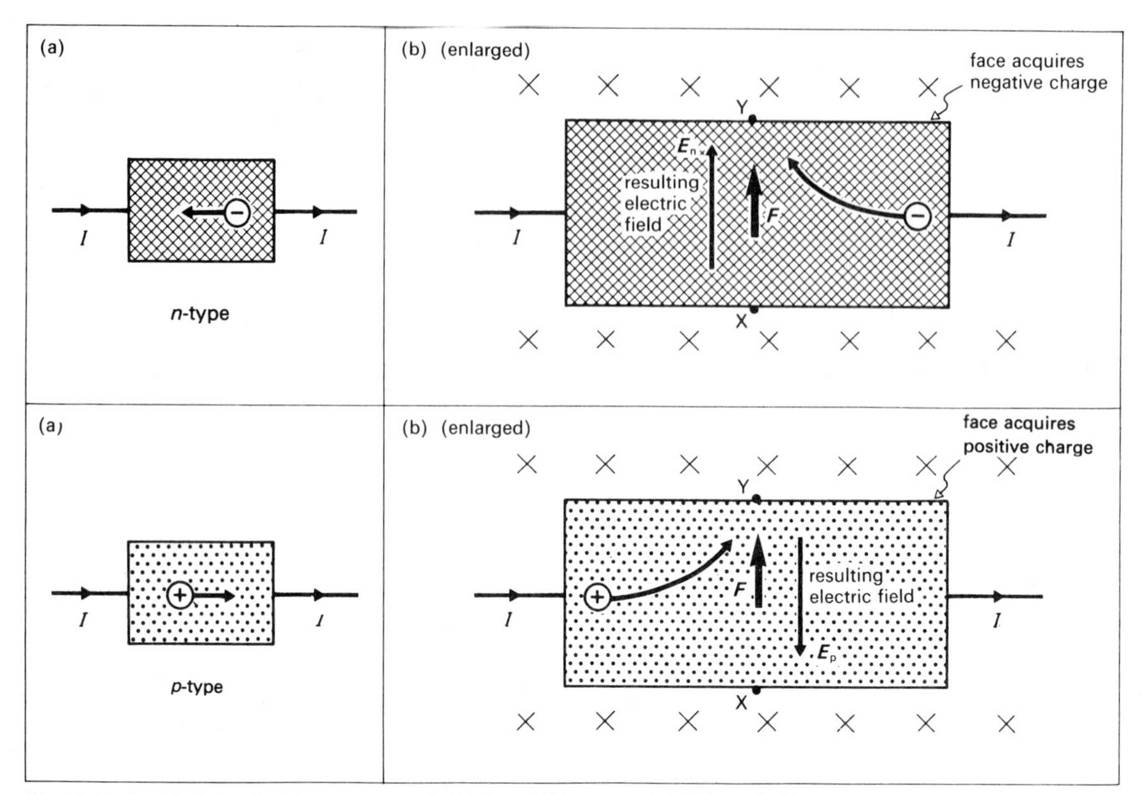

Fig. 52.8. The Hall effect: (a) zero magnetic field, and (b) magnetic field applied into paper

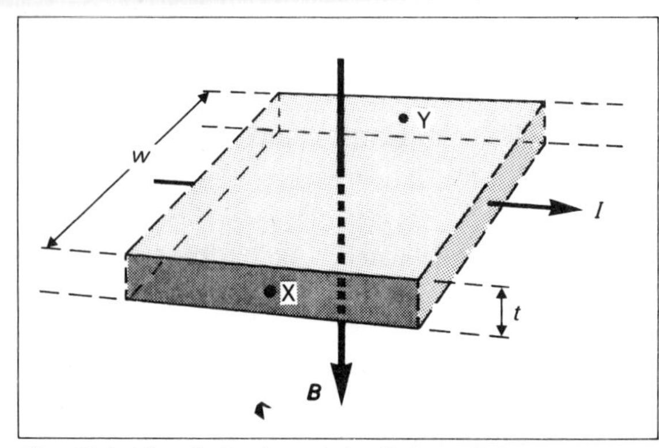

Fig. 52.9. The derivation of the Hall coefficient.

(3) When the resulting lateral charge movement has set up a potential gradient between X and Y, there will be an electric force $\boldsymbol{F}_e$ on the particles. This force is along the same line as $\boldsymbol{F}_m$, but is oppositely directed.

$$F_e = EQ$$
$$= \left(\frac{V_{XY}}{w}\right) e,$$

since the size of $\boldsymbol{E}$ equals the potential gradient.

(4) When equilibrium is established

$$\boldsymbol{F}_m = -\boldsymbol{F}_e$$
$$\frac{BJ}{n} = \left(\frac{V_H}{w}\right) e$$

$V_{XY} \equiv V_H$ is called the **Hall p.d.**

$$V_H = \frac{BJw}{ne} \quad \ldots (1)$$
$$= \frac{BI}{tne},$$

since $A = tw$.

Definition

For the situation of fig. 52.9 the* Hall coefficient *R_H is defined by the equation

$$\boxed{V_H = R_H(BJw)}$$

$$V_H\left[\frac{\text{N m}}{\text{C}}\right] = R_H\left[\frac{\text{m}^3}{\text{C}}\right] B\left[\frac{\text{N}}{\text{A m}}\right] J\left[\frac{\text{A}}{\text{m}^2}\right] w[\text{m}]$$

Notes.

(*a*) Comparison with equation (1) shows that

$$R_H = \frac{1}{ne},$$

which is immediately consistent with the unit $m^3\,C^{-1}$ for R_H.

(*b*) A material has a *negative* coefficient if the carriers primarily responsible for the *Hall* effect are *electrons* (since *e* is then negative). It is positive if they are positive holes (p. 351).

(*c*) When the *Hall* effect is being demonstrated in the laboratory the points X and Y are selected by experiment (using the method implied in fig. 52.10) to ensure that $V_{XY} = 0$ when $\boldsymbol{B} = 0$. The current through a specimen is a consequence of the potential gradient in the direction of $\boldsymbol{J}$.

Orders of Magnitude

The value of R_H enables us to calculate *n*, the number density of charge-carriers, as well as their sign. We can use this to find the number of charge-carriers provided by each parent atom.

	n/m^{-3}	$R_H/m^3\,C^{-1}$	*Typical* $V_H/\mu V$
Cu	10^{29}	-5×10^{-11}	1
p-type Ge	10^{20} (variable)	$+10^{-2}$ (variable)	10^4

V_H is much larger for semiconducting materials as both $R_H(=1/ne)$ and $\boldsymbol{v}_d$ are greater than for metals.

52.8 The Hall Probe

It was shown in the previous paragraph that the Hall p.d.

$$V_H \propto B$$

This result is used in the **Hall probe** to compare magnetic fields (fig. 52.10). The device is small and is mounted on the end of a long narrow handle for easy exploration of a field.

Procedure

(1) The slider is adjusted, in the absence of a $\boldsymbol{B}$-field, until X and Y are at the same potential.

(2) The p.d. subsequently recorded between the terminals X and Y by a sensitive voltmeter is a measure of the resolved part of $\boldsymbol{B}$ perpendicular to the wafer.

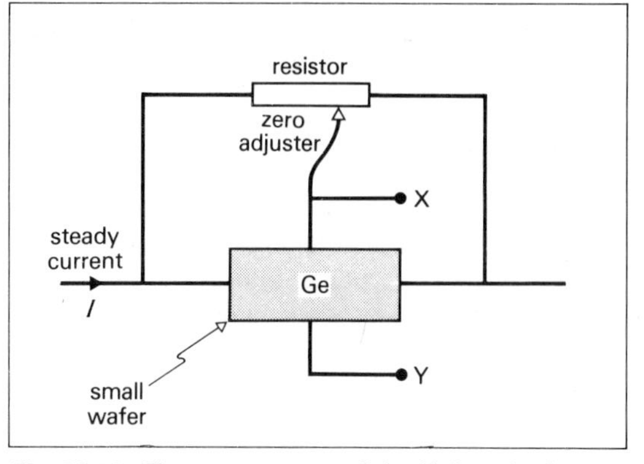

Fig. 52.10. The components of the Hall probe for comparing magnetic fields.

52.9 The Force on a Current-carrying Conductor

An electric current is an assembly of moving charges, and so we expect a current-carrying conductor to experience a (magnetic) force from the agent responsible for establishing any magnetic field round the conductor.

Refer to fig. 52.11.

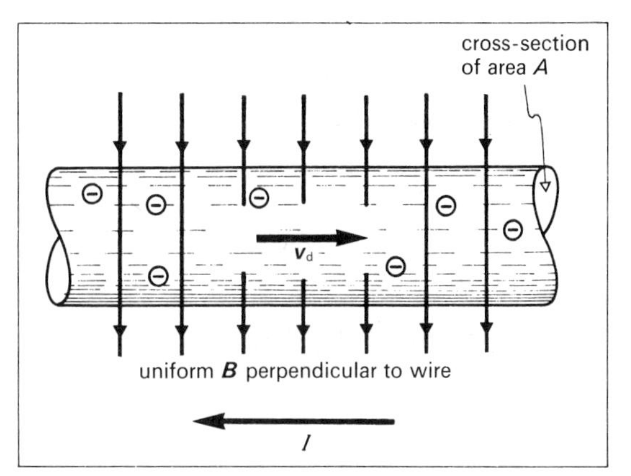

Fig. 52.11. Calculation of the magnetic force on a current-carrying conductor.

$n[m^{-3}]$ is the *number density* of conduction electrons. The size of the force F_e experienced by one electron is given by $F_e = Bev_d$, in which the drift speed

$$v_d = \frac{I}{nAe} \qquad \text{(p. 350)}$$

$$\therefore\ F_e = Be\left(\frac{I}{nAe}\right)$$

and is directed out of the paper.

In a length l of conductor, there are nAl conduction electrons. The electrons experience a total force $\boldsymbol{F}$, where

$$F = (nAl)Be\left(\frac{I}{nAe}\right) = BIl$$

This is also the force experienced by the conductor, as the electrons are not free to move outside the lattice, and experience a restraining force from the lattice.

The direction of the force acting on the wire can be found from *Fleming's left-hand motor rule*: it is perpendicular to the plane defined by $\boldsymbol{B}$ and the wire.

If the field makes an angle θ with the wire

$$\boxed{F = BIl \sin\theta}$$

$$F[\text{N}] = B\left[\frac{\text{N}}{\text{A m}}\right] I[\text{A}] l[\text{m}] \sin\theta$$

Note that this equation could have been used to define $\boldsymbol{B}$, but that it is more fundamental to use the force on a moving charge.

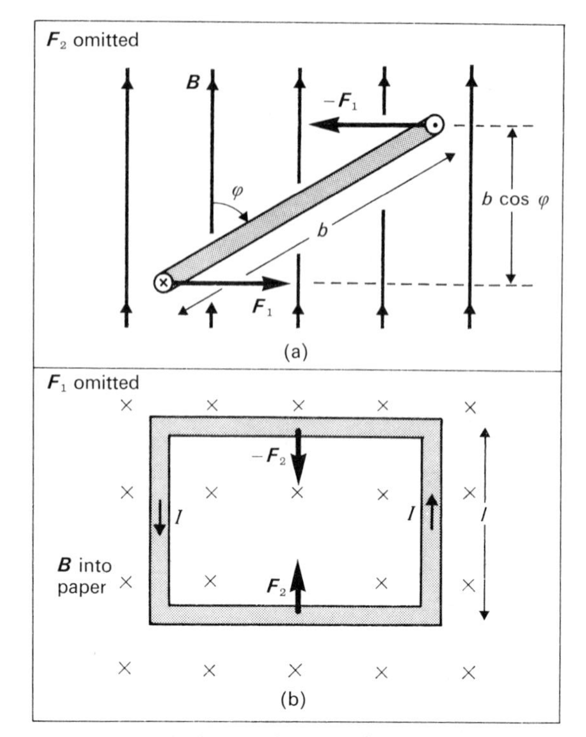

Fig. 52.12. Calculation of the resultant torque acting on a rectangular coil: (a) plan, and (b) section.

52.10 The Torque on a Current-carrying Coil

(a) Uniform Field

Consider a rectangular coil of length l and breadth b placed in a uniform magnetic field of flux density $\boldsymbol{B}$ (fig. 52.12).

The resultant of the forces $\boldsymbol{F}_2$ is zero, since they have the same line of action.

The forces $\boldsymbol{F}_1$ constitute a couple, since they are equal but antiparallel.

The torque acting on the coil

$$\begin{aligned} T &= F_1 \times (\text{arm of couple}) \qquad (\text{p. } 64) \\ &= BIl \times b\cos\phi \\ &= BI \times \text{area} \times \cos\phi \end{aligned}$$

Note that here ϕ is the angle between the *plane* of the coil and $\boldsymbol{B}$.

(b) Radial Field

If we use circular pole-pieces and a soft-iron cylindrical core, then the flux density in the air gap is radial (fig. 52.13). This means that the plane of the coil is always parallel to $\boldsymbol{B}$, so

$$\phi = 0 \quad \text{always}$$

and

$$\text{torque } T = IAB$$

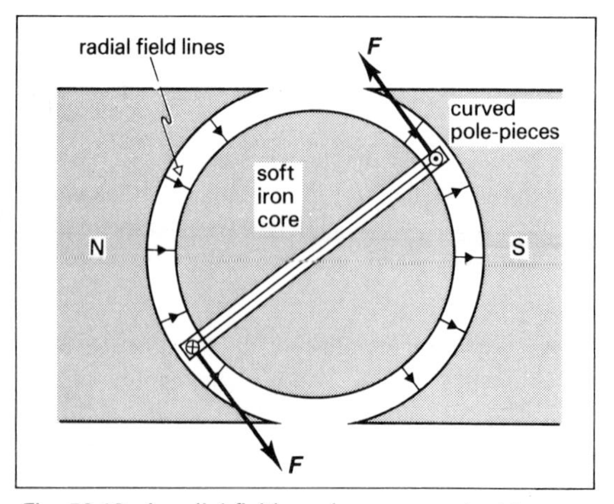

Fig. 52.13. A radial field produces a constant torque.

(c) Torque Acting on any Coil

It can be shown that for coils of any shape, carrying N turns of wire, and having area $\boldsymbol{A}$

$$T = NIAB \cos \phi$$

$$T[\mathrm{N\,m}] = NI[\mathrm{A}]A[\mathrm{m}^2]B\left[\frac{\mathrm{N}}{\mathrm{A\,m}}\right] \cos \phi$$

where ϕ is defined in fig. 52.12.

These results are applied in the moving-coil galvanometer (p. 388) and d.c. motor (p. 404).

52.11 Electromagnetic Moment *m*

(a) Current-carrying Coil

We will now work in terms of the angle θ (rather than ϕ), where $\theta = (90° - \phi)$, and is defined in fig. 52.14.

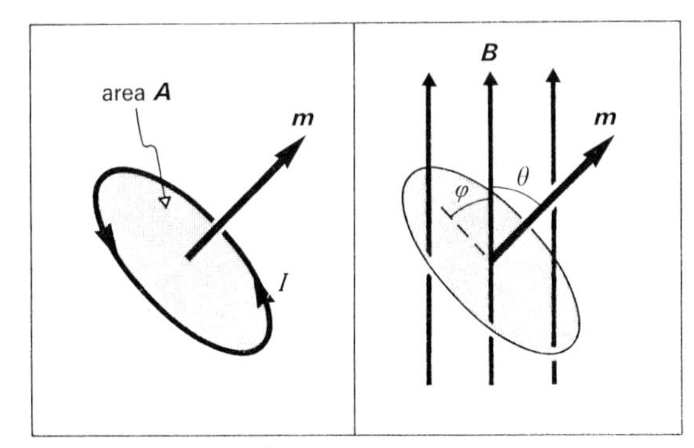

*Fig. 52.14. Electromagnetic moment **m** of a coil.*

Since

$$T = NIAB \sin \theta$$

$$T_{max} = NIAB.$$

The electromagnetic moment m of a current-carrying coil is defined by the equation

$$m = \frac{T_{max}}{B}$$

$$m[\mathrm{A\,m}^2] = \frac{T_{max}[\mathrm{N\,m}]}{B\left[\dfrac{\mathrm{N}}{\mathrm{A\,m}}\right]}$$

m is thus *numerically* equal to the torque required to hold the coil with its normal at right angles to a uniform flux density of size

$$B = 1\,\mathrm{T} = 1\,\mathrm{Wb\,m^{-2}} = 1\,\mathrm{N\,A^{-1}\,m^{-1}}$$

Notes.

(*i*) $\boldsymbol{m}$ is a *vector* whose direction is given in fig. 52.14. It is sometimes called simply **magnetic moment**.

(*ii*) Since $\boldsymbol{m} = NI\boldsymbol{A}$, we see at once that $\mathrm{A\,m^2}$ is a consistent unit for $\boldsymbol{m}$, which has the dimensions $[\mathrm{L^2\,I}]$.

(*iii*) The torque $\boldsymbol{T}$ acting on the coil tends to align $\boldsymbol{m}$ and $\boldsymbol{B}$, and has a magnitude

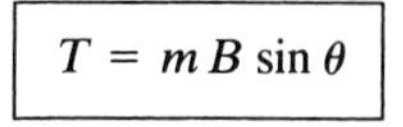

$$T = mB \sin \theta$$

where θ is the angle between $\boldsymbol{B}$ and the *normal* to the coil.

(*iv*) By defining $\boldsymbol{m}$ in terms of $\boldsymbol{B}$ rather than $\boldsymbol{H}$ (p. 432), we are following the *Sommerfeld* approach to electromagnetism. (A definition expressed in terms of $\boldsymbol{H}$ would follow the *Kennelly* approach.)

(b) Permanent Magnetic Dipole

A current-carrying coil and a permanent bar magnet have, to a first approximation, similar *external* fields. This suggests that we can extend the concept of electromagnetic moment to such a dipole (fig. 52.15).

The electromagnetic moment $\boldsymbol{m}$ of the dipole is again defined by the equation

$$m = \frac{T_{max}}{B}$$

Notes.

(*i*) $\boldsymbol{m}$ is a vector whose direction is given in fig. 52.15.

(*ii*) The torque $\boldsymbol{T}$ acting on the dipole tends to align $\boldsymbol{m}$ and $\boldsymbol{B}$, and is given by $T = mB \sin \theta$. Here θ is the angle between $\boldsymbol{B}$ and the dipole's *magnetic axis*.

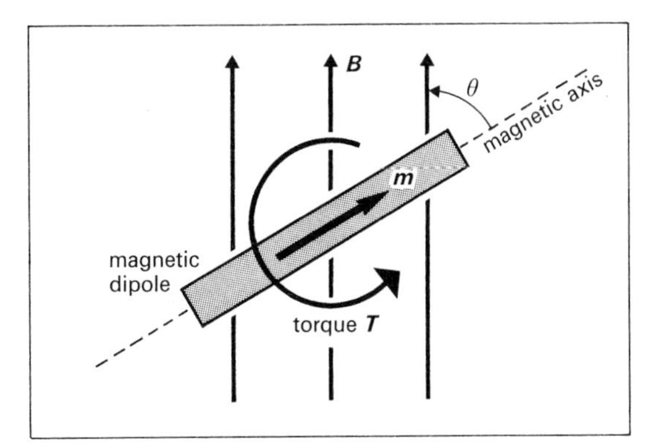

*Fig. 52.15. The electromagnetic moment **m** of a permanent magnetic dipole.*

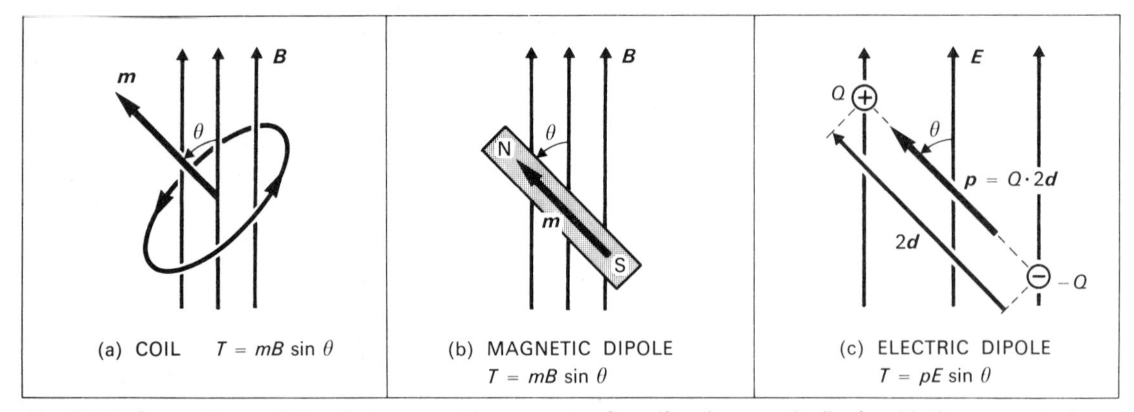

Fig. 52.16. Comparisons of the electromagnetic moments of a coil and magnetic dipole with the moment of an electric dipole.

(c) Comparison with Electric Dipole Moment

The reader should study fig. 52.16 and its equations very carefully.

To carry the analogy further, we can consider the magnet to have poles, called N and S, near the ends, and with which we can associate a particular magnetic pole strength. The direction of $\boldsymbol{m}$ is from S to N, and its magnitude would be proportional to

$$(\text{pole strength}) \times 2d,$$

where $2d$ is the pole separation. This idea is not developed in this book.

*52.12 Energy and Dipole Orientation

Suppose that in fig. 52.14 an external torque $\boldsymbol{T}$ increases the value of θ by $\delta\boldsymbol{\theta}$, the axis of $\boldsymbol{T}$ being parallel to that of $\delta\boldsymbol{\theta}$. The external agent does work

$$\begin{aligned}\delta W &= \boldsymbol{T}\cdot\delta\boldsymbol{\theta}\\ &= mB \sin\theta\,\delta\theta\end{aligned}$$

If we define the p.e. E_p of the system (the dipole and the currents responsible for the external field) such that $E_p = 0$ when $\theta = \pi/2$ rad, then

$$\begin{aligned}E_p &= \int_{\pi/2\,\text{rad}}^{\theta} mB \sin\theta\, d\theta\\ &= [-mB\cos\theta]_{\pi/2\,\text{rad}}^{\theta}\\ &= -mB\cos\theta\end{aligned}$$

Notes.

(*a*) The fact that E_p is negative when $\theta < \pi/2$ rad results from our arbitrary choice of the zero of p.e.

(*b*) When $\theta = \pi$ rad, the p.e. is a maximum. This occurs when the directions of $\boldsymbol{m}$ and $\boldsymbol{B}$ are antiparallel.

(*c*) An amount of energy $2mB$ is needed to change θ from 0 to π rad. This result is important in spectroscopy.

52.13 Measurement of Electromagnetic Moment

(*a*) Use the *defining equation*

$$T = mB \sin\theta$$

Measure values of $\boldsymbol{T}$, $\boldsymbol{B}$ and θ, from which $\boldsymbol{m}$ can be found.

(*b*) *Oscillation method*

Suspend the dipole so that it can rotate about its centre of mass. Allow it to oscillate about its equilibrium position in a uniform magnetic field $\boldsymbol{B}$ (such as that of the Earth).

The torque

$$T = -mB\sin\theta = I\ddot{\theta}$$

where the negative sign shows that the torque acts so as to decrease θ, and I is the moment of inertia of the suspended system.

For small oscillations

$$\ddot{\theta} = -\left(\frac{mB}{I}\right)\theta$$

The motion is simple harmonic, of time period

$$\frac{2\pi}{\omega} = 2\pi\sqrt{\frac{I}{mB}}$$

which is independent of the amplitude.

$\boldsymbol{m}$ can be calculated from a knowledge of the time period, I and $\boldsymbol{B}$. We assume that the suspension fibre exerts zero torque.

53 The Magnetic Field Related to its Cause

53.1 The Magnetic Field Created by Moving Charges

In chapter **52** we discussed the *effect* of a magnetic field on a moving charge. This effect was used to define the vector $\boldsymbol{B}$.

In this chapter we discuss the *cause* of the magnetic field, which is the movement of electric charge. One *moving* charge exerts a force on a second *moving* charge (and vice-versa) through their *magnetic interaction*.

(*a*) *Oersted* (1820) demonstrated that a magnetic field exists near a current-carrying wire. We infer that it is the moving charge in the wire that creates the field.

(*b*) *Rowland* (1878) demonstrated that such a field also exists close to an insulated charged disc which is rotated at high speed. This shows more fundamentally that electric charges in motion create the field. (Its magnitude was $\sim 10^{-7}$ tesla.)

In both experiments the existence of the field was shown by the torque experienced by a small magnetic dipole.

53.2 The Biot–Savart Law

Statement of the Law

Suppose a positive charge Q moves with velocity $\boldsymbol{v}$ as shown in fig. 53.1.

The magnitude of the magnetic field $\boldsymbol{B}$ at P is given by

$$B \propto \frac{Qv \sin\theta}{r^2}$$

where $\boldsymbol{r}$ and θ are defined in the diagram, which also shows the direction of $\boldsymbol{B}$.

The Permeability of Free Space μ_0

We write the constant of proportionality as $\mu_0/(4\pi)$ in the *rationalized* SI, leading to the equation

$$B = \left(\frac{\mu_0}{4\pi}\right) \cdot \frac{Qv \sin\theta}{r^2}$$

μ_0 is called the **permeability of free space**, or sometimes simply the **magnetic constant**.

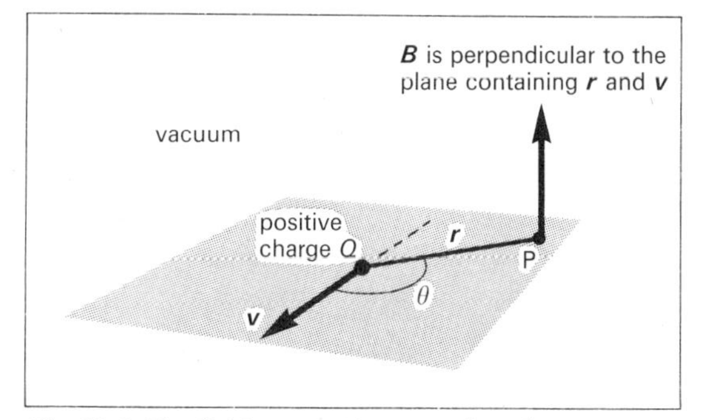

Fig. 53.1. The field created by a moving charge.

Notes.

(*a*) According to this defining equation μ_0 can have the unit Wb A^{-1} m^{-1}, but is more commonly expressed in the equivalent unit *henry* m^{-1} (p. 406). It has the dimensions $[MLT^{-2} I^{-2}]$.

(*b*) The *ampere* is defined (p. 10) in such a way that we *assign a value to* μ_0. (It *cannot,* therefore, be found by experiment.)

Thus

$$\mu_0 = 4\pi \times 10^{-7} \text{ Wb A}^{-1} \text{m}^{-1}$$

(*c*) The introduction of the 4π into this equation (*rationalization*) means that it does not appear in *Ampère's Law* (p. 386).

The Biot–Savart Law for a Current Element

A charge Q moving at velocity $\boldsymbol{v}$ sets up a magnetic field identical to that established by a current I in a conductor of length $\Delta \boldsymbol{l}$ (fig. 53.2), where

$$Qv = I\,\Delta \boldsymbol{l}$$

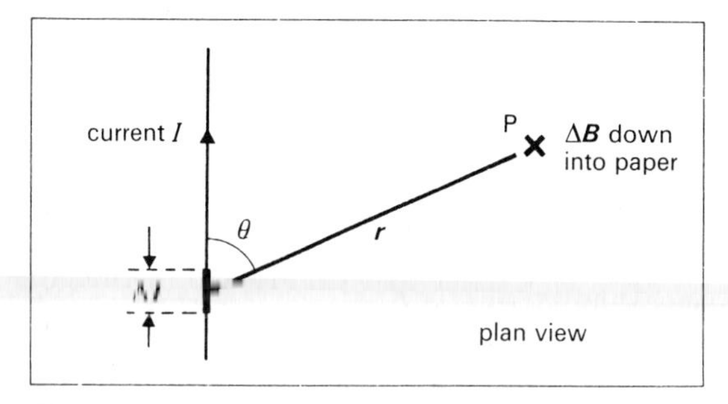

Fig. 53.2. The field set up by a current element $I\,\Delta \boldsymbol{l}$.

Then the contribution $\Delta \boldsymbol{B}$ of the **current element** $I\,\Delta \boldsymbol{l}$ to the total value of $\boldsymbol{B}$ at P is given by

$$\boxed{\Delta B = \left(\frac{\mu_0}{4\pi}\right)\cdot\frac{I\,\Delta l \sin\theta}{r^2}}$$

when $\Delta \boldsymbol{l}$ is very small.

Notes.

(*a*) This situation could have been used to quote the law, but it is more fundamental to use the field set up by a moving charge.

(*b*) The law can be tested experimentally only with difficulty, but is justified because all its *predictions* are in accord with experimental results.

(*c*) This equation is used to find $\boldsymbol{B}$ in simple situations (below).

The Direction of the Magnetic Field

We *defined* the sense and direction of $\boldsymbol{B}$ on page 373. We *work out* the sense and direction from either

(*a*) *Maxwell's right hand cork-screw rule,* or
(*b*) *the right hand grip rule.*

Figs. 53.1 and 53.2 both illustrate the result.

53.3 Comparisons of Three Important Fields

Field type	*Cause of field*	*Definition of field strength*	*Usual statement of inverse square law*	*Calculation of field from point cause*	*Gauss's law for a closed surface)*
gravitational	m	$\boldsymbol{g} = \dfrac{\boldsymbol{F}}{m}$	Newton $F = G\dfrac{m_1 m_2}{r^2}$	$g = G\dfrac{m}{r^2}$	$\left(\dfrac{1}{4\pi G}\right)\psi_g = -\sum m$
electric	Q	$\boldsymbol{E} = \dfrac{\boldsymbol{F}}{Q}$	Coulomb $F = \left(\dfrac{1}{4\pi\varepsilon_0}\right)\dfrac{Q_1 Q_2}{r^2}$	$E = \left(\dfrac{1}{4\pi\varepsilon_0}\right)\dfrac{Q}{r^2}$	$\varepsilon_0 \psi_E = \sum Q$
magnetic	$Qv \sin\theta$	$B = \dfrac{F}{Qv\sin\theta}$	Biot–Savart (right)	$B = \left(\dfrac{\mu_0}{4\pi}\right)\dfrac{Qv\sin\theta}{r^2}$	$\Phi = 0$

*53.4 The Magnetic Fields Established by Complete Circuits

Procedure

(1) Write down $\Delta \boldsymbol{B}$ in terms of a single variable, and
(2) sum or integrate between the limits applicable to the problem.

The results of this paragraph are summarized on page 387 for the student who wants to bypass the mathematics.

(*a*) $\boldsymbol{B}$ at the Centre of a Circular Coil

Suppose the coil has N turns each of radius r and carries a current I. The field at the centre is given by

$$B = \sum \Delta B$$

$$= \sum \left(\frac{\mu_0}{4\pi}\right) \frac{I\,\Delta l \sin\theta}{r^2}$$

$$= \left(\frac{\mu_0}{4\pi}\right) \frac{I \sin 90^\circ}{r^2} \sum \Delta l$$

$$= \frac{\mu_0}{4\pi} \cdot \frac{I}{r^2} \cdot N 2\pi r$$

$$\therefore \quad B = \frac{\mu_0 I N}{2} \cdot \frac{1}{r}$$

$$B\left[\frac{\text{Wb}}{\text{m}^2}\right] = \frac{\mu_0\left[\dfrac{\text{Wb}}{\text{A m}}\right] \cdot I[\text{A}] \cdot N}{2r[\text{m}]}$$

(*b*) $\boldsymbol{B}$ at any Point on the Axis of a Circular Coil

In fig. 53.3 we sum $\Delta \boldsymbol{B}$, taking its direction into account.

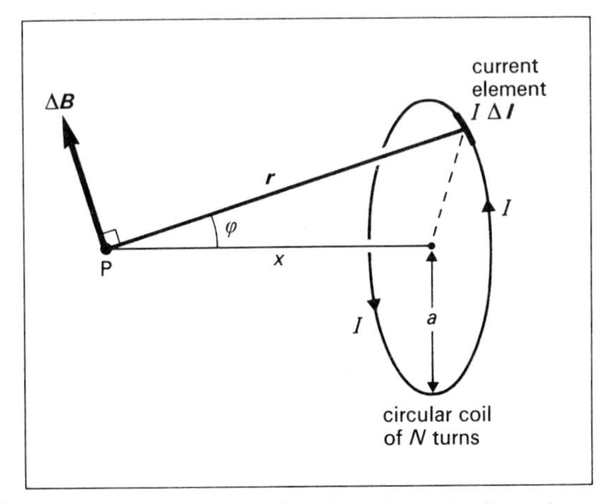

Fig. 53.3. Field at any point along the axis of a coil.

(*i*) By symmetry those components of $\Delta \boldsymbol{B}$ perpendicular to the axis sum to zero.

(*ii*) Each $\Delta \boldsymbol{B}$ has a component parallel to the axis, of magnitude

$$\Delta B \cdot \cos(90^\circ - \phi) = \left(\frac{\mu_0}{4\pi}\right) \cdot \frac{I\,\Delta l \sin 90^\circ}{r^2} \cdot \sin\phi$$

The magnetic field at P is given by

$$B = \sum \left(\frac{\mu_0}{4\pi}\right) \cdot \frac{I\,\Delta l}{r^2} \cdot \sin\phi$$

$$= \left(\frac{\mu_0}{4\pi}\right) \cdot \frac{2\pi a N I \sin\phi}{r^2}$$

$$= \frac{\mu_0 I N}{2} \cdot \frac{a^2}{(a^2 + x^2)^{3/2}}$$

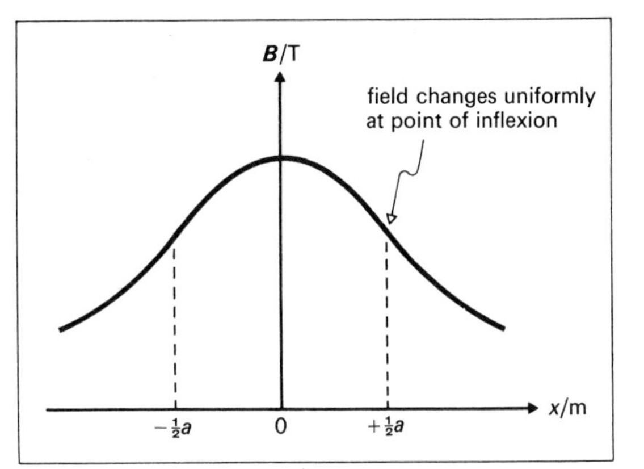

Fig. 53.4. Variation with distance x of the value of $\boldsymbol{B}$ *along the axis of a coil.*

This curve is plotted in fig. 53.4. It is nearly linear at the point of inflexion, given by $x = \pm(a/2)$. (See the *Helmholtz coil system*, overleaf.)

(*c*) $\boldsymbol{B}$ Along the Axis of a Solenoid

Using result (*b*), the contribution of $\delta \boldsymbol{B}$ of the *infinitesimal* element δx of fig. 53.5 is given by

$$\delta B = \frac{\mu_0 a^2 (n\,\delta x)\, I}{2(a^2 + x^2)^{3/2}}$$

where $(n\,\delta x)$ is the number of turns in the element.

Write all the variables in terms of θ:

$$x = a \cot\theta$$

$$\delta x = -a \operatorname{cosec}^2\theta\,\delta\theta$$

$$(x^2 + a^2) = a^2 \operatorname{cosec}^2\theta$$

Then $$\delta B = \frac{\mu_0 a^2 n(-a \operatorname{cosec}^2\theta\,\delta\theta) I}{2a^3 \operatorname{cosec}^3\theta}$$

$$= -\frac{\mu_0}{2} \cdot nI \sin\theta\,\delta\theta$$

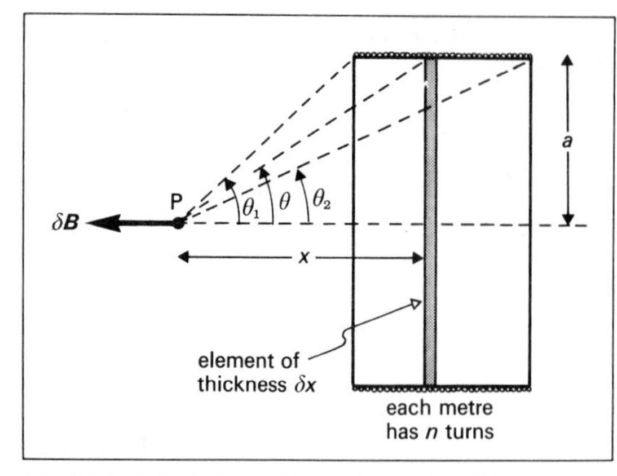

*Fig. 53.5. Calculation of **B** due to a solenoid.*

Note that the radius a cancels.

Let $\delta\theta \to 0$, and integrate

$$B = \int_{\theta_1}^{\theta_2} -\frac{\mu_0}{2} nI \sin\theta \, d\theta$$

$$= \frac{\mu_0}{2} nI (\cos\theta_2 - \cos\theta_1)$$

Special Cases

(*i*) When P is inside an *infinitely long* solenoid

$$\left.\begin{array}{l}\theta_2 = 0 \\ \theta_1 = \pi \text{ rad}\end{array}\right\} \text{ so } \quad B = \mu_0 nI$$

(*ii*) For P inside a toroidal solenoid (a **toroid**, such as that in fig. 57.2, p. 406), $B = \mu_0 nI$.

(*iii*) For P at one end of a long solenoid

$$\left.\begin{array}{l}\theta_2 = 0 \\ \theta_1 = \dfrac{\pi}{2} \text{ rad}\end{array}\right\} \quad B = \frac{\mu_0 nI}{2}$$

Note carefully that n has unit *turns* m^{-1}, whereas N was measured in *turns*.

It can be shown, using *Ampère's Law* (p. 386), that $\boldsymbol{B}$ is constant over the whole cross-section of a long straight solenoid.

(*d*) The Helmholtz Coil Arrangement

Two identical coils of radius a are placed a distance a apart. This achieves a magnetic field of exceptional uniformity over a large region. (See, for example, p. 438.)

At the mid-point 0, $x = a/2$.

Using result (*b*)

$$B = 2\frac{\mu_0 IN}{2} \cdot \frac{a^2}{\left[a^2 + \left(\frac{a}{2}\right)^2\right]^{3/2}}$$

$$= \frac{8\mu_0 NI}{\sqrt{125}\, a}$$

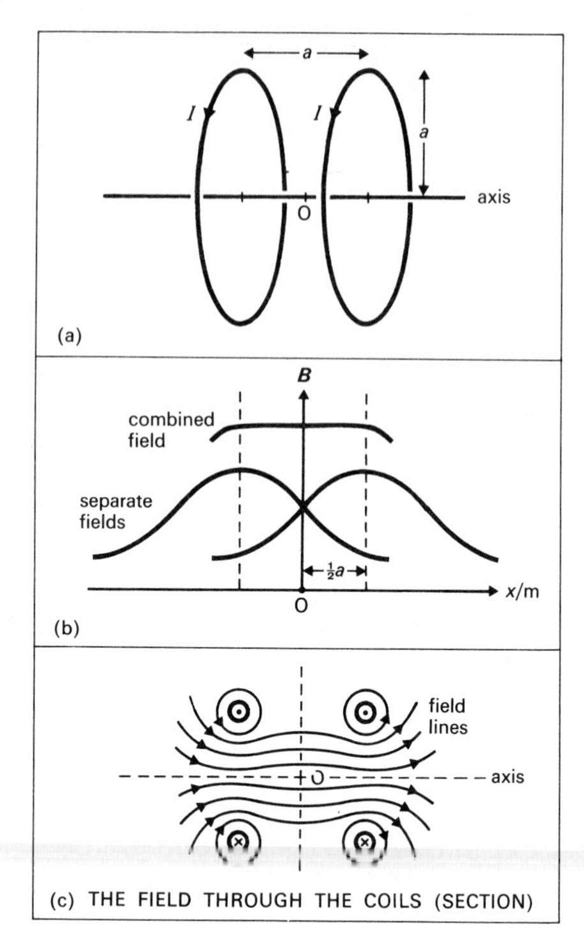

Fig. 53.6. The Helmholtz coils: (a) the arrangement, (b) and (c) the resulting field.

(*e*) The Field Due to a Long Straight Wire

In fig. 53.7 the contribution $\delta\boldsymbol{B}$ of the infinitesimal element $\delta\boldsymbol{l}$ is given by

$$\delta B = \left(\frac{\mu_0}{4\pi}\right) \cdot \frac{I\,\delta l \sin\theta}{r^2}$$

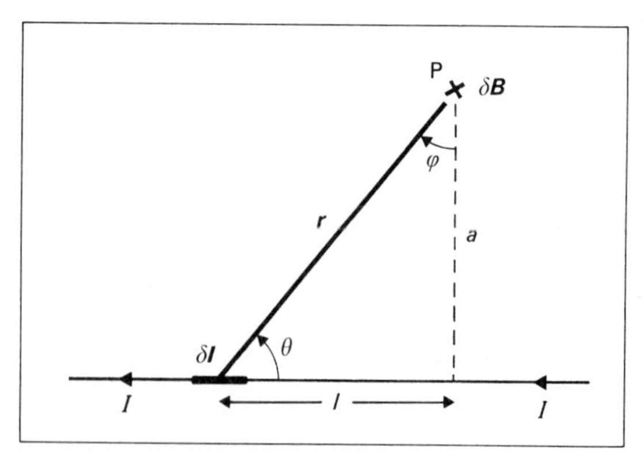

Fig. 53.7. Field due to a long straight wire.

Write all the variables in terms of ϕ:

$$l = a \tan \phi$$
$$\delta l = a \sec^2 \phi \, \delta\phi$$
$$r = a \sec \phi$$
$$\sin \theta = \cos \phi$$

Then

$$\delta B = \left(\frac{\mu_0}{4\pi}\right) \cdot \frac{Ia \sec^2 \phi \, \delta\phi \cos \phi}{a^2 \sec^2 \phi}$$
$$= \left(\frac{\mu_0}{4\pi}\right) \frac{I}{a} \cos \phi \, \delta\phi$$

Let $\delta\phi \to 0$, and integrate:

$$B = \left(\frac{\mu_0}{4\pi}\right) \cdot \frac{I}{a} \int_{\phi_1}^{\phi_2} \cos \phi \, d\phi$$
$$= \left(\frac{\mu_0}{4\pi}\right) \cdot \frac{I}{a} (\sin \phi_2 - \sin \phi_1)$$

For the special case of an *infinitely long* wire

$$\left.\begin{aligned} \phi_2 &= \frac{\pi}{2} \text{ rad} \\ \phi_1 &= -\frac{\pi}{2} \text{ rad} \end{aligned}\right\} B = \frac{\mu_0 I}{2\pi a}$$

53.5 Summary of Paragraph 53.4

SITUATION	SYMBOLS	RESULT
(*a*) Point at centre of circular coil	r = coil radius N = number of turns	$B = \frac{\mu_0 IN}{2r}$
(*b*) Any point on axis of circular coil	a = coil radius N = number of turns x = distance of point from coil centre	$B = \frac{\mu_0 IN}{2} \cdot \frac{a^2}{(a^2 + x^2)^{3/2}}$
(*c*) Point inside a very long solenoid or toroid Point at one end of a long solenoid	n = number of turns *per unit length*	$B = \mu_0 nI$ $B = \frac{\mu_0 nI}{2}$
(*d*) Point at centre of *Helmholtz* coil system	N = number of turns on each coil a = coil radius	$B = \frac{8\mu_0 NI}{\sqrt{125}a}$
(*e*) Point close to a very long straight wire	a = perpendicular distance from point to wire	$B = \frac{\mu_0 I}{2\pi a}$

In each result B can be written $\frac{(\text{dimensionless constant})\mu_0 I}{\text{length}}$. Note that $[n] = [\text{L}^{-1}]$.

53.6 Experimental Study of Magnetic Fields

The justification for the *Biot–Savart* law lies in the agreement between the predictions summarized in the table above and the values of $\boldsymbol{B}$ that are actually observed. This paragraph indicates how the fields are measured.

(*a*) ***Search coils.*** (*i*) If the circuit whose field is to be investigated carries an *alternating* current then it will establish an alternating field. If this field is made to thread a small search coil an alternating e.m.f. will be induced across the coil (p. 396). The e.m.f. can be measured by a c.r.o., and has a size proportional to the peak value of $\boldsymbol{B}$, and to the frequency.

(*ii*) If the circuit carries a *direct* current the field can be found by using a search coil in conjunction with a ballistic galvanometer in the way described on p. 399. Experimentally the method is not accurate, but it is the only one available in certain circumstances (see p. 427).

(*b*) ***Force on a wire.*** In principle the method used for the absolute measurement of current (p. 391) can be adapted for finding $\boldsymbol{B}$ if I is known, using $F = BIl$. In practice the method is not sensitive.

(*c*) ***The Hall probe*** (p. 377). This is probably the best way of comparing the fields in question. Allowance *is* necessary for the Earth's field, so the method is not suitable for very small fields. (Cf. (*a*)(*i*).)

*53.7 Ampère's Law

Demonstration of Ampère's Law

The field produced by an infinitely long straight conductor is given by

$$B = \frac{\mu_0 I}{2\pi a}$$

so

$$B(2\pi a) = \mu_0 I$$

This illustrates **Ampère's Law** which is

$$\oint \boldsymbol{B}\cdot d\boldsymbol{l} = \mu_0 I$$

where (*i*) $\oint \boldsymbol{B}\cdot d\boldsymbol{l}$ is the *line integral* round a closed loop, and

(*ii*) I is the current enclosed by the loop.

The law holds for any closed path about any configuration of current-carrying conductors in a vacuum (p. 432).

Uses of the Law

(*a*) For finding the magnetic field in situations which possess the necessary geometrical symmetry. It can be used for the long straight wire, toroid and solenoid.

(*b*) When applied to a long straight solenoid, it shows that $\boldsymbol{B}$ is constant over the whole cross-section (i.e. has the same value at any distance from the axis).

The law is an alternative to the *Biot–Savart Law.* (Cf. *Coulomb's Law* and *Gauss's Law.*)

53.8 The Magnetic Forces Exerted on each other by Current-carrying Conductors

At all points along the right wire of fig. 53.8 the field $\boldsymbol{B}$ created by the left wire is given by

$$B = \frac{\mu_0 I_1}{2\pi a}$$

Using $F = BIl$, the magnitude of the force $\boldsymbol{F}$ experienced by a length l of the right wire is given by

$$F = \left(\frac{\mu_0}{2\pi}\right)\cdot\frac{I_1 I_2}{a}\cdot l$$

Notes.

(*a*) When the currents are in the same direction, the conductors attract each other. When they are in opposite directions, they repel. This can be shown by *Fleming's left-hand motor rule.*

(*b*) *Newton's Third Law* tells us that two such *circuits* exert equal and opposite forces on each other.

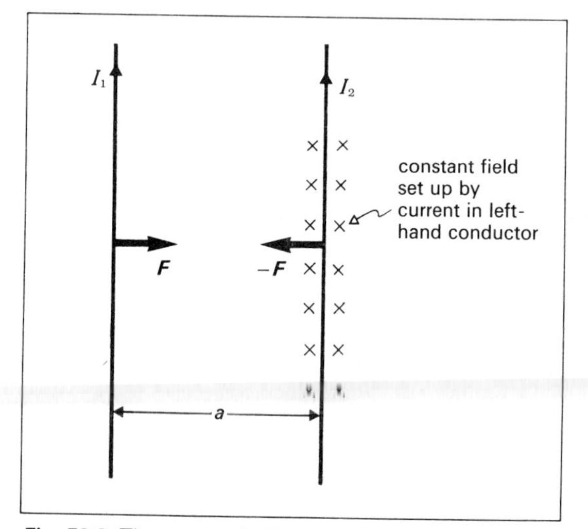

Fig. 53.8. The magnetic forces exerted by current-carrying conductors.

(*c*) The two wires of fig. 53.8 exert magnetic forces on each other even when

(*i*) they are both *electrically neutral,* and

(*ii*) they are both made of material which is *not ferromagnetic.*

This *fundamental magnetic effect* was first studied by *Ampère* (1822).

Correlation with the ampere

Suppose we assume the definition of the ampere:

*The **ampere** is that steady current which, flowing in two infinitely long straight parallel conductors of negligible circular cross-sectional area placed* 1 m *apart in a vacuum, causes each wire to exert a force of* 2×10^{-7} N *on each metre of the other wire.*

Then in

$$F = \left(\frac{\mu_0}{2\pi}\right)\cdot\frac{I_1 I_2}{a}\cdot l$$

we have

$$F = 2 \times 10^{-7}\,\text{N}$$

$$I_1 = I_2 = 1\,\text{A}$$

$$a = l = 1\,\text{m}$$

Substituting, we find

$$\mu_0 = 4\pi \times 10^{-7}\,\text{Wb A}^{-1}\,\text{m}^{-1}$$

(As an alternative procedure we could have taken the assigned value of μ_0, and shown that it leads to a result consistent with the definition of the ampere.)

The definition of the current standard of the SI

(1) *assigns* a value to μ_0,

(2) enables us to define the *coulomb*, which leads us

(3) to an *experimental* value for ε_0 (p. 311).

* The Relationship Between μ_0 and ε_0

(*a*) To define the ampere, we put

$$\mu_0 = 4\pi \times 10^{-7}\,\text{J s}^2\,\text{C}^{-2}\,\text{m}^{-1}$$

(*b*) By experiment, we find

$$\varepsilon_0 = 8.854 \times 10^{-12}\,\text{J}^{-1}\,\text{C}^2\,\text{m}^{-1}$$

So

$$\frac{1}{\mu_0\varepsilon_0} \approx 9 \times 10^{16}\,\text{m}^2\,\text{s}^{-2}$$

and

$$\frac{1}{\sqrt{\mu_0\varepsilon_0}} \approx 3 \times 10^{8}\,\text{m s}^{-1}$$

More advanced analysis predicts that the speed c of electromagnetic radiation in free space is given by

$$c = \frac{1}{\sqrt{\mu_0\varepsilon_0}}$$

The above calculation suggests that this agrees with the experimental evidence, and in fact this relationship is used for the accurate measurement of ε_0 (p. 333).

53.9 Summary of the Magnetic Field

	DEFINITIONS	LAWS	IMPORTANT RESULTS
$\boldsymbol{B}$ $\boldsymbol{m}$ $\boldsymbol{\Phi}$	$F = BQv\sin\theta$ $m = \dfrac{T_{max}}{B}$ $\Delta\Phi = B\cos\theta\,\Delta A$		$F = BIl$ $T = NIAB\sin\theta$ $T = mB\sin\theta$
μ_0	$B = \left(\dfrac{\mu_0}{4\pi}\right)\dfrac{Qv\sin\theta}{r^2}$	*Biot–Savart* $B \propto \dfrac{Qv\sin\theta}{r^2}$ *Ampère* $\oint B\,dl = \mu_0 I$	$\Delta B = \left(\dfrac{\mu_0}{4\pi}\right)\dfrac{I\,\Delta l\sin\theta}{r^2}$ coil $B = \dfrac{\mu_0 NI}{2r}$ long solenoid $B = \mu_0 nI$ straight wire $B = \dfrac{\mu_0 I}{2\pi a}$ $F = \left(\dfrac{\mu_0}{2\pi}\right)\cdot\dfrac{I_1 I_2}{a}\cdot l$

54 Electrical Measuring Devices

54.1 Survey of Instrument Types

Those marked † are described in this chapter.

(*a*) Those that Measure Steady or Alternating Quantities

(1) †Modified *moving-coil galvanometer*
(2) †*Current balance*
(3) †*Electrodynamometer*
(4) *Cathode ray oscillograph* (p. 441)
(5) *Electrostatic voltmeters*, such as

(*i*) the gold leaf electroscope, (p. 328)
(*ii*) the quadrant electrometer,
(*iii*) the attracted-disc electrometer. (p. 328)

They are not very sensitive instruments, and measure p.d.'s greater than about 50 V.

(6) *Diode valve voltmeter*
(7) *Hot-wire ammeter*
(8) *Thermocouple ammeter* (p. 412)
(9) *Moving-iron instruments*, which can be made attractive or repulsive. They

(*i*) have a non-uniform scale,
(*ii*) are hardly affected by the Earth's field,
(*iii*) need calibrating,
(*iv*) are ballistic in nature, and are usually damped (e.g. by a dashpot mechanism). Many of these instruments give a deflection which is proportional to the *square* of an alternating quantity. Their readings are discussed further on p. 412.

(*b*) Those only Suitable for Steady Quantities

(1) †*Moving-coil galvanometer*
(2) *Pivoted-magnet types*

A bar magnet suitably pivoted in a restoring magnetic field and a deflecting field gives a deflection that depends on the relative magnitudes of the fields.

This is the basis of the *tangent*, *sine*, *Helmholtz* and *astatic pair* galvanometers. They

(*i*) can be made absolute if the Earth's field is known,
(*ii*) are not very accurate,
(*iii*) have non linear scales,
(*iv*) are of academic interest only.

The applications of the d.c. amplifier are discussed in **46.9**.

54.2 The Moving-Coil Galvanometer

It is shown on page 379 that a coil of N turns each of area $\boldsymbol{A}$ and carrying a current I experiences a torque $\boldsymbol{T}$ given by

$$T = NIAB$$

when its plane is *parallel* to a field $\boldsymbol{B}$.

Such a coil is the basis of the moving-coil galvanometer (fig. 54.1).

The coil system is usually either

(*a*) pivoted on jewelled bearings for currents exceeding 5×10^{-5} A, or

(*b*) suspended from a torsion fibre for currents less than 5×10^{-5} A.

Galvanometer Sensitivity

Suppose the magnitude of the restoring torque $\boldsymbol{T}$ is given by $T = c\theta$ (p. 134) when the coil is deflected

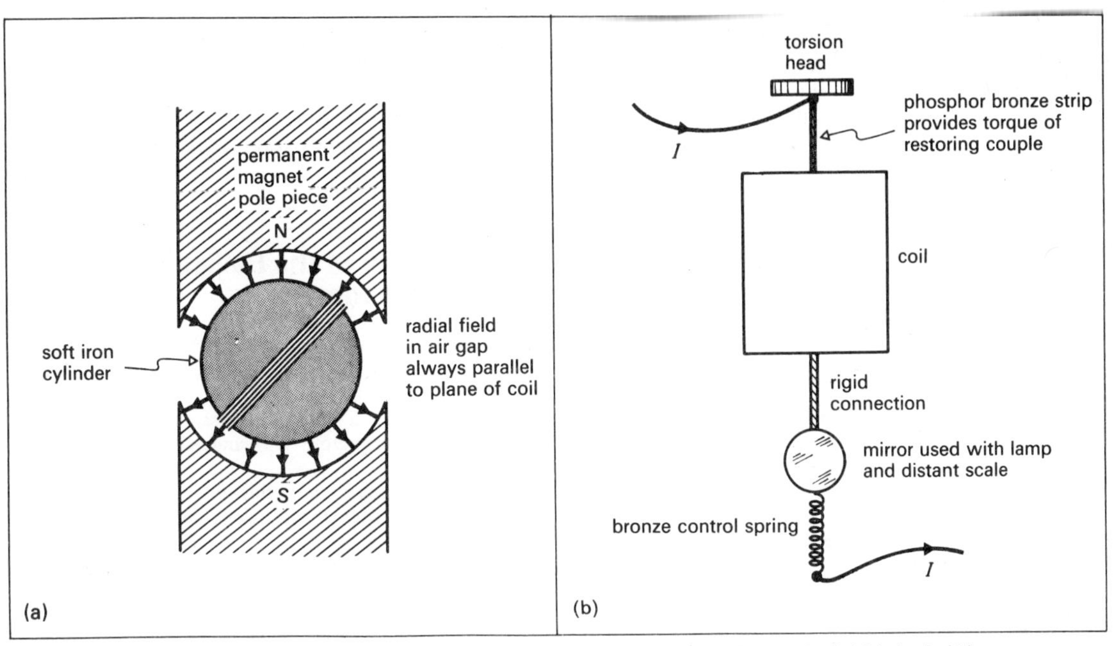

Fig. 54.1. A moving-coil galvanometer showing (a) the coil in the radial magnetic field (plan), (b) a suspension system (elevation).

through angle θ[rad].

$$T[\mathrm{N\,m}] = c\left[\frac{\mathrm{N\,m}}{\mathrm{rad}}\right]\theta[\mathrm{rad}]$$

When the coil is in equilibrium

$$NIAB = c\theta$$

$$\therefore \quad \boxed{\theta = \left(\frac{BAN}{c}\right)I}$$

The **current sensitivity** S *is defined by the equation*

$$S_I = \frac{\theta}{I}$$

$$S_I\left[\frac{\mathrm{rad}}{\mathrm{A}}\right] = \frac{\theta[\mathrm{rad}]}{I[\mathrm{A}]}$$

Although the unit of S is rad A^{-1}, in practice currents as large as 1 A are not passed through galvanometers of this type.

From the definition,

$$S_I = \frac{BAN}{c}$$

so we see that *high sensitivity* can be achieved by using

(*a*) a magnet which gives a *large value of* $\boldsymbol{B}$ (but its size must remain constant),

(*b*) a coil of *large area* $\boldsymbol{A}$ wound with a *large number of turns* N (but its size must be manageable),

(*c*) a restoring system of *small* c (consistent with the long period of swing that this would produce).

The sensitivity is sometimes quoted as (I/θ) [$\mathrm{A\,rad^{-1}}$]. Numerically this is the current required to give unit deflection. Its value might be $\sim 10^{-8}$ A for 1 mm deflection on a scale placed 1 m away.

High current sensitivity is desirable in circuits of high resistance.

Voltage sensitivity

Suppose that a p.d. V across the galvanometer produces a deflection θ. Then **voltage sensitivity** is defined by the equation

$$S_V = \frac{\theta}{V} = \frac{\theta}{IR} = \frac{S_I}{R} = \frac{BAN}{cR}$$

High voltage sensitivity is desirable in circuits of relatively low resistance, such as those of the *Wheatstone* bridge and potentiometer.

Notes on the Galvanometer

(*a*) $\theta \propto I$, so it cannot measure alternating values directly.

(*b*) The scale is linear when $\boldsymbol{B}$ is radial, so calibration is simple.

(*c*) External magnetic fields are relatively unimportant.

(*d*) It can be made very sensitive without loss of accuracy (but there is a lower limit imposed by *Brownian motion* (p.125)).

(*e*) A suspension system is delicate, and significantly affected by excessive vibration.

(*f*) It is very widely used.

54.3 Ways of Adapting the Galvanometer

(*a*) *As an ammeter*—add a **shunt** in parallel with a pivoted coil system (p. 366).

(*b*) *As a voltmeter*—add a high resistance bobbin in series with the coil (p. 366).

(*c*) *To measure alternating quantities*

(*i*) Put a rectifier in series with the coil. (N.B. This destroys the a.c. by converting it to a half-wave pulsating d.c. (p. 425)).

(*ii*) To obtain full-wave rectification *through the instrument* use the rectifier bridge of fig. 54.2.

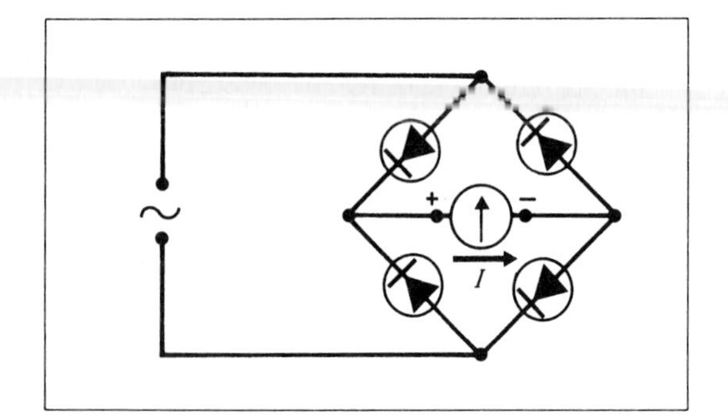

Fig. 54.2. Rectifier bridge for moving-coil galvanometer.

Dead-beat and Ballistic Use

(*a*) If the galvanometer is intended to attain a *steady value* quickly, damping is necessary to prevent oscillation. The galvanometer is **critically damped (dead-beat** movement) when oscillation is just prevented. Refer to fig. 11.4 (p. 89) for an illustrative diagram.

This is achieved by winding the coil on a *metallic frame*: the large induced currents in the frame give *electromagnetic damping*. (Refer to eddy currents, p. 400, for a discussion of the mechanism.)

(*b*) In a galvanometer adapted to measure charge, an impulsive torque sets the coil swinging. The damping is made as small as possible by placing a large resistance in series with the coil to reduce the induced current. The coil then swings freely (**ballistic** movement).

*54.4 Theory of the Ballistic Galvanometer

Principle

(1) Pass a charge Q through the coil in a time Δt which is $\ll \frac{1}{4}$ of the time period of swing.

(2) The suspended system receives an impulsive torque which is finished before the angular displacement is significant.

(3) The k.e. given to the coil is converted by rotation to p.e. stored in the system that provides the restoring torque.

Calculation

(*a*) The instantaneous torque

$$T = NIAB$$

The impulsive torque (angular impulse)

$$\int T\,\mathrm{d}t = \int_0^{\Delta t} NIAB\,\mathrm{d}t$$

$$= NAB\int_0^{\Delta t} I\,\mathrm{d}t$$

$$= NABQ$$

Put K = moment of inertia of coil system

ω_0 = initial angular velocity

Then

$$\text{angular impulse} = \text{change of angular momentum}$$

$$NABQ = K\omega_0$$

(*b*) The initial k.e. of the coil $= \frac{1}{2}K\omega_0^2$.

Put θ_m = final deflection of coil

Then the final p.e. stored by the suspension fibre is

$$\int_0^{\theta_m} c\theta\,\mathrm{d}\theta = \tfrac{1}{2}c\theta_m^2$$

Equating

$$\tfrac{1}{2}K\omega_0^2 = \tfrac{1}{2}c\theta_m^2$$

$$\tfrac{1}{2}K\left(\frac{NABQ}{K}\right)^2 = \tfrac{1}{2}c\theta_m^2$$

So

$$Q = \left(\frac{\sqrt{cK}}{NAB}\right)\theta_m$$

$$\boxed{Q \propto \theta_m}$$

Notes.

(*a*) (θ/Q) [rad C^{-1}] is called the **charge sensitivity**. It can be found by calculation or by calibration.

(b) Because neither air nor electromagnetic damping can be *eliminated*, the observed value of θ may have to be corrected to find the true θ_m by noting successive values of θ during the subsequent oscillation. (Again refer to fig. 11.4 on p. 89.)

(c) θ_m is frequently called the corrected **throw** of the galvanometer.

54.5 Uses of the Ballistic Galvanometer

The previous paragraph is summarized by the equation

$$Q = k\theta_m$$

where θ_m is *the corrected first throw*, and k is a constant.

(Note that where values of θ are to be *compared*, no correction is necessary.)

(*a*) Use on Open Circuit

Fig. 54.3 shows a circuit for comparing the capacitances of two capacitors.

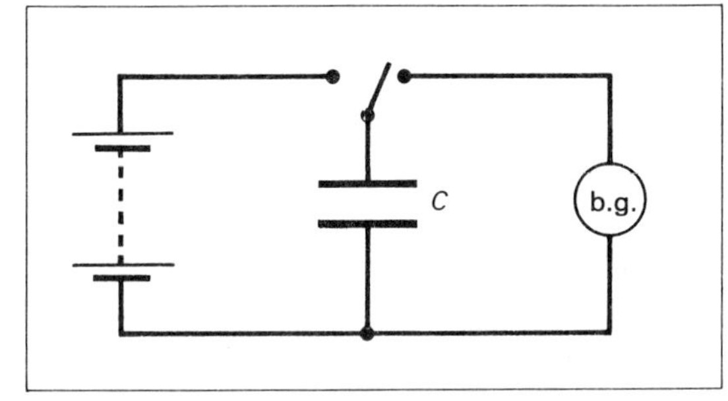

Fig. 54.3. The ballistic galvanometer on open circuit.

Connect two capacitors in turn, (1) to a battery which charges them to a p.d. V, and (2) to the ballistic galvanometer.

Then
$$Q_1 = C_1V = k\theta_{1,m}$$
$$Q_2 = C_2V = k\theta_{2,m}$$

Dividing
$$\frac{C_1}{C_2} = \frac{\theta_{1,m}}{\theta_{2,m}}$$

We do not need to know the e.m.f. of the battery.

(*b*) Calibration on Closed Circuit

When the b.g. is used on *closed* circuit, the damping is altered significantly. It must be calibrated in the circuit in which it is to be used (fig. 54.4).

To find k in $Q = k\theta_m$, a known Q is passed through the instrument by using the *standard mutual inductance* (p. 409). (A capacitor cannot be used because it would cause the circuit to revert to an open one.)

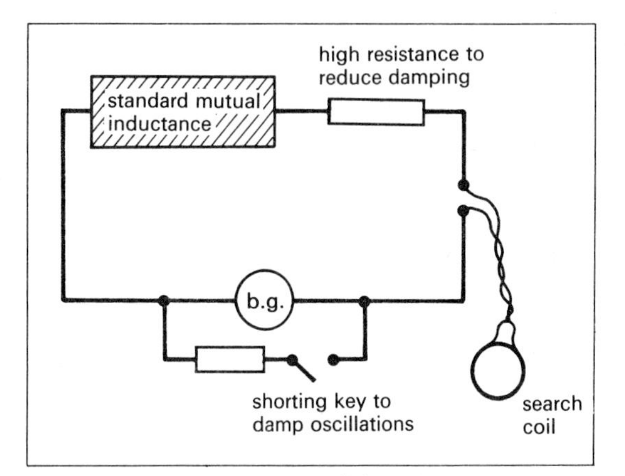

Fig. 54.4. Calibration of a ballistic galvanometer on closed circuit.

Apart from the comparison of capacitance (left), uses of the b.g. include

(i) measurement of *flux changes* $\Delta\Phi$ from

$$Q = \frac{\Delta\Phi}{R} \qquad \text{(p. 399)}$$

(ii) measurements on the *Earth's magnetic field* using the *Earth inductor*, and

(iii) the plotting of *hysteresis curves*. (p. 429)

54.6 A Simple Current Balance

A **current balance** is a device which measures the current in terms of the forces that current-carrying conductors exert on each other. It is an **absolute instrument** in the sense that the current is measured in terms of *mechanical* quantities, and is *directly related to the definition of the ampere*. No electrical measurements are needed, nor prior knowledge of any electrical quantity except the defined value of μ_0.

There are several versions: fig. 54.5 shows a simple one designed for the student laboratory. Note that the solenoid is in series with, and therefore carries the same current as, the wire.

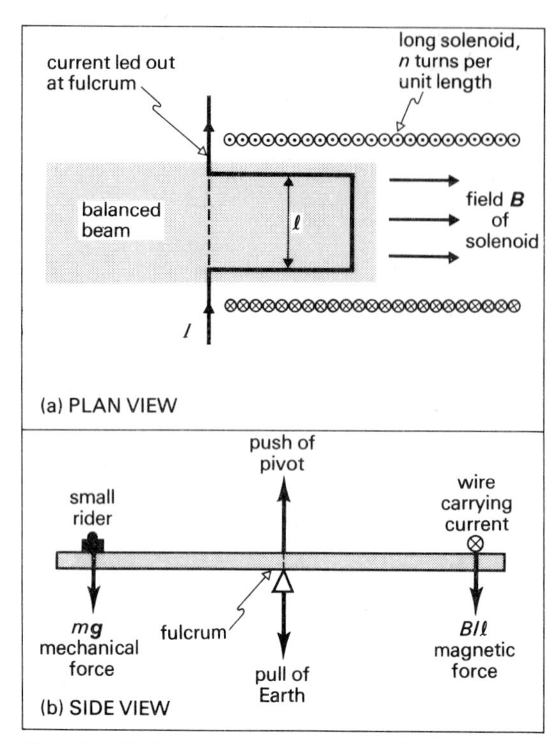

Fig. 54.5. A simple student-laboratory form of current balance.

If the beam is balanced through its centre of gravity

$$mg = BIl$$

in which $B = \mu_0 nI$ if we neglect corrections for the solenoid's finite length.

$$\therefore \; mg = (\mu_0 nI)Il$$
$$= (\mu_0 nl)I^2$$

Thus I can be calculated from a knowledge of μ_0 and the mechanical quantities m, $\boldsymbol{g}$, n and l.

Notes.

(*a*) The current balance represents a practical way of realizing the situation from which the ampere is defined (p. 10).

(*b*) The balance is primarily used to standardize other more convenient current-measuring devices.

(*c*) Since $mg \propto I^2$, the balance can be used to register the r.m.s. value of an alternating current.

54.7 The Electrodynamometer

This too, like the current balance, is an absolute instrument. It consists of a small pivoted coil carrying a current which experiences a torque when placed in the magnetic field at the centre of a second larger outer coil carrying the same current (fig. 54.6).

The principle is that of the moving-coil galvanometer, but the device is less sensitive because the deflecting field is much smaller than that established by a permanent magnet.

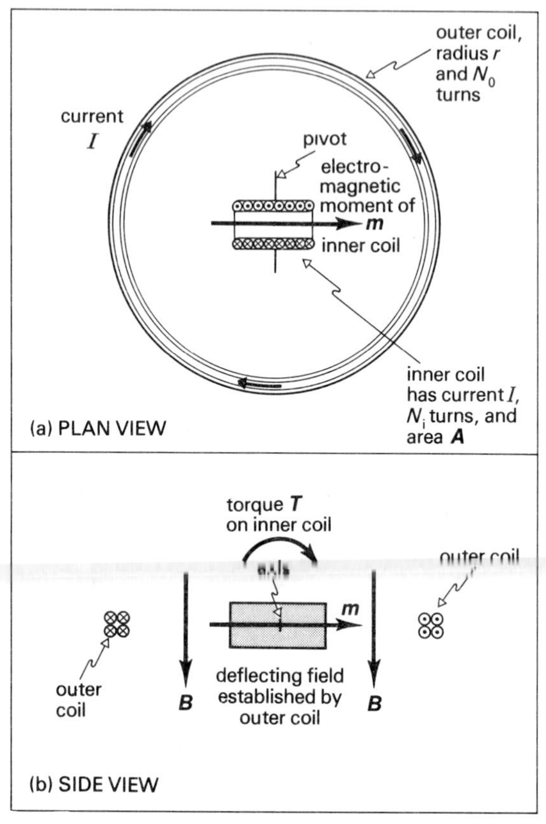

Fig. 54.6. Schematic arrangement of an electrodynamometer.

The deflecting torque on the inner coil is given by

$$T = mB \qquad \text{(p. 379)}$$

in which

$$m = N_i IA \qquad \text{(p. 379)}$$

and

$$B = \frac{\mu_0 N_0 I}{2r} \qquad \text{(p. 383)}$$

giving

$$T = \frac{\mu_0 N_i N_0 A I^2}{2r}$$

The sense of the torque is such as to align the electromagnetic moment $\boldsymbol{m}$ in the same direction as the deflecting field $\boldsymbol{B}$. Equilibrium can be attained by exerting an equal but opposite elastic torque, or by attaching the inner coil to a pivoted beam which can be loaded.

The notes of **54.6** apply equally here, and again I can be calculated from a knowledge of μ_0 and purely mechanical measurements.

54.8 The Wattmeter

Principle

The rate of conversion P of electrical energy by the load X in fig. 54.7 is given by

$$P = VI$$

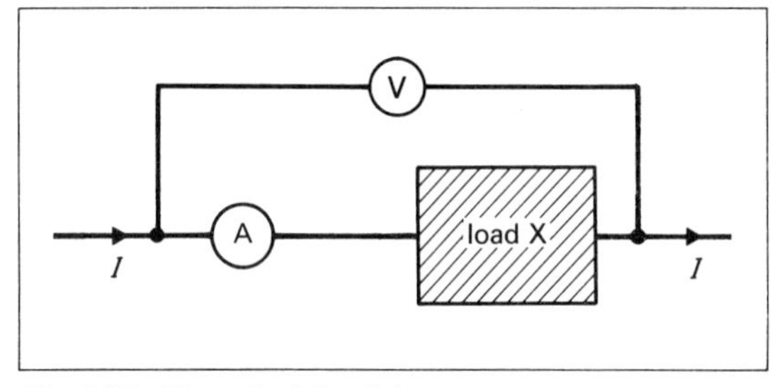

Fig. 54.7. The principle of the wattmeter.

The **wattmeter** combines the function of voltmeter and ammeter to record the *power* directly.

Adaptation of the Dynamometer

In fig. 54.8

(*a*) the *fixed coils* carry the same *current* I as the load X, and

(*b*) the *moving coil* carries a current which is proportional to the *potential difference* V across the load.

Deflecting couple

$\propto$ (fixed coil current) $\times$ (moving coil current)

$\propto VI$

$\propto P$, the instantaneous power supply.

$$\therefore \left(\begin{array}{c}\text{mean}\\ \text{deflection}\end{array}\right) \propto \left(\begin{array}{c}\text{mean}\\ \text{power supply}\end{array}\right).$$

The device can be used to measure the rate of conversion of electrical energy by a device carrying a.c. when V and I may be *out of phase* (p. 421).

The **joulemeter** (or watt-hour meter) incorporates a revolution counter to record *energy* consumption.

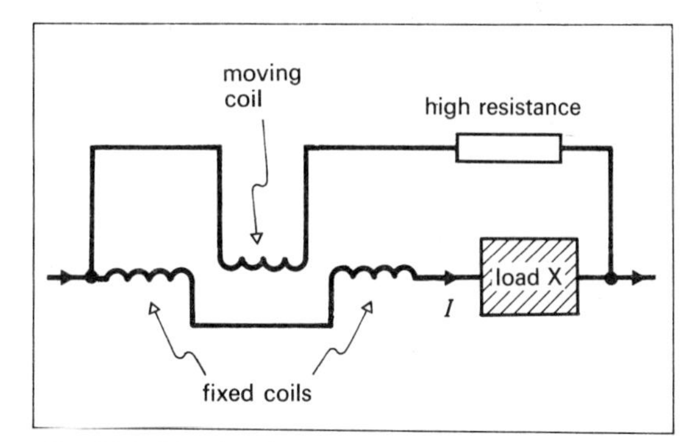

Fig. 54.8. The dynamometer as a wattmeter.

55 Principles of Electromagnetic Induction

55.1 Electromagnetic Induction

Fig. 55.1*a* shows a conductor being pulled by an external agent at velocity $\boldsymbol{v}$ through a magnetic field $\boldsymbol{B}$ directed into the paper.

Moving charge-carriers constitute a current: in a magnetic field they experience a (magnetic) force. By *Fleming's left hand motor rule* electrons experience a force *to the left*: because they are *free* they move to the left. (The positive ions of the lattice also experience the force, but are not free to move.)

The effects of their movement are shown in fig. 55.1*b*.

If one thinks of the conductor as a battery, the electric field *within it* (from positive to negative terminal) is in a direction opposite to that of the conventional e.m.f.

Notes.

(*a*) The *electric* field in the conductor is the *result* of the electron accumulation, and *not the cause* of the electron movement.

(*b*) The charge accumulation stops either when

(*i*) the relative movement stops, or

(*ii*) when the (electrical) forces that result from the charge accumulation are equal in magnitude but opposite in direction to the (magnetic) forces that result from the movement of the conductor.

(*c*) The e.m.f. results in a p.d. across the ends of the conductor so long as the movement is maintained. It is called an **induced** e.m.f.

(*d*) If the conductor is moved in the opposite direction, then the polarity of the e.m.f. is also reversed.

55.2 Magnetic Flux-Linkage $N\Phi$

The magnetic flux passing through a loop or coil is said to **thread** or **link** the coil.

Suppose a coil has N turns, and that the flux linking *each turn* is Φ.

***Then we define the* flux-linkage *to be* $N\Phi$.**

No particular symbol is used in this book.

The unit is the **weber-turn**, but $N\Phi$ has the same dimensions as the flux Φ.

55.3 Calculation of the Magnitude of the Induced e.m.f.

Fig. 55.2 shows an external agent exerting a force $\boldsymbol{F}$ so as to pull the rectangular coil at a *steady* velocity $\boldsymbol{v}$ through the field $\boldsymbol{B}$.

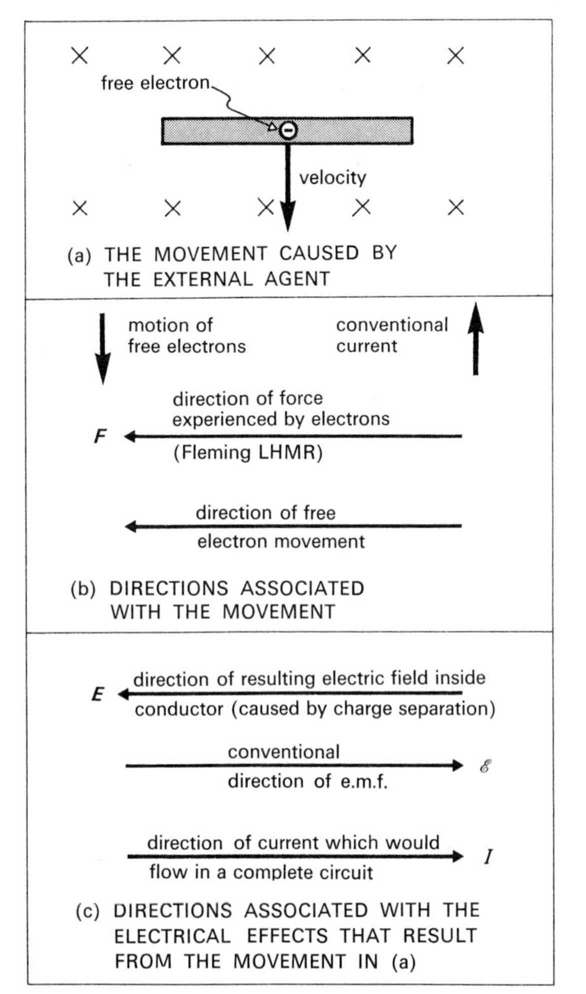

Fig. 55.1. Generation of a motional e.m.f.

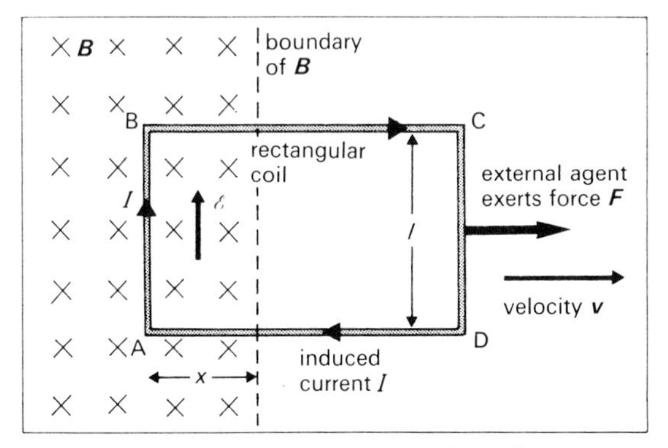

Fig. 55.2. Calculation of the motional e.m.f.

Suppose the e.m.f. $\mathscr{E}$ causes a current I. The (magnetic) force on the wire, by the motor rule,

$$= BIl \quad \text{to the left}$$

The wire is in equilibrium (since its velocity is constant), and so

$$F = BIl \quad \text{to the right}$$

$$\text{The rate of working of the agent } = Fv$$

$$\text{The rate of working of the seat of e.m.f. } = \mathscr{E}I$$

By the law of conservation of energy, if no other source (such as the cause of $\boldsymbol{B}$) provides energy, then these are equal.

$$\mathscr{E}I = (BIl)v$$
$$\mathscr{E} = Blv$$

Notes.

(a) If side CD had also been in the field, it would have been the seat of an equal e.m.f., which would have acted anticlockwise in the loop. The induced current would have been zero.

(*b*) The external agent does work, which is

(*i*) converted to electrical energy, but then

(*ii*) dissipated to internal energy by *Joule* heating.

The net effect is to increase the internal energy of the wire.

A Force Calculation of the e.m.f.

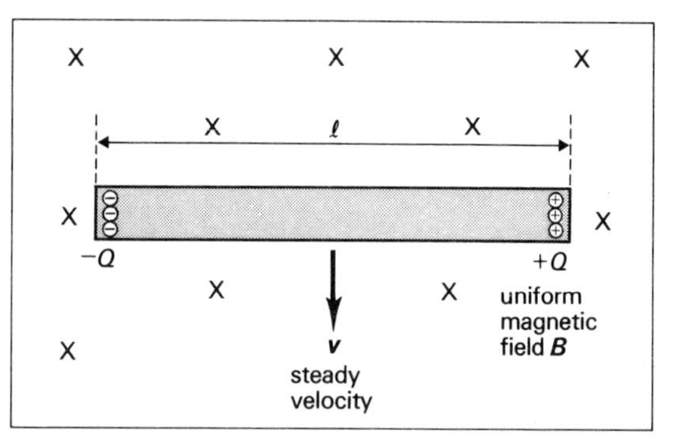

Fig. 55.3. Alternative calculation of the motional e.m.f.

Consider the equilibrium situation shown in fig. 55.3. The forces on the free charge $-Q$ are

(*a*) a magnetic force to the left, of size BQv

(*b*) an electric force to the right, of size EQ.

Putting $E = (\mathscr{E}/l)$ (p. 324) and equating, we have

$$(\mathscr{E}/l)Q = BQv$$
$$\mathscr{E} = Blv$$

This should be compared to the theory of the *Hall* effect (p. 375).

Ways of Expressing the Induced e.m.f.

(*a*) The side AB in fig. 55.2 is of length l. In time Δt it sweeps out an area $lv\,\Delta t$, through which there is a flux density $\boldsymbol{B}$. The flux $\Delta\Phi$ cut by the conductor in time Δt is given by

$$\Delta\Phi = Blv\,\Delta t$$

so

$$\mathscr{E} = Blv = \frac{\Delta\Phi}{\Delta t}$$

The induced e.m.f. is proportional to the rate at which the conductor cuts magnetic field lines.

(*b*) The flux through the coil of fig. 55.2 is given by $\Phi = Blx$. Since there is only one turn, the flux-linkage is also Blx. Then

$$\frac{d\Phi}{dt} = Bl\frac{dx}{dt} \quad \text{(since } \boldsymbol{B} \text{ and } l \text{ are constant)}$$

$$= Blv = \mathscr{E}$$

The induced e.m.f. is equal to the rate of change of flux-linkage.

This result is of more general use, since it can be used when the coil is stationary and the flux-linkage made to change by altering the magnitude of $\boldsymbol{B}$.

If there are N turns, then the total e.m.f.

$$\mathscr{E} = N\cdot\frac{d\Phi}{dt}$$

55.4 The Laws of Electromagnetic Induction

Faraday's First Law:

When the magnetic flux threading a circuit is changing, an e.m.f. is induced in the circuit.

The *magnitude* comes from

Faraday's Second Law:

The magnitude of an induced e.m.f. is proportional to the rate of change of flux-linkage.

This law is sometimes expressed in **Neumann's Equation**

$$\boxed{\mathscr{E} = -N\cdot\frac{d\Phi}{dt}}$$

$$\mathscr{E}[\text{volts}) = -N\cdot\frac{d\Phi[\text{Wb}]}{dt[\text{s}]}$$

On the left $\quad 1\,\text{V} = 1\,\text{J}\,\text{C}^{-1}$

On the right $\quad \dfrac{1\,\text{Wb}}{\text{s}} = \dfrac{1\,\text{N}\,\text{m}}{\text{A}\,\text{s}} = 1\,\text{J}\,\text{C}^{-1}$

The minus sign is discussed below.

The Direction of the Induced e.m.f.

This must be such that the law of conservation of energy is obeyed. **Le Chatelier's Principle** is a general statement:

When a constraint is applied to a physical system in equilibrium then, so far as it can, the system will adjust itself so as to oppose the constraint.

For electromagnetic induction this is expressed by

(1) **Lenz's Law**

The direction of an induced current (if one were to flow) is such that its effect would oppose the change in magnetic flux which gave rise to the current.

This is achieved by exerting forces of some kind, or establishing a magnetic field.

Or (2) **Fleming's Right-Hand Dynamo Rule**

This rule, which is particularly useful for straight conductors, is shown in fig. 55.4.

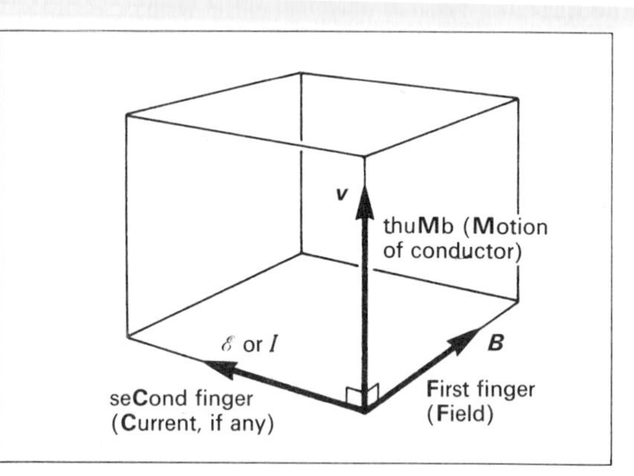

Fig. 55.4. Fleming's dynamo rule.

* The Electric Field Associated with an Induced e.m.f.

Fig. 55.5 shows the analogous situations in which

(*a*) a moving charge (current) establishes a magnetic field, whose sense and direction are given by *Maxwell's right-handed* corkscrew rule, and

(*b*) a time-varying magnetic flux gives rise to an electric field, whose sense and direction are given by a *left-handed* corkscrew rule.

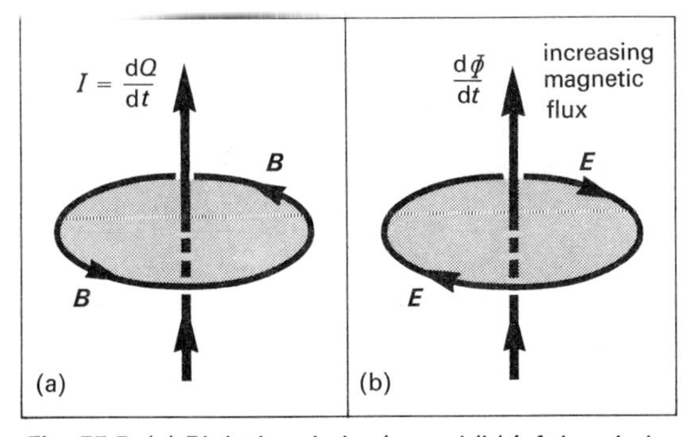

Fig. 55.5. (a) Right-handed rule, and (b) left-handed rule.

The *left-handedness* of the field $\boldsymbol{E}$ explains why *Neumann's equation* is given a *minus* sign.

The lines of $\boldsymbol{E}$ in fig. 55.5*b* are continuous: the electric field caused by the changing magnetic flux is *non-conservative*. (Compare with that established by static electric charges, p. 323.)

55.5 Further Examples of Electromagnetic Induction

In this paragraph we give examples of the conversion of *mechanical energy into electrical energy*, ((*a*) to (*d*)), or of the conversion of energy stored in a magnetic field into electrical energy (*e*).

In each example either

(*i*) the flux-linkage of a loop is changing, or
(*ii*) a conductor cuts across a magnetic field,

and this results in an electric field of the type discussed in paragraph **55.4**.

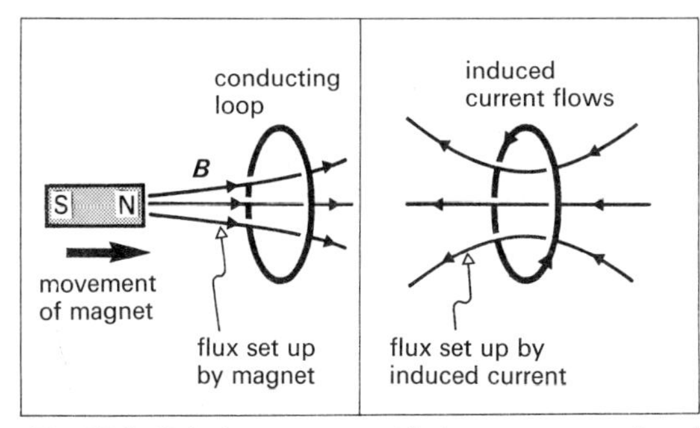

Fig. 55.6. Relative movement between a magnet and a conducting loop.

(*a*) If the magnet of fig. 55.6 is withdrawn, the sense of the induced current in the coil is reversed to preserve the *status quo*. The induced current is zero when their relative speed is zero.

Lenz's Law indicates that approach produces repulsion, and withdrawal produces attraction.

(*b*) The experiment of fig. 55.6 can be repeated using a loop carrying a current instead of a permanent magnetic dipole. The results are the same.

(*c*) Rotation of a loop in a magnetic field sets up an induced e.m.f. (and current).

This is the principle of the generator (p. 403). The conducting material of the loop cuts the magnetic field lines.

(*d*) If a loop is deformed in a magnetic field so that its area is changed, there is a change of flux-linkage, and a consequent e.m.f. An induced current causes forces that oppose further change of area.

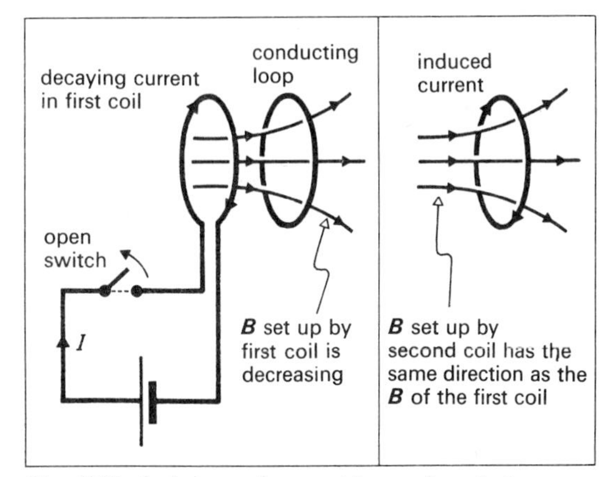

Fig. 55.7. A change of current in one loop induces a current in the other.

(*e*) In fig. 55.7 the induced current is in the same sense as the original current in the first coil. The decay of magnetic flux is delayed by the temporary current flow in the second coil. The energy to produce this current comes from the energy stored in the magnetic field. This effect (**mutual induction**) and a similar effect (**self-induction**) are discussed in chapter **57**.

55.6 The e.m.f. Induced in a Rotating Coil

Consider a conducting coil of N turns each of area $\boldsymbol{A}$ being made to rotate with angular velocity $\boldsymbol{\omega}$ in a *uniform* magnetic field $\boldsymbol{B}$ (fig. 55.8).

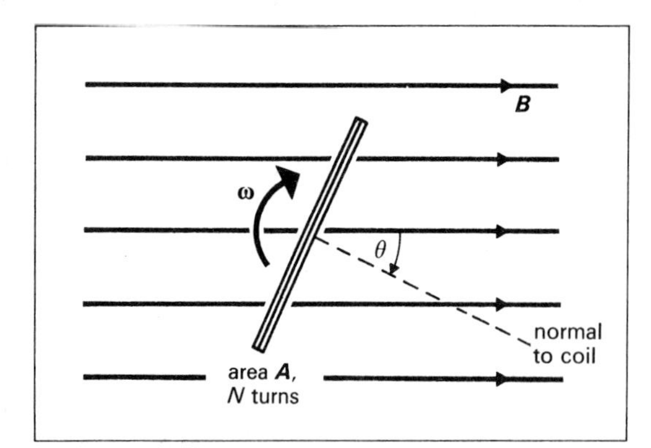

Fig. 55.8. Production of a sinusoidal e.m.f.

The flux through each turn is given by

$$\Phi = BA\cos\theta \qquad (\theta = \omega t)$$

The induced e.m.f. for each turn is

$$-\frac{d\Phi}{dt} = \omega AB\sin\omega t$$

For N turns

$$\mathscr{E} = N\omega AB\sin\omega t$$
$$= \mathscr{E}_0\sin\omega t$$

where the maximum value of $\mathscr{E}$ is $\mathscr{E}_0 = N\omega AB$, since $\sin\omega t$ cannot exceed 1.

The output from a simple generator is a **sinusoidally varying e.m.f.**

When the *plane* of the coil is

(a) perpendicular to $\boldsymbol{B}$, $\mathscr{E} = 0$
(b) parallel to $\boldsymbol{B}$, $\mathscr{E} = \mathscr{E}_0$

$\mathscr{E}_0$ can be made larger by increasing N, ω, A and B. The behaviour of a.c. is discussed in chapter **58**, p. 412. The dynamo (generator) is discussed on p. 403.

Motor Effect

In the generator a coil carrying a current is made to rotate in a magnetic field, and so experiences a torque which (by *Lenz's Law*) opposes the rotation. To produce electrical energy the external agent must expend mechanical energy (do work) against this opposing torque.

55.7 The e.m.f. Induced by a Changing Field

Suppose a coil of area $\boldsymbol{A}$ and N turns is placed normal to a changing magnetic field $\boldsymbol{B}$. For each turn $\Phi = BA$, so the e.m.f. induced across the coil is given by

$$\mathscr{E} = NA\left(\frac{dB}{dt}\right)$$

The a.c. Search Coil

The e.m.f. induced gives useful information about the field. If the field is established by a sinusoidal a.c. we can write

$$B = B_0\sin\omega t$$

so $$dB/dt = \omega B_0\cos\omega t$$

$$\therefore \quad \mathscr{E}_{max} \propto \omega B_0$$

The size of the induced e.m.f. can be measured by a c.r.o., and is proportional to the peak value of the field at that point. It is also proportional to the frequency (p. 403).

55.8 The Magnitude of an Induced Charge

Fig. 55.9 shows a search coil of N turns and area $\boldsymbol{A}$ connected into a circuit of resistance R.

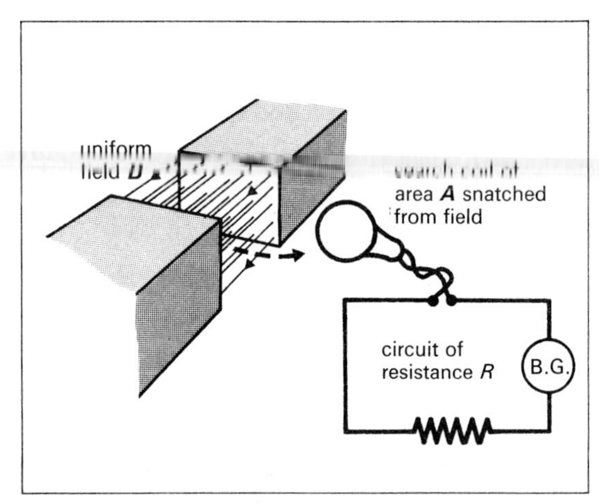

Fig. 55.9. Search coil.

Suppose the coil is totally immersed in the field $\boldsymbol{B}$ of the magnet, and then suddenly snatched away. Instantaneously

$$I = \frac{\mathscr{E}}{R}$$
$$= \frac{1}{R}\left[\frac{d}{dt}(N\Phi)\right] \qquad (\Phi = BA)$$
$$= \frac{NB}{R}\left(\frac{dA}{dt}\right)$$

(As we are not concerned here with the *sense* of the current, the minus sign has been dropped.)

As the area immersed in the field changes, so current flows in the circuit.

It is seen from fig. 55.10 that the *charge* passing

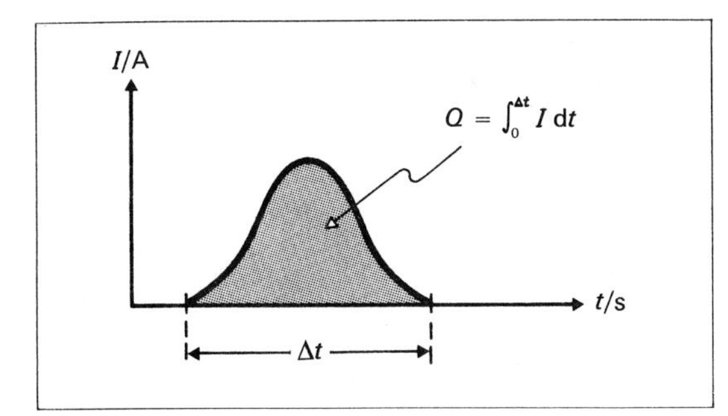

Fig. 55.10. Calculation of induced charge.

any cross-section of the circuit is *independent of the timing* Δt, being given by

$$Q = \int_0^{\Delta t} I \,\mathrm{d}t$$

$$= \frac{NB}{R}\int \frac{\mathrm{d}A}{\mathrm{d}t}\mathrm{d}t$$

$$= \frac{NBA}{R}$$

If the coil remains in the field, but is rotated through an angle π rad (half a revolution), then

$$Q = \frac{2NBA}{R}$$

More generally

$$\boxed{Q = \frac{\textit{change of flux-linkage}}{R}}$$

$$Q[\mathrm{C}] = \frac{N\,\Delta\Phi[\mathrm{Wb}]}{R[\Omega]}$$

Applications

(*a*) A search coil used in this way with a ballistic galvanometer can evaluate Q, $\Delta\Phi$ and hence $\boldsymbol{B}$, the flux density of a magnetic field. (See, for example, p. 391 and the *Rowland Ring* on p. 426.)

(*b*) The **Earth inductor** consists of a special search coil of a few hundred turns (and each of large area) mounted on a special wooden frame. It can measure the relatively weak magnetic field of the Earth.

If one measures the vertical and horizontal components $\boldsymbol{B}_{\text{ver}}$ and $\boldsymbol{B}_{\text{hor}}$ separately, one can calculate the **angle of dip** D from

$$\tan D = \frac{B_{\text{ver}}}{B_{\text{hor}}}.$$

56 Applications of Electromagnetic Induction

56.1 Eddy Currents

Any block of conducting material either

(*i*) moving in a magnetic field, or

(*ii*) placed in a changing magnetic field

experiences induced currents, called **eddy currents**.

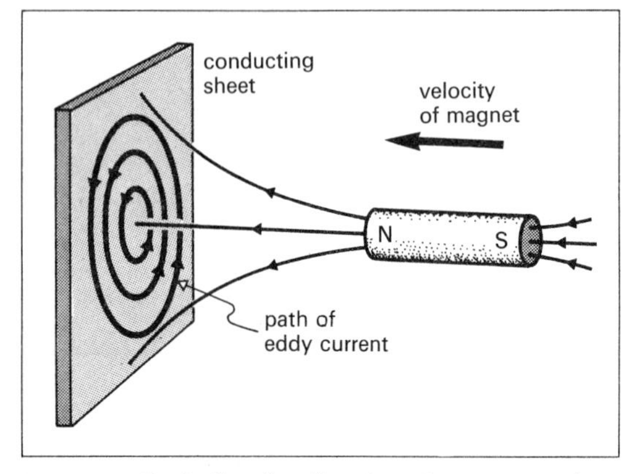

Fig. 56.1. Illustrating the direction of a typical eddy current.

They flow in whatever paths are dictated by the need to *oppose the change* that produced them (fig. 56.1). They show both heating and magnetic effects.

(*a*) A Heating Effect

(1) This can be useful, as in the **induction furnace**. *Joule heating* causes the melting of a *conductor* placed in the changing field, but insulators remain unaffected.

(2) This can be a nuisance in *iron-cored apparatus*, such as motors, dynamos and transformers. The *Joule heating* ($\propto V^2/R$) is reduced by increasing the resistance of the eddy current path by

(*i*) *laminating* the core,

(*ii*) using a bundle of iron wires, or

(*iii*) using a core packed with iron dust.

At extremely high frequencies one must use **ferrites**, materials of high resistivity whose magnetic properties are in some respects similar to those of iron (p. 429).

(*b*) A Magnetic Effect

A moving conductor will experience forces caused by the interaction of the external magnetic field with that set up by its own eddy currents. These will result in a *braking effect.*

Examples

(1) A copper plate swinging between the poles of a magnet is strongly damped unless it is slotted in the right direction.

(2) *Galvanometer damping* (see also p. 390). In

(*i*) the *dead-beat* movement the circuit resistance is just low enough to prevent oscillation (so that the induced current is relatively large), but not so low as to give over-damping,

(*ii*) the *ballistic* movement the induced current is as small as possible so that the coil swings freely.

56.2 The Transformer

Suppose fig. 56.2 represents an **ideal transformer** in which

(*a*) the primary resistance is zero,

(*b*) there is no flux leakage, so that we have the same flux through each turn of both the primary and secondary coils, and

(*c*) the secondary is on open circuit.

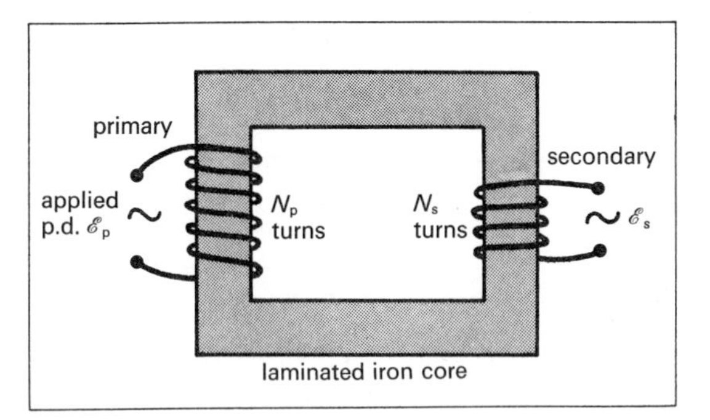

Fig. 56.2. An ideal transformer.

Suppose the current change in the primary changes the flux through the core at the rate $d\Phi/dt$.

Then the back e.m.f in the primary

$$= -N_p \frac{d\Phi}{dt}$$

and the induced e.m.f. in the secondary

$$\mathscr{E}_s = -N_s \frac{d\Phi}{dt}$$

Applying the loop equation to the primary

$$\mathscr{E}_p + \left(-N_p \frac{d\Phi}{dt}\right) = IR = 0 \qquad (R = 0)$$

$$\therefore \quad \mathscr{E}_p = N_p \frac{d\Phi}{dt}$$

from which $$\frac{\mathscr{E}_p}{\mathscr{E}_s} = -\frac{N_p}{N_s}$$

The minus sign indicates that the two e.m.f.'s are in antiphase.

The transformer *steps-up* the applied p.d. if $\mathscr{E}_s > \mathscr{E}_p$.

Secondary Not on Open Circuit

Suppose a *load resistance R* is connected across the secondary, so that the secondary current is I_s, and the primary current becomes I_p. If there is no energy loss

$$\text{power input} = \text{power output}$$

$$\mathscr{E}_p I_p = \mathscr{E}_s I_s$$

So $$\frac{I_p}{I_s} = \frac{\mathscr{E}_s}{\mathscr{E}_p} = \frac{N_s}{N_p}$$

In a step-up transformer $I_s < I_p$.

In practice a *non-ideal* transformer shows these losses:

(*a*) I^2R losses (*Joule heating*) in the copper windings,

(*b*) I^2R losses due to eddy currents in the iron core,

(*c*) hysteresis losses (the core dissipates energy on repeated magnetization, p. 429), and

(*d*) loss due to magnetic leakage.

* Reflected Resistance

Using the zero-loss equation

$$\frac{I_p}{I_s} = \frac{\mathscr{E}_s}{\mathscr{E}_p} = \frac{N_s}{N_p}$$

and $$I_s = \frac{\mathscr{E}_s}{R}$$

We have $$I_p = \left(\frac{N_s}{N_p}\right) \cdot \left(\frac{\mathscr{E}_s}{R}\right)$$

$$= \frac{\mathscr{E}_p}{R\left(\frac{N_p}{N_s}\right)^2}$$

A resistive load R across the secondary acts in the primary as though it were a resistance $R(N_p/N_s)^2$ which is called a **reflected resistance**.

Demonstration

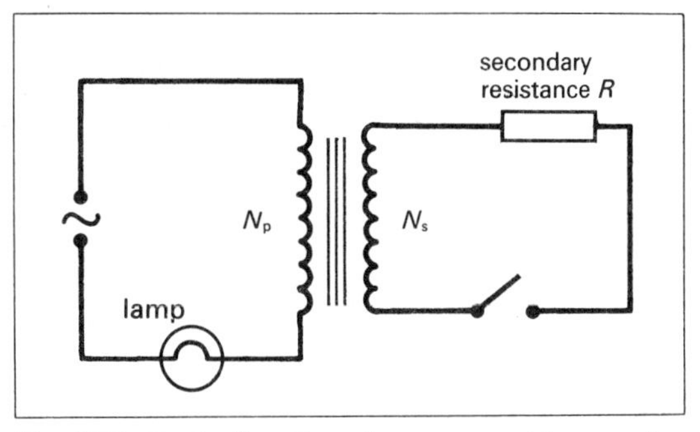

Fig. 56.3. Controlling the primary current by use of the secondary resistance.

When in fig. 56.3

(*i*) the switch is open, the lamp remains unlit,

(*ii*) the switch is closed, the resistance in the secondary *increases* the current through the *primary*, and this causes the lamp to shine.

56.3 Absolute Measurement of Resistance (Lorenz)

This method constitutes a determination of resistance without the need to make any *electrical* measurements. Because no knowledge of V or I is required, this is an **absolute** method.

Principle

Fig. 56.4 shows that there exists an e.m.f. (and hence a p.d.) between O and C when the copper disc is rotated in the magnetic field.

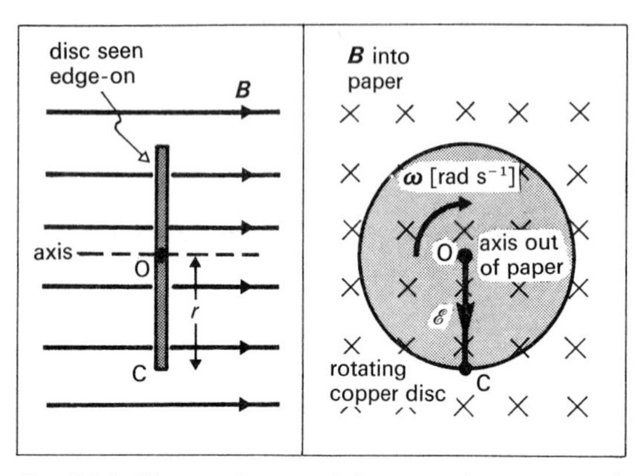

Fig. 56.4. There exists a p.d. between the centre and rim of the rotating disc.

Experiment

In fig. 56.5 the induced e.m.f. (which at balance equals the p.d. between the sliding contacts) is changed by altering the angular speed ω of the disc until G shows no deflection. Inside the long solenoid

$$B = \mu_0 nI$$

Any radius of the disc cuts flux at the rate BA in each revolution

$$\therefore \quad \frac{d\Phi}{dt} = \left(\frac{\omega}{2\pi}\right) BA$$

Thus the p.d. across the brushes

$$V_{oc} = \left(\frac{\omega}{2\pi}\right)(\mu_0 nI)(\pi r^2)$$

When ω is correctly adjusted,

$$V_{oc} = IR$$

Equating

$$R = \left(\frac{\mu_0}{2}\right)\omega n r^2$$

since *the current has cancelled.*

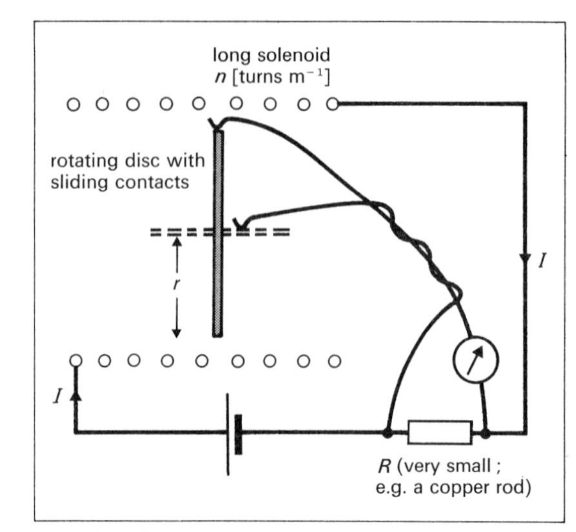

Fig. 56.5. The Lorenz experiment.

We need to measure r^2[m^2], n[m^{-1}] and ω[rad s^{-1}] so all readings are in terms of *mechanical* quantities.

Precautions

The induced e.m.f. is very small, so

(*a*) the *Earth's* field must be taken into account,
(*b*) contact p.d.'s must be avoided,
(*c*) allowance must be made for thermo-electric e.m.f.'s generated at the sliding contacts by frictional heating, and
(*d*) a very sensitive galvanoscope must be used.

*56.4 Absolute Measurement of Resistance (Belham)

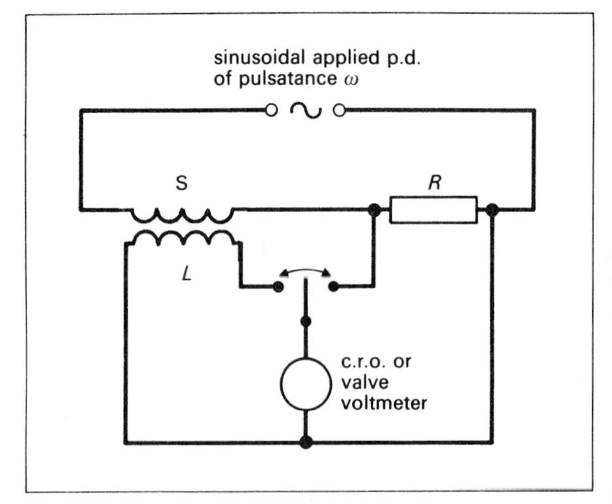

Fig. 56.6 The circuit of Belham's experiment.

Refer to fig. 56.6. Suppose a current $I_0 \sin \omega t$ flows through the long solenoid S and the resistor R. Let $\mathbf{A}$ be the solenoid's area of cross-section, n its number of turns per unit length, and N the number of turns on the search coil L. Inside the solenoid

$$B = \mu_0 n I_0 \sin \omega t$$

$$\frac{dB}{dt} = \mu_0 n I_0 \omega \cos \omega t$$

The maximum e.m.f. induced across L

$$\mathscr{E}_{max} = NA\mu_0 n I_0 \omega$$

and is proportional to ω. If R is non-inductive the maximum p.d. across it is

$$V_{max} = I_0 R$$

and is independent of ω. The experiment consists of adjusting ω until the detector shows that $\mathscr{E}_{max} = V_{max}$.

Then $\quad I_0 R = NA\mu_0 n I_0 \omega$

giving $\quad R = NA\mu_0 n \omega$

Notes.

(*a*) The method is absolute because the current amplitude cancels, and so need not be known.

(*b*) This is a more accurate experiment in the school laboratory than the *Lorenz* method, since the e.m.f. induced across L can easily be made as much as 1 volt. This enables R to be about 1 ohm, as compared to a few $\mu\Omega$ in the *Lorenz* method.

56.5 The Dynamo

The theory of the dynamo is given on p. 398. The induced e.m.f. causes a current to flow in the external circuit to which the dynamo is connected. This current is led away by **slip-rings** which rotate with the coil.

In the *d.c. dynamo* the current is made unidirectional by using a **split-ring commutator**: the output is then a unidirectional pulsating d.c., as shown in fig. 58.1 on page 411.

Dynamos can be

(*a*) **shunt-wound**, in which case the **field coils** (which energize the electromagnet) are in parallel with the **armature**, or

(*b*) **series-wound**, in which case they are connected in series with the armature.

In both types of winding the output p.d. depends on the current taken from the dynamo (the **load**). It can be made nearly constant if the dynamo is **compound-wound**.

The methods of winding are illustrated in fig. 56.7.

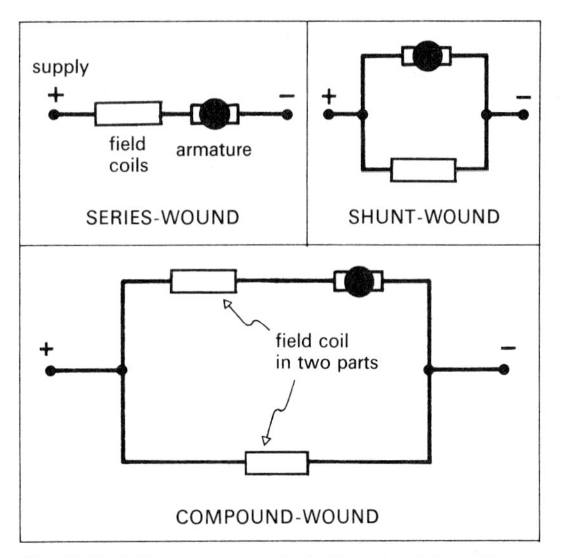

Fig. 56.7. Different ways of winding the field-coils and armature.

56.6 The Motor and Back e.m.f.

The motor is a device which converts electrical energy to mechanical energy. The principle is the same as that of the moving-coil galvanometer (p. 378), with the important difference that the rotating part (**armature**) consists of

(*a*) several coils: these experience torques whose average resultant value is steady, and

(*b*) a laminated soft-iron core, which provides both inertia (to maintain a steady angular speed), and a *radial field* (to provide a steady torque).

Because the armature is far more massive, it is supported on more robust bearings than those of the galvanometer.

D.C. Motor

In the **d.c. motor** the current sense is reversed automatically twice in each rotation by the split-ring commutator. This makes the sense of the torque constant.

Methods of winding the field-coils and the armature are given in fig. 56.7.

The characteristics of the different windings are given in fig. 56.8, and can be explained in terms of the back e.m.f.

The *series-wound* motor gives a large torque at slow speeds, and is used for pulling heavy loads (e.g. trains).

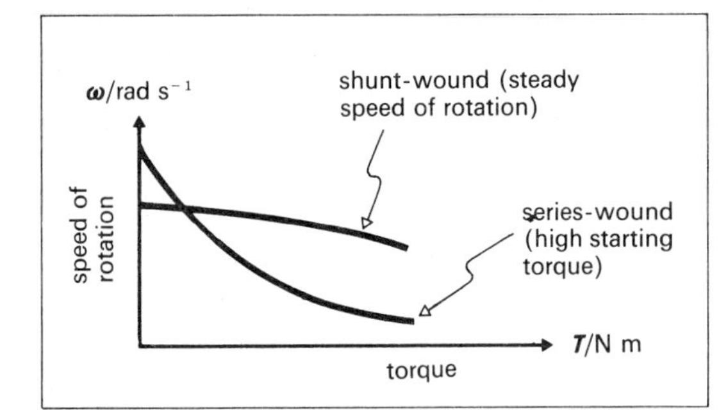

Fig. 56.8. Motor characteristics.

The *shunt-wound* motor has a steady current in the field coil, and so turns at a steady speed: it is used in situations where greater control is required (e.g. lathes and record-players).

In practice most motors are *compound-wound*, and have some of the advantages of both types.

A.C. Motor

In the **a.c. motor** one tries to reverse the sense of current flow in both the armature and the field coil at the same instant, so as to maintain a torque in the same sense.

(*i*) In the *series-wound* motor the electromagnet changes polarity in phase with the change of current in the armature, as required.

(*ii*) In the *shunt-wound* motor the large self-inductance (p. 404) of the field coil delays the establishment of the magnetic field until after the armature current has reached its maximum, and so such a motor cannot be used.

Back e.m.f. in the Motor

An e.m.f. $\mathscr{E}_B$ is induced in the rotating coil which tends to oppose the rotation of the coil.

Suppose V = p.d. applied to the coil
R = coil resistance
I = current through the coil

The loop equation (p. 362) gives

$$V - \mathscr{E}_B = IR$$

$$\therefore \quad VI = I^2R + \mathscr{E}_B I$$

(power supplied) = (power dissipated by Joule heating) + (rate of working of motor)

Suppose the coil has N turns, each of area $\boldsymbol{A}$, and that it rotates with an angular speed ω in a field $\boldsymbol{B}$. The torque acting on the coil is given by

$$T = NIAB \qquad \text{(p. 379)}$$

$$= NAB\left(\frac{V - \mathscr{E}_B}{R}\right)$$

If there were no load and no friction, the motor would accelerate until the resultant torque became zero. This would occur when $V = \mathscr{E}_B$, where $\mathscr{E}_B = N\omega AB$ (p. 398).

The Need for a Starting Resistance

When the motor is first switched on, $\omega = 0$, so $I_s = V/R$. This means that I_s is large if R is small.

When the motor is running, $\mathscr{E}_B$ increases, and so I decreases to its working value.

To prevent the armature burning out under a higher starting current, it is placed in series with a rheostat, whose resistance is decreased as the motor gathers speed.

56.7 The Alternator

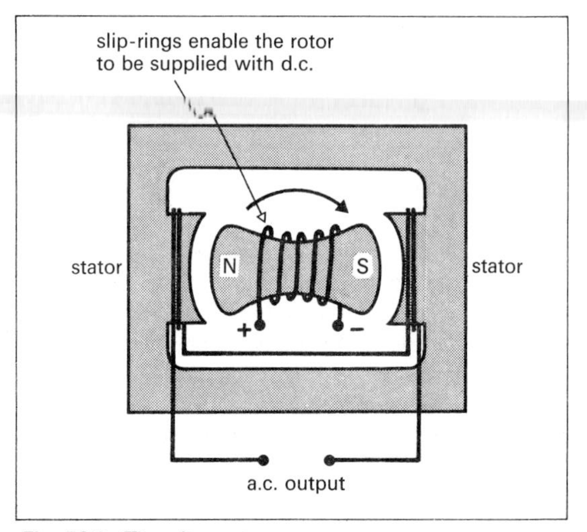

Fig. 56.9. The alternator.

The output current of the usual a.c. generator is carried by the slip-rings and brushes. In the **alternator** of fig. 56.9 the slip-rings carry only the relatively small current needed to energize the electromagnet.

This is achieved by rotating the magnet (the **rotor**) relative to the coil in which the current is induced (the **stator**).

57 Inductance

57.1 Definitions of Self-Inductance

Fig. 57.1 shows a single coil in which the current is being made to increase.

The flux threading the coil increases, and the *change of flux* induces an e.m.f. *in the coil itself*.

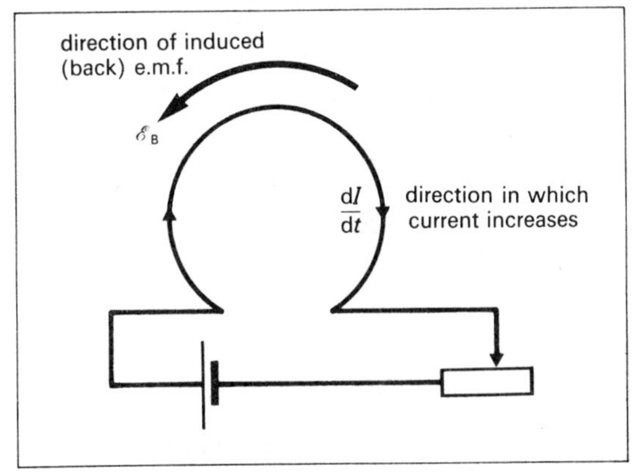

Fig. 57.1. Self-Inductance

By *Lenz's Law* the e.m.f. opposes the change: it tends to drive a current *against* the current that is *increasing*. If a current is *decreasing*, the induced e.m.f. tends to drive a current *in the same direction* as the original current to preserve the status quo.

For quantitative work we suppose no ferromagnetic materials are present. Then it follows from the *Biot–Savart Law* (p. 381) that $\Phi \propto I$.

We define the self-inductance L in two equivalent ways:

(a) Definition used to find L

Suppose a current I in a coil of N turns causes a flux Φ to thread each turn.

The self-inductance L is defined by the equation

$$\boxed{N\Phi = LI}$$

$$N\Phi[\text{Wb}] = L\left[\frac{\text{Wb}}{\text{A}}\right]I[\text{A}]$$

L is numerically equal to the flux-linkage of a circuit when unit current flows through it.

(b) Definition that describes the behaviour of an inductor in a circuit.

Using *Neumann's equation* on (p. 396), we can calculate the self-induced e.m.f. $\mathscr{E}$ from

$$\mathscr{E} = -\frac{d}{dt}N\Phi$$
$$= -L\frac{dI}{dt}$$

The self-inductance L may also be defined by the equation

$$\boxed{\mathscr{E} = -L\frac{dI}{dt}}$$

$$\mathscr{E}[\text{V}] = -L\left[\frac{\text{V s}}{\text{A}}\right]\frac{dI}{dt}\left[\frac{\text{A}}{\text{s}}\right]$$

L is numerically equal to the e.m.f. induced in the circuit when the current changes at the rate of 1 ampere in each second.

Notes.

(*a*) The units Wb A^{-1} and V s A^{-1} are equivalent. They occur so often that each is conveniently called a **henry** (H).

Thus 1 H = 1 V s A^{-1}. The mH and μH are useful subdivisions.

(*b*) The *permeability of free space* μ_0 is measured in Wb A^{-1} m^{-1}.

Since 1 H = 1 Wb A^{-1}, it follows that μ_0 may also be measured in H m^{-1}: this is the unit which is usually quoted.

(*c*) $[L] = [ML^2T^{-2}I^{-2}]$.

57.2 Calculation of Self-Inductance

Compare the defining equations

$$C = \frac{Q}{V} \quad \text{and} \quad L = \frac{N\Phi}{I}$$

In electrostatics we can calculate C if the geometry allows us to find V for a given Q. In electromagnetism we can find L if the geometry of a circuit allows us to find $N\Phi$ for a given I.

When a current flows in a *toroid*, the magnetic flux is entirely confined within the windings: outside $\boldsymbol{B}$ is self-cancelling. (Compare $\boldsymbol{E}$ inside and outside a parallel-plate capacitor.)

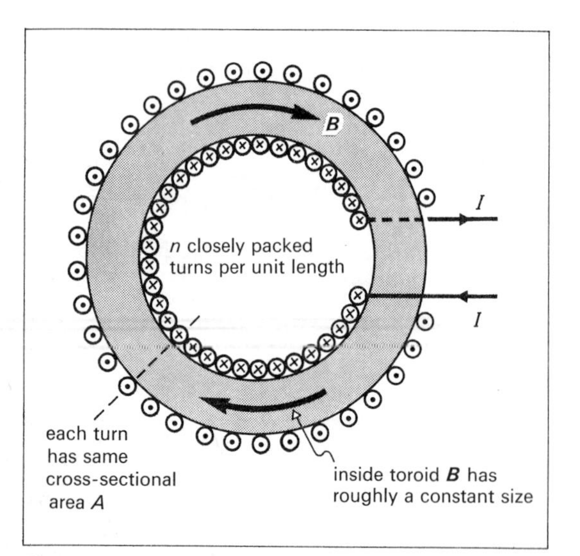

Fig. 57.2. Calculation of L for a toroid.

L for a Toroid

Consider a finite solenoid of length l(≫ the diameter) bent into the doughnut shape of fig. 57.2. The magnitude of $\boldsymbol{B}$ is nearly constant over the cross-section, and given by

$$B = \mu_0 nI \qquad \text{(p. 384)}$$

The flux-linkage $N\Phi = nlBA$.

So

$$L = \frac{N\Phi}{I} = \frac{nl(\mu_0 nI)A}{I} = \mu_0 n^2(Al) = \mu_0 n^2 V$$

where $V(= Al)$ is the volume of the toroid.

Notes

(*a*) The inductance of volume V of a long *straight* solenoid is also given by

$$L = \mu_0 n^2 V$$

near its centre.

(*b*) The inductance is $\propto n^2$. An increase in the number of turns per metre by a factor of k, increases

(*i*) the flux through *each* turn by k, and also

(*ii*) the number of turns to be linked by k.

The inductance then increases by a factor k^2. A coil of 1000 turns may have a relatively large self-inductance (~ mH) even when air-filled.

*57.3 The *LR* d.c. Circuit

(*a*) Growth of Current in the Inductor

When the switch of fig. 57.3 is connected across AC, the inductor provides an e.m.f. in the opposite direction to that of the battery. Using the loop equation (p. 362) clockwise.

$$\sum \mathscr{E} = \sum IR$$

$$\mathscr{E} - L \cdot \frac{dI}{dt} = IR$$

We can check by substitution that the differential equation has the solution

$$I = \frac{\mathscr{E}}{R}(1 - e^{-t/(L/R)})$$

Fig. 57.4 shows that the growth of current toward its final value $\mathscr{E}/R$ is exponential.

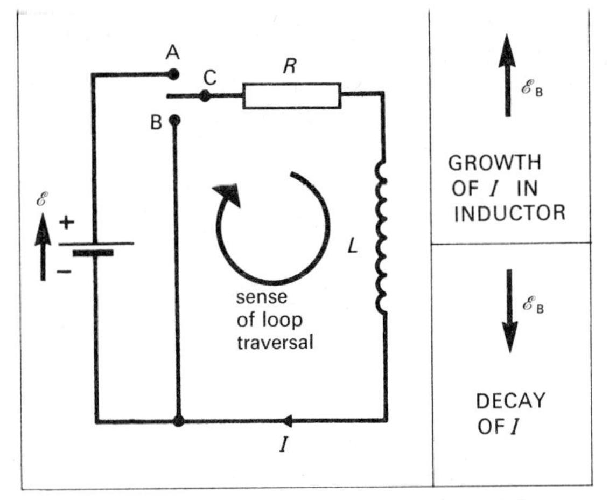

Fig. 57.3. Growth and decay of current in an LR circuit. The direction of the induced (back) e.m.f. is shown on the right.

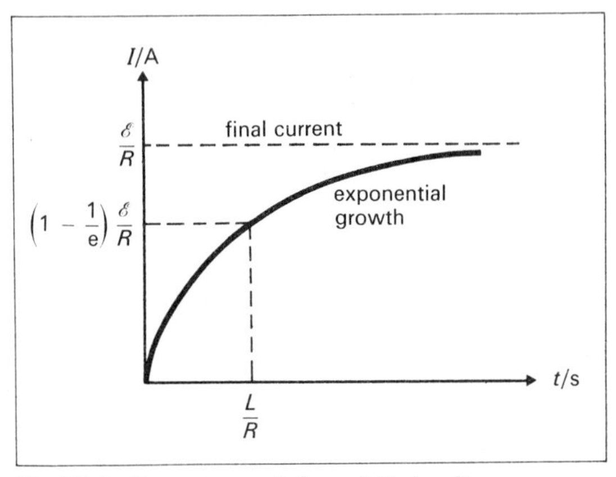

Fig. 57.4. Current growth in an LR circuit.

When

$$t = \left(\frac{L}{R}\right), \qquad I = \frac{\mathscr{E}}{R}\left(1 - \frac{1}{e}\right)$$

the current is smaller than its final value by 1/e or 37%. *(L/R) is called the* **inductive time constant** (τ_L).

$$\boxed{\tau_L = \frac{L}{R}}$$

$$\tau_L[\mathrm{s}] = \frac{L[\mathrm{V\,s/A}]}{R[\mathrm{V/A}]}$$

E.g. If $L = 2.0$ H and $R = 100\ \Omega$, then

$$\tau_L = \frac{L}{R} = 2.0 \times 10^{-2}\ \mathrm{s}$$

(b) Decay of Current

After a time $t(\gg L/R)$, connect the switch from AC to BC. The inductor now provides an e.m.f. in the same direction as was the battery's.

In the loop equation $\mathscr{E} = 0$, and $\mathrm{d}I/\mathrm{d}t$ is *negative*. (The back e.m.f. across the inductor reverses in direction.)

$$\therefore\ -L \cdot \frac{\mathrm{d}I}{\mathrm{d}t} = IR$$

This differential equation has the solution

$$I = \frac{\mathscr{E}}{R}\mathrm{e}^{-t/(L/R)}$$

as can be checked by substitution.

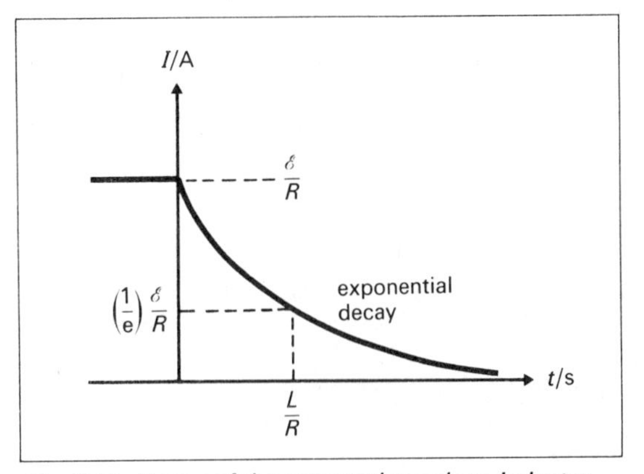

Fig. 57.5. Decay of the current through an inductor.

Fig. 57.5 shows that the current decays exponentially to zero.

When

$$t = \left(\frac{L}{R}\right) = \tau_L$$

the current has fallen to 1/e (or 37%) of its former value. This again is the inductive time constant, which has the same value for growth and decay.

* 57.4 The Energy Stored by an Inductor

An inductor carrying a current has energy associated with it. When the battery of a circuit is disconnected, the inductor becomes a seat of e.m.f. as it converts its stored magnetic energy to electrical energy.

We can write the loop equation

$$\Sigma\mathscr{E} = \Sigma V$$

as

$$\mathscr{E} = L\frac{\mathrm{d}I}{\mathrm{d}t} + IR,$$

in which $L(\mathrm{d}I/\mathrm{d}t)$ is regarded as a potential drop across the inductor.

$$\therefore \quad \mathscr{E}I = LI\frac{\mathrm{d}I}{\mathrm{d}t} + I^2R$$

input power from source	rate at which electrical energy is dissipated as internal energy

Using the law of conservation of energy we regard $LI(\mathrm{d}I/\mathrm{d}t)$ as *absorbed but non-dissipated power.* It represents the rate at which the inductor stores energy in its associated magnetic field.

When the current reaches its final value I_0 the energy W stored is given by

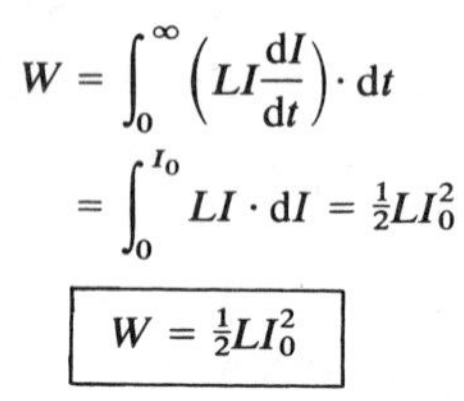

$$W = \int_0^{\infty}\left(LI\frac{\mathrm{d}I}{\mathrm{d}t}\right)\cdot \mathrm{d}t$$

$$= \int_0^{I_0} LI\cdot \mathrm{d}I = \tfrac{1}{2}LI_0^2$$

$$\boxed{W = \tfrac{1}{2}LI_0^2}$$

The energy storage in the magnetic field of an inductor is analogous to that in the electric field of a capacitor.

Compare $\quad W = \tfrac{1}{2}LI^2$

with $\quad W = \tfrac{1}{2}Q^2/C.$

* 57.5 Definitions of Mutual Inductance M

Fig. 57.6 shows a situation in which a current in the first circuit (the **primary**) causes a magnetic flux to thread the second circuit (the **secondary**). As on p. 405, we suppose no ferromagnetic materials are present.

Suppose a current I_p in the primary circuit causes a flux Φ_s to link each turn of the secondary circuit.

(1) The mutual inductance M is defined by the equation

$$\boxed{N_s\Phi_s = MI_p}$$

Using *Neumann's equation* (p. 396) we can calculate the e.m.f. $\mathscr{E}_s$, induced in the secondary by a changing current in the primary, from

$$\mathscr{E}_s = -\frac{\mathrm{d}}{\mathrm{d}t}(N_s\Phi_s)$$

$$= -M\frac{\mathrm{d}I_p}{\mathrm{d}t}$$

(2) The mutual inductance M may also be defined by the equation

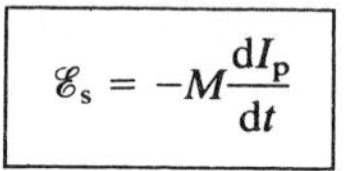

$$\boxed{\mathscr{E}_s = -M\frac{\mathrm{d}I_p}{\mathrm{d}t}}$$

Notes.

(*a*) It can be proved that the same value is obtained for M if one considers the flux threading the primary when a current flows in the secondary.

So
$$M = \frac{N_s\Phi_s}{I_p} = \frac{N_p\Phi_p}{I_s}$$

$$= -\frac{\mathscr{E}_s}{\mathrm{d}I_p/\mathrm{d}t} = -\frac{\mathscr{E}_p}{\mathrm{d}I_s/\mathrm{d}t}$$

(*b*) The two definitions for M, like those for L, are equivalent, and show that M is measured in the unit **henry**.

Expressed in words

(1) M is numerically equal to the flux-linkage in one circuit when unit current flows through the other.

We use this definition *to calculate M.*

(2) M is numerically equal to the e.m.f. induced in one circuit when the current changes in the other at the rate of 1 ampere in each second.

We use this definition *to describe the mutual behaviour* of two circuits.

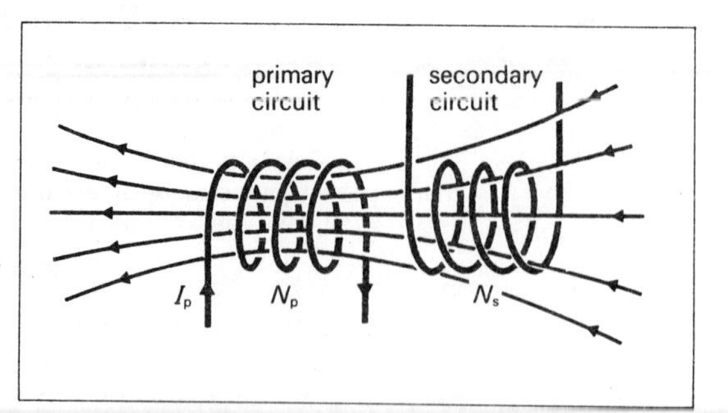

Fig. 57.6. Mutual induction.

*57.6 Calculation of Mutual Inductance

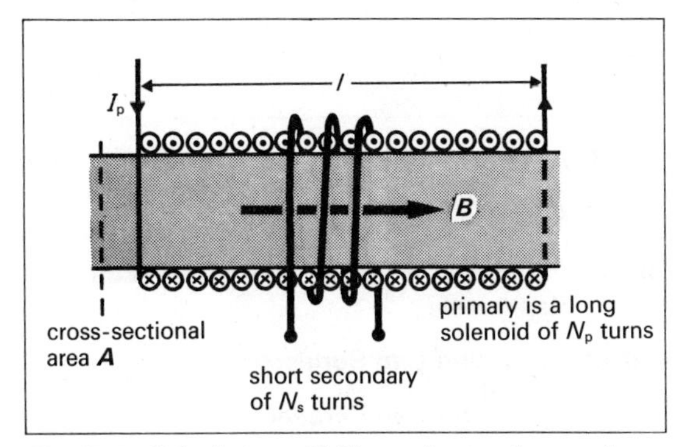

Fig. 57.7. Calculation of M for a short coil wound round a long solenoid.

The situation and symbols are defined in fig. 57.7.

The flux density $\boldsymbol{B}$ inside the long solenoid is given by

$$B = \mu_0\left(\frac{N_p}{l}\right)I_p$$

$$\text{The flux at the centre} = BA$$

$$= \mu_0 A\left(\frac{N_p}{l}\right)I_p$$

$$= \Phi_s$$

since *all* this flux links each turn of the secondary.

From the definition

$$M = \frac{N_s\Phi_s}{I_p}$$

$$= \frac{N_s[\mu_0 A(N_p/l)I_p]}{I_p}$$

$$= \frac{\mu_0 A N_p N_s}{l}$$

For an air-cored coil it is likely that $\Phi_s < BA$ because of magnetic flux leakage.

The idea of mutual inductance is used in the *transformer* (p. 401).

58 Alternating Current

58.1 Alternating Current and e.m.f.

An *alternating* current or e.m.f. is one whose magnitude and direction vary periodically with time.

The simplest a.c. is one which varies sinusoidally (in a simple harmonic manner) with time. It is of great importance

(*a*) because it can be produced by rotating a coil in a *uniform* magnetic field (p. 397), and

(*b*) because *any* periodic variation is equivalent to a combination of sinusoidal variations of appropriate frequencies, amplitudes and phase differences.

The *frequency* f of alternation, *angular frequency* (**pulsatance**) ω and *period* T are related by

$$\omega = 2\pi f = \frac{2\pi}{T}$$

Typical values for f are:

British mains	50 Hz ($50\ s^{-1}$)
Radio frequencies	10^5 to 2×10^{10} Hz
(Visible light	$\sim 10^{15}$ Hz)

Symbols

Throughout this chapter, this convention is adopted:

	Current	*e.m.f.*
instantaneous values	I	$\mathscr{E}$
root mean square (r.m.s.) values	$I_{\text{r.m.s.}}$	$\mathscr{E}_{\text{r.m.s.}}$
peak values	I_0	$\mathscr{E}_0$

Phase, Phase Difference, Lag and Lead

If $I = 0$ when $t = 0$, we write

$$I_1 = I_0 \sin \omega t$$

A second current I_2 may be given by

$$I_2 = I_0 \sin(\omega t + \phi)$$

The two currents have the same pulsatance ω, and the same amplitude I_0 (**peak** or maximum value), but differ in **phase**.

$$\left.\begin{array}{l} I_1 \text{ has a phase } \omega t \\ I_2 \text{ has a phase } (\omega t + \phi) \end{array}\right\}$$

The two currents are **out of phase**: their **phase difference** (or **shift**) is ϕ[rad], which corresponds to a time interval of ϕ/ω [seconds]. (See p. 98.)

I_2 **leads** I_1 by the angle ϕ,
I_1 **lags** I_2 by the angle ϕ.

Two *different* alternating quantities (such as p.d. and current) may also show a phase difference when they vary with the same frequency. Examples appear on p. 415, p. 416 and p. 418.

58.2 Average Value of a Sinusoidally Varying Current

By definition the *average value* of any periodic function f over its period T is given by

$$f_{av} = \frac{\int_0^T f \cdot dt}{T} \qquad \text{(p. 23)}$$

Consider $$I = I_0 \sin \omega t$$

(a) Whole cycle

By the symmetry of the sine curve, the average current $\langle I \rangle = 0$ over a whole number of cycles.

(b) Half cycle

$$\langle I \rangle = \frac{1}{T/2}\int_0^{T/2} I_0 \sin \omega t \cdot dt \qquad T = \frac{2\pi}{\omega}$$

$$= \frac{1}{\pi/\omega}\int_0^{\pi/\omega} I_0 \sin \omega t \cdot dt \qquad \frac{T}{2} = \frac{\pi}{\omega}$$

$$= \frac{I_0}{\pi/\omega}\left[-\frac{\cos \omega t}{\omega}\right]_0^{\pi/\omega} = \frac{I_0}{\pi/2}$$

Example. The pulsating d.c. of fig. 58.1 has an average value of $I_0/(\pi/2)$ over a whole number of half cycles.

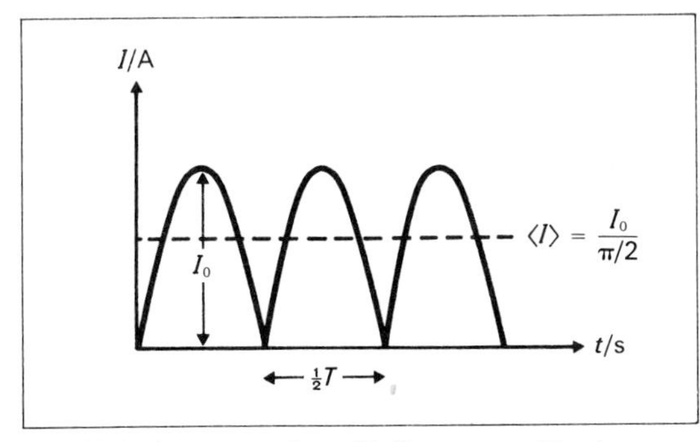

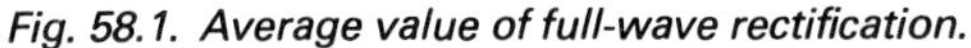

Fig. 58.1. Average value of full-wave rectification.

58.3 The Effective Value of Sinusoidally Varying Current

The **effective value** of an a.c. is that steady current which would convert electrical energy to other forms at the same rate as the a.c. (It is sometimes called the **virtual current**.)

The *instantaneous* rate of conversion

$$I^2R = I_0^2R \cdot \sin^2 \omega t$$

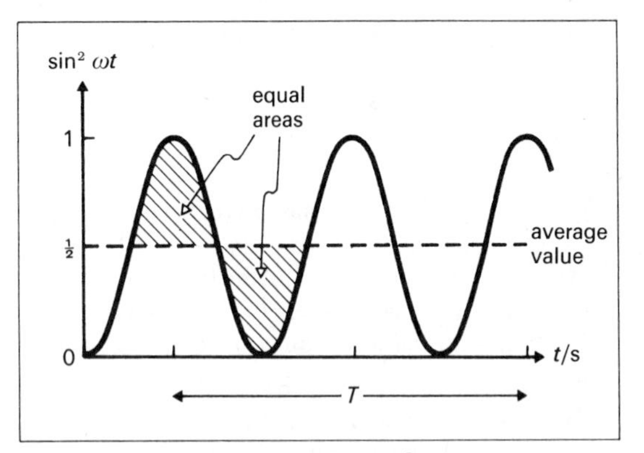

Fig. 58.2. Average value of the $\sin^2 \omega t$ curve.

The *average* rate of conversion

$$\langle I_0^2R \cdot \sin^2 \omega t\rangle = I_0^2R \cdot \langle \sin^2 \omega t \rangle$$
$$= \tfrac{1}{2}I_0^2R$$
$$= \left(\frac{I_0}{\sqrt{2}}\right)^2 R$$

Thus when the resistance is purely ohmic the effective value of current

$$= \sqrt{(\text{mean value of } I^2)}$$

and is called the **root mean square** current.

$$\boxed{I_{r.m.s.} = \frac{I_0}{\sqrt{2}}}$$

Example. For the British a.c. mains, we have

$$\mathscr{E} = \mathscr{E}_0 \sin \omega t,$$

in which $\mathscr{E}_{r.m.s.} = 240$ V

so

$$\mathscr{E}_0 = 240\sqrt{2}\ \text{V}$$
$$= 339\ \text{V}$$

$\mathscr{E}_{r.m.s.}$ is usually quoted.

58.4 Methods of Measuring Alternating Quantities

(a) Suppose an Instrument Gives a Deflection $\propto I$

Then if it has been calibrated by d.c., and a.c. is passed through it, it may read

(1) zero (e.g. an unadapted moving-coil galvanometer),
(2) $\frac{1}{2}I_0/(\pi/2)$, if half-wave rectification is used,
(3) $I_0/(\pi/2)$, if full-wave rectification is used (such as that in the bridge circuit of fig. 54.2).

(b) Suppose the Instrument Gives a Deflection $\propto I^2$

Then it may read

(4) $I_{\text{r.m.s.}} = I_0/\sqrt{2}$ if it has high inertia, and the frequency is high,
(5) I_0 if it has low inertia, and the frequency is low,
(6) a varying value between (4) and (5) if it tries to follow the variation, but is unable to keep up.

As a general rule, meters designed for a.c. use will read *root mean square values*. Examples include the current balance, electrodynamometer, hot-wire and thermocouple ammeter, and moving-iron instruments.

The diode valve voltmeter and c.r.o. both record *peak values*.

58.5 The Effects of Alternating Current

An alternating current differs from a direct current only in the *varying* magnitude and direction of the drift velocity imposed on the charge-carriers' random movement. The effects of a.c. are fundamentally the same as those of d.c., but this may not be apparent.

(a) Magnetic Effect

The a.c. gives a time-varying value of $\boldsymbol{B}$, which

(*i*) causes torques and forces of varying size to be exerted on coils and magnetic dipoles: when $\omega = 2\pi \times 50 \text{ rad s}^{-1}$, they may not respond significantly because of their inertia.

(*ii*) is part of an electromagnetic wave (the charge-carriers are accelerated).

(b) Chemical Effect

The observed effect depends on the frequency. At 50 Hz

(*i*) dilute $H_2SO_4 \rightarrow (2H_2 + O_2)$, a gaseous mixture at each platinum electrode, but

(*ii*) $CuSO_4$ gives zero *net* effect with Cu electrodes.

At higher frequencies the fact that an ion has a much greater mass and therefore smaller mobility than an electron may prevent sufficient movement to show any effect.

An a.c. can give an electric shock, i.e. does produce a *physiological effect.*

(c) Heating Effect

Joule heating is not dependent on the direction of the current. A filament lamp may show 100 brightnesses per second, since the *power* has twice the frequency of the current. The heat capacity of the filament controls the variation of brightness, but the filament may give a *stroboscopic effect.* (A discharge lamp has quicker response.)

The Thermocouple Ammeter

This uses the heating effect of the current, and is more satisfactory than the hot-wire ammeter.

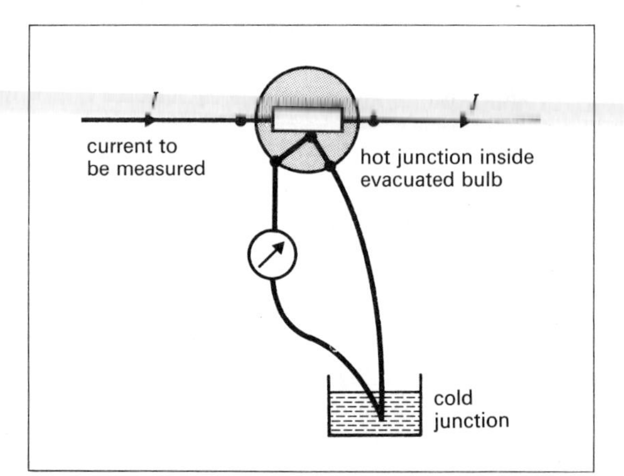

Fig. 58.3. A thermocouple ammeter.

The thermocouple (fig. 58.3) has the *net* effect of converting the a.c. to a d.c. which is measured by a sensitive moving-coil galvanometer.

58.6 Impedance

For d.c. we define *resistance* R by $R = V/I$. Suppose that for an a.c. the maximum value of the applied p.d.

across a device is V_0, and that the maximum current through the device is I_0.

The impedance Z of the device is defined by the equation

$$Z = \frac{V_0}{I_0} = \frac{V_{r.m.s.}}{I_{r.m.s.}}$$

$$Z[\Omega] = \frac{V_{r.m.s.}[V]}{I_{r.m.s.}[A]}$$

Notes.

(*a*) Z plays the same role in an a.c. circuit as R does in a d.c. circuit: it is the property of a component or a circuit that opposes the passage of a.c., and (together with V_0) determines its maximum value.

(*b*) Because V_0 and I_0 may occur at different instants of time, it is *not* true (in general) to say $Z = V/I$, where V and I are instantaneous values.

(*c*) The value of Z depends upon
 (*i*) the frequency of the a.c., and
 (*ii*) the values of R, L and C.

(*d*) $Z \geqslant R$ in series circuits.

Demonstration of Impedance

In the circuit of fig. 58.4, $V_{r.m.s.}$ is kept constant and $I_{r.m.s.}$ noted by a milliammeter.

Action	*Qualitative observation*	*Quantitative deduction*
increase ω	for R, $I_{r.m.s.}$ = constant	R does not depend on ω(†)
increase ω	for L, $I_{r.m.s.}$ decreases	$Z \propto \omega$
increase ω	for C, $I_{r.m.s.}$ increases	$Z \propto \frac{1}{\omega}$

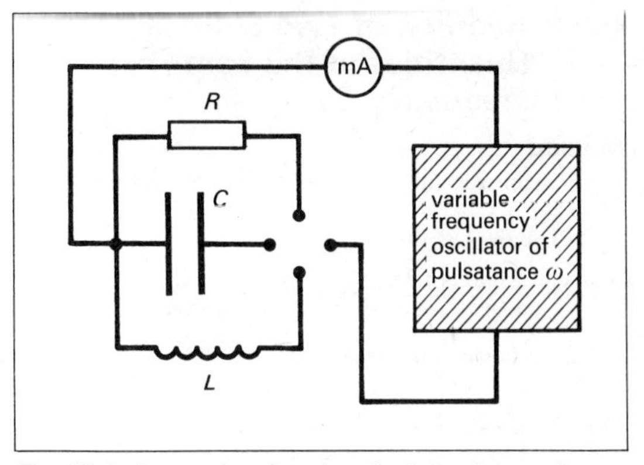

Fig. 58.4. Demonstration that the impedance of a device depends upon ω.

(† In practice R may be found to depend on ω because of the **skin effect** (when an a.c. is largely confined to surface layers), and because of various dissipative effects in inductive coils.)

The deductions are derived theoretically in this chapter.

58.7 Representation of a Sinusoidal Quantity by Rotating Vector

Suppose we have a vector of magnitude V_0 rotating in an anticlockwise sense at constant angular speed ω (fig. 58.5). It is referred to as a **phasor**.

The y-component of the vector at any instant has magnitude $V = V_0 \sin \omega t$.

Thus we can represent a sinusoidal quantity by the projection on to a fixed line of a rotating vector.

By convention the *anticlockwise* sense of rotation is taken to be *positive*.

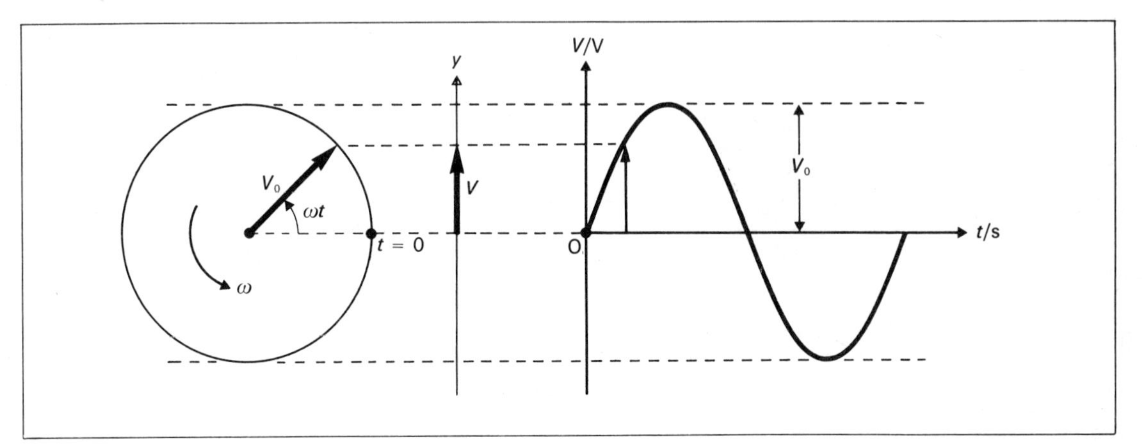

Fig. 58.5. $V = V_0 \sin \omega t$ represented by a rotating vector.

*58.8 Addition of Two Sinusoidal Quantities of the Same Frequency

Method

(1) Represent the two quantities by rotating vectors.

(2) Use the fact that

$$\begin{pmatrix}\text{projection of vector}\\ \text{sum of two vectors}\end{pmatrix} = \begin{pmatrix}\text{sum of projections}\\ \text{of two vectors}\end{pmatrix}$$

This is shown in fig. 58.6.

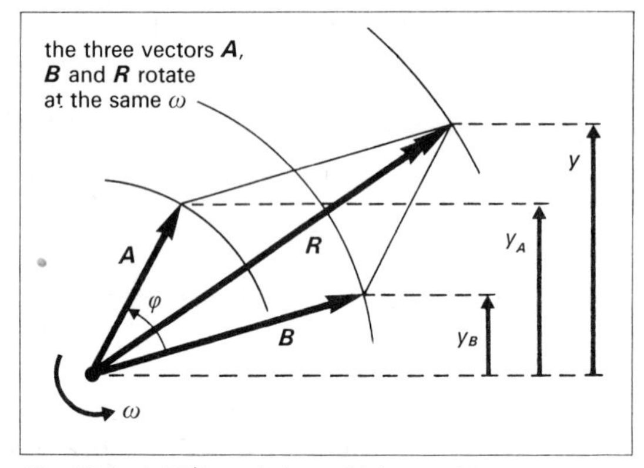

Fig. 58.6. Addition of sinusoidal quantities.

y is the projection of $\boldsymbol{R}$, where

(*i*) $\boldsymbol{R}$ = vector sum of $\boldsymbol{A}$ and $\boldsymbol{B}$, and

(*ii*) $y = y_A + y_B$

$$= A \sin \omega t + B \sin (\omega t - \phi).$$

The **phase lag** ϕ *of* y_B behind y_A corresponds to a **time lag** Δt, where

$$\frac{\Delta t}{T} = \frac{\phi}{2\pi}$$

so

$$\Delta t = \left(\frac{\phi}{2\pi}\right) \cdot T = \frac{\phi}{\omega}$$

Δt is *fixed* because $\boldsymbol{A}$ and $\boldsymbol{B}$ have the same angular frequency ω.

Notes

(*a*) The resultant y has the same frequency as y_A and y_B, but reaches its maximum at an instant different from either y_A or y_B (i.e. shows a *phase shift*).

(*b*) In general the *magnitude* of $\boldsymbol{R}$ is *not* equal to the sum of the magnitudes of $\boldsymbol{A}$ and $\boldsymbol{B}$.

(*c*) We apply these ideas to add p.d.'s across components in a.c. circuits. These p.d.'s have the *same frequency* but are *out of phase*.

Important Note

In paragraphs **58.9**, **58.10**, **58.11** *and* **58.12**, *the current is described by*

$$I = I_0 \sin \omega t$$

The phase of the applied p.d. is chosen appropriately to make this so.

58.9 A Purely Resistive Circuit

(a) Phase Relation for I and V

Let applied p.d. $V = V_0 \sin \omega t$.

Using the definition

$$R = \frac{V}{I}$$

we have

$$I = \frac{V}{R} = \frac{V_0}{R} \sin \omega t$$

$$= I_0 \sin \omega t$$

Fig. 58.7 shows

(*i*) I and V have the same frequency,

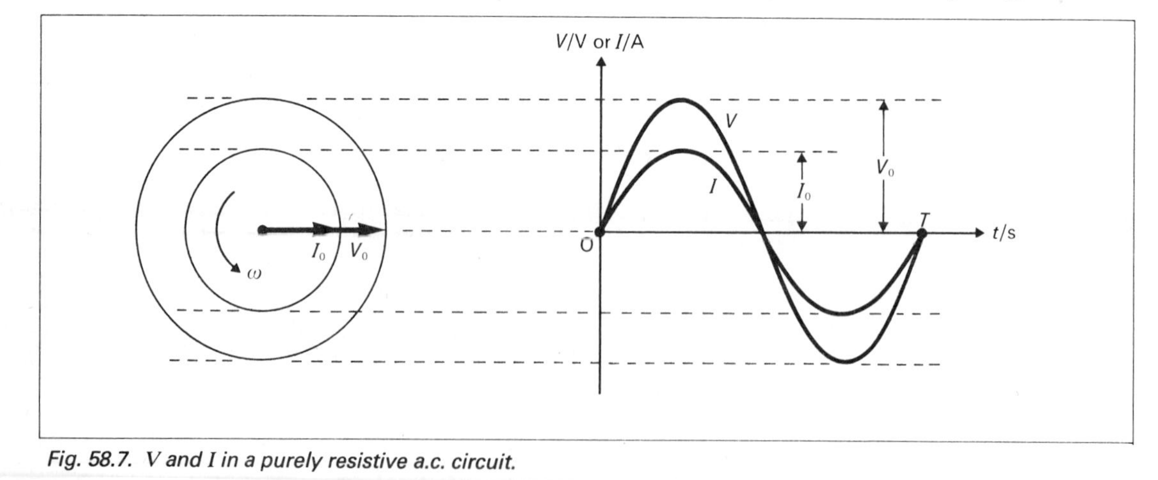

Fig. 58.7. V and I in a purely resistive a.c. circuit.

(*ii*) I and V are *in phase.*

$$R = \frac{V_0}{I_0}$$

(*b*) Power

Instantaneous power dissipation

$$P = VI$$
$$= (V_0 \sin \omega t) \times (I_0 \sin \omega t)$$

The power oscillates at an angular frequency 2ω, since

$$\sin^2 \omega t = \tfrac{1}{2}(1 - \cos 2\omega t)$$

Average power dissipation

$$\langle P \rangle = V_{\text{r.m.s.}} I_{\text{r.m.s.}}$$

which is always positive (fig. 58.2).

The resistor causes a *non-reversible* conversion of electrical energy to thermal energy.

58.10 A Purely Inductive Circuit

(*a*) Reactance and Phase Relation for I and V

Since we have assumed $R = 0$ in fig. 58.8, if we apply the loop equation $\sum \mathscr{E} = \sum IR$ in a clockwise sense,

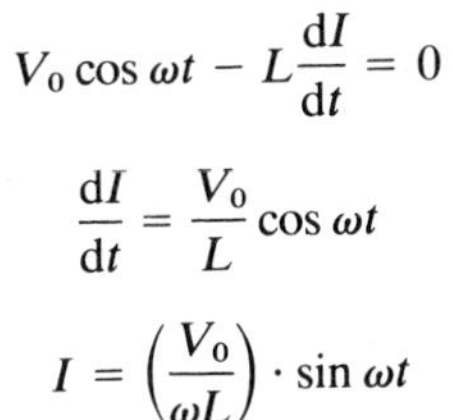

$$V_0 \cos \omega t - L\frac{dI}{dt} = 0$$

$$\frac{dI}{dt} = \frac{V_0}{L} \cos \omega t$$

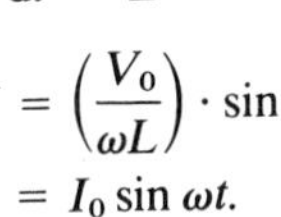

Integrating $$I = \left(\frac{V_0}{\omega L}\right) \cdot \sin \omega t$$
$$= I_0 \sin \omega t.$$

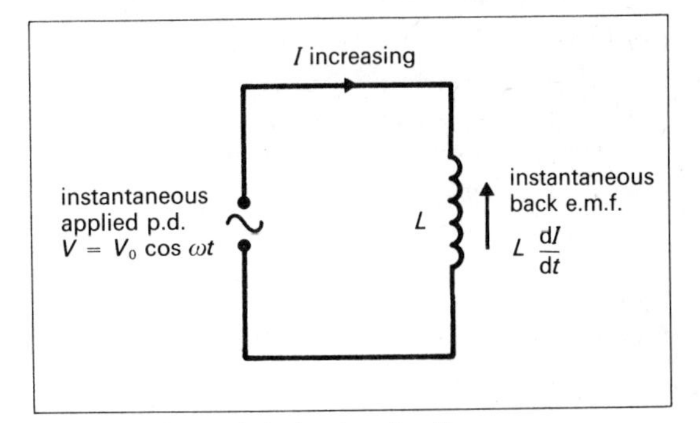

Fig. 58.8. A purely inductive circuit.

Fig. 58.9 shows

(*i*) I and V have the same frequency

(*ii*) the current lags behind the applied p.d. with a *phase lag* of $\pi/2$ rad: the current reaches its maximum value $T/4$ later in time.

We define the* inductive reactance X_L *by the equation

$$X_L = \frac{V_0}{I_0} = \frac{V_{\text{r.m.s.}}}{I_{\text{r.m.s.}}}$$

Since (above) $$I_0 = \frac{V_0}{\omega L}$$

$$\boxed{X_L = \omega L}$$

$$X_L[\Omega] = \omega\left[\frac{1}{\text{s}}\right] \cdot L\left[\frac{\text{V}}{\text{A/s}}\right]$$

ωL controls the current magnitude in the same way as

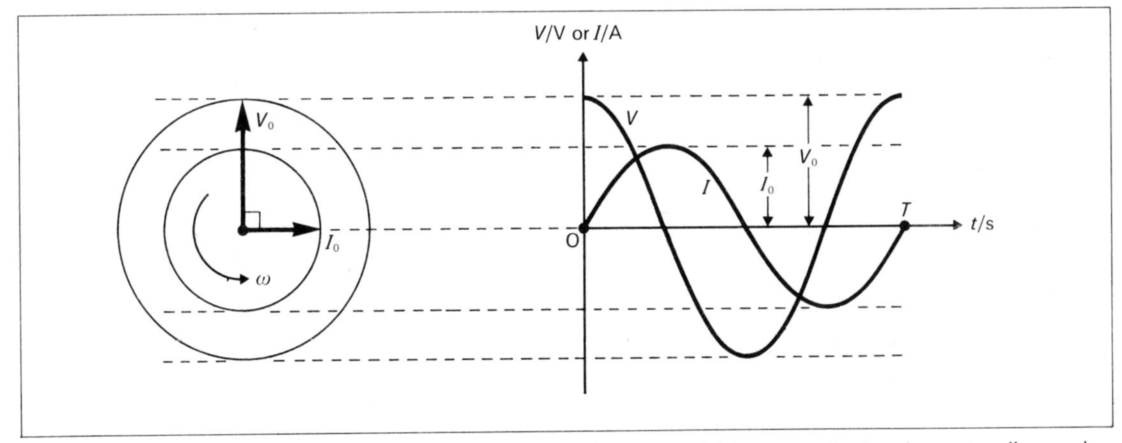

Fig. 58.9. V and I in an inductive circuit. (The symbol ↳ for L is helpful for remembering the vector diagram.)

resistance does. The inductive reactance is increased by

(*i*) increasing ω, and
(*ii*) increasing L.

See the *choke effect* (p. 426).

(*b*) Power

Instantaneous power conversion

$$P = VI$$
$$= (V_0 \cos \omega t) \times (I_0 \sin \omega t)$$

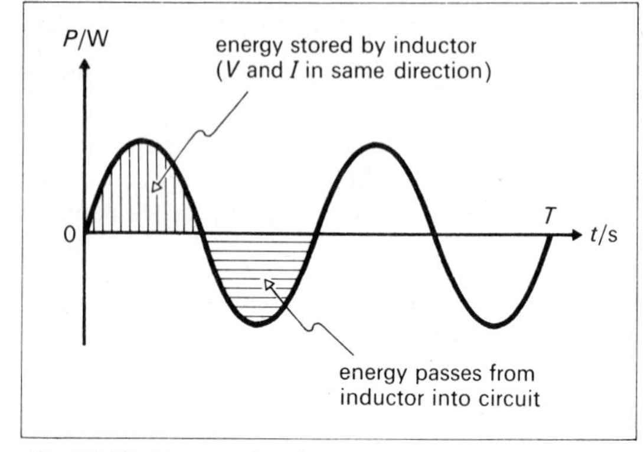

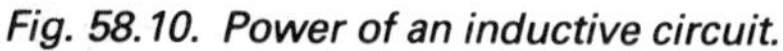
Fig. 58.10. Power of an inductive circuit.

The power oscillates at an angular frequency 2ω, as shown in fig. 58.10.

The *average* power (rate of energy conversion) $\langle P \rangle$ over a cycle is shown to be *zero*. Over a whole number of cycles there is no *net* energy transfer from the seat of e.m.f. to the magnetic field associated with the inductor. (In practice energy is always dissipated because R is finite.)

58.11 A Purely Capacitative Circuit

(*a*) Reactance and Phase Relation for I and V

For a capacitor $V = Q/C$ defines C.

The instantaneous current

$$I = \frac{dQ}{dt} = C\frac{dV}{dt}$$

We choose the applied p.d. $V = -V_0 \cos \omega t$ (for the reason on p. 414),

so
$$\frac{dV}{dt} = \omega V_0 \sin \omega t$$

and
$$I = (\omega C)\, V_0 \sin \omega t$$
$$= I_0 \sin \omega t$$

Fig. 58.11 shows

(*i*) I and V have the same frequency

(*ii*) the current leads the applied p.d. with a *phase lead* of $\pi/2$ rad: the current reaches its maximum value $T/4$ earlier in time.

Note that there is no real current flow *through the capacitor*, but the oscillation of charge through the circuit continually changes the sign of the charge on the capacitor plates.

We define the capacitative reactance X_C by the equation

$$X_C = \frac{V_0}{I_0} = \frac{V_{\text{r.m.s.}}}{I_{\text{r.m.s.}}}$$

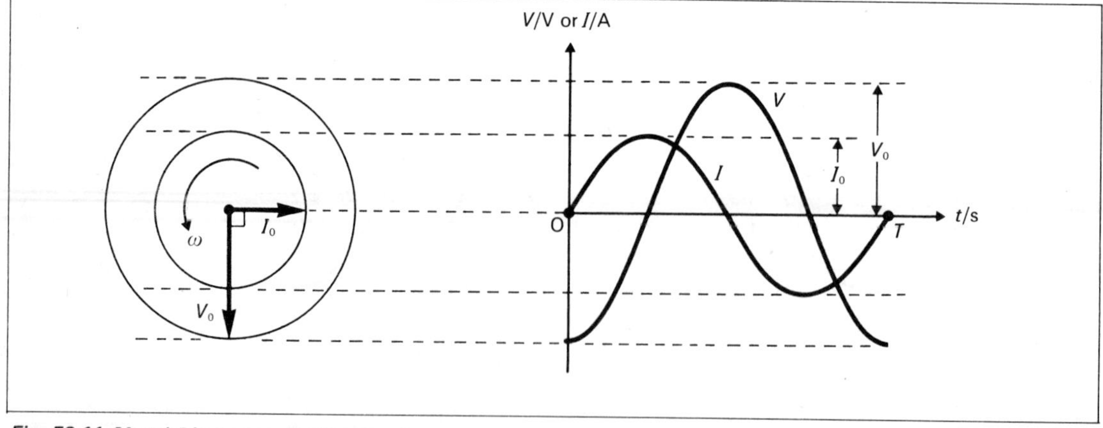

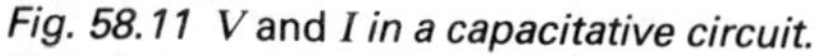
Fig. 58.11 V and I in a capacitative circuit.

Since (above)

$$I_0 = \frac{V_0}{(1/\omega C)}$$

$$\boxed{X_C = \frac{1}{\omega C}.}$$

$$X_C[\Omega] = \frac{1}{\omega[1/\text{s}]C[\text{A s/V}]}$$

The capacitative reactance is increased by

(*i*) reducing ω, and
(*ii*) reducing C.

(*b*) Power

Instantaneous power conversion

$$P = VI$$
$$= (-V_0 \cos \omega t) \times (I_0 \sin \omega t)$$

The power oscillates at an angular frequency 2ω, as shown in fig. 58.12.

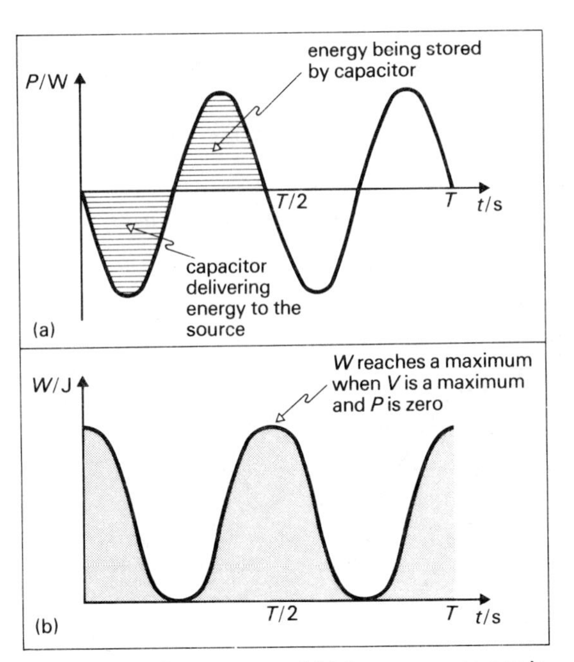

Fig. 58.12. (a) The power, and (b) the energy storage in a capacitative circuit.

The *average* power (rate of energy conversion) $\langle P \rangle$ over a cycle is zero, since the capacitor absorbs energy when the field between its plates is increasing, but returns it to the source as the field collapses. (In practice energy is always dissipated because R is finite.)

58.12 Summary of Circuit Elements taken Separately

Property of device	*Resistance or reactance X*	*V–I phase relation*	*Instantaneous energy storage*
R	R	in phase	nil
L	$X_L = \omega L$	V leads I by $\pi/2$ rad	$\frac{1}{2}LI^2$
C	$X_C = 1/(\omega C)$	V lags I by $\pi/2$ rad	$\frac{1}{2}CV^2$

This phase shift is independent of the frequency.

Fig. 58.13 shows the limiting values of ωL and $1/\omega C$ as $\omega \to 0$, and $\omega \to \infty$.

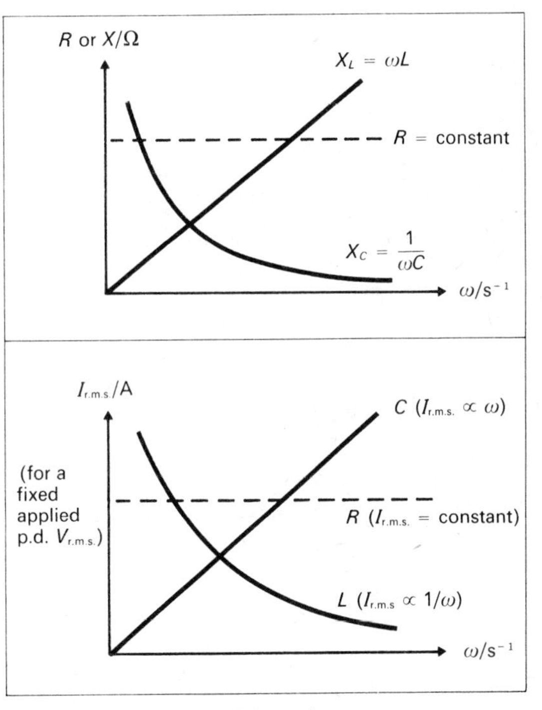

Fig. 58.13. The effect of the pulsatance ω on reactance and current.

*58.13 R, L and C in Series

Statement of the Problem

Refer to fig. 58.14. *Note* that $V = V_0 \sin \phi$ when $t = 0$.

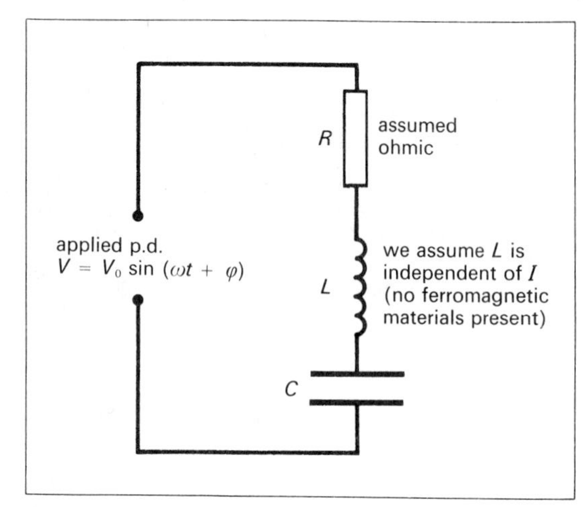

Fig. 58.14. The RLC series circuit.

We want to calculate

(1) ϕ (and Z) in terms of L, C and R (and ω).
(2) the net rate of transfer of energy from the source to the circuit (the power dissipation).

Principles

(*a*) The law of *charge conservation* tells us that the instantaneous value of I through R, L, C and the source must be the same.

(*b*) The law of *energy conservation* gives us the loop equation. *At any instant* $V = V_R + V_L + V_C$.

Analysis by Differential Equation

The loop equation (above) can be written

$$V_0 \sin(\omega t + \phi) = IR + L\frac{dI}{dt} + \frac{Q}{C}$$

$$= L\frac{d^2Q}{dt^2} + R\frac{dQ}{dt} + \frac{Q}{C}$$

since $\quad I = \dfrac{dQ}{dt}$.

The solution of this equation gives Q, and hence I, as a function of t. We will not adopt this method.

Solution by Rotating Vector Method

Since V_R is in phase with I, we can relate the phases of V_R, V_L and V_C as in fig. 58.15.

By choice $V_R = 0$ when $t = 0$. To evaluate

$$V = V_R + V_L + V_C$$

we can either

(*i*) add their instantaneous values, or
(*ii*) add $V_{R,0} + V_{L,0} + V_{C,0}$ by *vector* addition and calculate the projection of their resultant.

This is done in fig. 58.16.

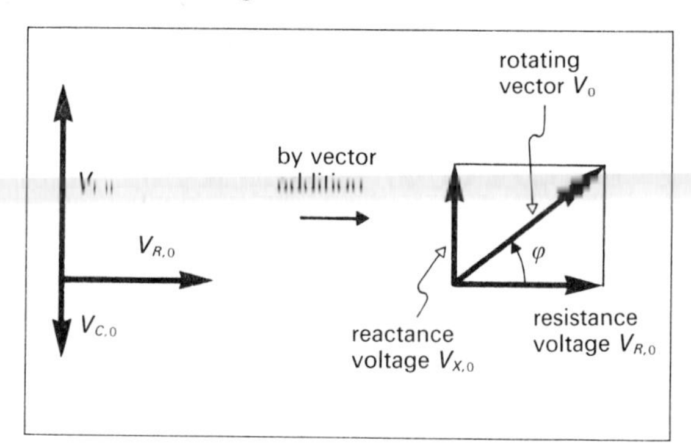

Fig. 58.16. Vector addition of maximum p.d's.

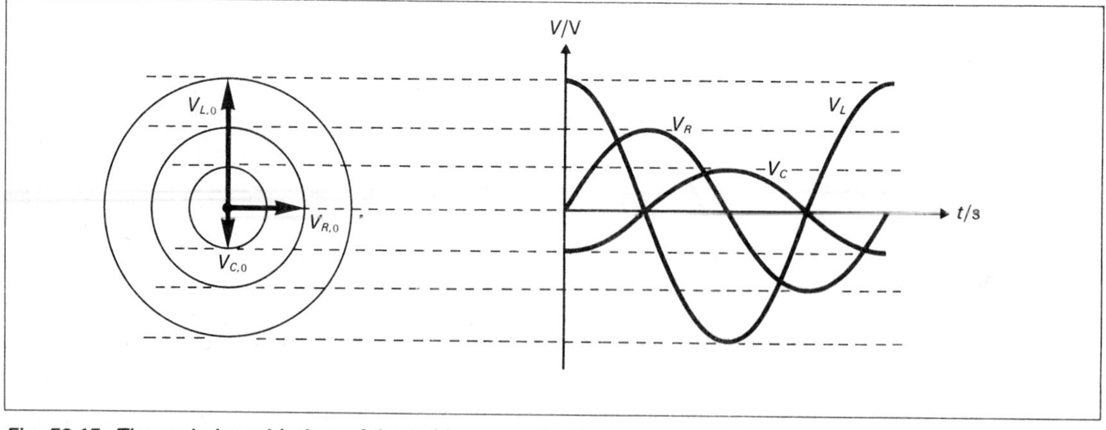

Fig. 58.15. The variation with time of the p.d.'s across the three components.

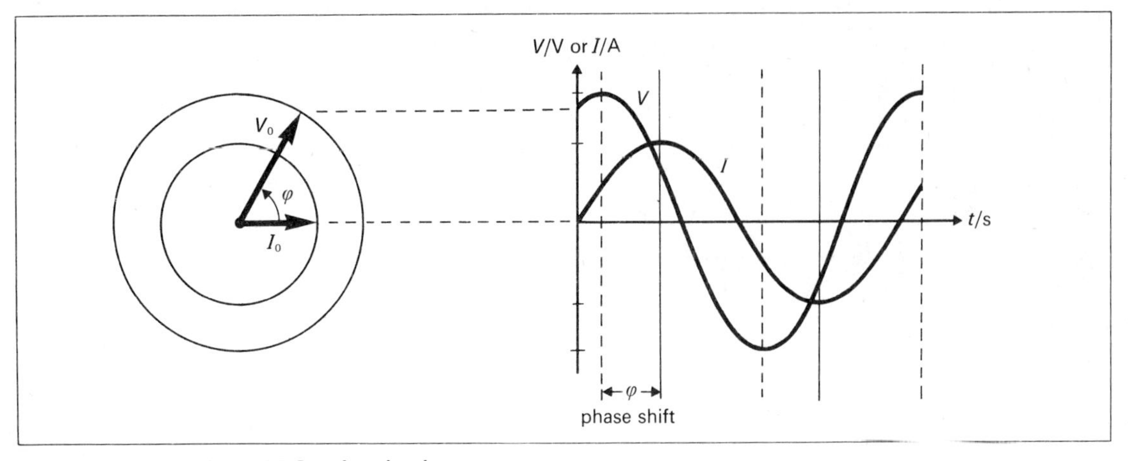

Fig. 58.17. V and I in an RLC series circuit.

The phase shift between V_0 and I_0 is given by

$$\tan \phi = \frac{V_{X,0}}{V_{R,0}} = \frac{V_{L,0} - V_{C,0}}{V_{R,0}}$$

The relationship between V and I is given in fig. 58.17. This shows how

(*i*) I and V have the same frequency,
(*ii*) the applied p.d. leads the current (in this example) by a *phase angle* ϕ. ϕ is negative when $V_{C,0} > V_{L,0}$.

Summary Diagram

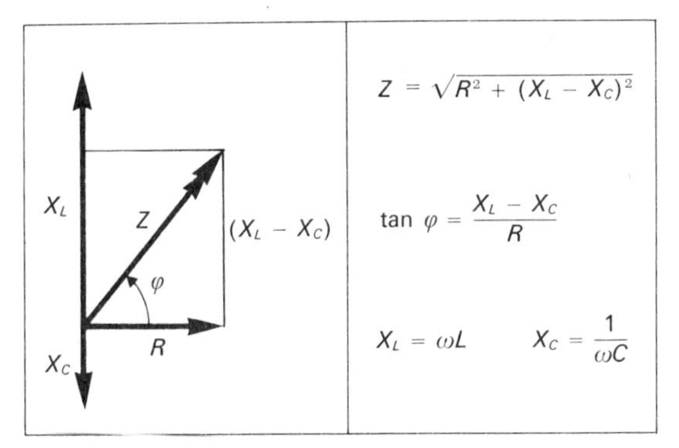

Fig. 58.18. How to calculate impedance in an RLC series circuit.

Fig. 58.18 shows that

$$V_0 = \sqrt{[V_{R,0}^2 + (V_{L,0} - V_{C,0})^2]}$$

Dividing each side by I_0, we have

$$Z = \frac{V_0}{I_0} = \sqrt{[R^2 + (X_L - X_C)^2]}$$

Remember that current is common to all elements of the *series* circuit.

To investigate

(*a*) an RC circuit, let $L \to 0$,
(*b*) an LR circuit, let $C \to \infty$, or
(*c*) an LC circuit, let $R \to 0$.

Power dissipation is discussed on page 420.

*58.14 *R, L and C* in Parallel

The parallel circuit in fig. 58.19 is different from a series circuit in that

(1) the instantaneous p.d. V across both branches is the same,
(2) the instantaneous current I is equal to the vector sum of the instantaneous currents I_1 and I_2, because
(3) the currents I_1 and I_2 are out of phase (I_1 leads I_2).

The equivalent impedance of the circuit can be found by using the rotating vector method (fig. 58.20).

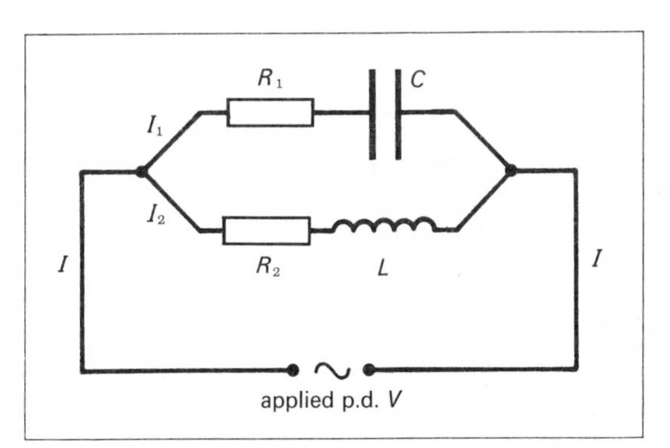

Fig. 58.19. An RLC parallel circuit.

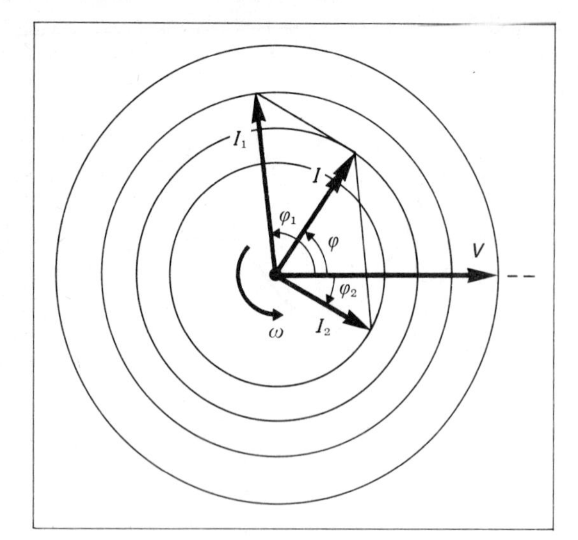

Fig. 58.20. Rotating vector diagram for the circuit of Fig. 58.19.

From fig. 58.20 it can be seen that:

(i) I_1 leads V by ϕ_1 where

$$\phi_1 = \arctan \frac{X_C}{R_1}$$

(ii) I_2 lags V by ϕ_2 where

$$\phi_2 = \arctan \frac{X_L}{R_2}$$

(iii) I leads V by ϕ.

The equivalent impedance of the circuit is given by $Z = V/I$, where V is known and I must be calculated.

The evaluation of I involves the calculation of I_1 and I_2, their phase angles, and some trigonometry.

The final expression for the equivalent impedance will obviously be complicated and is not given here, although it can be readily derived from fig. 58.20.

*58.15 Power in a.c. Circuits

The *RLC* Series Circuit

Instantaneous power

Refer to fig. 58.21.

$$\begin{aligned} P &= VI \\ &= (V_0 \sin(\omega t + \phi)) \times (I_0 \sin \omega t) \\ &= V_0 I_0 \sin \omega t (\sin \omega t \cos \phi + \cos \omega t \sin \phi) \\ &= V_0 I_0 \cos \phi \sin^2 \omega t + V_0 I_0 \sin \phi \sin \omega t \cos \omega t \end{aligned}$$

This is shown in fig. 58.22.

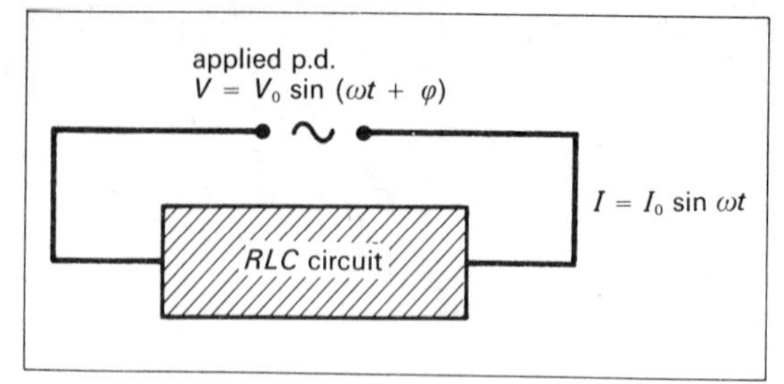

Fig. 58.21. Calculation of power in the *RLC* series circuit.

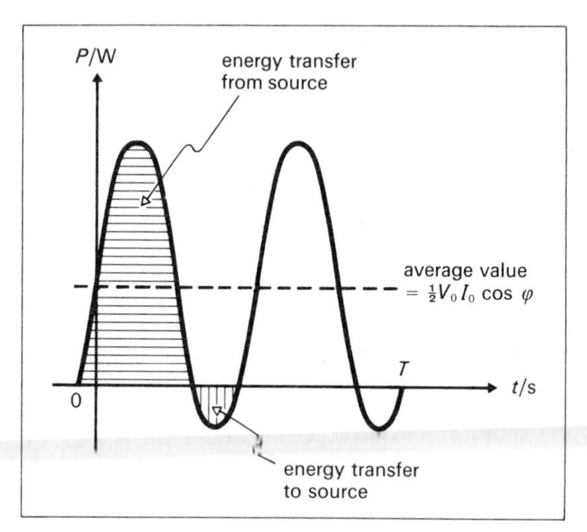

Fig. 58.22. Power of an *RLC* series circuit.

The average power ⟨P⟩ over one cycle

$$\begin{aligned} \langle P \rangle &= \langle V_0 I_0 \cos \phi \sin^2 \omega t \rangle + 0 \\ &= \tfrac{1}{2} V_0 I_0 \cos \phi \\ &= V_{\text{r.m.s.}} I_{\text{r.m.s.}} \cos \phi & \left(\frac{V_{\text{r.m.s.}}}{I_{\text{r.m.s.}}} = Z\right) \\ &= I_{\text{r.m.s.}}^2 Z \cos \phi & (R = Z \cos \phi) \\ &= I_{\text{r.m.s.}}^2 R \end{aligned}$$

$$\boxed{\langle P \rangle = I_{\text{r.m.s.}}^2 Z \cos \phi = I_{\text{r.m.s.}}^2 R}$$

We define the power factor of an RLC series circuit to be

$$\frac{\text{average power input}}{V_{\text{r.m.s.}} I_{\text{r.m.s.}}}$$

Thus the power factor $= \cos \phi = \dfrac{R}{Z}$

Notes

(i) When $\phi = \pm \pi/2$ rad, the *net* energy conversion is zero, and the current is said to be **wattless**.

(*ii*) The resistance R is solely responsible for the *net* conversion of electrical energy, which it does at the rate

$$I^2_{r.m.s.} R = \left(\frac{I_0}{\sqrt{2}}\right)^2 R$$

Measurement of a.c. Power

(1) For the *ammeter/voltmeter method* measure $V_{r.m.s.}$ and $I_{r.m.s.}$. Then $V_{r.m.s.} I_{r.m.s.}$ is the *apparent* power, and $V_{r.m.s.} I_{r.m.s.} \cos\phi$ the *real* power.

(2) *In the dynamometer wattmeter* (p. 393) the instantaneous couple on the moving coil

$$\propto VI \text{ (instantaneous values)}$$

and is always in the same sense.

The reading is proportional to the *real* power because the power factor $\cos\phi$ is taken into account automatically.

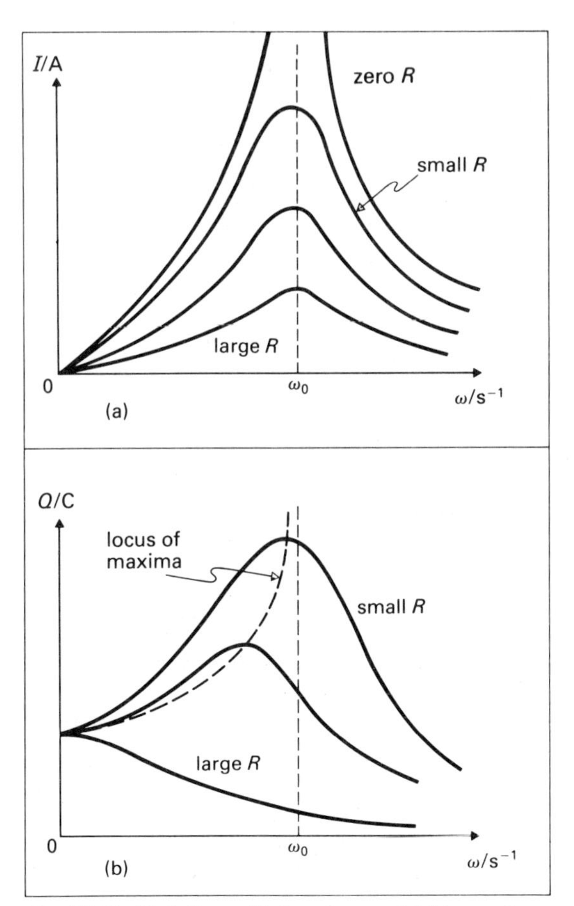

Fig. 58.23. Variation with frequency of (a) current, and (b) charge in the RLC series circuit. (Compare with fig. 11.5.)

*58.16 Resonance

(*a*) *RLC* Series Circuit

$$I_0 = \frac{V_0}{Z} = \frac{V_0}{\sqrt{R^2 + \left(\omega L - \frac{1}{\omega C}\right)^2}}$$

For a fixed value of V_0, I_0 reaches a maximum value when $\omega L = 1/(\omega C)$ and thus when V and I are *in phase*.

The occurrence of a maximum current at $\omega_0 = 1/\sqrt{LC}$ is known as **current resonance** (cf. *velocity* resonance in a mechanical system).

Charge resonance corresponds to mechanical *amplitude* resonance. The frequency of charge resonance, that of current resonance and the natural frequency of the circuit approach each other as R tends to zero (fig. 58.23).

(*b*) *RLC* Parallel Circuit

Resonance occurs at the frequency at which the current and potential difference are in phase (fig. 58.24*b*).

In fig. 58.24*a*, the current is the same in each branch: one current leads V while the other lags V by the same angle.

If X_L and X_C are much greater than R, the phase angle is large and therefore I is very small. Thus the equivalent impedance of the circuit is much greater than that of either branch.

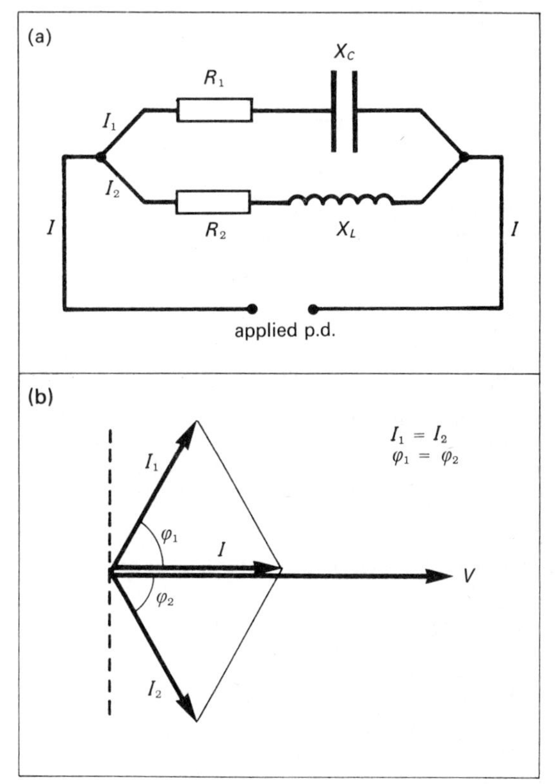

Fig. 58.24. (a) A parallel circuit, and (b) its corresponding vector diagram for resonance.

At *resonance* (fig. 58.25), there is maximum equivalent impedance and *minimum current.*

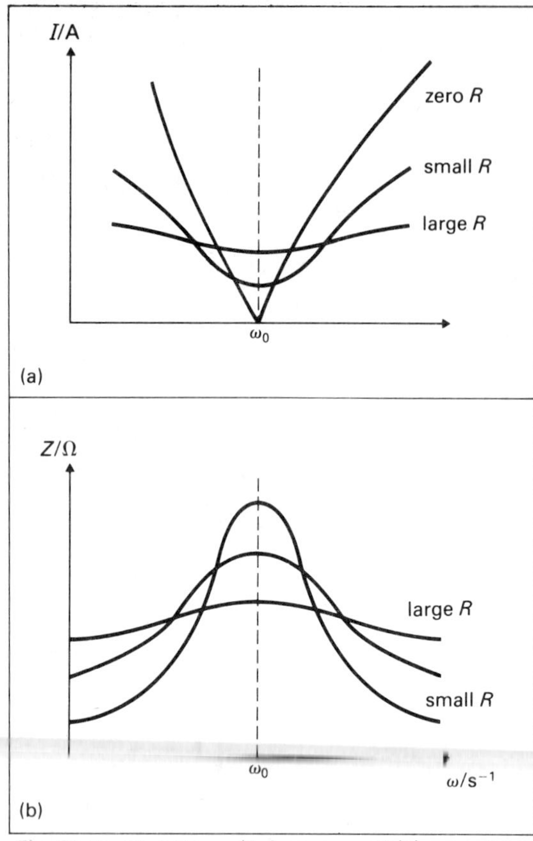

Fig. 58.25. Variation with frequency of (a) current and (b) impedance, in the parallel RLC circuit of Fig. 58.24.

Use of Resonant Circuits

(*a*) The sharp maximum response of a *series* tuned circuit makes it suitable for selecting a small band of frequencies from a wide range.

(*b*) The *parallel* circuit has a sharp maximum impedance at resonance and in situations where the amplification of a circuit depends on an impedance, the parallel tuned circuit gives high amplification over only a small range of frequencies.

*58.17 Electrical Oscillations

An electronic oscillator produces an output signal which continuously repeats the same pattern (e.g. sine wave, square wave, saw-tooth, etc.) of current and p.d. variations with time. Applications include

(*a*) sine-wave oscillators which provide the **carriers** for transmission of sound and picture information, and which are also used for reception,

(*b*) induction and dielectric heating in industry, and

(*c*) timing and triggering in computers and oscilloscopes.

The *LC* Oscillatory Circuit

Fig. 58.26 shows the type of circuit used for radio-frequency generation. The circuit tends to oscillate at its resonant frequency when a pulse of energy is put into it. The energy stored in the electric field of the charged capacitor is converted into energy stored in the magnetic field of the coil. During build-up a back e.m.f. is induced, and this causes the capacitor to be recharged.

When $X_L = X_C$, this oscillatory process occurs at a frequency f_0, where

$$2\pi f_0 L = \frac{1}{2\pi f_0 C},$$

so

$$f_0 = \frac{1}{2\pi\sqrt{LC}}$$

Notes. (*a*) For f_0 to be high, L and C must be small.
(*b*) To alter f_0, either L or C may be varied.

These oscillations will be damped by the resistive losses in the coil and wiring, and by dielectric losses in the capacitor. (Cf. damping by frictional forces p. 89.)

The oscillations can be maintained by using **feedback** which continuously puts energy into the circuit in phase with its natural oscillations (p. 453).

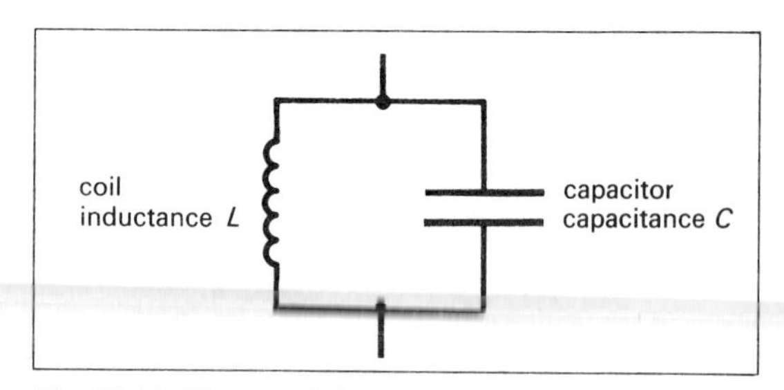

Fig. 58.26. The parallel LC resonant circuit.

*58.18 An Analogue Between Mechanical and Electrical Oscillations

Analogues are models which utilize some (supposed) similarity between physical processes. They can be misleading if used incautiously, and some give solutions which are qualitative rather than quantitative.

For *direct current* useful analogues exist using concepts of

(*a*) *heat flow* (p. 217), and also

(*b*) *water flow*.

The latter is particularly useful for an elementary introduction, since the ideas of water pumps, hydrostatic pressure differences and volume-flow rates of (visible) water are easier to visualize than those of sources of e.m.f., electric p.d. and rates of flow of (invisible) electric charge.

A *mechanical* system is found most useful for visualizing *electrical oscillations*, and the diagram and the table opposite outline one such system. (Other self-consistent systems exist: one, for example, compares ***F*** and *I* rather than ***F*** and *V*.)

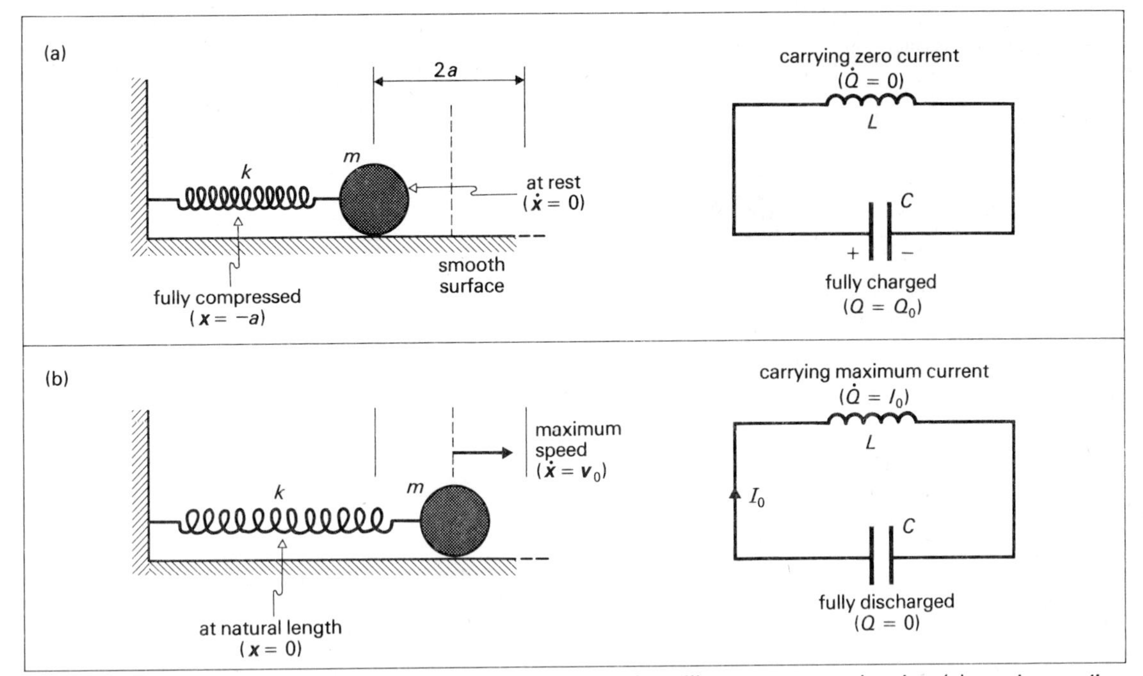

Fig. 58.27. Examples of undamped electrical and mechanical oscillatory systems showing (a) maximum displacement, and (b) zero displacement.

Undamped Free Oscillations

To compare the two systems shown in fig. 58.27 we can use the analogous quantities and ideas in the table below.

Mechanical		*Electrical*	
mass	m	L	inductance
displacement	$\boldsymbol{x}$	Q	charge
velocity	$\boldsymbol{v} = \dot{\boldsymbol{x}}$	$I = \dot{Q}$	current
acceleration	$\ddot{\boldsymbol{x}}$	$\ddot{Q} = \dfrac{\mathrm{d}I}{\mathrm{d}t}$	rate of change of current
force (Newton II)	$\boldsymbol{F} = m\ddot{\boldsymbol{x}}$	$V = L\dfrac{\mathrm{d}I}{\mathrm{d}t} = L\dfrac{\mathrm{d}^2Q}{\mathrm{d}t^2}$	p.d. (following from definition of L)
elastance (stiffness or spring constant)	$k = F/x$	$1/C = V/Q$	
compliance	$1/k = x/F$	$C = Q/V$	definition of capacitance
elastic p.e.	$\frac{1}{2}kx^2$	$\frac{1}{2}(Q^2/C)$	electric field energy
kinetic energy	$\frac{1}{2}m\dot{x}^2$	$\frac{1}{2}LI^2$	magnetic field energy
natural frequency of oscillation	$\dfrac{1}{2\pi}\sqrt{\dfrac{k}{m}}$	$\dfrac{1}{2\pi}\sqrt{\dfrac{1}{LC}}$	natural frequency of oscillation

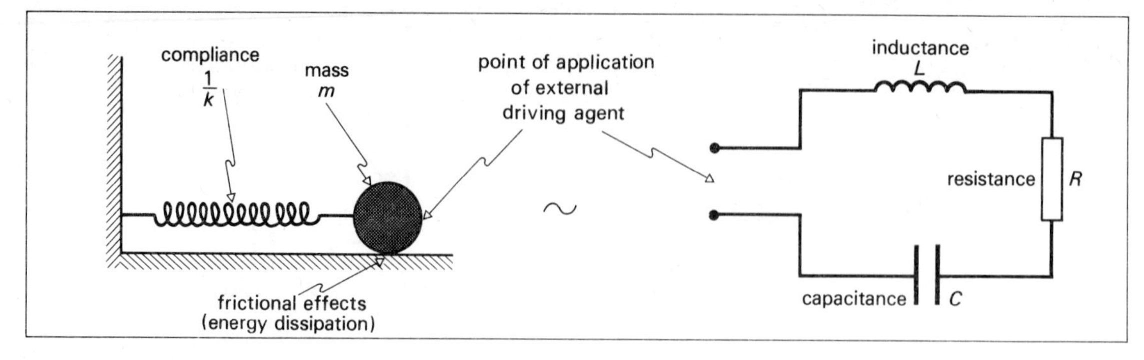

Fig. 58.28. Analogues for forced damped oscillations.

Forced Damped Oscillations

Fig. 58.28 introduces two new factors.

(*a*) A *resistance* in the electrical circuit causes energy dissipation. $R = V/I$ defines R, and has a mechanical analogue $\rho = F/\dot{x}$, in which ρ is a measure of frictional or viscous drag.

(*b*) An *external agent* (periodic force or alternating p.d.) is required to supply power if the amplitude of the system is to be maintained. During each cycle an amount of energy is provided to replace that which has been converted into internal energy.

Each system has a Q-factor, a measure of the selectivity of the resonant system, defined by

$$Q = \frac{2\pi\,(\text{total energy of system})}{\left(\begin{array}{c}\text{energy supplied to the}\\ \text{system during one cycle}\end{array}\right)}$$

58.19 Rectification of a.c.

The need for a.c.: Power loss in transmission is reduced by using large p.d.'s (p. 358). Such p.d.'s are obtained using a transformer (p. 401) which needs a *varying* current.

One needs d.c. for

(*i*) electronic apparatus (such as the c.r.o. and transistors), and

(*ii*) some electrolytic processes.

Rectification is the process by which one converts a.c. to d.c. It is achieved by a device which has a characteristic similar to fig. 58.29.

A **rectifier** impedes current flow in one direction far more than in the reverse direction. Devices which have an effective *one-way property* are called **valves**.

Examples include

(*i*) the metal rectifier,

(*ii*) the semiconductor junction-diode (p. 446), and

(*iii*) the thermionic diode.

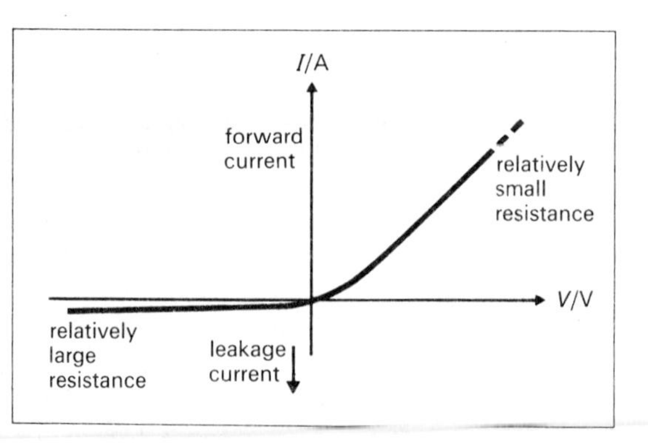

Fig. 58.29. A typical rectifier characteristic.

Suppose we have a sinusoidal input to a rectifier. Fig. 58.30 shows the circuit and output waveform for (*a*) **half-wave** and (*b*) **full-wave** rectification. The latter is also given by the *rectifier bridge circuit* of fig. 54.2 on page 390.

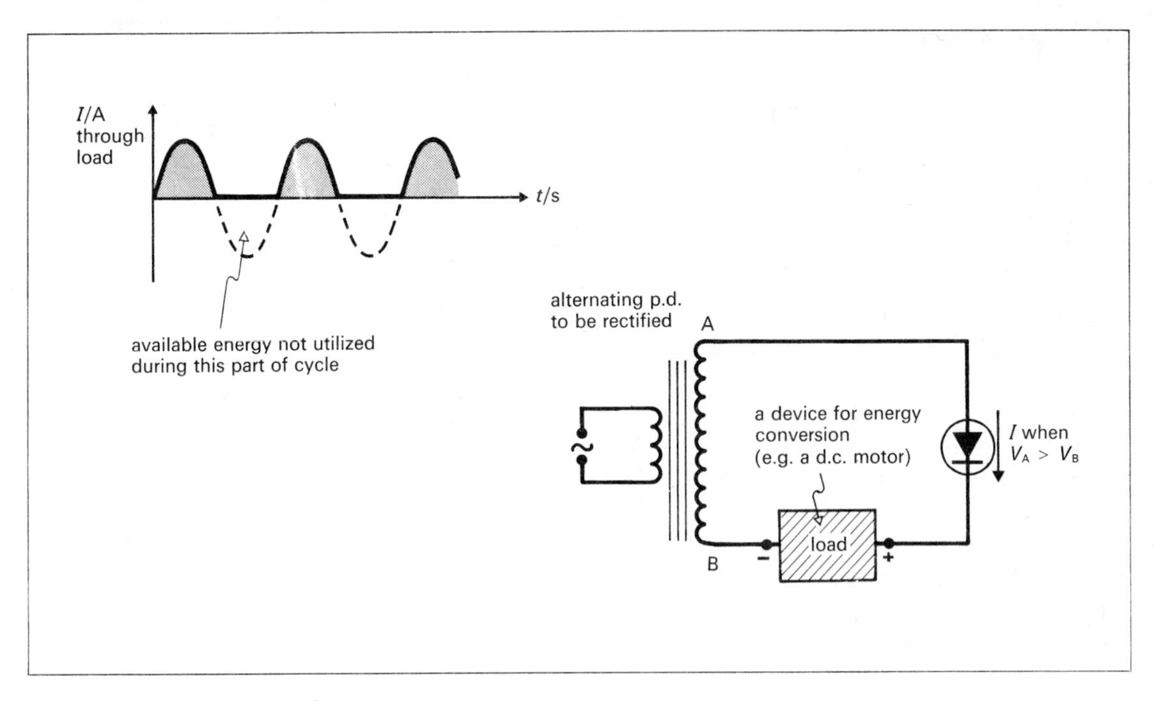

Fig. 58.30(a). Half-wave rectification.

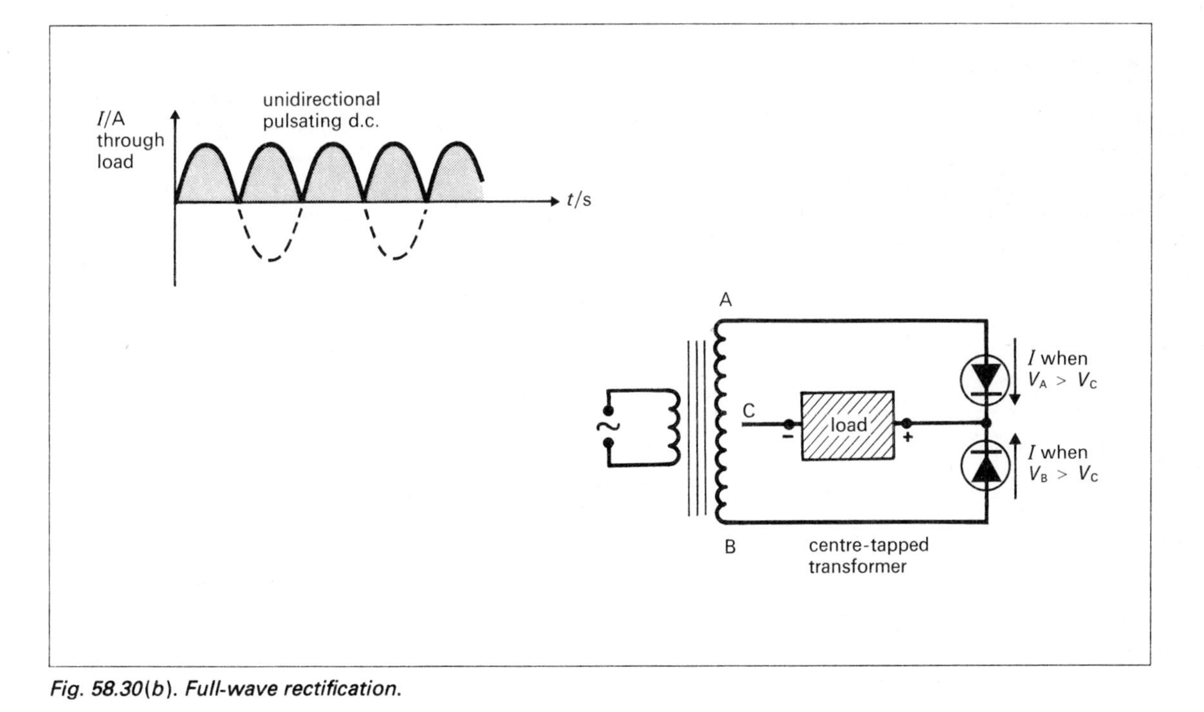

Fig. 58.30(b). Full-wave rectification.

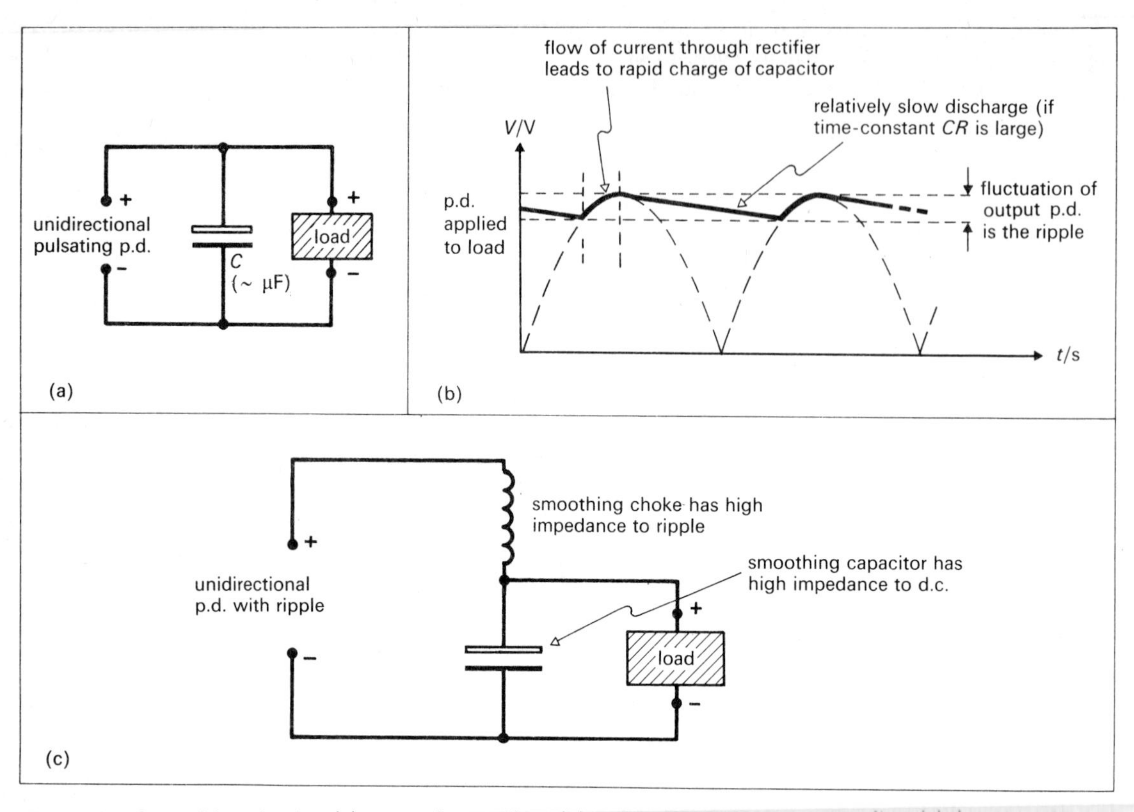

Fig. 58.31. Smoothing circuits: (a) reservoir capacitor, (b) action of the reservoir capacitor, (c) the smoothing filter suppressing the ripple.

58.20 Smoothing Circuits

Rectification gives a *unidirectional* but *pulsating* p.d. or current. To achieve a *steady* p.d. or current we use

(1) a **reservoir capacitor**, whose action in smoothing the pulsating d.c. is explained by fig. 58.31*a* and *b*, together with

(2) a **smoothing filter**: this consists of a *smoothing choke* and *smoothing capacitor* (fig. 58.31*c*), which reduce the effect of the *ripple*.

A **power pack** comprises a rectifier, reservoir capacitor and smoothing filter built into a single unit.

59 Magnetic Properties of Matter

59.1 The Rowland Anchor Ring

It was explained on page 399 that if the flux through a search coil of N turns changes by $\Delta\Phi$, the charge Q that flows past a cross-section of the circuit is given by

$$Q = \frac{N\,\Delta\Phi}{R}$$

where R is the resistance of the circuit.

If we use a given search coil with a particular galvanometer, the throw

$$\theta \propto Q$$
$$\propto \Delta\Phi$$

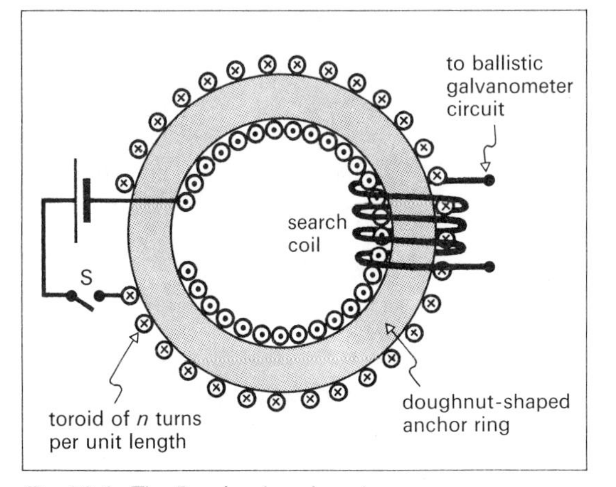

Fig. 59.1. The Rowland anchor ring.

In fig. 59.1. the toroidal coil is wound onto a **Rowland anchor ring.** Because $\boldsymbol{B}$ is effectively constant over the cross-section of the toroid, the flux Φ through the search coil $\propto B$, the average flux density over a cross-section.

If the switch is opened or closed, then

the ballistic throw $\theta \propto B$, *the flux density.*

To investigate the magnetic properties of a material, we measure

(*a*) $\boldsymbol{B}_0$, the flux density using a toroid filled with air (which approximates to a vacuum), and
(*b*) $\boldsymbol{B}$, the flux density when the toroid is *filled* by the material.

Materials are called

(*i*) **diamagnetic** if $\boldsymbol{B} < \boldsymbol{B}_0$, and
(*ii*) **paramagnetic** if $\boldsymbol{B} > \boldsymbol{B}_0$. For some materials $\boldsymbol{B} \gg \boldsymbol{B}_0$; they are called **ferromagnetic**.

59.2 Definitions

We write $\boldsymbol{B} = \boldsymbol{B}_0 + \boldsymbol{B}_M$, where

$\boldsymbol{B}$ = the resultant flux density,
$\boldsymbol{B}_0$ = the flux density when the toroid is empty (i.e. that set up by the current alone),
$\boldsymbol{B}_M$ = the additional flux density set up by the material of the core. ($\boldsymbol{B}_M$ is negative for a diamagnetic material.)

Dividing by B_0

$$\frac{B}{B_0} = 1 + \frac{B_M}{B_0}$$

or

$$\boxed{\mu_r = 1 + \chi_m}$$

(*a*) We have defined the **relative permeability** μ_r by the equation

$$\boxed{\mu_r = \frac{B}{B_0}}$$

For a ferromagnetic material μ_r is not a constant, but depends on $\boldsymbol{B}_0$. It can be considered as the permeability of a material relative to that of a vacuum.

Thus

$$\boxed{\mu_r = \frac{\mu}{\mu_0}}$$

where μ is called the **absolute permeability**. μ_r is a pure number.

(*b*) We have put $B_M/B_0 = \chi_m$, the **magnetic susceptibility**, although it is not defined in this way in more advanced work. It is a parameter which measures the response of the material to the external magnetic field. χ_m is a pure number.

Since $B_0 = \mu_0 nI$, where I is the current in the toroid, we can measure corresponding values of $\boldsymbol{B}$ and $\boldsymbol{B}_0$ by using different values of I, and hence evaluate μ_r for *ferromagnetic materials*. For other materials

$$|\chi_m| < 10^{-3}$$

and the method is not suitable in practice.

59.3 Explanation of Magnetism by Equivalent Surface Currents

In a first study of magnetism, the facts of permanent magnetism are sometimes explained by the concept of the *elementary magnetic dipole*.

We replace that idea here by the concept of an *elementary coil circuit* (fig. 59.2).

The resultant **Amperian surface currents** set up a flux density which is superimposed on $\boldsymbol{B}_0$. The equation

$$\boldsymbol{B} = \boldsymbol{B}_0 + \boldsymbol{B}_M$$

has an equivalent in

$$I = I_r + I_M$$

where I = effective current responsible for the resultant $\boldsymbol{B}$,
I_r = real current (through the windings) responsible for $\boldsymbol{B}_0$, and
I_M = the surface current on the magnetized body responsible for $\boldsymbol{B}_M$.

Using this idea we explain the effects of a permanent magnetic dipole in terms of surface currents similar to those carried by a solenoid.

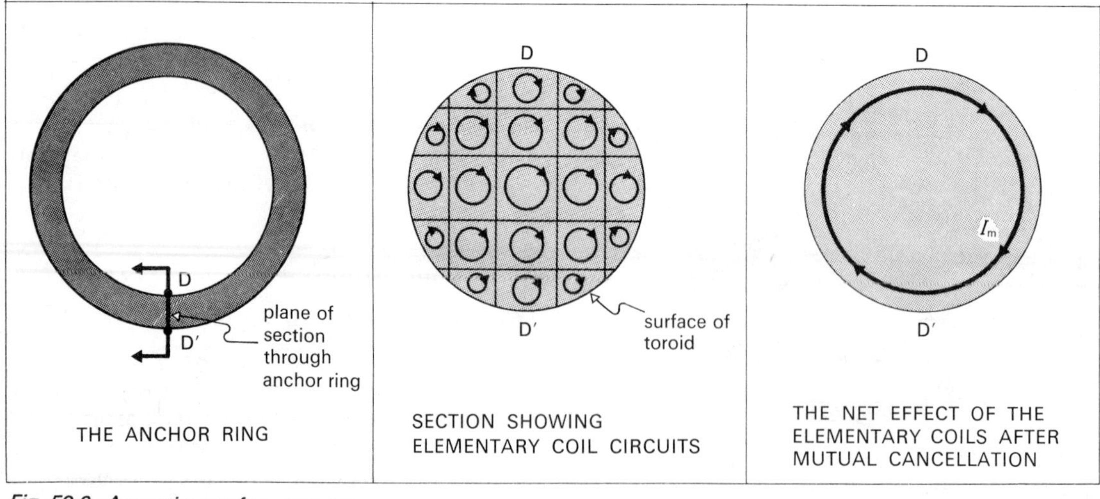

Fig. 59.2. Amperian surface currents.

59.1 Magnetization Curves and Hysteresis Loops

The curve which shows $\boldsymbol{B}$ plotted against $\boldsymbol{B}_0$ for a ferromagnetic material is called a **hysteresis loop** (fig. 59.3).

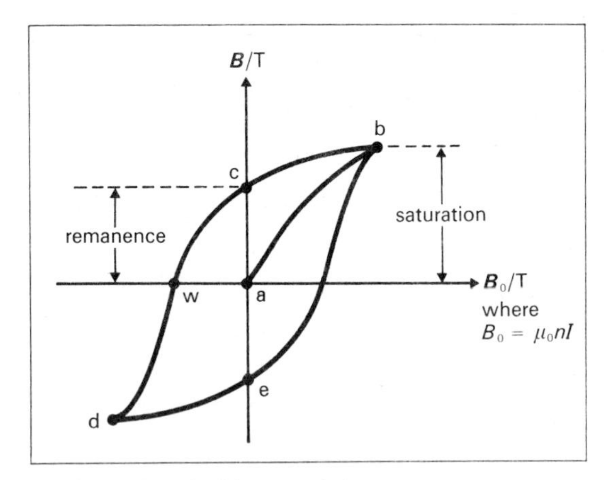

Fig. 59.3. A typical hysteresis loop.

The line ab represents the **magnetization curve**. The different parts of the curve represent the following:

Part of Curve	*Operation*
a to b	increase of toroid current I, using an unmagnetized ferromagnetic core
b to c	I is reduced to zero
c to d	I is reversed in direction, and its magnitude increased to a maximum
d to e	I is again reduced to zero
e to b	I is again increased to its forward maximum

The hysteresis loop shows the following features:

(*i*) The lack of retraceability is called **hysteresis** ('lagging behind').

(*ii*) At the points c and e the core is magnetized, even though $I = 0$. This exemplifies **permanent magnetism**.

(*iii*) ac is a measure of the **remanence** or **retentivity** of the material.

(*iv*) aw shows the reversed value of $\boldsymbol{B}_0$ required to reduce $\boldsymbol{B}$ to zero after the specimen has been **saturated** in the reverse direction.

aw $\propto$ **coercive force**. Coercive force has the dimensions of B_0/μ_0.

(*v*) The area under the loop is proportional to the energy dissipated as internal energy in unit volume of the material when it is taken once round the cycle.

(*vi*) The loop shows that $\mu_r (= B/B_0)$ is *not* a constant.

(*vii*) At saturation the hysteresis loop has a gradient of 1, but the different scales for $\boldsymbol{B}$ and $\boldsymbol{B}_0$ disguise this.

A specimen which is withdrawn along the axis of a solenoid carrying an alternating current is taken through a succession of ever-decreasing hysteresis cycles, and is thus demagnetized.

Requirements of Magnetic Materials

(*a*) *Permanent magnets* are made of **hard** magnetic material which has a large remanence (to make the magnet 'strong'), and a large coercive force (to prevent easy demagnetization by stray fields (fig. 59.4). Such materials can be made from

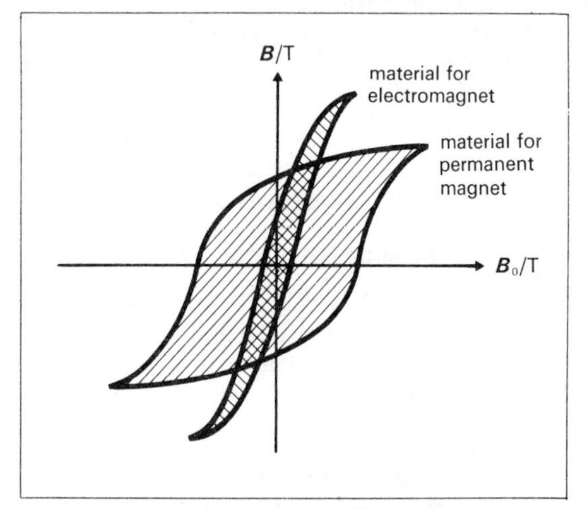

Fig. 59.4. Hysteresis loops for different types of ferromagnetic materials. (Note that the scale for $\boldsymbol{B}$ *is perhaps one hundred times that for* $\boldsymbol{B}_0$*.)*

(*i*) *alloys* (e.g. *Ticonal* and *Alcomax*), or

(*ii*) *ceramic* materials (e.g. *Magnadur*), which have the mechanical properties of (say) china, the magnetic properties of alloys, and yet high resistivity. (See *ferrites* on p. 400).

(*b*) *Transformer cores* have a very narrow loop to prevent overheating by **hysteresis loss**.

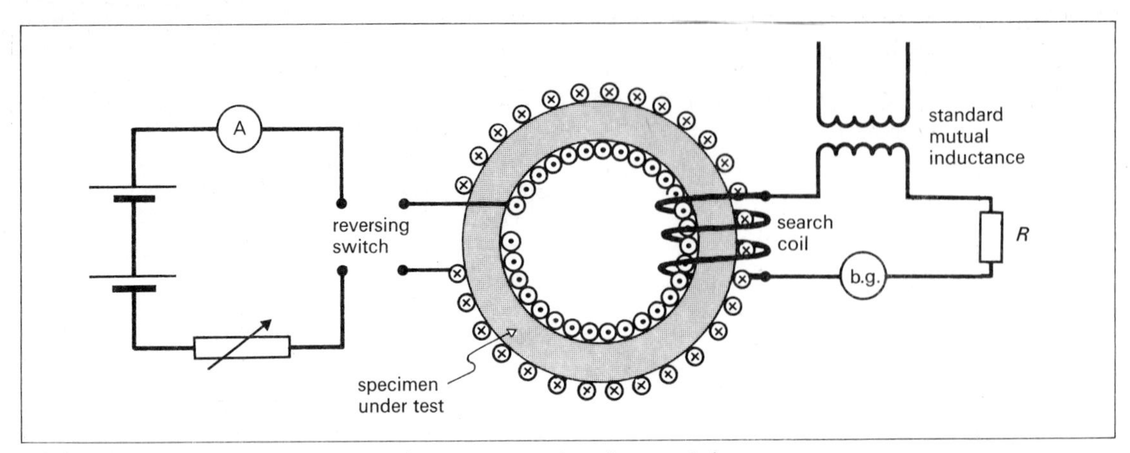

Fig. 59.5. Circuit for using a ballistic galvanometer to plot a hysteresis loop.

(*c*) *An electromagnet* is made of a **soft** magnetic material which has a small remanence and a small coercive force. It can thus be made a 'strong' but temporary magnet.

(*d*) *An inductor* (such as a choke) is often provided with a laminated core of high permeability to increase its self-inductance.

59.5 Plotting a Hysteresis Loop

(*a*) Using a Ballistic Galvanometer

Use the circuit of fig. 59.5.

(1) Using a pre-determined toroid current I, throw the reversing switch, and note the galvanometer throw θ.

(2) Plot $\theta(\propto B)$ along the y-axis, and $I(\propto B_0)$ along the x-axis.

The procedure is slow and complex because of the necessity of taking the ring through the hysteresis cycle *systematically*.

(*b*) Using a Cathode Ray Oscilloscope

This simple quick method can be used with an oscilloscope whose mu-metal screen can be removed to enable an external magnetic field to deflect the electron beam (fig. 59.6).

(1) The solenoid current I magnetizes the specimen. The $\boldsymbol{B}$ that results causes a deflection of the electron beam in the y-direction (according to the motor rule), which is proportional to B.

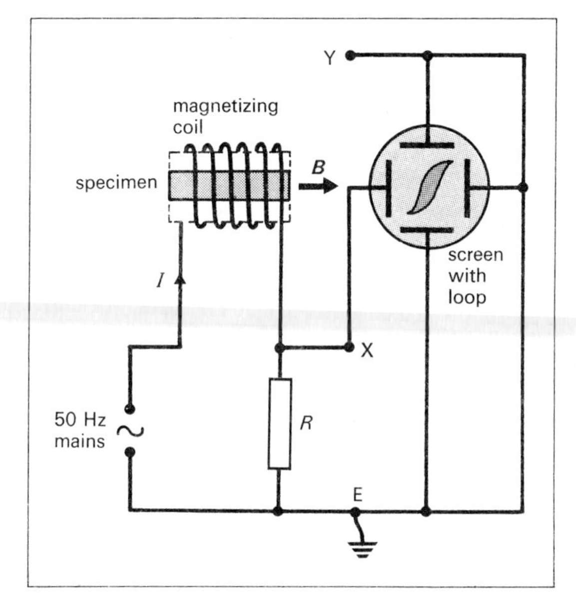

Fig. 59.6. Circuit for using a cathode ray oscilloscope to display a hysteresis loop.

(2) The p.d. across $R \propto I$, which is $\propto B_0$. This p.d. is applied across the X-plates, and gives a deflection in the x-direction $\propto B_0$.

Notes.

(*i*) I_{max} must be large enough to produce saturation.

(*ii*) If the conditions are not changed, the method can be used to compare the loops of different materials.

(*iii*) This is an example of *coordinate representation* on the oscilloscope (p. 86).

*59.6 Classification of Magnetic Materials

All substances display some response to a magnetic field.

(*a*) Diamagnetic Substances

(*i*) When freely suspended in a uniform magnetic field $\boldsymbol{B}_0$, they align themselves *across* the field. ($\boldsymbol{B}_0$ is the flux density *before* the specimen is placed in the field.)

(*ii*) They become magnetized in a direction opposite to that of $\boldsymbol{B}_0$. This means that $\boldsymbol{B}_M$ (p. 427) is *negative*, and inside the specimen $\boldsymbol{B} < \boldsymbol{B}_0$.

(*iii*) We draw the field lines as though they avoided passing through the specimen: it experiences a repulsion from the stronger part of the field to the weaker (fig. 59.7*a*).

(*iv*) Bismuth and copper are diamagnetic.

(*v*) Diamagnetics align themselves parallel to the field if it is non-uniform.

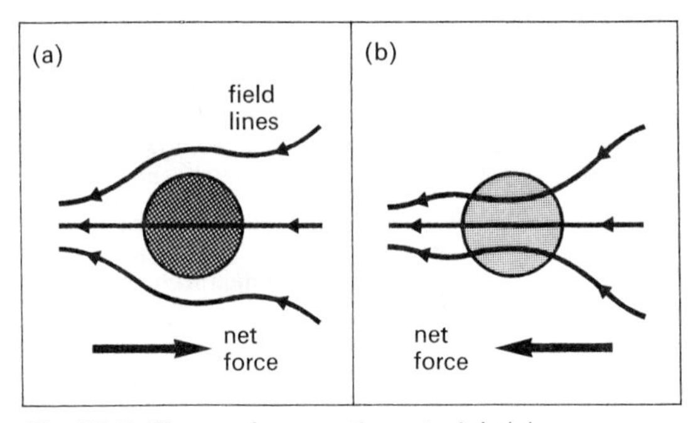

Fig. 59.7. Types of magnetic material: (a) diamagnetic, (b) paramagnetic.

(*b*) Paramagnetic Substances

(*i*) When freely suspended in a uniform magnetic field $\boldsymbol{B}_0$, they align themselves *parallel* to the field.

(*ii*) They become magnetized in the same direction as $\boldsymbol{B}_0$. $\boldsymbol{B}_M$ is *positive*.

(*iii*) The field lines are drawn as though being attracted into the specimen (as with a magnetic dipole): it experiences an attraction toward the stronger part of the field (fig. 59.7*b*).

(*iv*) Platinum and aluminium are paramagnetic.

(*c*) Ferromagnetic Substances

These display paramagnetic behaviour, but much more strongly. They may be physically moved to the stronger part of the field.

*59.7 Explanation of Magnetic Behaviour

When matter is subject to an external magnetic field, there are two basic effects:

(1) the electronic motion is *distorted*, and
(2) any permanent magnetic dipoles become partially *oriented*.

(1) The **distortion effect** shows the response of the moving charges to the magnetic force. The movement that results from the distortion establishes (according to *Lenz's Law*) a magnetic field in *opposition* to the applied field. The magnetic field in the specimen is *reduced*, corresponding to diamagnetism.

(2) The **orientation effect** will be shown by materials whose atoms have, before the field is applied, an electromagnetic moment. Such atoms include those whose **electron spins** are not paired. The external field $\boldsymbol{B}_M$ supports $\boldsymbol{B}_0$, so the field in the specimen is *increased*, corresponding to paramagnetism.

All substances have diamagnetic properties, which are nearly independent of temperature. Except at high temperatures, they may be swamped by paramagnetic properties (if present). Paramagnetism decreases with rising temperature, but some metals show a *weak* paramagnetism which is temperature-independent.

Ferromagnetism

Quantum mechanics explains how, in some substances, the unpaired electron spins become **coupled**, so that $\sim 10^{10}$ atoms may combine to form a **domain** (of linear dimension $\sim 10^{-7}$ m) which has a large electromagnetic moment. When the domains are randomly arranged, the specimen as a whole is unmagnetized.

When an external magnetic field is applied to a specimen, the domains that are suitably oriented grow in preference to

Summary

Effect	χ_m	μ_r	*Origin* (paragraph **59.7**)
diamagnetism	about -10^{-5}	< 1	distortion, according to *Lenz's Law*
paramagnetism	about $+10^{-3}$	> 1	orientation of magnetic dipoles in material
ferromagnetism	depend on $\boldsymbol{B}_0$, but can be very large ($\sim 10^{+3}$)		magnetic domains

others. In soft magnetic specimens the domain boundaries (*Bloch walls*) move easily, and weak fields can produce alignment. In hard specimens the boundaries are more restricted, and greater fields are needed for magnetization.

Evidence for Domains

(*i*) **Bitter photographs**, which show visible surface effects,
(*ii*) **Barkhausen effect**, in which the jumpy orientation of domains is demonstrated by clicks in a loudspeaker, and
(*iii*) **Magnetostriction**, the change of body size on magnetization.

The Curie Temperature

When the temperature exceeds the **Curie temperature** the coupling effect disappears, and ferromagnetic materials become paramagnetic.

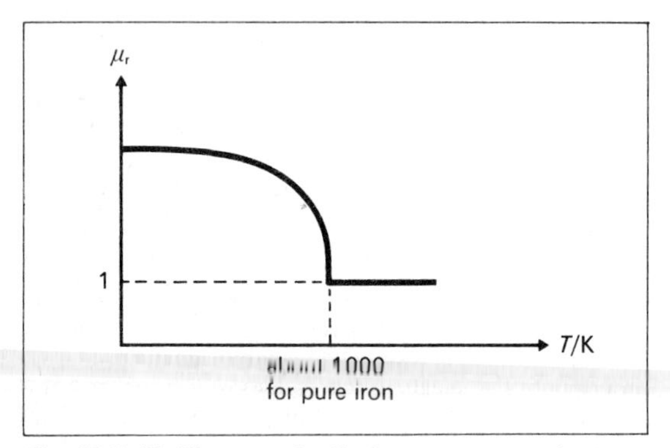

Fig. 59.8. To illustrate the Curie temperature. (μ_r is plotted for a given value of $\boldsymbol{B}_0$.)

Fig. 59.8 shows how μ_r decreases as T increases. The *Curie temperature* depends upon both $\boldsymbol{B}_0$, and the material.

*59.8 The Magnetic Circuit

A magnetic circuit consists of a closed loop of magnetic flux. There is a useful analogy between (*i*) the electric circuit equation and (*ii*) a magnetic circuit equation.

(*i*) (electromotive force) = (current) × (total circuit resistance)

$$\mathscr{E} = I \times \Sigma\left(\frac{l}{\sigma A}\right)$$

(*ii*) (magnetomotive force) = (magnetic flux) × (total circuit reluctance R_m)

$$F_m = \Phi \times \Sigma\left(\frac{l}{\mu A}\right)$$

The **magnetomotive force** is the **source** of magnetic effects. For example a toroid of N turns each of which carries a current I has an m.m.f. of NI. The unit of m.m.f. is the ampere (turn). (See below.)

Notes.

(*a*) The concept of a magnetic circuit is not very useful if there is appreciable flux leakage. (Electric charge leakage does not occur in a well-insulated electric circuit.)

(*b*) The unit of **reluctance** R_m is that of $l/\mu A$, i.e. $A\ Wb^{-1}$, which is consistent with that of F_m being A.

(*c*) There are no loose ends to either flux lines or electric current paths. Charged particles flow round an electric circuit whereas nothing flows in a magnetic circuit, although some atoms become reoriented.

(*d*) μ is a function of I, $\boldsymbol{B}$ and hence Φ in ferromagnetic cores; that is they show non-linear behaviour. On the other hand σ does not vary with I in an ohmic resistor unless the temperature changes. The range of values of σ in various materials is many times greater than that of μ.

(*e*) If a part of a magnetic circuit does not contain ferromagnetic material then the reluctance of this part is virtually constant, i.e. independent of flux density. (This statement is analogous to *Ohm's* law.)

(*f*) Flux can be made to pass easily through air and free space, though any air gap effectively controls the total reluctance of a circuit. No meaning can be attached to the terms a good magnetic insulator or a magnetic short circuit.

(*g*) *Reluctances in series and parallel* can often be combined in the same way as resistances, as long as care is taken to establish the boundaries between different components.

(*h*) Current density $J = I/A$ appears as an analogy to magnetic flux density $B = \Phi/A$.

(*i*) When *Ampère's* law is applied round a closed loop (i.e. a magnetic circuit) which encloses N turns of wire, we obtain

$$\frac{1}{\mu_0}\oint B_0 \cdot dl = \Phi\Sigma\frac{l}{\mu A} = NI$$

From this we deduce that $F_m = NI$, which was implied above.

*59.9 Ampère's Law Modified

(This paragraph is not entirely rigorous. Its aim is to introduce the magnetic field vector $\boldsymbol{H}$.)

On page 386 *Ampère's Law* was quoted by the equation

$$\oint B \cdot dl = \mu_0 I$$

which applies in a vacuum.

When magnetic material is present, we write

$$\oint B \cdot dl = \mu_0(I_r + I_M)$$

in which I_M was introduced on p. 428.

To rewrite the law in a simpler form, we can proceed as follows.

The equation

$$\boldsymbol{B} = \boldsymbol{B}_0 + \boldsymbol{B}_M$$

can be rewritten

$$\boldsymbol{B} = \mu_0\boldsymbol{H} + \mu_0\boldsymbol{M}$$

in which

(1) It is convenient to write $\boldsymbol{B}_0 = \mu_0\boldsymbol{H}$ in a vacuum. $\boldsymbol{H}$ is called the **magnetic field strength**, and is a *vector* quantity. It has the unit $\mathrm{A\,m^{-1}}$. (It is sometimes called the *magnetizing force*.)

(2) $\boldsymbol{M}$ stands for the **magnetization**, which is *defined* as the electromagnetic moment per unit volume. It is also a *vector*, and has unit

$$\frac{\mathrm{A\,m^2}}{\mathrm{m^3}} = \mathrm{A\,m^{-1}}$$

Rearrangement of the last equation gives

$$\boldsymbol{H} = \frac{\boldsymbol{B} - \mu_0\boldsymbol{M}}{\mu_0}$$

which may be regarded as a more formal *defining equation for* $\boldsymbol{H}$.

For dia- and paramagnetic materials we have the empirical equations

$$\boldsymbol{B} = \mu_r\mu_0\boldsymbol{H} = \mu\boldsymbol{H}$$

in which μ_r is a constant.

For ferromagnetic materials μ_r is *not* a constant.

$\boldsymbol{H}$ is not a new quantity of fundamental significance, but it enables us to rewrite *Ampère's Law* in a simpler form. It becomes

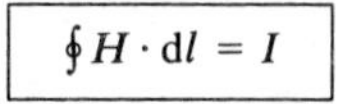

$$\oint H \cdot \mathrm{d}l = I$$

where I represents the sum of the *true currents* (as opposed to *Amperian* currents) enclosed by the loop.

A proper study of the vectors $\boldsymbol{B}$, $\boldsymbol{H}$ and $\boldsymbol{M}$ is beyond the scope of this book: in this paragraph we merely indicate that $\boldsymbol{H}$ is a useful quantity.

XI Electronic, Atomic and Nuclear Physics

60 The Free Electron

60.1 Sources of Electrons

Small negatively charged particles can be obtained in the following ways:

(*a*) *By applying a potential difference* of a few kilovolts across a gas maintained at a pressure of about 10 Pa. The tube containing the gas is called a **discharge tube**. The particles seem to come from the cathode, and are called **cathode rays** (see below).

(*b*) *By heating a conductor.* **Thermions** are emitted from the surface. The process is called **thermionic emission** (p. 440).

(*c*) *By irradiating a metal surface* with electromagnetic radiation of appropriate frequency. **Photoelectrons** are emitted from the surface. The process is called **photoelectric emission** (p. 455).

(*d*) *From certain radioactive nuclei.* β^-**-particles** are emitted (p. 474). The process is called β-**emission**.

(*e*) *By imposing an intense electric field* ($\sim 10^8\ \mathrm{V\,m^{-1}}$) at the surface of a metal. The process is called **field emission.**

(*f*) *By bombarding a metal surface* with primary (incident) electrons. The process is called **secondary emission**, and the particles are called **secondary electrons**.

Their Identification

The **specific charge** q of a particle is its ratio

$$\frac{\text{charge}}{\text{mass}}$$

and can be measured by experiments like *Thomson's* (p. 438). Such experiments indicate that *all these negative particles have the same specific charge*, and *all* are believed to be **electrons**.

Properties

The electron has the following properties:

(*i*) It carries a *charge*

$$e = -1.6 \times 10^{-19}\ \mathrm{C}$$

(*Millikan's* experiment—p. 440). This is the fundamental quantum of charge.

(*ii*) It has a *specific charge* q given by

$$q = \frac{e}{m_e} = -1.76 \times 10^{11}\ \mathrm{C\,kg^{-1}}$$

(*iii*) It has a *rest mass* 9.1×10^{-31} kg, which can be deduced from e and e/m_e, but which cannot be measured directly.

(*iv*) It is common to all matter.

**Relativistic Mass*

Because an electron has such a small mass, it is easily accelerated by (electrical) forces to speeds approaching that of light in a vacuum. Its mass m is then given by

$$m = \frac{m_e}{\sqrt{1-(v/c)^2}}$$

where v is the speed of the electron relative to the observer measuring the mass. (An equation of the above form applies to any moving particle, not just an electron.)

Unless otherwise stated, in this book we consider the **rest mass** m_e, rather than the relativistic mass.

Observation	*Deduction*
(*a*) They cast a shadow of an obstacle (such as a *Maltese Cross*) on the wall of the tube.	(*a*) They travel in nearly straight paths despite the transverse force (their weight) acting on them, and so are moving *very fast*.
(*b*) They can, indirectly, transfer their energy so as to set a small paddle wheel into rotation.	(*b*) They probably consist of fast moving *particles*, whose k.e. is being converted to rotational energy.
(*c*) They can pass through aluminium foil.	(*c*) They must, if charged, have a very small charge (and hence (from e/m_e), a very small (sub-atomic) mass).
(*d*) They are deflected by a magnetic field in a direction opposite to that in which positive particles with the same velocity would be.	(*d*) According to *Fleming*'s left hand motor rule, they are *negatively* charged.
(*e*) In an electric field they are deflected towards a positive plate.	(*e*) This confirms that their charge is negative.
(*f*) They can be collected in a metal cylinder. An electroscope then shows that the cylinder acquires a negative charge (*Perrin's* experiment).	(*f*) This associates them with the negative charge produced in electrostatics experiments. (The rays were deflected into the cylinder by a magnetic field, to check that the effect was not caused by any electromagnetic radiation emitted by the hot cathode.)
(*g*) They affect a photographic plate.	(*g*) This confirms that they have energy.

60.2 Properties of Cathode Rays

Although cathode rays were first observed in a discharge tube, it is convenient to demonstrate their properties using thermions, electrons emitted from a heated cathode. See the table above.

60.3 Principles of Electron Deflection

Care must be taken not to confuse the following situations:

(*a*) Magnetic Field *B* only

In fig. 60.1*a* the electron velocity $\boldsymbol{v}$ is perpendicular to $\boldsymbol{B}$. Hence $F = Bev$, and $\boldsymbol{F}$ is perpendicular to the plane containing $\boldsymbol{B}$ and $\boldsymbol{v}$.

Since $\boldsymbol{F}$ is perpendicular to $\boldsymbol{v}$, the speed remains constant, and the path is *circular*.

Using $$\boldsymbol{F} = m\boldsymbol{a}$$

we have $$Bev = \frac{m_e v^2}{r}$$

where v^2/r is the electron's centripetal acceleration.

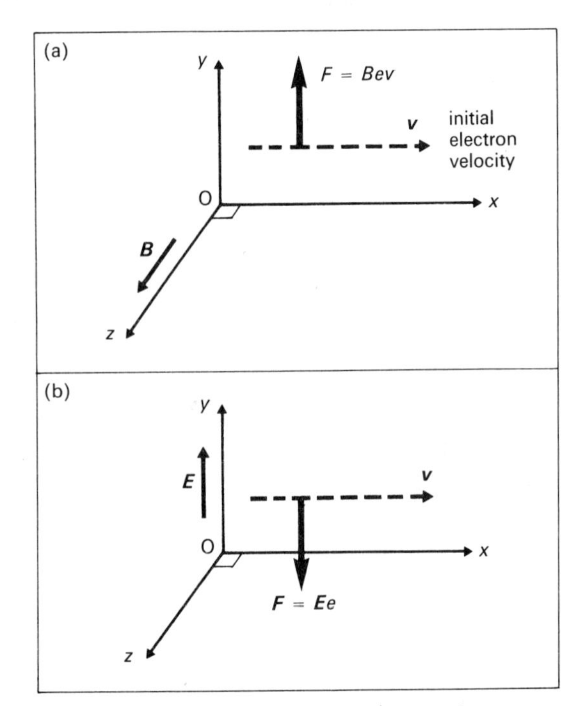

*Fig. 60.1. The initial force that acts on an electron moving through uniform fields of **B** and **E**.*

(b) Electric Field E only

In fig. 60.1*b*, the electron moves in a *parabolic* path. Refer to page 315.

(c) E and B Perpendicular (crossed)

Using the fields shown in fig. 60.1, the magnetic *force* is in the positive y-direction, and the electric *force* in the negative y-direction. Their magnitudes can be made equal, so that the trajectory becomes a straight line.

Then $$Bev = eE$$

so $$v = \frac{E}{B}$$

$$v\left[\frac{\text{m}}{\text{s}}\right] = \frac{E[\text{N/C}]}{B[\text{N s/C m}]}$$

The method is used to *calculate the speed of the electrons* in a beam. It depends for its action on the fact that the electric force is independent of the electron speed, while the magnetic force is proportional to that speed.

Application

Measurement of (e/m_e) for *electrons*.

(d) E and B Parallel

Suppose we have a situation in which E and B are both parallel to Oy. Then the electric and magnetic *forces are perpendicular*, and the electrons in the beam experience simultaneous deflections

(*i*) along yO (electric) and
(*ii*) along zO (magnetic).

A screen placed some distance along Ox, and parallel to the y–z plane would show a *parabolic trace* if the electrons were travelling at different velocities.

Application

Measurement of specific charge q for *positive ions* (p. 461).

60.4 Thomson's Measurement of the Electron's Specific Charge

Fig. 60.2 shows a modernized version of *Thomson's* apparatus.

Procedure

(1) With E and B both zero, note position O of the spot on the screen.

(2) Using a fixed value of B (alone), deflect the spot to a measured position S.

(3) Adjust the value of E so that the spot returns to its undeflected position O.

B can be established using a *Helmholtz* coil arrangement (p. 384).

Calculation

$$Bev = \frac{m_e v^2}{r} \qquad \text{(p. 374)}$$

so $$\frac{e}{m_e} = \frac{v}{Br}$$

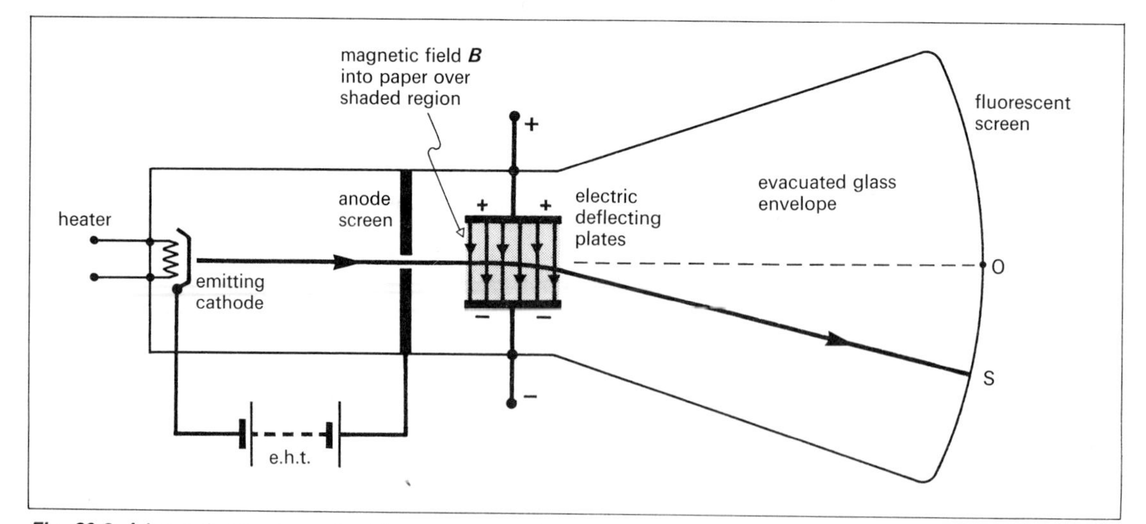

Fig. 60.2. A heated cathode method for measuring e/m_e.

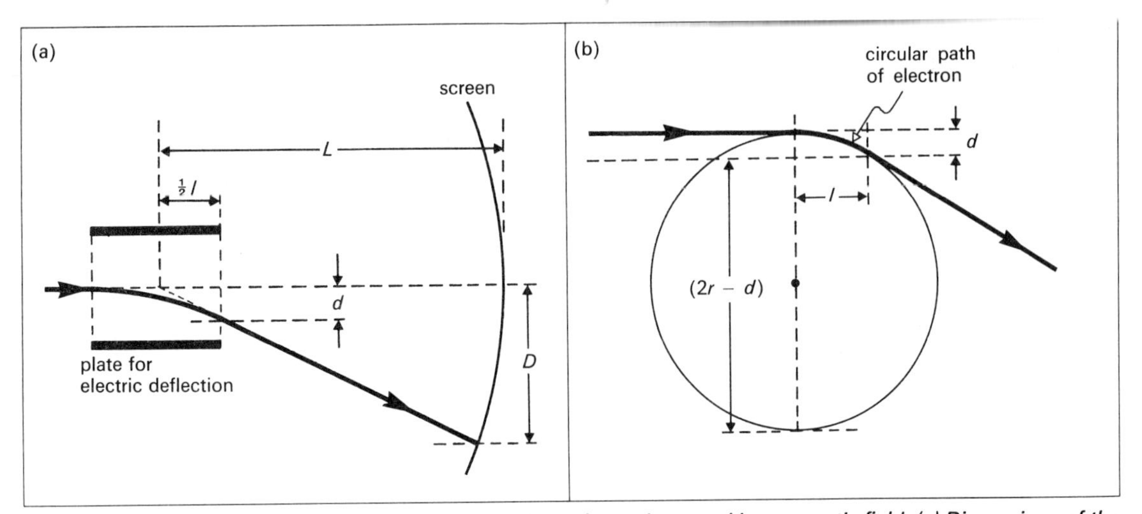

Fig. 60.3. Geometry for measurement of r, radius of circular path caused by magnetic field. (a) Dimensions of the apparatus, and (b) the electron path.

in which

(*a*) v is found from $v = E/B$ (left), and
(*b*) r is found as follows.

Measurement of r

From fig. 60.3

(*a*) By similar triangles

$$\frac{d}{D} = \frac{l/2}{L}$$

(*b*) By the intersecting chords theorem (p. 23)

$$d(2r - d) = l^2$$

d is calculated from (*a*), and substitution into (*b*) gives r.

Notes.

(*a*) *Thomson's* result (1897) was 30% inaccurate because not all the electrons travelled at the same speed, and because the exact extent of the fields was not well defined. *Dunnington's* method (1933) is more accurate.

(*b*) The experiment

(*i*) confirms that the cathode ray speed is less than that of electromagnetic waves (i.e. that the rays, if waves, are not *electromagnetic* waves),

(*ii*) gives a value for e/m_e about 2000 times that found for H^+ from electrolysis.

Millikan's oil-drop measurement of the elementary charge (below) confirms that this is because the electron has a smaller mass (rather than a larger charge).

60.5 Millikan's Measurement of the Electron's Charge

Procedure (fig. 60.4)

(1) Earth both plates, and observe the fall of *a particular drop* under gravity. It acquires a steady downward terminal velocity $\boldsymbol{v}_d$.

(2) Charge the drop by a burst of X-rays. Apply an electric field $\boldsymbol{E}$ by connecting an e.h.t. battery across the plates. Suppose that the same drop, because it experiences an electric force upwards acquires a new (upward) terminal velocity $\boldsymbol{v}_u$.

Calculation

(1) Since the drop is in equilibrium, the resultant force acting on it is zero.

$$\begin{pmatrix}\text{Archimedean}\\ \text{upthrust}\end{pmatrix} + \begin{pmatrix}\text{viscous}\\ \text{force}\end{pmatrix} = (\text{weight})$$

$$U \quad + \quad F_v \quad = W$$

$$\frac{4}{3}\pi r^3 \sigma g + 6\pi\eta r v_d = \frac{4}{3}\pi r^3 \rho g$$

where ρ, σ are oil and air densities, and r is the drop radius. (See p. 148 for *Stoke's Law* $F_v = 6\pi\eta r v$.)

Millikan did a separate experiment to find η, and so could calculate r.

(2) For equilibrium

$$(\text{upthrust}) + \begin{pmatrix}\text{electric}\\ \text{force}\end{pmatrix} = (\text{weight}) + \begin{pmatrix}\text{viscous}\\ \text{force}\end{pmatrix}$$

$$U \quad + \quad F_E \quad = \quad W \quad + \quad F_v$$

$$\frac{4}{3}\pi r^3 \sigma g + EQ = \frac{4}{3}\pi r^3 \rho g + 6\pi\eta r v_u$$

where Q is the charge carried by the drop.

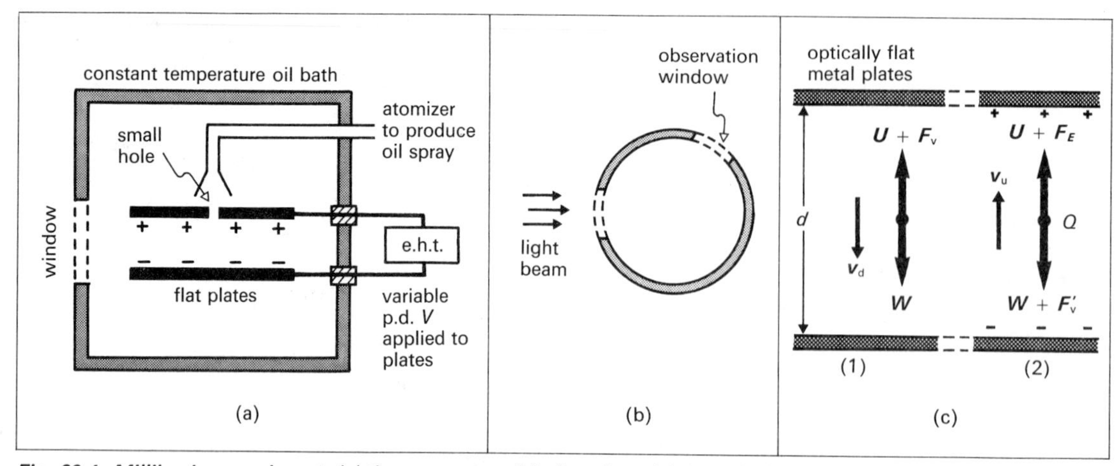

Fig. 60.4. Millikan's experiment: (a) the apparatus, (b) plan view, (c) the calculation.

Since r has been found, and $E = V/d$, Q is now the only unknown, and can be calculated.

Notes.

(*a*) *Millikan* took the following precautions:

(*i*) He used a special oil to reduce evaporation.

(*ii*) He carried out a separate experiment which resulted in his using a modified form of *Stokes's Law* (the oil drop is falling through a non-continuous medium).

(*iii*) The applied field $E = V/d$ was measured with reference to standard cells.

(*b*) *Millikan* repeated his measurements many times and the graph plotted in fig. 60.5 shows how his results might have appeared. It can be seen from the graph that Q does not have every conceivable value, but only whole multiples of one particular value x.

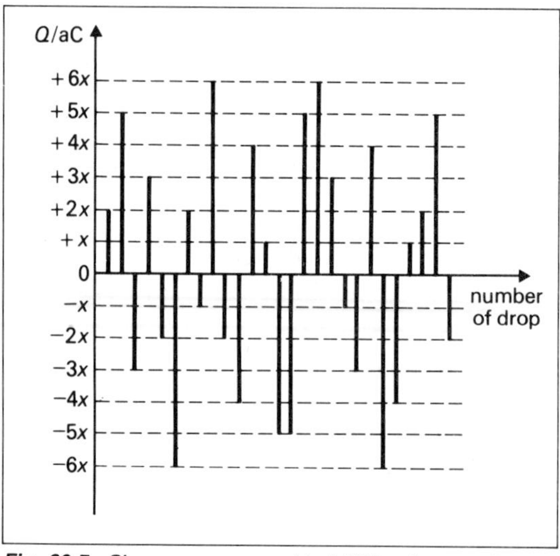

Fig. 60.5. Charges measured in Millikan's experiment.

The interpretation of this observation, which is the fundamentally important contribution of the experiment, is that all electric charges can be only whole multiples of the smallest discrete unit of charge. This basic quantum of charge is of the same size as that associated with an electron. Evidence from other experimental work shows that the charge of an electron is negative, so that in *Millikan's* experiment an excess of electrons gives the oil drop a negative charge, and a deficit gives the oil drop a positive charge. *Millikan* found that the size of x was 0.16 aC, and we write this as $e = -0.16$ aC. The general expression for any charge is $Q = \pm Ne$, where N is a whole number; no fraction of e has ever been observed.

60.6 Thermionic Emission

Electrons are free to migrate within the boundary of a metal, but if an electron tries to leave the surface, it experiences an attractive force from the resultant positive charge it leaves behind.

When an electron is removed from a metal, an external agent does work against that attractive force.

*The **work function energy** W is the least energy that must be supplied to remove an electron from the surface of the metal.*

The work function potential Φ is defined by

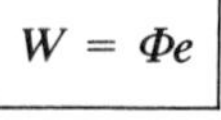

$$W = \Phi e$$

where e is the magnitude of the electronic charge. (Refer also to the photoelectric effect p. 456.)

Examples

Metal	Φ/V	$W/10^{-19}$J
caesium	1.8	2.9
nickel	5.0	8.0
tungsten	4.5	7.2

The low value of Φ (and hence W) for caesium explains why it is used in the photoelectric emission cell (p. 458).

Space Charge

When a metal is heated, its electrons may acquire sufficient thermal energy to leave the surface. This process is **thermionic emission**. The greater the temperature, the greater the rate of emission from unit area of surface. There is a good analogy with *latent heat* and *evaporation*.

As with evaporation a dynamic equilibrium is attained when the region around the metal (the **space charge**) returns electrons to the surface as fast as they leave it.

An **emission current** flows when a positive plate (an **anode**) withdraws electrons from the space charge. The metal then becomes a **cathode**. This is the principle of the **thermionic diode valve**.

Types of Cathode

Because thermionic emission occurs close to a metal's m.p., only some metals are suitable.

(*a*) A thoriated tungsten filament has Φ = 2.6 V, and emits copiously at 1900 K.

(*b*) A metal coated with barium oxide has Φ = 1 V, and can be operated at about 1300 K.

(*c*) As heating is frequently achieved by using a.c., the **indirectly-heated** cathode is used: it has the advantage that its surface is at a *uniform potential.*

60.7 Construction of the Cathode Ray Oscilloscope

Refer to fig. 60.6.
The oscilloscope has three basic components.

(1) The Electron Gun

This consists of

(*a*) the indirectly heated *cathode* to emit electrons,

(*b*) the *grid*, whose potential relative to the cathode controls the electric field at the cathode, and so the number of electrons passing through the grid per second: the grid is the *brightness* control,

(*c*) the *electron lens*, made up of

(*i*) a *focussing anode*, and

(*ii*) an *accelerating anode*.

The potential of the focussing anode is altered by the *focus* control.

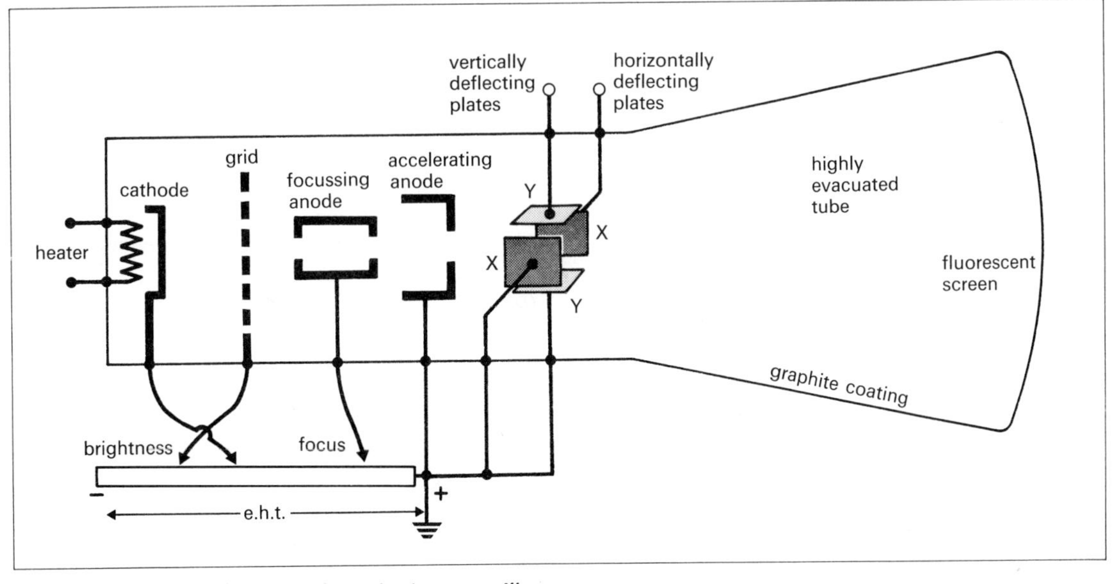

Fig. 60.6. The basic elements of a cathode ray oscilloscope.

Typical potentials for the gun of a small tube might be

(*a*) cathode −800 V,
(*b*) grid −800 V to −850 V,
(*c*) (*i*) focussing anode −500 V to −600 V,
(*ii*) accelerating anode zero (earthed).

The figures represent p.d. relative to the accelerating anode.

The change of k.e. of an accelerated electron can be found from

$$eV_{CA} = \tfrac{1}{2}m_e v^2 - \tfrac{1}{2}m_e u^2$$

where it is accelerated from a speed u to speed v because of the p.d. V_{CA} between anode and cathode. Usually $\tfrac{1}{2}m_e u^2 \ll \tfrac{1}{2}m_e v^2$, and can be considered negligible.

(2) The Deflecting System

This consists of two pairs of parallel plates arranged so that their respective electric fields produce horizontal and vertical deflection of the electron beam.

(*a*) The *X-shift* and *Y-shift* controls change the undisplaced location of the spot on the screen.

(*b*) The *sensitivity* control changes the displacement produced by a given change of p.d. applied to the input (p. 315).

One of each pair of the deflecting plates is at the same (earth) potential as the accelerating anode.

(3) The Display System

This consists of a *screen* coated with a *phosphor* such as zinc sulphide, whose molecules absorb energy from the incident electron beam, and re-radiate it as visible light. The re-radiation is

(*a*) *fluorescence* if there is no afterglow, and
(*b*) *phosphorescence* if the effect continues for a short while after impact.

The screen potential remains constant because the electrons return to earth through the internal graphite coating.

The Time Base

A time-base consists of a system which applies a saw-tooth p.d. of the type in fig. 60.7 across the X-plates.

Along AB the spot traverses the screen at a steady finite speed. Along BC it flies back (almost instantaneously) to the starting point.

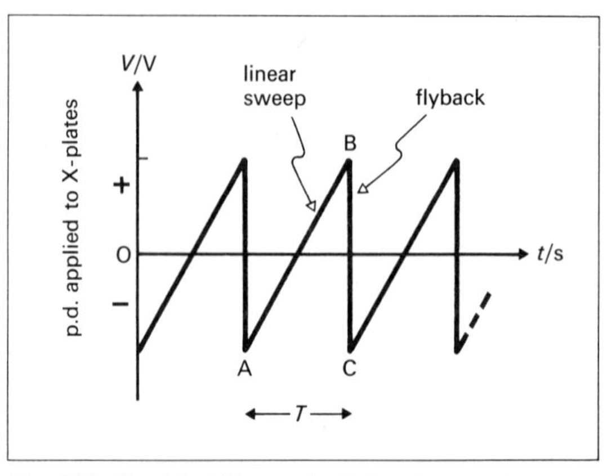

Fig. 60.7. The ideal (theoretical) time-base p.d.

The time period T can be changed by coarse and fine controls on the front of the c.r.o.

The time-base is frequently *synchronized* with the signal applied to the Y-plates.

60.8 Uses of the C.R.O.

The c.r.o. is most versatile. The following are among the uses to which it is put.

(a) Use as a Voltmeter

(*i*) D.C. Connect the p.d. to be measured across the X- or Y-plates, without using the time-base. The spot deflection can be calibrated in mm/V by using a known p.d. The c.r.o. has a very high impedance, and is not damaged by overloading (which merely causes the spot to be deflected off the screen).

(*ii*) A.C. Connect the p.d. to be measured as in (*i*). The spot deflection represents 2 × (*peak voltage*), because the inertia of the electron beam is negligible.

The c.r.o. can be used as a voltmeter over a wide range of frequencies. (Cf. the moving-coil galvanometer which reads zero at high frequency.)

(b) The Display of Waveforms

Connect the signal to be displayed to the Y-plates. Fig. 60.8 gives two examples. The fly-back trace is usually suppressed.

This method can be used to display

(*i*) half-wave rectification (p. 425),
(*ii*) full-wave rectification (p. 425),
(*iii*) the effects of smoothing and filtering (p. 426),
(*iv*) the waveforms of musical notes.

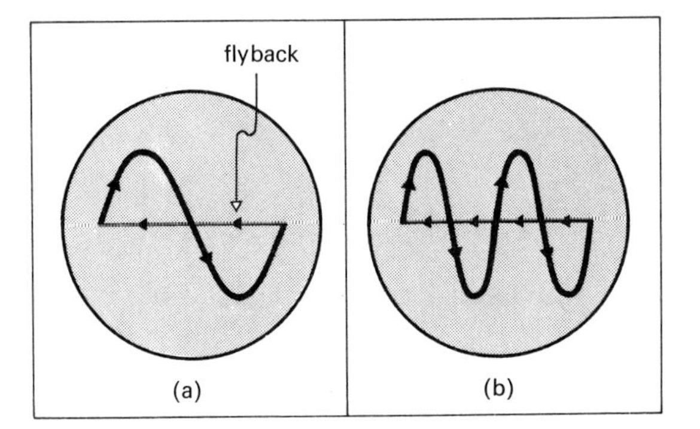

Fig. 60.8. British a.c. mains waveform (50 Hz).
(a) Time-base frequency 50 Hz,
(b) time-base frequency 25 Hz.

(c) The Display of Phase Relationships

Suppose V_X and V_Y are sinusoidal p.d.'s of the same frequency. Then if they are of equal amplitude, and connected to the X- and Y-plates, one would see

(*i*) a straight line at 45° to the axis if they are in phase,
(*ii*) a circle if the phase difference is $\pi/2$ rad,
(*iii*) an ellipse for other phase differences.

The method is useful for demonstrating phase relationships in a.c. circuits (p. 417), and the theory is discussed on p. 97.

(d) To Compare Frequencies

Suppose the two p.d.'s whose frequencies are to be compared are connected across the deflecting plates.

When the frequency ratio is appropriate a steady pattern (called a **Lissajous figure**) appears on the screen. The form of the pattern enables the ratio to be calculated (see p. 86).

(e) To Display Hysteresis Loops

(See page 430.)

61 Solid State Electronics

61.1 The Principle of the p–n Junction Diode

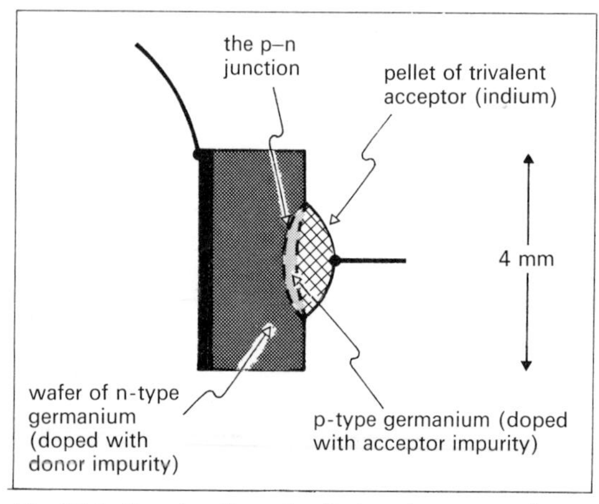

Fig. 61.1. Construction of a diode using germanium.

The whole p–n junction (fig. 61.1) must be part of a single continuous crystal of semiconductor material to avoid the high rates of recombination that would otherwise occur at the junction of two separate crystals. For protection the junction diode is enclosed in a light-tight capsule.

Its Action

(a) Zero Bias (no Applied Field) (fig. 61.2)

Notes.

(*i*) The thermal diffusion of holes into the n-region and electrons into the p-region causes a net migration of charge across the boundary or junction.

(*ii*) The resulting p.d. is called the **contact potential**, and represents a *barrier* of a fraction of a volt. The resulting field tends to prevent further diffusion of charge-carriers.

(*iii*) The height of the barrier (the p.d.) is determined by the width of the **depletion layer** or **transition region**. In this region the total carrier number density is less than one thousandth of its value in either the p- or n-regions. This depletion is not due to recombination. It is because the sum of the densities of the holes and electrons is least when the densities of each are the same.

(*iv*) The size of the contact potential decreases as the temperature rises. It also decreases with increasing density of doping impurities.

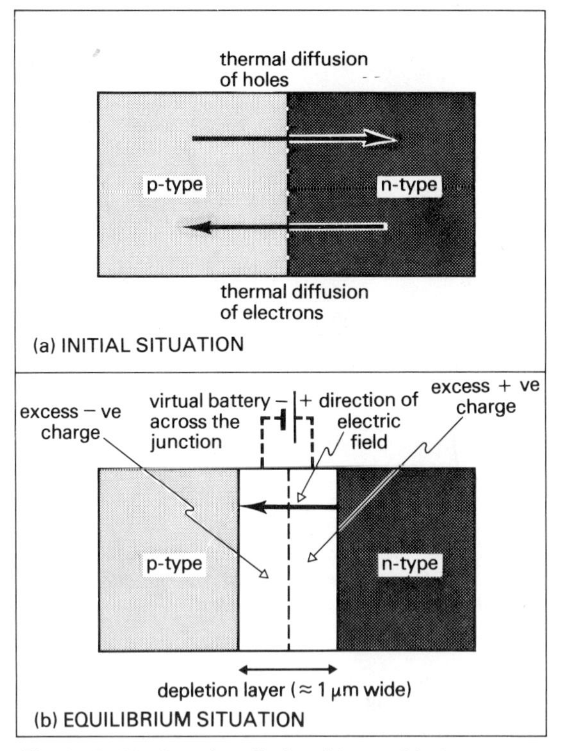

Fig. 61.2. The junction diode with zero bias.

(*v*) No current would flow if the two regions were connected by a wire, since equal and opposite contact potentials would be set up at the two metal/semiconductor contacts.

(*vi*) The p–n junction behaves like a capacitor of capacitance equal to

(permittivity) × (junction area) ÷ (junction width)

(b) Reverse Bias (a Field Applied) (fig. 61.3)

Notes.

(*i*) Reverse bias means that the p-region is made negative w.r.t. the n-region. The zero-bias contact potential across the boundary is increased (approximately in proportion to the square root of the applied p.d.) and the flow of majority carriers is further inhibited.

(*ii*) The electron number density in the p-region and the hole density in the n-region are both greatly reduced. Therefore the minority-carrier densities are also reduced on either side of the depletion layer.

(*iii*) The typical current flow for a small diode is a few μA. It is due to the *minority* carriers being pushed across the junction, and constitutes a **leakage current**.

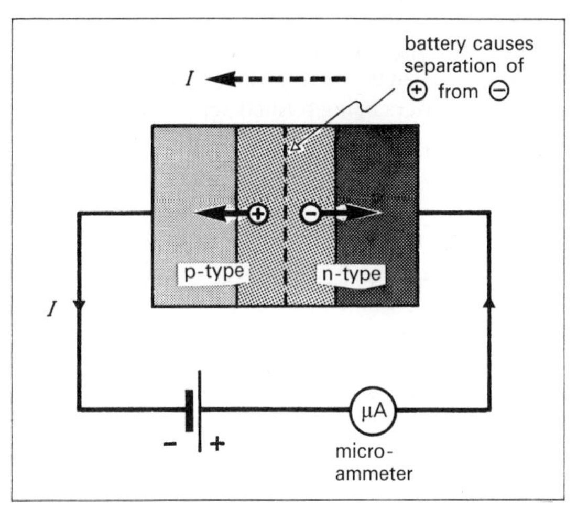

Fig. 61.3. The junction diode with reverse bias.

(*iv*) As the junction width is increased, the capacitance is decreased.

(c) Forward Bias (fig. 61.4)

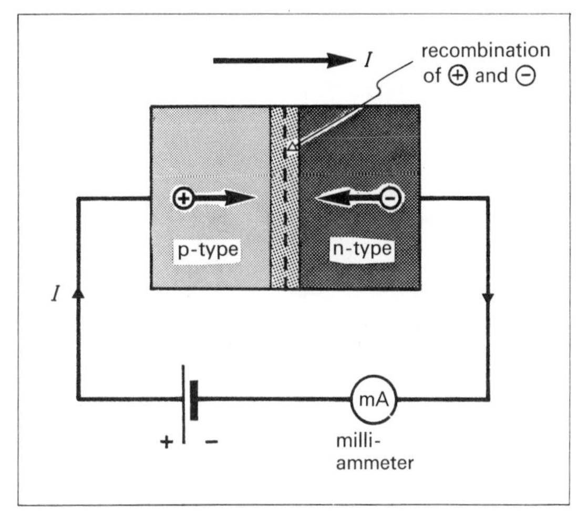

Fig. 61.4. The junction diode with forward bias.

Notes.

(*i*) Forward bias means that the p-region is made positive relative to the n-region. The zero-bias p.d. across the junction is overcome. Charge-carriers of both signs can move and recombine at the junction. The width of the depletion layer is reduced.

(*ii*) The minority-carrier densities have increased relative to their equilibrium values on either side of the depletion layer.

(*iii*) The typical current flow for a small diode is a few mA, but can be made much larger. It is due to the *majority* carriers being pushed across the junction.

(*iv*) The circuit symbol for the junction diode is shown in fig. 61.5.

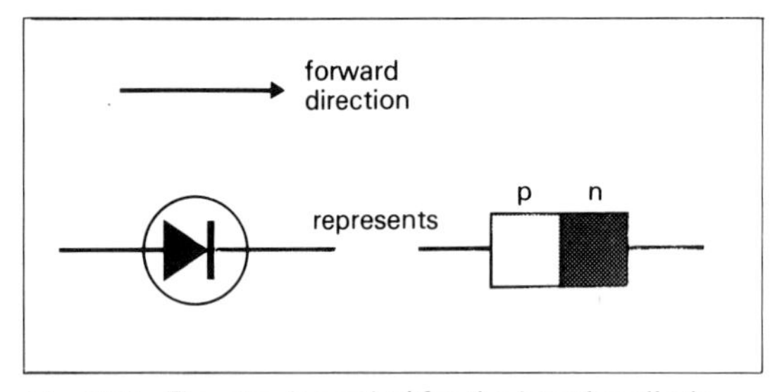

Fig. 61.5. The circuit symbol for the junction diode.

61.2 Characteristic of the p–n Junction Diode

Circuits (fig. 61.6)

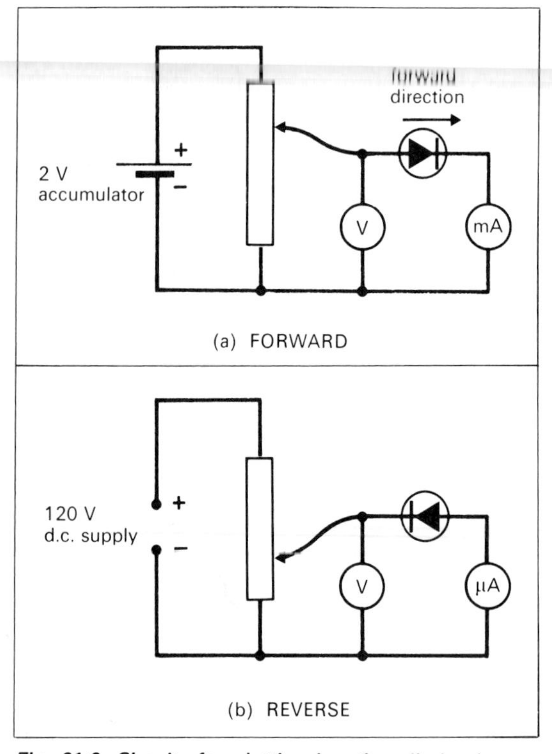

Fig. 61.6. Circuits for plotting junction diode characteristics.

Results (fig. 61.7)

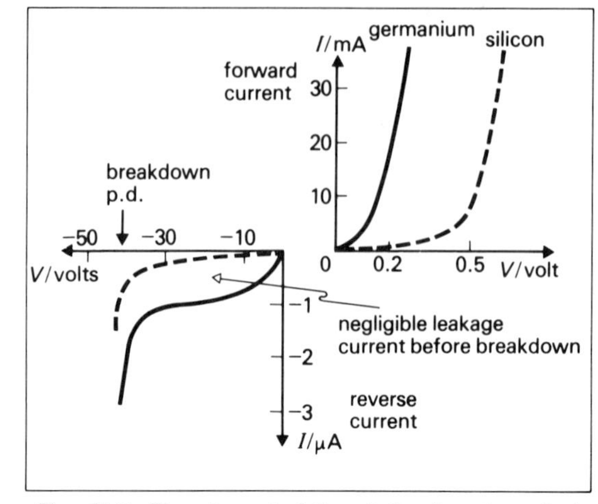

Fig. 61.7. The characteristic of the junction diode. (Note the different current scales.)

Notes.

(*a*) The graph of fig. 61.7 holds provided the temperature is kept below a certain value (usually by a blackened sheet of copper).

(*b*) The leakage current due to minority carriers remains small for quite a large range of reverse p.d.'s.

(*c*) The junction diode can therefore be used for *rectification* as the forward currents continue to increase with increase in applied p.d.

(*d*) The device has a *maximum peak inverse voltage*: a greater reverse field damages the crystal structure as the current increases rapidly.

(*i*) The **Zener effect** (or *tunnelling*) occurs when the applied electric field pulls electrons directly out of their bonds, giving an increased current.

(*ii*) The **avalanche effect** occurs when high energy electrons passing through the crystal release further electrons from the bonds by collisions. These extra electrons in turn dislodge others causing an *avalanche*.

Voltage regulator diodes (or *Zener* diodes) make use of these breakdown effects.

61.3 Introduction to Transistors

Their Uses

Both field effect and bipolar junction transistors are described in this chapter and, although the student can often regard them as alternatives, their particular properties must be considered in practical applications.

Transistor type	f.e.t (unipolar)	b.j.t. (bipolar)
Input impedance/Ω	10^9 to 10^{18}	1 to 10^6
Effect of incident radiation	negligible	damage caused
Effect of high temperatures	negligible	damage caused
Quantity amplified directly	p.d.	current

For the general-purpose transmission of electrical signals, bipolar transistors are preferred because of their greater amplification, higher frequency response and higher power capability. Unipolar transistors are often more suitable in voice-frequency amplifiers with their very small control current and in certain radio receivers where they can reduce interference and signal distortion. In logic or computing circuits the choice is not obvious and requires detailed investigation: unipolar types involve only one diffusion to make both source and drain, and are popular in integrated circuits in computer memories; whereas bipolar types are frequently more appropriate owing to their higher amplification and lower operating power requirement.

The Distinction between Them

(*a*) In **unipolar** transistors the current is carried by the drift of majority carriers of *a single type*, and is controlled by the electric fields within the semiconducting materials. Types include

(*i*) *the junction gate field effect transistor* (j.f.e.t.) (**61.4**), and

(*ii*) *the insulated gate field effect transistor* (i.g.f.e.t.) (**61.6**).

Unipolar transistors are often referred to as *majority-carrier devices*.

(*b*) In **bipolar** transistors (such as the *junction transistor* b.j.t. of **61.8**) the current is carried by the drift of *both* electrons and positive holes; they are often referred to as *minority-carrier devices*.

61.4 The Junction Gate Field Effect Transistor (j.f.e.t.)

Figs. 61.8 and 61.9 show the construction and bias p.d.'s of an n-channel j.f.e.t.

Notes.

(*a*) A *p-channel* j.f.e.t. has an n-type gate, a p-type channel and an n-type substrate. The circuit symbol has the arrow in the opposite direction.

(*b*) The current in the channel is controlled by reverse-biasing the p–n junctions formed by the gate-source and the gate-drain interfaces. When reverse-biased, the depletion layer widens, leaving a narrower channel.

(*c*) *Electrons* flow through the n-channel when the drain is made positive with respect to the source. For a p-channel j.f.e.t. the polarities of V_{DS} and V_{GS} would be reversed and *holes* would flow through the channel.

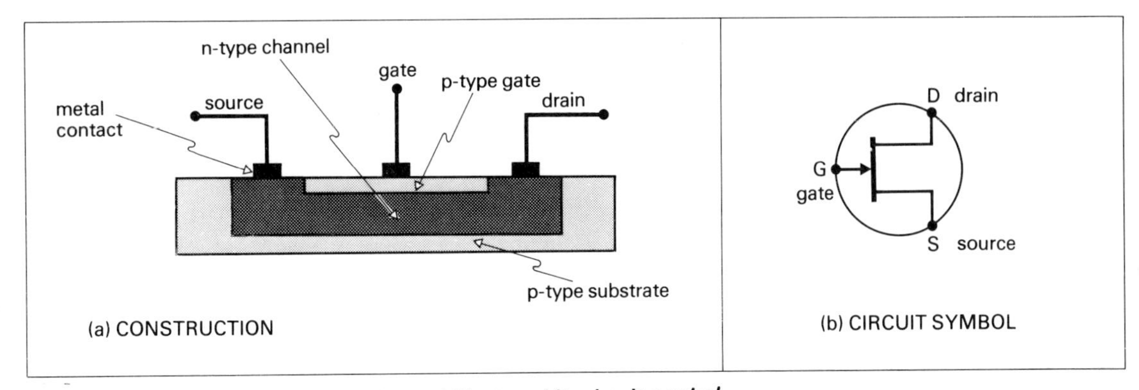

Fig. 61.8. The construction of an n-channel j.f.e.t. and its circuit symbol.

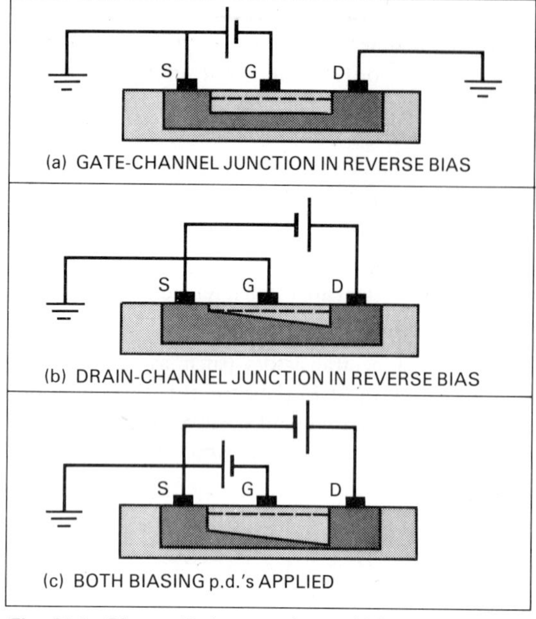

Fig. 61.9. Bias p.d.'s in an n-channel j.f.e.t.

(*d*) The potential drop along the channel causes a variation in the reverse bias so that the gate becomes deeper at the drain end than at the source end.

(*e*) At **pinch-off** the gate reaches across the channel making it very narrow. The drain current I_D does not now show any significant increase with increase of V_{DS}. This is because any increase in I_D predicted from an increase in V_{DS} is balanced owing to the narrowing of the channel. If V_{GS} has a small negative value, pinch-off will occur at a lower value of V_{DS} and of I_D.

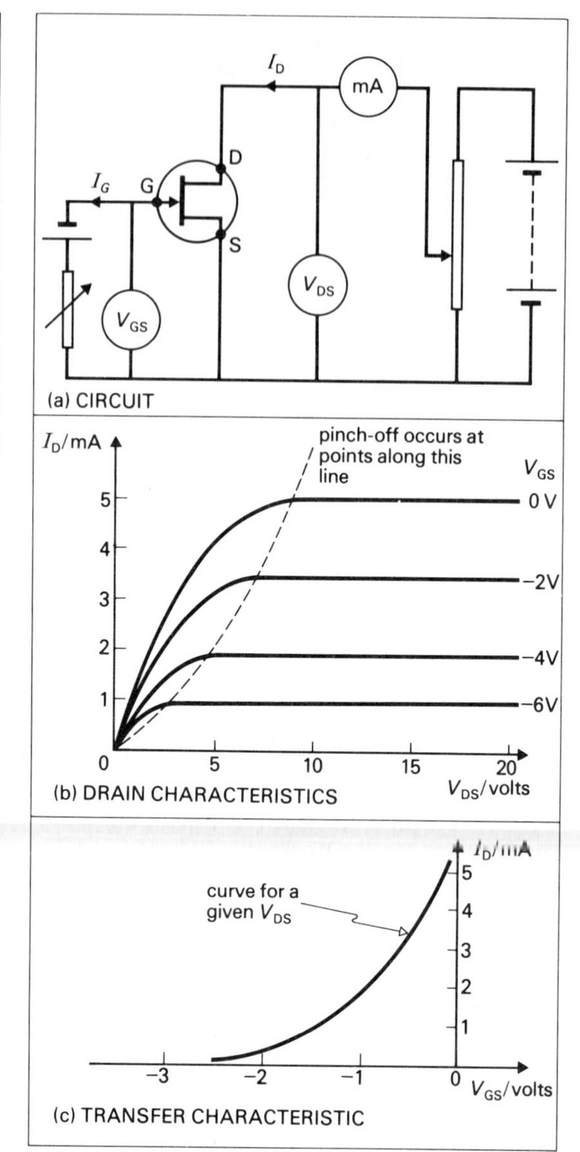

Fig. 61.10. Characteristics for an n-channel j.f.e.t.

61.5 Characteristics of n-channel j.f.e.t.'s

The circuit with the **drain** and **transfer characteristics** that can be obtained for an n-channel j.f.e.t. are shown in fig. 61.10.

Notes.

(*a*) The dotted line in fig. 61.10(*b*) is the locus of the points at which pinch-off occurs.

(*b*) (*i*) The *triode region* of operation, below pinch-off, is where the drain current I_D is strongly affected by V_{DS} as well as by V_{GS}. In this region some j.f.e.t.'s are used as voltage-controlled *resistors*.

(*ii*) The *pentode* or *pinched-off region* is where the drain current is largely independent of V_{DS}. In this region j.f.e.t.'s are used as *amplifiers*.

(*c*) The **pinch-off p.d.** is the value of V_{DS} when the value of I_D becomes effectively constant. It depends on the value of V_{GS}.

(*d*) The gate current I_G is negligibly small (~nA) and is independent of V_{DS}. If the junction becomes forward-biased then I_G becomes large, and the j.f.e.t. is no longer a field effect device.

(*e*) The input impedance is very high ($>10^9\ \Omega$).

(*f*) The **mutual conductance**

$$g_m = \frac{\Delta I_D}{\Delta V_{GS}}$$

and the **drain** or **output conductance**

$$g_D = \frac{\Delta I_D}{\Delta V_{DS}} = \frac{1}{r_D}$$

where r_D = drain resistance. g_D is usually very small compared with g_m and is correspondingly less important. (E.g. at $V_{DS} = 15$ V, $g_m \approx 10^{-3}$ S, $g_D \approx 10^{-5}$ S.)

61.6 The Insulated Gate Field Effect Transistor (i.g.f.e.t.)

Figs. 61.11 and 61.12 show the construction of, and application of drain and gate p.d.'s to, a p-channel i.g.f.e.t. These devices are also referred to as *metallic oxide semiconductor transistors* (*m.o.s.t.'s*), as the insulating layer is usually a metallic oxide.

Notes.

(*a*) Both source and drain form p–n junctions with the channel. The channel material is n-type but the device is a **p-channel** i.g.f.e.t. as it is the induced *holes* that carry the current. The gate is a metal film separated from the silicon by SiO_2 or some other good insulator.

(*b*) Holes are induced in the channel after V_{GS} has become more negative than that necessary for the threshold p.d.

(*c*) If the drain is made negative relative to the source, holes flow towards it owing to the electric field. As the drain becomes more negative, the potential difference between gate and drain decreases and, when it falls below the threshold p.d., the channel is pinched off.

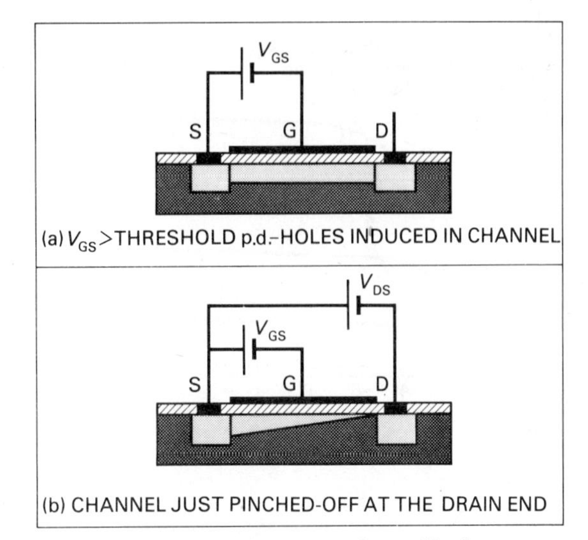

Fig. 61.12. P.d.'s applied to a p-channel i.g.f.e.t.

(*d*) An **n-channel** i.g.f.e.t. would have p-type substrate and *electrons* would be induced in it using a positive gate/source p.d. The circuit symbol for the n-channel type has the arrow reversed.

61.7 Characteristics of p-channel i.g.f.e.t.'s

Fig. 61.13 shows the drain characteristics for a p-channel i.g.f.e.t.

Notes.

(*a*) The student should sketch a circuit diagram suitable for investigating these characteristics, using fig. 61.10(a) as a guide.

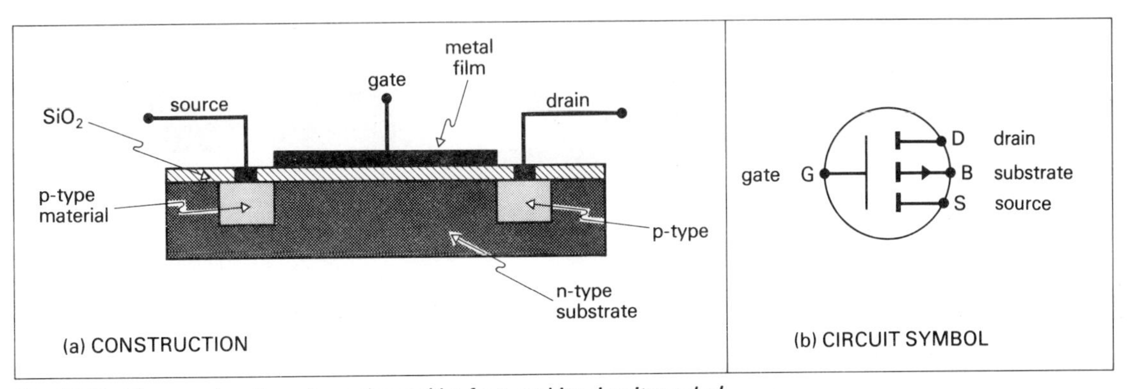

Fig. 61.11. The construction of a p-channel i.g.f.e.t. and its circuit symbol.

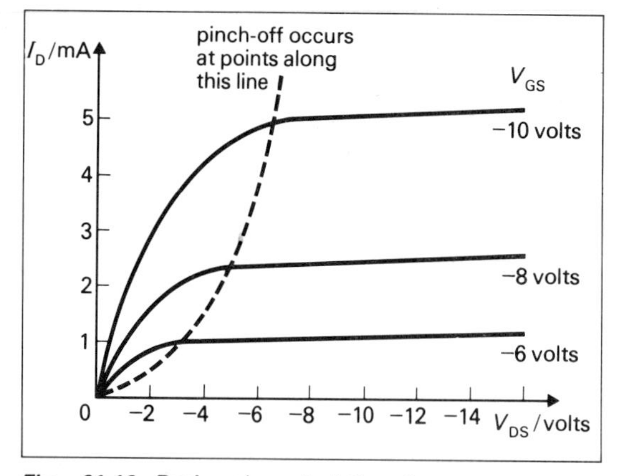

Fig. 61.13. Drain characteristics for a p-channel i.g.f.e.t.

(*b*) The broken line shows the locus of the points where the channel pinches-off. This occurs when the drain/gate potential difference drops below the threshold value. At all points on this line

$$V_{DS} - V_{GS} = V_{threshold}$$

(*c*) I_G is very small ($\approx 10^{-14}$ A) because the gate is insulated.

(*d*) The i.g.f.e.t. described here is an *enhancement-mode version* as current can be enhanced only by the gate potential. (A less common device is the *depletion mode* type in which the channel exists (because of appropriate impurity concentrations) before any gate potential is applied—the p.d. is then used to reduce the conductivity of the channel.)

61.8 The Bipolar or Junction Transistor (b.j.t.)

Bipolar transistors consist of two p–n junctions very close together in a single piece of semiconducting material. Fig. 61.14 shows the construction of n–p–n and p-n-p transistors and their circuit symbols.

Notes.

(*a*) The arrowhead in the symbol indicates the conventional direction of current flow. The n–p–n types are now more commonly used and are described here. (The action of the p–n–p type can be understood by reversing the polarity of the applied potential differences and remembering that the majority carriers have the opposite sign of charge.)

(*b*) The transistor is enclosed in a light-tight capsule for protection.

(*c*) In the n–p–n transistor, the terms **emitter** and **collector** indicate what is happening to the *electrons.*

The Action of the Junction Transistor

Notes.

(*a*) Fig. 61.15 shows a circuit in which the emitter is common to base and collector circuits. This is called a **common-emitter** connection. The emitter is frequently earthed or **grounded**. Common-base and common-collector are other methods of connection which can be used.

(*b*) In both n–p–n and p–n–p transistors the collector-base junction must be reverse-biased: in the absence of $\mathscr{E}_2$ no current would flow across it.

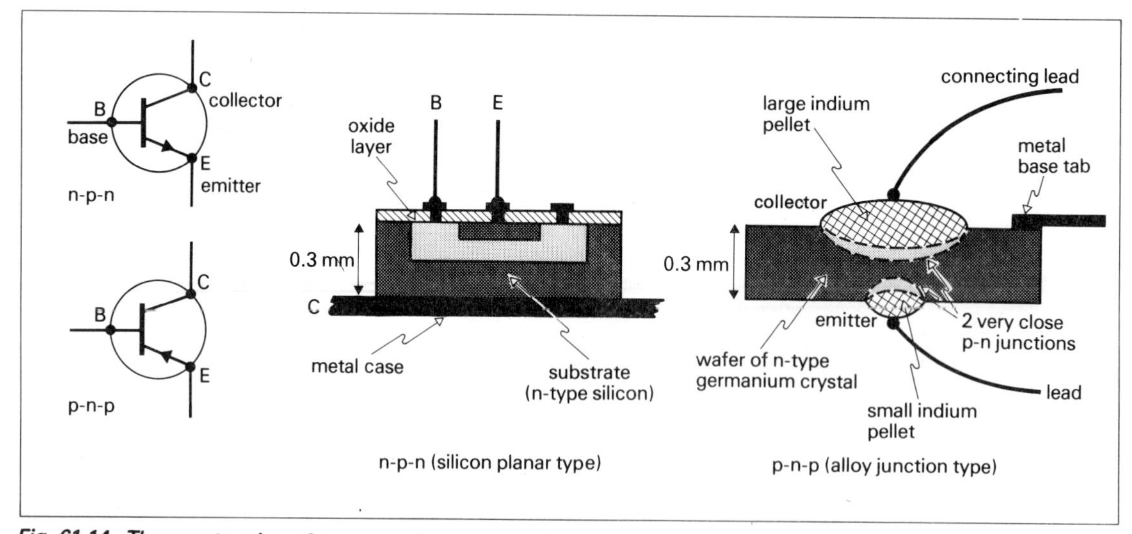

Fig. 61.14. The construction of n–p–n and p–n–p transistors and their circuit symbols.

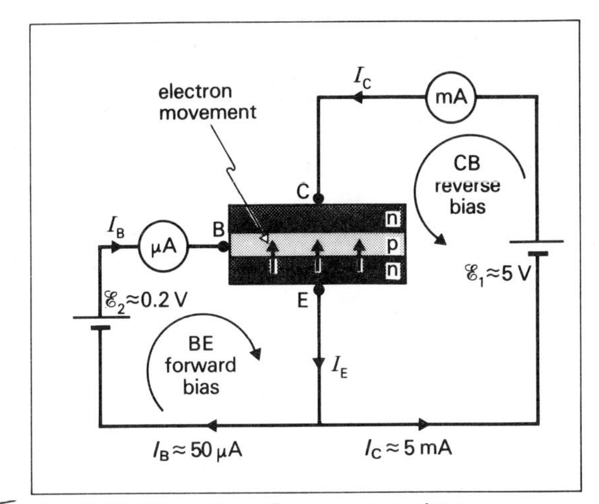

Fig. 61.15. The action of a n–p–n transistor.

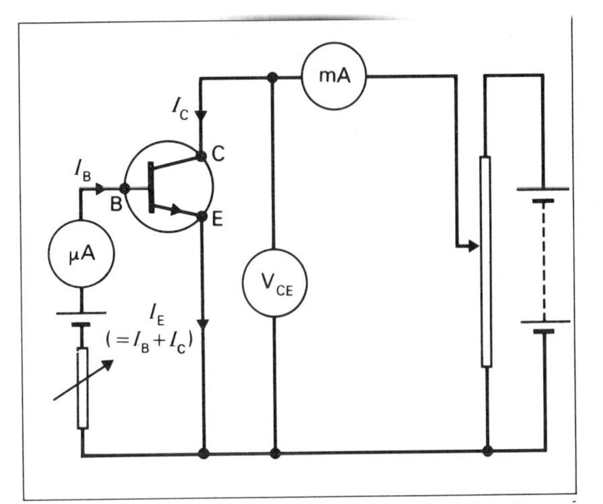

Fig. 61.16. Circuit for plotting characteristics for an n–p–n transistor.

The minority-carrier number density in the base next to the collector junction is reduced.

(*c*) Similarly, the emitter-base junction must always be forward-biased. This raises the minority-carrier density in the base next to the emitter junction.

(*d*) Electrons are swept across the base-emitter junction into the very thin base region and,

(*i*) about 99 per cent of the *electrons* are attracted straight through into the collector where they constitute the collector current I_C, but

(*ii*) only about one per cent combine with holes in the base causing a small base current I_B to flow.

Thus the value of I_C may be perhaps 100 times I_B.

(*e*) In the p–n–p transistor, *holes* are swept across the p–n junction. A small proportion combine with electrons in the base to give a small base current. The electron flow in the base is maintained by supply from the external circuit.

61.9 Characteristics of the Junction Transistor

For the circuit of fig. 61.16, which uses the common-emitter connection, we are interested in the interdependence of

(*i*) V_{CE}, the p.d. between collector and emitter

(*ii*) I_B, the base (input) current, and

(*iii*) I_C, the collector (output) current.

These relationships can be displayed by the **output** and **transfer** characteristics.

(*a*) The Output Characteristic

This shows the dependence of I_C on V_{CE} for fixed values of I_B. The results can be seen in fig. 61.17.

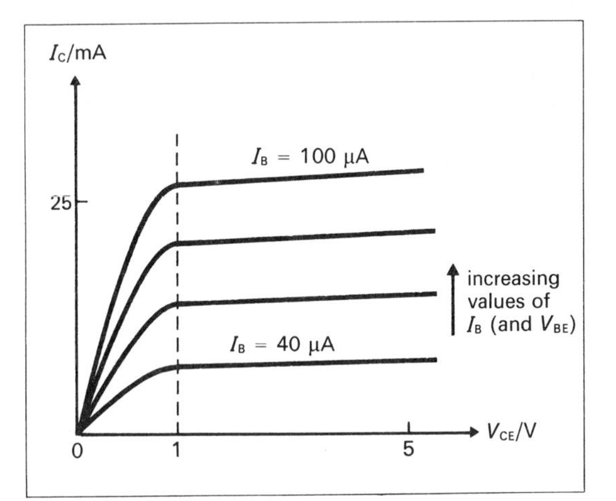

Fig. 61.17. Output characteristics for an n–p–n transistor.

The transistor is usually operated beyond the sharp change of slope. For this region we define the

$$\textbf{output resistance}\,[\Omega] = \frac{\Delta V_{CE}[\text{V}]}{\Delta I_C[\text{A}]}$$

for a fixed I_B.

It is the inverse of the slope of the characteristic. The collector-base junction is reverse-biased, which makes the output resistance of the order of 1 MΩ.

(b) The Transfer Characteristic

This shows the dependence of I_C on I_B for fixed values of V_{CE}. The results can be seen in fig. 61.18.

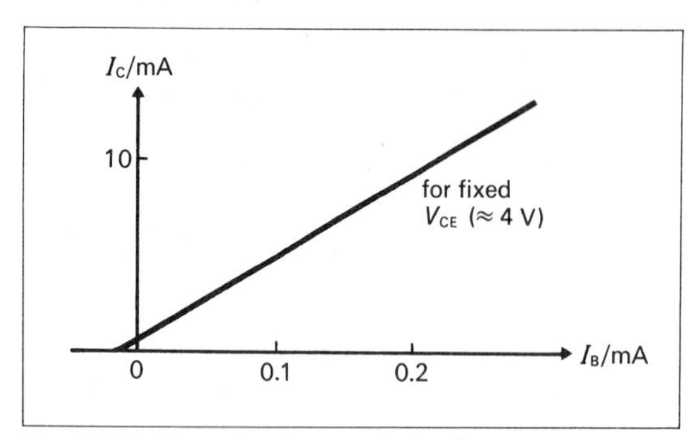

Fig. 61.18. Transfer characteristics for an n–p–n transistor.

The transfer characteristic is linear. For a fixed value of V_{CE} the quantity I_C/I_B is called the **static value of the forward current transfer ratio**, symbol h_{FE} (it used to be called the *current amplification factor*, α').

We define the* small signal forward current transfer ratio h_{fe} *by the equation

$$h_{fe} = \frac{\Delta I_C}{\Delta I_B}$$

The subscript $_e$ denotes the use of a common-emitter connection. h_{fe} is approximately equal to h_{FE}, being the slope of the characteristic. Typically it has a value ≈100 (no unit).

Note that the polarities and currents would be reversed in the circuit used to obtain the characteristics for a p–n–p transistor.

61.10 The Transistor as an Amplifier

(a) The f.e.t. amplifier (fig. 61.19)

Notes

(*a*) A p-channel device would have the circuit symbol arrow in the opposite direction and the polarities would be reversed.

(*b*) The alternating input signal changes the potential difference between gate and source, which causes an alternating current to be superimposed on the otherwise steady drain current I_D. The result is a variation in the p.d. across R_D, and so an alternating

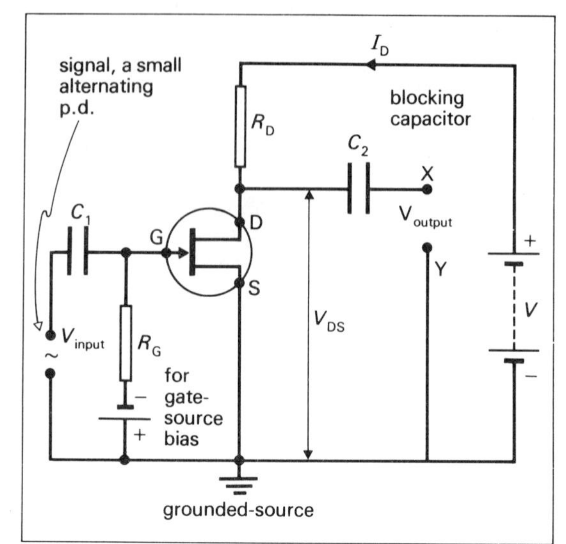

Fig. 61.19. A n-channel f.e.t. amplifier.

p.d. appears superimposed on the direct p.d. between drain and source. C_2 blocks the steady component so that the output p.d. V_{XY} is entirely alternating. The f.e.t. is a **voltage-controlled** device.

(*c*) C_1 blocks any direct component in the input signal, and the gate resistor R_G ensures that the correct value of V_{GS} is applied to the f.e.t. The arrangement is known as **resistor-capacitor coupling**.

(*d*) The source of the applied p.d. V_{app} supplies the power necessary for amplification.

$$V_{app.} = (V_R)_D + V_{XY} = \text{constant}$$

Since V_{XY} decreases when $(V_R)_D$ (and hence V_{input}) increases, the input signal and the output are *in antiphase*.

(*e*) The **amplification of p.d.** or **gain** equals $\frac{\Delta V_{out}}{\Delta V_{in}}$. This can be shown to be approximately equal to $g_m R_D$, where g_m is the mutual conductance (p. 449).

(b) The b.j.t. amplifier

The circuit is basically the same as that for the f.e.t. amplifier (fig. 61.19), apart from the transistor symbol. Figs. 61.16 and 61.18 illustrate how the small current I_B in the base circuit controls the relatively large current I_C in the collector circuit. The b.j.t. is thus a **current-controlled** device as the current is amplified directly, whereas the f.e.t. is *voltage-controlled*. The input signal causes an alternating current to be superposed on the steady base current.

This results in an alternating current (h_{fe} times as great) superposed on the steady state collector current. Variations in the current through the collector resistor R_C cause variations in the p.d. across R_C, and so in the potential of X relative to Y.

(c) The practical amplifier (fig. 61.20)

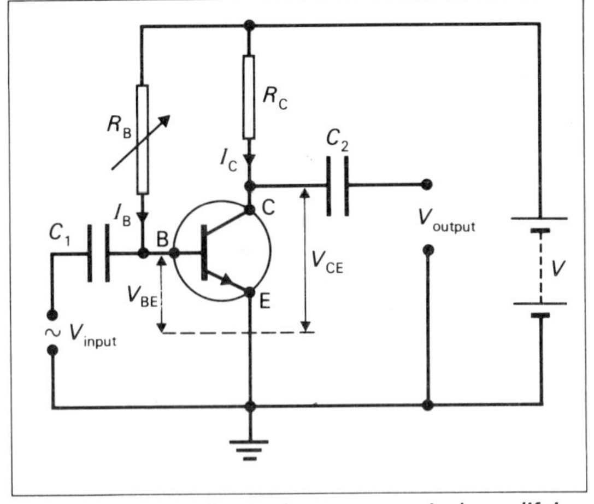

Fig. 61.20. A simple single-stage practical amplifying circuit.

The correct operating p.d.'s for a b.j.t. amplifier (V_{CE} and V_{BE}) are obtained from a single supply using the circuit of fig. 61.20. The base-bias resistor R_B has large resistance so that the steady bias current flowing into the base is very small. The base is thus held at a potential of the *same polarity* as that of the collector but much smaller in magnitude. (Note that in the f.e.t. the gate is held at a potential *opposite* from the polarity of the drain.) For greater amplification, several stages are connected *in cascade* (the output from one stage providing the input of the next.)

61.11 The Transistor as an Oscillator

The tuned circuit L_1C_1 in fig. 61.21 is connected to the collector and changes in current through L_1 are fed back to the base-emitter circuit by L_2. The battery makes the emitter negative relative to the base and a high resistance R is used to provide the positive bias for the collector relative to the base. C_2 prevents the base and emitter from being short-circuited, so that a steady bias is maintained between them.

The circuit acts as an amplifier supplying its own input. If a small portion of the amplified p.d. is fed

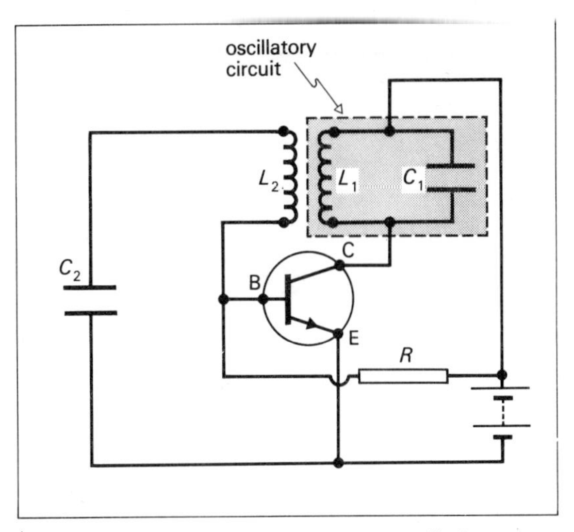

Fig. 61.21. Circuit for generating a.c. oscillations.

back in the correct phase, enough energy will be replaced in the circuit L_1C_1 to overcome the resistive and dielectric energy losses.

Note that the coil/capacitor circuit oscillates, and the transistor merely provides the necessary energy from the battery at the right time.

61.12 The Transistor as a Switch

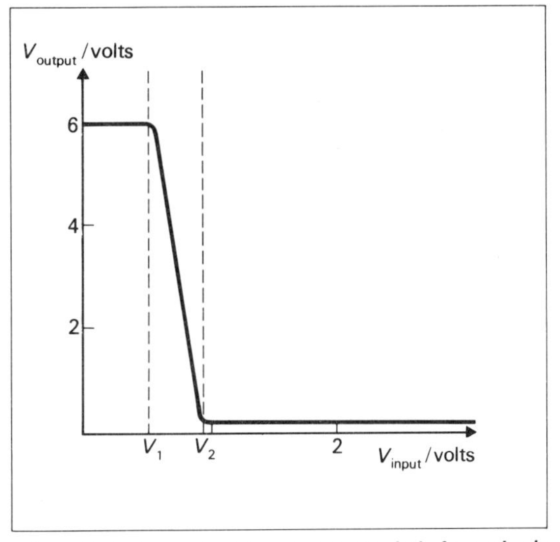

Fig. 61.22. The output-input characteristic for a single stage transistor amplifier.

The output-input characteristic of fig. 61.22 (which refers to fig. 61.23) shows that when V_{input} is below V_1, no current flows through the transistor so that

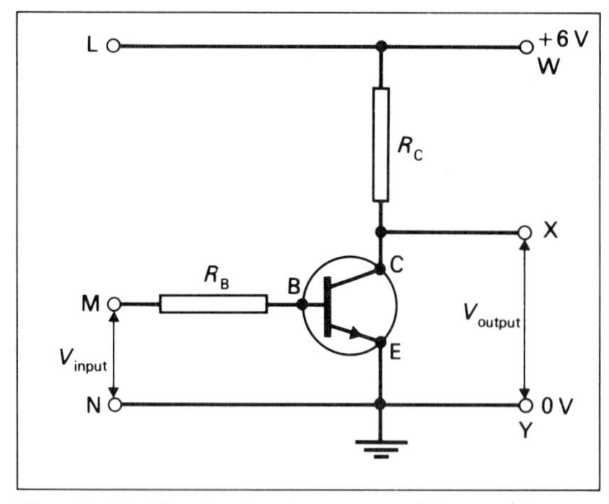

Fig. 61.23. Circuit showing a transistor used as a switch.

V_{output} is +6 V. When V_{input} is above V_2, the collector current is large enough for most of the input p.d. to appear across the collector resistor; V_{output} now remains near to 0 V as V_{input} is increased. The circuit of fig. 61.23 shows how a transistor can be used as a **switch**. If L and M are connected, the potential of X will be close to that of Y, so that V_{output} is near to 0 V. When N and M are connected, the potential of X will be the same as that of W so that V_{output} is +6 V. I.e. when

$$V_{input} = +6\text{ V}, \qquad V_{output} \approx 0\text{ V}$$

and when

$$V_{input} = 0\text{ V}, \qquad V_{output} = +6\text{ V}$$

V_{input} is normally provided by another circuit and switching operations can be executed very rapidly. Switches with several inputs (and therefore various outputs) are called **gates**, and in general they are more complicated versions of the simple transistor switch.

61.13 Other Uses of the Transistor

The transistor also finds application in

(*i*) the *thermistor* (p. 355),
(*ii*) the *phototransistor*, a type of photocell (p. 348) with automatic current amplification,
(*iii*) the *solid detector* (p. 471), and
(*iv*) *integrated circuits.* Each circuit is a small piece of silicon (≈0.2 mm thick) in which perhaps thousands of transistors and other electrical components have been formed and interconnected. They are of fundamental importance in electronics because they are

(*a*) small and light,
(*b*) more reliable than conventional circuits,
(*c*) mass-produced and therefore cheap, and
(*d*) used in circuits with special properties and applications.

61.14 Thermionic and Semiconductor Devices Compared

Junction diodes and transistors have largely replaced thermionic diodes and triodes because they

(*i*) are cheaper,
(*ii*) can be made much smaller and more resistant to damage from impact,
(*iii*) do not need heating to energize them (and so the transistor needs less connections), and
(*iv*) use applied p.d.'s about one-tenth of those of thermionic devices.

On the other hand they may

(*i*) suffer breakdown if the temperature becomes too high, and
(*ii*) be damaged permanently by using connections of wrong polarity, or p.d.'s that are too high.

62 The Photoelectric Effect

62.1 Description of the Photoelectric Effect

The first observations were made by *Hertz* in 1887.

The effect consists of the emission of electrons from the surface of a metal when electromagnetic waves of high enough frequency fall on the surface.

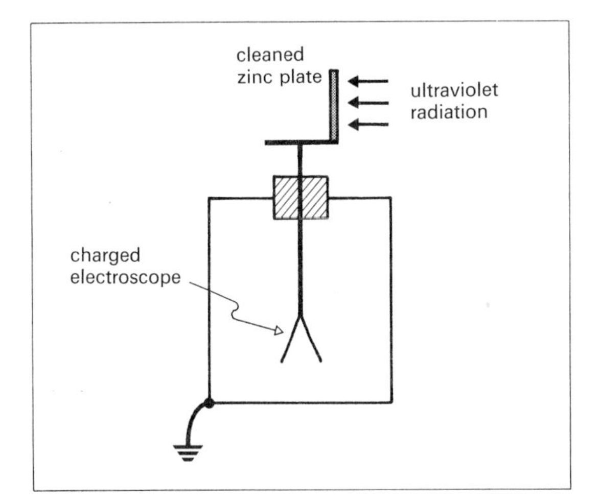

Fig. 62.1. To demonstrate the photoelectric effect for zinc.

In fig. 62.1,

(*a*) if the electroscope is *negatively* charged, the leaves collapse steadily, but

(*b*) if it is *positively* charged, the leaves show no change of deflection.

In 1899 *Thomson* and *Lenard* showed that the liberated negative ions (**photoelectrons**) have the same (e/m_e) as any other electrons.

Fig. 62.2 shows the apparatus used to study the effect quantitatively. The galvanometer is a very

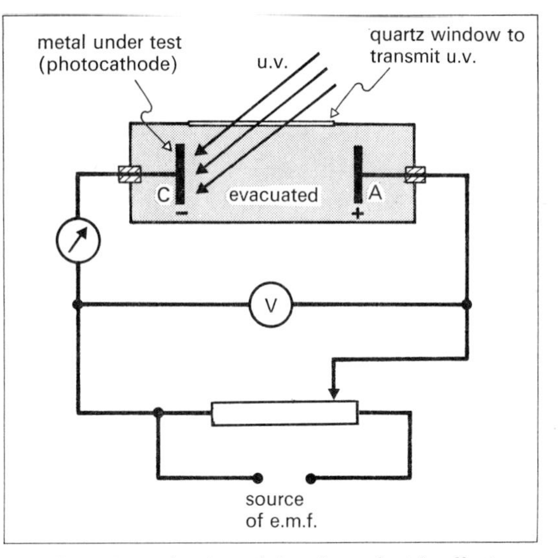

Fig. 62.2. Investigation of the photoelectric effect.

sensitive ammeter (such as the d.c. electrometer described on p. 336).

The results of such experiments are sometimes called the **Laws of Photoelectric Emission**.

I When the incident light is monochromatic, the *number of photoelectrons emitted per second* is proportional to the light *intensity* I'. Such an emission occurs effectively *instantaneously*.

II The k.e. of the emitted electrons varies from 0 to a maximum value. This *definite maximum* and the electron energy distribution both depend only on the *frequency* of the light, and not on its intensity.

III Electrons are not emitted when the light has a frequency lower than a certain *threshold* value ν_0. The value of ν_0 varies from metal to metal.

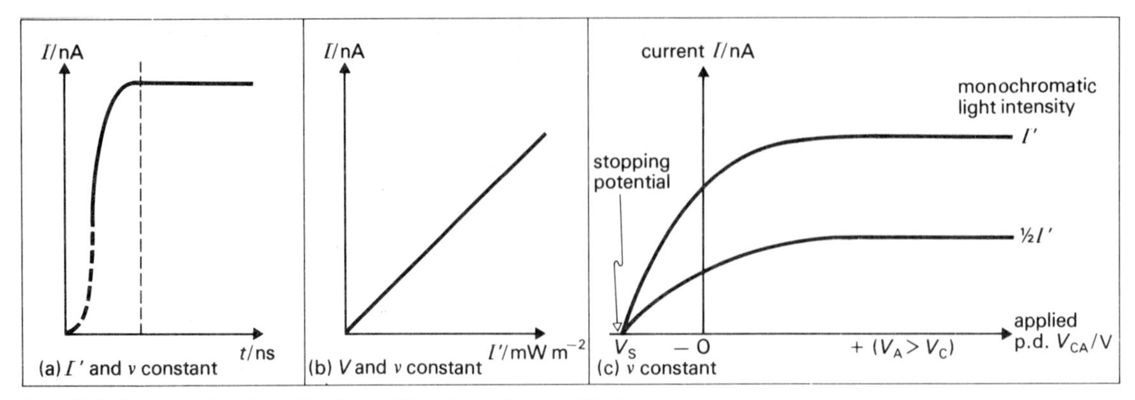

Fig. 62.3. Results of an investigation of the photoelectric effect.

Some of these results are demonstrated by the graphs of fig. 62.3.

The curved part of fig. 62.3(*c*) indicates that the photoelectrons are emitted with variable speeds.

Difficulties of a Wave Theory

If we consider light to be *wholly wave-like*, we would expect a uniform distribution of energy over the whole area on which the light is incident. We then cannot explain why

(*a*) The effect is only observed when the light frequency reaches a value ν_0, regardless of its intensity.

(*b*) The electron emission is instantaneous even if very faint light is used. (On the assumption that wave energy is uniformly distributed amongst the free electrons we would predict time delays of the order of 10^3 s.)

(*c*) The maximum speed of the emitted electrons is independent of the intensity of the incident radiation.

62.2 Einstein's Photoelectric Equation

Photons and the Quantum Theory

(*a*) *Planck* (1900), in an attempt to explain the distribution of energy in the black-body spectrum (p. 225), suggested that when radiation was emitted or absorbed, the emitting or absorbing oscillator always showed a *discrete sudden change of energy* ΔE. ΔE is related to the radiation frequency ν by

$$\boxed{\Delta E = h\nu}$$

$$\Delta E[\text{J}] = h[\text{J s}]\cdot\nu\left[\frac{1}{\text{s}}\right]$$

h is called the **Planck constant**, and has the value 6.6×10^{-34} J s. It has dimensions of $[\text{ML}^2\text{T}^{-1}]$.

For visible light the **quantum** of energy, or **photon**, carries energy

$$\begin{aligned}\Delta E &= h\nu \\ &= (6.6 \times 10^{-34}\ \text{J s}) \times (10^{15}\ \text{s}^{-1}) \\ &\approx 10^{-19}\ \text{J}\end{aligned}$$

The exact value of ΔE depends on the frequency (colour) of the light, but is so small that for many purposes we regard energy emission and absorption as continuous processes.

Experiments on black-body radiation enabled *Planck* to calculate a value for h.

(*b*) *Einstein* (1905) extended *Planck*'s original idea by suggesting that electromagnetic waves were not only *generated* discontinuously, but that they could exhibit particle behaviour while being *absorbed.*

Explanation of the Photoelectric Effect

Suppose such a photon collides directly with, and imparts *all* its energy to, an electron in a metal *surface*. Using the law of conservation of energy

$$\begin{pmatrix}\text{energy of}\\ \text{photon}\end{pmatrix} = \begin{pmatrix}\text{energy needed to}\\ \text{remove electron}\\ \text{from metal}\end{pmatrix} + \begin{pmatrix}\text{k.e. imparted}\\ \text{to ejected}\\ \text{electron}\end{pmatrix}$$

$$h\nu = W + \tfrac{1}{2}m_e v_m^2$$

v_m is the maximum speed of ejection. Using one simple model, we imagine those electrons that are in the surface to have the least values of W, the *work function energy* (p. 440). This is illustrated in fig. 62.4.

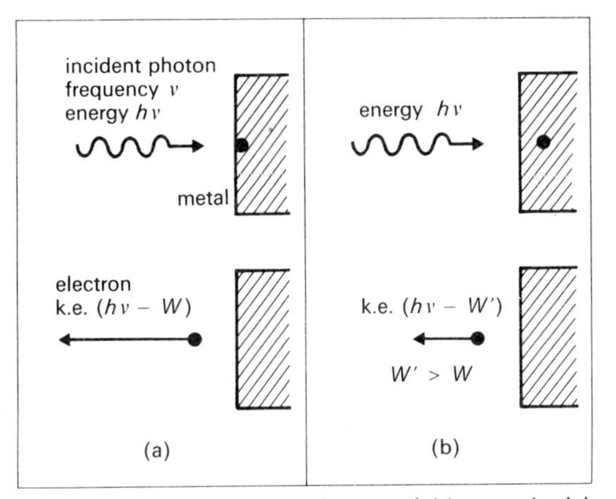

Fig. 62.4. Ejected electrons have variable speeds: (a) surface electron, (b) electron in metal. (Note that in more advanced study a radically different model is used.)

For the surface electrons of a given metal, W always has the same value. (Note that in more advanced work this simple model is replaced by one using more sophisticated concepts.)

We explain the difficulties encountered by the wave theory (above) as follows:

(*a*) $\frac{1}{2}m_e v_m^2 = h\nu - W$

$= h\nu - h\nu_0 \qquad (W = h\nu_0)$

When $\nu = \nu_0$, no individual photon imparts enough energy to an individual electron to cause ejection. (An increase in light intensity does not cause an increase in photon energy.)

(*b*) Electron emission is the result of a direct collision between an electron and a photon, so there is no time delay before emission starts. (A photon cannot collide with more than *one* electron. This is because it has zero rest mass, and so according to relativity theory it ceases to exist once it no longer moves with the speed of light.)

(*c*) An increase in light intensity results in an increase in the number of photons arriving per second, but not in the energy of individual photons.

$$\boxed{\frac{1}{2}m_e v_m^2 = h\nu - W}$$

is called **Einstein's photoelectric equation.**

*These ideas lead to the concept of the **wave-particle duality of light**—under some circumstances it behaves as a wave motion, and under others like a particle (or photon) (p. 261).

Matter shows analogous behaviour (p. 265).

62.3 Millikan's Experiment to Verify the Equation

If a metal has a positive potential, more energy must be given to an electron to enable it to escape. If the metal potential is V, this additional energy is eV. This situation is illustrated in fig. 62.5.

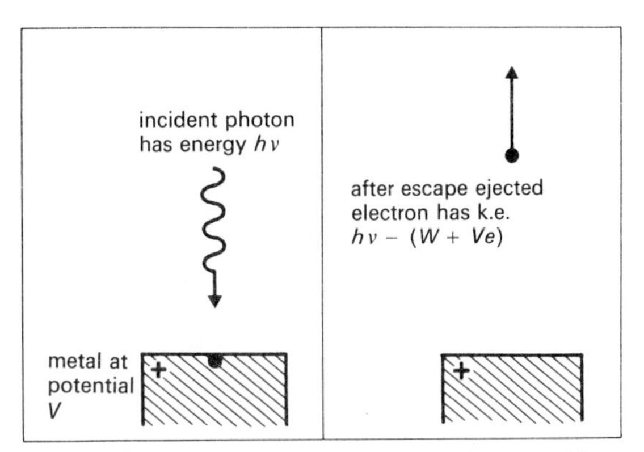

Fig. 62.5. The effect of raising the metal to a positive potential.

The **stopping potential** V_s is the potential to which the metal must be raised so that when electrons are ejected from the surface by suitable radiation, they are all just attracted back again.

Since in general $\frac{1}{2}m_e v_m^2 = h\nu - (W + Ve)$

$v_m = 0$ when $eV_s = h\nu - W$

When $V_s = 0$, $\quad h\nu_0 = W$

where ν_0 is the **threshold frequency**.

When photoelectric emission has been occurring, the loss of electrons will itself cause the potential of an insulated metal to become positive: emission will cease when the stopping potential V_s has been reached.

Using apparatus similar to that of fig. 62.2 (but with the terminals reversed) *Millikan* measured the potential difference V_{CA} such that the current was just reduced to zero for different values of ν.

The stopping potential

$$V_s = V_C - V_A \qquad (V_C > V_A)$$

Gain of electric p.e. (QV) equals loss of electron k.e.

$$eV_s = h\nu - h\nu_0$$

$$V_s = \left(\frac{h}{e}\right)\cdot\nu - \left(\frac{h}{e}\right)\cdot\nu_0$$

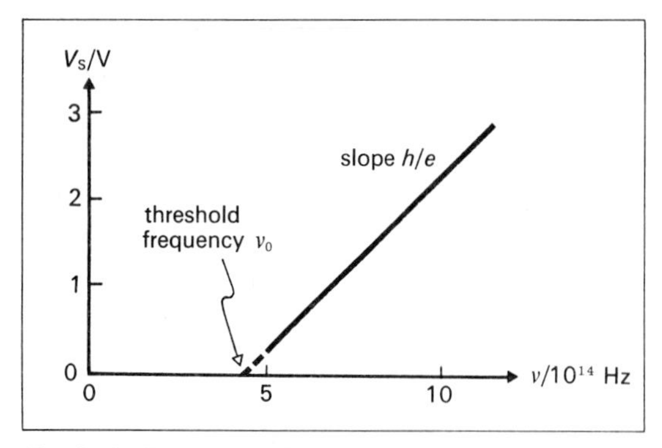

Fig. 62.6. The result of Millikan's experiment when a freshly-cut sodium surface is used.

If we plot a graph of V_s against ν (fig. 62.6) we find a straight line whose

(a) slope $= h/e$, and
(b) intercept $= \nu_0$ (the value of ν when $V_s = 0$).

Einstein's photoelectric equation was confirmed by

(a) the straight line graph, and
(b) the fact that the slope h/e, and the intercept $\nu_0(= W/h)$ on the ν-axis, both enabled h to be calculated. (*Millikan* had already measured e.)

The determined value of h agreed with that found from *Planck's radiation formula* (p. 225).

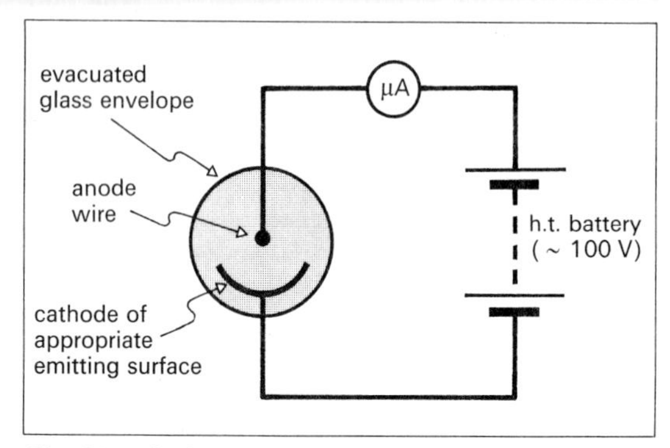

Fig. 62.7. A photoemission cell in a circuit.

62.4 Photocells

(a) Photoelectric emission cell

In fig. 62.7 the material of the cathode can be chosen to respond to the frequency range over which the cell will operate. The response can be made proportional to the light intensity.

Uses

The cell can be used in any situation where a beam of light which was falling on a cell is broken, e.g.

(i) to count vehicles or items on a factory belt,
(ii) as a burglar alarm, or
(iii) to open doors.

(b) Photoconductive cell

An **internal photoelectric effect** may free charge-carriers in a material that is otherwise an insulator, and thereby increase its conductivity by as much as 10^4 when it is illuminated. A current will flow if the material is in a circuit with a source of e.m.f.

Uses

(i) Detection and measurement of infra-red radiation ($\lambda \approx 10^{-6}$ m).
(ii) Relays, for switching on artificial lighting, such as street lighting.

(c) Photovoltaic Cell

This may consist of a sandwich of copper, copper oxide and a thin layer of translucent gold. An e.m.f. capable of giving a current of 1 mA can be generated when the gold is illuminated.

Uses

The cell needs no source of e.m.f., and is frequently used as an exposure meter in photography.

63 X-rays and the Atom

63.1 Production of X-Rays

X-rays were discovered by *Röntgen* in 1895: they were produced by electrons colliding with the walls or electrodes of a discharge tube.

The Coolidge Tube

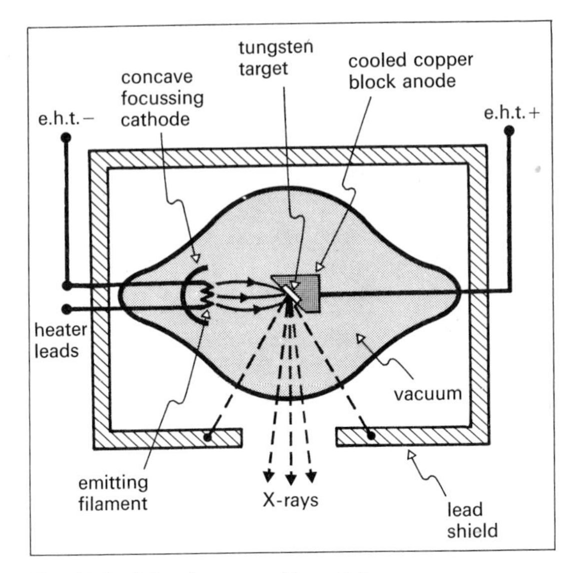

Fig. 63.1. A hard vacuum X-ray tube.

In fig. 63.1 the electrons emitted from the heated filament are attracted to and focussed on the cooled anode. (Less than $\frac{1}{2}$% of the electron k.e. goes to X-rays. The remainder is converted to the internal energy of the anode.) Their rapid deceleration results in the emission of **X-rays**, an *electromagnetic radiation.*

(*a*) The *intensity* of the X-ray beam is controlled by the filament current, which determines the number of electrons hitting the anode per second.

(*b*) *The penetrating power* (**quality**) of the beam is controlled by the p.d. between the anode and cathode. This may be 100 kV or higher.

63.2 Properties of X-Rays

General

(*a*) They *travel in straight lines*, and are not deflected by electric or magnetic fields. (They are not charged particles.)

(*b*) They *produce ionization* of a gas when they pass through it. An electroscope charged either positively *or* negatively is rapidly discharged if X-rays make the surrounding air conducting.

(*c*) They *expose photographic plates* in the same way as visible light waves.

(*d*) They cause *fluorescence.*

(*e*) They cause *photoelectric emission.*

(*f*) They *penetrate matter.* Three reasons for absorption are

(*i*) **Compton effect** (in which the quantum is scattered by an electron), $\propto$ number of electrons present

(*ii*) photoelectric effect (for quantum energies up to $\sim 1.6 \times 10^{-14}$ J), roughly $\propto Z^4$
(*iii*) pair production (p. 463) (for quantum energies $> 1.6 \times 10^{-13}$ J), roughly $\propto Z^2$.

The absorption is thus highest for elements of high *atomic number* (dense materials). This is the basis of radiography (see below).

The Wave Nature of X-Rays

Because X-rays are electromagnetic waves, they possess the usual properties of such waves (p. 263). The typical wavelength is $\sim 10^{-10}$ m, which makes the wave properties difficult to demonstrate.

(*a*) *Von Laue* (*1912*) demonstrated that X-rays give a typical *diffraction pattern* when made to pass through a crystal. The pattern is recorded on a photographic plate (p. 287).

(*b*) *The Braggs* (*1913*) developed a simpler method of analysing the 'reflected' diffraction pattern from a crystal. Considering each ion in a crystal lattice to act as a secondary source, they developed an equation (the **Bragg equation** p. 288) which enabled the wavelength λ of the X-rays to be calculated in terms of the separation d of the ion planes.

X-Ray Spectra

Fig. 63.2 shows a typical X-ray spectrum.

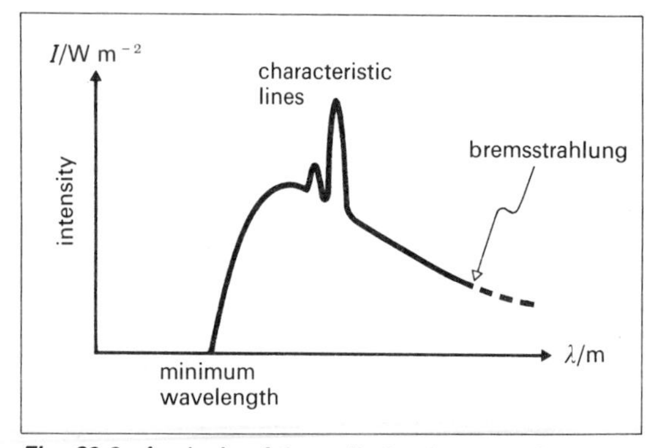

Fig. 63.2. Analysis of the radiation from an X-ray tube.

(*a*) *The continuous spectrum* shows a well defined minimum wavelength (maximum frequency). This corresponds to an electron losing all its energy in a single collision with a target atom. The longer wavelengths (smaller energies) correspond to a more gradual loss of energy, which happens when the electron experiences several deflections and collisions and so is slowed down more gradually. All or some of the k.e. of the electron is converted into the energy of the photon(s). This radiation is called **bremsstrahlung** (braking radiation). All targets show this continuous spectrum.

(*b*) The k.e. of a bombarding electron $= Ve$ where V is the accelerating p.d.

$$\therefore \quad Ve = h\nu_{\max} = h\frac{c}{\lambda_{\min}}$$

where $\nu_{\max}$ is the frequency of the most energetic photon (possessing all the initial k.e. of the colliding electron).

(*c*) *The line spectrum* is characteristic of the element used for the target in the X-ray tube. It corresponds to the quantum of radiation emitted when an electron changes energy levels very close to a nucleus. (See p. 459 and p. 467.)

Application of X-rays

(*a*) **Medicine**

(*i*) *Radiography*—dense materials such as bones absorb the radiation and this produces shadows on the film which can be analysed. This technique requires no lenses or focussing.

(*ii*) *Radiotherapy*—destruction of cancer cells.

(*b*) **Industrial radiography**—location of internal imperfections (e.g. in a casting).

(*c*) **X-ray crystallography**—analysis of structure of organic molecules. This can be extended to analyse fine powders containing minute crystals. X-ray spectrometer measurements (p. 288) lead to the most accurate determination of N_A.

(*d*) A single crystal can be used as a *monochromator* for X-rays—it can diffract an intense beam of *given* λ into a particular direction.

63.3 Positive Rays

Cathode rays (p. 436) are the streams of negative particles (electrons) that seem to come *from* the cathode of the discharge tube. **Positive** (**canal**) rays are the streams of positive ions that pass *through* a perforated cathode—see fig. 63.3.

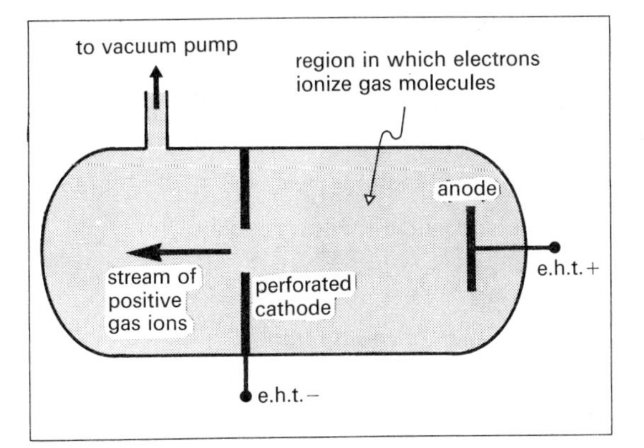

Fig. 63.3. Positive rays in the discharge tube. The pressure in the tube is ~0.4 kPa.

Properties of Positive Rays

(*a*) They cause fluorescence and affect a photographic plate.

(*b*) They *are* related to the gas in the tube. (Cathode rays are *not*.)

(*c*) They are deflected in magnetic and electric fields, but much less than cathode rays. (The positive ions are more massive.)

(*d*) Their velocities are variable, since they are accelerated only from the points at which they are formed. (These vary in position between the anode and the cathode.)

Thomson's Experiment

Thomson applied parallel fields of $\boldsymbol{B}$ and $\boldsymbol{E}$ which resulted in parabolic traces on a suitably placed photographic plate (p. 438). He deduced that

(*a*) the value of the specific charge $q = e/m_p$ for the hydrogen ion (a proton) is about 1840 times smaller than e/m_e for the electron, so

(*b*) if the hydrogen ion and electron carry charges of the same magnitude, the mass of the hydrogen ion is about 1840 times that of the electron.

He was also able to measure the masses of other positive ions by using different gases in the tube. This led to the discovery of **isotopes** (below).

63.4 Definitions

Models of the atom are discussed on p. 465. Consider a model in which the atom is composed of

(*a*) a nucleus, which contains neutrons and protons (**nucleons**) only, and

(*b*) an encircling cloud of electrons.

Then we can define the following terms.

A **nuclide** is a particular **species** of atom (one whose nucleus contains a specified number of protons, and a specified number of neutrons).

The **atomic number** Z of an *element* is the number of protons contained in the nucleus. This is equal to the number of electrons in a neutral atom of the element, and so determines the *chemical nature* of an element.

The **mass number** A of an *atom* or nuclide is the number of nucleons in the nucleus. It is by definition an integer.

If two different nuclides have

(*a*) the same atomic number Z but different mass numbers A, they are called **isotopes**

(*b*) the same mass number A but different atomic numbers Z, they are called **isobars**.

(*Isotope* implies *same place* in the periodic table. *Isobar* implies *same weight*, meaning same total mass.) Isotopes are chemically identical (p. 479).

A **unified atomic mass unit** (u) is defined such that the mass of the carbon-12 atom is 12 u. By experiment

$$m_u = 1\,\text{u} \approx 1.66 \times 10^{-27}\,\text{kg}$$

(*N.B.* $m_u = 1$ u is not really a *unit*, but we find it convenient to express atomic masses relative to this *physical quantity*.)

The **relative atomic mass** is the ratio

$$\frac{\text{mass of an atom}}{\text{mass of a } {}^{12}_{6}\text{C atom}} \times 12$$

The *mass* of an atom is usually quoted in unified atomic mass units, and is then numerically equal to its relative atomic mass. (It is often incorrectly called its *atomic weight*.) Atoms have masses which are very close to an integral number times m_u. If we assume that the proton and neutron each have a mass very close to 1 u, this is consistent with our simple model of the atom.

In practice this model is oversimplified. As can be seen below, the masses of nucleons are nearly one per cent greater than 1 u, and since some nuclei have more than 200 nucleons, the nearly integral values for atomic masses can be explained only in terms of binding energy (p. 462). The rest masses and rest energies of electrons and nucleons are:

$$\begin{aligned} \text{electron } m_e &= 9.109\,53 \times 10^{-31}\,\text{kg} \equiv 81.872\,4\,\text{fJ} \\ \text{proton } m_p &= 1.672\,65 \times 10^{-27}\,\text{kg} \\ &\equiv 0.150\,330\,\text{nJ} \\ \text{neutron } m_n &= 1.674\,95 \times 10^{-27}\,\text{kg} \\ &\equiv 0.150\,537\,\text{nJ} \end{aligned}$$

Representation by Symbol

A given nuclide is represented by

$$^A_Z\mathrm{X}$$

in which

(*a*) the superscript A is the mass number,
(*b*) the subscript Z is the atomic number, and
(*c*) X is the **chemical symbol**.

X and Z present the same information. Some examples are given on p. 476.

63.5 Mass-Energy Equivalence

The masses of atomic particles are small. For a given energy, these particles may be accelerated to such high speeds that relativistic mechanics must be used. The mass of a particle m moving at speed v relative to an observer would be measured by him to be

$$m = \frac{m_0}{\sqrt{1-(v/c)^2}}$$

where m_0 is its rest mass. Its relativistic momentum is therefore

$$\boldsymbol{p} = m\boldsymbol{v} = \frac{m_0\boldsymbol{v}}{\sqrt{1-(v/c)^2}}$$

The theory of relativistic mechanics gives the kinetic energy of a particle to be

$$E_k = (m - m_0)c^2$$

(Note that this is *not* equal to the classical value $\frac{1}{2}mv^2$.) If we write the total energy as E, then

$$mc^2 = E_k + m_0c^2$$
$$= E$$

The *Einstein* relationships are

$E = mc^2$	**total energy**
$E_0 = m_0c^2$	**rest energy**

The laws of energy and mass conservation are thus combined in the *law of conservation of mass-energy*.

The relationship between total energy and momentum is

$$E^2 = E_0^2 + (pc)^2$$

Some Implications

These relativistic equations show that

(*i*) when $\boldsymbol{v} \ll \boldsymbol{c}$, so that *Newtonian* mechanics is a good approximation,

$$m \approx m_0, \qquad p \approx m_0v, \qquad E_k \approx \tfrac{1}{2}m_0v^2, \qquad E_k \ll E_0$$

(*ii*) when $\boldsymbol{v}$ approaches $\boldsymbol{c}$, so that relativistic mechanics is necessary,

$$m \gg m_0 \qquad E \gg E_0, \qquad p \approx \frac{E}{c}, \qquad E_k \approx E$$

and

(*iii*) for a particle of zero rest mass,

$$m_0 = 0, \qquad E = pc, \qquad E_k = E, \qquad \boldsymbol{v} = \boldsymbol{c}$$

Unbound and Bound Systems

(*a*) In an **unbound** system, the rest mass of the composite system is greater than the sum of the rest masses of the separated particles by an amount equal to the k.e. of the amalgamating particles at combination.

(*b*) In a **bound** system, the rest mass of the composite system is less than the sum of the rest masses of the separated particles by an amount called the **binding energy** E_b. If a system of rest mass M_0 is split into two particles of rest mass m_0' and m_0'' by adding energy equal to E_b, then

$$E_b = (m_0' + m_0'')c^2 - M_0c^2$$

A measurable mass difference is obtained only when one is dealing with nuclear forces. The total mechanical energy E_m of a system of particles that have mutual attraction, is taken by convention to be zero when the particles are at rest and infinitely separated. Thus when particles are bound, E_m becomes negative; that is energy would have to be added to the system to separate the particles again completely, and thus to increase the energy to zero.

63.6 Photon-Electron Interactions

In all such interactions the laws of conservation of charge, mass-energy and relativistic momentum can be applied, and the particle-like nature of electromagnetic radiation is emphasized. The photons have energy $h\nu$, momentum h/λ and effective mass $h/c\lambda$. These interactions usually involve high-energy photons and electrons.

(*a*) ***The photoelectric effect*** (p. 455). A photon is annihilated on colliding with a bound electron. Most of the photon's energy is transferred to the electron which is ejected, whereas most of the photon's momentum is transferred to the object to which the electron was bound. (This effect cannot, therefore, take place with a free electron.)

(*b*) ***The Compton effect.*** A photon collides with a free or lightly-bound electron, giving the electron kinetic energy and causing it to recoil. A second (scattered) photon of lower energy and therefore greater wavelength is created.

(*c*) ***Pair production.*** A photon passes near a massive nucleus and its energy is converted into matter. This cannot happen spontaneously in free space where it is not possible to satisfy simultaneously the conservation laws of mass-energy, momentum and electric charge. The photon energy is converted into

(*i*) the rest mass of the electron-positron pair, and
(*ii*) the k.e. of the particles so formed.

The equation for this can be written

$$h\nu = 2m_0c^2 + [E_k^+ + E_k^-]$$

The minimum energy of the photon for pair production is 1.64×10^{-13} J, and it can therefore be achieved only by γ-photons or X-ray photons.

(*d*) ***X-ray production*** (p. 459). An electron loses kinetic energy through collisions and deflections near massive particles. Some of the energy is converted into the energy of one or more photons in the production of bremsstrahlung. (Most of the k.e. is converted into the internal energy of the target.)

(*e*) ***Pair annihilation.*** A positron loses its kinetic energy by successive ionization, comes to rest and combines with a **negatron** (negative electron). Their total mass is converted into two oppositely directed photons (**annihilation radiation**), and the process is thus the reverse of pair production. As $h\nu_{min} = m_0c^2$, the total energy available is 1.64×10^{-13} J and, to conserve momentum, each quantum has energy 8.2×10^{-14} J. They move off in opposite directions.

63.7 The Nucleus

The radius of an atom is $\sim 10^{-10}$ m. The radius of the nucleus of an atom $\sim 10^{-15}$ m.

The electron cloud shields the nucleus, so that its behaviour is generally independent of the environment (temperature, pressure, chemical combination, etc.). Information about the nucleus comes from

(*a*) *ejected particles*, such as those emitted during radioactivity,
(*b*) *injected particles*, as when the nucleus is bombarded by artificially accelerated ions, and
(*c*) *spectra* of all kinds.

Moseley's Experiments on X-Ray Spectra (1913)

Using a *Bragg* X-ray spectrometer, *Moseley* measured the frequency ν of corresponding characteristic line spectra for different elements. He plotted the graph of $\sqrt{\nu}$ against Z (which we now call the *atomic number*) as shown in fig. 63.4.

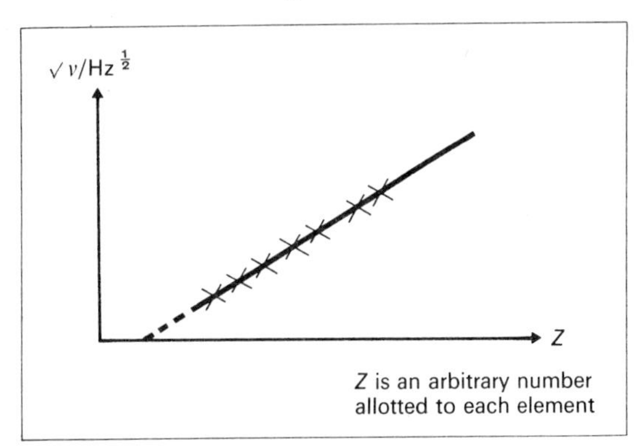

Fig. 63.4. The graph which gives the clue to atomic number.

The graph is described by

$$\sqrt{\nu} = (\text{constant})(Z - b)$$

where b is a constant.

Deductions

(*a*) Each nucleus has a *characteristic number Z* in addition to its mass number A, which increases by 1 from one element to the next.

(*b*) Gaps in *Moseley's* graph were explained by elements not then discovered.

(*c*) The number Z measured by this method follows the numbering of elements in the periodic table. Anomalies arising from an atomic *mass* order disappear.

(*d*) Comparison with *Bohr's theory* of the hydrogen atom shows that Z is numerically equal to the positive charge on the nucleus.

Moseley's experiment is a *direct way of measuring atomic number.*

Rutherford's Experiment

Evidence for the existence of the nucleus comes from scattering experiments carried out for *Rutherford* by *Geiger* and *Marsden*. (See fig. 63.5.)

Detailed measurements of the proportion of α-particles that suffer a deviation θ

(*i*) confirm that each atom has a nucleus in which its mass is concentrated,

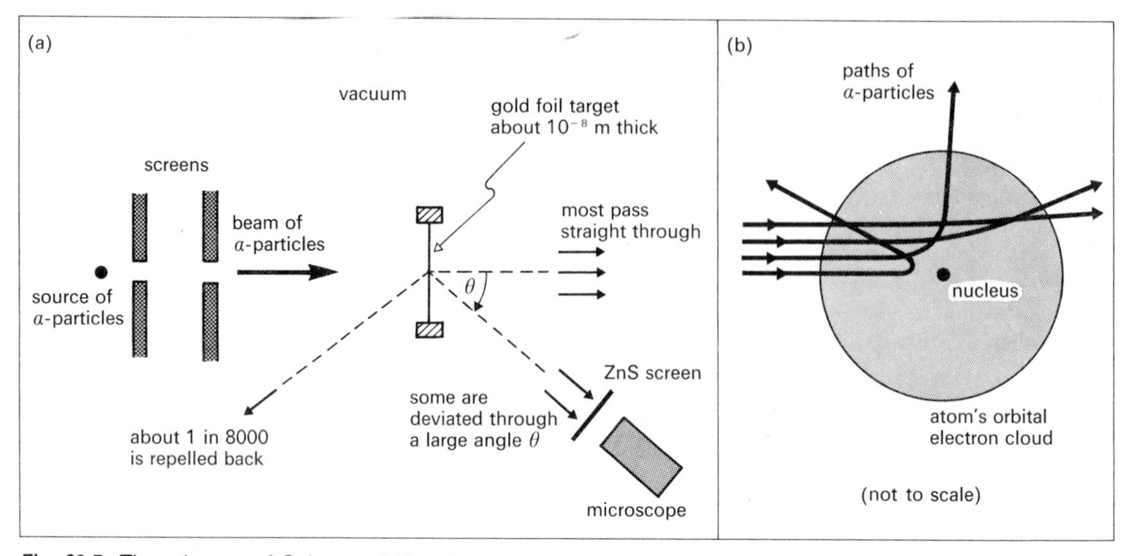

Fig. 63.5. The schemes of Geiger and Marsden's experiment: (a) the general idea, and (b) what happens on a microscopic scale.

(*ii*) enable the size and charge of the nucleus to be measured.

The experiment also provides confirmatory evidence that *Coulomb's Law* (which is used for the calculation) holds for atomic dimensions.

Nuclear Stability

The ***n***uclear mass M_n is always less than the sum of the rest masses of the constituent nucleons because the p.e. associated with the (strong) interactions between nucleons is so great. If a nucleus has Z protons and N neutrons ($N = A - Z$), then

$$M_n < Zm_p + Nm_n$$

As it is the masses of neutral ***a***toms M_a that are measured, then

$$M_a < Zm_p + Nm_n + Zm_e$$

or $$M_a < ZM_H + Nm_n \qquad \dagger$$

When Z protons and N neutrons combine to form a nucleus, they give up energy which is revealed as a mass reduction—this is the binding energy E_b of the nucleus (p. 462).

$$M_a + \frac{E_b}{c^2} = ZM_H + Nm_n$$

† This step is justified because the strong (nuclear) interaction between protons and neutrons is about 10^6 times as strong as the electromagnetic interaction between protons and electrons.

The greater the energy that must be supplied to unbind the system (i.e. to split it into its constituent particles), the more stable is that system.

The binding energy *per nucleon* E_b/A is a more useful measure of stability. It corresponds to the atomic ionization potential as it is the energy needed to remove an average nucleon from the nucleus.

$$\frac{E_b}{A} = \left(\frac{ZM_H + Nm_n - M_a}{A}\right)c^2$$

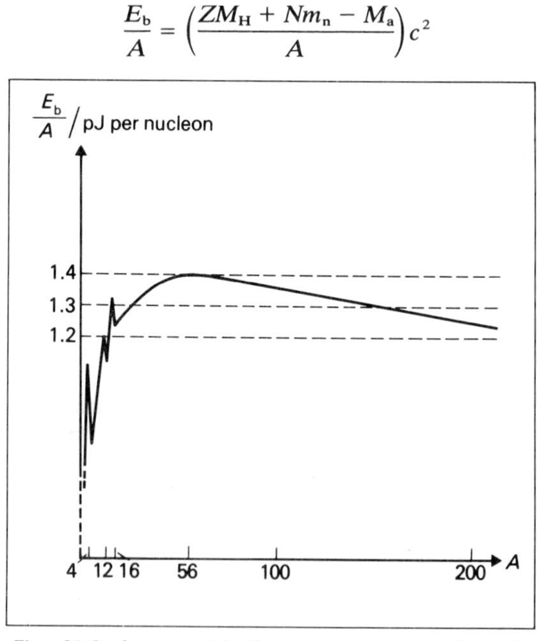

Fig. 63.6. Average binding energy per nucleon for naturally occurring nuclides.

Fig. 63.6 shows that, after the initial irregularities for nuclei with few nucleons (note the stability peaks for $^4_2$He, $^{12}_6$C and $^{16}_8$O), the curve reaches a maximum at about $^{56}_{26}$Fe and then drops gradually towards the massive nuclei. When $A > 20$, the total binding energy is roughly proportional to A. ($E_b/A \approx$ 1.3 pJ/nucleon, on average.)

Nuclear forces tend to give constant E_b/A and to make the value of $Z = N$. Nevertheless some deviation is caused because

(a) electrostatic repulsion between protons in heavy nuclei tends to reduce E_b/A and to make $Z < N$, and

(b) surface energy in light nuclei tends to reduce E_b/A.

All nuclear changes happen in an attempt to achieve maximum stability by increasing E_b/A. They therefore tend to minimize the mass of the system, since A = constant.

* 63.8 The Bainbridge Mass Spectrograph

(a) The Velocity Selector (fig. 63.7a)

Suppose the ions carry a charge Q, and have mass m. Only those ions which experience zero net force will emerge from the last slit. These ions all have the *same speed* v such that

$$EQ = B'Qv$$

$$v = \frac{E}{B'}$$

(b) The Magnetic Deflection (fig. 63.7b)

For the semi-circular path of radius r

$$\boldsymbol{F} = m\boldsymbol{a}$$

gives
$$BQv = \frac{mv^2}{r}$$

so
$$r = \frac{m}{Q} \cdot \frac{v}{B}$$

Since all the positive ions move at the same speed v in the same magnetic field $\boldsymbol{B}$

$$r \propto \frac{m}{Q}$$

For singly charged ions

$$r \propto m$$

The ions form dark lines on the plate. Their positions enable m to be calculated once the spectrum has been calibrated by admitting gases which produce ions of known mass.

When ions of isotopes are present several lines will be produced in slightly different positions owing to the different masses of the nuclei. From the degree of exposure of the photographic plate the relative abundance of isotopes can be deduced.

63.9 Models of the Atom

For a discussion of the term *model*, see p. 4.

(a) Thomson's plum pudding model (1904)

Thomson's model suggested that the atom's positive charge was spread throughout its whole volume, and

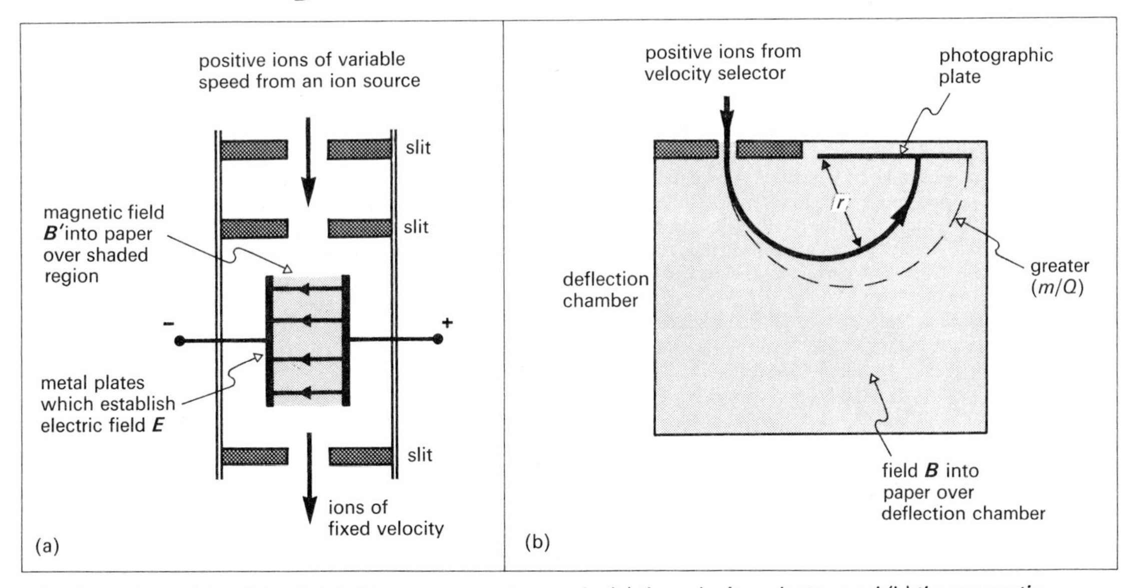

Fig. 63.7. Essentials of the Bainbridge mass spectrograph: (a) the velocity selector, and (b) the magnetic deflection through π rad.

that the electrons vibrated about fixed centres embedded in the sphere. It gave a qualitative explanation of dielectric polarization (p. 338).

(b) Rutherford's nuclear atom (1911)

Rutherford's model resulted from *Geiger* and *Marsden's* experiment: it confined the positive charge to a central core (**nucleus**) of radius $\sim 10^{-15}$ m, around which the electrons were supposed to move in a spherical volume of radius 10^{-10} m. The electrons in orbit account for the atom's volume.

Orbiting electrons are accelerated electric charges, and by classical theory should lose energy by radiating electromagnetic waves: they should spiral into the nucleus.

(c) Bohr's atom (1913)

This is discussed more fully opposite. *Bohr* applied some of the quantum theory ideas to the *Rutherford* atom, and this enabled him to make quantitative predictions for hydrogen and singly ionized helium.

His model correlates well with

(*i*) the *Franck* and *Hertz* experiment (right),
(*ii*) *Moseley's* work on X-ray spectra (p. 463), and in addition can be used to predict
(1) some spectroscopic results with an accuracy of 1 part in 10^6, and
(2) the correct value for the most probable location of the electron orbit in the hydrogen atom.

The *Bohr* atom has now been replaced by the ideas that come from wave mechanics.

**(d) The wave mechanics atom*

The wave mechanics model is too difficult for detailed study in this book. It has the advantage of not making arbitrary assumptions, but can only be described adequately using mathematics. The procedure is (1) to set up a *wave equation* for a particular situation, (2) to solve that equation where possible, and (3) to deduce from the solution the *probability* of finding an electron at a particular distance from the nucleus.

The idea of orbits is replaced by that of an **electron cloud** of particular density, but the energy levels of the *Bohr* atom remain (p. 467).

63.10 The Orbital Electrons

(*a*) Barkla's Experiment (1911)

Barkla measured the number of electrons in an atom by observations on the scattering of X-rays. He concluded that

$$(\text{number of electrons}) \sim \tfrac{1}{2}(\text{mass number})$$

This raised the question of an atomic property Z as distinct from A.

(*b*) Bohr's Theory (1913)

Bohr's Theory gave a quantitative prediction of the energy of an orbital electron in the hydrogen atom.

If an electron is to change from an allowed energy level E_2 to a lower level E_1 (fig. 63.8), then the energy difference appears as an electromagnetic wave of frequency ν, where

$$(E_2 - E_1) = h\nu \qquad \text{(p. 456)}$$

The theory leads to a detailed account of the spectral lines of hydrogen, and the ionization energy.

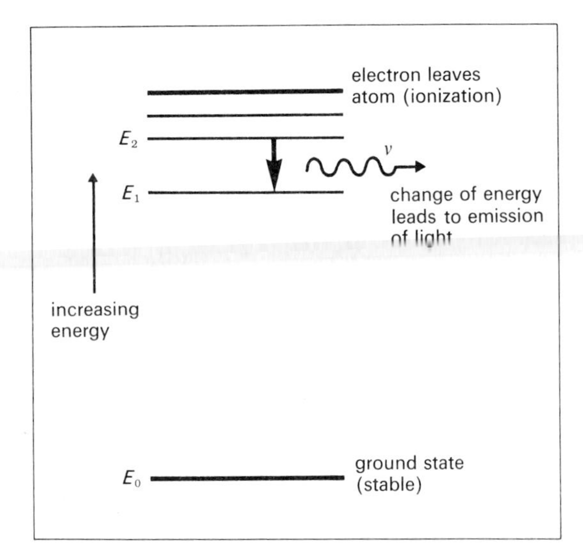

Fig. 63.8. Energy level diagram for hydrogen.

This is indirect evidence for the existence of energy levels in the orbital electrons.

(*c*) The Franck and Hertz Experiment (1916)

The experiment provided further evidence for *discrete energy levels* within the atom. (See fig. 63.9.)

Procedure

(1) Electrons emitted by the cathode are attracted to and pass through the grid. They just reach the anode because $V_{GC} > V_{GA}$.

(2) Increase V_{GC} and plot a graph as shown in fig. 63.9*b*.

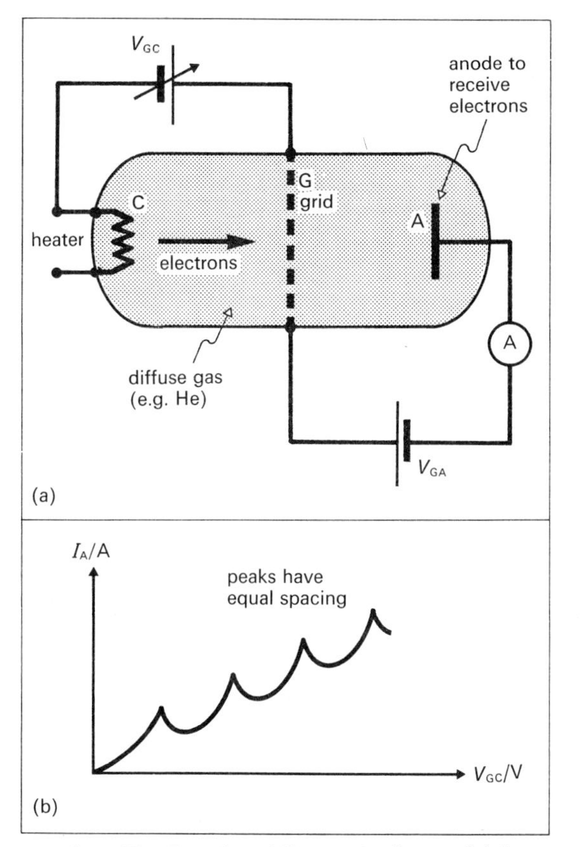

Fig. 63.9. The Franck and Hertz experiment: (a) the apparatus, and (b) the results.

Interpretation

Electrons colliding with He atoms do not lose energy unless the He atoms are excited (unless the orbital electrons are raised to the next energy level). When they do, those electrons do not reach the anode A, and the current drops. Further increase in V_{GC} leads to an increased current, until another peak is reached, corresponding to a second excitation. Each peak represents an *inelastic collision* with energy exchange between the free electron and the atom.

*(*d*) The Wave Mechanics Energy Levels

Solution of the wave equation predicts the number of electrons that can fit into each **shell** of an atom, and describes, using **quantum numbers**, how much energy each electron may have.

There is exact agreement between the predictions of wave mechanics and empirical methods such as those based on chemical behaviour and the periodic table.

*63.11 The Bohr Atom

The Spectrum of Atomic Hydrogen

Hydrogen *atoms* in a discharge tube emit characteristic spectral lines whose frequencies ν fit the general formula

$$\nu = cR_H\left(\frac{1}{n_1^2} - \frac{1}{n_2^2}\right)$$

c is the speed of light, R_H is called the **Rydberg constant** (for hydrogen), and n_1 and n_2 are integers, such that $n_1 < n_2$.

By experiment $R_H = 1.096\,78 \times 10^7\ \text{m}^{-1}$.

Spectral Series

(*a*) When $n_1 = 1, n_2 \geqslant 2$, the frequencies correspond to the **Lyman series**, which is in the *ultra-violet.*

(*b*) When $n_1 = 2, n_2 \geqslant 3$, the frequencies correspond to the **Balmer series**, which is in the *visible region.*

(*c*) When $n_1 = 3, n_2 \geqslant 4$, the frequencies correspond to the **Paschen series**, which is in the *infra-red.*

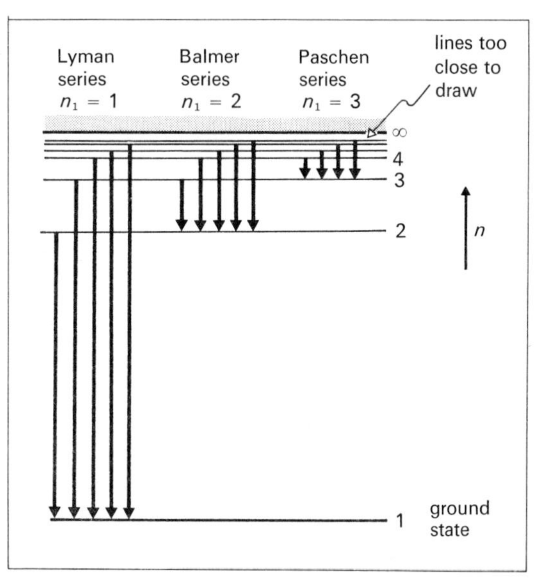

Fig. 63.10. The three main spectral series of atomic hydrogen.

These spectral series are illustrated in fig. 63.10 and they provide convincing evidence for the existence of discrete energy levels in atoms.

Classical Physics could not explain the general equation above.

Bohr's Analysis of the Hydrogen Atom

Consider an electron of mass m_e and charge $(-e)$ in circular motion of radius r around a proton of much greater mass and charge $(+e)$.

Bohr postulated

(1) That when the electron had an angular momentum given by

$$m_e v r = \frac{nh}{2\pi} \tag{1}$$

then the orbit was stable: although the electron was accelerated, it did not radiate electromagnetic waves.

(2) That when the electron changed from an allowed orbit in which the system had energy E_2 to a different allowed orbit of lower energy E_1, then it radiated electromagnetic waves of frequency ν such that

$$E_2 - E_1 = h\nu$$

where h is the *Planck* constant.

For the circular orbit, the centripetal force is provided by the *Coulomb* force

$$\left(\frac{1}{4\pi\varepsilon_0}\right)\cdot\frac{e^2}{r^2} = \frac{m_e v^2}{r} \tag{2}$$

Eliminating the speed v from (1) and (2)

$$r = \frac{\varepsilon_0 n^2 h^2}{\pi m_e e^2} \tag{3}$$

The total energy E of the system is given by

$$E = E_p + E_k$$

where k.e. $E_k = \frac{1}{2} m_e v^2$

$$= \left(\frac{1}{4\pi\varepsilon_0}\right)\cdot\frac{e^2}{2r} \quad \text{(Equ. 2)}$$

and p.e. $E_p = -\left(\frac{1}{4\pi\varepsilon_0}\right)\cdot\frac{e^2}{r}$

So total energy

$$E = -\left(\frac{1}{4\pi\varepsilon_0}\right)\cdot\frac{e^2}{2r}$$

$$= -\frac{m_e e^4}{8\varepsilon_0^2 n^2 h^2}$$

using the value of r from equation (3).

This equation shows

(*a*) that the system always has *negative* energy—this is because the electron is *bound* to the proton, and

(*b*) that the energy of an orbit varies as $-1/n^2$, i.e. becomes greater for larger values of n. Since $r \propto n^2$, this means for larger values of r.

The frequency ν radiated by a transition from

$$n = n_2 \quad \text{to} \quad n = n_1 \qquad (n_2 > n_1)$$

is given by

$$h\nu = E_2 - E_1$$

$$= \frac{m_e e^4}{8\varepsilon_0^2 h^2}\left(\frac{1}{n_1^2} - \frac{1}{n_2^2}\right)$$

which can be written

$$\nu = cR_H\left(\frac{1}{n_1^2} - \frac{1}{n_2^2}\right) \tag{4}$$

where $R_H = \dfrac{m_e e^4}{8\varepsilon_0^2 c h^3}$

Notes.

(*a*) The values obtained for ν from this expression are in excellent agreement with those found empirically.

(*b*) Using values of ε_0, m_e, e and h found by other methods, R_H can be found, and agrees with the values found spectroscopically.

(*c*) By putting $n = 1$ in equation (3), we can calculate the *electron orbital radius* (the *Bohr* radius) in the ground state a_0 to be 0.53×10^{-10} m. Methods based on gas kinetic theory give an atomic diameter $\sim 10^{-10}$ m, which is in agreement.

(*d*) By putting $n_1 = 1$ and $n_2 = \infty$ in equation (4) we can calculate the energy E required to ionize atomic hydrogen to be about 2.2×10^{-18} J, which agrees with experiment. E is called the **ionization energy**.

Using $E = VQ = V_i e$, we can show that the *ionization potential*

$$V_i = 13.6 \text{ V}$$

(*e*) All energy states above the ground state are called *excited states*. The difference between the energy of an excited state and the energy of the ground state is called the **excitation energy**.

The **wave mechanics atom**, which is mentioned on page 466, has replaced the *Bohr* atom for the following reasons:

(*i*) The *Bohr theory* cannot account for the spectra of atoms or ions having more than one electron.

(*ii*) The postulate that the angular momentum takes values $nh/2\pi$ turns out to be an over-simplification.

(*iii*) High resolution spectroscopy indicates that each spectral line predicted by the theory consists in fact of many closely-spaced lines.

64 Radioactivity

64.1 General Description of Radioactivity

History

In 1896 *Becquerel* noted that radiation from uranium affected photographic plates.

In 1898 *Curie* noted that thorium and radium showed the same effect.

In 1899 *Rutherford* distinguished

(1) a low penetration radiation (α-particles),
(2) a radiation of higher penetration (β-particles).

In 1900 *Villard* discovered a third type of radiation of even greater penetration (γ-rays).

Radioactivity

Radioactivity is the spontaneous disintegration of the nucleus of an atom, from which may be emitted some or all of the following:

(1) α-particles,
(2) β-particles,
(3) γ-rays,
(4) other particles or types of electromagnetic radiation (less important at this level).

The process represents an attempt by an unstable nucleus to become more stable, and is not affected by

(*a*) chemical combination, or
(*b*) any change in physical environment other than nuclear bombardment.

We suspect that it is the *nucleus which is concerned* with the effect, but not the orbital electrons.

A source of radiation continuously emits *energy* in large quantities as

(*a*) k.e. of ejected particles, and/or
(*b*) γ-radiation.

Natural sources of radiation include cosmic rays, rocks (especially granite), luminous paint and television screens. These contribute to *background* radiation which must be taken into account in the measurement of activities.

Radioactive Series

Naturally occurring radioactive nuclides can be classified into three groups, in each of which a series of successive disintegrations produces eventually a different lead isotope. For example

(*a*) $^{238}_{92}U$ produces $^{206}_{82}Pb$ (uranium series)
(*b*) $^{232}_{90}Th$ produces $^{208}_{82}Pb$ (thorium series)
(*c*) $^{235}_{92}U$ produces $^{207}_{82}Pb$ (actinium series)

At any instant some of all the intermediate elements produced by disintegration will be present. For laboratory work it is often convenient to choose a radioactive element which can be isolated (e.g. chemically), and which emits *only one type of radiation*. For example

(*a*) $^{241}_{95}Am$ for α-particles only
(*b*) $^{90}_{38}Sr$ for β^--particles only
(The β-particles also produce bremsstrahlung (p. 460).)
(*c*) $^{60}_{27}Co$ for γ-rays only.
(β-particles are also emitted but these have low energy and are relatively easily absorbed.)

Activity of Sources

The **activity** A of a source is the number of disintegrations it undergoes per unit time. The SI unit of activity is called the **becquerel** (Bq). Thus $1 \text{ Bq} = 1 \text{ s}^{-1}$.

Laboratory sources have activities of about 10^4 Bq, whereas some medical sources might have an activity of about 10^{14} Bq.

64.2 Methods of Detection

Most detectors indicate the arrival of *energy*, which may produce

(*i*) *ionization*, as in
  (*a*) the ionization chamber,
  (*b*) the Geiger Müller tube,
  (*c*) the cloud chamber, and similar chambers,
  (*d*) the spark counter,
(*ii*) *exposure* of a photographic emulsion,
(*iii*) *fluorescence* in the phosphor of a scintillation counter, and
(*iv*) *mobile charge-carriers* in a semiconductor.

Some of these detectors are discussed in this book but details are not included.

(*i*) Ionization Methods

Ionization is the removal of one or more of the extra-nuclear electrons from an atom, thus creating an **ion-pair**, consisting of *the removed electron*, negatively charged, and *the massive remainder of the atom*, positively charged. All radiations can produce ionization of the atoms of solids, liquids and gases.

α-particles have the greatest ionizing power of the three types of radiation, because

(1) they have twice the charge of a proton which enables them to attract electrons strongly, and
(2) they travel comparatively slowly and consequently remain a relatively long time in the vicinity of each atom near their path.

Ionization needs energy, and each ion-pair formed results in the ionizing agent losing that amount of kinetic energy. α- and β-particles decrease in speed in the process, and if γ-quanta are not wholly absorbed, their frequency is considerably decreased.

The distances travelled by α- and β-particles of given energy in a given medium are fairly constant, as these particles are slowed down in a large number (10^3–10^5) of very small steps, *in each of which* an ion-pair is formed.

X- or γ-rays are absorbed exponentially as a quantum usually loses all its energy (as in pair production) or much of it (as in the *Compton* effect) *in a single event*.

The X- and γ-rays cause ionization by ejecting electrons (secondary radiation).

(*a*) Ionization Chamber

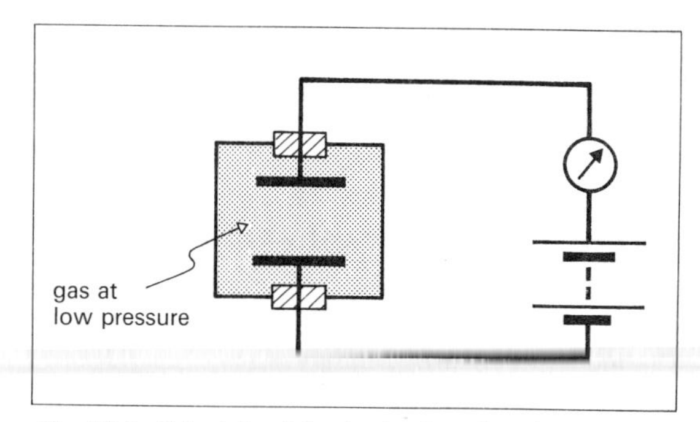

Fig. 64.1. Principle of the ionization chamber.

In fig. 64.1, radiation which enters the chamber ionizes gas molecules by collision, and the applied p.d. causes an electrical pulse to pass through the circuit. The pulse can be used to activate

(1) an amplifier and loudspeaker,
(2) a d.c. amplifier (p. 336),
(3) an electronic counting device called a **scaler**, or
(4) a **ratemeter**, which records the average rate of pulse arrival.

(*b*) The G.M. tube

This is a very sensitive type of ionization chamber and is shown in fig. 64.2.

γ-rays easily penetrate the walls of the tube but less penetrating radiation enters through the mica window. For detection of α-particles and slow β-particles the window has to be extremely thin.

(*c*) The Cloud and Similar Chambers

A charged particle travelling at speed leaves behind a trail of ions. If these occur in a region in which there is

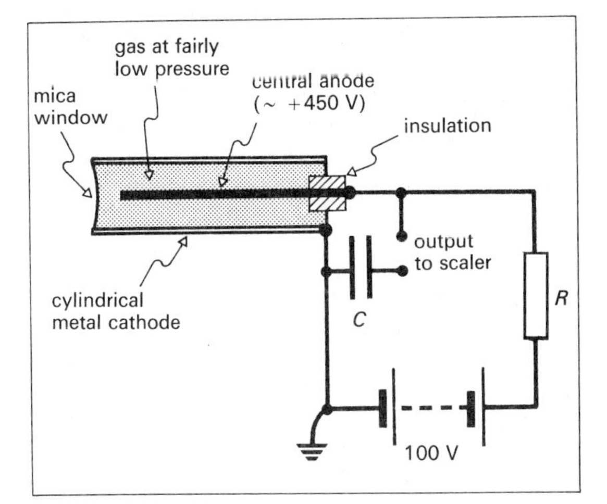

Fig. 64.2. A typical G.M. tube.

vapour about to condense, or liquid about to boil, then the ions will encourage the formation of droplets and vapour bubbles respectively. Either will leave a trail to show the path of the particle, which may be photographed if the chamber is brightly lit from the side. These principles are used in

(1) *The Wilson cloud chamber,* in which a sudden adiabatic expansion of the right amount cools water vapour so as to produce condensation on ions of either positive or negative charge.

(2) *The diffusion cloud chamber,* in which super-saturated alcohol vapour condenses. The pattern seen in the chamber is characteristic of the nature of the radiation which causes the ionization.

(3) *The bubble chamber,* in which the charged particle leaves a trail of bubbles in liquid hydrogen. This has the advantage that any photograph will not be distorted by expansion processes.

(d) The Spark Counter

This uses the ionization produced by particles to trigger a spark discharge in a region where air is close to breakdown in a strong electric field.

(*ii*) Photographic Emulsion Methods

A primary ion colliding with an emulsion gives a line of silver grains on processing. This enables a well-defined permanent record to be made, which can be made three-dimensional by stacking the plates. The emulsion used for detecting radiation has a high density of silver halide which results in the particle tracks being short.

(*iii*) Scintillation Methods

The impact of radiation on a suitable material (which can be matched to the radiation) causes the emission of a minute flash of light—a scintillation. In the **spinthariscope** α-particles cause zinc sulphide (a **phosphor**) to scintillate. A single crystal of sodium iodide is used for counting γ-rays.

A **scintillation counter** consists of an appropriate phosphor combined with a photomultiplier tube, which enables the weak flashes produced by β-particles and γ-rays to be detected.

Scintillation detectors have several advantages:

(1) The energies of the particles detected can be measured.

(2) They can deal with very high counting rates with pulse durations as short as 1 ns (10^{-9} s).

(3) Their efficiency in counting γ-rays is nearly 100%.

(*iv*) The Solid State Detector

If electrons and holes are suddenly created at the junction of a junction diode (p. 444) by ionizing radiation, an applied p.d. will cause a very short pulse to pass through the circuit. Amplification and counting by scaler or ratemeter now enables α-, β-, γ- and X-radiation, as well as protons and neutrons, to be detected.

64.3 Measurements with a d.c. Amplifier

The d.c. amplifier (**46.9**) can be used to detect and measure very small ionization currents by using the resistance ranges and connecting a milliammeter at the output (fig. 64.3). The working potential difference for an ionization chamber can be established by using an α-source and the d.c. amplifier to plot a graph of ionization current against applied p.d. V.

(*a*) The d.c. amplifier can be used to investigate the *range of radiation* by selecting a source which is known to emit only one kind of radiation and placing it at different distances d from the ionization chamber. A graph can then be plotted of the corresponding ionization currents I against the distances d (fig. 64.4), from which the range of that particular radiation in air can be deduced. The experiment can be repeated using layers of different materials.

(*b*) The d.c. amplifier can also be used for *determination of half-life* (of for example thoron gas pumped

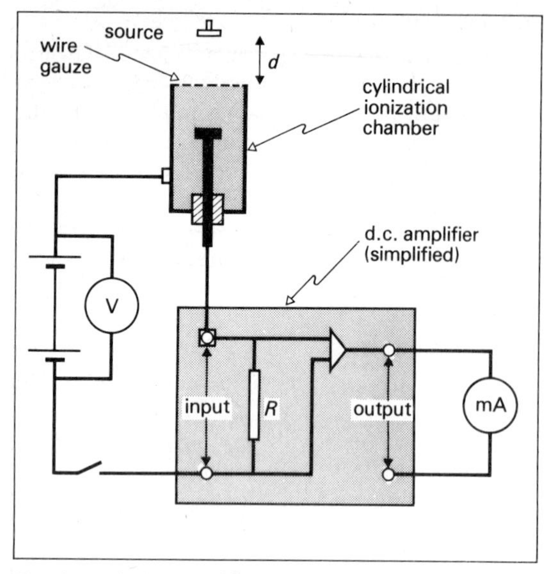

Fig. 64.3. A d.c. amplifier being used to measure the current in ionization chamber.

into the (closed) ionization chamber), since the ionization currents are directly related to the activities (p. 470). The half-life (p. 477) can be deduced from a graph of $I:t$ or $\log I:t$ (fig. 64.7).

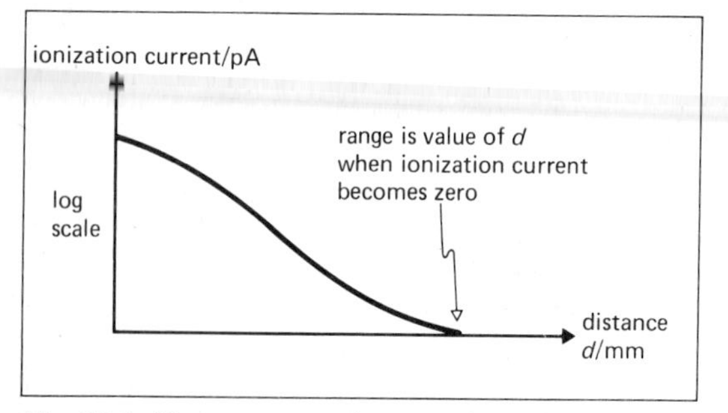

Fig. 64.4. Measurement of range of radiation.

64.4 Identification of the Radiations by Selective Absorption

Effect of different Radiations

(*a*) *α-particles* penetrate a few tens of mm of air or a thin foil of Al. For a particular disintegrating nuclide all the particles have about the same speed—there is often a line spectrum of nearly equal energies.

They have a small range, because they may produce about 10^4 ion-pairs *per* mm *of path* in air at atmospheric pressure.

(*b*) *β-particles* are emitted with very variable speeds even for a particular disintegrating nuclide. Low energy β-particles have a smaller penetration than α-particles, but typically they have about ten times the range. The ionization they produce per unit length of path is about one hundredth that of α-particles.

(*c*) *γ-rays* have variable energies which fall into several distinct monoenergetic groups from any particular emitter. The most energetic radiation may penetrate up to several tens of mm of lead, whereas the least energetic can pass through only very thin foils.

These measurements are usually done with a G.M. tube since γ-rays typically produce 10^{-4} times as many ion-pairs per unit length as do α-particles.

Air acts as an almost perfectly transparent medium to γ-rays and the intensity (rate of energy arrival per unit area) of γ-rays emanating from a point source varies inversely as the square of the distance from the source. The inverse square law does not hold for α- and β-particles in air because it absorbs them. The law would hold for beams of these particles when travelling through a vacuum, provided they emanated from a point source.

Effect of Different Absorbing Materials

Ability to absorb depends on

(*i*) the material *density*, and
(*ii*) on material *thickness*.

The atomic number of the material has little effect on the absorption of α- and β-particles, but, for γ-rays in high and low energy ranges there is very much higher absorption in materials of high atomic number.

It is convenient to quote data in terms of *mass per unit area*.

Suppose an α-particle could penetrate 4×10^{-2} m of air of density 1.3 kg m^{-3}. Then the mass per unit area to stop these particles is 5.2×10^{-2} kg m^{-2}. If we used aluminium of density 2700 kg m^{-3}, we would require a thickness of 2×10^{-5} m.

Distinguishing two Simultaneous Radiations

We can plot a graph of ionization currents against different thicknesses of absorbing material. A graph like fig. 64.5 which has an abrupt change of slope shows the range of the less penetrating radiation.

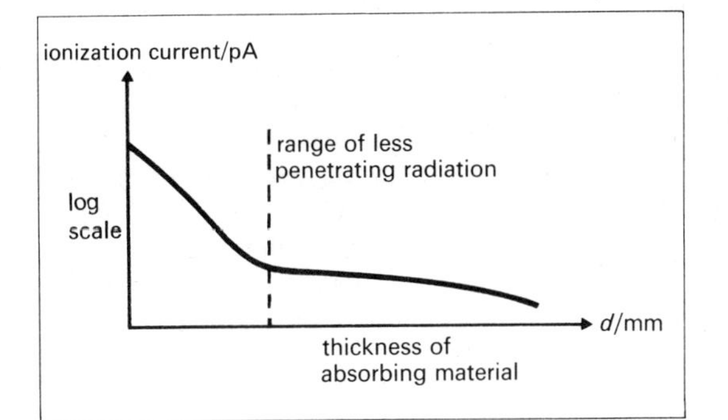

Fig. 64.5. The form of an absorption curve for two simultaneous radiations.

64.5 The Effect on the Radiations of *E* and *B*

(*a*) Electric Field *E*

(*i*) *α-particles* are deflected as positive charge would be deflected,

(*ii*) *β-particles* are deflected as negative charge would be deflected,

(*iii*) *γ-rays* are undeflected and have no charge.

(*b*) Magnetic Field *B*

(*i*) *α-particles* show a small deflection (fig. 64.6), which suggests they may be *particles* of relatively large mass,

(*ii*) *β-particles* show a large deflection—they are *particles* of small mass. They also show considerable dispersion, which demonstrates that they are emitted with variable speeds.

(*c*) Crossed *E* and *B*

This enables the *specific charge* and *speed* of the particles to be measured (p. 438).

64.6 Summary of the Properties of the Radiations

(*a*) α-particles

Their specific charge is half that of the proton. When they acquire electrons which neutralize them, they show the spectral lines typical of helium, so we deduce that they are *helium nuclei* (*Rutherford* and *Royds*—1909). Each α-particle consists of four nucleons—two protons and two neutrons. This is a particularly stable arrangement as each energy level in the shell model of the nucleus can accommodate 2p and 2n.

α-radiation is the result of the ejection at high speed of α-particles from a heavy nucleus. Many α-emitters are naturally occurring nuclides whose atomic number is between that of lead (82) and uranium (92). One very important emitter, plutonium-239, is manufactured in nuclear reactors.

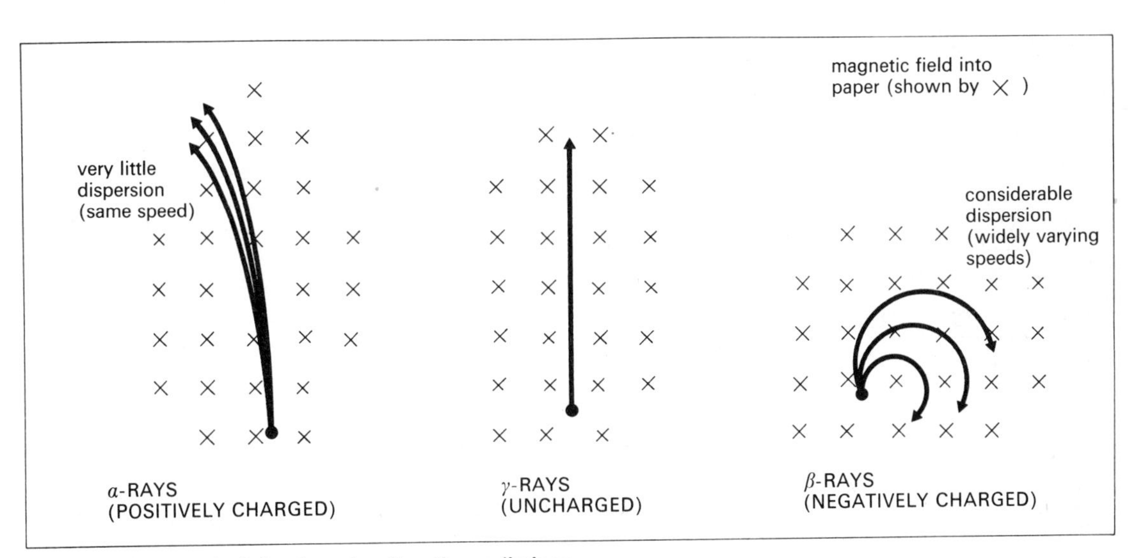

Fig. 64.6. Schematic deflection of radioactive radiations.

(b) β-particles

(*i*) ***Negative β-decay.*** In this decay the nucleus attains greater stability by the conversion of a neutron to a proton. The energy released is the k.e. of an ejected *electron* plus the energy of the **antineutrino**—a particle with zero rest mass and zero charge. Since the rest mass is zero it has an energy pc only by virtue of travelling at speed c.

$$\mathrm{n} \rightarrow \mathrm{p}^+ + \mathrm{e}^- + \bar{\nu}$$

β^--emission is characteristic of nuclei having a large proportion of neutrons. (The ejected electrons come from the *nucleus* rather than from the electron cloud, and yet do not exist as electrons while in the nucleus.)

(*ii*) ***Positive β-decay.*** Positron decay is possible if the mass of the parent atom exceeds that of the daughter atom by at least $2m_e$ (or its energy equivalent). Any extra energy appears as the k.e. of the daughter nucleus, the positron and the **neutrino**—a particle with its sense of spin opposite from that of the antineutrino.

$$\mathrm{p}^+ \rightarrow \mathrm{n} + \mathrm{e}^+ + \nu$$

β^+-emission is characteristic of nuclides with low atomic number which have a high proportion of protons.

If a nucleus has insufficient energy for positron decay, it may attain the same result by **electron capture** from one of the innermost shells. The energy released is that of a neutrino and there will also be X-radiation owing to the vacancy in the electron shell. A particular species of nuclide often decays by either positive β-decay or electron capture.

Quite a few β-emitters are found in nature (e.g. potassium-40), but most are prepared in the laboratory by irradiating matter with neutrons.

All the decays mentioned above involve a change in Z and N with A remaining constant. If a graph of *mass defect* of nuclides is plotted against Z for an isobar of odd number A, the result is a parabola. The most stable position is at the lowest point of the curve. The β-decay of other nuclei will help them to reach this point. For an isobar of even number A, two parabolae are obtained.

(c) γ-rays

These (being electromagnetic waves) have properties similar to X-rays (p. 459), but originate *in the nucleus* and generally have a smaller wavelength.

When a nucleus is excited it can return to its ground state by emitting a γ-ray. It is often found that α- and β-emitters also emit γ-rays because nuclei are frequently produced in excited states as a result of their decay. However, many radioactive substances do not emit γ-rays as the resultant nuclei are produced in the ground state. The wavelength, λ, is measured by crystal diffraction methods.

Using

(*i*) $c = \nu\lambda$, and
(*ii*) $E = h\nu$

their energies E may be measured.

*Antiparticles

The positron is said to be the antiparticle of the negatron, and the antineutrino is the antiparticle of the neutrino. All

Table of Properties

Property	α	β^-	γ
fluorescence on a phosphor	yes	less	very little but is observed with NaI
effect on photographic plate	yes	yes	yes
number of ion-pairs per mm in air	$\sim 10^4$	$\sim 10^2$	~ 1
absorption: stopped by up to	$\sim 10^{-2}$ mm Al ~tens of mm air	~mm Al ~m air	~100 mm Pb
deflection by ***E*** and ***B***	as + ve charge	as − ve charge	nil
speeds	$\sim 10^7\ \mathrm{m\,s^{-1}}$	$\sim 10^8\ \mathrm{m\,s^{-1}}$ but variable	$3 \times 10^8\ \mathrm{m\,s^{-1}}$

elementary particles have such counterparts (e.g. proton-antiproton and neutron-antineutron pairs) but these are not often detected as a large amount of energy is required to generate such pairs. The annihilation process is often the most effective method of detecting antiparticles, and occurs when an antiparticle meets its corresponding particle.

64.7 Radiation Hazards

(*a*) A source should never be present *inside the body*, since at all times its radiation will damage the living cells. Therefore protective clothing should be worn to prevent the contamination of the body by radioactive substances. For example, strontium-90 is absorbed into the bones where its radiation damages the bone marrow.

(*b*) An *external source* is less dangerous, because it will cause damage only while it is close to the body. Even while it is there

(*i*) the α-radiation cannot penetrate the skin, but care must still be taken because some α-emitters create radioactive gaseous daughter products which can be absorbed into the body

(*ii*) the other radiations can be screened: e.g. β-radiation by a sheet of Perspex, and γ-radiation by tens of mm of lead or a few metres of concrete.

As a general rule materials will not become radioactive on exposure to radiation from other nuclei, but they may do so when used as a target in a particle accelerator.

Radiation Dose

(a) General

The dose of an ionizing radiation absorbed by an irradiated material is measured in **grays** (Gy). 1 Gy = 1 J kg^{-1}.

Sometimes one distinguishes between the total energy absorbed, and that absorbed from X- and γ-radiation.

(b) The Biological Effect

Equal doses of different ionizing radiations provide the same amount of energy in a given absorber, but may produce different biological effects. Thus

$$\begin{pmatrix}\text{effective}\\ \text{dose}\end{pmatrix} = \begin{pmatrix}\text{total}\\ \text{dose}\end{pmatrix} \times \begin{pmatrix}\text{relative biological}\\ \text{effectiveness (r.b.e.)}\end{pmatrix}$$

where the r.b.e. is a factor controlled by the nature of the radiation.

The effective dose for those engaged on radiation work must not exceed a specified maximum in a given time.

64.8 Radioactive Decay

The Russell, Fajans and Soddy Rules (1913)

(*a*) *In α-decay*, an α-particle is emitted, so

(*i*) A decreases by 4, and

(*ii*) Z decreases by 2.

If P represents the parent nuclide and D the daughter, then

$${}^{A}_{Z}\mathrm{P} \rightarrow {}^{A-4}_{Z-2}\mathrm{D} + {}^{4}_{2}\alpha$$

(*b*) *In β-decay*

(*i*) A does not change since no nucleon is emitted, and

(*ii*) Z increases by 1(β^--decay) or decreases by 1(β^+-decay).

$$\beta^-:\ {}^{A}_{Z}\mathrm{P} \rightarrow {}_{Z+1}^{A}\mathrm{D} + {}^{0}_{-1}\mathrm{e}$$

$$\beta^+:\ {}^{A}_{Z}\mathrm{P} \rightarrow {}_{Z-1}^{A}\mathrm{D} + {}^{0}_{+1}\mathrm{e}$$

(*c*) *A γ-ray* represents the emission of energy from a nucleus which is returning to its ground state (p. 467). Both A and Z remain unaffected.

$$(\text{excited nucleus}) \rightarrow (\text{more stable nucleus}) + \gamma$$

When the emission of a γ-ray accompanies the emission of an α-particle or a β-particle, rule (*a*) or (*b*) will apply.

Examples of Radioactive Decay

We represent the changes by equations that show what happens *to the nucleus.* They must be distinguished from equations that represent chemical reactions (which are concerned with the bonding of an *atom's outermost electrons*). In all such equations, the sums of

(*a*) the mass numbers (representing the *number of nucleons*), and

(*b*) the atomic numbers (representing electric *charge*)

must be the same on each side of the equation if it is correct. (The number of nucleons and charge are both *conserved.*)

α-decay ${}^{226}_{88}\mathrm{Ra} \rightarrow {}^{222}_{86}\mathrm{Rn} + {}^{4}_{2}\alpha$ (radium → radon)

β^--decay ${}^{12}_{5}\mathrm{B} \rightarrow {}^{12}_{6}\mathrm{C} + {}^{0}_{-1}\mathrm{e}$ (boron → carbon)

β^+-decay ${}^{12}_{7}\mathrm{N} \rightarrow {}^{12}_{6}\mathrm{C} + {}^{0}_{+1}\mathrm{e}$ (nitrogen)

64.9 Nuclear Reactions

A nuclear reaction is an induced change involving a spontaneous decay of a nucleus from a highly excited state. It requires the bombardment of a nucleus by high energy particles (such as those from a particle accelerator) or by γ-radiation which results in their capture by, and the subsequent decay of, a compound nucleus.

The symbolic way of expressing a nuclear reaction is:

$$\text{initial nuclide}\left(\begin{array}{cc}\textit{incoming} & \textit{outgoing}\\ \textit{quanta or} & \textit{quanta or}\\ \textit{particles,} & \textit{particles}\end{array}\right)\text{final nuclide}$$

Reactions are normally classified in terms of the quantities enclosed by the brackets. The equations which follow are examples of what can take place.

(a) A (d, p) Reaction

$$^{23}_{11}\text{Na}\left(^{2}_{1}\text{H},\ ^{1}_{1}\text{p}\right)^{24}_{11}\text{Na}$$

A stable nucleus absorbs a deuterium nucleus which leads to the ejection of a proton and the formation of a new radioactive nucleus.

(b) An (α, p) Reaction

$$^{14}_{7}\text{N}\left(^{4}_{2}\alpha,\ ^{1}_{1}\text{p}\right)^{17}_{8}\text{O}$$

This reaction was carried out by *Rutherford* (1919), and was the first artificial transmutation. The presence of the oxygen isotope was shown spectroscopically.

(c) A (p, α) Reaction

$$^{7}_{3}\text{Li}\left(^{1}_{1}\text{p},\ ^{4}_{2}\alpha\right)^{4}_{2}\text{He}$$

This reaction was carried out by *Cockcroft* and *Walton* (1932), and was the first transmutation achieved using artificially accelerated ions (the protons).

(d) An (α, n) Reaction

$$^{9}_{4}\text{Be}\left(^{4}_{2}\alpha,\ ^{1}_{0}\text{n}\right)^{11}_{6}\text{C}$$

This was carried out by *Chadwick* (1932) and led to the discovery of the **neutron**. This is an example of a *neutron source*.

(e) An (n, α) Reaction

$$^{10}_{5}\text{B}\left(^{1}_{0}\text{n},\ ^{4}_{2}\alpha\right)^{7}_{3}\text{Li}$$

This is one method of detecting neutrons.

(f) An (n, γ) Reaction

$$^{27}_{13}\text{Al}\left(^{1}_{0}\text{n},\ \gamma\right)^{28}_{13}\text{Al}$$

The unstable isotope of the target nucleus is formed by neutron capture.

(g) A (γ, p) Reaction

$$^{25}_{12}\text{Mg}\left(\gamma,\ ^{1}_{1}\text{p}\right)^{24}_{11}\text{Na}$$

This reaction is known as *photodisintegration.*

64.10 The Law of Radioactive Decay

Individual nuclei disintegrate independently and so the decay is a *random process.* This is well illustrated by the spark counter. The number of nuclei present in a small sample is *so great that we can treat the process statistically*, and determine the rate at which a **parent** element disintegrates into its **daughter** element.

Nevertheless, strictly only the *probability* of disintegration in a particular time interval can be stated.

The Radioactive Decay Constant λ

Suppose at any instant there are N active nuclei present in a sample. We assume

$$\left(\begin{array}{c}\text{rate of}\\ \text{disintegration}\\ \text{or } \textbf{activity}\end{array}\right) \propto \left(\begin{array}{c}\text{number of}\\ \text{nuclei available}\\ \text{to disintegrate}\end{array}\right)$$

$$A \propto N \qquad \text{or} \qquad -\frac{dN}{dt} \propto N$$

We define the* radioactive decay constant λ *by the equation

$$\boxed{A = -\frac{dN}{dt} = \lambda N}$$

$$A[\text{Bq}] = -\frac{dN}{dt}\left[\frac{1}{\text{s}}\right] = \lambda\left[\frac{1}{\text{s}}\right]\cdot N$$

Suppose at $t = 0$ we have N_0 active nuclei, and after a time t we have N active nuclei. Then

$$\int_{N_0}^{N}\frac{dN}{N} = -\lambda\int_{0}^{t} dt$$

$$\ln\left(\frac{N}{N_0}\right) = -\lambda t$$

$$\boxed{N = N_0\, e^{-\lambda t}}$$

The number of parent nuclei present decays *exponentially.*

The Half-Life Period $T_{\frac{1}{2}}$

Suppose $T_{\frac{1}{2}}$ is the time taken for the number of active nuclei to be reduced to $N_0/2$. Then

$$\frac{N}{N_0} = e^{-\lambda T_{\frac{1}{2}}} = \tfrac{1}{2}$$

$$e^{\lambda T_{\frac{1}{2}}} = 2$$

$$\therefore \quad T_{\frac{1}{2}} = \frac{\ln 2}{\lambda} = \frac{0.693}{\lambda}$$

The value of $T_{\frac{1}{2}}$ depends on the emitting nuclide, being 10^9 years for uranium-238 and 10^{-6} s for some products of radium decay.

Fig. 64.7 illustrates the concepts of $T_{\frac{1}{2}}$ and exponential decay. The second graph is a semilogarithmic plot and the slope of the line measures the decay constant.

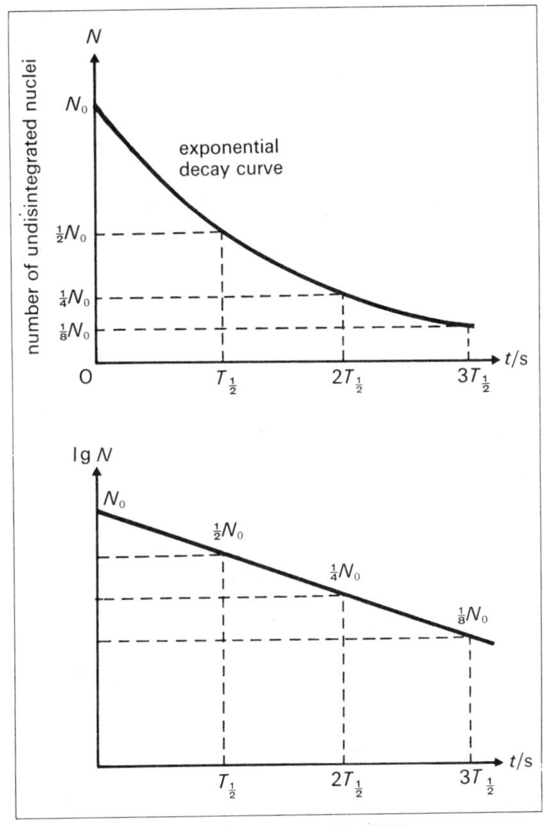

Fig. 64.7. Exponential decay and half life.

The decay of radioactive *parent* atoms and the growth in the number of *daughter* atoms (assumed stable) are illustrated graphically in fig. 64.8 as functions of time.

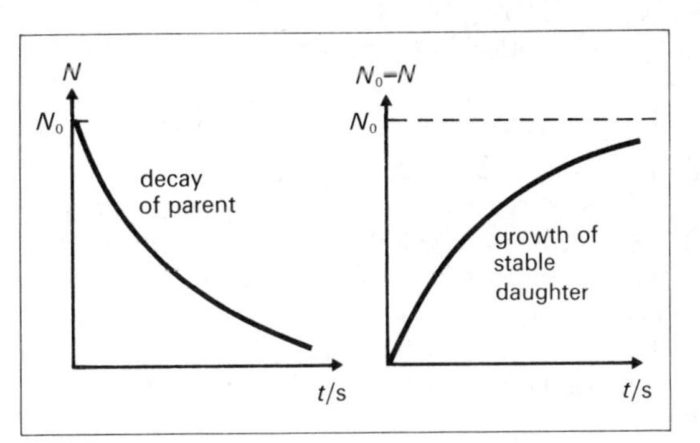

Fig. 64.8. Decay of parent atoms and growth in number of daughter atoms.

Radioactive Equilibrium

If a sample of the first member of a radioactive series has been left for some time, all members of the series will be present together. The *relative* amounts of the various nuclides will be constant at radioactive equilibrium, which occurs when each nuclide is decaying at the rate at which it is formed.

Therefore $\quad A_1 = A_2 = A_3$

which means that $\quad \lambda_1 N_1 = \lambda_2 N_2 = \lambda_3 N_3$

and so $\quad \dfrac{N_1}{(T_{\frac{1}{2}})_1} = \dfrac{N_2}{(T_{\frac{1}{2}})_2} = \dfrac{N_3}{(T_{\frac{1}{2}})_3}$

64.11 Energy Changes in Radioactive Decay and Nuclear Reactions

The *disintegration* or *reaction energy* (Q value) is the energy released in a radioactive decay or a nuclear reaction per disintegrating nucleus.

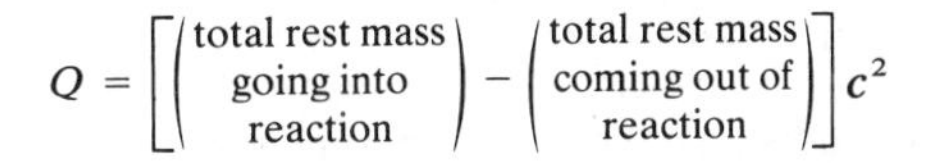

$$Q = \left[\begin{pmatrix}\text{total rest mass}\\ \text{going into}\\ \text{reaction}\end{pmatrix} - \begin{pmatrix}\text{total rest mass}\\ \text{coming out of}\\ \text{reaction}\end{pmatrix}\right] c^2$$

If $Q > 0$, nuclear energy is released—an **exothermic** reaction.

If $Q < 0$, nuclear energy is absorbed—an **endothermic** reaction.

(*a*) In α-decay the α-particles have discrete energies which lie in the range 0.6 to 1.6 pJ.

(*b*) In β-decays the negatrons or positrons have a continuous energy spectrum up to the calculated maximum value, since the remainder of the energy is carried away by the antineutrino or the neutrino which is also emitted. Examples:

(*i*) β^--decay of ${}^{12}_{5}B$ to ${}^{12}_{6}C$ has $Q = +2.1$ pJ per disintegration.

(*ii*) β^+-decay of ${}^{12}_{7}N$ to ${}^{12}_{6}C$ has $Q = +2.6$ pJ per disintegration.

(*c*) In γ-decay the γ-photon energies cover the range 0 to about 1 pJ.

High-energy nuclear reactions involve bombarding particles of energies ≈ 10 nJ; *low-energy* reactions involve particle energies up to ≈ 3 pJ. For example, in the *Cockcroft* and *Walton* (p, α) reaction, $Q = +2.8$ pJ, whereas in the *Rutherford* (α, p) reaction, $Q = -0.19$ pJ.

Nuclear Fission

Nuclear fission, which is the disintegration of a massive nucleus into two fragments with comparable masses, is a special type of low-energy nuclear reaction. Sufficient excitation energy for the nucleus to split may be provided spontaneously or by particle bombardment. Oscillations of the excited nucleus cause a deformation of the spherical shape which allows the *Coulomb* repulsion forces to overcome the supposed nuclear surface tension. The two nuclei produced will lie in the middle of the curve of fig. 63.6 so that the final value of E_b/A is greater than its original value. A nucleus undergoing fission does not split into the same product nuclei each time. For uranium-235 the product masses have values of A between 70 and 165. An example is given below:

$$ {}^{235}_{92}\mathrm{U}\left({}^{1}_{0}\mathrm{n},\ 3{}^{1}_{0}\mathrm{n}\right){}^{141}_{56}\mathrm{Ba},\ {}^{92}_{36}\mathrm{Kr} $$

neutron bombardment   barium   krypton

The products have too many neutrons to be stable and so these are either released simultaneously or they change into protons via β^-- and γ-decays. The total energy released per disintegration is very high compared with a typical exothermic nuclear reaction and can be calculated from either

(*i*) the loss of mass, or

(*ii*) the change in the value of E_b/A.

A typical energy release is 32 pJ per disintegration, of which about 85 per cent is the k.e. of the fission fragments. The rest of the energy is shared by antineutrinos, γ-rays, β-particles and neutrons. The fission process provides a great source of energy as it can result in a *self-sustaining* **chain reaction**. The requirement for this is that for every uranium atom split there must be at least one neutron, out of the average of 2.5 produced, that will split another uranium atom. The flux inside the *reactor* may be $\approx 10^{19}$ neutrons $\mathrm{m^{-2}\,s^{-1}}$. Loss of neutrons is minimized by

(*i*) having a large nuclear reactor,

(*ii*) slowing down the neutrons using a **moderator** (such as beryllium or graphite)—the aim being to increase the fission *cross-section* but to avoid neutron capture, and

(*iii*) arranging the uranium fuel rods appropriately.

The rate at which fission reactions occur is regulated by control rods which can easily absorb neutrons. When each fission produces *less than one* further fission the reactor is **subcritical**. When it produces *at least one* further fission the reactor becomes **critical** and the reaction is self-sustaining. The fuel in a *breeder* reactor consists of

(*i*) a *fissionable* material, e.g. plutonium-239, and

(*ii*) a *fertile* material, e.g. uranium-238.

The fertile material can be converted in the reactor into fissionable material.

Nuclear Fusion

Light elements can fuse together to increase their binding energy per nucleon. This mass reduction results in the release of large amounts of energy (of the order of pJ per fusion). A maintained temperature of 10^8 to 10^9 K is needed to give the nuclei enough energy to overcome the initial electrostatic repulsion and to come together (hence '**thermonuclear**' reaction). The radiation of solar and stellar energy is caused by cycles of thermonuclear reactions, in which the overall effect is the formation of helium from hydrogen. For example,

$$ {}^{3}_{1}\mathrm{H}\left({}^{2}_{1}\mathrm{H},\ {}^{1}_{0}\mathrm{n}\right){}^{4}_{2}\mathrm{He} $$

tritium   deuterium

64.12 Some Uses of Radioactivity

(*a*) Radioisotopes

A radioactive isotope is called a **radioisotope**. Radioisotopes are most conveniently made within a nuclear reactor where a large flux of neutrons is available.

Uses

(*i*) Cobalt-60 emits γ-rays which are used for treating cancer, and for sterilization.

(*ii*) As **tracer elements**. A radioisotope will experience the same chemical treatment as its non-active isotope, but its activity enables it to be detected even in minute quantities.

(*iii*) The thickness of a material can be measured from the decrease in the intensity of the radiation that has passed through it.

(*b*) Atomic Energy

(*i*) **Nuclear fission** has been described in **64.11** as the basis for the atomic reactor within which the k.e. of the fission fragments is a source of thermal energy. When each fission reaction produces *well above the one* further fission, the reactor is said to be **supercritical**. An example of this is the *atomic bomb.*

(*ii*) **Nuclear fusion** as described in **64.11** is the basis of the *hydrogen bomb*, which needs a fission bomb to provide the required starting temperature. A controlled thermonuclear reaction would provide an enormous source of energy but, as yet, research work has been unsuccessful.

(*c*) Dating Methods

(*i*) *Geological*

Uranium has a convenient half-life ($\sim 4.5 \times 10^9$ years). The age of rocks can be measured from observations on the relative amounts of ^{232}Th, ^{238}U, ^{208}Pb (from Th), ^{206}Pb (from ^{238}U) and ^{207}Pb (from ^{235}U) in the rocks. Another method, useful for younger rocks, uses the ratio $^{40}K/^{40}Ar$.

(*ii*) *Archaeological*

Radioactive ^{14}C has a convenient half-life ($\sim 5.7 \times 10^3$ years), and also has a constant ratio to ^{12}C in *living matter.* Measurement of their proportions in dead organisms enables an estimate to be made of the time that has elapsed since death.

Appendix A. Special Names and Symbols for Derived SI Units

The quantities in this table are measured in derived units to which special names have been allotted.

Physical quantity	*Symbol for quantity*	*Name of SI unit*	*Symbol for unit*	*Immediate definition of unit*	*Definition in terms of basic units*
frequency	f or ν	hertz	Hz	s^{-1}	s^{-1}
force	$\mathbf{F}$	newton	N	$kg\ m\ s^{-2}$	$kg\ m\ s^{-2}$
work or energy	W	joule	J	N m	$kg\ m^2\ s^{-2}$
power	P	watt	W	$J\ s^{-1}$	$kg\ m^2\ s^{-3}$
pressure	p	pascal	Pa	$N\ m^{-2}$	$kg\ m^{-1}\ s^{-2}$
electrical charge	Q	coulomb	C	A s	A s
electrical potential difference	V	volt	V	$J\ C^{-1}$	$kg\ m^2\ s^{-3}\ A^{-1}$
electrical capacitance	C	farad	F	$C\ V^{-1}$	$A^2\ s^4\ kg^{-1}\ m^{-2}$
electrical resistance	R	ohm	Ω	$V\ A^{-1}$	$kg\ m^2\ s^{-3}\ A^{-2}$
electrical conductance	G	siemens	S	Ω^{-1}	$kg^{-1}\ m^{-2}\ s^3\ A^2$
magnetic flux density	$\mathbf{B}$	tesla	T	$N\ A^{-1}\ m^{-1}$	$kg\ s^{-2}\ A^{-1}$
magnetic flux	Φ	weber	Wb	$T\ m^2$	$kg\ m^2\ s^{-2}\ A^{-1}$
inductance	L or M	henry	H	$Wb\ A^{-1} = V\ A^{-1}\ s$	$kg\ m^2\ s^{-2}\ A^{-2}$
activity	A	becquerel	Bq	s^{-1}	s^{-1}
absorbed dose	[illegible]	gray	Gy	$J\ kg^{-1}$	$m^2\ s^{-2}$

Appendix B. Recommended Values of Selected Physical Constants

These are given to only four significant figures, as at this level we seldom need to use more. Nevertheless, the reader should examine the last column carefully.

Physical quantity	*Symbol*	*Value*	*Approximate uncertainty in 1973 in parts per million*
speed of light in a vacuum	c	2.998×10^{8} m s^{-1}	0.004
permeability of a vacuum	μ_0	$4\pi \times 10^{-7}$ H m^{-1} exactly	–
the permittivity constant	ε_0	8.854×10^{-12} F m^{-1}	0.01
mass of proton	m_p	1.673×10^{-27} kg	5
mass of neutron	m_n	1.675×10^{-27} kg	5
mass of electron	m_e	9.110×10^{-31} kg	5
charge of proton	e	1.602×10^{-19} C	3
Boltzmann constant	k	1.381×10^{-23} J K^{-1}	32
Planck constant	h	6.626×10^{-34} J s	5
Stefan–Boltzmann constant	σ	5.670×10^{-8} W m^{-2} K^{-4}	125
Rydberg constant	R_∞	1.097×10^{7} m^{-1}	0.08
Avogadro constant	N_A	6.022×10^{23} mol^{-1}	5
molar gas constant	R	8.314 J mol^{-1} K^{-1}	31
Faraday constant	F	9.648×10^{4} C mol^{-1}	3
gravitational constant	G	6.672×10^{-11} N m^{2} kg^{-2}	615

Bibliography

(**a**) The following books provide careful and detailed discussion of many topics from school physics at a relatively elementary level:

PSSC Physics (Heath), 1965.
College Physics (Raytheon), 1968.

(**b**) *Practical Physics*, by Squires (McGraw-Hill) 1976, is not a catalogue of experiments, but rather sets out to develop a critical approach to experimental work.

(**c**) *Essential Principles of Physics* needs to be amplified by more detailed texts. The authors recommend the following as a complementary set. Each aims at about the same standard as *E.P.O.P.*, and together they cover the same ground in a more explanatory manner.

Mechanics, Vibrations and Waves, by Akrill and Millar (John Murray), 1974.
The Mechanical and Thermal Properties of Materials, by Collieu and Powney (Arnold), 1973.
Principles of Light and Optics, by Wheadon (Longmans), 1968.
Electricity and Modern Physics, by Bennet (Arnold), 1974.

(**d**) The following, listed in approximate order of increasing sophistication, comprise complete texts which lead up to first and second year standard of English universities. They discuss fundamentals very carefully, and this makes them invaluable for bridging what is sometimes a difficult gap between school and degree physics. Of the many texts available, these reflect very much the authors' personal choice.

College Physics, by Sears and Zemansky (Addison-Wesley), 1974.
University Physics, by Sears and Zemansky (Addison-Wesley), 1976.
Fundamentals of Physics, by Halliday and Resnick (Wiley), 1970.
Physics, by Alonso and Finn (Addison-Wesley), 1970.
Elementary Classical Physics (2 vols), 1973, and
Elementary Modern Physics, 1968, by Weidner and Sells (Allyn and Bacon).
Physics (2 parts), by Halliday and Resnick (Wiley), 1978.
Fundamental University Physics (3 vols), by Alonso and Finn (Addison-Wesley), 1967.
Berkeley Physics Course (5 vols), (McGraw-Hill), 1965 to 1973.
Feynman Lectures on Physics (3 vols), by Feynman, Leighton and Sands (Addison-Wesley), 1963.

(**e**) Useful books which do not form part of a complete text:

Gases, Liquids and Solids, by Tabor (Penguin Books), 1969.
Properties of Matter, by Flowers and Mendoza (Wiley), 1970.
Particles, by Powles (Addison-Wesley), 1968.
Newtonian Mechanics, by French (Nelson), 1971.
Vibrations and Waves, by French (Nelson), 1971.
Special Relativity, by French (Nelson), 1968.
Forces and Particles, by Pippard (Macmillan), 1972.
Optics, by Hecht and Zajac (Addison-Wesley), 1974.

(**f**) Further suggestions for more general reading will be found in the authors' *Questions on Principles of Physics* (John Murray), 1979.

To that list should be added the publications of the **Open University**, details of which may be obtained from

The Marketing Division
The Open University
P.O. Box 81
Milton Keynes, MK7 6AT.

Index

Key **316** refers to a main entry 217*d* refers to a definition 184*t* refers to a table

E

F

J

K

L

T